Handbook of Experimental Pharmacology

Volume 179

Editor-in-Chief

K. Starke, Freiburg i. Br.

Editorial Board

G.V.R. Born, London
S. Duckles, Irvine, CA
M. Eichelbaum, Stuttgart
D. Ganten, Berlin
F. Hofmann, München
W. Rosenthal, Berlin
G. Rubanyi, San Diego, CA

Prof. Dr. Veit Flockerzi
Experimentele und Klinische
Pharmakologie und Toxikologie
der Universität des Saarlandes
Gebäude 46
66421 Homburg
Deutschland
veit.flockerzi@uniklinikum-saarland.de

Prof. Dr. Bernd Nilius
Laboratorium voor Fysiologie
Katholieke Universiteit Leuven
Campus Gasthuisberg
Herestraat 49
3000 Leuven
België
Bernd.Nilius@med.kuleuven.be

With 55 Figures and 19 Tables

ISSN 0171-2004

ISBN-10 3-540-34889-1 Springer Berlin Heidelberg New York

ISBN-13 978-3-540-34889-4 Springer Berlin Heidelberg New York

This work is subject to copyright. All rights reserved, whether the whole or part of the material is concerned, specifically the rights of translation, reprinting, reuse of illustrations, recitation, broadcasting, reproduction on microfilm or in any other way, and storage in data banks. Duplication of this publication or parts thereof is permitted only under the provisions of the German Copyright Law of September 9, 1965, in its current version, and permission for use must always be obtained from Springer. Violations are liable for prosecution under the German Copyright Law.

Springer is a part of Springer Science+Business Media
springer.com

© Springer-Verlag Berlin Heidelberg 2007

The use of general descriptive names, registered names, trademarks, etc. in this publication does not imply, even in the absence of a specific statement, that such names are exempt from the relevant protective laws and regulations and therefore free for general use.

Product liability: The publishers cannot guarantee the accuracy of any information about dosage and application contained in this book. In every individual case the user must check such information by consulting the relevant literature.

Editor: Simon Rallison, London
Desk Editor: Susanne Dathe, Heidelberg
Cover design: *design & production* GmbH, Heidelberg, Germany
Typesetting and production: LE-TeX Jelonek, Schmidt & Vöckler GbR, Leipzig, Germany
Printed on acid-free paper   27/3100-YL - 5 4 3 2 1 0

Transient Receptor Potential (TRP) Channels

Contributors

J. Abramowitz, G.P. Ahern, I.S. Ambudkar, K.S. Authi,
G.J. Barritt, D.J. Beech, R.J.M. Bindels, L. Birnbaumer,
M. Bödding, K.A. del Carmen, A. Cavalié, C.K. Colton,
A. Dietrich, P. Eder, J. Eisfeld, A. Fleig, V. Flockerzi,
J. García-Añoveros, D.L. Gill, D. Griesemer, K. Groschner,
T. Gudermann, L.-P. He, T. Hewavitharana, J.G.J. Hoenderop,
M. Hoth, D.A. Jacobson, J.-Y. Kim, K. Kiselyov, L. Lemonnier,
W. Liedtke, E.R. Liman, P. Luncsford, A. Lückhoff, J.A. Matta,
A.R. Mensenkamp, Y. Mori, S. Muallem, K. Nagata,
B.A. Niemeyer, B. Nilius, T. Numaga, J. Oberwinkler,
G. Owsianik, R.L. Patterson, R. Penner, S.E. Philipp,
L.H. Philipson, S.C. Pingle, T.D. Plant, M. Poteser, J.W. Putney,
F. Qin, A. Quintana, D. van Rossum, G.Y. Rychkov,
E.C. Schwarz, D.M. Shin, J.T. Smyth, J. Soboloff, M. Spassova,
A. Stokes, R. Strotmann, M. Tominaga, S. Tonner, M. Trebak,
H. Turner, G. Vazquez, K. Venkatachalam, R. Vennekens,
T. Voets, M. Wakamori, A.S. Wenning, U. Wissenbach,
R. Witzgall, M.-J. Wolfs, W. Xu, X. Yao, E. Yildirim, J. Young
Kim, J.P. Yuan, M.X. Zhu

Editors

Veit Flockerzi and Bernd Nilius

We dedicate this volume of the Handbook of Experimental Pharmacology to Franz Hofmann who has made, and continues to make outstanding contributions to the field of ion channels and molecular pharmacology

Preface

Only a decade has passed since the first mammalian homologues of the transient receptor potential (TRP) gene from *Drosophila melanogaster* were discovered. Today the superfamily of mammalian TRP genes comprises 28 members. Most if not all of them encode ion channel proteins, and these TRP cation channels display an extraordinary assortment of selectivities and activation mechanisms, some of which represent previously unrecognized modes for regulating ion channels. The biological roles of TRP channels appear to be equally diverse. Some TRPs are involved in sensory functions such as smell, taste, primitive forms of vision, pheromone sensing and pain perception. Found in skin, hair follicles and nerve cells, the "thermo"-TRPs are responsive to heat or cold and may help regulate body temperature and avoidance of tissue-damaging noxious temperatures. Other TRPs are also sensitive to the active ingredients of oregano, thyme and clove, allicin, mustard, horseradish and wasabi, cinnamon oil and chili peppers. Some of those natural compounds or their ingredients have been in medical (and kitchen) use for centuries. Other TRPs appear to be intimately connected to G protein-coupled receptors, thereby linking chemical signals such as hormones and neurotransmitters to membrane excitability in the cardiovascular system. TRPs may even be involved in mechanosensation and regulation of neurite length and growth cone morphology. At least three TRP proteins are distinguished from other known ion channels in that they consist of enzyme domains linked to the C termini of ion channel domains. The result of this marriage of channel and enzyme has been baptized "chanzyme". We could continue in this way, and actually, a quick click for "TRP channel" in PubMed will immediately reveal additional novel and surprising TRP features. Apparently we are just beginning to understand the TRP functions and their biological roles.

In this fast-moving field the main goal of this volume is to provide up-to-date information on the molecular and functional properties and pharmacology of mammalian TRP channels. Leading experts in the field have written 35 chapters which describe properties of a single TRP protein/channel or portray more general principles of TRP function and important pathological situations linked to mutations of TRP genes or their altered expression. Thereby, this volume on TRP channels provides valuable information for readers with different expectations and backgrounds, for those who are approaching this field of research as well as for those wanting to make a trip to TRPs, from pharmacologists and physiologists to medical doctors, other scientists, students and lecturers. It fills a gap between pharmacology textbooks and the latest manuscripts in scientific periodicals.

All contributing authors as well as the editors have taken great care to provide up-to-date information. However, inconsistencies or errors may remain, for which we assume full responsibility. We are indebted to our colleagues for their excellent contributions. It has been a great experience, both personally and scientifically, to interact with and learn from the 81 contributing authors. We would also like to thank Mrs. Inge Vehar, for her excellent and invaluable secretarial assistance during all stages of the project. Within Springer, we are grateful to Ms. Susanne Dathe for successfully managing this volume and for her encouraging support. It has been a pleasure to work with her.

Homburg,
September 2006   Veit Flockerzi, Bernd Nilius

List of Contents

Subject Index XIII

An Introduction on TRP Channels 1
   V. Flockerzi

Part I. TRPC Channel Subfamily

TRPC1 Ca^{2+}-Permeable Channels in Animal Cells 23
   G. Rychkov, G. J. Barritt

TRPC2: Molecular Biology and Functional Importance 53
   E. Yildirim, L. Birnbaumer

TRPC3: A Multifunctional, Pore-FormingSignalling Molecule 77
   P. Eder, M. Poteser, K. Groschner

Ionic Channels Formed by TRPC4 93
   A. Cavalié

Canonical Transient Receptor Potential 5 109
   D. J. Beech

TRPC6 125
   A. Dietrich, T. Gudermann

TRPC7 143
   T. Numaga, M. Wakamori, Y. Mori

Part II. TRPV Channel Subfamily

Capsaicin Receptor: TRPV1A Promiscuous TRP Channel 155
   S. C. Pingle, J. A. Matta, G. P. Ahern

2-Aminoethoxydiphenyl Borate as a Common Activator
of TRPV1, TRPV2, and TRPV3 Channels 173
   C. K. Colton, M. X. Zhu

TRPV4 189
    T. D. Plant, R. Strotmann

TRPV5, the Gateway to Ca^{2+} Homeostasis 207
    A. R. Mensenkamp, J. G. J. Hoenderop, R. J. M. Bindels

TRPV6 221
    U. Wissenbach, B. A. Niemeyer

Part III. TRPM Channel Subfamily

TRPM2 237
    J. Eisfeld, A. Lückhoff

TRPM3 253
    J. Oberwinkler, S. E. Philipp

Insights into TRPM4 Function, Regulation and Physiological Role . . . 269
    R. Vennekens, B. Nilius

TRPM5 and Taste Transduction 287
    E. R. Liman

TRPM6: A Janus-Like Protein 299
    M. Bödding

The Mg^{2+} and Mg^{2+}-Nucleotide-Regulated
Channel-Kinase TRPM7 313
    R. Penner, A. Fleig

TRPM8 329
    T. Voets, G. Owsianik, B. Nilius

Part IV. Other TRP Channels

TRPA1 347
    J. García-Añoveros, K. Nagata

TRPP2 Channel Regulation 363
    R. Witzgall

Part V. TRP Proteins and Specific Cellular Functions

Know Thy Neighbor: A Survey of Diseases and Complex
Syndromes that Map to Chromosomal Regions
Encoding TRP Channels 379
    J. Abramowitz, L. Birnbaumer

List of Contents

TRP Channels of the Pancreatic Beta Cell 409
    D. A. Jacobson, L. H. Philipson

TRP Channels in Platelet Function 425
    K. S. Authi

TRP Channels in Lymphocytes 445
    E. C. Schwarz, M.-J. Wolfs, S. Tonner, A. S. Wenning, A. Quintana,
    D. Griesemer, M. Hoth

Link Between TRPV Channels and Mast Cell Function 457
    H. Turner, K. A. del Carmen, A. Stokes

TRPV Channels' Role in Osmotransduction
and Mechanotransduction 473
    W. Liedtke

Nociception and TRP Channels 489
    M. Tominaga

Part VI. TRP Proteins – Integrators of Multiple Inputs

Regulation of TRP Ion Channels
by Phosphatidylinositol-4,5-Bisphosphate 509
    F. Qin

TRPC, cGMP-Dependent Protein Kinases and Cytosolic Ca^{2+} 527
    X. Yao

Trafficking of TRP Channels: Determinants of Channel Function 541
    I. S. Ambudkar

TRPC Channels: Interacting Proteins 559
    K. Kiselyov, D. M. Shin, J.-Y. Kim, J. P. Yuan, S. Muallem

TRPC Channels: Integrators of Multiple Cellular Signals 575
    J. Soboloff, M. Spassova, T. Hewavitharana, L.-P. He, P. Luncsford,
    W. Xu, K. Venkatachalam, D. van Rossum, R. L. Patterson, D. L. Gill

Phospholipase C-Coupled Receptors and Activation
of TRPC Channels 593
    M. Trebak, L. Lemonnier, J. T. Smyth, G. Vazquez, J. W. Putney Jr.

Subject Index 615

Subject Index

Abramowitz, J. , 379
Ahern, G. P. , 155
Ambudkar, I. S. , 541
Authi, K. S. , 425

Barritt, G. J. , 23
Beech, D. J. , 109
Bindels, R. J. M. , 207
Birnbaumer, L. , 53, 379
Bödding, M. , 299

Carmen, K. A. del , 457
Cavalié, A. , 93
Colton, C. K. , 173

Dietrich, A. , 125

Eder, P. , 77
Eisfeld, J. , 237

Fleig, A. , 313
Flockerzi, V. , 1

García-Añoveros, J. , 347
Gill, D. L. , 575
Griesemer, D. , 445
Groschner, K. , 77
Gudermann, T. , 125

He, L.-P. , 575
Hewavitharana, T. , 575
Hoenderop, J. G. J. , 207
Hoth, M. , 445

Jacobson, D. A. , 409

Kim, J.-Y. , 559
Kiselyov, K. , 559

Lemonnier, L. , 593
Liedtke, W. , 473
Liman, E. R. , 287
Lückhoff, A. , 237
Luncsford, P. , 575

Matta, J. A. , 155
Mensenkamp, A. R. , 207
Mori, Y. , 143
Muallem, S. , 559

Nagata, K. , 347
Niemeyer, B. A. , 221
Nilius, B. , 269, 329
Numaga, T. , 143

Oberwinkler, J. , 253
Owsianik, G. , 329

Patterson, R. L. , 575
Penner, R. , 313
Philipp, S. E. , 253
Philipson, L. H. , 409
Pingle, S. C. , 155
Plant, T. D. , 189
Poteser, M. , 77
Putney, J. W. , 593

Qin, F. , 509
Quintana, A. , 445

Rossum, D. van , 575

Rychkov, G., 23

Schwarz, E. C., 445
Shin, D. M., 559
Smyth, J. T., 593
Soboloff, J., 575
Spassova, M., 575
Stokes, A., 457
Strotmann, R., 189

Tominaga, M., 489
Tonner, S., 445
Trebak, M., 593
Turner, H., 457

Vazquez, G., 593

Venkatachalam, K., 575
Vennekens, R., 269
Voets, T., 329

Wakamori, M., 143
Wenning, A. S., 445
Wissenbach, U., 221
Witzgall, R., 363
Wolfs, M.-J., 445

Xu, W., 575

Yao, X., 527
Yildirim, E., 53
Yuan, J. P., 559

Zhu, M. X., 173

An Introduction on TRP Channels

V. Flockerzi

Experimentelle und Klinische Pharmakologie und Toxikologie,
Universität des Saarlandes, 66421 Homburg, Germany
veit.flockerzi@uniklinikum-saarland.de

1	The TRP Roots	1
2	What Makes a TRP a TRP?	2
3	Biological Functions and Pharmacology	4
4	TRPs and Store-Operated Channels	6
References		15

Abstract The transient receptor potential (TRP) ion channels are named after the role of the channels in *Drosophila* phototransduction. Mammalian TRP channel subunit proteins are encoded by at least 28 genes. TRP cation channels display an extraordinary assortment of selectivities and activation mechanisms, some of which represent previously unrecognized modes of regulating ion channels. In addition, the biological roles of TRP channels appear to be equally diverse and range from roles in thermosensation and pain perception to Ca^{2+} and Mg^{2+} absorption, endothelial permeability, smooth muscle proliferation and gender-specific behaviour.

Keywords TRP channels · TRP pharmacology · TRP gene mutation · Thermo TRPs · Store-operated channel

1
The TRP Roots

Transient receptor potential (TRP) channels constitute a superfamily of cation permeable channels. The founding member of this superfamily was identified as a *Drosophila* gene product required for visual transduction, which in the fruit fly is a phospholipase C-dependent process (Hardie and Minke 1993). The name *transient receptor potential* is based on the transient rather than sustained response to light of the flies carrying a mutant in the *trp* locus (Montell and Rubin 1989). Furthermore, *trp* mutants display a defect in light-induced Ca^{2+} influx. The predicted structure of TRP and the related protein, TRPL, and a variety of in vitro expression studies supported the proposal that TRP and TRPL (Minke and Parnas 2006) were cation influx channels. In addition, the studies of *Drosophila* phototransduction indicated that TRP is

Table 1 The seven subfamilies of TRP genes. In yeast (*Saccharomyces cerevisiae*) a single TRP-related gene has been identified (*yvc1*), which cannot be assigned to any of the subfamilies. The TRPC2-gene appears to be a pseudogene in humans; this is the reason why the human TRPC-family comprises six members, the mouse family seven members (TRPC1 through TRPC7). The three TRPP genes in humans and mice comprise TRPP2 (PC2, PKD2), TRPP3 (PKD2L1, polycystin L, PCL) and TRPP5 (PKD2L2). The much larger polycystin-1 (PC1, PKD1, sometimes called TRPP1), polycystin-REJ, polycystin-1L1 and polycystin-1L2 proteins are 11-transmembrane proteins that contain a C-terminal 6-transmembrane TRP-like channel domain; they are not explicitly included in the TRP channel family. The genes indicated have been identified in the genomes of *Drosophila melanogaster* (fly), *Caenorhabditis elegans* (worm), *Ciona intestinalis* (sea squirt) and *Fugu rubripes* (fish). Different from *Fugu*, two members of TRPA and one of TRPN have been identified in *Danio rerio* (zebrafish)

TRP subfamilies	Fly	Worm	Sea squirt	Fish	Mouse	Human
TRPC	3	3	8	8	7	6
TRPV	2	5	2	4	6	6
TRPM	1	4	2	6	8	8
TRPA	4	2	4	1	1	1
TRPN	1	1	1	-	-	-
TRPP	4	1	9	2	3	3
TRPML	1	1	1	4	3	3
Total	16	17	27	25	28	27

permeable to Ca^{2+} and is the target of a phosphoinositide cascade, leading to the suggestion that phototransduction in *Drosophila* might be analogous to the general and widespread process of phosphoinositide-mediated Ca^{2+} influx in other cells (Hardie 2003).

On the basis of similarity to *Drosophila* TRP and TRPL protein sequences, the complementary DNA (cDNA) of 28 mammalian TRP related proteins has been cloned in recent years using database searches of expressed sequence tags (EST), RT-PCR or expression-cloning strategies. As it turns out, there are not only mammalian homologues of TRP, but a panoply of TRP-related channels are conserved in flies, worms, fish and tunicates (Table 1); a single TRP-related gene, *yvc1*, has been identified in yeast, but no TRP-related genes have been found in plants so far.

2
What Makes a TRP a TRP?

The crystal structures of TRP proteins are not yet available, but all TRP protein sequences comprise at least six predicted transmembrane domains (6TM) and thereby resemble voltage-gated K^+ channels in overall transmembrane

An Introduction on TRP Channels

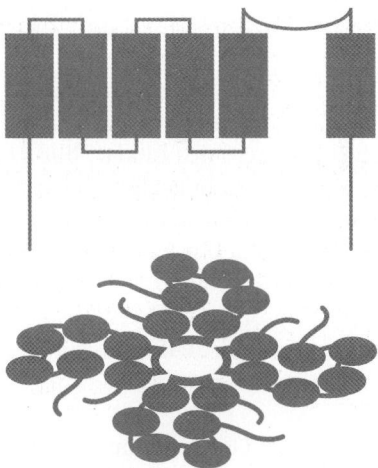

Fig. 1 A TRP protein contains six domains, predicted to cross the cell membrane, and a pore loop within the extracellular linker separating the fifth and sixth transmembrane domains (*top*). Four TRP proteins are assumed to form homo-oligomeric and hetero-oligomeric channels (*bottom*)

architecture (Fig. 1). But despite these topographic similarities, TRPs show limited conservation of the S4 positive charges and P loop sequences that are the hallmarks of the voltage-gated channel families (Clapham et al. 2005); accordingly, TRPs are only distantly related to these channels. Voltage-gated K^+ channels are tetramers, and a similar tetrameric architecture is assumed for TRP channels. Although it may be a good guess, direct biochemical evidence for such an architecture of TRP channels in native tissues is scarce (Kedei et al. 2001; Hoenderop et al. 2003a; Erler et al. 2004).

Based on their amino acid sequence similarities, the TRP-related proteins fall into seven subfamilies (Table 1): the classical TRPs (TRPCs), which display greatest similarity to *Drosophila* TRP; the vanilloid receptor TRPs (TRPVs); the melastatin TRPs (TRPMs); the mucolipins (TRPMLs); the polycystins (TRPPs); and ankyrin transmembrane protein 1 (TRPA1). The seventh subfamily, TRPN, comprises proteins in tunicates, flies and worms; members of this subfamily have not been identified yet in mammals (Table 1). The degree of amino acid sequence similarity among members of one subfamily approaches up to more than 90%, but a corresponding similarity is hardly detectable among members of different subfamilies. Accordingly, only members of one subfamily may account for heteromeric TRP channels in vivo—that is, in native tissues and primary cells. But it is still unclear at present whether functional channels require heteromultimerization with other TRPs or whether TRPs can only form homomeric ion channels.

At present, the overall 6TM architecture/homology and the cationic permeability are the major features which define proteins of the TRP family. Whereas all TRP channels are permeable for cations, only two TRP channels are impermeable for Ca^{2+} (TRPM4, TRPM5), and two others are highly Ca^{2+} permeable (TRPV5, TRPV6). By showing that an aspartate-to-alanine mutation at position D542 of TRPV5 abolishes Ca^{2+} permeation whereas permeation of monovalent cations basically remains intact, Nilius and colleagues demonstrate directly that the TRPV5 is responsible for channel activity and that the TM5–TM6 linker is part of the TRPV5 channel pore (Nilius et al. 2001). Since then similar strategies have been applied to TRPV6, TRPV1, TRPM4, TRPM5 and TRPM8 channels (Owsianik et al. 2006). In the case of TRPM3 it was shown that alternative splicing of its TM5–TM6 linker switches the divalent cation selectivity (Oberwinkler et al. 2005); this finding unequivocally proves that TRPM3 itself forms functional channels and that its TM5–TM6 linker is part of the channel pore.

Other structural features of TRP proteins include ankyrin repeats and coiled-coil regions within the N-terminal and C-terminal cytoplasmatic domains of TRPCs, TRPVs, TRPMs and TRPA1 but not in TRPPs or TRPMLs (Fig. 2). In addition, four short amino acid sequence motifs, M1–M4, located at comparable positions are present in the proteins of some TRP subfamilies but not in those of others. Motif M1 consists of 38 amino acids upstream of TM1 of TRPC1 to TRPC7 (Fig. 3). An alternative M1 motif of 52 to 63 amino acid residues is detectable at the corresponding position of TRPV1 to TRPV6 (Fig. 3). The amino acid sequence motif M2 resides within the cytosolic loop between transmembrane segments 4 and 5 of TRPC1 to TRPC7 and is fairly conserved in the TRPVs and TRPMs. Motif M3 (consensus sequence ILLLNM-LIAMM) is part of TM6 of all TRPCs, TRPVs and TRPMs (Fig. 3). Finally, the highly conserved 26-amino-acid motif M4, the TRP domain (Fig. 2) containing a "TRP-box" (WKFQR) follows immediately transmembrane segment 6 (Fig. 3) and is located within the C-terminal cytosolic tail of most TRPs. It is followed by proline-rich sequences in some TRPs. These motifs do not resemble any other amino acid sequence of known function and might represent typical features of TRP proteins. For example, the cluster of positive charged residues in motif 4 (the TRP domain) has been implicated to determine the modulation of TRPM8 channel activity (Rohacs et al. 2005) by phosphatidylinositol 4,5-bisphosphate (PIP_2), although other groups showed that neutralization of critical residues in this domain did not modulate or only weakly modulated TRPM4 activation by PIP_2 (Zhang et al. 2005b; Nilius et al. 2006).

3
Biological Functions and Pharmacology

Of paramount importance is the identification of the biological functions of this heterogeneous family of proteins. A selective TRP channel pharmacology

An Introduction on TRP Channels

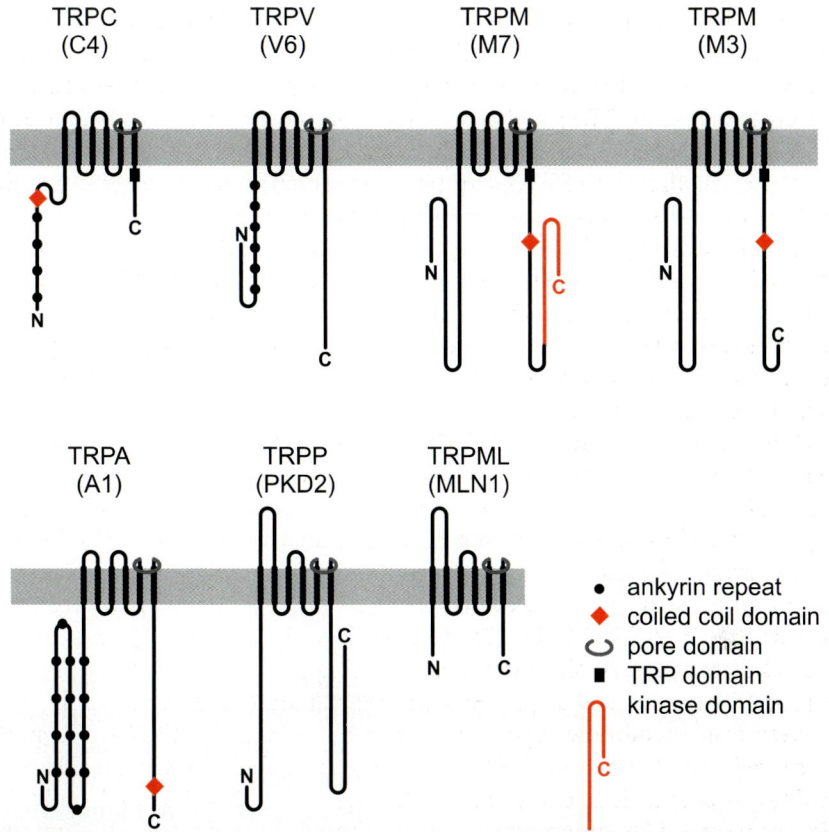

Fig. 2 The seven mammalian TRP subfamilies. Representatives of the subfamilies are indicated in *brackets*. Several protein domains are indicated: ankyrin repeats, coiled-coil domain, protein kinase domain (TRPM6/7 only) and the TRP domain. Ankyrin repeats are common modular protein interaction motifs. Coiled-coil structures are 2- to 5-stranded bundles of α-helices which are stabilized by hydrophobic and other interactions; they are widely used to connect different or like subunits in oligomeric proteins

to unravel roles of TRP channels does not yet exist. The best compounds, both blocking and activating, are available for TRPV1 and TRPM8 (Table 2). Both channels are prototypes of the thermoTRPs, a subset of TRPs which are activated by distinct temperatures and are involved in converting thermal information into chemical and electrical signals (Table 2). In addition, genetic approaches in worms, flies and mice (Table 3) demonstrate a role for many TRP proteins in an assortment of sensory processes ranging from thermosensation to osmosensation, olfaction, taste, mechanosensation, vision and pain perception. Several recent studies have begun to unravel roles for TRP proteins in non-excitable cells, including a requirement in endothelium-dependent vasorelaxation (Freichel et al. 2001; Tiruppathi et al. 2002) and smooth muscle-dependent contractility (Dietrich et al. 2005), as indicated by analyses of TRPC4 and TRPC6 knock-out mice (Table 3). Finally, to date, mutations in five different TRPs have been linked to human disease (Table 4). Studies in *Caenorhabditis elegans* indicate that at least one TRPM member may have role in cell-cycle control (West et al. 2001). Moreover, TRPM1 has been suggested to be a tumour suppressor (Duncan et al. 1998), and a decrease in expression of TRPM1 appears to be a prognostic marker for metastasis in patients with localized malignant melanoma (Duncan et al. 2001). In addition, TRPC6 was reported to be down-regulated in a murine autocrine tumour model (Buess et al. 1999), whereas expression of TRPM8 (Tsavaler et al. 2001) and TRPV6 (Wissenbach et al. 2001) seems to be up-regulated in prostate cancers and may represent new markers for such cancer (Wissenbach et al. 2004).

4
TRPs and Store-Operated Channels

Many cells including exocrine gland cells and vascular smooth muscle cells show a co-ordinated regulation of an intracellular Ca^{2+} release mechanism and the entry of Ca^{2+} across the plasma membrane. This Ca^{2+} entry is the basis for sustained $[Ca^{2+}]_i$ elevations, which are important for various cellular functions including gene expression, secretion and cell proliferation. A link between Ca^{2+} release and Ca^{2+} entry into cells was proposed by Putney (1977). His model described an initial emptying of the intracellular Ca^{2+} stores by inositol 1,4,5-trisphosphate (IP_3), followed by entry of Ca^{2+} into the cytosol and refilling of the stores. This type of Ca^{2+} entry is referred to as capacitative or store-operated Ca^{2+} entry. The first electrical measurement of current through

Fig. 3 Positions of amino acid sequence motifs M1, M2, M3 and M4 relative to TM1–TM6 (*bottom right*) and sequence alignments of amino acid sequence motifs M1, M2, M3 and M4 of the mouse TRPCs, TRPVs and TRPMs. Identical amino acid residues (*ident*) and conservative substitutions are indicated by *bold letters*. Conservative substitutions are defined as pairs of residues belonging to one of the following groups: S, T, P, A and G; N, D, E and Q; H, R and K; M, I, L and V; F, Y and W (Dayhoff et al. 1981). GenBank accession numbers: *TRPC1*, NM 011643; *TRPC2*, NM 011644; *TRPC3*, NM 019510; *TRPC4*, NM 016984; *TRPC5*, NM 009428; *TRPC6*, NM 013838; *TRPC7*, NM 012035; *TRPV1*, NM 001001445; *TRPV2*, BC005415; *TRPV3*, NM 145099; *TRPV4*, NM 022017; *TRPV5*, XM 149866, *TRPV6*, AJ542487; *TRPM1*, NM 018752; *TRPM2*, NM 138301, TRPM3, AJ 544534; *TRPM4*, AJ575814; *TRPM5*, CAE05940; *TRPM6*, NM 153417; *TRPM7*, NM 021450; *TRPM8*, NM 134252

An Introduction on TRP Channels

M1

```
mTRPC4  DLARLKLAIKYRQKEFVAQPNCQQLLASRWYDEFPGWR  322
mTRPC5  DLAKLKVAIKYHQKEFVAQPNCQQLLATLWYDGFPGWR  323
mTRPC1  NLSRLKLAIKYNQKEFVSQSNCQQFLNTVWFGQMSGYR  359
mTRPC3  SLSRVKLAIKYEVKKFVAHPNCQQQLLTIWYENLSGLR  330
mTRPC7   LSRIKLAIKYEVKKFVAHPNCQQQLLTMWYENLSGLR  344
mTRPC6  NLSRLKLAIKDEVKKFVAHPNCQQQLLSIWYENLSGLR  398
mTRPC2  NLARLRLAVNYNQKQFVAHPICQQVLSSIWCGNLAGWR  621
ident        L    A   KFV  CQQ  L    W     GR
```

```
mTRPV1  EELTNKKGLTPLALAASSGKIGVLAYILQREIHEPECRHLSRKFTEWAYGPVHSSLYDLSCID  389
mTRPV3  ETMRNNDGLTPLQLAAKMGKAEILKYILSREIKEKPLRSLSRKFTDWAYGPVSSLYDLTNVD  396
mTRPV4  ETVLNNDGLSPLMMAAKTGKIGVFQHIIRREVTDEDTRHLSRKFKDWAYGPVYSSLYDLSSLD  425
mTRPV2  EDICNHQGLTPLKLAAKEGKIEIFRHILQREFSG-LYQPLSRKFTEWCYGPVRVSLYDLSSVD  344
mTRPV5  ELVPNNQGLTPFKLAGVEGNTVMFQHLMQ----------KRKRIQWSFGPLTSSLYDLTEID  277
mTRPV6  ELVPNNQGLTPFKLAGVEGNIVMFQHLMQ----------KRKHIQWTYGPLTSTLYDLTEID  283
ident      E   NGL PA  G                        RK  W  GP    LYDL  D
```

M2

```
mTRPV1  MGIYAVMIEKMILRDLCRF   581
mTRPV2  TGIYSVMIQKVILRDLLRF   535
mTRPV4  TGTYSIMIQKILFKDLFRF   617
mTRPV3  MGMYSVMIQKVILHDVLKF   590
mTRPV5  LGPFTIMIQKMIFGDLLRF   486
mTRPV6  LGPFTIMIQKMILLIAMM    492
mTRPM2  LGPKIIIVKRMMK-DVFFF   935
mTRPM8  LGPKIIMLQRMLI-DVFFF   870
mTRPM4  LGPKIVIVSKMMK-DVFFF   929
mTRPM5  LGPKIVIVSKMMK-DVFFF   929
mTRPM1  LGPYVMMIGKMMI-DMLYF   898
mTRPM3  LGPYVMMIGKMMI-DMMYF  1013
mTRPM6  AGPYVTMIAKMAA-NMFYI   986
mTRPM7  AGPYVMMIGKMVA-NMFYI  1003
mTRPC4  LGPLQISLGRMLL-DILKF   519
mTRPC5  LGPLQISLGRMLL-DILKF   520
mTRPC1  LGPLQISMGQMLQ-DFGKF   556
mTRPC3  FGPLQISLGRTVK-DIFKF   568
mTRPC7  FGPLQISLGRTVK-DIFKF   582
mTRPC6  FGPLQISLGRTVK-DIFKF   636
mTRPC2  LGTLQISIGKMID-DMIRF   834
ident       G
```

M3

```
mTRPV1  ILLLNMLIALM   683
mTRPV2  VLLLNMLIALM   640
mTRPV4  VLLLNMLIALM   718
mTRPV3  VLLLNMLIALM   677
mTRPV5  LLMLNLFIAMM   571
mTRPV6  LLMLNLLIAMM   577
mTRPM2  ILLLNLLIAMF  1046
mTRPM8  ILLVNLLVAMF   979
mTRPM4  ILLLNLLIAMF  1039
mTRPM5  ILLLNLLIAMF  1038
mTRPM1  ILLVNLLIAVF  1004
mTRPM3  ILLVNLLIAVF  1126
mTRPM6  IIMVNLLIACF  1075
mTRPM7  IIMVNLLIAFF  1096
mTRPC4  VVLLNMLIAMM   620
mTRPC5  VVLLNMLIAMM   624
mTRPC1  IVLTKLLVAML   661
mTRPC3  VVLLNMLIAMI   658
mTRPC7  VVLLNMLIAMI   672
mTRPC6  IVLLNMLIAMI   726
mTRPC2  IVLLNMLIAMI   920
ident        A
```

M4

```
mTRPV1  KNIWKLQRAITILDTEKSFLKCMRKA     720
mTRPV2  WSIWKLQKAISVLEMENGYWWCRRKR     677
mTRPV4  KHIWKLQWATTILDIERSFPVFLRKA     755
mTRPV3  ERIWRLQRARTILEFEKMLPEWLRSR     714
mTRPV5  DELWRAQVVATTVMLERKMPRFLWPR     608
mTRPV6  DELWRAQVVATTVMLERKLPRCLWPR     614
mTRPM2  DQIWKFQRHDLIEEYHGRP-PAPPPL    1083
mTRPM8  DQVWKFQRYFLVQEYCNRL-NIPFPF    1015
mTRPM4  DLYWKAQRYSLIREFHSRP-ALAPPL    1075
mTRPM5  DLYWKAQRYSLIREFHSRP-ALAPPL    1075
mTRPM1  NQVWKFQRYQLIMTFHDRP-VLPPPM    1040
mTRPM3  NQVWKFQRYQLIMTFHERP-VLPPPL    1162
mTRPM6  NKLWKYNRYRYIMTYHQKP-WLPPPF    1111
mTRPM7  NIVWKYQRYHFIMAYHEKP-VLPPPL    1132
mTRPC4  DIEWKFARTKLWMSYFEEGGTLPTPF     657
mTRPC5  DIEWKFARTKLWMSYFDEGGTLPPPF     661
mTRPC1  DKEWKFARAKLWLSYFDDKCTLPPPF     698
mTRPC3  DVEWKFARSKLWLSYFDDGKTLPPPF     695
mTRPC7  DVEWKFARAKLWLSYFDEGRTLPAPF     709
```

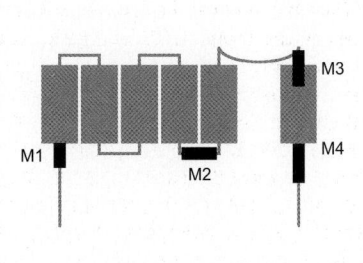

Table 2 ThermoTRPs: mammalian thermal-sensitive channels and their pharmacology

Channel	Threshold [°C]	Agonists	Antagonists

TRPV1	≥ 43	Vanilloid compounds
		  Capsaicin (EC$_{50}$, 0.7 µM)
		  Resiniferatoxin
		  Olvanil
		  Piperine
		Endocannabinoid lipids
		  Anandamide
		  Arachidonylethanolamide (AEA)
		  2 Arachidonoyl-glycerol (2-AG)
		Eicosanoids
		  12-(S)-HPETE
		  15-(S)-HPETE
		  5-(S)-HETE
		  Leukotriene B4
		Extracellular protons
		2-APB
		Camphor
		Eugenol
		Allicin
			Ruthenium red;
			Capsazepine

TRPV2	≥ 52	IGF-I 2-APB Carvacrol	Ruthenium red; La^{3+}; SKF96365
TRPV3	$\geq 30/\leq 39$	Camphor 2-APB Carvacrol Thyme Eugenol	Ruthenium red
TRPV4	$\geq 24/\leq 32$	4-α Phorbol Eicosanoid 5',6'epoxyeicosatrienoic acid (5',6'-EET)	Ruthenium red; Gadolinium; La^{3+}
TRPM4	$\geq 15/\leq 37$	[Ca]$_{ic}$	Intracellular free ATP^{4-} (1.7 µM); Spermine
TRPM5		[Ca]$_{ic}$	

Table 2 continued

Channel	Threshold [°C]	Agonists	Antagonists
TRPM8	23–28	Menthol (EC$_{50}$, 70 µM)	Capsazepine
		Icilin (EC$_{50}$, 360 nM)	
		Linalool	
		Geraniol	
		Eucalyptol	
		Hydroxycitronellal	
		PIP$_2$	
TRPA1[b]	17	Cinnamaldehyde[b]	Ruthenium red (5 µM)
		Allyl isothiocyanate[a]	
		Garlic (allicin)	
		Δ^9-Tetrahydrocannabinol	
		Icilin (EC$_{50}$, 100 µM)	

[a]Isothiocyanate is the pungent component of mustard, horseradish and wasabi
[b]The issue of whether noxious cold activates TRPA1 is not without controversy (compare Table 3)

a store-operated channel (SOC) was achieved in mast cells and this current was termed Ca^{2+} release-activated Ca^{2+} channel (CRAC) current (Hoth and Penner 1992). The physiological hallmark of the underlying channel is a high selectivity for Ca^{2+} over other cations. The desire to identify the molecular nature of this SOC was and still is the motivation for many workers to characterize mammalian TRP channels. However, the pore properties of most TRP channels that have been studied in detail, including TRPV6 and TRPV5, appear not to match the pore properties of CRAC channels (Owsianik et al. 2006). Although it may turn out that one ore more of the TRP proteins participate in store-operated Ca^{2+} entry, the focus has been switched on new players in this field, the stromal interaction molecule (STIM) (Liou et al. 2005; Roos et al. 2005; Zhang et al. 2005a; Spassova et al. 2006) and Orai or CRACM (Feske et al. 2006; Vig et al. 2006; Yeromin et al. 2006; Prakriya et al. 2006), which appear to be essential components and regulators of SOC/CRAC channel complexes (Fig. 4).

In this volume we try to give a concise survey on the members of the TRP subfamilies TRPC, TRPV, TRPM and TRPA; in addition, their integration in physiological systems is described. We are still at the very beginning of understanding their impact on very diverse cell and organ functions and their roles in disease. TRP pharmacology is still not at an advanced stage—with the exceptions indicated in Table 2—but needless to say, one of the main challenges in this field will be to define in more detail physiological TRP channel functions and properties of TRP channels as targets for novel drugs.

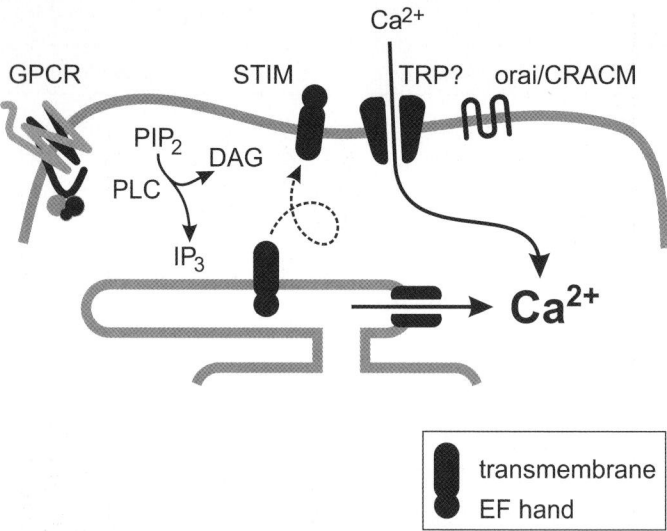

Fig. 4 STIM, Orai/CRACM and SOC activation. Activation begins with an agonist binding to a plasma membrane receptor (*GPCR*) coupled to phospholipase C (*PLC*) through a G protein mechanism. PLC activation leads to breakdown of phosphatidylinositol 4,5-bisphosphate (*PIP$_2$*) and production of diacylglycerol (DAG) and inositol 1,4,5-trisphosphate (*IP$_3$*). IP$_3$ in turn activates the IP$_3$ receptor, causing release of Ca^{2+} from an intracellular Ca^{2+}-containing compartment which is part of the endoplasmic reticulum (ER). The fall in ER Ca^{2+} then signals to plasma membrane store-operated channels through a mechanism that involves STIM1 (*STIM*) in the ER, plasma membrane or both. The structure of STIM includes a single transmembrane domain and a NH$_2$-terminal single EF hand which would face the lumen of the ER and may be involved in sensing ER Ca^{2+} levels; upon Ca^{2+} store depletion it may translocate to the plasma membrane to activate store-operated channels, especially Ca^{2+} release-activated calcium (CRAC) channels. Orai/CRACM, a 4TM plasma membrane protein, could be the essential subunit of the channel or a component of the machinery required to activate the channel. Most TRP channels are not gated by the usual processes defined as activating store-operated Ca^{2+} entry nor do they reveal the pore properties of CRAC channels. Nevertheless it may turn out that one or more of these channels participate in this process

Table 3 Biological relevance of TRP channels identified in transgenic mouse models or in mice with spontaneous mutations in TRP genes: chromosomal location of mouse TRP protein genes, mutation and described phenotype

Gene	Chromosomal locus	Type of mutation	Phenotype	Reference(s)
TRPC1	9 E4		n.r.	
TRPC2	7 F1	Targeting (exons 8–11)	Defects in pheromone perception, sex discrimination and gender-specific social behaviour, lack of male-male aggression, TRPC2$^{-/-}$ males initiate sexual and courtship behaviours towards both males and females	Stowers et al. 2002
		Targeting (exons 6–11)	Defects in pheromone perception and pheromone-evoked male-male aggression, attenuated aggressive behaviour in lactating females, increased sexual behaviour towards other males	Leypold et al. 2002
TRPC3	3 B		n.r.	
TRPC4	3 D		Impaired endothelial-dependent vasorelaxation, endothelial permeability, 5-HT induced GABA-release from thalamic interneurons depends on TRPC4	Freichel et al. 2001; Tiruppathi et al. 2002; Munsch et al. 2003
TRPC5	X F2		n.r.	
TRPC6	9 A1		Increased agonist-induced and smooth muscle-dependent contractility of tracheal and vessel rings, elevated blood pressure	Dietrich et al. 2003; Dietrich et al. 2005
TRPC7	13 B2		n.r.	
TRPV1	11 B4	Targeting (exon 13)	Decreased sensitivity to vanilloids, protons, mild noxious heat; impaired thermal hyperalgesia in the setting of tissue injury; no effects on nerve injury-induced thermal hyperalgesia	Caterina et al. 2000
			Diminished stretch-evoked and hypo-osmolarity-induced ATP release and impaired bladder contraction	Birder et al. 2002
		Targeting (exon 10 and 11)	Lack of carrageenan-induced thermal hyperalgesia, decreased sensitivity to vanilloids, protons, mild noxious heat	Davis et al. 2000

Table 3 continued

Gene	Chromosomal locus	Type of mutation	Phenotype	Reference(s)
TRPV2	11 B2	Transgenic (cardiac-α-MHC promoter)	Cardiomyopathy: cardiac hypertrophy affecting all four chambers, with disorganized myocyte arrangement and interstitial fibrosis	Iwata et al. 2003
TRPV3	11 B4	Targeting (exons 14 and 15)	Impaired thermosensation: defective responses to innocuous and noxious heat but not in other sensory modalities; decreased sensitivity to camphor in keratinocytes	Moqrich et al. 2005
TRPV4	5 F	Targeting (exon 4)	Impaired osmotic sensation	Mizuno et al. 2003
			Impaired pressure sensitivity and acidic nociception	Suzuki et al. 2003
			Impaired thermal hyperalgesia in the setting of tissue injury	Todaka et al. 2004
			Impaired thermal hyperalgesia under injury-free conditions	Lee et al. 2005
			Delayed-onset hearing loss	Tabuchi et al. 2005
		Targeting (exon 12)	Impaired osmotic sensation	Liedtke and Friedman 2003
TRPV5	6 B2	Targeting (exon 13)	Diminished active Ca^{2+} re-absorption causing severe hypercalciuria and compensatory hyperabsorption of dietary Ca^{2+}, reduced trabecular and cortical bone thickness	Hoenderop et al. 2003b
			Disrupted osteoclastic bone resorption	van der Eerden et al. 2005
TRPV6	6 B2		n.r.	
TRPM1	7 C		n.r.	
TRPM2	10 C1		n.r.	
TRPM3	19 C1		n.r.	
TRPM4	7 B4	Targeting (exons 15 and 16)	Increased IgE-mediated Ca^{2+} entry due to lack of Ca^{2+} activated non-selective cation channels in mast cells	Vennekens et al. 2006

Table 3 continued

Gene	Chromosomal locus	Type of mutation	Phenotype	Reference(s)
TRPM5	7 F5	Targeting (exons 15–19)	Abolished sweet, amino acid, and bitter taste reception (no impact on sour or salty tastes)	Zhang et al. 2003
		Targeting (promoter and exons 1–4)	Bitter taste response: reduced avoidance, normal (NG) chorda tympani (CT), and diminished glossopharyngeal response Sweet taste response: no licking response, diminished preference ratio, reduced nerve responses; Umami taste response: no licking response, diminished preference ratio, normal NG response, and diminished CT response	Damak et al. 2006
			Increasing temperature (15–35°C) enhanced the gustatory nerve response to sweet compounds in wild-type but not in knock-out mice	Talavera et al. 2005
TRPM6	19 B		n.r.	
TRPM7	2 F2		n.r.	
TRPM8	1 C5		n.r.	
TRPA1	1 A3	Targeting (exon 23)	Impaired mustard oil- and garlic-induced activation of primary afferent nociceptors and subsequent production of inflammatory pain; deficits in bradykinin-evoked nociceptor excitation and pain hypersensitivity, normal cold sensitivity and unimpaired auditory function	Bautista et al. 2006
		Targeting (exon 21–25)	Behavioural deficits in response to mustard oil, to cold (0°C), and to punctate mechanical stimuli Hair-cell transduction not impaired	Kwan et al. 2006

Table 3 continued

Gene	Chromosomal locus	Type of mutation	Phenotype	Reference(s)
TRPP2 (PKD2)	5 E4	Targeting (exon 1)	Polycystic kidney and liver lesions	Wu et al. 1998
			Embryonic lethality between E13.5 and parturition; defective cardiac septation	Wu et al. 2000
		Targeting (exon 1)	Cyst formation in nephrons and pancreatic ducts in homozygous embryos and heterozygous adult mice	
			Embryonic lethality between E12.5 and parturition; right pulmonary isomerism, randomization of embryonic turning, heart looping, and abdominal situs	Pennekamp et al. 2002
TRPP3 (PKD2L1)	19 D1	Spontaneous mutation ("Krd")	Hypoplastic and cystic kidneys	Nomura et al. 1998
TRPP5 (PKD2L2)	18 B3		n.r.	
TRPML1	8 A1.1		n.r.	
TRPML2	3 H3		n.r.	
TRPML3	3 H3	Spontaneous mutation ("varitint-waddler")	Early-onset hearing loss, vestibular defects, pigmentation abnormalities and perinatal lethality	Di Palma et al. 2002

n.r., Not reported

Table 4 TRP to disease: summary of TRP genes linked to human diseases and of TRP gene expression linked to human cancers

Association with disease	TRP channel gene	Chromosomal locus (human)	Disease
Mutations linked to human disease	TRPC6	11q21–22	Focal segmental glomerular sclerosis
	TRPM6	9q21.13	Hypomagnesia with secondary hypocalcemia
	TRPM7	15q21	Guamanian amyotrophic lateral sclerosis and parkinsonism dementia
	TRPML1	19p13.3–13.2	Mucolipidosis type IV
	TRPP2 (PKD2, PC2)	4q21–23	Autosomal dominant polycystic kidney disease
Correlation of disease with abnormal expression of TRP genes	TRPV6	7q33–34	Up-regulated in prostate cancer
	TRPM1	15q13-q14	Down-regulated in malignant melanomas
	TRPM8	2q37.1	Up-regulated in cancers of the prostate, breast, colon, lung, skin

References

Bautista DM, Jordt SE, Nikai T, Tsuruda PR, Read AJ, Poblete J, Yamoah EN, Basbaum AI, Julius D (2006) TRPA1 mediates the inflammatory actions of environmental irritants and proalgesic agents. Cell 124:1269–1282

Birder LA, Nakamura Y, Kiss S, Nealen ML, Barrick S, Kanai AJ, Wang E, Ruiz G, De Groat WC, Apodaca G, Watkins S, Caterina MJ (2002) Altered urinary bladder function in mice lacking the vanilloid receptor TRPV1. Nat Neurosci 5:856–860

Buess M, Engler O, Hirsch HH, Moroni C (1999) Search for oncogenic regulators in an autocrine tumor model using differential display PCR: identification of novel candidate genes including the calcium channel mtrp6. Oncogene 18:1487–1494

Caterina MJ, Leffler A, Malmberg AB, Martin WJ, Trafton J, Petersen-Zeitz KR, Koltzenburg M, Basbaum AI, Julius D (2000) Impaired nociception and pain sensation in mice lacking the capsaicin receptor. Science 288:306–313

Clapham DE, Julius D, Montell C, Schultz G (2005) International Union of Pharmacology. XLIX. Nomenclature and structure-function relationships of transient receptor potential channels. Pharmacol Rev 57:427–450

Damak S, Rong M, Yasumatsu K, Kokrashvili Z, Perez CA, Shigemura N, Yoshida R, Mosinger B Jr, Glendinning JI, Ninomiya Y, Margolskee RF (2006) Trpm5 null mice respond to bitter, sweet, and umami compounds. Chem Senses 31:253–264

Davis JB, Gray J, Gunthorpe MJ, Hatcher JP, Davey PT, Overend P, Harries MH, Latcham J, Clapham C, Atkinson K, Hughes SA, Rance K, Grau E, Harper AJ, Pugh PL, Rogers DC, Bingham S, Randall A, Sheardown SA (2000) Vanilloid receptor-1 is essential for inflammatory thermal hyperalgesia. Nature 405:183–187

Dayhoff MO, Schwartz RM, Chen HR, Hunt LT, Barker WC, Orcutt BC (1981) Data bank. Nature 290:8

Di Palma F, Belyantseva IA, Kim HJ, Vogt TF, Kachar B, Noben-Trauth K (2002) Mutations in Mcoln3 associated with deafness and pigmentation defects in varitint-waddler (Va) mice. Proc Natl Acad Sci U S A 99:14994–14999

Dietrich A, Gollasch M, Chubanov V, Mederos y Schnitzler M, Dubrovska G, Herz U, Renz H, Gudermann T, Birnbaumer L (2003) Studies on TRPC6 deficient mice reveal its non-redundant role in the regulation of smooth muscle tone. Naunyn Schmiedebergs Arch Pharmacol 367 (Suppl 1):R95

Dietrich A, Mederos YS, Gollasch M, Gross V, Storch U, Dubrovska G, Obst M, Yildirim E, Salanova B, Kalwa H, Essin K, Pinkenburg O, Luft FC, Gudermann T, Birnbaumer L (2005) Increased vascular smooth muscle contractility in TRPC6$^{-/-}$ mice. Mol Cell Biol 25:6980–6989

Duncan LM, Deeds J, Hunter J, Shao J, Holmgren LM, Woolf EA, Tepper RI, Shyjan AW (1998) Down-regulation of the novel gene melastatin correlates with potential for melanoma metastasis. Cancer Res 58:1515–1520

Duncan LM, Deeds J, Cronin FE, Donovan M, Sober AJ, Kauffman M, McCarthy JJ (2001) Melastatin expression and prognosis in cutaneous malignant melanoma. J Clin Oncol 19:568–576

Erler I, Hirnet D, Wissenbach U, Flockerzi V, Niemeyer BA (2004) Ca^{2+}-selective transient receptor potential V channel architecture and function require a specific ankyrin repeat. J Biol Chem 279:34456–34463

Feske S, Gwack Y, Prakriya M, Srikanth S, Puppel SH, Tanasa B, Hogan PG, Lewis RS, Daly M, Rao A (2006) A mutation in Orai1 causes immune deficiency by abrogating CRAC channel function. Nature 441:179–185

Freichel M, Suh SH, Pfeifer A, Schweig U, Trost C, Weissgerber P, Biel M, Philipp S, Freise D, Droogmans G, Hofmann F, Flockerzi V, Nilius B (2001) Lack of an endothelial store-operated Ca^{2+} current impairs agonist-dependent vasorelaxation in TRP4$^{-/-}$ mice. Nat Cell Biol 3:121–127

Hardie RC (2003) Regulation of TRP channels via lipid second messengers. Annu Rev Physiol 65:735–759

Hardie RC, Minke B (1993) Novel Ca^{2+} channels underlying transduction in Drosophila photoreceptors: implications for phosphoinositide-mediated Ca^{2+} mobilization. Trends Neurosci 16:371–376

Hoenderop JG, Voets T, Hoefs S, Weidema F, Prenen J, Nilius B, Bindels RJ (2003a) Homo- and heterotetrameric architecture of the epithelial Ca^{2+} channels TRPV5 and TRPV6. EMBO J 22:776–785

Hoenderop JG, van Leeuwen JP, van der Eerden BC, Kersten FF, van der Kemp AW, Merillat AM, Waarsing JH, Rossier BC, Vallon V, Hummler E, Bindels RJ (2003b) Renal Ca^{2+} wasting, hyperabsorption, and reduced bone thickness in mice lacking TRPV5. J Clin Invest 112:1906–1914

Hoth M, Penner R (1992) Depletion of intracellular calcium stores activates a calcium current in mast cells. Nature 355:353–356

Iwata Y, Katanosaka Y, Arai Y, Komamura K, Miyatake K, Shigekawa M (2003) A novel mechanism of myocyte degeneration involving the Ca^{2+}-permeable growth factor-regulated channel. J Cell Biol 161:957–967

Kedei N, Szabo T, Lile JD, Treanor JJ, Olah Z, Iadarola MJ, Blumberg PM (2001) Analysis of the native quaternary structure of vanilloid receptor 1. J Biol Chem 276:28613–28619

Kwan KY, Allchorne AJ, Vollrath MA, Christensen AP, Zhang DS, Woolf CJ, Corey DP (2006) TRPA1 contributes to cold, mechanical, and chemical nociception but is not essential for hair-cell transduction. Neuron 50:277–289

Lee H, Iida T, Mizuno A, Suzuki M, Caterina MJ (2005) Altered thermal selection behavior in mice lacking transient receptor potential vanilloid 4. J Neurosci 25:1304–1310

Leypold BG, Yu CR, Leinders-Zufall T, Kim MM, Zufall F, Axel R (2002) Altered sexual and social behaviors in trp2 mutant mice. Proc Natl Acad Sci U S A 99:6376–6381

Liedtke W, Friedman JM (2003) Abnormal osmotic regulation in trpv4−/− mice. Proc Natl Acad Sci U S A 100:13698–13703

Liou J, Kim ML, Heo WD, Jones JT, Myers JW, Ferrell JE Jr, Meyer T (2005) STIM is a Ca^{2+} sensor essential for Ca^{2+}-store-depletion-triggered Ca^{2+} influx. Curr Biol 15:1235–1241

Minke B, Parnas M (2006) Insights on trp channels from in vivo studies in Drosophila. Annu Rev Physiol 68:649–684

Mizuno A, Matsumoto N, Imai M, Suzuki M (2003) Impaired osmotic sensation in mice lacking TRPV4. Am J Physiol Cell Physiol 285:C96–101

Montell C, Rubin GM (1989) Molecular characterization of the Drosophila trp locus: a putative integral membrane protein required for phototransduction. Neuron 2:1313–1323

Moqrich A, Hwang SW, Earley TJ, Petrus MJ, Murray AN, Spencer KS, Andahazy M, Story GM, Patapoutian A (2005) Impaired thermosensation in mice lacking TRPV3, a heat and camphor sensor in the skin. Science 307:1468–1472

Munsch T, Freichel M, Flockerzi V, Pape HC (2003) Contribution of transient receptor potential channels to the control of GABA release from dendrites. Proc Natl Acad Sci U S A 100:16065–16070

Nilius B, Vennekens R, Prenen J, Hoenderop JG, Droogmans G, Bindels RJ (2001) The single pore residue Asp542 determines Ca^{2+} permeation and Mg^{2+} block of the epithelial Ca^{2+} channel. J Biol Chem 276:1020–1025

Nilius B, Mahieu F, Prenen J, Janssens A, Owsianik G, Vennekens R, Voets T (2006) The Ca^{2+}-activated cation channel TRPM4 is regulated by phosphatidylinositol 4,5-biphosphate. EMBO J 25:467–478

Nomura H, Turco AE, Pei Y, Kalaydjieva L, Schiavello T, Weremowicz S, Ji W, Morton CC, Meisler M, Reeders ST, Zhou J (1998) Identification of PKDL, a novel polycystic kidney disease 2-like gene whose murine homologue is deleted in mice with kidney and retinal defects. J Biol Chem 273:25967–25973

Oberwinkler J, Lis A, Giehl KM, Flockerzi V, Philipp SE (2005) Alternative splicing switches the divalent cation selectivity of TRPM3 channels. J Biol Chem 280:22540–22548

Owsianik G, Talavera K, Voets T, Nilius B (2006) Permeation and selectivity of trp channels. Annu Rev Physiol 68:685–717

Pennekamp P, Karcher C, Fischer A, Schweickert A, Skryabin B, Horst J, Blum M, Dworniczak B (2002) The ion channel polycystin-2 is required for left-right axis determination in mice. Curr Biol 12:938–943

Prakriya M, Feske S, Gwack Y, Srikanth S, Rao A, Hogan PG (2006) Orai1 is an essential pore subunit of the CRAC channel. Nature 443:230–233

Putney JW Jr (1977) Muscarinic, alpha-adrenergic and peptide receptors regulate the same calcium influx sites in the parotid gland. J Physiol 268:139–149

Rohacs T, Lopes CM, Michailidis I, Logothetis DE (2005) PI(4,5)P2 regulates the activation and desensitization of TRPM8 channels through the TRP domain. Nat Neurosci 8:626–634

Roos J, DiGregorio PJ, Yeromin AV, Ohlsen K, Lioudyno M, Zhang S, Safrina O, Kozak JA, Wagner SL, Cahalan MD, Velicelebi G, Stauderman KA (2005) STIM1, an essential and conserved component of store-operated Ca^{2+} channel function. J Cell Biol 169:435–445

Spassova MA, Soboloff J, He LP, Xu W, Dziadek MA, Gill DL (2006) STIM1 has a plasma membrane role in the activation of store-operated Ca(2+) channels. Proc Natl Acad Sci U S A 103:4040–4045

Stowers L, Holy TE, Meister M, Dulac C, Koentges G (2002) Loss of sex discrimination and male-male aggression in mice deficient for TRP2. Science 295:1493–1500

Suzuki M, Mizuno A, Kodaira K, Imai M (2003) Impaired pressure sensation in mice lacking TRPV4. J Biol Chem 278:22664–22668

Tabuchi K, Suzuki M, Mizuno A, Hara A (2005) Hearing impairment in TRPV4 knockout mice. Neurosci Lett 382:304–308

Talavera K, Yasumatsu K, Voets T, Droogmans G, Shigemura N, Ninomiya Y, Margolskee RF, Nilius B (2005) Heat activation of TRPM5 underlies thermal sensitivity of sweet taste. Nature 438:1022–1025

Tiruppathi C, Freichel M, Vogel SM, Paria BC, Mehta D, Flockerzi V, Malik AB (2002) Impairment of store-operated Ca^{2+} entry in TRPC4(−/−) mice interferes with increase in lung microvascular permeability. Circ Res 91:70–76

Todaka H, Taniguchi J, Satoh J, Mizuno A, Suzuki M (2004) Warm temperature-sensitive transient receptor potential vanilloid 4 (TRPV4) plays an essential role in thermal hyperalgesia. J Biol Chem 279:35133–35138

Tsavaler L, Shapero MH, Morkowski S, Laus R (2001) Trp-p8, a novel prostate-specific gene, is up-regulated in prostate cancer and other malignancies and shares high homology with transient receptor potential calcium channel proteins. Cancer Res 61:3760–3769

van der Eerden BC, Hoenderop JG, de Vries TJ, Schoenmaker T, Buurman CJ, Uitterlinden AG, Pols HA, Bindels RJ, van Leeuwen JP (2005) The epithelial Ca^{2+} channel TRPV5 is essential for proper osteoclastic bone resorption. Proc Natl Acad Sci U S A 102:17507–17512

Vennekens R, Olausson R, Meissner M, Nilius B, Flockerzi V, Freichel M (2006) Conditional inactivation of the TRPM4 gene in the mouse. Naunyn Schmiedebergs Arch Pharmacol 372[Suppl 7]:60

Vig M, Peinelt C, Beck A, Koomoa DL, Rabah D, Koblan-Huberson M, Kraft S, Turner H, Fleig A, Penner R, Kinet JP (2006) CRACM1 is a plasma membrane protein essential for store-operated Ca^{2+} entry. Science 312:1220–1223

West RJ, Sun AY, Church DL, Lambie EJ (2001) The C. elegans gon-2 gene encodes a putative TRP cation channel protein required for mitotic cell cycle progression. Gene 266:103–110

Wissenbach U, Niemeyer BA, Fixemer T, Schneidewind A, Trost C, Cavalie A, Reus K, Meese E, Bonkhoff H, Flockerzi V (2001) Expression of CaT-like, a novel calcium-selective channel, correlates with the malignancy of prostate cancer. J Biol Chem 276:19461–19468

Wissenbach U, Niemeyer B, Himmerkus N, Fixemer T, Bonkhoff H, Flockerzi V (2004) TRPV6 and prostate cancer: cancer growth beyond the prostate correlates with increased TRPV6 Ca^{2+} channel expression. Biochem Biophys Res Commun 322:1359–1363

Wu G, D'Agati V, Cai Y, Markowitz G, Park JH, Reynolds DM, Maeda Y, Le TC, Hou H Jr, Kucherlapati R, Edelmann W, Somlo S (1998) Somatic inactivation of Pkd2 results in polycystic kidney disease. Cell 93:177–188

Wu G, Markowitz GS, Li L, D'Agati VD, Factor SM, Geng L, Tibara S, Tuchman J, Cai Y, Park JH, van Adelsberg J, Hou H Jr, Kucherlapati R, Edelmann W, Somlo S (2000) Cardiac defects and renal failure in mice with targeted mutations in Pkd2. Nat Genet 24:75–78

Yeromin AV, Zhang SL, Jiang W, Yu Y, Safrina O, Cahalan MD (2006) Molecular identification of the CRAC channel by altered ion selectivity in a mutant of Orai. Nature 443:226–229

Zhang SL, Yu Y, Roos J, Kozak JA, Deerinck TJ, Ellisman MH, Stauderman KA, Cahalan MD (2005a) STIM1 is a Ca^{2+} sensor that activates CRAC channels and migrates from the Ca^{2+} store to the plasma membrane. Nature 437:902–905

Zhang Y, Hoon MA, Chandrashekar J, Mueller KL, Cook B, Wu D, Zuker CS, Ryba NJ (2003) Coding of sweet, bitter, and umami tastes: different receptor cells sharing similar signaling pathways. Cell 112:293–301

Zhang Z, Okawa H, Wang Y, Liman ER (2005b) Phosphatidylinositol 4,5-bisphosphate rescues TRPM4 channels from desensitization. J Biol Chem 280:39185–39192

Part I
TRPC Channel Subfamily

TRPC1 Ca^{2+}-Permeable Channels in Animal Cells

G. Rychkov[1] · G. J. Barritt[2] (✉)

[1]School of Molecular and Biomedical Science, University of Adelaide,
5005 Adelaide, South Australia, Australia

[2]Department of Medical Biochemistry, School of Medicine, Flinders University,
GPO Box 2100, 5001 Adelaide, South Australia, Australia
greg.barritt@flinders.edu.au

1	Introduction	24
2	Structures and Properties of the TRPC1 Genes and Proteins	25
2.1	Coding Sequence, Gene Sequence, Protein Structure	25
2.2	Likely Quaternary Structure of the TRPC1 Channel	26
2.3	Proteins Which May Interact with TRPC1	29
2.4	Antibodies Against TRPC1	29
3	Expression of TRPC1 Protein in Animal Cells	31
3.1	Expression of TRPC1 in Tissues and Cells	31
3.2	Intracellular Localisation of TRPC1	32
4	Ion Channel Properties of TRPC1	33
5	Modes of Activation of Channels Comprising TRPC1	35
5.1	Activation by Depletion of Intracellular Ca^{2+} Stores	35
5.2	Interaction of the TRPC1 Polypeptide with IP$_3$ Receptors	37
5.3	Mechanical: Stretch-Activated	42
5.4	Conclusions About the Role and Mechanisms of Activation of the TRPC1 Protein	43
5.5	Modulation of TRPC1 Function	43
6	Pharmacology	44
7	Biological Relevance and Emerging Biological Roles for TRPC1	44
7.1	Immediate Consequences of Na$^+$ and Ca^{2+} Entry Through TRPC1 Channels	44
7.2	Downstream Events	45
8	Conclusions	46
	References	47

Abstract The full-length transient receptor (TRPC)1 polypeptide is composed of about 790 amino acids, and several splice variants are known. The predicted structure and topology is of an integral membrane protein composed of six transmembrane domains, and a cytoplasmic C- and N-terminal domain. The N-terminal domain includes three ankyrin repeat motifs. Antibodies which recognise TRPC1 have been developed, but it has been difficult to obtain antibodies which have high affinity and specificity for TRPC1. This has made studies of the cellular functions of TRPC1 somewhat difficult. The TRPC1 protein is widely expressed in different types of animal cells, and within a given cell is found at the plasma membrane and at intracellular sites. TRPC1 interacts with calmodulin, caveolin-1, the InsP$_3$ receptor, Homer, phospholipase C and several other proteins. Investigations of the biological roles and mechanisms of action of TRPC1 have employed ectopic (over-expression or heterologous expression) of the polypeptide in addition to studies of endogenous TRPC1. Both approaches have encountered difficulties. TRPC1 forms heterotetramers with other TRPC polypeptides resulting in cation channels which are non-selective. TRPC1 may be: a component of the pore of store-operated Ca^{2+} channels (SOCs); a subsidiary protein in the pathway of activation of SOCs; activated by interaction with InsP$_3$R; and/or activated by stretch. Further experiments are required to resolve the exact roles and mechanisms of activation of TRPC1. Cation entry through the TRPC1 channel is feed-back inhibited by Ca^{2+} through interaction with calmodulin, and is inhibited by Gd^{3+}, La^{3+}, SKF96365 and 2-APB, and by antibodies targeted to the external mouth of the TRPC1 pore. Activation of TRPC1 leads to the entry to the cytoplasmic space of substantial amounts of Na$^+$ as well as Ca^{2+}. A requirement for TRPC1 is implicated in numerous downstream cellular pathways. The most clearly described roles are in the regulation of growth cone turning in neurons. It is concluded that TRPC1 is a most interesting protein because of the apparent wide variety of its roles and functions and the challenges posed to those attempting to elucidate its primary intracellular functions and mechanisms of action.

Keywords TRPC1 · Non-selective cation channel · Store-operated Ca^{2+} channel · Inositol 1, 4, 5 triphosphate receptor · Antibody · Na$^+$

1
Introduction

The TRPC1 (canonical transient receptor potential) polypeptide forms non-selective cation channels that are expressed in a wide variety of cell types and tissues (Vasquez et al. 2004; Beech 2005; Ramsey et al. 2006; Ambudkar 2006). It is likely that TRPC1 plays a significant part in intracellular Na$^+$ and Ca^{2+} homeostasis. However, its exact role is not entirely clear. TRPC1 is one of seven members of the TRPC sub-family of non-selective cation channels (Ramsey et al. 2006). This sub-family is, in turn, a member of the broad transient receptor potential (TRP) family, which also includes the TRPV, TRPM, TRPA, TRPP and TRPML sub-families. Following the first identification of the TRP and TRPL channels in the *Drosophila* photoreceptor cell and the idea that there are likely to be animal cell homologues of TRP and TRPL, TRPC1 was one of the first of these homologues to be discovered (Wes et al. 1995; Zhu et al. 1995).

Having established the nature and some properties of the TRPC1 gene, messenger RNA (mRNA) and proteins, the pathway towards elucidation of the intracellular locations and mechanisms of activation of TRPC1, and its biological functions, has not been easy (Ramsey et al. 2006). Much attention has been focussed on TRPC1 as a Ca^{2+}-permeable channel in the plasma membrane, and whether or not the TRPC1 polypeptide comprises store-operated Ca^{2+} channels (SOCs). The difficulties encountered in elucidation of the molecular details of TRPC1 function have been due, in part, to the unavailability of high-affinity selective TRPC1 inhibitors, and inherent technical difficulties in designing effective strategies to selectively ablate the TRPC1 protein, including the generation of a TRPC1 gene knock-out mouse. Contributing to the trials and tribulations encountered by those seeking to understand TRPC1 are:

- The complexities surrounding the definition of SOCs and in designing effective strategies for investigating the mechanism(s) of activation of SOCs (Parekh and Putney 2005; Ramsey et al. 2006);
- Some confusion created by using an unknown pathway (SOC activation) to try to elucidate the function and mechanism of activation of another unknown (TRPC1);
- The need to often rely heavily on results obtained with ectopically (over-) expressed TRPC1 rather than the endogenous protein.

The aims of this chapter are to summarise the main features of TRPC1 structure, mechanisms of activation and function. Some emphasis has been given to a description of the proteins thought to interact with TRPC1, anti-TRPC1 antibodies, elucidation of the roles of TRPC1 as a plasma membrane non-selective cation channel and its mechanisms of activation, since they are important current issues. It is hoped that the information presented will assist in providing a base for further studies of TRPC1 function, and in understanding its future use as a target for pharmaceutical intervention.

2
Structures and Properties of the TRPC1 Genes and Proteins

2.1
Coding Sequence, Gene Sequence, Protein Structure

The human gene encoding TRPC1 is located on chromosome 3q22–3q24 (Clapham et al. 2005). Expression of this gene yields two splice variants, α and β, of TRPC1 mRNA (Sakura and Ashcroft 1997; Engelke et al. 2002). The longest (unspliced) form (*TRPC1α*) encodes a peptide of 793 amino acids (Clapham et al. 2005). Mouse TRPC1 mRNA exhibits four splice variants (α, β, γ and δ) but two of these (γ and δ) do not appear to be translated and hence are likely non-functional (Sakura and Ashcroft 1997). The predicted sizes of the mouse α (full-length) and β isoforms are 91.2 and 87.6 kDa, respectively (Sakura and Ashcroft 1997; Sinkins et al. 1998; Wang et al. 1999). However, the actual sizes of the TRPC1 peptides extracted from cells may be greater than the predicted sizes as TRPC1 can be *N*-glycosylated (Birnbaumer et al. 1996; Boulay et al. 1997; Zhu et al. 1998). There are a number of similarities between the TRPC1, TRPC4 and TRPC5 so that TRPC1 is considered to be a member of the TRPC1, TRPC4 and TRPC5 sub-group in the TRPC sub-family (Ramsey et al. 2006). The other two sub-groups comprise the TRPC3, TRPC6 and TRPC7 proteins and the TRPC2 protein.

The predicted structure of the TRPC1 polypeptide when inserted in the plasma membrane consists of a cytoplasmic N-terminal domain, six transmembrane (TM) domains, an S5–S6 pore-forming domain containing the LFW motif, and a cytoplasmic C-terminal domain (Dohke et al. 2004; Beech, 2005; Ramsey et al. 2006) (Fig. 1). This predicted structure is based on the TRPC1 amino acid sequence, analysis of hydropathy plots, and comparison of the amino acid sequence of the TRPC1 polypeptide with the sequences and predicted structures of other TRP polypeptides and other cation channels, especially the bacterial K^+ (KcsA) channel and mammalian voltage-operated K^+ channel (Dohke et al. 2004; Beech 2005; Ramsey et al. 2006). As discussed in Sect. 3.2 below, TRPC1 is also localised in intracellular membranes (Wang et al. 1999; Goel et al. 2002, 2006; Chen and Barritt, 2004; Rao et al. 2006) where it likely exhibits a topology similar to that shown in Fig. 1 with cytoplasmic N- and C-terminal domains (Dohke et al. 2004).

Within the cytoplasmic N-terminal domain of TRPC1, three ankyrin repeat motifs, an N-terminal coiled-coil domain, a dimerisation domain and a binding site for caveolin-1 have been identified (reviewed in Vazquez et al. 2004; Beech 2005; Ramsey et al. 2006; Fig. 1). In the cytoplasmic C-terminal domain, a TRP box (EWKFAR), a C-terminal coiled-coil domain and binding sites for some proteins which interact with TRPC1 have been identified (Fig. 1).

2.2
Likely Quaternary Structure of the TRPC1 Channel

The composition of endogenous functional cation channels involving the TRPC1 polypeptide in the plasma membrane and in the intracellular mem-

Fig. 1 Topology of TRPC1. A schematic representation of the TRPC1 polypeptide as an integral protein in the plasma membrane. It is hypothesised that the endogenous non-selective cation channels formed from TRPC1 polypeptides are heterotetramers composed of a mixture of TRPC1 and one or more other TRPC polypeptides. The LFW hydrophobic domain of TRPC1 contributes to the channel pore. Features of the membrane-spanning domains and sequences with known functions are shown. References are given in the text. *ANK 1–3*, ankyrin-like repeats 1 to 3; *CC-N* and *CC-C*, N- and C- terminal coiled-coil domains, respectively; *CIRB*, calmodulin/InsP$_3$ receptor (IP$_3$R) binding region; *LFW*, amino acid motif conserved in the hydrophobic putative pore region

TRPC1 Ca^{2+}-Permeable Channels in Animal Cells

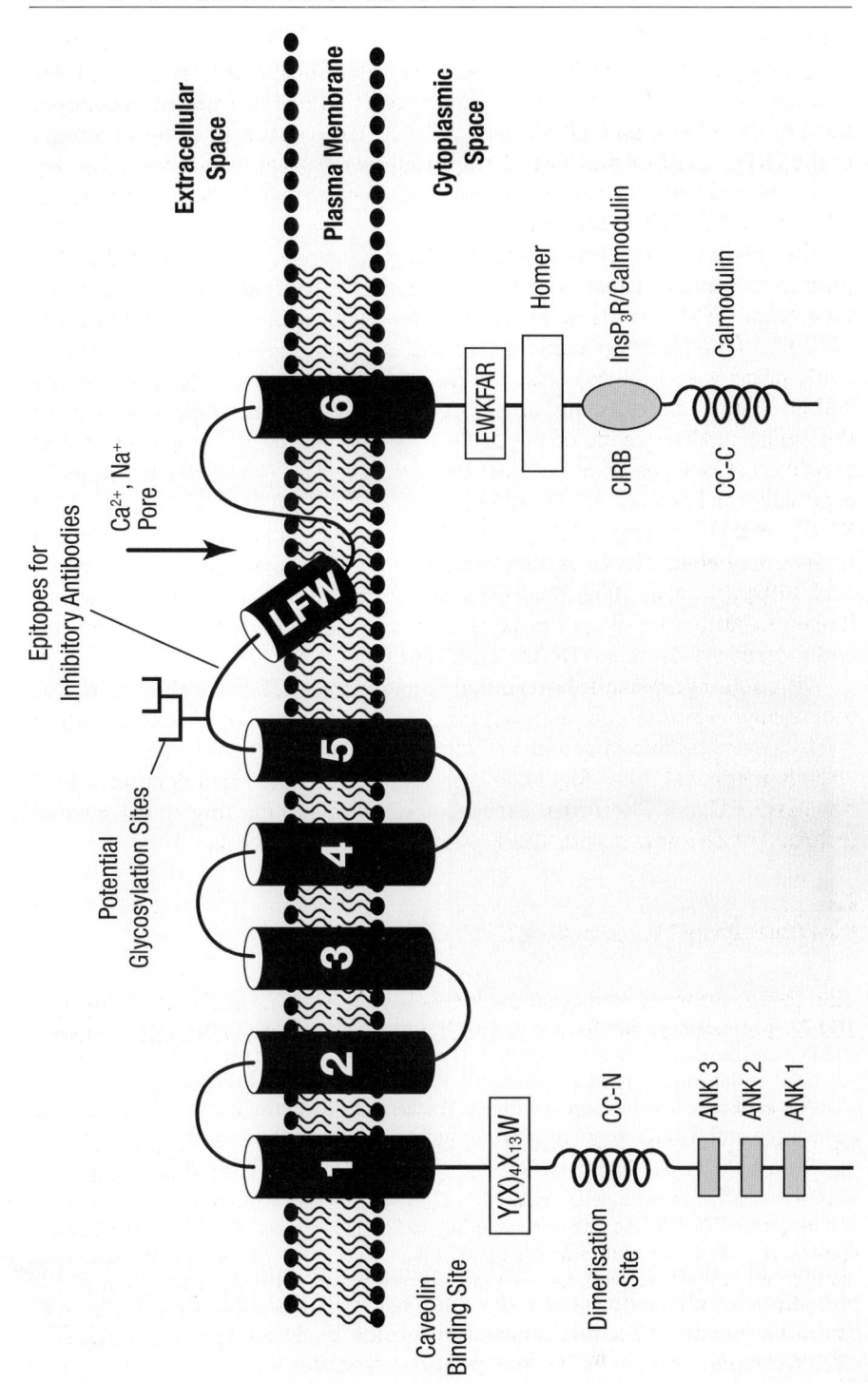

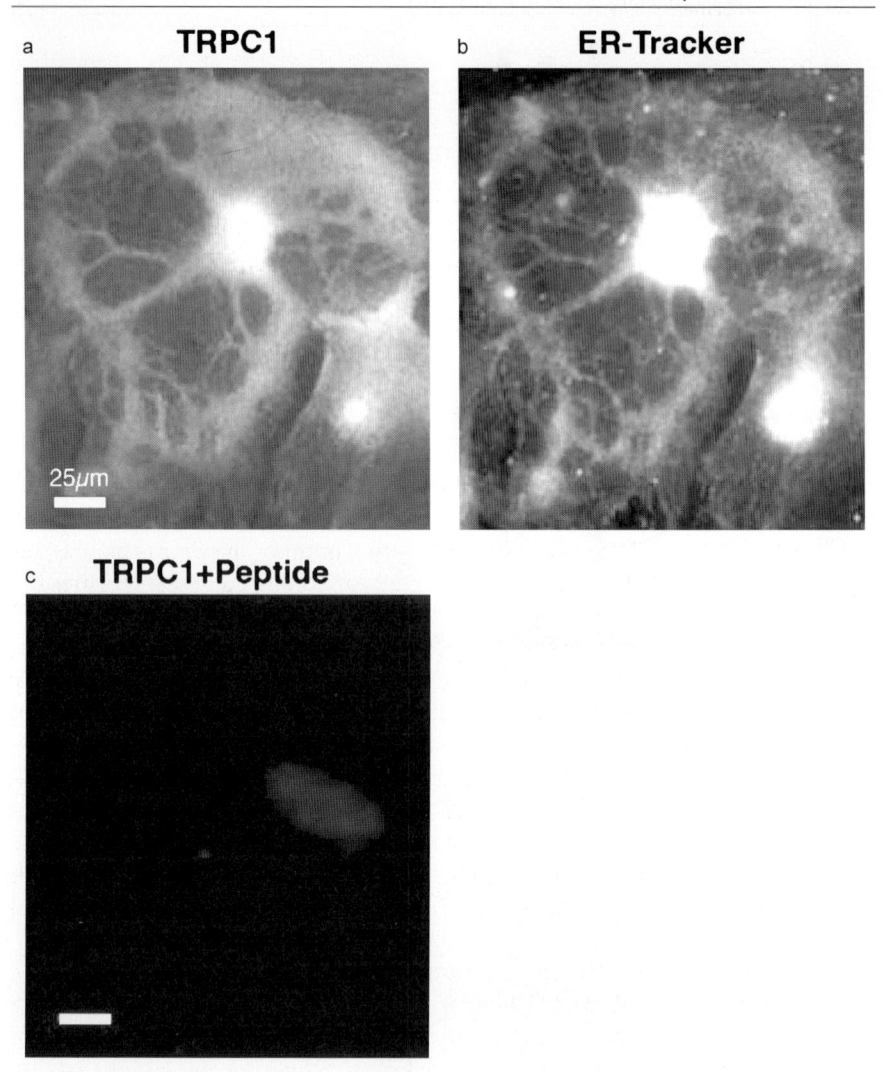

Fig. 2 a–c Intracellular distribution of TRPC1. Immunofluorescence images of endogenous TRPC1 and the endoplasmic reticulum (ER) marker ER-Tracker in permeabilised astrocytes showing co-location of TRPC1 with the ER. Panel **a**, cells cross-reacted with polyclonal anti TRPC1 antibody; **b**, cells stained with ER-Tracker; and **c**, control immunofluorescence in which the anti TRPC1 antibody was pre-incubated with TRPC1 peptide. The *scale bar* represents 25 µm. Reproduced from Golovina (2005) by copyright permission of The Physiological Society

branes of animal cells is not yet well defined. Co-immunoprecipitation and functional studies with ectopically expressed TRPC1 and other TRPC polypeptides have provided some information on the likely tetrameric structure of TRPC1 channels. The TRPC1 polypeptide associates with TRPC4 and TRPC5

(Strubing et al. 2001, 2003; Hofmann et al. 2002; Goel et al. 2002; Brownlow and Sage 2005) and with TRPC2, TRPC3, TRPC6 and TRPC7 (Lintschinger et al. 2000; Tsiokas et al. 1999; Wu et al. 2004; Liu et al. 2005 Zagranichnaya et al. 2005). Taken together, these results suggest that functional TRPC1 channels are heterotetramers comprising the TRPC1 polypeptide together with one or more other TRPC polypeptides. Homomeric interactions between TRPC1 may be mediated by N-terminal coiled-coil domains (Engelke et al. 2002), and heteromeric interactions between TRPC1 and TRPC3 may involve the first ankyrin repeat of TRPC1 (Liu et al. 2005).

2.3
Proteins Which May Interact with TRPC1

Several different experimental approaches have provided evidence which indicates that in addition to the likely formation of heterotetramers with other TRPC polypeptides, TRPC1 can interact with a number of other proteins (Table 1). For some of these, further experiments are required to test whether the protein interacts directly or indirectly (via other proteins) with TRPC1 and whether the interaction occurs with endogenous TRPC1 in vivo. This is particularly important where the experiments have been performed with ectopically expressed proteins, and where the principal or sole evidence for an interaction is co-immunoprecipitation from cell extracts. The data in Table 1 have been obtained in different cell types, and an interaction between a given protein and TRPC1 may not occur in all cell types. For inositol 1,4,5-trisphosphate receptor (IP$_3$R), Homer, calmodulin and caveolin-1, where the interaction with TRPC1 has been studied in more detail, the domains of TRPC1 likely to be involved in the interaction have been identified (Fig. 1) and are discussed in more detail later (see below for references).

The functions served by the interaction of TRPC1 with other proteins are only partly understood. Some protein–TRPC1 interactions may be part of the TRPC1 activation pathway, some are likely to be involved in the regulation of TRPC1 by Ca^{2+}, and some are likely to determine the intracellular localisation of TRPC1 (Table 1). It has been suggested that TRPC1 forms part of a protein complex ("signalplex") which may link TRPC1 to the cytoskeleton and possibly to intracellular organelles (Beech et al. 2003; Beech 2005; Ambudkar 2006). This signalling complex may allow signalling both to and from TRPC1.

2.4
Antibodies Against TRPC1

A number of polyclonal (Wang et al. 1999; Bobanovic et al. 1999; Brereton et al. 2000; Rosado and Sage 2000; Xu and Beech 2002; Goel et al. 2002; Crousillac et al. 2003; Goel et al. 2006) and monoclonal (Tsiokas et al. 1999) antibodies which recognise the TRPC1 polypeptide have been developed and employed in studies

Table 1 Proteins which have been shown to interact with the TRPC1 polypeptide[a]

Protein	Hypothesised role of interaction	Reference(s)
Glutamate metabotropic receptor 1α (G protein-coupled receptor)	Component of activation pathway and/or regulation of intracellular location	Kim et al 2003
Basic fibroblast growth factor receptor-1 (tyrosine kinase-coupled receptor)	Component of activation pathway and/or regulation of intracellular location	Fiorio-Pla et al 2005
Gq/11 α	Component of activation pathway and/or regulation of intracellular location	Lockwich et al 2000
PLCβ	Component of activation pathway and/or regulation of intracellular location	Lockwich et al 2000
IP$_3$R	Component of activation pathway	Lockwich et al 2000; Rosado and Sage 2000; Yuan et al 2003
Homer	Component of activation pathway	Yuan et al 2003
RhoA	Component of activation pathway and/or plasma membrane localisation	Mehta et al 2003
Plasma membrane (Ca^{2+}+Mg^{2+}) ATP-ase (PMCA)	Regulation by Ca^{2+} and/or regulation of intracellular location	Singh et al 2002; Kiselyov et al 2005
Sarcoplasmic-endoplasmic (Ca^{2+}+Mg^{2+}) ATP-ase (SERCA)	Regulation by Ca^{2+} and/or regulation of intracellular location	Golovina 2005
Calmodulin	Feedback-inhibition by Ca^{2+}	Tang et al 2001; Singh et al 2002
TRP Polycystin 2	Intracellular localisation	Tsiokas et al 1999
PKC α	Regulation of channel activity	Ahmmed et al 2004
Caveolin-1	Plasma membrane localisation and insertion into the membrane	Lockwich et al 2000; Brazer et al 2003
Inhibitor of myogenic family (I-mfa)		Ma et al 2003
MxA (member of dynamin superfamily)	Link to scaffolding proteins and cytoskeleton	Lussier et al 2005
β-Tubulin	Plasma membrane localisation	Bollimuntha et al 2005

[a] This table combines data obtained using a number of different techniques and cell types, and endogenous and ectopically expressed TRPC1. Further experiments are required to confirm that, for some proteins in the list, the interaction with TRPC1 occurs in vivo, to elucidate the functional consequences of this interaction, and the cell types in which this occurs

of TRPC1 function. The polyclonal antibodies include those raised against peptide sequences in an extracellular site of the putative pore region (Wang et al. 1999; Bobanovic et al. 1999; Xu and Beech 2002) and the cytoplasmic C-terminal domain (Bobanovic et al. 1999; Brereton et al. 2000). The monoclonal antibody recognises the N-terminal domain (Tsiokas et al. 1999). In Western blot and/or immunofluorescence experiments, these antibodies appear to bind to the TRPC1 polypeptide with varying degrees of affinity and specificity. Convincing evidence that a given antibody recognises the TRPC1 polypeptide under the experimental conditions employed has not always been available. It has often been difficult to generate anti-TRPC1 antibodies with sufficient specificity and affinity for TRPC1 (Ong et al. 2002; Hassock et al. 2002; Parekh and Putney 2005).

Glycosylation of TRPC1 (Birnbaumer et al. 1996; Boulay et al. 1997; Zhu et al. 1998), which may vary with the cell type and method of extraction, has made it difficult to use the observed apparent molecular weight of TRPC1 in Western blots as a criterion that a given antibody specifically recognises TRPC1 (Ong et al. 2002). In an attempt to increase the specificity of detection of the TRPC1 protein, some groups have employed the immunoprecipitation of TRPC1 with one antibody followed by Western blot using a second anti-TRPC1 antibody to detect TRPC1 proteins in cell extracts (Ong et al. 2000; Rosado et al. 2002).

3
Expression of TRPC1 Protein in Animal Cells

3.1
Expression of TRPC1 in Tissues and Cells

Expression of endogenous TRPC1 in cells and tissues has been studied using Northern blot analysis, RT-PCR, in situ hybridisation, Western blot analysis and immunofluorescence. In contrast to some other members of the TRP family, TRPC1 is very widely expressed in animal cells. This suggests that TRPC1 may serve a range of specific functions in different types of cells. Cells and tissues in which TRPC1 has been detected by one or more of these methods include brain, heart, testis, ovaries, smooth muscle, endothelium, salivary glands and liver (reviewed in Beech 2005). For some cells and tissues the only evidence for expression of TRPC1 is provided by the detection of TRPC1 mRNA using RT-PCR. It has either not been possible to detect the TRPC1 protein by Western blot or immunofluorescence, or this has not been attempted. This raises the possibility that there are some cell types for which a very low level of TRPC1 mRNA is expressed and, while this is detectable by PCR, there may be no functional TRPC1 protein expressed. The results of several studies, chiefly employing immunofluorescence, have identified those cell types in a particular organ in which TRPC1 is expressed. These include neurons (Goel et al. 2002; Glazebrook et al. 2005; von Bohlen-Und-Halbach et al. 2005) the retina (Crousillac et al. 2003) and kidney (Goel et al. 2006).

3.2
Intracellular Localisation of TRPC1

The intracellular localisation of TRPC1 has been studied both in cells expressing endogenous TRPC1 and in cells in which TRPC1 is ectopically expressed. The results of studies of endogenous TRPC1 are likely to provide the most reliable data on the in vivo intracellular localisation of TRPC1. Several factors could lead to the abnormal intracellular localisation of ectopically expressed TRPC1. The expression of mRNA encoding TRPC1 from a given species in a host cell of a different species (e.g. expression of rat TRPC1 in a human cell line or expression of very high, possibly supra-physiological, levels of TRPC1) could lead to altered trafficking of the TRPC1 polypeptide. Notwithstanding these reservations, analysis of the intracellular distribution of ectopically expressed TRPC1 suggests patterns of TRPC1 distribution similar to those observed for endogenous TRPC1 (Wang et al. 1999; Brereton et al. 2000; Chen and Barritt 2003). Moreover, the results of studies with ectopically expressed TRPC1 can provide valuable information on the intracellular trafficking of TRPC1 and its functions at specific intracellular localisations.

In many cell types, TRPC1 is found both at the plasma membrane and at intracellular sites, including the endoplasmic reticulum (ER) and Golgi apparatus (Wang et al. 1999; Goel et al. 2002, 2006; Chen and Barritt 2003; Uehara 2005; Rao et al. 2006). In cells in which TRPC1 is ectopically expressed, the movement of TRPC1 to the plasma membrane is dependent on, or enhanced by, co-expression with TRPC4 and/or TRPC5 (Hofmann et al. 2002; Yuan et al. 2003). Immunofluorescence images of endogenous TRPC1 are often punctate, suggesting that TRPC1 is concentrated in sub-regions of organelles (Wang et al. 1999; Xu and Beech 2001). The localisation of TRPC1 at intracellular sites may reflect, in part, a normal intracellular function of TRPC1, such as the facilitation of Ca^{2+} release from intracellular stores (see Sect. 5.1 below), its role in forming signalling complexes, and/or TRPC1 polypeptide being trafficked from the ER through the Golgi apparatus to the plasma membrane (Brereton et al. 2000; Dohke et al. 2004).

In the plasma membrane, TRPC1 appears to be confined to specific regions of the membrane (Lockwich et al. 2000; Bergdahl et al. 2003; Brownlow et al. 2004; Dalrymple et al. 2002; Golovina 2005). The application of immunofluorescence and other techniques suggests that in smooth muscle cells and astrocytes, TRPC1 in the plasma membrane is localised in regions adjacent to the underlying ER (Fig. 2; Dalrymple et al. 2002; Golovina 2005). The results of immunofluorescence, sub-cellular fractionation and Western blot experiments conducted with sub-mandibular gland cells (Bergdahl et al. 2003; Brazer et al. 2003; Lockwich et al. 2000), platelets (Brownlow et al. 2004), smooth muscle cells (Bergdahl et al. 2003) and endothelial cells (Rao et al. 2006) indicate that in the plasma membrane TRPC1 is associated with cholesterol-rich lipid rafts and with caveolin-1. Correlation of these observations with TRPC1 function suggests that the localisation of TRPC1 in lipid rafts and its interaction with cholesterol are required for the activation of Ca^{2+} inflow in sub-mandibular gland cells (Lockwich et al. 2000) and for the contraction of smooth muscle cells (Bergdahl et al. 2003).

4
Ion Channel Properties of TRPC1

There are limited data available from patch clamp experiments that unequivocally assign a specific current to a channel that is either a homotetramer of endogenous TRPC1 polypeptides or a heterotetramer containing endogenous TRPC1. What is known about currents mediated by TRPC1 has mainly come from the ectopic expression of the TRPC1 polypeptide in different cell lines. In salivary gland cells, current through the endogenous SOCs was indistinguishable in its properties from the membrane current activated by the depletion of intracellular Ca^{2+} stores in cells in which TRPC1 was ectopically expressed. Both are inwardly rectifying currents through non-selective cation channels of 20 pS conductance (Liu et al. 2003). Moreover, transfection with antisense TRPC1 decreased the activity of the endogenous SOCs in salivary gland cells.

In CHO cells, over-expression of TRPC1 resulted in a linear non-selective current activated by store-depletion. This was slowly inhibited by extracellular Ca^{2+} (1.2 mM) and Gd^{3+} (20 µM) (Zitt et al. 1996). The estimated single channel conductance of TRPC1 in CHO cells was 16 pS. Membrane currents with similar properties could be recorded in Sf9 insect cells infected with baculovirus containing complementary DNA (cDNA) encoding TRPC1 (Sinkins et al. 1998). In Sf9 cells, however, TRPC1 current was constitutively active and insensitive to store depletion.

In H4IIE liver cells, current that was attributed to ectopically expressed TRPC1 could be activated by maitotoxin, but not by store depletion. The channel underlying this current did not discriminate between Na^+ and Ca^{2+} and was inhibited by Gd^{3+} (50 µM) (Brereton et al. 2001). A store-independent non-selective cation current activated by 1-oleoyl-2 acetyl-*sn* glycerol (OAG) was observed in HEK293 cells in which TRPC1 was ectopically expressed (Lintschinger et al. 2000). This current had a linear current–voltage relationship and was completely inhibited by 2 mM extracellular Ca^{2+} or 50 µM La^{3+}. In some experiments where TRPC1 has been expressed ectopically, no detectable current attributable to TRPC1 could be observed (Strubing et al. 2001).

In any experiment where the TRPC1 peptide is ectopically expressed, the nature of the current attributable to the presence of the TRPC1 is likely to depend on the nature of the host cell. This may specify how the TRPC1 polypeptide is trafficked to the plasma membrane, the nature of other TRPC peptides that could form a putative tetrameric pore with TRPC1, and the nature of the pathways that may regulate the channel. Another issue in studying TRPC1-mediated currents is to distinguish these from background currents due to other channels present in the host cell.

The notion that the nature of the current attributable to ectopically expressed TRPC1 may depend on the presence of the endogenous TRPC polypeptides is well demonstrated by co-expression studies. Co-expression of TRPC1 with TRPC3 in HEK293 cells generated membrane current that was different from the currents produced by either TRPC1 or TRPC3 alone, in its sensitivity to external and internal Ca^{2+} and the cation selectivity (Lintschinger et al. 2000).

Co-expression of TRPC1 with TRPC5 in HEK293-M1 cells produced a non-selective cation conductance which was activated by carbachol but insensitive to store depletion (Strubing et al. 2001). The characteristics of the TRPC1/TRPC5 current were different from those attributable to TRPC5 alone, while expression of TRPC1 alone did not produce any currents. The TRPC1/TRPC5 heterotetramers displayed an outwardly rectifying cation current and a single channel conductance of 5 pS. Similar results were obtained when TRPC1 was co-expressed with TRPC4 (Strubing et al. 2001). In addition to forming heterotetramers with TRPC1, TRPC4 and TRPC5 may promote the trafficking of TRPC1 to the plasma membrane (Hofmann et al. 2002).

Studies by Ambudkar's group of the biogenesis and topology of the TRPC1 polypeptide expressed in HEK293 cells have provided information on some features of the likely pore formed by TRPC1 polypeptides (Dohke et al. 2004). Their results show that while eight hydrophobic (potentially transmembrane) α-helices can be identified in the TRPC1 polypeptide, it is likely that only six of these span the membrane. Moreover, they provide evidence that one of the non-membrane spanning α-helices (the region TM5–TM6) partly enters the membrane. This evidence was obtained, in part, by preparing a mutant form of the TRPC1 polypeptide in which all seven negatively charged residues in the region between TM5 and TM6 were neutralised (D–N and E–Q). This resulted in changes in the selectivity of the channel, and a decrease in the Ca^{2+} current through the pore presumably formed by the TRPC1 polypeptide. Surprisingly, they found no change in the magnitude of the Na^+ current. On the basis of these results, and using the more extensive structural and topological studies of the K^+ channel, KcsA, and TRPV5 and TRPV6 channels as models, it was suggested that the TM5–TM6 region in TRPC1 may constitute the pore of the channel formed by the TRPC1 polypeptide (Dohke et al. 2004; Owsianik et al. 2006). The results obtained by Dohke et al. (2004) suggest that the negatively charged amino acids in this TM5–TM6 region, which are likely located close to but outside the pore, regulate ion flow through the pore.

5
Modes of Activation of Channels Comprising TRPC1

5.1
Activation by Depletion of Intracellular Ca^{2+} Stores

Although the subject of many published studies, the mechanism of activation of Ca^{2+} and Na$^+$ entry through plasma membrane channels comprising TRPC1 has not been well established. There is evidence to indicate that TRPC1 is activated by store depletion, G protein- or tyrosine kinase-coupled receptors independent of store depletion, and stretch (reviewed in Putney 2005; Parekh and Putney 2005; Ramsey et al. 2006).

Before summarising the evidence upon which these conclusions are based, it is useful to clarify the definitions of the main types of plasma membrane non-voltage operated Ca^{2+}-permeable channels. These are defined here as ligand-gated, receptor-activated, store-operated (SOC) and stretch-activated Ca^{2+}-permeable channels. In ligand-gated channels, the binding site for agonist and the channel pore are on the same oligomeric protein. Receptor-activated channels (as defined here) comprise an agonist-receptor complex coupled to the channel by an intracellular messenger(s) and/or by protein–protein interaction. Store depletion is not a causative step in the activation of receptor-activated Ca^{2+}-permeable channels. A SOC is defined as a channel which requires a decrease in Ca^{2+} in an intracellular store, most likely the ER (Ca^{2+} store depletion), as a necessary and causative step in its activation (Parekh and Putney 2005). Physiologically, activation of SOCs is initiated by the binding of an agonist to a G protein- or tyrosine kinase-coupled receptor, the activation of phospholipase C, the generation of IP$_3$ and Ca^{2+} release mediated by IP$_3$Rs and ryanodine receptors. Artificially, Ca^{2+} store depletion can be induced by inhibition of the sarco-endoplasmic reticulum (Ca^{2+}+Mg^{2+})ATPase Ca^{2+} pump (SERCA) by thapsigargin, 2,5-di(*tert*-butyl)-1,4-hydroquinone (DBHQ) or other agents. The mechanism(s) or pathway(s) by which the consequences of store depletion are signalled to the SOC channel in the plasma membrane are not well understood. It is not even known whether there is a unique pathway for SOC activation or several different pathways and corresponding different arrangements of the component proteins. Stretch-activated channels are thought to be activated by changes in the lipid membrane or in the cytoskeleton initiated by a mechanical signal such as stretching or a change in cell volume.

Ectopic expression of the TRPC1 polypeptide leads to either the appearance of a Ca^{2+} entry pathway which can be activated by a SERCA inhibitor (Delmas et al. 2002) or to the enhancement of an existing Ca^{2+} entry pathway activated by a SERCA inhibitor (Liu et al. 2000; Kunichika et al. 2004; Fiorio-Pla et al. 2005). This phenomenon is not observed in all cell types subjected to this experimental regime (Strubing et al. 2001). The failure to detect Ca^{2+} entry in response to SERCA inhibitor in some cell types in which TRPC1 is ectopically expressed may be due to a cellular environment in which TRPC1 is not correctly localised in the plasma membrane and/or to the presence of, and presumably interference or masking by, endogenous SOCs (Beech 2006). The latter would be activated by the same experimental strategies used to activate TRPC1. In some experiments, ectopic expression of TRPC1 has led to enhanced Ca^{2+} entry which appears to be independent of store depletion (Sinkins et al. 1999). The absence of more extensive and convincing data showing SERCA-initiated Ca^{2+} entry in cells ectopically expressing TRPC1 is puzzling. It suggests that TRPC1 is not activated by store depletion and/or the function of TRPC1 in the plasma membrane and its activation mechanism are more complex than required for a SOC.

A reduction in TRPC1 expression using antisense constructs or small interfering RNAs (siRNA) targeted to TRPC1 has been found to be associated with a decrease in SERCA inhibitor-initiated Ca^{2+} entry in a wide variety of cells (Tomita et al. 1998; Ahmmed et al. 2004; Brough et al. 2001; Vaca and Sampieri 2002; Rosado and Sage 2002; Wu et al. 2000, 2004; Bergdahl et al. 2005; Lin et al. 2000, 2003, 2004; Sweeny et al. 2002; Brazer et al. 2003; Golovina 2005; Mori et al. 2002; Tu et al. 2005; Vandebrouck et al. 2002; Chen and Barritt 2003; Cai et al. 2006; Brueggemann et al. 2006). Another approach to test the participation of TRPC1 as a component of SOCs has employed polyclonal antibodies directed against a peptide on the extracellular loop of TRPC1 close to the putative cation pore. In vascular smooth muscle cells (Xu and Beech 2001) and platelets (Rosado et al. 2002; Kunzelmann-Marche et al. 2002), such a polyclonal antibody, when applied to the extracellular medium, inhibited SERCA-initiated Ca^{2+} entry. In the case of smooth muscle cells, the degree of inhibition was small.

While many authors have interpreted the observations described above as indicating that TRPC1 is a component of the SOC pore (shown schematically in Fig. 2a), there could be other explanations. (1) The Ca^{2+} entry measured may not be through SOCs. The use of an intracellular fluorescent Ca^{2+} sensor to detect increases in $[Ca^{2+}]_{cyt}$ and hence Ca^{2+} entry does not allow characterisation of the Ca^{2+} entry channels activated by a SERCA inhibitor. (2) SERCA inhibitors can initiate the activation of Ca^{2+} entry through non-SOCs, including non-selective cation channels activated by Ca^{2+} or metabolites of arachidonic acid, formed by Ca^{2+}-activated phospholipase A_2, (reviewed in Parekh and Putney 2005; Eder et al. 2005). (3) Thapsigargin has been shown to cause an increase in IP_3 (Nishida et al. 2003) which could activate a putative IP_3R-coupled channel (see below) independent of store depletion. (4) The activation of TRPC1 leads to Na^+ inflow which, in turn, could lead to Ca^{2+} entry through the Na^+–Ca^{2+} exchange working in reverse mode and, in excitable cells, to depolarisation of the plasma membrane and Ca^{2+} entry through voltage-operated Ca^{2+} channels (Eder et al. 2005). (5) The magnitudes of some observed changes in TRPC1 expression and the rate of Ca^{2+} entry are small (Chen and Barritt 2003). (6) The relationship between decreased TRPC1 expression and decreased Ca^{2+} entry may only be an indirect one. As pointed out by Ramsey et al. (2006), suppression of expression of the cyclic nucleotide-gated channel, which has no known or anticipated connection with SOCs, can alter SERCA-initiated Ca^{2+} inflow (Zhang et al. 2002).

Another consideration in the hypothesis that TRPC1 is a component of the pore of a SOC is ion selectivity. The SOCs which have so far been most clearly defined and characterised are the CRAC (Ca^{2+} release-activated Ca^{2+}) channels in lymphocytes, mast cells, hepatocytes and megakaryocytes (reviewed in Parekh and Putney 2005). All these have a high selectivity for Ca^{2+}. By contrast, as indicated above, channels putatively assigned to TRPC1 are nonselective. Thus it is unlikely that TRPC1 polypeptides are components of these well-characterised Ca^{2+}-selective SOCs (Parekh and Putney 2005). It has been suggested that there are several types of SOCs with different selectivities for Ca^{2+} and that TRPC1 polypeptides are components of a non-selective SOC (Beech 2005; Ambudkar 2006).

The observation that considerable TRPC1 polypeptide is detected at intracellular locations (Wang et al. 1999; Goel et al. 2002; Chen and Barritt 2003; Rao et al. 2006; Uehara 2005) raises the possibility that TRPC1 could be involved in mediating and/or regulating the release of Ca^{2+} from intracellular stores, and hence could play an indirect role in the activation of SOCs without being a component of the SOC pore itself (Parekh and Putney 2005). Some evidence for this possibility has come from the essentially complete ablation of TRPC1 expression in chicken DT40B lymphocytes using targeted genetic disruption of the TRPC1 gene (Mori et al. 2002). This led to a decrease in SERCA-initiated Ca^{2+} release and, in 80% of cells tested, to the absence of current through CRAC channels, while normal currents were observed in the other 20% of cells. The authors concluded that TRPC1 is required, in some way, in the mechanism of activation of CRAC channels, but is unlikely to be a component of the CRAC channel pore. This general idea of TRPC1 as an accessory protein is shown schematically in Fig. 3b.

5.2
Interaction of the TRPC1 Polypeptide with IP$_3$ Receptors

Several experimental approaches have provided evidence which suggests that TRPC1 interacts with IP$_3$Rs (Rosado and Sage 2000; Lockwich et al. 2000; Yuan et al. 2003). In extracts of platelets, IP$_3$R2 and TRPC1 co-immunoprecipitate, suggesting that in the native platelet environment these two proteins interact (Rosado and Sage 2000; 2001). The co-immunoprecipitation of TRPC1 and IP$_3$R2 was enhanced by prior treatment of platelets with SERCA inhibitors or thrombin, a platelet agonist, and was inhibited by xestospongin C an inhibitor of IP_3-induced Ca^{2+} release from the ER via IP_3R (Rosado and Sage 2000; Brownlow and Sage 2003; Rosado et al. 2002). The effect of thrombin in promoting TRPC1-IP_3R2 co-immunoprecipitation could be detected within 1 s of thrombin addition. The authors interpreted these and other results to indicate that (1) TRPC1 is a component of platelet SOCs, (2) the activation of platelet SOCs requires interaction of TRPC1 with IP_3R2, and (3) this interaction is a necessary component of the mechanism by which platelet SOCs are activated (Fig. 2a, sub-part b; Rosado et al. 2002; Brownlow and Sage 2003). Another study involving sub-cellular fractionation and Western blot analysis with a different anti-TRPC1 antibody concluded that TRPC1 in platelets is only located at intracellular sites (Hassock et al. 2002). Such a localisation is not consistent with a role for TRPC1 as a component of the pore of a SOC (Parekh and Putney 2005). As suggested below for other cell types, further experiments, possibly requiring high-resolution localisation and more specific anti-TRPC1 antibodies, are required to understand the role of TRPC1 in platelets.

Lockwich et al. (2000) have shown that in membrane extracts of submandibular gland cells, an anti-TRPC1 antibody immunoprecipitates IP_3R3 as well as caveolin-1 and $G_{\alpha q|11}$. They also showed that TRPC1 is part of

Fig. 3 a–d Hypothesised localisations, functions and mechanisms of activation of TRPC1. **a** TRPC1 may be a component of the pore-forming sub-units of a SOC in the plasma membrane. The TRPC1 SOC may be activated by a decrease in Ca^{2+} in intracellular stores (ER) involving either (*a*) an unknown mechanism (e.g. a mobile intracellular messenger such as Ca^{2+} influx factor, CIF) or (*b*) interaction of the IP_3R protein with the TRPC1 protein. This mechanism is a necessary step in TRPC1 activation. **b** TRPC1 may be required for the activation of a SOC located in the plasma membrane. It is hypothesised that TRPC1 is not a component of the pore-forming sub-units of the SOC, but is a component of a necessary step in the SOC activation pathway. Two possibilities are shown: (*a*) an hypothesised role of TRPC1 at the plasma membrane as an accessory protein interacting with the SOC; (*b*) an hypothesised role of TRPC1 in the ER as an accessory protein and/or involved in IP_3-initiated Ca^{2+} release from the ER. **c** TRPC1 may be a component of the pore-forming sub-units of a plasma membrane Ca^{2+}-permeable channel (not an SOC) which may be activated by interaction between TRPC1 and an IP_3R protein (a necessary step in activation). In this case, it is hypothesised that the TRPC1 channel is not activated by a decrease of Ca^{2+} in intracellular stores, although this event may be associated with the activation. Two possibilities are shown: (*a*) the relevant IP_3R may be located in the plasma membrane with the IP_3 binding site facing the cytoplasmic space; (*b*) the relevant IP_3R may be located in the ER. **d** TRPC1 is a component of the pore-forming sub-units of a plasma membrane Ca^{2+}-permeable channel which may be activated by stretch (mechanical means). Two possibilities are shown: (*a*) an hypothesised stretch activation pathway mediated solely by physical changes in lipids in the plasma membrane bilayer; (*b*) an hypothesised stretch-activation pathway mediated by the interaction of TRPC1 with the cytoskeleton and other scaffolding proteins

a Triton S-100 insoluble complex in the plasma membrane. The results of co-immunoprecipitation experiments conducted with astrocytes indicate that TRPC1 can interact with IP$_3$R1 and with SERCA 2B. Moreover, the analysis of high-resolution immunofluorescence images indicated that TRPC1, IP$_3$R1 and SERCA 2B are co-located in astrocytes (Golovina 2005). The re-

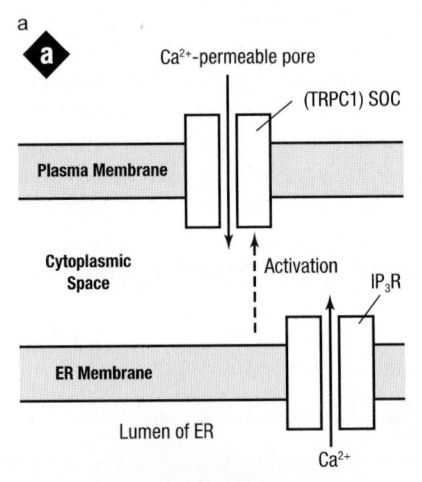

a

a A decrease in ER [Ca^{2+}] is required to initiate SOC (TRPCl) activation via an unknown activation pathway

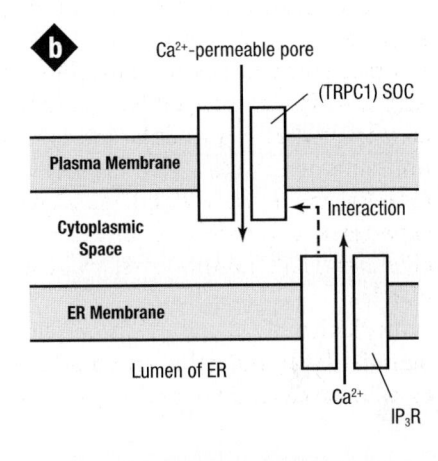

b A decrease in ER [Ca^{2+}] is required to initiate SOC (TRPC1) activation via TRPC1-InsP$_3$R interaction

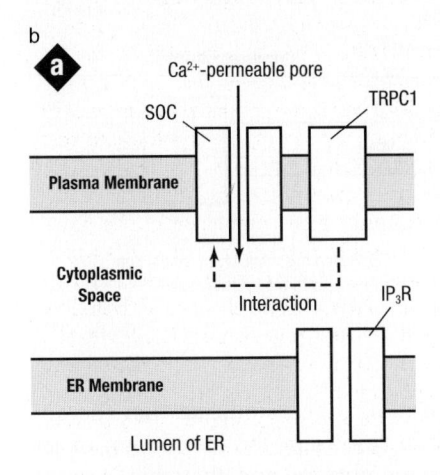

b

a A decrease in ER [Ca^{2+}] is required to initiate SOC activation. TRPC1 (in plasma membrane) is a necessary acessory protein

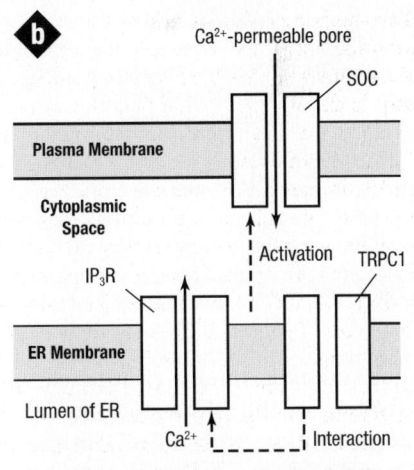

b A decrease in ER [Ca^{2+}] is required to initiate SOC activation. TRPC1 (in the ER) is a necessary acessory protein

c

a The TRPC1 channel (not a SOC) is activated by interaction with the IP$_3$R protein located in the plasma membrane

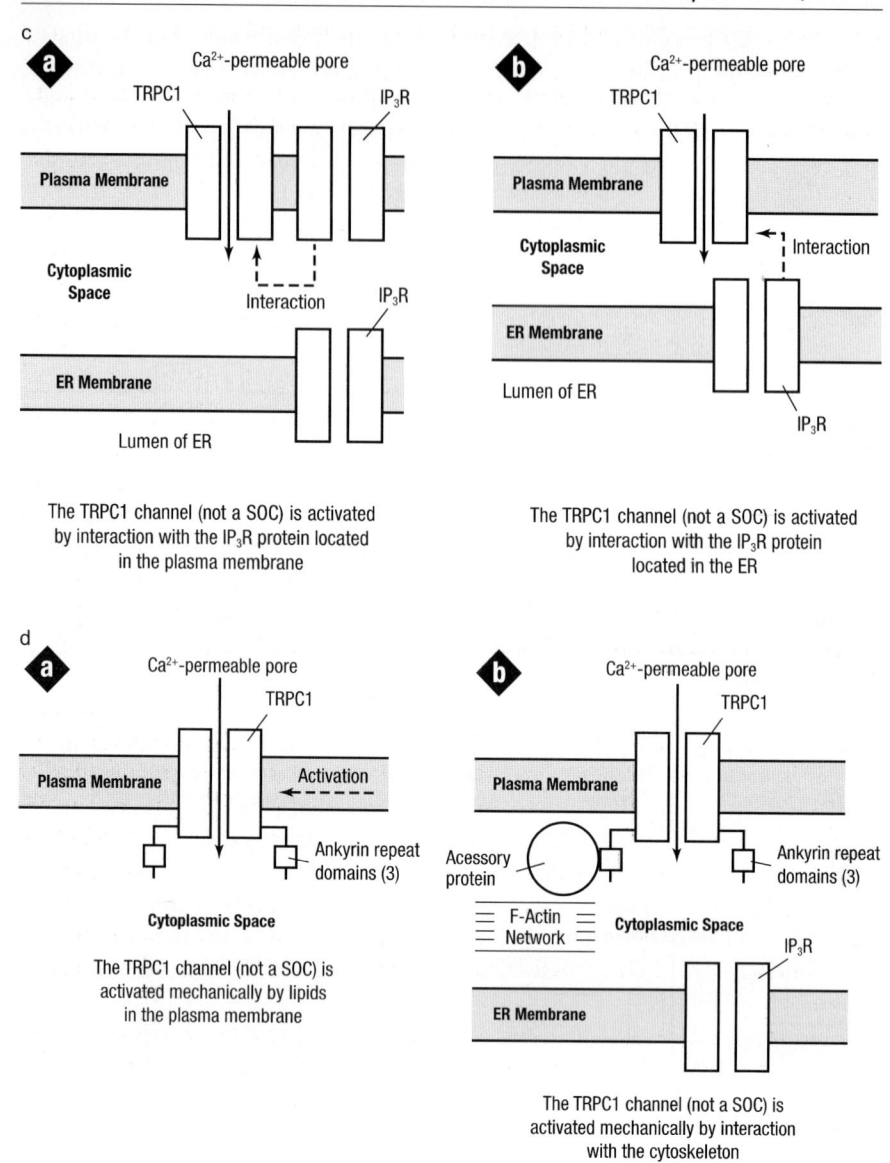

b The TRPC1 channel (not a SOC) is activated by interaction with the IP$_3$R protein located in the ER

d

a The TRPC1 channel (not a SOC) is activated mechanically by lipids in the plasma membrane

b The TRPC1 channel (not a SOC) is activated mechanically by interaction with the cytoskeleton

Fig. 3 continued

sults of experiments conducted with endothelial cells indicate that the activation of the RhoA pathway by thrombin results in the interaction of an IP$_3$R subtype with TRPC1 at the plasma membrane, and that this is associated with the activation of SERCA inhibitor-initiated Ca^{2+} inflow (Mehta et al. 2003). The observation that latrunculin, which inhibits actin polymerisation, inhibited the RhoA-facilitated association of TRPC1 and IP$_3$R and

Ca^{2+} entry suggested that this process is dependent on F-actin polymerisation.

Yuan et al. (2003) have shown, chiefly using co-immunoprecipitation techniques, that in HEK293 cells, ectopically expressed TRPC1 interacts with the adapter protein, Homer, and with IP$_3$R1. TRPC1 and Homer 1 were also shown to co-immunoprecipitate in brain extracts. The results of further experiments in which mutants of Homer were ectopically expressed in HEK293 cells suggest that Homer modulates the interaction between TRPC1 and IP$_3$R1 which, in turn, affects Ca^{2+} entry. Treatment of the cells with thapsigargin reduced the amount of TRPC1 which co-immunoprecipitated with Homer. In HEK293 cells in which TRPC1 and mutants of TRPC1 were expressed ectopically, it was found that mutants of TRPC1 which cannot bind Homer exhibit increased Ca^{2+} entry. Moreover, ectopic expression of Homer 1 in submandibular gland cells, or ablation of Homer 1 in pancreatic acinar cells by construction of a Homer 1-negative mouse, also altered Ca^{2+} entry. Yuan et al. (2003) suggest that Homer regulates the interaction of TRPC1 with IP$_3$R1 such that, in the resting state, the TRPC1–Homer–IP$_3$R1 complex is formed, whereas store depletion leads to disassociation of the complex and enhanced Ca^{2+} entry [cf. the conclusions for platelets above where it is proposed that store-depletion enhances the IP$_3$R2–TRPC1 interaction (Rosado and Sage 2000)].

The results described above have generally been interpreted as providing evidence that TRPC1 interacts with IP$_3$Rs, and this protein–protein interaction is a necessary part of the mechanism by which at least one type of SOC, composed of TRPC1 polypeptides, is activated [Fig. 2a (part b)]. However, there are questions which require further investigation: first, whether TRPC1 does, indeed, constitute the pore of some SOCs; second, whether the TRPC1-IP$_3$R1 interaction occurs in vivo (i.e. in the environment of the native cell); and third, whether the TRPC1 and IP$_3$R1 proteins involved in this interaction are localised in the plasma membrane or in an intracellular membrane.

Bolotina and Csutora (2005) have suggested that one interpretation of these and other results is that in some cell types there are plasma membrane Ca^{2+}-permeable channels, including those composed of TRPC1, which are activated by the binding of IP$_3$ to IP$_3$R in a pathway involving the direct interaction of the IP$_3$R protein with the channel protein, but not involving a requirement for Ca^{2+} release from intracellular stores (Fig. 2c). They called these "IP$_3$R-operated Ca^{2+}-permeable channels". Thus it is possible that TRPC1 constitutes IP$_3$R-operated Ca^{2+} permeable channels in the plasma membrane where the mechanism of TRPC1 activation involves IP$_3$ and IP$_3$R leading to channel activation. This would not require store depletion as a causative step, although store depletion may be associated with channel activation.

5.3
Mechanical: Stretch-Activated

There is some evidence to indicate that TRPC1 can be activated by mechanical means. Studies with *Xenopus laevis* oocytes suggest that TRPC1 constitutes a mechanosensitive cation channel (MscCa) channel (Maroto et al. 2005). MscCa, which is also present in some mammalian cells, is thought to be involved in the regulation of cell volume and locomotion. Maroto et al. (2005) prepared extracts of oocyte plasma membrane proteins and reconstituted these into liposomes. They showed, by patch-clamp recording, that MscCa activity can be observed in these liposomes. Sub-fractionation of the membrane proteins yielded a fraction which gave high MscCa activity in liposomes and contained TRPC1. Ectopic expression of human TRPC1, which is homologous to *Xenopus* TRPC1, in oocytes resulted in a large increase in a current attributed to MscCa. In other experiments, administration of antisense RNA directed against TRPC1 abolished the endogenous *Xenopus* MscCa current. Transfection of CHO-K1 cells, which do not normally express mechanosensitive channel activity, with DNA encoding hTRPC1 resulted in significantly increased mechanosensitive channel expression.

In addressing the mechanism by which TRPC1 might be activated by stretch, Maroto et al. (2005) observed that MscCa activity is detected both in oocyte membrane vesicles apparently deficient in the cytoskeleton and when partially purified membrane proteins, including TRPC1, are reconstituted in liposomes. These preparations should contain no elements of the cytoskeleton. The authors proposed that TRPC1 can be activated solely by tension developed in the lipid bilayer without a requirement for direct protein–protein interaction. Such a mechanism, shown schematically in Fig. 2d (part *a*) is responsible for the activation of stretch-activated channels in bacteria (Maroto et al. 2005). As described in Sect. 2.1, TRPC1 also contains three N-terminal cytoplasmic ankyrin domains and binding sites for numerous regulatory proteins. Therefore it is also possible that a pathway initiated by stretch but involving the cytoskeleton network, scaffolding proteins, and protein–protein interaction with TRPC1 [shown schematically in Fig. 2d (*b*)] either alone or in combination with the membrane lipid pathway, is responsible for the mechanosensitive activation of TRPC1 in vivo.

Ablation of TRPC1 expression in H4IIE liver cells using siRNA was found to be associated with an enhancement of volume recovery after exposure to hypertonic and hypotonic media (Chen and Barritt 2003). This observation provides some evidence to suggest that TRPC1 may be involved in the regulation of liver cell volume (Chen and Barritt 2003) possibly through a stretch-activated pathway. Further experiments are required to test this idea, and to investigate whether TRPC1 constitutes, or is a component of, stretch-activated non-selective cation channels in many cell types, and to further elucidate the mechanism of activation by stretch (Ramsey et al. 2006).

5.4
Conclusions About the Role and Mechanisms of Activation of the TRPC1 Protein

The results of all experiments with TRPC1 reported to date provide evidence to indicate that TRPC1:

- May be a component of a SOC activated by either an unknown mechanism(s) or by direct interaction between TRPC1 and IP$_3$R (Fig. 2a);

- May be an accessory protein required for the activation of a SOC which is not, itself, composed of TRPC1 polypeptides (Fig. 2b);

- May have a role as an intracellular Ca^{2+}-permeable channel (Fig. 2b);

- May be a component of a non-SOC plasma membrane Ca^{2+}-permeable channel activated by direct interaction with IP$_3$R (Fig. 2c);

- May be a component of a non-SOC plasma membrane Ca^{2+}-permeable channel activated by stretch (mechanical pathway; Fig. 2d).

At present none of these hypotheses for the function and activation of TRPC1 has been eliminated. It is possible that there is one role and mechanism of activation for TRPC1 channels in the plasma membrane which underlies the various and sometimes apparently diverse observations on the nature and activation of TRPC1. For example, if TRPC1 were a Ca^{2+}-permeable channel located in the plasma membrane activated by stretch, it is possible that one of the consequences of the action of SERCA inhibitors and/or store depletion is to modify the shape of the ER near the plasma membrane, alter the conformation of the cytoskeleton and initiate activation of TRPC1 by a "mechanical" pathway which may also involve IP$_3$R (cf. Beech 2006).

5.5
Modulation of TRPC1 Function

While the mechanism of activation of TRPC1 is not yet clearly defined, it remains difficult to distinguish the pathway or pathways central to the activation mechanism from those pathways which may be involved in modulation or regulation of the TRPC1 channel. The activation and/or activity of TRPC1 is modulated by Ca^{2+} and calmodulin (Singh et al. 2002), protein kinase C (Ahmmed et al. 2004; Larsson et al. 2005) and other protein kinases, (reviewed in Beech 2005). Like many of the other TRPC proteins, TRPC1 is inhibited by Ca^{2+} and calmodulin, and this is thought to be part of a feedback inhibition mechanism (Tang et al. 2001; Singh et al. 2002). Two calmodulin binding sites, aa719–751 and aa758–793, in the C-terminus of TRPC1 have been identified (Singh et al. 2002). Deletion of aa758–793, but not deletion of the aa719–751 site, was found to reduce Ca^{2+}-dependent inhibition of SERCA inhibitor-initiated Ca^{2+} inflow in sub-mandibular gland cells, suggesting that aa758–793 is involved in Ca^{2+}/calmodulin-mediated inhibition of TRPC1.

6
Pharmacology

As is the case for most other members of the TRPC channel family, the absence of known high-affinity and selective inhibitors of TRPC1 has hampered studies of the cellular roles and functions of TRPC1, its broader biological roles, its mechanism of activation and its use as a potential pharmaceutical target. There are no known small-molecule high-infinity inhibitors of either the activation pathway for TRPC1 or of cation entry through the channel pore (channel blockers). Cation entry through TRPC1 is blocked by 10 µM Gd^{3+} (Zitt et al. 1996), 50 µM La^{3+} (Litschinger et al. 2000), 100 µM 2-aminoethoxydiphenyl borate (2APB) (Zagranichnaya et al. 2005) and 10 µM SKF96365 (Zagranichnaya et al. 2006). However, these agents are not specific for TRPC1 and inhibit a number of other TRP channels and some members of other Ca^{2+}-permeable channels (reviewed in Parekh and Putney 2005; Putney 2005). In the case of the organic inhibitors, further experiments to test their effects on defined TRPC1 currents are required.

As described in Sect. 5.1, two polyclonal antibodies raised against peptides which constitute part of the external component of the putative pore-forming hydrophobic sequence of TRPC1 inhibit SERCA inhibitor-stimulated Ca^{2+} entry in vascular smooth muscle cells (Xu and Beech 2001) and platelets (Rosado et al. 2002). The mechanisms by which these anti-TRPC1 antibodies inhibit Ca^{2+} entry are not fully understood, but it is proposed that the antibodies may block the pore. These results suggest that antibodies directed against the extracellular domain of the pore of TRPC1 may provide promising pharmacological inhibitors of this and other channels (Xu et al. 2005).

7
Biological Relevance and Emerging Biological Roles for TRPC1

7.1
Immediate Consequences of Na^+ and Ca^{2+} Entry Through TRPC1 Channels

In discussing present knowledge of the biological functions of TRPC1, it is useful to consider first the immediate consequences of Na^+ and Ca^{2+} entry through open TRPC1 channels, and then the likely downstream events initiated by changes in the intracellular concentrations of these ions. On the basis of studies to date, chiefly involving ectopic expression of TRPC1 where the composition of the resulting channel is not precisely known, it can be predicted that TRPC1 is a non-selective cation channel which facilitates Na^+ and Ca^{2+} entry across the plasma membrane (Eder et al. 2005). Therefore, Na^+ entry as well as Ca^{2+} entry through TRPC1 is likely to be physiologically significant (Eder et al. 2005). By contrast, CRAC (SOC) channels in lymphocytes, mast cells and hepatocytes, with a high selectivity for Ca^{2+}, admit predominantly Ca^{2+} (Parekh and Putney 2005).

The concentration of Na^+ in the cytoplasmic space ($[Na^+]_{cyt}$) is normally maintained at about 10 mM. In the excitable cells, Na^+ entry through TRPC1 would lead to an increase in $[Na^+]_{cyt}$, membrane depolarisation and, among other things, to the activation of voltage-operated Ca^{2+} channels and further Ca^{2+} entry. In non-excitable cells, Na^+ entry through TRPC1 is likely to drive membrane potential more positive, to modify cell volume and affect other cell functions. Na^+ is involved in cation transporters, including (Na^++K^+) ATP-ase, Na^+-anion co-transporters and plasma membrane Na^+–Ca^{2+} exchange. Thus, an increase in $[Na^+]_{cyt}$ would be expected to activate these transporters and exchangers including, as mentioned above, the possibility of Ca^{2+} inflow through the Na^+–Ca^{2+} exchange working in reverse mode (Eder et al. 2005).

Ca^{2+} entry through TRPC1 may have several intracellular functions, depending in part on the spatial location of the TRPC1 channels (Beech 2005). One of these may be to replenish Ca^{2+} in intracellular stores since, as discussed above, there is some evidence to indicate that in some cell types TRPC1 is activated by store depletion. Further experiments are required to test whether Ca^{2+} which enters cells through TRPC1 refills intracellular stores, although some studies suggest that non-selective cation channels are less effective than Ca^{2+}-selective SOCs for this purpose (Gregory et al. 2003).

7.2
Downstream Events

Experiments involving the ablation of TRPC1 expression or physiological interventions to increase or decrease TRPC1 expression have provided evidence that Na^+ and Ca^{2+} inflow through TRPC1 is involved in the initiation, or modulation, of a number of intracellular processes including smooth muscle contraction, stem cell differentiation, endothelium-induced arterial contraction, endothelial cell permeability, salivary gland secretion, glutamate-mediated neurotransmission, growth cone movement, neuroprotection, neuronal differentiation and the regulation of liver cell volume (reviewed in Beech 2005; Ambudkar 2006). To date, there are no reports of the successful construction of a TRPC1 knock-out mouse which may provide a useful experimental model for further studies of the downstream biological functions of TRPC1.

One of the most clearly defined roles of TRPC1 is its involvement in the regulation of movement of nerve growth cones. Wang and Poo (2005) and Shim et al. (2005) have provided evidence to indicate that in *Xenopus* brain a normal function of TRPC1 in nerve cells is in the pathway by which netrin 1, brain-derived neurotrophic factor, and myelin-associated glycoprotein (extracellular guidance factors) regulate the extension and turning of growth cones. Wang and Poo (2005), using in part the non-selective inhibitor SKF96365 to inhibit cation entry through TRPC1, and a morpholino oligonucleotide targeted to TRPC1 to ablate TRPC1 expression, obtained evidence to suggest that activation of TRPC1 by netrin-1 is a necessary step in the mechanisms by which this extracellular signal guides the movement of nerve growth cones. They provided evidence to suggest that Ca^{2+} entry through TRPC1 leads to partial depolarisation of the plasma membrane, subsequent activation of voltage-operated Ca^{2+} channels and Ca^{2+} entry, which, in turn, increases $[Ca^{2+}]_{cyt}$ and leads to altered growth cone tuning.

Shim et al. (2005) also employed morpholino oligonucleotides to ablate TRPC1 expression, as well as the non-selective inhibitors SKF96365 and La^{3+} to inhibit TRPC1 function. They demonstrated a requirement for Ca^{2+} entry through TRPC1 for both growth cone tuning in response to neurotrophic factors, and for midline guidance of axons of commissural interneurons in the developing spinal cord. On the assumption that TRPC1 forms SOCs, they speculated that the mechanism of activation of TRPC1 by netrin-1 involves the binding of netrin-1 to its receptor, induction of the phosphorylation of tyrosine residues on the cytoplasmic domain of the receptor, activation of phospholipase C_δ, the generation of IP_3, release of Ca^{2+} from the ER and activation of TRPC1 (Shim et al. 2005).

8
Conclusions

The results of studies reported to date have defined the main features of the structure and topology of TRPC1 as a pore-forming integral membrane protein, and have identified many of the proteins which can, under various circumstances, interact with TRPC1. The TRPC1 protein is particularly interesting in that it appears to have diverse biological functions which are only partly understood. Moreover, elucidation of the immediate cellular localisations of TRPC1, its functions, mechanisms of action and role in SOCs continues to present considerable challenges. Further investigations are required to provide a more complete understanding of these properties. These investigations will be aided by the development of additional specific high-affinity anti-TRPC1 antibodies, and further studies of the structure of the TRPC1 polypeptide which, among other benefits, will aid in a search for inhibitors with high affinity which could be used for pharmaceutical intervention.

Acknowledgements Research conducted in the authors' laboratories on TRPC1 in liver cells was supported by grants from the National Health and Medical Research Council, the Australian Research Council and the Flinders Medical Centre Research Foundation. Permission of Dr. V. Golovina and The Physiological Society to reproduce Fig. 1b is gratefully acknowledged.

References

Ahmmed GU, Mehta D, Vogel S, Holinstat M, Paria BC, Tiruppathi C, Malik AB (2004) Protein kinase Calpha phosphorylates the TRPC1 channel and regulates store-operated Ca^{2+} entry in endothelial cells. J Biol Chem 279:20941–20949

Ambudkar IS (2006) Ca^{2+} signalling microdomains:platforms for the assembly and regulation of TRPC channels. Trends Pharmacol Sci 27:25–32

Beech D (2005) TRPC1: store-operated channel and more. Pflugers Arch 451:53–60

Beech DJ, Xu SZ, McHugh D, Flemming R (2003) TRPC1 store-operated cationic channel subunit. Cell Calcium 33:433–440

Bergdahl A, Gomez MF, Dreja K, Xu SZ, Adner M, Beech DJ, Broman J, Hellstrand P, Sward K (2003) Cholesterol depletion impairs vascular reactivity to endothelin-1 by reducing store-operated Ca^{2+} entry dependent on TRPC1. Circ Res 93:839–847

Bergdahl A, Gomez MF, Wihlborg AK, Erlinge D, Eyjolfson A, Xu SZ, Beech DJ, Dreja K, Hellstrand P (2005) Plasticity of TRPC expression in arterial smooth muscle: correlation with store-operated Ca^{2+} entry. Am J Physiol 288:C872–880

Birnbaumer L, Zhu X, Jiang MS, Boulay G, Peyton M, Vannier B, Brown D, Platani D, Sadeghi H, Stefani E, Birnbaumer M (1996) On the molecular basis and regulation of cellular capacitative calcium entry: roles for Trp proteins. Proc Natl Acad Sci U S A 93:15195–15202

Bobanovic L, Laine M, Peterson C, Bennett D, Berridge M, Lipp P, Ripley S, Bootman M (1999) Molecular cloning and immunolocalization of a novel vertebrate trp homologue from Xenopus. Biochem J 340:593–599

Bollimuntha S, Cornatzer E, Singh BB (2005) Plasma membrane localization and function of TRPC1 is dependent on its interaction with beta-tubulin in retinal epithelium cells. Vis Neurosci 22:163–170

Bolotina V, Csutora P (2005) CIF and other mysteries of the store-operated Ca^{2+}=entry pathway. Trends Biochem Sci 30:378–387

Boulay G, Zhu X, Peyton M, Jiang MS, Hurst R, Stefani E, Birnbaumer L (1997) Cloning and expression of a novel mammalian homolog of the Drosphilia transient receptor potential (trp) involved in calcium entry secondary to activation of receptors coupled by the Gq class of G protein. J Biol Chem 272:29672–29680

Brazer SC, Singh BB, Liu X, Swaim W, Ambudkar IS (2003) Caveolin-1 contributes to assembly of store-operated Ca^{2+} influx channels by regulating plasma membrane localization of TRPC1. J Biol Chem 278:27208–27215

Brereton H, Harland M, Auld A, Barritt G (2000) Evidence that the TRP-1 protein is unlikely to account for store-operated Ca^{2+} inflow in Xenopus laevis oocytes. Mol Cell Biochem 214:63–74

Brereton HM, Chen J, Rychkov G, Harland ML, Barritt GJ (2001) Maitotoxin activates an endogenous non-selective cation channel and is an effective initiator of the activation of the heterologously expressed hTRPC-1 (transient receptor potential) non-selective cation channel in H4-IIE liver cells. Biochim Biophys Acta 1540:107–126

Brough GH, Wu S, Cioffi D, Moore TM, Li M, Dean N, Stevens T (2001) Contribution of endogenously expressed Trp1 to a Ca^{2+}-selective, store-operated Ca^{2+} entry pathway. FASEB J 15:1727–1738

Brownlow S, Sage S (2003) Rapid agonist-evoked coupling of type II Ins(1,4,5)P3 receptor with human transient receptor potential (hTRPC1) channels in human platelets. Biochem J 375:697–704

Brownlow S, Sage S (2005) Transient receptor potential protein subunit assembly and membrane distribution in human platelets. Thromb Haemost 94:839–845

Brownlow SL, Harper AG, Harper MT, Sage SO (2004) A role for hTRPC1 and lipid raft domains in store-mediated calcium entry in human platelets. Cell Calcium 35:107–113

Brueggemann LI, Markun DR, Henderson KK, Cribbs LL, Byron KL (2006) Pharmacological and electrophysiological characterization of store-operated currents and capacitative Ca^{2+} entry in vascular smooth muscle cells. J Pharmacol Exp Ther 317:488–499

Cai S, Fatherazi S, Presland RB, Belton CM, Roberts FA, Goodwin PC, Schubert MM, Izutsu KT (2006) Evidence that TRPC1 contributes to calcium-induced differentiation of human keratinocytes. Pflugers Arch 452:43–52

Chen J, Barritt GJ (2003) Evidence that TRPC1 (transient receptor potential canonical 1) forms a Ca^{2+}-permeable channel linked to the regulation of cell volume in liver cells obtained using small interfering RNA targeted against TRPC1. Biochem J 373:327–336

Clapham D, Julius D, Montell C, Schultz G (2005) International Union of Pharmacology. XLIX. Nomenclature and structure-function relationships of transient receptor potential channels. Pharmacol Rev 57:427–450

Crousillac S, Lerouge M, Rankin M, Gleason E (2003) Immunolocalization of TRPC channel subunits 1 and 4 in the chicken retina. Vis Neurosci 20:453–463

Dalrymple A, Slater DM, Beech D, Poston L, Tribe RM (2002) Molecular identification and localization of Trp homologues, putative calcium channels, in pregnant human uterus. Mol Hum Reprod 8:946–951

Delmas P, Wanaverbecq N, Abogadie FC, Mistry M, Brown DA (2002) Signaling microdomains define the specificity of receptor-mediated InsP3 pathways in neurons. Neuron 34:209–220

Dohke Y, Oh YS, Ambudkar IS, Turner RJ (2004) Biogenesis and topology of the transient receptor potential Ca^{2+} channel TRPC1. J Biol Chem 279:12242–12248

Eder P, Poteser M, Romanin C, Groschner K (2005) Na^+ entry and modulation of Na^+/Ca^{2+} exchange as a key mechanism of TRPC signalling. Pflugers Arch 451:99–104

Engelke M, Friedrich O, Budde P, Schafer C, Niemann U, Zitt C, Jungling E, Rocks O, Luckhoff A, Frey J (2002) Structural domains required for channel function of the mouse transient receptor potential protein homologue TRP1beta. FEBS Lett 523:193–199

Fiorio-Pla A, Maric D, Brazer SC, Giacobini P, Liu X, Chang YH, Ambudkar IS, Barker JL (2005) Canonical transient receptor potential 1 plays a role in basic fibroblast growth factor (bFGF)/FGF receptor-1-induced Ca^{2+} entry and embryonic rat neural stem cell proliferation. J Neurosci 25:2687–2701

Glazebrook PA, Schilling WP, Kunze DL (2005) TRPC channels as signal transducers. Pflugers Arch 451:125–130

Goel M, Sinkins WG, Schilling WP (2002) Selective association of TRPC channel subunits in rat brain synaptosomes. J Biol Chem 277:48303–48310

Goel M, Sinkins W, Zuo CD, Estacion M, Schilling W (2006) Identification and localization of TRPC channels in rat kidney. Am J Physiol 290:F1241–F1252

Golovina VA (2005) Visualization of localized store-operated calcium entry in mouse astrocytes. Close proximity to the endoplasmic reticulum. J Physiol 564:737–749

Gregory RB, Sykiotis D, Barritt GJ (2003) Evidence that store-operated Ca^{2+} channels are more effective than intracellular messenger-activated non-selective cation channels in refilling rat hepatocyte intracellular Ca^{2+} stores. Cell Calcium 34:241–251

Hassock S, Zhu M, Trost C, Flockerzi V, Authi K (2002) Expression and role of TRPC proteins in human platelets: evidence that TRPC6 forms the store-independent calcium entry channel. Blood 100:2801–2811

Hofmann T, Schaefer M, Schultz G, Gudermann T (2002) Subunit composition of mammalian transient receptor potential channels in living cells. Proc Natl Acad Sci USA 99:7461–7466

Kim SJ, Kim YS, Yuan JP, Petralia RS, Worley PF, Linden DJ (2003) Activation of the TRPC1 cation channel by metabotropic glutamate receptor mGluR1. Nature 426:285–291

Kiselyov K, Kim JY, Zeng W, Muallem S (2005) Protein-protein interaction and function TRPC channels. Pflugers Arch 451:116–124

Kunichika N, Yu Y, Remillard CV, Platoshyn O, Zhang S, Yuan JX (2004) Overexpression of TRPC1 enhances pulmonary vasoconstriction induced by capacitative Ca^{2+} entry. Am J Physiol Lung Cell Mol Physiol 287:L962–L969

Kunzelmann-Marche C, Freyssinet JM, Martinez MC (2002) Loss of plasma membrane phospholipid asymmetry requires raft integrity. J Biol Chem 277:19876–19881

Larsson KP, Peltonen HM, Bart G, Louhivuori LM, Penttonen A, Antikainen M, Kukkonen JP, Akerman KE (2005) Orexin-A-induced Ca^{2+} entry: evidence for involvement of trpc channels and protein kinase C regulation. J Biol Chem 280:1771–1781

Lin MJ, Leung GP, Zhang WM, Yang XR, Yip KP, Tse CM, Sham JS (2004) Chronic hypoxia-induced upregulation of store-operated and receptor-operated Ca^{2+} channels in pulmonary arterial smooth muscle cells: a novel mechanism of hypoxic pulmonary hypertension. Circ Res 95:496–505

Lintschinger B, Balzer-Geldsetzer M, Baskaran T, Graier WF, Romanin C, Zhu MX, Groschner K (2000) Coassembly of Trp1 and Trp3 proteins generates diacylglycerol- and Ca^{2+}-sensitive cation channels. J Biol Chem 275:27799–27805

Liu X, Wang W, Singh BB, Lockwich T, Jadlowiec J, OConnell B, Wellner R, Zhu MX, Ambudkar IS (2000) Trp1, a candidate protein for the store-operated Ca^{2+} influx mechanism in salivary gland cells. J Biol Chem 275:3403–3411

Liu X, Singh BB, Ambudkar IS (2003) TRPC1 is required for functional store-operated Ca^{2+} channels. Role of acidic amino acid residues in the S5-S6 region. J Biol Chem 278:11337–11343

Liu X, Bandyopadhyay BC, Singh BB, Groschner K, Ambudkar IS (2005) Molecular analysis of a store-operated and 2-acetyl-sn-glycerol-sensitive non-selective cation channel. Heteromeric assembly of TRPC1-TRPC3. J Biol Chem 280:21600–21606

Lockwich TP, Liu X, Singh BB, Jadlowiec J, Weiland S, Ambudkar IS (2000) Assembly of Trp1 in a signaling complex associated with caveolin-scaffolding lipid raft domains. J Biol Chem 275:11934–11942

Lussier MP, Cayouette S, Lepage PK, Bernier CL, Francoeur N, St-Hilaire M, Pinard M, Boulay G (2005) MxA, a member of the dynamin superfamily interacts with the ankyrin-like repeat domain of TRPC. J Biol Chem 280:19393–19400

Ma R, Rundle D, Jacks J, Koch M, Downs T, Tsiokas L (2003) Inhibitor of myogenic family, a novel suppressor of store-operated currents through an interaction with TRPC1. J Biol Chem 278:52763–52772

Maroto R, Raso A, Wood TG, Kurosky A, Martinac B, Hamill OP (2005) TRPC1 forms the stretch-activated cation channel in vertebrate cells. Nat Cell Biol 7:179–185

Mehta D, Ahmmed GU, Paria BC, Holinstat M, Voyno-Yasenetskaya T, Tiruppathi C, Minshall RD, Malik AB (2003) RhoA interaction with inositol 1,4,5-trisphosphate receptor and transient receptor potential channel-1 regulates Ca^{2+} entry. Role in signaling increased endothelial permeability. J Biol Chem 278:33492–33500

Mori Y, Wakamori M, Miyakawa T, Hermosura M, Hara Y, Nishida M, Hirose K, Mizushima A, Kurosaki M, Mori E, Gotoh K, Okada T, Fleig A, Penner R, Iino M, Kurosaki T (2002) Transient receptor potential 1 regulates capacitative Ca^{2+} entry and Ca^{2+} release from endoplasmic reticulum in B lymphocytes. J Exp Med 195:673–681

Nishida M, Sugimoto K, Hara Y, Mori E, Morii T, Kurosaki T, Mori Y (2003) Amplification of receptor signalling by Ca^{2+} entry-mediated translocation and activation of PLCγ2 in B lymphocytes. EMBO J 22:4677–4688

Ong HL, Chen J, Chataway T, Brereton H, Zhang L, Downs T, Tsiokas L, Barritt G (2002) Specific detection of the endogenous transient receptor potential (TRP)-1 protein in liver and airway smooth muscle cells using immunoprecipitation and Western-blot analysis. Biochem J 364:641–648

Owsianik G, Talavera K, Voets T, Nilius B (2006) Permeation and selectivity of TRP channels. Annu Rev Physiol 68:685–717

Parekh A, Putney J (2005) Store-operated calcium channels. Physiol Rev 85:757–810

Putney J (2005) Physiological mechanisms of TRPC activation. Pflugers Arch 451:29–34

Ramsey I, Delling M, Clapham D (2006) An introduction to TRP channels. Annu Rev Physiol 68:619–647

Rao J, Platoshyn O, Golovina V, Liu L, Zou T, Marasa B, Turner D, Yuan J, Wang JY (2006) TRPC1 functions as a store-operated Ca^{2+} channel in intestinal epithelial cells and regulates early mucosal restitution after wounding. Am J Physiol 290:G782–792

Rosado J, Sage S (2001) Activation of store-mediated calcium entry by secretion-like coupling between the inositol 1,4,5-triphosphate receptor type II and human transient receptor potential (hTrp1) channels in human platelets. Biochem J 356:191–198

Rosado JA, Sage SO (2000) Coupling between inositol 1,4,5-triphosphate receptors and human transient receptor potential channel 1 when intracellular Ca^{2+} stores are depleted. Biochem J 350:631–635

Rosado JA, Brownlow SL, Sage SO (2002) Endogenously expressed Trp1 is involved in store-mediated Ca^{2+} entry by conformational coupling in human platelets. J Biol Chem 277:42157–42163

Sakura H, Ashcroft FM (1997) Identification of four trp1 gene variants murine pancreatic β-cells. Diabetologia 40:528–532

Shim S, Goh E, Ge S, Sailor K, Yuan J, Roderick H, Bootman M, Worley P, Song H, Ming GL (2005) XTRPC1-dependent chemotropic guidance of neuronal growth cones. Nat Neurosci 8:730–735

Singh BB, Liu X, Tang J, Zhu MX, Ambudkar IS (2002) Calmodulin regulates Ca^{2+}-dependent feedback inhibition of store-operated Ca^{2+} influx by interaction with a site in the C terminus of TrpC1. Mol Cell 9:739–750

Sinkins W, Estacion M, Schilling W (1998) Functional expression of TrpC1: a human homologue of the Drosophila Trp channel. Biochem J 331:331–339

Sinkins WG, Goel M, Estacion M, Schilling WP (2004) Association of immunophilins with mammalian TRPC channels. J Biol Chem 279:34521–34529

Strubing C, Krapivinsky G, Krapivinsky L, Clapham DE (2001) TRPC1 and TRPC5 form a novel cation channel in mammalian brain. Neuron 29:645–655

Strubing C, Krapivinsky G, Krapivinsky L, Clapham DE (2003) Formation of novel TRPC channels by complex subunit interactions in embryonic brain. J Biol Chem 278:39014–39019

Sweeney M, Yu Y, Platoshyn O, Zhang S, McDaniel SS, Yuan JX (2002) Inhibition of endogenous TRP1 decreases capacitative Ca^{2+} entry and attenuates pulmonary artery smooth muscle cell proliferation. Am J Physiol 283:L144–L155

Tang J, Lin Y, Zhang Z, Tikunova S, Birnbaumer L, Zhu MX (2001) Identification of common binding sites for calmodulin and inositol 1,4,5-trisphosphate receptors on the carboxyl termini of trp channels. J Biol Chem 276:21303–21310

Tomita Y, Kaneko S, Funayama M, Kondo H, Satoh M, Akaike A (1998) Intracellular Ca^{2+} store-operated influx of Ca^{2+} through TRP-R, a rat homologue of TRP, expressed in Xenopus oocytes. Neurosci Lett 248:195–198

Tsiokas L, Arnould T, Zhu C, Kim E, Walz G, Sukhatme VP (1999) Specific association of the gene product of PKD2 with the TRPC1 channel. Proc Natl Acad Sci USA 96:3934–3939

Tu CL, Chang W, Bikle DD (2005) Phospholipase cgamma1 is required for activation of store-operated channels in human keratinocytes. J Invest Dermatol 124:187–197

Uehara K (2005) Localization of TRPC1 channel in the sinus endothelial cells of rat spleen. Histochem Cell Biol 123:347–356

Vaca L, Sampieri A (2002) Calmodulin modulates the delay period between release of calcium from internal stores and activation of calcium influx via endogenous TRP1 channels. J Biol Chem 277:42178–42187

Vandebrouck C, Martin D, Colson-Van Schoor M, Debaix H, Gailly P (2002) Involvement of TRPC in the abnormal calcium influx observed in dystrophic (mdx) mouse skeletal muscle fibers. J Cell Biol 158:1089–1096

Vazquez G, Wedel B, Aziz O, Trebak M, Putney J (2004) The mammalian TRPC cation channels. Biochim Biophys Acta 1742:21–36

von-Bohlen-Und-Halbach O, Hinz U, Unsicker K, Egorov AV (2005) Distribution of TRPC1 and TRPC5 in medial temporal lobe structures of mice. Cell Tissue Res 322:201–206

Wang GX, Poo MM (2005) Requirement of TRPC channels in netrin-1-induced chemotropic turning of nerve growth cones. Nature 434:898–904

Wang W, OConnell B, Dykeman R, Sakai T, Delporte C, Swaim W, Zhu X, Birnbaumer L, Ambudkar IS (1999) Cloning of Trp1beta isoform from rat brain: immunodetection and localization of the endogenous Trp1 protein. Am J Physiol 276:C969–C979

Wes PD, Chevesich J, Jeromin A, Rosenberg C, Stetten G, Montell C (1995) TRPC1, a human homolog of a Drosophila store-operated channel. Proc Natl Acad Sci USA 92:9652–9656

Wu X, Babnigg G, Villereal ML (2000) Functional significance of human trp1 and trp3 in store-operated Ca^{2+} entry in HEK-293 cells. Am J Physiol 278:C526–C536

Wu X, Zagranichnaya TK, Gurda GT, Eves EM, Villereal ML (2004) A TRPC1/TRPC3-mediated increase in store-operated calcium entry is required for differentiation of H19-7 hippocampal neuronal cells. J Biol Chem 279:43392–43402

Xu SZ, Beech DJ (2001) TrpC1 is a membrane-spanning subunit of store-operated Ca^{2+} channels in native vascular smooth muscle cells. Circ Res 88:84–87

Xu SZ, Zeng F, Lei M, Li J, Gao B, Xiong C, Sivaprasadarao A, Beech D (2005) Generation of functional ion-channel tools by E3 targeting. Nat Biotechnol 23:1234–1235

Yuan JP, Kiselyov K, Shin DM, Chen J, Shcheynikov N, Kang SH, Dehoff MH, Schwarz MK, Seeburg PH, Muallem S, Worley PF (2003) Homer binds TRPC family channels and is required for gating of TRPC1 by IP3 receptors. Cell 114:777–789

Zagranichnaya T, Wu X, Villereal M (2005) Endogenous TRPC1, TRPC3 and TRPC7 proteins combine to form native store-operated channels in HEK-293 cells. J Biol Chem 280:29559–29569

Zhang J, Xia SL, Block E, Patel J (2002) No upregulation of a cyclic nucleotide-gated channel contributes to calcium elevation in endothelial cells. Am J Physiol Cell Physiol 283:1080–1089

Zhu X, Chu PB, Peyton M, Birnbaumer L (1995) Molecular cloning of a widely expressed human homologue for the Drosophila trp gene. FEBS Lett 373:193–198

Zhu X, Jiang MS, Birnbaumer L (1998) Receptor-activated Ca^{2+} influx via human Trp3 stably expressed in human embryonic kidney (HEK)293 cells. J Biol Chem 273:133–142

Zitt C, Zobel A, Obukhov AG, Harteneck C, Kalkbrenner F, Luckhoff A, Schultz G (1996) Cloning and functional expression of a human Ca^{2+}-permeable cation channel activated by calcium store depletion. Neuron 16:1189–1196

… 53

TRPC2: Molecular Biology and Functional Importance

E. Yildirim[*] · L. Birnbaumer (✉)

Division of Intramural Research, National Institute of Environmental Health Sciences,
National Institutes of Health, Department of Health and Human Services,
111 T.W. Alexander Drive, Research Triangle Park NC, 27709, USA
birnbau1@niehs.nih.gov
*Present address: Howard Hughes Medical Institute, Department of Molecular Biology,
Massachusetts General Hospital, Boston MA, 02114 USA

1	The TRP Family of Cation Channels	54
2	The TRPC2 Channel	56
2.1	Expression, Structure, and Regulation	56
2.2	Functional Roles	58
2.2.1	The Acrosome Reaction	58
2.2.2	Gender-Specific Sexual Behavior	62
3	Concluding Remarks: One Channel with two Distinct Roles?	68
	References	70

Abstract TRPC (canonical transient receptor potential) channels are the closest mammalian homologs of *Drosophila* TRP and TRP-like channels. TRPCs are rather nonselective Ca^{2+} permeable cation channels and affect cell functions through their ability to mediate Ca^{2+} entry into cells and their action to collapse the plasma membrane potentials. In neurons the latter function leads to action potentials. The mammalian genome codes for seven TRPCs of which TRPC2 is the largest with the most restricted pattern of expression and has several alternatively spliced variants. Expressed in model cells, TRPC2 mediates both receptor- and store depletion-triggered Ca^{2+} entry. TRPC2 is unique among TRPCs in that its complete gene has been lost from the Old World monkey and human genomes, in which its remnants constitute a pseudogene. Physiological roles for TRPC2 have been studied in mature sperm and the vomeronasal sensory system. In sperm, TRPC2 is activated by the sperm's interaction with the oocyte's zona pellucida, leading to entry of Ca^{2+} and activation of the acrosome reaction. In the vomeronasal sensory organ (VNO), TRPC2 was found to constitute the transduction channel activated through signaling cascade initiated by the interaction of pheromones with V1R and V2R G protein-coupled receptors on the dendrites of the sensory neurons. V1Rs and V2Rs, the latter working in conjunction with class I MHC molecules, activate G_i- and G_o-type G proteins which in turn trigger activation of TRPC2, initiating an axon potential that travels to the axonal terminals. The signal is then projected to the glomeruli of the auxiliary olfactory bulb from where it is carried first to the amygdala and then to higher cortical cognition centers. Immunocytochemistry and gene deletion studies have shown that (1) the V2R-G_o-MHCIb-β2m pathway mediates male aggressive behavior in response to pheromones; (2) the V1R-G$_{i2}$ pathway mediates mating partner recognition, and (3) these differences have an anatomical correlate in that these functional components are located in anatomically distinct compartments of the VNO. Interestingly, these anatomically segregated signaling pathways use a common transduction channel, TRPC2.

Keywords TRP · TRPC2 · G protein · Acrosome reaction · Vomeronasal organ

1
The TRP Family of Cation Channels

Activation of the Gq-phospholipase C (PLC) signaling results in hydrolysis of phosphoinositol bisphosphate (PIP$_2$) into two second messengers—diacylglycerol (DAG) and inositol 1,4,5-trisphosphate (IP$_3$)—followed by IP$_3$-induced Ca^{2+} release from intracellular stores. Further, the depletion of intracellular Ca^{2+} stores activates Ca^{2+} permeable cation channels in the plasma membrane.

Mammalian homologs of the fly transient receptor potential (*trp*) channel, canonical TRPs (TRPCs), have been postulated as the pore-forming molecules through which store depletion-activated Ca^{2+} entry takes place (Birnbaumer et al. 1996). To date, seven subfamilies of TRP and TRP-like channels have been identified, and are found to be activated by an astounding set of diverse mechanisms and ligands. Figure 1 illustrates the phylogenetic relations among the TRP family of ion channels and lists some of signals and factors that

Fig. 1 (*Left*) Phylogenetic tree of TRP channels. The ion channel domains of TRP channels (reviewed in Birnbaumer et al. 2003) were analyzed by the GrowTree routine of the Wisconsin GCG Molecular Biology Package. Individual trees were constructed for the TRP C, M, V, P, ML and A subfamilies. Separately, a tree was constructed with the ion channel domains of one member of each of the subfamilies [TRPC3, TRPM2, TRPV1, PKD2 (TRPP2), TRPML1, TRPA1 (ANKTM1)], and used as anchoring branches for the families. Human sequences were used for all except the TRPC2, for which the mouse TRP channel domain sequence was used. The years that members of TRP subfamily were discovered are indicated in *bold* next to the channel names. (*Right*) The diverse regulatory inputs for different TRP channels. References (Ref): 1, Caterina et al. 1997; 2, Zygmunt et al. 1993; 3, Prescott and Julius 2003; 4, Caterina et al. 1999; 5, Kanzaki et al. 1999; 6, Peier et al. 2002; 7, Smith et al. 2002; 8, McKemy et al. 2003; 9, Story et al. 2003; 10, Guler et al. 2002; 11, Grimm et al. 2003; 12, Hoenderop et al. 1999; 13, Montilh-Zoller et al. 2003; 14, Runnels et al. 2002; 15, Perraud et al. 2001; 16, Wehage et al. 2002; 17, Hara et al. 2002; 18, Launay et al. 2002; 19, Nilius et al. 2006; 20, Hofmann et al. 2003; 21, Talavera et al. 2005

regulate their activity. While we have a somewhat better understanding of the physiological relevance of TRP-like channels (V, M, P, ML and A subfamilies), the role and the mechanism of activation of the TRPC class of channels is, for the most part, still incompletely understood. One exception is TRPC2, which was proposed to be the candidate cation entry channel that plays major roles in two distinct pathways: store-operated Ca^{2+} entry during acrosome reaction and pheromone-induced vomeronasal signaling. In this chapter, we review the properties of the TRPC2 channel and the current understanding of its physiological significance.

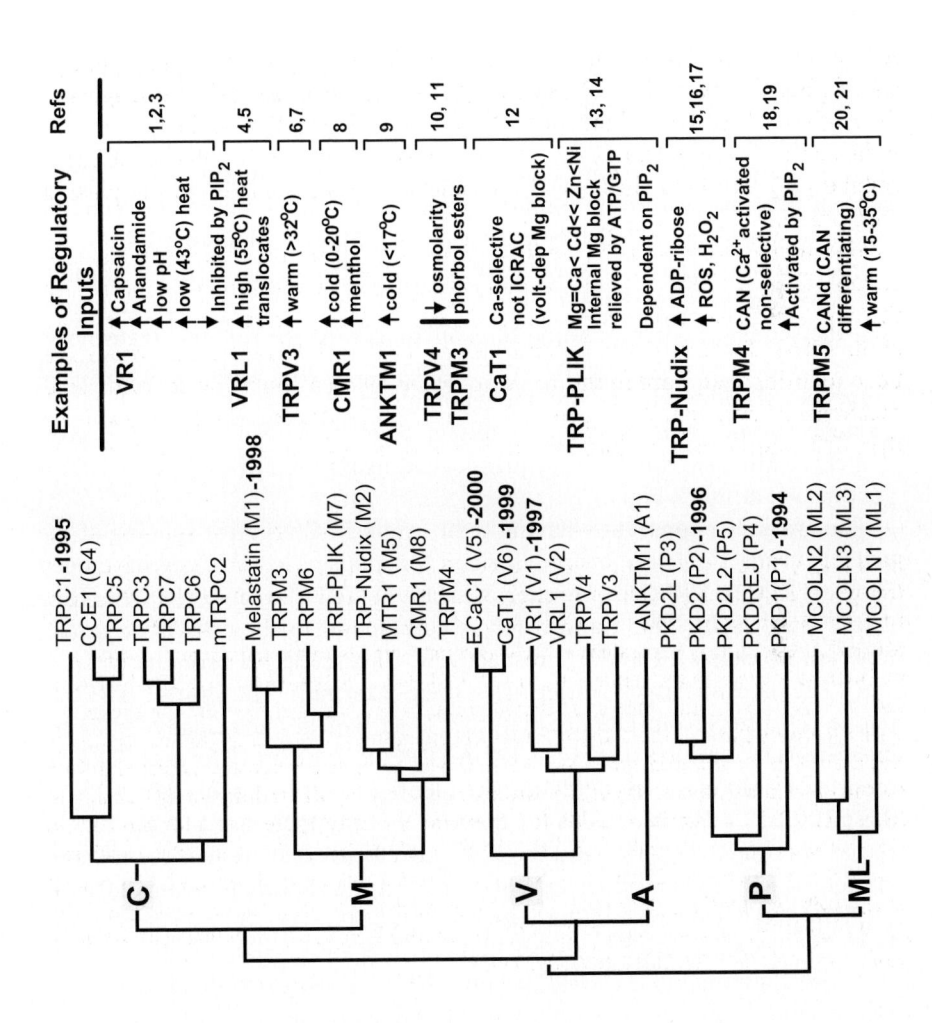

2
The TRPC2 Channel

2.1
Expression, Structure, and Regulation

The existence of the mouse TRPC2 gene was first reported by Zhu et al. (1996). This was followed by the cloning of bovine (Wissenbach et al. 1998), mouse (Vannier et al. 1999), and rat (Liman et al. 1999) complementary DNAs (cDNAs). Phylogenetic analysis of mammalian TRPC2s showed that the genes encode uninterrupted open reading frames (ORFs) with the potential of forming active channels in all vertebrates up to New World monkeys but no longer encode functional ORFs in Old World monkeys, apes and humans, in which the TRPC2 is a pseudogene (Liman and Innan 2003; Yildirim et al. 2003).

The mouse TRPC2 gene is found on chromosome 7 in a region that is syntenic to human chromosome 11p15.3–15.4. Rather than finding a full TRPC2 gene in this location of the human chromosome 11, we found the genomic region to encode a TRPC2-like transcript. Mapping the human cDNA to the human chromosome 11 sequence revealed an intron–exon structure similar to that of the mouse gene spanning from exons 14 to 21, with the exception that the sequence corresponding to exon 16, coding for the fifth transmembrane segment and half of the pore region, is absent. A search for TRPC2 sequences elsewhere in the human genome was negative, except for locating sequences similar to those comprising a fusion of mouse exons 2 and 3 approximately 70 Mb upstream of the sequences coding for the TRPC2-like transcript (Yildirim et al. 2003; Fig. 2).

Four variants, apparently having their origin in alternative splicing of the primary transcript, have been described for mouse TRPC2. They have been named TRPC2a, TRPC2b (Vannier et al. 1999), TRPC2α, and TRPC2β (Hofmann et al. 2000). The a and b forms correspond to the cDNAs originally reported as clones 14 and 17, respectively. They code for proteins of 1,172 and 1,072 aa, respectively, differing in their N termini. The shorter, TRPC2b (clone 17), differs from TRPC2a in its first 11 aa, with amino acids 12–1,072 being identical to amino acids 112–1,172 of TRPC2a. The α and β forms of TRPC2, reported by Hofmann et al. (2000), are identical to amino acids 287–1,172 of the TRPC2a. Codon 283 codes for the first P of the PQP motif located at the 5′-end of exon 10 (Fig. 3). Thus, the α, β, and a forms share sequences encoded in exons 10–21, differing upstream of exon 10. Amino acids upstream of PQP in TRPC2α and β are Met and Met-Asp-Pro-Leu-Ser (MDPLS), respectively. The TRPC2α cDNA was reported with 1,004 nt of 5′-untranslated sequence. The sequence alignment of the TRPC2a and TRPC2α cDNAs revealed that the first 713 nt of the 5′-untranslated sequence of TRPC2α are identical to those sections of the TRPC2a cDNA that are derived from exons 3–8, the remaining 291 nt just before the initiator ATG of TRPC2α are identical to the 5′-end of intron H

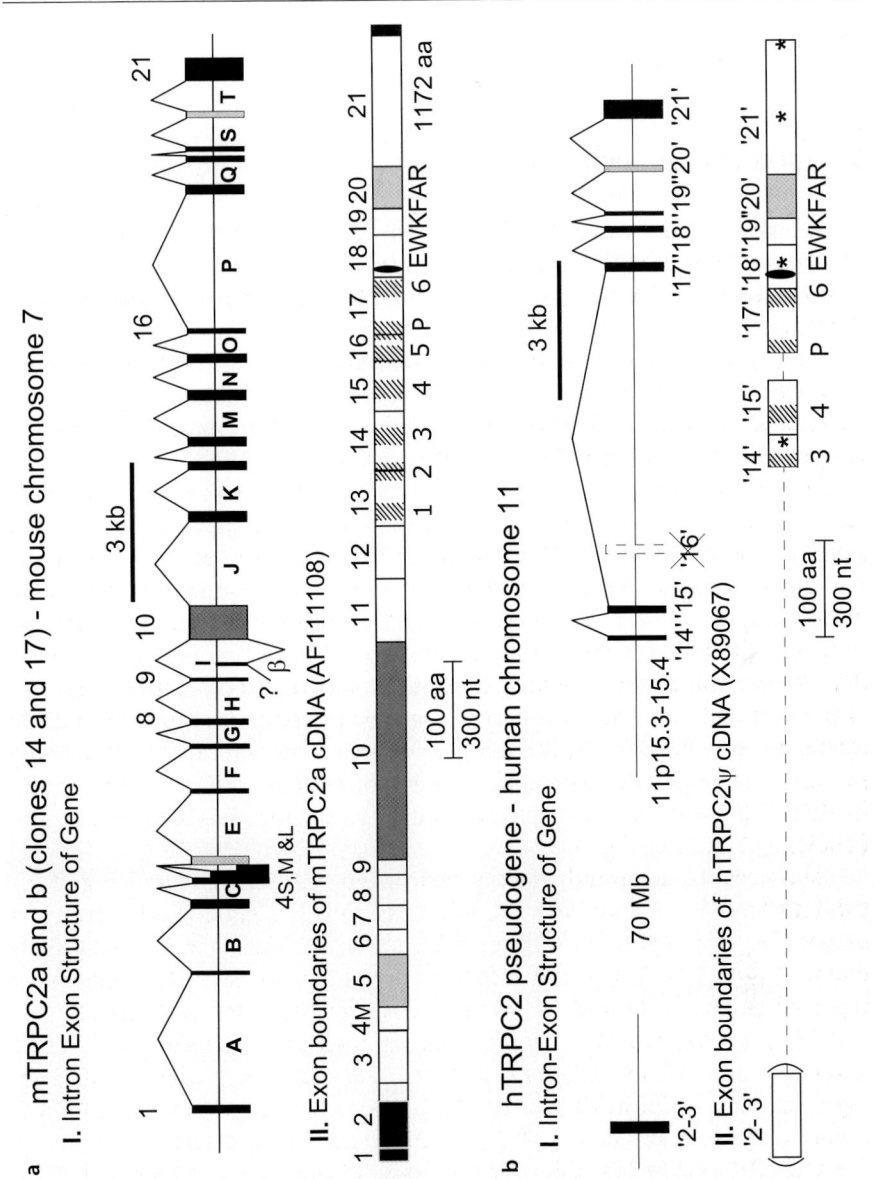

TRPC2: Molecular Biology and Functional Importance

Fig. 2 a,b Comparison of the mouse chromosome 7 region harboring the TRPC2 gene to the syntenic region of human chromosome 11 harboring the TRPC2 pseudogene. **a** Intron–exon distribution along the chromosome is shown *above* the diagram of the cDNA with its exon boundaries. **b** Intron–exon distribution along the human chromosome is shown *above* the diagram of the human cDNA. Note that it leaves as undecided (?) the existence of an additional exon with a separate promoter located 5′ to exon β (between exons 9 and 10). The human Trpc2 sequences can be found in the complements to regions 63,880 to 74,537 of genomic contig NT_035090 and 2,272,990 to 2,336,104 of contig NT_033927 (reprinted from Yildirim et al. 2003 with permission)

between exons 8 and 9 (Fig. 3aII). It is noteworthy that the mouse TRPC2β is nearly identical to the rat TRPC2 cDNA, expressed in the vomeronasal sensory organ (VNO) (Liman et al. 1999). The ORF of the rat TRPC2β cDNA codes for a protein that is almost identical to TRPC2a from amino acids 289–1,172, beginning with MDPLSP, where the P corresponds to the second P of the PQP motif of mouse exon 10. As is the case for mouse TRPC2β, the upstream nucleotides reported for rat TRPC2 are wholly represented within the genomic sequence upstream of exon 10, giving no further information as to whether the cDNA is derived from a bona fide mRNA or not. Further tests will be required to settle this issue.

Kyte–Doolittle analysis of the amino acid sequence of TRPC2 protein identifies seven hydrophobic regions with the potential to form six transmembrane segments. The long N-terminus of TRPC2 was shown to contain ankyrin (reviewed in Birnbaumer et al. 2003), calmodulin (CaM) (Yildirim et al. 2003), and enkurin (Sutton et al. 2004) binding domains. In analogy to all other TRPCs, TRPC2 also has a CaM binding site on its C-terminus (Tang et al. 2001; Yildirim et al. 2003). In addition, the mouse TRPC2 C-terminus has binding site for junctate, an IP$_3$R-associated protein (Treves et al. 2004; Stamboulian et al. 2005). The role of these domains in channel function is not clear at this time.

Upon expression in model cells, the mouse TRPC2 variants a and b are activated by both mere store depletion and activation of a G$_q$-coupled receptor (Vannier et al. 1999; Fig. 4). It is also noteworthy that neither the murine TRPC2α nor the murine TRPC2β cDNAs encode a protein capable of forming active channels. Instead, when observed by confocal microscopy, TRPC2α and TRPC2β fail to reach the plasma membrane (Hofmann et al. 2000).

2.2
Functional Roles

2.2.1
The Acrosome Reaction

In mammals, oocyte fertilization by a sperm cell involves a preparatory acrosome reaction triggered by the interaction of the sperm head with zona pellucida proteins surrounding the oocyte (Fig. 5a). The acrosome reaction

Fig. 3 a Diagrams of ORF in mature TRPC2 transcripts based on RT-PCR results reported by Yildirim et al. (2003) and on α and β cDNAs reported by Hofmann et al. (2000). **b** Alignment of deduced amino acid sequences of ORFs encoded in the TRPC2 transcripts with exons 4S, 4M, and 4L. *Top line* of alignment is master sequence; -, amino acids identical to TRPC2a in TRPC2b, peptide L and peptide M, gap; *, stop; |, exon boundaries (reprinted from Yildirim et al. 2003 with permission)

a
Predicted mTRPC2 ORFs based on cDNA and RT-PCR Analysis
I. formed by exons 1 through 21
using either exon 4S (86 nt), 4M (165 nt) or 4L (273 nt)

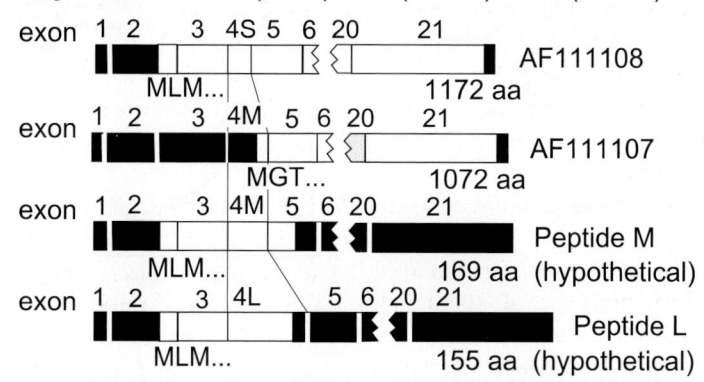

exon 1 2 3 4S 5 6 20 21 AF111108
MLM... 1172 aa

exon 1 2 3 4M 5 6 20 21 AF111107
MGT... 1072 aa

exon 1 2 3 4M 5 6 20 21 Peptide M
MLM... 169 aa (hypothetical)

exon 1 2 3 4L 5 6 20 21 Peptide L
MLM... 155 aa (hypothetical)

II. formed by exons α (1007 nt) plus 10 through 21,
or exons β (72 nt) plus 10 through 21

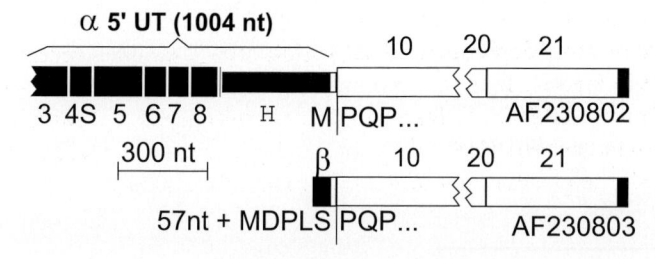

α 5' UT (1004 nt)
10 20 21
3 4S 5 6 7 8 H M PQP... AF230802
300 nt
β 10 20 21
57nt + MDPLS PQP... AF230803

b
Deduced amino acid sequence of putative peptides M (exon 4M) and L(exon 4L)

```
           exon 2              |exon 3
TRPC2a  MLMSRTDSKS GKNRSGVRMF KD|GDFLTPAS GESWDRLRLT CSQPFTRHQS   50
Pept.M  ---------- ---------- --|-------- ---------- ----------   50
Pept.L  ---------- ---------- --|-------- ---------- ----------   50
                                          |exon 4
TRPC2a  FGLAFLRVRS SLGSLADPVV DPSAPGSSGL NQ|NSTDVLES DPRPWLTNPS  100
Pept.M  ---------- ---------- ---------- --|-------- ----------  100
Pept.L  ---------- ---------- ---------- --|-------- ----------  100

TRPC2a  IRRTFFPDPQ T.......... .......... .......... ..........  111
Pept.M  ---------- -YVPAVDISQ GQGHIAHGHK NPSRGPL... ..........  137
Pept.L  ---------- -YVPAVDISQ GQGHIAHGHK NPSRGPLVSH LRSKHVRMAN  150
TRPC2b                                                  MGTK      4
                |exon 5
TRPC2a  .......|SKT EISELKGMLK QLQPGPLGRA ARMVLSAARK (+1028aa) 1172
Pept.M  .......|EHQ GNFRAQGYVE AVAARASGAG SPHGAFCCP*            169
Pept.L  YVSFK*                                                  155
TRPC2b  THPVVPW|--- ---------- ---------- ---------  (+1028aa) 1072
```

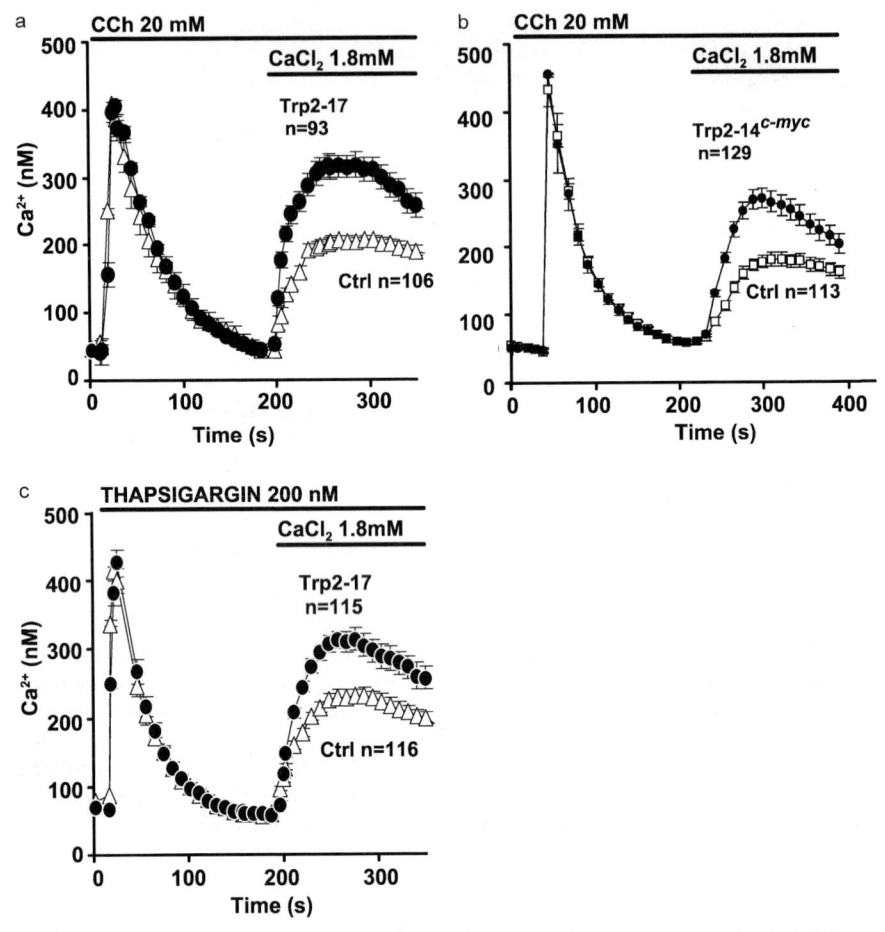

Fig. 4 a–c Expression of TRPC2 in COS-M6 cells enhances receptor-operated and store-operated Ca^{2+} entry. Ca^{2+} transients induced by carbachol (*CCh*) in the presence of extracellular Ca^{2+} induces capacitative calcium entry (CCE) that is significantly enhanced in COS-M6 cells expressing either **a** TRP2-17 or **b** TRP2-14 tagged at its C terminus with the epitope EQKLISEEDL that is recognized by the 9E10 anti-*myc* monoclonal antibody. **c** Thapsigargin-induced store depletion stimulates CCE in *TRP2-17* expressing COS-M6 cells. For details see Vannier et al. 1999 (adapted from Vannier et al. 1999 with permission)

is an exocytotic reaction that depends on Ca^{2+} influx from the extracellular milieu culminating with the release the acrosome's content, i.e., of enzymes that facilitate the penetration of the sperm head through the zona pellucida and injection of the male genome into the oocyte's cytoplasm. The zona pellucida is a glycoprotein envelope composed of three main glycoproteins termed ZP1, ZP2, and ZP3 (reviewed in Primakoff and Myles 2002; Jovine et al. 2005). Of these, ZP3 is the main activator of the acro-

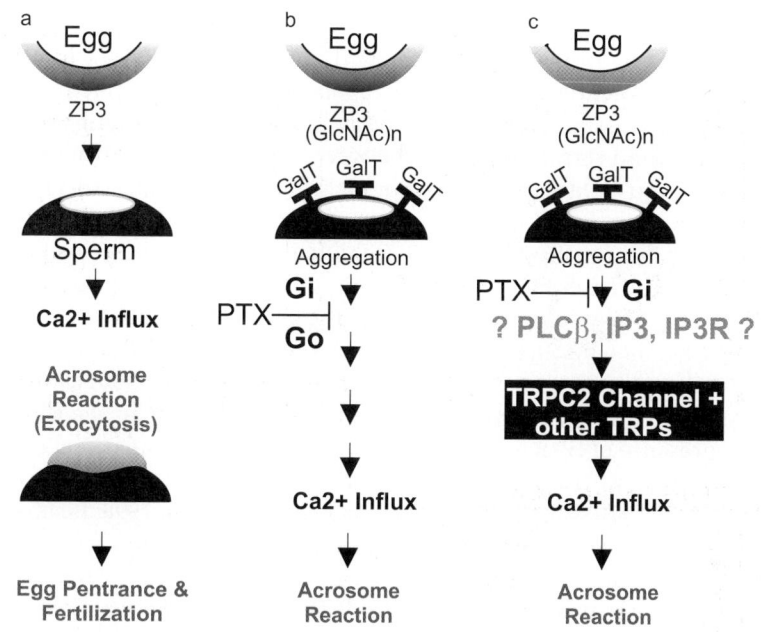

Fig. 5 a Interaction of sperm head with ZP3 protein of zona pellucida of the egg triggers Ca^{2+} influx and an exocytotic, acrosome reaction. **b** ZP3 of zona pellucida interacts with galactosyltransferase (GalT) promoting its clustering and aggregation and resulting in a pertussis toxin (PTX)-sensitive Ca^{2+} influx, which implicates involvement of G_i/G_o proteins in acrosome reaction (Endo et al. 1987; Miller et al. 1992). **c** The PTX-sensitive step is mediated by G_i (no G_o was detected in sperm heads), which activates a TRPC2 channel presumably through involvement of PLCβ, DAG, and IP_3–IP_3R (Jungnickel et al. 2001)

some reaction (Bleil and Wassarman 1983; Fig. 5b). A range of intracellular signaling cascades is triggered during the acrosome reaction. These include aggregation of sperm surface galactosyltransferase (GalT), activation of pertussis toxin-sensitive G proteins (Wilde et al. 1992; Ward et al. 1992; Ward and Kopf 1993), and activation of a voltage-independent cation channel that induces membrane depolarization leading to activation of a voltage-dependent T-type calcium channel that causes a transient increase in intracellular Ca^{2+} levels (Florman et al. 1992; Arnoult et al. 1996, 1999; Wassarman 1999). The transient phase is followed by a sustained phase of elevated intracellular Ca^{2+} (Florman et al. 1989; Lee and Storey 1989; Shirakawa and Miyazaki 1999). The mechanism leading to the sustained Ca^{2+} phase during the acrosome reaction was analyzed in capacitated sperm by Fura-2 Ca^{2+} imaging experiments in which application of thapsigargin caused store-operated Ca^{2+} entry into sperm. This suggested a role for store-operated Ca^{2+} entry channels in ZP3-induced acrosome reaction (O'Toole et al. 2000; Fig. 5b).

Among the members of TRPC channels, TRPC2 transcripts were detected in mouse testis by Northern blot analysis (Vannier et al. 1999), which made this channel a potential candidate for regulation of store-operated Ca^{2+} entry during the acrosome reaction (Fig. 5c). In 2001, using an antibody raised against the second extracellular loop between transmembrane domains 4 and 5 of TRPC2 (anti-RDAS antibody), Jungnickel et al. (2001) located TRPC2 by immunofluorescence on the anterior part of sperm heads. They also provided evidence that the anti-RDAS antibody inhibited Ca^{2+} entry into fluo3-loaded sperm cells induced either by thapsigargin or ZP3. The active participation of TRPC2 in the physiologic activation of the acrosome reaction was further supported by showing that the anti-RDAS antibody not only blocked Ca^{2+} entry but also the acrosome reaction itself (Jungnickel et al. 2001). Walensky and Snyder (1995) had shown earlier the presence of IP_3Rs in the acrosome cap. The presence of IP_3R and TRPC2 in close proximity is consistent with the possibility that TRPC2 may be activated according to the protein–protein interaction model of TRPC channel activation (Kiselyov et al. 1998; Birnbaumer et al. 2000). However, when the role of intracellular Ca^{2+} pools was analyzed by applying Ca^{2+} ionophores, ionomycin, and A23187, the Ca^{2+} released from intracellular stores was not sufficient to drive the acrosome reaction and required extracellular Ca^{2+} (O'Toole et al. 2000). While no doubt important, TRPC2 is not essential under in vivo conditions, as TRPC2-deficient mice are fertile (Stowers et al. 2002; Leypold et al. 2002). This last observation suggests that other TRPCs or TRP-related channels are involved in modulating Ca^{2+} entry during the acrosome reaction (Wissenbach et al. 1998; Trevino et al. 2001; Castellano et al. 2003). Thus, the mechanism of how ZP3 signaling induces store depletion still needs to be elucidated.

2.2.2
Gender-Specific Sexual Behavior

2.2.2.1
Transduction of Pheromone Signals

The mammalian olfactory system contains sensory neurons that are found in two different locations; the main olfactory epithelium of the nose (MOE) and the VNO. These two types of sensory neurons belong to two neuronal networks that differ in their function and their molecular makeup. From the functional viewpoint, MOE neurons respond to volatile odors, whereas VNO neurons respond to nonvolatile "odors" including pheromones, and are important for sexual behavior (reviewed in Buck 2000). The olfactory signals perceived by the MOE are sent to the main olfactory bulb (MOB) where axons of cells expressing a given olfactory receptor converge in glomeruli (Ressler et al. 1994; Vassar et al. 1994). From there, mitral cells project to the olfactory cortex leading to various cognitive and emotional responses as well as measured thoughts and behaviors.

In contrast, the axons from neurons of the VNO are projected to the glomeruli of the accessory olfactory bulb (AOB), from where projections extend to the amygdala and the hypothalamus in which behaviors such as social recognition, neuroendocrine function, and reproductive behavior are conceived (Keverne 1999; Dulac 2000). The detection of pheromones by the VNO neurons occurs via two classes of G protein-coupled receptors (GPCRs), V1Rs, that are restricted to the apical zone of the VNO and are co-expressed with $G_{i2}\alpha$, and V2Rs that are expressed in the basal zone together with $G_o\alpha$, class Ib major histocompatibility complex (MHC1b) proteins and β2 microglobulin (β2m) (Dulac and Axel 1995; Herrada and Dulac 1997; Berghard and Buck 1996; Jia and Halpern 1996; Saito et al. 1998). Immunohistochemical analysis of VNO sections and immunoaffinity chromatography analysis of VNO extracts showed that V2Rs and MHC1b molecules interact (Ishii et al. 2003; Loconto et al. 2003), leading to the proposal that these two types of molecules might function together during chemosensory perception (Loconto et al. 2003). In support, Leinders-Zufall and colleagues (2004) showed that peptides that bind to MHC molecules function as signals of genetic individuality that are required for mate recognition and thus influence social behavior.

Studies on the olfactory transduction cascade in the MOE neurons provided evidence that odors induce an inward cationic current that causes membrane depolarization and generate an action potential (Trotier and MacLeod 1983; Firestein and Werblin 1989; Firestein et al. 1990). The components of this cascade include a family of 500–1,000 odorant receptors that activate the heterotrimeric G protein G_{olf} (Buck and Axel 1991; Raming et al. 1993), G_{olf} (Pace and Lancet 1986; Jones and Reed 1989), the G_{olf} G protein, adenylyl cyclase III [activated by $G_{olf}\alpha$ (Pace et al. 1985; Sklar et al. 1986; Bakalyar and Reed 1990)], and the olfactory nucleotide (cAMP)-gated channel (oCNG) (Nakamura and Gold 1987; Firestein et al. 1991; Liman and Buck 1994; Fig. 6). In contrast, VNO neurons use a different set of signaling molecules for the transduction of chemical stimuli (Berghard et al. 1996; Liman and Corey 1996), in which activation of the V1R or V2R GPCRs activate $G_{i2}\alpha$ or $G_o\alpha$, respectively (Dulac and Axel 1995; Herrada and Dulac 1997; Berghard and Buck 1996; Jia and Halpern 1996; Saito et al. 1998; Fig. 7a). V1Rs belong to the rhodopsin or class I type of GPCRs, while V2Rs belong to the class 3 family of GPCRs, to which metabotropic glutamate receptors (mGluRs), GABA-B receptors and Ca^{2+} sensing GPCRs belong. Activated G_i or G_o, either through GTP–Gα or Gβγ, in turn stimulate the PLCβ pathway causing synthesis of the DAG and IP_3 second messengers.

2.2.2.2
TRPC2: The Candidate Transduction Channel Responsible for Transducing Vomeronasal Signaling

The evidence for the identity of the ion channels that mediate this signaling pathway with involvement of PLC in VNO neurons came from Dulac and colleagues in 1999. These authors cloned the rat TRPC2 (rTRPC2) cDNA from rat VNO and characterized its spatial expression pattern both at the mRNA and protein levels (Liman et al. 1999; Fig. 7b, 3–5). These analyses showed that rTRPC2 was exclusively expressed in both apical and basal zones of VNO, particularly at the dendritic tips of the vomeronasal sensory neurons (VSNs) suggesting a role for rTRPC2 in sensory signal transduction. A follow-up study by Menco et al. (2001) supported these findings and showed by light and electron microscopy that the rTRPC2 gene product localizes to the microvilli of rat vomeronasal receptor cells. These are the sites of VNO cells that interact with odorants and pheromones. These authors also showed that the sites of TRPC2 expression are enriched in $G_{i2}\alpha$ and $G_o\alpha$ (Menco et al. 2001), with $G_{i2}\alpha$ being found in the apical zone where V1R receptors are found, and $G_o\alpha$ being found in the basal zone where V2R receptors and the MHC1b molecules had been found (Fig. 7b, 6–9).

The TRPC2 gene was inactivated by homologous recombination independently by two groups in order to investigate the significance of the channel in VNO-mediated sensory responses and attendant behaviors (Leypold et al. 2002; Stowers et al. 2002). As listed in Table 1, analyses of the phenotype of TRPC2$^{-/-}$ mice provided evidence that TRPC2 is a component of the VNO signaling pathway(s) that detects male-specific pheromones, mediates gender-specific behavior, and has a role in male–male aggression (Leypold et al. 2002; Stowers et al. 2002). These data and the analysis of $\beta 2m^{-/-}$ mice has shed light on the structural components that make up the functional units in VNO (Loconto et al. 2003; Fig. 8).

The findings from the phenotypic analysis of TRPC2$^{-/-}$ mice made TRPC2 a candidate ion channel that contributes to the VSN signaling. To elucidate this concept at the molecular level, Zufall and colleagues investigated the existence and characteristics of PLC-activated cation channels by performing inside-out recordings from plasma membranes of the dendritic tips of VSNs in the presence of a plasma membrane permeable DAG analog 1-stearoyl-2-arachidonoyl-*sn*-glycerol (SAG) in the bath solution (Lucas et al. 2003). They

Fig. 6 a,b Signal transduction in the olfactory epithelium. **a** Odor (*circle*) stimulates olfactory receptor (*OR*), which activates olfactory G protein (*G*_{*olf*}) triggering the adenylyl cyclase (*AC*)-cAMP ON signal. cAMP activates the olfactory cyclic nucleotide gated channels (*cNGC*) which results in two responses: an action potential and entry of Ca^{2+} which via Ca^{2+}/CaM and CaMKII inhibits the adenylyl cyclase, thereby terminating the signaling process. **b** Localization of the messenger RNA (mRNA) of signal transduction components of the olfactory epithelium; odorant receptors (*OR*) (adapted from Ressler et al. 1993 with permission), G_{olf}, oCNC1, oCNC2, ACIII (adapted from Berghard et al. 1996 with permission). *OE*, olfactory epithelium; *OB*, olfactory bulb; *VNO*, vomeronasal organ; *AOB*, accessory olfactory bulb

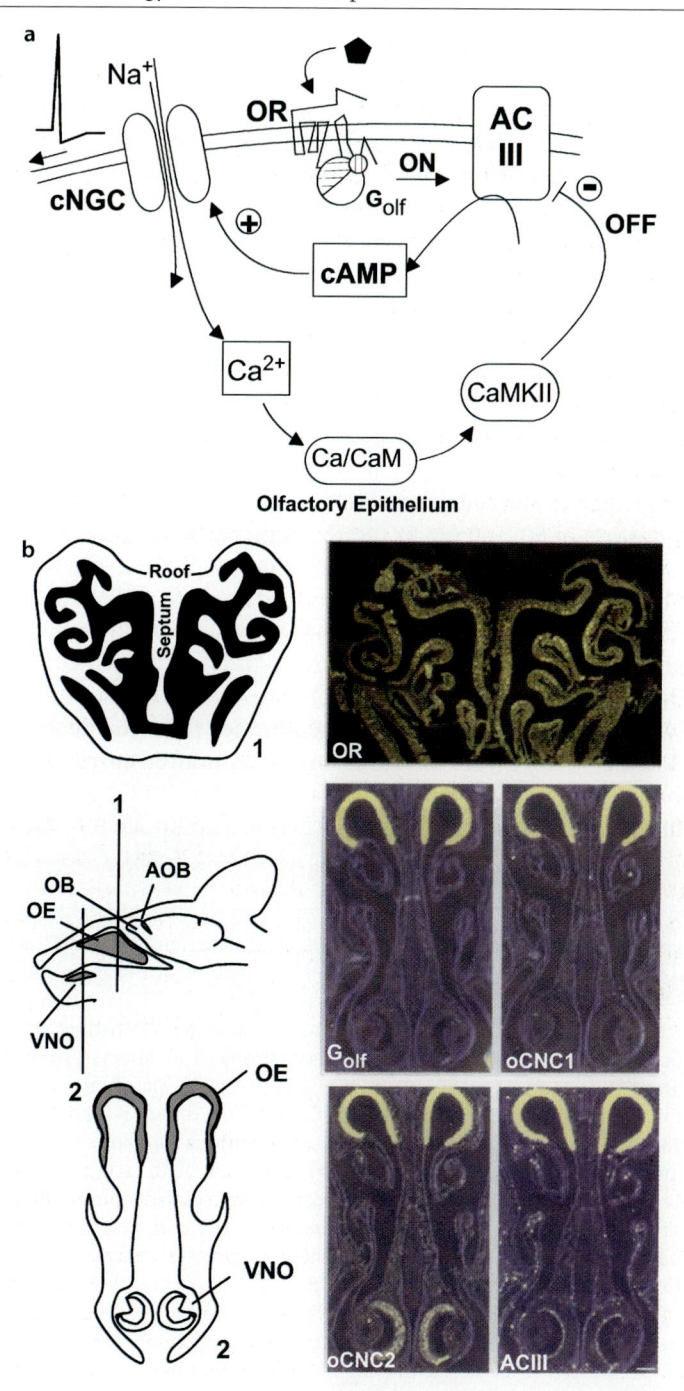

TRPC2: Molecular Biology and Functional Importance 65

Table 1 Phenotype of the TRPC2$^{-/-}$ mice. (From studies of Stowers et al. 2002; Leypold et al. 2002)

Abolished sensory response to pheromone cues in urine
Abolished pheromone-evoked male–male aggression
Abolished maternal aggression against intruders
Males display loss of gender discrimination for mating
Males display sexual behavior toward male intruders
Males display defects in territorial markings
Normal testosterone production by males
Normal male–female mating behavior

found a sustained inward current, which could not be activated by Ca^{2+} alone or by application of 50–100 µM IP$_3$ to the bath, suggesting that the activation pathway is not through IP$_3$R signaling or store-depletion. The SAG-induced currents had a conductance of 42 pS, and extracellular Ca^{2+}, Mg^{2+}, ATP, or GTP was not necessary for the development of this current. The characteristics of this cation channel are listed in Table 2.

In order to test whether the TRPC2 channel is involved in the generation of this DAG-induced current, plasma membranes from the dendritic tips of VSNs of TRPC2$^{-/-}$ mice were used to record SAG-induced currents (Lucas et al. 2003). These recordings showed that ablation of the TRPC2 gene reduced the SAG-induced current, indicating that TRPC2 is a component of the signaling

Fig. 7 a,b Signal transduction in the VNO. **a** In the apical zone (*left*), the TRPC2 transduction channel is activated by the V1R-G$_{i2}\alpha$-triggered stimulation of PLCβ. In the basal zone (*right*), the V2R.MHCIb (M10).β2m complex stimulates PLC via activation of G$_o\alpha$. This results in the activation of the TRPC2 vomeronasal transduction channel. **b** Localization of the signal transduction components of the vomeronasal olfactory system. *B1–2*, V1R (adapted from Dulac and Axel 1995) and G$_{i2}\alpha$ (adapted from Matsunami and Buck 1997) colocalize in the VNO apical zone. *B3* Rat TRPC2 mRNA is found in both apical and basal zones of the VNO (adapted from Liman et al. 1999 with permission). *B4–5*, Rat TRPC2 and G$_{i2}\alpha$ protein localizes to the dendritic tips of the VNO neurons (adapted from Liman et al. 1999 with permission). *B6–9*, V2R, G$_o\alpha$ (adapted from Matsunami and Buck 1997 with permission), M10, and β2m (adapted from Ishii et al. 2003 with permission) mRNAs are found in basal zone of the VNO. *VNO*, vomeronasal organ; *AOB*, accessory olfactory bulb; *D*, dorsal, *P*, posterior, *V*, ventral, *A*, anterior

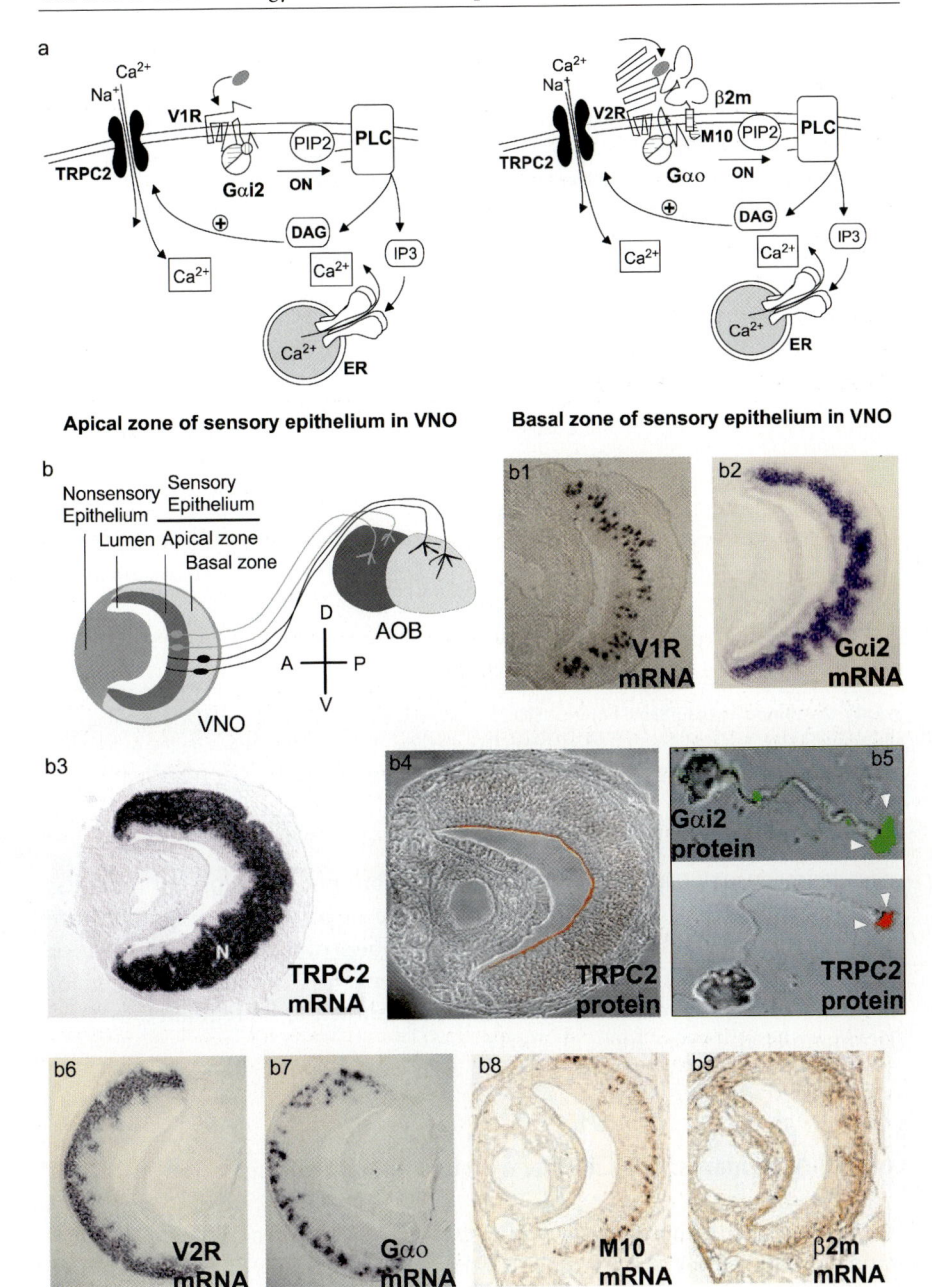

TRPC2: Molecular Biology and Functional Importance 67

Apical zone of sensory epithelium in VNO   **Basal zone of sensory epithelium in VNO**

Table 2 Properties of the PLC-activated cation channel at the dendritic tips of VSNs. (These observations were reported by Lucas et al. 2003)

Conductance	Ion permeability	Cation current can be induced by	Pheromone-induced cation current is insensitive to	Pheromone/SAG-induced cation current is sensitive to
42 pS	Na^+, Ca^{2+}, Cs^+ But not NMDG	Pheromone cues in urine DAG analogs SAG, OAG, DOG But not by OG, IP_3, store depletion Cyclooxygenase inhibitor, indomethacin (50 µM) Lipoxygenase inhibitor, nordihydroguaiaretic acid (100 or 500 µM)	PLC antagonist, U-73122 2-ABP	DAG lipase inhibitor, RHC-80267 PLA_2 inhibitor, isotetrandrine PI3 kinase inhibitor, LY-29002 PKC inhibitors, staurosporine or calphostin C

2-ABP, 2-Aminoethoxydiphenyl borate; DOG, 1,2-dioctanoyl-*sn*-glycerol; NMDG, *N*-methyl-D-glucamine; OAG, 1-oleoyl-2-acetyl-*sn*-glycerol; OG, 1-oleoyl-glycerol; PLA_2, phospholipase A_2; SAG, 1-stearoyl-2-arachidonoyl-*sn*-glycerol; VSN, vomeronasal sensory neurons

cascade in sensory transduction. A reduction, but not a full loss, of the SAG- or urine-induced current is indicative that one or more cation channels are involved in this process, presumably other members of the TRPC family, which might interact with TRPC2 to regulate Ca^{2+} entry. A candidate partner for TRPC2 is the TRPC6 channel, which has been shown to be regulated by DAG (Hofmann et al. 1999; Chu et al. 2004).

3
Concluding Remarks: One Channel with two Distinct Roles?

The findings from the analysis of the sperm acrosome reaction, the sensory transduction in the VNO neurons and the characterization of $TRPC2^{-/-}$ mice

Fig. 8 Basal and apical vomeronasal zones mediate distinct behavioral responses using separate receptor/G protein systems but a common TRPC2 transduction channel (adapted from Dulac and Torello 2003 with permission)

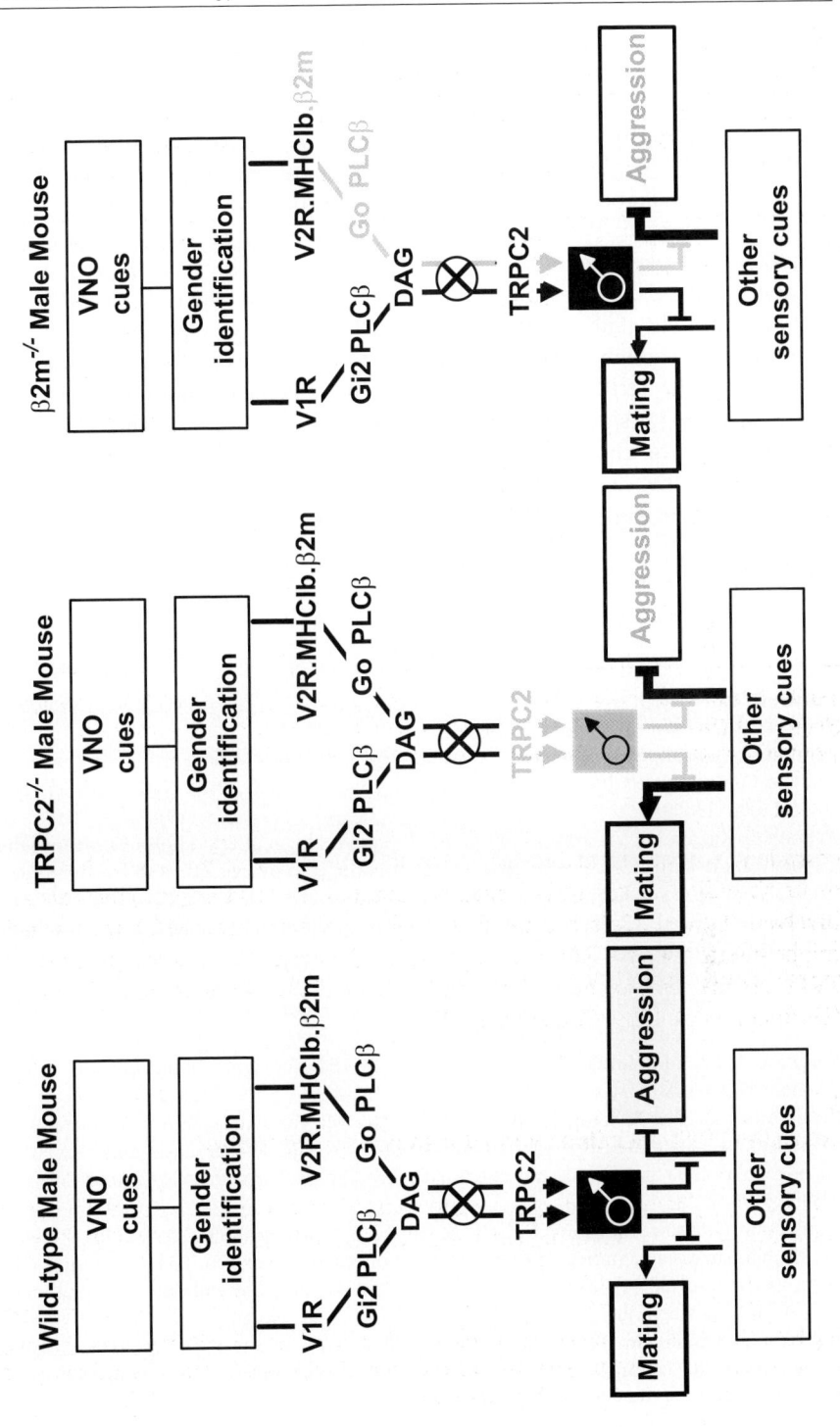

TRPC2: Molecular Biology and Functional Importance

provided evidence that the TRPC2 channel is important for both sperm function and vomeronasal pheromone sensing. Yet the main role in these two important functions appears to differ. In the acrosome reaction the channel shows characteristics of a store-operated Ca^{2+} entry channel supplying the Ca^{2+} required to trigger exocytosis. In sensory neurons, on the other hand, the channel shows characteristics of a depolarizing cation channel that triggers an action potential. Whether Ca^{2+} entering into the neurons during depolarization also plays an additional role is not known. In neurons the TRPC2 channel appears to be activated by DAG in response to stimulation of PLC activity (Jungnickel et al. 2001; Lucas et al. 2003). In sperm, DAG did not induce an acrosome reaction (Stamboulian et al. 2005), suggesting that the mode of activation of the TRPC2 channel might be cell-specific and might require different partners for signaling to proceed. Thus, while advances are being made in identifying the components that make up different signaling pathways in different cells, the molecular details of the process by which TRPC2 is being activated remain in need of further clarification.

Acknowledgements Supported by the Intramural Research Program of the NIH, and NIEHS.

References

Arnoult C, Cardullp RA, Lemos JR, Florman HM (1996) Activation of mouse sperm T-type Ca^{2+} channels by adhesion to the egg zona pellucida. Proc Natl Acad Sci USA 93:13004–13009

Arnoult C, Kazam IG, Visconti PE, Kopf GS, Villaz M, Florman HM (1999) Control of the low voltage-activated calcium channel of mouse sperm by egg ZP3 and by membrane hyperpolarization during capacitation. Proc Natl Acad Sci USA 96:6757–6762

Bakalyar HA, Reed RR (1990) Identification of a specialized adenylyl cyclase that may mediate odorant detection. Science 250:1403–1406

Berghard A, Buck LB (1996) Sensory transduction in vomeronasal neurons: evidence for G alpha o, G alpha i2, and adenylyl cyclase II as major components of a pheromone signaling cascade. J Neurosci 16:909–918

Berghard A, Buck LB, Liman ER (1996) Evidence for distinct signaling mechanisms in two mammalian olfactory sense organs. Proc Natl Acad Sci USA 93:2365–2369

Birnbaumer L, Zhu X, Jiang M, Boulay G, Peyton M, Vannier B, Brown D, Platano D, Sadeghi H, Stefani E, Birnbaumer M (1996) On the molecular basis and regulation of cellular capacitative calcium entry: roles for Trp proteins. Proc Natl Acad Sci USA 93:15195–15202

Birnbaumer L, Boulay G, Brown D, Jiang M, Dietrich A, Mikoshiba K, Zhu X, Qin N (2000) Mechanism of capacitative Ca^{2+} entry (CCE): interaction between IP3 receptor and TRP links the internal calcium storage compartment to plasma membrane CCE channels. Recent Prog Horm Res 55:127–162

Birnbaumer L, Yildirim E, Abramowitz J (2003) A comparison of the genes coding for canonical TRP channels and their M, V, and P relatives. Cell Calcium 33:419–432

Bleil JD, Wassarman PM (1983) Sperm–egg interactions in the mouse: sequence of events and induction of the acrosome reaction by a zona pellucida glycoprotein. Dev Biol 95:317–324

Buck L, Axel R (1991) A novel multigene family may encode odorant receptors: a molecular basis for odor recognition. Cell 65:175–187

Buck LB (2000) The molecular architecture of odor and pheromone sensing in mammals. Cell 100:611–618

Castellano LE, Trevio CL, Rodriguez D, Serrano CJ, Pacheco J, Tsutsumi V, Felix R, Darszon A (2003) Transient receptor potential (TRPC) channels in human sperm: expression, cellular localization and involvement in the regulation of flagellar motility. FEBS Lett 541:69–74

Caterina MJ, Schumacher MA, Tominaga M, Rosen TA, Levine JD, Julius D (1997) The capsaicin receptor: a heat-activated ion channel in the pain pathway. Nature 389:816–824

Caterina MJ, Rosen TA, Tominaga M, Brake AJ, Julius D (1999) A capsaicin-receptor homologue with a high threshold for noxious heat. Nature 398:436–441

Chu X, Tong Q, Cheung JY, Wozney J, Conrad K, Mazack V, Zhang W Stahl R, Barber DL, Miller BA (2004) Interaction of TRPC2 and TRPC6 in erythropoietin modulation of calcium influx. J Biol Chem 279:10514–10522

Devary O, Heichal O, Blumenfeld A, Cassel D, Suss E, Barash S, Rubinstein CT, Minke B, Selinger Z (1987) Coupling of photoexcited rhodopsin to inositol phospholipid hydrolysis in fly photoreceptors. Proc Natl Acad Sci USA 84:6939–6943

Dulac C (2000) Sensory coding of pheromone signals in mammals. Curr Opin Neurobiol 10:511–518

Dulac C, Axel R (1995) A novel family of genes encoding putative pheromone receptors in mammals. Cell 83:195–206

Dulac C, Torello AT (2003) Molecular detection of pheromone signals in mammals: from genes to behaviour. Nat Rev Neurosci 4:551–562

Endo Y, Lee MA, Kopf GS (1987) Evidence for the role of a guanine nucleotide-binding regulatory protein in the zona pellucida-induced mouse sperm acrosome reaction. Dev Biol 119:210–216

Firestein S, Werblin F (1989) Odor-induced membrane currents in vertebrate-olfactory receptor neurons. Science 244:79–82

Firestein S, Shepherd GM, Werblin FS (1990) Time course of the membrane current underlying sensory transduction in salamander olfactory receptor neurons. J Physiol 430:135–158

Firestein S, Zufall F, Shepherd GM (1991) Single odor-sensitive channels in olfactory receptor neurons are also gated by cyclic nucleotides. J Neurosci 11:3565–3572

Florman HM, Tombes RM, First NL, Babcock DF (1989) An adhesion-associated agonist from the zona pellucida activates H-protein-promoted elevations of internal Ca^{2+} and pH that mediate mammalian sperm acrosomal exocytosis. Dev Biol 135:133–146

Florman HM, Corron ME, Kim TD, Babcock DF (1992) Activation of voltage-dependent calcium channels of mammalian sperm is required for zona pellucida-induced acrosomal exocytosis. Dev Biol 152:304–314

Grimm C, Kraft R, Sauerbruch S, Schultz G, Harteneck C (2003) Molecular and functional characterization of the melastatin-related cation channel TRPM3. J Biol Chem 278:21493–21501

Guler AD, Lee HS, Iida T, Shimizu I, Toinaga M, Caterina M (2002) Heat-evoked activation of the ion channel, TRPV4. J Neurosci 22:6408–6414

Hara Y, Wakamori M, Ishii M, Maeno E, Nishida M, Yoshida T, Yamada H, Shimizu S, Mori E, Kudoh J, Shimizu N, Kurose H, Okada Y, Imoto K, Mori Y (2002) LTRPC2 Ca^{2+}-permeable channel activated by changes in redox status confers susceptibility to cell death. Mol Cell 9:163–173

Herrada G, Dulac C (1997) A novel family of putative pheromone receptors in mammals with a topographically organized and sexually dimorphic distribution. Cell 90:763–773

Hoenderop JG, van der Kemp AW, Hartog A, van de Graaf SF, van Os CH, Willems PH, Bindels RJ (1999) Molecular identification of the apical calcium channel in 1,25-dihydroxy-vitamin D3-responsive epithelia. J Biol Chem 274:8375–8378

Hofmann T, Obukhov AG, Schaefer M, Harteneck C, Gudermann T, Schultz G (1999) Direct activation of human TRPC6 and TRPC3 channels by diacylglycerol. Nature 397:259–263

Hofmann T, Schaefer M, Schultz G, Gudermann T (2000) Cloning, expression and subcellular localization of two novel splice variants of mouse transient receptor potential channel 2. Biochem J 351:115–122

Hofmann T, Chubanov V, Guderman T, Montell C (2003) TRPM5 is a voltage-modulated and Ca^{2+}-activated monovalent selective cation channel. Curr Biol 13:1153–1158

Hotta Y, Benzer S (1970) Genetic dissection of the Drosophila nervous system by means of mosaics. Proc Natl Acad Sci USA 67:1156–1163

Ishii T, Hirota J, Mombaerts P (2003) Combinatorial coexpression of neural and immune multigene families in mouse vomeronasal sensory neurons. Curr Biol 13:394–400

Jia C, Halpern M (1996) Subclasses of vomeronasal receptor neurons: differential expression of G proteins (G_{i2}-alpha and G_o-alpha) and segregated projections to the accessory olfactory bulb. Brain Res 719:117–128

Jones DT, Reed RR (1989) Golf: an olfactory neuron specific-G protein involved in odorant signal transduction. Science 244:790–795

Jovine L, Darie CC, Litscher ES, Wassarman PM (2005) Zona pellucida domain proteins. Annu Rev Biochem 74:83–114

Jungnickel MK, Marrero H, Birnbaumer L, Lemos JR, Florman HM (2001) TRP2 is essential for Ca^{2+} entry into mouse sperm triggered by egg ZP3. Nat Cell Biol 3:499–502

Kanzaki M, Zhang YQ, Mashima H, Li L, Shibata H, Kojima I (1999) Translocation of a calcium-permeable cation channel induced by insulin-like growth factor I. Nat Cell Biol 1:165–170

Keverne EB (1999) The vomeronasal organ. Science 286:716–720

Kiselyov K, Xu X, Mozhayeva G, Kuo T, Pessah I, Mignery G, Zhu X, Birnbaumer L, Muallem S (1998) Functional interaction between InsP3 receptors and store-operated Htrp3 channels. Nature 396:478–482

Launay P, Fleig A, Perraud AL, Scharenberg AM, Penner R, Kinet JP (2002) TRPM4 is a Ca^{2+}-activated nonselective cation channel mediating membrane depolarization. Cell 109:397–407

Lee MA, Storey BT (1989) Endpoint of first stage of zona pellucida-induced acrosome reaction in mouse spermatozoa characterized by acrosomal H+ and Ca^{2+} permeability: population and single cell kinetics. Gamete Res 24:303–326

Leinders-Zufall T, Brennan P, Widmayer P, S PC, Maul-Pavicic A, Jager M, Li XH, Breer H, Zufall F, Boehm T (2004) MHC class I peptides as chemosensory signals in the vomeronasal organ. Science 306:1033–1037

Leypold BG, Yu CR, Leinders-Zufall T, Kim MM, Zufall F, Axel R (2002) Altered sexual and social behaviors in trp2 mutant mice. Proc Natl Acad Sci USA 99:6376–6381

Liman ER, Buck LB (1994) A second subunit of the olfactory cyclic nucleotide-gated channel confers high sensitivity to cAMP. Neuron 13:611–621

Liman ER, Corey DP (1996) Electrophysiological characterization of chemosensory neurons from the mouse vomeronasal organ. J Neurosci 16:4625–4637

Liman ER, Innan H (2003) Relaxed selective pressure on an essential component of pheromone transduction in primate evolution. Proc Natl Acad Sci USA 100:3328–3232

Liman ER, Corey DP, Dulac C (1999) TRP2: a candidate transduction channel for mammalian pheromone sensory signaling. Proc Natl Acad Sci USA 96:5791–5796

Loconto J, Papes F, Chang E, Stowers L, Jones EP, Takada T, Kumanovics A, Fischer Lindahl K, Dulac C (2003) Functional expression of murine V2R pheromone receptors involves selective association with the M10 and M1 families of MHC class Ib molecules. Cell 112:607–618

Lucas P, Ukhahov K, Leinders-Zufall T, Zufall F (2003) A diacylglycerol-gated cation channel in vomeronasal neuron dendrites is impaired in TRPC2 mutant mice: mechanism of pheromone transduction. Neuron 40:551–561

Matsunami H, Buck LB (1997) A multigene family encoding a diverse array of putative pheromone receptors in mammals. Cell 90:775–784

McKemy DD, Neuhauser WM, Julius D (2002) Identification of a cold receptor reveals a general role for TRP channels in thermosensation. Nature 416:52–58

Menco BP, Carr VM, Ezeh PI, Liman ER, Yankova MP (2001) Ultrastructural localization of G-proteins and the channel protein TRP2 to microvilli of rat vomeronasal receptor cells. J Comp Neurol 438:468–489

Miller DJ, Macek MB, Shur BD (1992) Complementarity between sperm surface beta-1,4-galactosyltransferase and egg-coat ZP3 mediates sperm-egg binding. Nature 357:589–593

Monteilh-Zoller MK, Hermosura MC, Nadler MJ, Scharenberg AM, Penner R, Fleig A (2003) TRPM7 provides an ion channel mechanism for cellular entry of trace metal ions. J Gen Physiol 121:49–60

Nakamura T, Gold GH (1987) A cyclic nucleotide-gated conductance in olfactory receptor cilia. Nature 325:442–444

Nilius B, Mahieu F, Prenen J, Janssens A, Owsianik G, Vennekens R, Voets T (2006) The Ca^{2+}-activated cation channel TRPM4 is regulated by phosphatidylinositol 4,5-bisphosphate. EMBO J 25:467–478

O'Toole CM, Arnoult C, Darszon A, Steinhardt RA, Florman HM (2000) Ca^{2+} entry through store-operated channels in mouse sperm is initiated by egg ZP3 and drives the acrosome reaction. Mol Biol Cell 11:1571–1584

Pace U, Lancet D (1986) Olfactory GTP-binding protein: signal-transducing polypeptide of vertebrate chemosensory neurons. Proc Natl Acad Sci USA 83:4947–4951

Pace U, Hanski E, Salomon Y, Lancet D (1985) Odorant-sensitive adenylate cyclase may mediate olfactory reception. Nature 316:255–258

Pak WL, Grossfield J, Arnold K (1970) Mutant of the visual pathway of Drosophila melanogaster. Nature 227:518–520

Peier AM, Reeve AJ, Andersson DA, Moqrich A, Earley TJ, Hergarden AC, Story GM, Colley S, Hogenesch JB, McIntyre P, Bevan S, Patapoutian A (2002) A heat-sensitive TRP channel expressed in keratinocytes. Science 296:2046–2049

Perraud AL, Fleig A, Dunn CA, Bagley LA, Launay P, Schimitz C, Stokes AJ, Zhu Q, Bessman MJ, Penner R, Kinet JP, Scharenberg AM (2001) ADP-ribose gating of the calcium-permeable LTRPC2 channel revealed by Nudix motif homology. Nature 411:595–599

Prescott ED, Julius D (2003) A modulator PIP2 binding site as a determinant of capsaicin receptor sensitivity. Science 300:1284–1288

Primakoff P, Myles P (2002) Penetration, adhesion, and fusion in mammalian sperm-egg interaction. Science 296:2183–2185

Raming K, Krieger J, Strotmann J, Boekhoff I, Kubick S, Baumstark C, Breer H (1993) Cloning and expression of odorant receptors. Nature 361:353-356

Ressler KJ, Sullivan SL, Buck LB (1993) A zonal organization of odorant receptor gene expression in the olfactory epithelium. Cell 73:597-609

Ressler KJ, Sullivan SL, Buck LB (1994) Information coding in the olfactory system: evidence for a stereotyped and highly organized epitope map in the olfactory bulb. Cell 79:1245-1255

Runnels LW, Yue L, Clapham DE (2002) The TRPM7 channel is inactivated by PIP2 hydrolysis. Nat Cell Biol 4:329-336

Saito H, Mimmack ML, Keverne EB, Kishimoto J, Emson PC (1998) Isolation of mouse vomeronasal receptor genes and their co-localization with specific G-protein messenger RNAs. Brain Res Mol Brain Res 60:215-227

Selinger Z, Minke B (1988) Inositol lipid cascade of vision studied in mutant flies. Cold Spring Harb Symp Quant Biol 53:333-341

Shirakawa H, Miyazaki S (1999) Spatiotemporal characterization of intracellular Ca^{2+} rise during acrosome reaction of mammalian spermatozoa induced by zona pellucida. Dev Biol 208:70-78

Sklar PB, Anholt RR, Snyder SH (1986) The odorant-sensitive adenylate cyclase of olfactory receptor cells: differential stimulation by distinct classes of odorants. J Biol Chem 261:15538-15543

Smith GD, Gunthrope MJ, Kelsell RE, Hayes PD, Reilly P, Facer P, Wright JE, Jerman JC, Walhin JP, Ooi L, Egerton J, Charles KJ, Smart D, Randall AD, Anand D, Davis JB (2002) TRPV3 is a temperature sensitive vanilloid receptor-like protein. Nature 418:186-190

Story GM, Peier AM, Reeve AJ, Eid SR, Mosbacher J, Hricik TR, Earley TJ, Hergarden AC, Andersson DA, Hwang SW, McIntyre P, Jegla T, Bevan S, patapoutian A (2003) ANKTM1, a TRP-like channel expressed in nociceptive neurons, is activated by cold temperatures. Cell 112:819-829

Stowers L, Holy TE, Meister M, Dulac C, Koentges G (2002) Loss of sex discrimination and male-male aggression in mice deficient for TRP2. Science 295:1493-1500

Stamboulian S, Moutin MJ, Treves S, Pochon N, Grunwald D, Zorzato F, De Waard M, Ronjat M, Arnoult C (2005) Junctate, an inositol 1,4,5-trisphosphate receptor associated protein, is present in rodent sperm and binds TRPC2 and TRPC5 but not TRPC1 channels. Dev Biol 286:326-337

Sutton KA, Jungnickel MK, Wang Y, Cullen K, Lambert S, Florman HM (2004) Enkurin is a novel calmodulin and TRPC channel binding protein in sperm. Dev Biol 274:426-435

Talavera K, Yasumatsu K, Voets T, Droogmans G, Shigemura N, Ninomiya Y, Margolskee RF, Nilius B (2005) Heat activation of TRPM5 underlies thermal sensitivity of sweet taste. Nature 438:1022-1025

Tang J, Lin Y, Zhang Z, Tikunova S, Birnbaumer L, Zhu MX (2001) Identification of common binding sites for calmodulin and inositol 1,4,5-trisphosphate receptors on the carboxyl termini of trp channels. J Biol Chem 276:21303-21310

Treves S, Franzini-Armstrong C, Moccagatta L, Arnoult C, Grasso C, Schrum A, Ducreux S, Zhu MX, Mikoshiba K, Girard T, Smida-Rezgui S, Ronjat M, Zorzato F (2004) Junctate is a key element in calcium entry induced by activation of IP3 receptors and/or calcium store depletion. J Cell Biol 166:537-548

Trevino CL, Serrano CJ, Beltran C, Felix R, Darszon A (2001) Identification of mouse trp homologs and lipid rafts from spermatogenic cells and sperm. FEBS Lett 509:119-125

Trotier D, MacLeod P (1983) Intracellular recordings from salamander olfactory receptor cells. Brain Res 268:225-237

Vannier B, Peyton M, Boulay G, Brown D, Qin N, Jiang M, Zhu X, Birnbaumer L (1999) Mouse trp2, the homologue of the human trpc2 pseudogene, encodes mTrp2, a store depletion-activated capacitative Ca^{2+} entry channel. Proc Natl Acad Sci USA 96:2060–2064

Vassar R, Chao SK, Sitcheran R, Nunez JM, Vosshall LB, Axel R (1994) Topographic organization of sensory projections to the olfactory bulb. Cell 79:981–991

Walensky LD, Snyder SH (1995) Inositol 1,4,5-trisphosphate receptors selectively localized to the acrosomes of mammalian sperm. J Cell Biol 130:857–869

Ward CR, Kopf GS (1993) Molecular events mediating sperm activation. Dev Biol 158:9–34

Ward CR, Storey BT, Kopf GS (1992) Activation of a G_i protein in mouse sperm membranes by solubilized proteins of the zona pellucida, the egg's extracellular matrix. J Biol Chem 267:14061–14067

Wassarman PM (1999) Mammalian fertilization: molecular aspects of gamete adhesion, exocytosis, and fusion. Cell 96:175–183

Wehage E, Eisfeld J, Heiner I, Eberhard J, Zitt C, Luckhoff A (2002) Activation of the cation channel long transient receptor potential channel 2 (LTRPC2) by hydrogen peroxide. J Biol Chem 277:23150–23156

Wilde MW, Ward CR, Kopf GS (1992) Activation of a G-protein in mouse sperm by the zona pellucida, an egg-associated extracellular matrix. Mol Reprod Dev 31:297–306

Wissenbach U, Schroth G, Philipp S, Flockerzi V (1998) Structure and mRNA expression of a bovine trp homologue related to mammalian trp2 transcripts. FEBS Lett 429:61–66

Yildirim E, Dietrich A, Birnbaumer L (2003) The mouse C-type transient receptor potential 2 (TRPC2) channel: alternative splicing and calmodulin binding to its N terminus. Proc Natl Acad Sci USA 100:2220–2225

Zhu X, Jiang M, Peyton MJ, Boulay G, Hurst R, Stefani E, Birnbaumer L (1996) trp, a novel mammalian gene family essential for agonist-activated capacitative Ca^{2+} entry. Cell 85:661–671

Zygmunt PM, Petersson J, Andersson DA, Chuang H, Sorgard M, Di Marzo V, Julius D, Hogestatt ED (1999) Vanilloid receptors on sensory nerves mediate the vasodilator action of anandamide. Nature 400:452–457

TRPC3: A Multifunctional, Pore-Forming Signalling Molecule

P. Eder · M. Poteser · K. Groschner (✉)

Institute of Pharmaceutical Sciences, Pharmacology and Toxicology,
Karl-Franzens-University of Graz, Universitaetsplatz 2, 8010 Graz, Austria
klaus.groschner@uni-graz.at

1	TRPC3 Basic Features	78
1.1	Gene Products and Expression Pattern	78
1.2	Domain Structure and Membrane Topology of TRPC3	79
1.3	Subunit Assembly and Multimerization	80
2	**Channel Properties**	81
2.1	The Pore	81
2.2	The Gating: Primary Modes of Activation	82
2.3	Subcellular Targeting and Cellular Regulation	84
2.4	Regulatory Phosphorylation	86
3	**Pharmacology**	87
4	**Biological Relevance and Emerging Biological Roles of TRPC3**	88
	References	89

Abstract TRPC3 represents one of the first identified mammalian relatives of the *Drosophila trp* gene product. Despite intensive biochemical and biophysical characterization as well as numerous attempts to uncover its physiological role in native cell systems, this channel protein still represents one of the most enigmatic members of the transient receptor potential (TRP) superfamily. TRPC3 is significantly expressed in brain and heart and likely to play a role in both non-excitable as well as excitable cells, being potentially involved in a wide spectrum of Ca^{2+} signalling mechanisms. Its ability to associate with a variety of partner proteins apparently enables TRPC3 to form different cation channels in native cells. TRPC3 cation channels display unique gating and regulatory properties that allow for recognition and integration of multiple input stimuli including lipid mediators and cellular Ca^{2+} gradients as well as redox signals. The physiological/pathophysiological functions of this highly versatile cation channel protein are as yet barely delineated. Here we summarize current knowledge on properties and possible signalling functions of TRPC3 and discuss the potential biological relevance of this signalling molecule.

Keywords TRPC3 · Cation channel subunit · Cellular regulation · Ca^{2+} signalling · Lipid sensor

1
TRPC3 Basic Features

1.1
Gene Products and Expression Pattern

Human TRPC3 (hTRPC3) cDNA was originally cloned using a library derived from HEK293 cells and the expressed sequence tag (EST) sequence R34716 as a probe (Zhu et al. 1996). The human TRPC3 gene consists of 11 exons located on chromosome 4. TRPC3 messenger RNA (mRNA) comprises roughly 4 kb and was found abundantly expressed in brain (Riccio et al. 2002). In peripheral tissues, substantial expression of TRPC3 mRNA has been detected in pituitary gland, and somewhat lower levels in heart and lung (Riccio et al. 2002). Notably, TRPC3 appears to be expressed predominantly in embryonic and developing tissues (Strubing et al. 2003). The human TRPC3 protein comprises 848 amino acids (aa) and shares 96.41% homology with mouse TRPC3 (mTRPC3; 836 aa) and 94% with rat TRPC3 rTRPC3 828 aa; Preuss et al. 1997). Human, mouse and rat genomes contain one additional exon that gives rise to the expression of an N-terminally extended splice variant of TRPC3 (Yildirim et al. 2005; hTRPC3a: 921 aa; mTRPC3a: 911 aa; rTRPC3a: 910 aa). A short splice variant

Fig. 1 a Proposed membrane topology and domain structure of TRPC3. The N-terminus of TRPC3 contains four ankyrin domains, a coiled-coil domain and a hydrophobic domain preceding the first transmembrane segment. A second coiled-coil domain is located in the C-terminus downstream of the TRP box and the proline-rich region. A putative glycosylation site is located in the first extracellular loop. Transmembrane 5 (TM5), TM6 and the connecting pore loop are proposed to form the central cation-conducting pore. TRPC3 splice variants exhibiting different N-termini, TRPC3sv (short variant) lacking two ankyrin domains (736 aa) and TRPC3a with an extended N-terminus (921 aa) are shown. **b** Detection of three TRPC3 splice variants in rat heart by immunoblotting. Rat cardiomyocytes were lysed and 150 µg of the resulting protein was subjected to sodium dodecyl sulphate polyacrylamide gel electrophoresis (SDS-PAGE) and Western blotting. A custom made TRPC3-specific antibody detected three proteins (∼84 kDa, ∼100 kDa, ∼120 kDa) potentially corresponding to the TRPC3 splice forms reported so far. It is of note that the 84 kDa protein was detected only occasionally. **c** Hypothetical organization of a tetrameric TRPC3 pore structure. **d** Inhibition of exo HA-tagged TRPC3 channels by an anti-HA antibody demonstrates the existence of homotetramers. *Left*: Time course of FURA-2 Ca^{2+}-measurement in HEK293 cells co-transfected with YFP-TRPC3 and exo HA-TRPC3 (DNA ratio 1/1). Incubation with HA-antibody significantly inhibits carbachol (CCh)-induced Ca^{2+} entry in a Ca^{2+} re-addition protocol. *Right*: Maximum Ca^{2+} entry derived from Ca^{2+} re-addition experiments as described above (average values±SEM, $n=40$). Dependency of inhibition on cDNA ratio (tagged/non-tagged=1/5, *black columns* and 1/1, *grey columns*) is displayed. Expected values for different order multimers ($n=1-4$), based on the concept that incorporation of a single HA-tagged protein is sufficient to confer antibody sensitivity, are shown

TRPC3: A Multifunctional, Pore-Forming Signalling Molecule

of TRPC3, termed Trp3sv, has been isolated from a rat heart complementary DNA cDNA) library encoding a 736-aa protein with a truncated N-terminus (Ohki et al. 2000; Fig. 1a, b).

1.2
Domain Structure and Membrane Topology of TRPC3

TRPC3 has been supposed as an integral membrane protein with seven membrane-spanning hydrophobic regions (H1–H7), of which six helical stretches (H2–H7) form the transmembrane core domain (TM1–TM6), which is flanked by an intracellular N- and C-terminal domain (Vannier et al. 1998; Zhu et al.

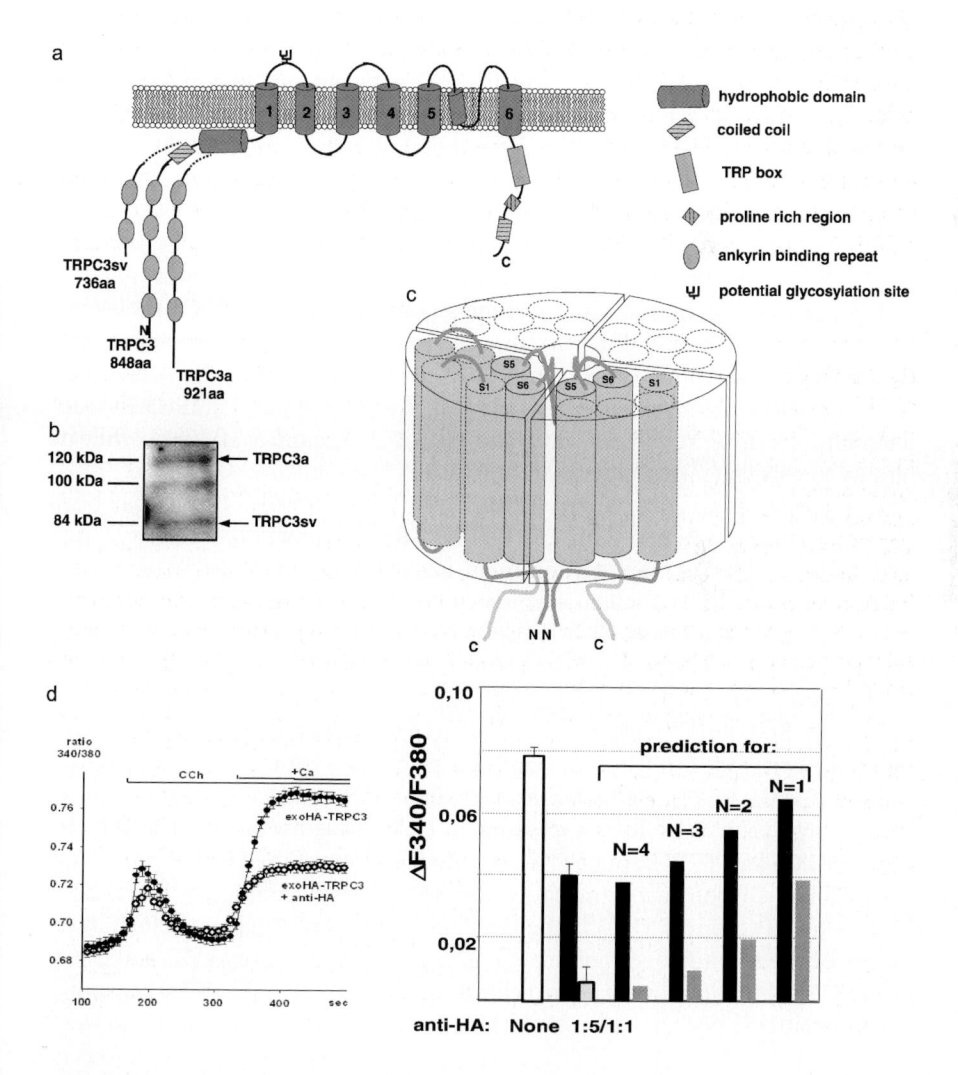

1996). The first hydrophobic region, H1, has been proposed as an intracellular, membrane-associated segment, and based on the similarity of the TRPC membrane topology to that of voltage-gated K$^+$ channels (K$_V$), transmembrane segments TM5, TM6 and their connecting loop were designated as the putative pore region (Vannier et al. 1998; Fig. 1a).

1.3
Subunit Assembly and Multimerization

According to the general concept of a tetrameric pore structure in cation channels with a 6TM membrane topology, TRPC3 is assumed to form tetrameric channel complexes as illustrated in Fig. 1c. It is important to note that the existence of native homotetrameric TRPC3 channels has not definitely been proved and the subunit composition and stoichiometry of native TRPC3 heterotetramers remains elusive. Nonetheless, heterologous overexpression of TRPC3 is likely to generate homomultimeric channels. The stoichiometry of pore complexes may be tested in cells expressing defined mixtures of blocking sensitive mutants and wild-type proteins (Kosari et al. 1998). Characterization of a TRPC3 mutant that contains an exohemagglutinin (exo-HA) tag, which confers sensitivity to block by anti-HA antibody (Poteser et al. 2006), substantiated the concept of TRPC3 being able to assemble in homotetramers (Fig. 1d). Inhibition was consistent with multimers comprising four TRPC3 molecules ($n = 4$) but significantly different from the inhibition predicted for $n = 1$ or 2.

Analysis of the potential of other TRPC isoforms to associate with TRPC3 upon heterologous overexpression suggested a preference of TRPC3 to associate with its closer relatives TRPC6 and TRPC7 (Hofmann et al. 2002), while its ability to associate with more distant relatives was controversially interpreted (Lintschinger et al. 2000; Liu et al. 2005). In HEK293 cells, TRPC3 has been demonstrated to interact with TRPC6 and TRPC7 to form receptor-operated channels. Co-transfection of a dominant-negative mutant of TRPC6 (dnTRPC) and TRPC3 resulted in a decreased receptor activated Mn^{2+} influx. Moreover, association of TRPC3 and TRPC6 fusion proteins was clearly demonstrated by fluorescence resonance energy transfer (FRET) experiments, while minimal FRET signals were detected in cells co-expressing fluorescent protein fusions of TRPC3 and TRPC1, TRPC4 or TRPC5 (Hofmann et al. 2002). In epithelial cells, the existence of TRPC3/6 complexes associated with Ca^{2+} signalling proteins of the G$_q$-PI-PLC pathway has been demonstrated (Bandyopadhyay et al. 2004). Similarly, in brain synaptosomes, TRPC3 has been shown to interact with TRPC6 and TRPC7, but not with other members of the TRPC family (Goel et al. 2002). By contrast, Lintschinger et al. observed functional interactions between TRPC1 and TRPC3 in the HEK293 expression system, and such interactions have recently been confirmed in hippocampal neurons (Wu et al. 2004) as well as in human salivary gland cells in which TRPC3 and TRPC1 were shown to associate via N-terminal interactions to form heteromeric store-operated channels (Liu et al. 2005). Moreover, the association of TRPC3 with TRPC4 was recently demonstrated in a study using HEK293 as well as native vascular endothelial cells (Poteser et al. 2006).

In summary, TRPC3 displays a remarkable potential to form divergent types of cation channels by multimerization with other TRPC proteins. Assembly of TRP homo- and heterotetramers is likely based on interactions between the N-terminal domains of the pore-forming subunits as illustrated in Fig. 1c. Assembly of TRP pore complexes via N-terminal interactions was suggested by demonstration of the ability of N-terminal domains to associate as well as by marked dominant-negative effects of N-terminal fragments (Engelke et al. 2002; Groschner et al. 1998; Liu et al. 2005). Nonetheless, stable assembly of tetramers is likely to involve additional interactions between other domains in hydrophobic segments (Xu et al. 1997) or in the C-terminus (Poteser et al. 2006).

2
Channel Properties

2.1
The Pore

In analogy to K_V channels, the short hydrophobic segments located between transmembrane segments five and six of the individual subunits are considered to line the central ion conducting pathway of TRP channels (see Clapham et al. 2001). Evidence for the contribution of a membrane protein or a protein domain to an ion channel pore may be obtained by different experiments, with the most convincing proof coming from mutation of the putative pore-forming region that results in altered pore properties (Hofmann et al. 2002; Strubing et al. 2003). Alternatively, the blocking activity of antibodies directed against epitopes within motives that putatively contribute to the outer vestibule of the channel pore may be taken as evidence in support of a particular pore structure concept. To date, such evidence is sparse for TRPC channels, and particular information on the structure of the pore region and the selectivity filter of TRPC3 channels is currently lacking. Nonetheless, both approaches yielded evidence in support of a role of the S5–S6 linker in the permeation pathway of a related protein, TRPC1 (Liu et al. 2003). Hence, it appears reasonable to consider this structure essential for the formation of the ion-conducting pathway of TRPC3 channels.

Overexpression of TRPC3 in classical expression systems generates a cation-conducting pathway, which is most likely based on the formation of multimeric pore complexes that allow cation permeation without appreciable selectivity (Hurst et al. 1998; Lintschinger et al. 2000). A thorough investigation of TRPC3 channels in the specific environment of endothelial cells was performed by Kamouchi et al. (1999) and revealed a permeability ratio (pCa/pNa) of about 1.6 (Kamouchi et al. 1999). As TRPC3 is likely to associate with other TRPC species and/or other not-yet-identified transmembrane proteins, multiple populations of channels containing TRPC3 as pore-forming subunits may coexist at variable ratios depending on TRPC expression levels. This concept has been put forward to explain TRPC3-dependent divalent entry phenomena that displayed divergent regulatory properties and different sensitivity to block by the lanthanide Gd^{3+}, indicating the existence of TRPC3 channels with different pore structures (Trebak et al. 2002). Single TRPC3 channels generated by heterologous overexpression have repeatedly been characterized, and a unitary conductance of 60–66 pS (Hurst et al. 1998; Kamouchi et al. 1999; Kiselyov et al. 1998; Poteser et al. 2006; Zitt et al. 1997), as well as a 17-pS sub-conductance (Kiselyov et al. 1999), has been reported.

As Ca^{2+} permeation through TRPC3 has been demonstrated in classical ion substitution experiments (Kamouchi et al. 1999), promotion of cellular Ca^{2+} entry in response to overexpression of TRPC3 has been generally attributed to Ca^{2+} permeation through the TRPC3 pore. However, recent evidence indicates a potential role of TRPC3-mediated Na^+ entry and membrane depolarization as essential determinants of cellular Ca^{2+} homeostasis, and it points to a more complex and indirect link between TRPC3 channel activity and cellular Ca^{2+} signals (Eder et al. 2005; Rosker et al. 2004).

2.2
The Gating: Primary Modes of Activation

Several studies suggest that TRPC3 channels display significant constitutive activity (Dietrich et al. 2003; Hurst et al. 1998) exceeding that of other related TRPC species such as TRPC6. Basal activity of TRPC channels, specifically of TRPC3/6/7 channels, has been related to the glycosylation status, mimicking the higher glycosylation state of TRPC6 and resulting in a reduced constitutive activity (Dietrich et al. 2003). Typically, TRPC3 currents display only little voltage dependence as illustrated in Fig. 2a.

There is general agreement in that cellular TRPC3 channel activity is enhanced in response to stimulation of cellular phospholipase C (PLC) activity (Fig. 2a), and solid evidence has been presented for a rather direct activation of TRPC3 currents by the lipid mediator diacylglycerol (DAG) (Hofmann et al. 1999; Lintschinger et al. 2000; McKay et al. 2000). Hence, TRPC3 as well as its closest relatives TRPC6 and TRPC7 are considered as a unique family of lipid-sensitive cation channels. Besides direct activation of TRPC3 currents by PLC-derived DAG, control of TRPC3 channels by the filling state of intracellular Ca^{2+} stores has repeatedly been suggested (Kiselyov et al. 1998; Vazquez et al. 2003). The ability of TRPC3 to contribute to store-operated Ca^{2+} entry is likely dependent on the expression pattern of Ca^{2+} signalling molecules such as TRPC heteromerization partners or regulatory channel subunits. TRPC1 has recently been identified as one potential heteromerization partner that enables formation of store-operated TRPC3 cation channels (Liu et al. 2005). In essence, both the binding of lipid mediators to TRPC3 channels complexes and an ill-defined stimulus generated by reduction of the filling state of intracellular Ca^{2+} stores are considered as gating mechanisms for TRPC3 channels. Recently, channels generated by the long TRPC3 splice variant TRPC3a were found sensitive to both activation via the G_q–PLC pathway and to direct store depletion (Yildirim et al. 2005). Signalling proteins potentially involved in gating of TRPC3 channels are the inositol 1,4,5-trisphosphate (IP_3) receptor (IP_3R) and calmodulin (CaM) (Zhang et al. 2001), which bind to a combined interaction domain in the C-terminus of TRPC3 that has been termed CIRB (CaM–IP_3 receptor-binding site) (Fig. 2c; Tang et al. 2001). IP_3R-mediated gating of TRPC3 channels was suggested as a mechanism that confers sensitivity of TRPC3 channels to the filling state of intracellular Ca^{2+} stores in terms of a "conformational coupling model" (Berridge 1995; Irvine 1990). This concept was furthered along by the identification of another integral membrane protein resident in the membrane of the endoplasmic reticulum, junctate, which was found to associate with both TRPC3 and the IP_3R (Treves et al. 2004). However, a key role of IP_3Rs was questioned by other studies demonstrating the presence of PLC-dependent as well as store-operated function of TRPC3 in expression systems lacking all three isoforms of the IP_3R (Wedel et al. 2003). Nonetheless, Ca^{2+}-dependent binding of CaM to the CIRB region was shown to interfere with the IP_3R–TRPC3 interaction and to inhibit TRPC channel activity (Zhang et al. 2001). Intracellular Ca^{2+} has repeatedly been demonstrated as a key regulator of TRPC3 channels, and it governs TRPC3 conductances in a rather complex manner, as both permissive (Zitt et al. 1997) and inhibitory effects of intracellular Ca^{2+} have been demonstrated (Lintschinger et al. 2000).

Importantly, classical gating processes, which determine open probability, as well as processes that control plasma membrane presentation of channels, are considered essential for cellular regulation of TRPC3 conductances. Rapid enhancement of TRPC3 channel density in the plasma membrane may explain activation of a TRPC3 conductance without a classical gating process. As recently reported for TRPC5 (Bezzerides et al. 2004), a substantial fraction of TRPC3 appears to be targeted to a pool of highly mobile, plasma membrane-associated vesicles as evident from high-resolution fluorescence microscopy experiments (TIRFM, Fig. 2b). Rapid insertion and retrieval of TRPC3 channels via regulated exo- and endocytosis may therefore be considered as one mechanism regulating cellular TRPC3 conductances. However, a recent study suggests that activation of TRPC3 channels via the G_q–PLC pathway involves a process that is independent of membrane recruitment of channels (Smyth et al. 2006). Further analysis of the contribution of gating mechanisms and mechanisms that merely govern membrane insertion/retrieval of TRPC3 channels in conductance activation will require a thorough analysis at the single channel level.

2.3
Subcellular Targeting and Cellular Regulation

TRPC3 has been suggested as part of a multimolecular signalling complex containing proteins of the Gq–PLC pathway, proteins of the endoplasmic reticulum (ER) membrane and scaffolds and adaptor proteins such as ezrin and caveolin-1 (Cav-1) (Lockwich et al. 2001). Interaction between TRPC3 and other proteins have been identified that may as well be essential for correct targeting or activation of TRPC3 conductances, such as immunophilins, which interact with TRPC3 via the C-terminal proline-rich region (Sinkins et al. 2004). Appropriate assembly of TRPC3 pore complexes with scaffolds and adaptor proteins is likely to enable particular mechanisms of cellular regulation of TRPC3 conductances. Structural motifs in TRPC3 that are involved in such interactions are highlighted in Fig. 2c. Cytoskeletal rearrangements were found to trigger internalization of TRPC3 complexes, reminiscent of the internalization of caveolae, and Cav-1 was demonstrated to co-localize and associate with TRPC3 (Lockwich et al. 2001). Consequently, a caveolin-binding motif has been identified between aa 324–351 (Brazer et al. 2003) in the N-terminus

Fig. 2 a Current-to-voltage relation of basal and carbachol-stimulated TRPC3 conductances in HEK293 cells. Typical whole-cell membrane currents recorded during voltage-ramp protocols in TRPC3-expressing HEK293 cells (T3–9) under basal conditions and after challenge with carbachol (CCh, 200 µM) are shown. Bath solution contained 137 mM NaCl, 5.4 mM KCl, 10 mM hydroxyethylpiperazine ethanesulphonic acid (HEPES), 10 mM glucose, 1 mM MgCl (nominally Ca^{2+} free). Patch pipettes contained 120 mM Cs-methanesulphonate, 20 mM CsCl, 10 mM HEPES, 1 mM $MgCl_2$, 1 mM ethyleneglycoltetraacetic acid (EGTA; pH adjusted to 7.4). Voltage-clamp protocols (depolarizing ramps from −100 to +80 mV/0.6 V/s, 0.2 Hz, holding potential −70 mV) were controlled by pClamp software (Axon Instruments, Foster City). Experiments were performed at room temperature. **b** Rapid membrane insertion and retrieval of TRPC3 in HEK293 cells visualized by total internal reflection fluorescent microscopy (TIRFM). *Left*: Epifluorescence image illustrating the cellular distribution of YFP-TRPC3 in HEK293 cells. *Right*: TIRFM images illustrating clustered localization of YFP-TRPC3 in the cell membrane, with a series of small images representing a sequence illustrating the time course of the fluorescence within a defined area, as indicated (*white square*). The *arrow* highlights the appearance of a TRPC3 cluster probably due to a vesicle fusion event, and its disappearance within 3 s. **c** Putative sites relevant for regulatory phosphorylation of TRPC3 and for protein–protein interactions. Sites relevant for regulatory phosphorylation by the cyclic guanosine monophosphate (cGMP)-dependent kinase protein kinase G (*PKG*) and *PKC* are indicated along with domains involved in the interaction of TRPC3 with phospholipase Cγ1 (*PLCγ1*), caveolin-1 (*Cav-1*), vesicle-associated membrane protein (*VAMP*), FK506 binding protein 12 (*FKBP12*), calmodulin (*CaM*), and inositol-tris-phosphate receptor (IP_3R). The domain for mutually exclusive binding of CaM or the IP_3R designated as CIRB is shown

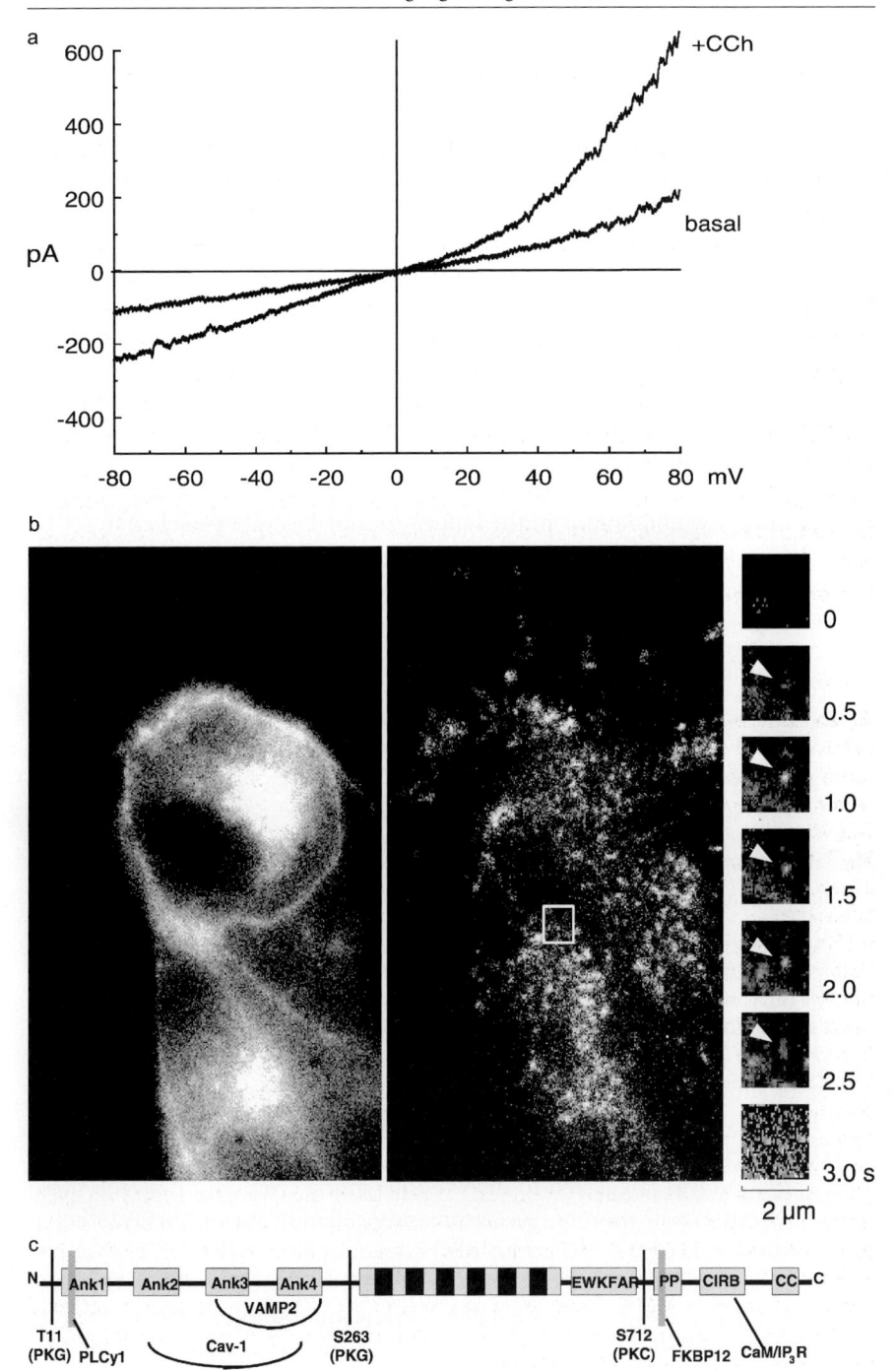

TRPC3: A Multifunctional, Pore-Forming Signalling Molecule 85

of TRPC3. It remains to be clarified if targeting of TRPC3 to the specific membrane environment of caveolae is essential for channel gating by lipids, or if caveolin-TRPC3 association is involved in cellular trafficking of TRPC3 complexes. Importantly, oxidative modification of membrane lipids, specifically of cholesterol, was found to markedly promote TRPC3 activity, a mechanism that may enable TRPC3 to serve as a sensor for the cellular redox state as proposed for channels in vascular endothelial cells (Balzer et al. 1999; Groschner et al. 2004). Such redox-dependent lipid regulation of TRPC3 may well be related to targeting of TRPC3 channels to the cholesterol-rich environment of caveolae. Consistently, cholesterol has recently been found to promote membrane presentation of TRPC3 (Graziani et al. 2006).

The cytoplasmic N-terminus of TRPC3 (aa 123–221) contains a site for interaction with a protein that most likely governs vesicular trafficking of TRPC3, the vesicle-associated membrane protein VAMP2 (Singh et al. 2004). Moreover, correct plasma membrane targeting of TRPC3 has been shown to involve the ankyrin domains in the N-terminus of TRPC3 (Wedel et al. 2003), as well as a unique interaction between TRPC3 and PLCγ1, which generates a composite PH (pleckstrin homology) domain (van Rossum et al. 2005). This bimolecular domain comprising two incomplete lipid binding structures represented by the very N-terminus of TRPC3 and PH-c of PLCγ1 was found to bind PIP_2 and sphingosine-1-phosphate (van Rossum et al. 2005) and has been suggested as a structure essential for plasma membrane targeting and the function of TRPC3 channel complexes. Plasma membrane targeting appears to be governed in addition by the CIRB (761–795 aa) region in the C-terminus of TRPC3 (Wedel et al. 2003).

2.4
Regulatory Phosphorylation

TRPC3 displays several potential sites for regulatory phosphorylation in both the N-and C-terminal cytoplasmic domain (Fig. 2c). Protein kinase C (PKC) has been implicated in down-regulation of TRPC3 activity via phosphorylation of S712 (Trebak et al. 2005), and cyclic GMP (cGMP)-dependent phosphorylation has been shown to suppress TRPC3-mediated store-operated Ca^{2+} entry in HEK293 cells mediated by phosphorylation of TRPC3 at positions T11 and S263 (Kwan et al. 2004). Notably, suppression of TRPC3 activity in response to PKC activation was suggested to involve PKG and a cross-talk between these phosphorylation pathways in PKG-expressing cell systems and in endothelial cells (Kwan et al. 2005). Moreover, PLC-dependent activation of TRPC3 has been demonstrated to require regulatory phosphorylation involving the non-receptor tyrosine kinase Src. Pharmacological inhibition of Src, as well as a dominant negative Src, suppressed TRPC3 activation (Vazquez et al. 2004).

3
Pharmacology

Useful blockers with appreciable selectivity for TRPC3 channels are still missing. Nonetheless, a number of non-selective Ca^{2+} entry blockers were identified as inhibitors of TRPC3 channel activity. Such non-selective inhibitors include the organoborane, 2-aminoethoxydiphenyl borate (2APB), and the imidazole derivative SKF96365, as well as flufenamate and lanthanides. Since 2APB was initially described as a membrane-permeant inhibitor of IP_3Rs, the observed suppression of TRPC3-mediated Ca^{2+} entry has been attributed to effects on IP_3Rs (Ma et al. 2000). However, a more recent report demonstrated direct inhibitory effects of 2APB on TRPC3 channels. IP_3-independent, DAG-triggered TRPC3 activity was inhibited by about 50% with 2APB (30 µM) in HEK293 cells expressing TRPC3 (Lievremont et al. 2005). The apparent incomplete inhibition of lipid mediator-induced TRPC3 channel activity is in contrast to the reported complete block of store-operated channel activity. This discrepancy may be taken as support for the concept of multiple TRPC3 pore structures. A similar phenomenon was reported for the sensitivity of TRPC3 to the lanthanides Gd^{3+} and La^{3+}. When expressed in HEK293 cells, TRPC3 channels were completely inhibited by 150 µM La^{3+} (Zhu et al. 1998), while for channels expressed in COS-1 cells complete block required 10 µM of La^{3+} (Preuss et al. 1997). Gd^{3+} was found to inhibit receptor/PLC-regulated TRPC3 in HEK293 cells completely at 200 µM but lacked any effect on this TRPC3 pathway up to 10 µM concentrations, which suppressed an endogenous Ca^{2+} entry pathway in this cell line (Zhu et al. 1998). TRPC3, expressed at limited levels in DT40 chicken B lymphocytes, was reported to generate a store-operated Ca^{2+} entry pathway that is sensitive to Gd^{3+} in the low micromolar range (Vazquez et al. 2003), again indicating that lanthanides may discriminate different TRPC3 containing pore structures.

Lanthanides have been shown to block TRPC3 currents in CHO cells at rather low levels from the cytoplasmic side (EC_{50} 0.02 µM; Halaszovich et al. 2000). Thus intracellular accumulation of lanthanides may contribute at a variable degree to block of TRPC3 channels, depending on the experimental conditions.

SKF96365 (1-(β-[3-(4-methoxy-phenyl)propoxy]-4-methoxyphenethyl)-1H-imidazole hydrochloride) and flufenamate represent inhibitors that have been widely employed for inhibition of TRPC cation channels. Like 2APB and lanthanides, these compounds effectively suppress membrane transport pathways in a rather non-selective manner with remarkable inhibitory potency for channels outside the TRP superfamily, such as voltage-gated Ca^{2+} channels (Merritt et al. 1990) as well as Cl^- channels (Sergeant et al. 2001). Thus, none of the currently available inhibitors is suitable as a tool to identify native TRPC conductances. The mechanisms by which the organic compounds suppress TRPC3 are largely elusive, and indirect mechanisms such as interference with regulatory components of TRPC3 complex may be considered.

4
Biological Relevance and Emerging Biological Roles of TRPC3

Considering the predominant expression of TRPC3 in specific regions of the brain (Li et al. 2005; Riccio et al. 2002) and in the heart (Eder et al. 2006; Goel et al. 2006), along with particular high expression in embryonic tissues (Strubing et al. 2003), one might speculate about a function of TRPC3 in development of neuronal and cardiac tissue, including a possible role as a determinant of cell proliferation and differentiation. So far, information from a knock-out animal model is lacking, and alternative approaches in cellular model systems—such as dominant-negative suppression, small interfering RNA (siRNA)-mediated knock-down of expression or selective block of channels by isoform specific antibodies—have so far barely been exploited to investigate the role of TRPC3 in neuronal or cardiac cells. Nonetheless, such experiments have provided evidence for functional expression and physiological significance of TRPC3 channels in classical non-excitable tissues such as epithelial and endothelial cells (Bandyopadhyay et al. 2004; Groschner et al. 1998; Liu et al. 2005) as well as immune cells (Philipp et al. 2003).

The role of TRPC3 channels in excitable tissues is so far barely understood. Importantly, the function of TRPC3 channels in excitable cells may critically depend on the functional link of TRPC3 to voltage-gated channels and other prominent ion transport systems in these cells. One example for such a signalling partnership is the recently reported association of TRPC3 channels to the cardiac-type Na^+/Ca^{2+} exchanger NCX1 (Rosker et al. 2004; Eder et al. 2006). TRPC3 may govern Ca^{2+} homeostasis in NCX1-expressing cells, not only via Ca^{2+} permeation through its pore structure but also by translation of TRPC3-mediated Na^+ entry into Ca^{2+} signals. A similar partnership may exist between TRPC3 and voltage-gated Ca^{2+} channels. In view of the significant basal activity of TRPC3 channels, which may be a property of native TRPC3 channels (Albert et al. 2006), it appears reasonable to speculate that enhanced expression of TRPC3 by itself might result in relevant reduction of cell membrane potential and profound alterations of the function of excitable cells. An interesting observation, with respect to this concept, is the recently observed association of pulmonary artery smooth muscle dysfunction with increased expression of TRPC3 (Yu et al. 2004). Elevated density of TRPC3 is expected to cause membrane depolarization of vascular smooth muscle due to enhanced constitutive TRPC3 activity, leading to increased, pathophysiologically relevant Ca^{2+} entry through CaV1.2 (L-type) Ca^{2+} channels.

In summary, our current knowledge suggests TRPC3 as a multifunctional and versatile sensor molecule of particular physiological and pathophysiological relevance.

References

Albert AP, Pucovsky V, Prestwich SA, Large WA (2006) TRPC3 properties of a native constitutively active Ca^{2+}-permeable cation channel in rabbit ear artery myocytes. J Physiol 571:361–369

Balzer M, Lintschinger B, Groschner K (1999) Evidence for a role of Trp proteins in the oxidative stress-induced membrane conductances of porcine aortic endothelial cells. Cardiovasc Res 42:543–549

Bandyopadhyay BC, Swaim WD, Liu X, Redman R, Patterson RL, Ambudkar IS (2004) Apical localization of a functional TRPC3/TRPC6-Ca^{2+}-signaling complex in polarized epithelial cells: role in apical Ca^{2+} influx. J Biol Chem 280:12908–12916

Berridge MJ (1995) Capacitative calcium entry. Biochem J 312:1–11

Bezzerides VJ, Ramsey IS, Kotecha S, Greka A, Clapham DE (2004) Rapid vesicular translocation and insertion of TRP channels. Nat Cell Biol 6:709–720

Brazer SC, Singh BB, Liu X, Swaim W, Ambudkar IS (2003) Caveolin-1 contributes to assembly of store-operated Ca^{2+} influx channels by regulating plasma membrane localization of TRPC1. J Biol Chem 278:27208–27215

Clapham DE, Runnels LW, Strubing C (2001) The TRP ion channel family. Nat Rev Neurosci 2:387–396

Dietrich A, Mederos y Schnitzler M, Emmel J, Kalwa H, Hofmann T, Gudermann T (2003) N-linked protein glycosylation is a major determinant for basal TRPC3 and TRPC6 channel activity. J Biol Chem 278:47842–47852

Eder P, Poteser M, Romanin C, Groschner K (2005) $Na(^+)$ entry and modulation of $Na(^+)/Ca(2+)$ exchange as a key mechanism of TRPC signaling. Pflugers Arch 451:99–104

Eder P, Probst D, Rosker C, Poteser M, Wolinski H, Kohlwein SD, Romanin C, Groschner K (2006) Phospholipase C-dependent control of cardiac calcium homeostasis involves a TRPC3-NCX1 signaling complex. Cardiovasc Res in press

Engelke M, Friedrich O, Budde P, Schafer C, Niemann U, Zitt C, Jungling E, Rocks O, Luckhoff A, Frey J (2002) Structural domains required for channel function of the mouse transient receptor potential protein homologue TRP1beta. FEBS Lett 523:193–199

Goel M, Sinkins WG, Schilling WP (2002) Selective association of TRPC channel subunits in rat brain synaptosomes. J Biol Chem 277:48303–48310

Goel M, Zuo CD, Sinkins WG, Schilling WP (2006) TRPC3 channels co-localize with the Na^+Ca^{2+} exchanger and the Na^+ pump in the axial component of the transverse-axial-tubular system (TATS) of rat ventricle. Am J Physiol Heart Circ Physiol 0:00785.2006v1

Graziani A, Rosker C, Kohlwein SD, Zhu MX, Romanin C, Sattler W, Groschner K, Poteser M (2006) Cellular cholesterol controls TRPC3 function: evidence from a novel dominant negative knock-down strategy. Biochem J 396:147–155

Groschner K, Hingel S, Lintschinger B, Balzer M, Romanin C, Zhu X, Schreibmayer W (1998) Trp proteins form store-operated cation channels in human vascular endothelial cells. FEBS Lett 437:101–106

Groschner K, Rosker C, Lukas M (2004) Role of TRP channels in oxidative stress. Novartis Found Symp 258:222–266

Halaszovich CR, Zitt C, Jungling E, Luckhoff A (2000) Inhibition of TRP3 channels by lanthanides. Block from the cytosolic side of the plasma membrane. J Biol Chem 275:37423–37428

Hofmann T, Obukhov AG, Schaefer M, Harteneck C, Gudermann T, Schultz G (1999) Direct activation of human TRPC6 and TRPC3 channels by diacylglycerol. Nature 397:259–263

Hofmann T, Schaefer M, Schultz G, Gudermann T (2002) Subunit composition of mammalian transient receptor potential channels in living cells. Proc Natl Acad Sci U S A 99:7461–7466

Hurst RS, Zhu X, Boulay G, Birnbaumer L, Stefani E (1998) Ionic currents underlying HTRP3 mediated agonist-dependent Ca^{2+} influx in stably transfected HEK293 cells. FEBS Lett 422:333–338

Irvine RF (1990) 'Quantal' Ca^{2+} release and the control of Ca^{2+} entry by inositol phosphates—a possible mechanism. FEBS Lett 263:5–9

Kamouchi M, Philipp S, Flockerzi V, Wissenbach U, Mamin A, Raeymaekers L, Eggermont J, Droogmans G, Nilius B (1999) Properties of heterologously expressed hTRP3 channels in bovine pulmonary artery endothelial cells. J Physiol 518:345–358

Kiselyov K, Xu X, Mozhayeva G, Kuo T, Pessah I, Mignery G, Zhu X, Birnbaumer L, Muallem S (1998) Functional interaction between InsP3 receptors and store-operated Htrp3 channels. Nature 396:478–482

Kiselyov K, Mignery GA, Zhu MX, Muallem S (1999) The N-terminal domain of the IP3 receptor gates store-operated hTrp3 channels. Mol Cell 4:423–429

Kosari F, Sheng S, Li J, Mak DO, Foskett JK, Kleyman TR (1998) Subunit stoichiometry of the epithelial sodium channel. J Biol Chem 273:13469–13474

Kwan HY, Huang Y, Yao X (2004) Regulation of canonical transient receptor potential isoform 3 (TRPC3) channel by protein kinase G. Proc Natl Acad Sci U S A 101:2625–2630

Kwan HY, Huang Y, Yao X (2005) Protein kinase C can inhibit TRPC3 channels indirectly via stimulating protein kinase G. J Cell Physiol 207:315–321

Li Y, Jia YC, Cui K, Li N, Zhang ZY, Wang YZ, Yuan XB (2005) Essential role of TRPC channels in the guidance of nerve growth cones by brain-derived neurotrophic factor. Nature 437:894–898

Lievremont JP, Bird GS, Putney JW Jr (2005) Mechanism of inhibition of TRPC cation channels by 2-aminoethoxydiphenylborane. Mol Pharmacol 68:758–762

Lintschinger B, Balzer-Geldsetzer M, Baskaran T, Graier WF, Romanin C, Zhu MX, Groschner K (2000) Coassembly of Trp1 and Trp3 proteins generates diacylglycerol- and Ca^{2+}-sensitive cation channels. J Biol Chem 275:27799–27805

Liu X, Singh BB, Ambudkar IS (2003) TRPC1 is required for functional store-operated Ca^{2+} channels. Role of acidic amino acid residues in the S5-S6 region. J Biol Chem 278:11337–11343

Liu X, Bandyopadhyay BC, Singh BB, Groschner K, Ambudkar IS (2005) Molecular analysis of a store-operated and OAG sensitive non-selective cation channel: heteromeric assembly of TRPC1-TRPC3. J Biol Chem 280:21600–21606

Lockwich T, Singh BB, Liu X, Ambudkar IS (2001) Stabilization of cortical actin induces internalization of transient receptor potential 3 (Trp3)-associated caveolar Ca^{2+} signaling complex and loss of Ca^{2+} influx without disruption of Trp3-inositol trisphosphate receptor association. J Biol Chem 276:42401–42408

Ma HT, Patterson RL, van Rossum DB, Birnbaumer L, Mikoshiba K, Gill DL (2000) Requirement of the inositol trisphosphate receptor for activation of store-operated Ca^{2+} channels. Science 287:1647–1651

McKay RR, Szymeczek-Seay CL, Lievremont JP, Bird GS, Zitt C, Jungling E, Luckhoff A, Putney JW Jr (2000) Cloning and expression of the human transient receptor potential 4 (TRP4) gene: localization and functional expression of human TRP4 and TRP3. Biochem J 351:735–746

Merritt JE, Armstrong WP, Benham CD, Hallam TJ, Jacob R, Jaxa-Chamiec A, Leigh BK, McCarthy SA, Moores KE, Rink TJ (1990) SK&F 96365, a novel inhibitor of receptor-mediated calcium entry. Biochem J 271:515–522

Ohki G, Miyoshi T, Murata M, Ishibashi K, Imai M, Suzuki M (2000) A calcium-activated cation current by an alternatively spliced form of Trp3 in the heart. J Biol Chem 275:39055–39060

Philipp S, Strauss B, Hirnet D, Wissenbach U, Mery L, Flockerzi V, Hoth M (2003) TRPC3 mediates T-cell receptor-dependent calcium entry in human T-lymphocytes. J Biol Chem 278:26629–26638

Poteser M, Graziani A, Rosker C, Eder P, Derler I, Kahr H, Zhu MX, Romanin C, Groschner K (2006) Trpc3 and trpc4 associate to form a redox-sensitive cation channel-evidence for expression of native trpc3/trpc4 heteromeric channels in endothelial cells. J Biol Chem 281:13588–13595

Preuss KD, Noller JK, Krause E, Gobel A, Schulz I (1997) Expression and characterization of a trpl homolog from rat. Biochem Biophys Res Commun 240:167–172

Riccio A, Medhurst AD, Mattei C, Kelsell RE, Calver AR, Randall AD, Benham CD, Pangalos MN (2002) mRNA distribution analysis of human TRPC family in CNS and peripheral tissues. Brain Res Mol Brain Res 109:95–104

Rosker C, Graziani A, Lukas M, Eder P, Zhu MX, Romanin C, Groschner K (2004) Ca^{2+} signaling by TRPC3 involves Na+ entry and local coupling to the Na^+/Ca^{2+} exchanger. J Biol Chem 279:13696–13704

Sergeant GP, Hollywood MA, McHale NG, Thornbury KD (2001) Spontaneous Ca^{2+} activated Cl− currents in isolated urethral smooth muscle cells. J Urol 166:1161–1166

Singh BB, Lockwich TP, Bandyopadhyay BC, Liu X, Bollimuntha S, Brazer SC, Combs C, Das S, Leenders AG, Sheng ZH, Knepper MA, Ambudkar SV, Ambudkar IS (2004) VAMP2-dependent exocytosis regulates plasma membrane insertion of TRPC3 channels and contributes to agonist-stimulated Ca^{2+} influx. Mol Cell 15:635–646

Sinkins WG, Goel M, Estacion M, Schilling WP (2004) Association of immunophilins with mammalian TRPC channels. J Biol Chem 279:34521–34529

Smyth JT, Lemonnier L, Vazquez G, Bird GS, Putney JW Jr (2006) Dissociation of regulated trafficking of TRPC3 channels to the plasma membrane from their activation by phospholipase C. J Biol Chem 281:11712–11720

Strubing C, Krapivinsky G, Krapivinsky L, Clapham DE (2003) Formation of novel TRPC channels by complex subunit interactions in embryonic brain. J Biol Chem 278:39014–39019

Tang J, Lin Y, Zhang Z, Tikunova S, Birnbaumer L, Zhu MX (2001) Identification of common binding sites for calmodulin and inositol 1,4,5-trisphosphate receptors on the carboxyl termini of trp channels. J Biol Chem 276:21303–21310

Trebak M, Bird GS, McKay RR, Putney JW Jr (2002) Comparison of human TRPC3 channels in receptor-activated and store-operated modes. Differential sensitivity to channel blockers suggests fundamental differences in channel composition. J Biol Chem 277:21617–21623

Trebak M, Hempel N, Wedel BJ, Smyth JT, Bird GS, Putney JW Jr (2005) Negative regulation of TRPC3 channels by protein kinase C-mediated phosphorylation of serine 712. Mol Pharmacol 67:558–563

Treves S, Franzini-Armstrong C, Moccagatta L, Arnoult C, Grasso C, Schrum A, Ducreux S, Zhu MX, Mikoshiba K, Girard T, Smida-Rezgui S, Ronjat M, Zorzato F (2004) Junctate is a key element in calcium entry induced by activation of InsP3 receptors and/or calcium store depletion. J Cell Biol 166:537–548

van Rossum DB, Patterson RL, Sharma S, Barrow RK, Kornberg M, Gill DL, Snyder SH (2005) Phospholipase Cgamma1 controls surface expression of TRPC3 through an intermolecular PH domain. Nature 434:99–104

Vannier B, Zhu X, Brown D, Birnbaumer L (1998) The membrane topology of human transient receptor potential 3 as inferred from glycosylation-scanning mutagenesis and epitope immunocytochemistry. J Biol Chem 273:8675–8679

Vazquez G, Wedel BJ, Trebak M, Bird GS, Putney JW Jr (2003) Expression level of TRPC3 channel determines its mechanism of activation. J Biol Chem 278:21649–21654

Vazquez G, Wedel BJ, Kawasaki BT, Bird GS, Putney JW Jr (2004) Obligatory role of Src kinase in the signaling mechanism for TRPC3 cation channels. J Biol Chem 279:40521–40528

Wedel BJ, Vazquez G, McKay RR, St JBG, Putney JW Jr (2003) A calmodulin/inositol 1,4,5-trisphosphate (IP3) receptor-binding region targets TRPC3 to the plasma membrane in a calmodulin/IP3 receptor-independent process. J Biol Chem 278:25758–25765

Wu X, Zagranichnaya TK, Gurda GT, Eves EM, Villereal ML (2004) A TRPC1/TRPC3-mediated increase in store-operated calcium entry is required for differentiation of H19-7 hippocampal neuronal cells. J Biol Chem 279:43392–43402

Xu XZ, Li HS, Guggino WB, Montell C (1997) Coassembly of TRP and TRPL produces a distinct store-operated conductance. Cell 89:1155–1164

Yildirim E, Kawasaki BT, Birnbaumer L (2005) Molecular cloning of TRPC3a, an N-terminally extended, store-operated variant of the human C3 transient receptor potential channel. Proc Natl Acad Sci U S A 102:3307–3311

Yu Y, Fantozzi I, Remillard CV, Landsberg JW, Kunichika N, Platoshyn O, Tigno DD, Thistlethwaite PA, Rubin LJ, Yuan JX (2004) Enhanced expression of transient receptor potential channels in idiopathic pulmonary arterial hypertension. Proc Natl Acad Sci U S A 101:13861–13866

Zhang Z, Tang J, Tikunova S, Johnson JD, Chen Z, Qin N, Dietrich A, Stefani E, Birnbaumer L, Zhu MX (2001) Activation of Trp3 by inositol 1,4,5-trisphosphate receptors through displacement of inhibitory calmodulin from a common binding domain. Proc Natl Acad Sci U S A 98:3168–3173

Zhu X, Jiang M, Peyton M, Boulay G, Hurst R, Stefani E, Birnbaumer L (1996) trp, a novel mammalian gene family essential for agonist-activated capacitative Ca^{2+} entry. Cell 85:661–671

Zhu X, Jiang M, Birnbaumer L (1998) Receptor-activated Ca^{2+} influx via human Trp3 stably expressed in human embryonic kidney (HEK)293 cells. Evidence for a non-capacitative Ca^{2+} entry. J Biol Chem 273:133–142

Zitt C, Obukhov AG, Strubing C, Zobel A, Kalkbrenner F, Luckhoff A, Schultz G (1997) Expression of TRPC3 in Chinese hamster ovary cells results in calcium-activated cation currents not related to store depletion. J Cell Biol 138:1333–1341

Ionic Channels Formed by TRPC4

A. Cavalié

Pharmakologie und Toxikologie, Universität des Saarlandes, 66421 Homburg, Germany
adolfo.cavalie@uniklinikum-saarland.de

1	Molecular Properties of TRPC4	94
1.1	Binding Sites for Calmodulin and IP$_3$ Receptors	96
1.2	Protein 4.1-Binding Domain	97
1.3	PDZ-Binding Site	97
1.4	Autoinhibitory Domains	98
1.5	Structural Features Common to TRPC5	98
1.5.1	N-Terminal Coiled-Coil Domain	98
1.5.2	The LFW Motif in the Pore Region	98
1.5.3	Amino Acid Residues Responsible for La^{3+} Facilitation	99
1.5.4	Amino Acid Residues Responsible for the Outward Rectification	99
2	Ionic Channel Properties	100
3	Activation Mechanisms	102
4	Pharmacology	102
5	Biological Relevance of TRPC4	103
References		106

Abstract TRPC4 (transient receptor potential canonical 4) is a member of the TRPC sub-family and, within this sub-family, TRPC4 is most closely related to TRPC5. A number of splice variants of TRPC4 have been identified, whereby TRPC4α and TRPC4β appear to be the most abundant isoforms in various species. TRPC4α comprises six transmembrane segments and the N- and C-termini are located intracellularly. Additionally, TRPC4α shares other structural features with members of the TRPC sub-group, including ankyrin-like repeats, coiled-coil regions and binding sites for calmodulin and IP$_3$ receptors. Three calmodulin-binding domains have been identified in the C-terminus of TRPC4α. TRPC4β lack 84 amino acids in the C-terminus, which correspond to the last two calmodulin-binding sites of TRPCα. The first and last calmodulin-binding domains of TRPC4α overlap with binding sites for the N- and C-termini of IP$_3$ receptors. The ionic channels formed by TRPC4 appear to be Ca^{2+}-permeable, although there is a considerably discrepancy in the degree of Ca^{2+} selectivity. Studies with mice lacking TRPC4 ($TRPC4^{-/-}$) suggest an important role for TRPC4 in supporting Ca^{2+} entry. The defect in Ca^{2+} entry in $TRPC4^{-/-}$ mice appears to be associated with a reduction of the vasorelaxation of arteries, vascular permeability in the lung and neurotransmitter release from thalamic dendrites.

Keywords TRPC4 · Calmodulin · IP3 receptors · Ca^{2+} selectivity · TRPC4$^{-/-}$

1
Molecular Properties of TRPC4

The first full-length transient receptor potential canonical 4 (TRPC4) complementary DNA (cDNA) was originally cloned from bovine adrenal gland, and subsequent studies reported various orthologue genes from species including mouse, rat and human (Philipp et al. 1996; Mori et al. 1998; Mizuno et al. 1999; McKay et al. 2000). The corresponding proteins (bTRPC4α, mTRPC4α, rTRPC4α, hTRPC4α) comprise 974 to 981 amino acids (aa), whereby the longest one is bTRPC4α. Based on computational sequence analysis (Philipp et al. 2000b), it has been proposed that TRPC4α comprises six transmembrane segments (S1–S6) with a putative pore-forming region (P) between S5 and S6 (Fig. 1). In this model, the N- and C-termini are located in the intracellular side. Besides TRPC4α, ten different splice variants of TRPC4 have been so far reported. One group of splice variants shows deletions in the C-terminal region but still contains the six transmembrane segments (Plant and Schaefer 2003). Particularly important is the variant TRPC4β that lacks a domain of 84 aa in the C-terminal region (Fig. 1), which corresponds to putative binding sites for calmodulin (CaM) and inositol 1,4,5-trisphosphate (IP_3) receptors in TRPC4α (see Sect. 1.1). Another group of the TRPC4 splice variants might encode proteins truncated at the second transmembrane segment (Plant and Schaefer 2003). However, the most abundant transcripts in the mouse, rat and humans appear to be TRPC4α and TRPC4β (Mery et al. 2001; Schaefer et al. 2002). So far, only TRPC4α and TRPC4β function has been tested in over-expression systems (see Sect. 2) and it remains to be seen whether the truncated variants have negative dominant effects in the formation of ionic channels.

TRPC4 shows four protein motifs (M1–M4) characteristic for the TRPC sub-family (Philipp et al. 2000b). M1 and M4 are localised in the N-terminal and C-terminal regions of the protein, respectively (Fig. 1). M2 forms the intracellular loop between S4 and S5, and M3 is part of the last transmembrane segment S6 (Fig. 1). M4 contains the so-called TRP box, the highly conserved motif EWKFAR (Fig. 1) that is a hallmark of all TRP channel proteins (Montell et al. 2002). TRPC4 also shares other structural features with members of the TRPC sub-group. These include ankyrin-like repeats, a coiled-coil region and

Fig. 1 General structural features of TRPC4. Based on the amino acid sequence of the murine TRPC4α isoform (mTRPC4α), the diagram shows the location of binding sites as well as important protein domains and amino acid residues identified either in TRPC4 orthologues or TRPC5 (see text). *aa*, amino acids; *ANK1–4*, ankyrin-like repeats; *CCD*, coiled-coil domain; *M1–4*, TRPC protein motifs; *NCB*, N-terminal calmodulin-binding site; *S1–6*, transmembrane segments; *P*, pore region; *P4.1B*, protein 4.1-binding domain; *IP₃RB1–2*, IP_3 receptor-binding sites; *CCB1–3*, C-terminal calmodulin-binding sites; *CIRB*, calmodulin- and IP_3 receptor-binding site; *PDZB*, PDZ-binding site

Ionic Channels Formed by TRPC4

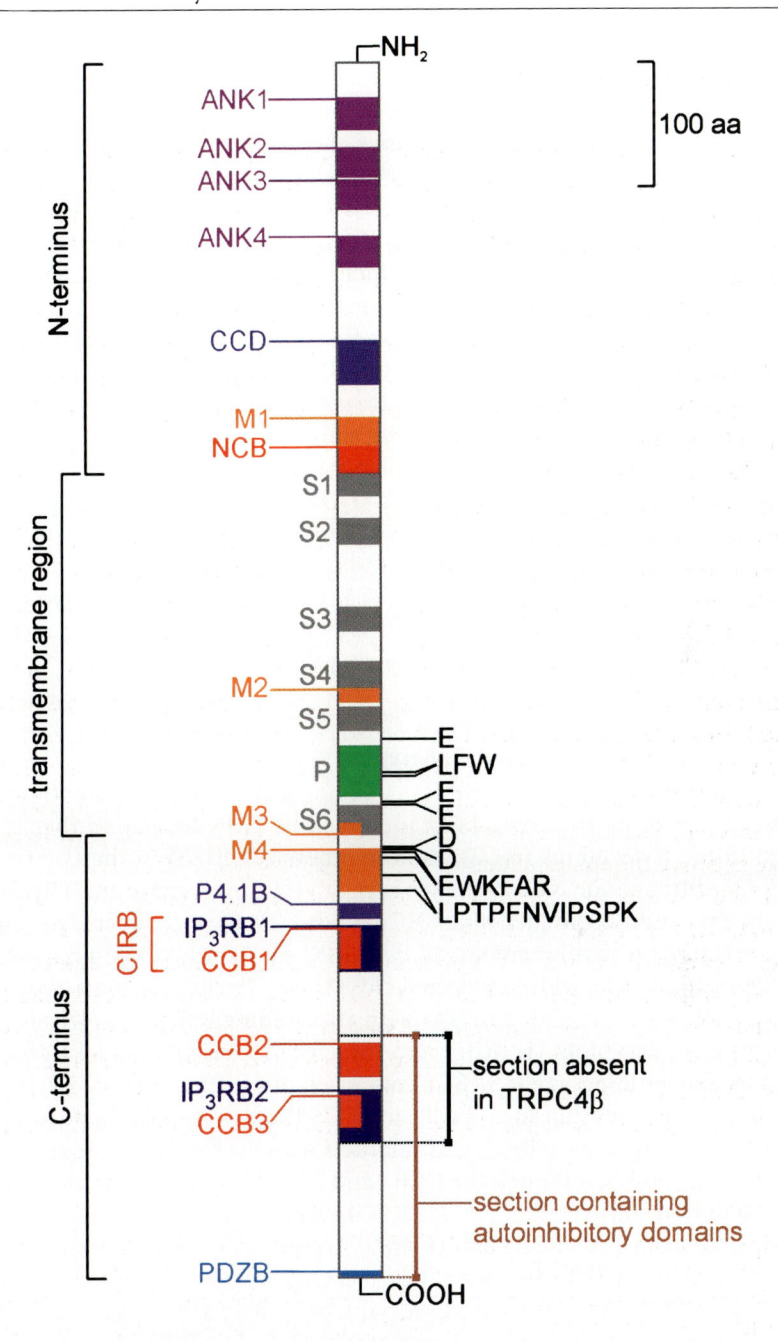

M1: LARLKLAIKYRQKEFVAQPNCQQLL
M2: GPLQISLGRM
M3: VLLNMLIAM
M4: DIEWKFARTKLWMSYFEEGGTLPTPFNVIPSPK

a caveolin-binding region in the N-terminus as well as a proline-rich region, binding sites for CaM and IP$_3$ receptors and a second coiled-coil domain in the C-terminus (Vazquez et al. 2004). Although these protein domains have been studied in great detail for various members of the TRPC sub-family, their precise amino acid sequences are not always well-defined in TRPC4, and our knowledge about their function in TRPC4 is also still limited in some cases. For instance, four ankyrin-like repeats have been predicted for members of the TRPC sub-family (Fig. 1), which—including TRPC4—interact with MxA, a member of the dynamin super-family (Lussier et al. 2005). Apparently, MxA interacts with the second ankyrin-like repeat of TRPC6 and modulates TRPC6 in a guanosine triphosphate (GTP)-dependent manner, but little is known about their action on TRPC4. Structure–function studies specifically with TRPC4 have demonstrated binding sites for CaM, IP$_3$ receptors and protein 4.1, as well a PDZ-binding site, which is unique to TRPC4 and TRPC5 (see Sects. 1.1, 1.2, and 1.3). Since TRPC4 and TRPC5 are the most closely related isoforms within the TRPC sub-family (Montell et al. 2002; Vazquez et al. 2004), insights gained in structure–function studies of TRPC5 also provide important hints to understand the function of TRPC4. The following sections summarise the structural features specifically studied in TRPC4 as well as others common to TRPC5 (Fig. 1).

1.1
Binding Sites for Calmodulin and IP$_3$ Receptors

Four CaM-binding sites have been identified in TRPC4α, one of them is located in the N-terminus (NCB) and the others (CCB1–3) in the C-terminus (Fig. 1). Although the NCB sequence is relatively conserved in the TRPC sub-family, the conditions under which NCB binds CaM have not been determined (Zhu 2005). As with all members of the TRPC sub-family, TRPC4 has a CaM-binding site (CCB1), which is located close to S6 (Trost et al. 2001; Tang et al. 2001). Importantly, CCB1 overlaps with the binding site for the N-terminal region of IP$_3$ receptors (IP$_3$RB1) and therefore represents a dual CaM- and IP$_3$ receptor-binding site (CIRB) (Tang et al. 2001; Zhang et al. 2001). Two additional CaM-binding sites (CCB2 and CCB3) are present in the C-terminus of TRPC4α (Tang et al. 2001). CCB3 interacts with the C-terminal region of IP$_3$ receptors (IP$_3$RB2), although the CCB3 and IP$_3$RB2 sequences do not overlap (Zhu 2005). As illustrated in Fig. 1, TRPC4β lacks a section of the C-terminus, which contains CCB2, CCB3 and IP$_3$RB2. Accordingly, it has been shown that the C-terminus of TRPC4α, but not of TRPC4β, binds to IP$_3$ receptors (Mery et al. 2001). All CCBs appear to interact with CaM in a Ca^{2+}-dependent manner, with half-maximal binding occurring at 9.1–27.9 µM (CCB1), 1.0 µM (CCB2) and 16.6 µM (CCB3) (Trost et al. 2001; Tang et al. 2001). Furthermore, CaM and IP$_3$ receptors compete for binding to CIRB (Tang et al. 2001; Zhang et al. 2001). Thus, it is likely that IP$_3$ receptors are bound to CIRB at low Ca^{2+} concentrations and, conversely, CaM is preferentially bound to CIRB at high Ca^{2+} concentrations. The functional implications of the Ca^{2+}-dependent competition of CaM and IP_3 receptors for binding to TRPC4 is not well-understood but the presence of CIRB might be crucial, since deletion of CIRB in TRPC5 results in non-functional channels (Ordaz et al. 2005). By contrast, deletion of the second CaM-binding site of TRPC5, which has low homology with CCB3 (Zhu 2005), only attenuates the CaM-dependent modulation of channel activation (Ordaz et al. 2005). Further studies with deletions of CCB2, CCB3 or both are needed to clarify the function of these CaM-binding sites in TRPC4α. Similarly, point mutations in CIRB might be useful to analyse the effects of the binding of CaM and IP_3 receptors to TRPC4.

1.2
Protein 4.1-Binding Domain

Actin and spectrin are principally found in the cellular membrane and cytoskeleton, whereas protein 4.1 and ankyrin link transmembrane proteins to the underlying cytoskeleton. In a recent study, it was shown that protein 4.1 binds to TRPAC4 likely within the conserved protein 4.1-binding domain THLCKKKMRRK (Cioffi et al. 2005), which is located in the C-terminus between M4 and CIRB as illustrated in Fig. 1. This region is apparently present in both TRPC4α and TRPC4β. The same study showed that expression of a TRPC4 mutant, in which the protein 4.1-binding domain has been deleted, reduces the activation of store-operated calcium (SOC) currents in pulmonary endothelial cells. Similarly, over-expression of a peptide corresponding to the protein 4.1-binding domain and the adjacent proline-rich region L*PTP*FNVI*PSP*K, which is part of M4 (see Fig. 1), abolished SOC current activation in these endothelial cells (Cioffi et al. 2005).

1.3
PDZ-Binding Site

PDZ domains are named after the three proteins in which they were first identified (PSD-95, dlg, ZO1) and represent important protein–protein interaction sites involved in clustering and organisation of signalling components including ion channels and transporters. TRPC4 interacts with NHERF (Na^+/H^+ exchanger regulatory factor) and EBP50 (ezrin/moesin/radixin-binding phosphoprotein 50), which as PDZ domain proteins also interact with members of the phospholipase Cβ (PLCβ) family and membrane-cytoskeletal adaptors of the ERM (ezrin/radixin/moesin) family (Tang et al. 2000; Mery et al. 2002). Tang et al. (2000) mapped the PDZ-binding site to the last three C-terminal amino acids (TRL) of TRPC4 (see Fig. 1). Within the TRPC sub-family, this TRL motif is unique to TRPC4 and TRPC5. TRPC4 mutants lacking the TRL motif accumulate into cell outgrowths and their overall cell surface expression is reduced by approximately 60%, indicating that the TRL motif is important for the cell surface delivery of TRPC4 that most probably involves interactions with the ERM-bound scaffolding EBP50 (Mery et al. 2002).

1.4
Autoinhibitory Domains

In a comparative study of TRPC4α and TRPC4β, it has been shown that TRPC4α acts as a dominant negative regulator in heteromultimeric channels formed by TRPC4α and TRPC4β (Schaefer et al. 2002). Additionally, homomeric channels formed by TRPC4α were poorly activated upon stimulation of G protein-coupled receptors. Deletion of the last 192 aa, which comprise CCB2, CCB3, IP$_3$RB2 and PDZB (Fig. 1), fully restored the G protein-mediated activation of TRPC4α to the same levels of TRPC4β (Schaefer et al. 2002). Furthermore, the same study showed that deletion of the last 135 aa partially recovers the G protein-mediated activation of TRPC4α. Since the surface expression of these mutants was expected to be low due to the missing PDZB (Tang et al. 2000; Mery et al. 2002), the study by Schaefer et al. (2002) suggests that the last part of the C-terminus (Fig. 1) contains domains with autoinhibitory effects in the formation of TRPC4 channels.

1.5
Structural Features Common to TRPC5

1.5.1
N-Terminal Coiled-Coil Domain

Ionic channels formed by TRPC1 and TRPC5 appear to be confined to the cell body and proximal processes of neurons, whereas TRPC5 is transported to growth cones to form homomeric channels (Greka et al. 2003). In the same study, it is reported that the N-terminus of TRPC5 interacts with members of the stathmin family of phosphoproteins and, specifically, the interaction with stathmin 2 appears to be important for the transport of TRPC5 to the growth cones. Using serial deletions, Greka et al. (2003) showed that the N-terminal coiled-coil domain represent the stathmin-interacting domain of TRPC5, indicating that the interaction between the channels and stathmins is mediated through their common α-helical domains. The authors also showed that the N-terminus of TRPC4 interacts with stathmin 2. Since the N-termini of TRPC4 and TRPC5 are highly homology, it is likely the N-terminal coiled-coil domain of TRPC4 (Fig. 1) is important for the interactions with members of the stathmin family.

1.5.2
The LFW Motif in the Pore Region

The pore helix of members of the TRPC sub-family is not well defined (Owsianik et al. 2006). Nevertheless, changes in the LFW motif, which is contained in the pore region of TRPC5, to AAA resulted in dominant-negative mutants (Hofmann et al. 2002; Strübing et al. 2003). As for other members of the TRPC sub-family (Owsianik et al. 2006), the pore region of TRPC4 also contains the LFW motif (Fig. 1), which may be important for the formation of the pore helix of TRPC4 channels.

1.5.3
Amino Acid Residues Responsible for La^{3+} Facilitation

Previous studies reported that La^{3+} potentiates the ionic currents through TRPC4 and TRPC5 (Schaefer et al. 2000; Strübing et al. 2001). In a systematic study of the negatively charged residues located in the extracellular loops of TRPC5, it has been shown that neutralisation of three out of ten residues results not only in a loss of the La^{3+} potentiation but instead in inhibition of ionic currents by La^{3+} (Jung et al. 2003). These residues correspond to glutamates (E543, E595 and E598), which are very close to the segments S5 and S6 of TRPC5 and are also conserved in TRPC4 (Fig. 1). Mutation of the glutamate residue (E570) located in the central part of the loop between S5 and S6 had no effect on the La^{3+} potentiation (Jung et al. 2003), suggesting that this glutamate residue of TRPC5 is not accessible to external polyvalent cations. The effect of La^{3+} on the mutants E543Q and E595Q/E598Q of TRPC5 were mimicked by increasing external Ca^{2+} from 2 mM to 20 mM (Jung et al. 2003). Thus, the negatively charged residues close to S5 and S6 of TRPC4 and TRPC5 might not only react with La^{3+} but might also serve as sensors for extracellular cations such as Ca^{2+} and Mg^{2+} (see Sect. 4).

1.5.4
Amino Acid Residues Responsible for the Outward Rectification

Channels formed by TRPC5 display a strong outward rectification at positive potentials (Plant and Schaefer 2003). Substitution of the aspartate residue at position 633 by alanine abolished the outward rectification of TRPC5 channels, indicating that D633 is the determinant of the blockade of TRPC5 channels by internal Mg^{2+} (Obukhov and Nowycky 2005). By contrast, the exchange of the adjacent aspartate residue at position 636 had only minor effects on the conductivity of ionic channels formed by TRPC5 (Obukhov and Nowycky 2005). These aspartate residues are conserved in TRPC4 and are located intracellularly between M3 and M4 in TRPC4 (Fig. 1). Since the current–voltage relationship of TRPC4 channels is not well defined (see Sect. 2), it remains to be seen whether these aspartate residues, which may form a ring of negatively charges around the channel axis, mediate TRPC4 channel blockade by intracellular Mg^{2+} or enhance the channel conductivity by attracting positively charged ions.

2
Ionic Channel Properties

The number of TRPC sub-units needed to form an ionic channel has not been determined yet but, as for other well-known ionic channels such as voltage-dependent potassium channels, it is believed that TRP channels are composed of four sub-units. Following this idea, the diversity of native TRPC channels might depend on whether TRPC sub-units are able to form homomultimeric or heteromultimeric (or both types of) ionic channels. Furthermore, the question arises concerning which TRPC sub-units can combine to form a heteromultimeric channel. Within the TRPC sub-family, TRPC4 is more closely related to TRPC5 (Montell et al. 2002; Vazquez et al. 2004). Accordingly, associations of TRPC4 and TRPC5 have been shown both in over-expression system and brain synaptosome and microsome preparations—that is to say, in vivo (Hofmann et al. 2002; Goel et al. 2002; Strübing et al. 2003). Additionally, the same studies showed interactions between TRPC4 and TRPC1. Associations of TRPC4 with TRPC3, TRPC6 and TRPC7 were not detected in adult brain but are likely during embryonic development (Strübing et al. 2003). All these results together support the idea that TRPC4 might form homomultimeric channels as well as heteromultimeric channels mainly with TRPC1, TRPC5 or both, at least in adult cells. Furthermore, heteromultimeric and homomeric channels might co-exist in the same cell, as has been demonstrated for TRPC5 and TRPC1 (Greka et al. 2003). In over-expression systems, it is assumed that the main fraction of recombinant channels represent homomeric forms of the transfected TRPC4 sub-unit, although this assumption has never been proved systematically. Due to the fact that almost all cells express TRPC1, however, it is likely that a fraction of the recombinant channels might contain both TRPC1 and TRPC4 sub-units. Accordingly, the biophysical properties of ionic channels formed by TRPC4 might differ, depending on whether the study is conducted on native or over-expressed channels.

Strübing et al. (2001) showed that the current–voltage relationships (I–V) in cells over-expressing TRPC4 differ strongly from those observed in cells co-transfected with TRPC1 and TRPC4. The ionic currents reversed their polarity between +10 and +20 mV in both cases. At membrane potentials more positive than the reversal potential, the outward currents were nearly constant up to +60 mV in cells transfected with TRPC4 but displayed a strong outward rectification in cells expressing TRPC1+TRPC4. Below the reversal potential, the inward current amplitudes were larger at slightly negative membrane potentials than at strongly negative potentials both in cells expressing TRPC4 and TRPC1+TRPC4. The corresponding I–Vs of TRPC4- and TRPC1+TRPC4-expressing cells displayed a complex shape, which has been often described as doubly-rectifying and outwardly-rectifying, respectively (Plant and Schaefer 2003). In order to compare the different studies on the current–voltage relationships of ionic channels formed by TRPC4, the following sections focus independently on inward currents, outward currents and reversal potential.

Reversal potentials have been used to determine the relative permeability of Ca^{2+} over Na^+ ions (P_{Ca}/P_{Na}) and, therefore, to measure the Ca^{2+}-selectivity of ionic channels (Owsianik et al. 2006). In the first study of TRPC4α (Philipp et al. 1996), the reversal potential was close to 0 mV with nearly physiological cation concentrations. The estimated P_{Ca}/P_{Na} was 7. Subsequent studies with TRPC4α and TRPC4β (McKay et al. 2000; Schaefer et al. 2000, 2002; Strübing et al. 2001) also reported reversal potentials close to 0 mV under comparable ionic conditions. The calculated P_{Ca}/P_{Na} was 0.9–1.05 in the latter studies (McKay et al. 2000; Schafer et al. 2000), indicating that TRPC4 forms a non-selective cationic channel.

In most of the measurements of reversal potentials in cells over-expressing TRPC4, Ca^{2+} concentrations of 1–2 mM have been used. At higher Ca^{2+} concentrations, the reversal potentials appear to shift into strongly positive potentials (Warnat et al. 1999). A comparative study with Ca^{2+} and Ba^{2+} at various concentrations showed that the reversal potential increased above +10 mV when the cationic concentrations were increased to 10 mM (Philipp et al. 1996). In line with this observation, the ionic currents, which are absent in mouse aortic endothelial cells (MAECs) obtained form mice lacking TRPC4 ($TRPC4^{-/-}$), displayed reversal potentials above +40 mV and the calculated P_{Ca}/P_{Na} was 159.7 with 5 mM Ca^{2+} in the corresponding wild-type cells (Freichel et al. 2001). Similarly, the currents, which are abolished by TRPC4 antisense cDNA in adrenal cells, also displayed strongly positive reversal potentials with 10 mM Ca^{2+} in the control cells (Philipp et al. 2000a). Thus, it appears that ionic channels formed by TRPC4 are Ca^{2+}-permeable channels, and the permeation of Ca^{2+} ions becomes more evident at high Ca^{2+} concentrations. It remains to be proved, however, whether the apparent different selectivity of over-expressed and native channels can be explained by the different Ca^{2+} concentrations used in these studies.

Consistent with the high P_{Ca}/P_{Na}, the inward currents attributed to TRPC4 in MAECS and adrenal cells (Philipp et al. 2000a; Freichel et al. 2001) showed strong inward rectification. By contrast, the shape of the I–V below the reversal potentials appears to be very variable in cells over-expressing TRPC4. In some experiments, inward rectification has been observed independently of the reversal potential (Philipp et al. 1996; Warnat et al. 1999). Other studies with over-expressed TRPC4 reported ionic currents with a flat region at negative potentials (Schaefer et al. 2000, 2002; Strübing et al. 2001). The reasons for this variability is not well understood, but it appears that the shape of the I–V depends on the extent of channel activation in cells over-expressing TRPC4 (Plant and Schaefer 2003). The flat region of the I–V extends to potentials as high as −40 mV when the inward current amplitudes are in the picoampere range, and this flat region tends to disappear as the inward current amplitudes reach the nanoampere range. Intriguing is also the fact that outward currents were small or absent in some studies of native and over-expressed TRPC4. Probably, the reason is that the I–V of cells over-expressing TRPC4 also displays a flat region between 0 and +50 mV (Schaefer et al. 2000, 2002; Strübing et al. 2001); accordingly, small outward currents are expected above +50 mV. In studies of TRPC5, this flat region of the I–V at positive potentials has been attributed to a blockade by internal Mg^{2+} (Obukhov and Nowycky 2005). Since the amino acids involved in the Mg^{2+} blockade of ionic channels formed by TRPC5 are also conserved in TRPC4 (see Sect. 1; Fig. 1), it is likely that ionic channels formed by TRPC4 are also blocked by internal Mg^{2+}. Accordingly, the outward rectification of the I–V of ionic channels formed by TRPC4 might depend strongly on the internal Mg^{2+} concentration, but this issue to date has not been addressed experimentally for TRPC4.

3
Activation Mechanisms

Almost all studies with cells over-expressing TRPC4 have shown that the activation of the recombinant channels requires activation of G proteins (Philipp et al. 1996; Schaefer et al. 2000, 2002; Strübing et al. 2001). Furthermore, the activation of PLCβ or PLCγ (or both) is required for activation of recombinant channels formed by TRPC4 (Schaefer et al. 2000). There are, however, considerable discrepancies on the precise mechanism leading to channel activation behind PLC (see Plant and Schaefer 2003). Nevertheless, the ionic currents that are attributed to TRPC4 in MAECs and adrenal cells appear to be activated by depletion of internal Ca^{2+} stores (Freichel et al. 2001; Philipp et al. 2000a). As for TRPC5 (Zeng et al. 2004), it might be that the ionic channels formed by TRPC4 are activated by a multiplicity of signals. For instance, it has been shown that both external and internal Ca^{2+} ions potentiate the activation of ionic currents induced by stimulation of G proteins in cells over-expressing TRPC4 (Schaefer et al. 2000). This observation is consistent with the existence of CaM-binding sites in TRPC4 (see Sect. 1) but it also suggests that, besides Ca^{2+}, another intracellular molecule triggers the activation of TRPC4 channels. Future studies are required to address the question concerning whether a single intracellular signal triggers the channel activation or the coincidence of two or more signals is needed for the activation of ionic channels formed by TRPC4.

4
Pharmacology

Specific blockers or activators of TRPC4 channels have not been found yet, with the probable exception of lanthanides. Generally, La^{3+} and Gd^{3+} have been used for a long time to block a number of ionic conductances. However, micromolar concentrations of lanthanides appear to potentiate the ionic currents through recombinant TRPC4 and TRPC5 channels (Schaefer et al. 2000; Strübing et al. 2001). In a single-channel study of TRPC5, Jung et al. (2003) showed that lanthanides both enhance the open probability and inhibit the single-channel currents in a dose-dependent manner. At micromolar concentrations, the potentiation of single-channel activity is the dominant effect of lanthanides, and at millimolar concentrations the inhibitory effects on the single-channel conductance become dominant. In the analysis of whole-cell currents, Jung et al. (2003) found potentiation of whole-cell currents at La^{3+} and Gd^{3+} concentrations below 100 µM. The inhibitory effects were seen at La^{3+} and Gd^{3+} concentrations higher than 100 µM. The specificity of lanthanide effects on TRPC4 and TRPC5 was underlined by the observation that the lanthanides inhibited currents through recombinant TRPC6 channels with an apparent IC_{50} of 6.1 µM. Additionally, Jung et al. (2003) showed that three glutamate residues flanking the pore region of TRPC5 are involved in the potentiating effect of lanthanides (see also Sect. 1). Since these glutamate residues are conserved in TRPC4 (Fig. 1), a similar mechanism can be assumed for the effects of lanthanides on TRPC4. Concerning lanthanide effects, however, there is also a discrepancy between over-expressed and native channels formed by TRPC4. For instance, the ionic currents missing in cells of mice lacking TRPC4 are not potentiated, but are inhibited, by lanthanides in the micromolar range (Freichel et al. 2001; see also the following section).

5
Biological Relevance of TRPC4

Seminal experiments with mice lacking TRPC4 ($TRPC4^{-/-}$) suggested an important role for TRPC4 in the vascular endothelium (Freichel et al. 2001; Tiruppathi et al. 2002). The Ca^{2+} entry in macrovascular aortic endothelial cells and the vasorelaxation of aortic rings induced by acetylcholine was markedly reduced in $TRPC4^{-/-}$ mice (Freichel et al. 2001). Primary cultured MAECs express TRPC4 transcripts and proteins. In these cells, Ca^{2+}-permeable channels can be activated by depletion of internal Ca^{2+} stores and represent, therefore, SOC, which presumably supports the Ca^{2+} entry induced by acetylcholine. Since the MAECs from $TRPC4^{-/-}$ mice lack the expression of TRPC4 and also SOC currents, the experiments with $TRPC4^{-/-}$ mice indicate that TRPC4 is an indispensable component of SOC channels in vascular endothelial cells and these channels directly provide a Ca^{2+}-entry pathway essential contributing to the regulation of blood vessel tone (Freichel et al. 2001). Similarly, lung vascular endothelial cells (LECs) isolated from $TRPC4^{-/-}$ mice showed a reduced Ca^{2+} entry upon stimulation with thrombin and PAR-1 peptide (Tiruppathi et al. 2002). Although SOC currents were not measured in LEC cells, it is likely that

TRPC4 also forms a Ca^{2+}-entry pathway in these microvascular endothelial lung cells. The defect in Ca^{2+} entry in the LEC cells of $TRPC4^{-/-}$ was associated with a lack of actin-stress fibre formation and reduced cell retraction in response to thrombin stimulation. Experiments with isolated-perfused mouse lungs demonstrated that the increase in vascular permeability induced by the PAR-1 peptide was also less pronounced in $TRPC4^{-/-}$ mice, suggesting that the Ca^{2+}-entry pathway formed by TRPC4 is essential for the modulation of the microvessel permeability in intact lungs (Tiruppathi et al. 2002). The Ca^{2+} entry supported by TRPC4 both in MAEC and LEC cells was blocked by 1 µM La^{3+} (Freichel et al. 2001; Tiruppathi et al. 2002). Similarly, 1 µM La^{3+} was sufficient to block the SOC currents supported by TRPC4 in MAEC cells (Freichel et al. 2001).

A role for TRPC4 in the brain is also indicated by studies with $TRPC4^{-/-}$ mice (Munsch et al. 2003). Thalamocortical relay neurons receive γ-aminobutyric acidergic (GABA-ergic) inputs from interneurons that are modulated by extrathalamic systems acting on various metabotropic receptors. Application of serotonin (5-hydroxytryptamine, 5-HT) and dihydroxyphenylglycol (DHPG) to these cells increases the spontaneous inhibitory post-synaptic current (sIPSC) activity via metabotropic 5-HT (type 2) and glutamate (type 1, mGluRI) receptors, respectively. By contrast, the sIPSC activity in the thalamic relay neurons is reduced by β-methylcholine (MCh) through action on metabotropic cholinergic receptors. While the DHPG and MCh effects were similar in wild-type and $TRPC4^{-/-}$ mice, the 5-HT effects on sIPSC activity were strongly reduced in $TRPC4^{-/-}$ mice, indicating that TRPC4 proteins critically support the GABA release from interneuronal dendrites onto thalamic relay neurons (Munsch et al. 2003). Since neither voltage-dependent Ca^{2+} channels nor Ca^{2+} release from internal Ca^{2+} stores was involved in the 5-HT effects on sIPSC activity, TRPC4 appeared to be critically involved in the Ca^{2+} entry pathway that is needed for GABA release from thalamic interneuronal dendrites (Munsch et al. 2003). Interestingly, 100 µM La^{3+} and 50 µM Gd^{3+} caused a slow increase in sIPSC activity and occluded the responses to 5-HT, which apparently depend on the intracellular PLC signalling pathway.

Thus, studies with $TRPC4^{-/-}$ mice have demonstrated so far that TRPC4 plays a central role in supporting the Ca^{2+} entry which is required for the function of macrovascular and microvascular endothelial cells (Freichel et al. 2001; Tiruppathi et al. 2002) as well as for neurotransmitter release from thalamic interneuronal dendrites (Munsch et al. 2003). In line with this suggestion, transfection experiments with mouse adrenal chromaffin and PC12 cells have shown that TRPC4 can provide a sufficient Ca^{2+} influx to trigger a robust secretory response in neurosecretory cells (Obukhov and Nowycky 2002). Considering that previous studies suggested that TRPC4 channels have a reversal potential close to 0 mV and support not only Ca^{2+} currents but also other cationic currents (see Sect. 2), a distinct possibility is that TRPC4 also contributes to mechanisms underlying membrane depolarisation by primarily allowing a Na$^+$ influx at potentials close to the resting potential. Although not yet conclusively demonstrated, a number of studies provide hints for a possible role of TRPC4 in generating depolarising currents in neurons and smooth muscles.

Stimulation of G protein-coupled metabotropic glutamate receptors leads to activation of non-selective cationic channels in neurons, for instance, in pyramidal neurons of the CA3 region of hippocampus, midbrain dopaminergic neurons and cerebellar Purkinje neurons (e.g. Bengtson et al. 2004; Gee et al. 2003; Kim et al. 2003). In the pre-frontal cortex, G protein-coupled muscarinic receptors also activate cationic non-selective channels (e.g. Haj-Dahmane and Andrade 1999). The ionic currents activated via metabotropic glutamate receptors in some of these studies resembled the double rectifying currents observed after co-expression of TRPC4 and TRPC1 (see Sect. 2). So far, it has been shown that manipulations that interfere with TRPC1 block the excitatory post-synaptic potentials (EPSPs) induced by mGluRI activation in Purkinje cells (Kim et al. 2003). Since TRPC1 might form heteromultimeric ionic channels with TRPC4 and TRPC5 rather than homomultimeric TRPC1 channels (see Sect. 1), the distinct possibility remains that TRPC4 or TRPC5 (or the combination of both channel proteins) is involved in the generation of EPSPs induced by activation of mGluRI in neurons. Since metabotropic glutamate receptors are located in the periphery of post-synaptic membranes and require of burst of activity rather than single synaptic volleys for activation, such ionic currents activated via metabotropic glutamate receptors might contribute to slow EPSPs and to increase neuronal activity.

In the gastrointestinal (GI) muscles, the rhythmic oscillations in membrane potential known as slow waves are generated by the interstitial cells of Cajal (ICC) which represent GI pacemaking cells. Such slow waves propagate within ICC networks, conduct into smooth muscle and initiate phasic contractions via activation of Ca^{2+} entry through voltage-dependent Ca^{2+} channels. The pacemaker mechanisms of ICC appear to involve rhythmic oscillations in the intracellular Ca^{2+} concentration and periodic activation of non-selective pacemaker channels. Comparison of the properties of these native currents in ICC and ionic currents of heterologously expressed TRPC4 revealed strongly similar current-voltage relationships (Walker et al. 2002), indicating that TRPC4 is a strong candidate in the search for membrane proteins that form pacemaker channels in ICC. Supporting this suggestion, the presence of TRPC4 in the caveolae of ICC has been demonstrated (Torihashi et al. 2002). In intestinal smooth muscle, activation of G-coupled muscarinic acetylcholine receptors also evokes a membrane depolarisation that is supported by non-selective cationic currents. In these cells, the membrane depolarisation activates voltage-dependent calcium channels which allow the Ca^{2+} influx and, in this way, contribute to the rise in the intracellular Ca^{2+} concentration needed for the contractile response. Since transcripts of TRPC4 were detected in this tissue (Walker et al. 2001;

Beech et al. 2004) and the ionic currents recorded in GI cells (e.g. Inoue and Isenberg 1990; Zholos et al. 2004) resemble the ionic currents observed some studies with heterologously expressed TRPC4 (see Sect. 2), it is likely that the depolarising currents activated by acetylcholine in smooth muscle might be supported by ionic channels formed by TRPC4.

References

Beech DJ, Muraki K, Flemming R (2004) Non-selective cationic channels of smooth muscle and the mammalian homologues of Drosophila TRP. J Physiol 559:685–706

Bengtson CP, Tozzi A, Bernardi G, Mercuri NB (2004) Transient receptor potential-like channels mediate metabotropic glutamate receptor EPSCs in rat dopamine neurones. J Physiol 555:323–330

Cioffi DL, Wu S, Alexeyev M, Goodman SR, Zhu MX, Stevens T (2005) Activation of the endothelial store-operated ISOC Ca^{2+} channel requires interaction of protein 4. 1 with TRPC4. Circ Res 97:1164–1172

Freichel M, Suh SH, Pfeifer A, Schweig U, Trost C, Weissgerber P, Biel M, Philipp S, Freise D, Droogmans G, Hofmann F, Flockerzi V, Nilius B (2001) Lack of an endothelial store-operated Ca^{2+} current impairs agonist-dependent vasorelaxation in TRP4$^{-/-}$ mice. Nat Cell Biol 3:121–127

Gee CE, Benquet P, Gerber U (2003) Group I metabotropic glutamate receptors activate a calcium-sensitive transient receptor potential-like conductance in rat hippocampus. J Physiol 546:655–664

Goel M, Sinkins WG, Schilling WP (2002) Selective association of TRPC channel subunits in rat brain synaptosomes. J Biol Chem 277:48303–48310

Greka A, Navarro B, Oancea E, Duggan A, Clapham DE (2003) TRPC5 is a regulator of hippocampal neurite length and growth cone morphology. Nat Neurosci 6:837–845

Haj-Dahmane S, Andrade R (1999) Muscarinic receptors regulate two different calcium-dependent non-selective cation currents in rat prefrontal cortex. Eur J Neurosci 11:1973–1980

Hofmann T, Schaefer M, Schultz G, Gudermann T (2002) Subunit composition of mammalian transient receptor potential channels in living cells. Proc Natl Acad Sci U S A 99:7461–7466

Inoue R, Isenberg G (1990) Acetylcholine activates nonselective cation channels in guinea pig ileum through a G protein. Am J Physiol 258:C1173–C1178

Jung S, Muhle A, Schaefer M, Strotmann R, Schultz G, Plant TD (2003) Lanthanides potentiate TRPC5 currents by an action at extracellular sites close to the pore mouth. J Biol Chem 278:3562–3571

Kim SJ, Kim YS, Yuan JP, Petralia RS, Worley PF, Linden DJ (2003) Activation of the TRPC1 cation channel by metabotropic glutamate receptor mGluR1. Nature 426:285–291

Lussier MP, Cayouette S, Lepage PK, Bernier CL, Francoeur N, St-Hilaire M, Pinard M, Boulay G (2005) MxA, a member of the dynamin superfamily, interacts with the ankyrin-like repeat domain of TRPC. J Biol Chem 280:19393–19400

McKay RR, Szymeczek-Seay CL, Lievremont JP, Bird GS, Zitt C, Jungling E, Luckhoff A, Putney JW Jr (2000) Cloning and expression of the human transient receptor potential 4 (TRP4) gene: localization and functional expression of human TRP4 and TRP3. Biochem J 351:735–746

Mery L, Magnino F, Schmidt K, Krause KH, Dufour JF (2001) Alternative splice variants of hTrp4 differentially interact with the C-terminal portion of the inositol 1,4,5-trisphosphate receptors. FEBS Lett 487:377–383
Mery L, Strauss B, Dufour JF, Krause KH, Hoth M (2002) The PDZ-interacting domain of TRPC4 controls its localization and surface expression in HEK293 cells. J Cell Sci 115:3497–3508
Mizuno N, Kitayama S, Saishin Y, Shimada S, Morita K, Mitsuhata C, Kurihara H, Dohi T (1999) Molecular cloning and characterization of rat trp homologues from brain. Brain Res Mol Brain Res 64:41–51
Montell C, Birnbaumer L, Flockerzi V (2002) The TRP channels, a remarkably functional family. Cell 108:595–598
Mori Y, Takada N, Okada T, Wakamori M, Imoto K, Wanifuchi H, Oka H, Oba A, Ikenaka K, Kurosaki T (1998) Differential distribution of TRP Ca^{2+} channel isoforms in mouse brain. Neuroreport 9:507–515
Munsch T, Freichel M, Flockerzi V, Pape HC (2003) Contribution of transient receptor potential channels to the control of GABA release from dendrites. Proc Natl Acad Sci U S A 100:16065–16070
Obukhov AG, Nowycky MC (2002) TRPC4 can be activated by G-protein-coupled receptors and provides sufficient Ca^{2+} to trigger exocytosis in neuroendocrine cells. J Biol Chem 277:16172–16178
Obukhov AG, Nowycky MC (2005) A cytosolic residue mediates Mg^{2+} block and regulates inward current amplitude of a transient receptor potential channel. J Neurosci 25:1234–1239
Ordaz B, Tang J, Xiao R, Salgado A, Sampieri A, Zhu MX, Vaca L (2005) Calmodulin and calcium interplay in the modulation of TRPC5 channel activity. Identification of a novel C-terminal domain for calcium/calmodulin-mediated facilitation. J Biol Chem 280:30788–30796
Owsianik G, Talavera K, Voets T, Nilius B (2006) Permeation and selectivity of trp channels. Annu Rev Physiol 68:685–717
Philipp S, Cavalie A, Freichel M, Wissenbach U, Zimmer S, Trost C, Marquart A, Murakami M, Flockerzi V (1996) A mammalian capacitative calcium entry channel homologous to Drosophila TRP and TRPL. EMBO J 15:6166–6171
Philipp S, Trost C, Warnat J, Rautmann J, Himmerkus N, Schroth G, Kretz O, Nastainczyk W, Cavalié A, Hoth M, Flockerzi V (2000a) TRP4 (CCE1) protein is part of native calcium release-activated Ca^{2+}-like channels in adrenal cells. J Biol Chem 275:23965–23972
Philipp S, Wissenbach U, Flockerzi V (2000b) Molecular biology of calcium channels. In: Putney JW Jr (ed) Calcium signaling. CRC Press, Boca Raton, pp 321–342
Plant TD, Schaefer M (2003) TRPC4 and TRPC5: receptor-operated Ca^{2+}-permeable nonselective cation channels. Cell Calcium 33:441–450
Plant TD, Schaefer M (2005) Receptor-operated cation channels formed by TRPC4 and TRPC5. Naunyn Schmiedebergs Arch Pharmacol 371:266–276
Schaefer M, Plant TD, Obukhov AG, Hofmann T, Gudermann T, Schultz G (2000) Receptor-mediated regulation of the nonselective cation channels TRPC4 and TRPC5. J Biol Chem 275:17517–17526
Schaefer M, Plant TD, Stresow N, Albrecht N, Schultz G (2002) Functional differences between TRPC4 splice variants. J Biol Chem 277:3752–3759
Strübing C, Krapivinsky G, Krapivinsky L, Clapham DE (2001) TRPC1 and TRPC5 form a novel cation channel in mammalian brain. Neuron 29:645–655

Strübing C, Krapivinsky G, Krapivinsky L, Clapham DE (2003) Formation of novel TRPC channels by complex subunit interactions in embryonic brain. J Biol Chem 278:39014–39019

Tang J, Lin Y, Zhang Z, Tikunova S, Birnbaumer L, Zhu MX (2001) Identification of common binding sites for calmodulin and inositol 1,4,5-trisphosphate receptors on the carboxyl termini of trp channels. J Biol Chem 276:21303–21310

Tang Y, Tang J, Chen Z, Trost C, Flockerzi V, Li M, Ramesh V, Zhu MX (2000) Association of mammalian trp4 and phospholipase C isozymes with a PDZ domain-containing protein, NHERF. J Biol Chem 275:37559–37564

Tiruppathi C, Freichel M, Vogel SM, Paria BC, Mehta D, Flockerzi V, Malik AB (2002) Impairment of store-operated Ca^{2+} entry in TRPC4$^{-/-}$ mice interferes with increase in lung microvascular permeability. Circ Res 91:70–76

Torihashi S, Fujimoto T, Trost C, Nakayama S (2002) Calcium oscillation linked to pacemaking of interstitial cells of Cajal: requirement of calcium influx and localization of TRP4 in caveolae. J Biol Chem 277:19191–19197

Trost C, Bergs C, Himmerkus N, Flockerzi V (2001) The transient receptor potential, TRP4, cation channel is a novel member of the family of calmodulin binding proteins. Biochem J 355:663–670

Vazquez G, Wedel BJ, Aziz O, Trebak M, Putney JW Jr (2004) The mammalian TRPC cation channels. Biochim Biophys Acta 1742:21–36

Walker RL, Hume JR, Horowitz B (2001) Differential expression and alternative splicing of TRP channel genes in smooth muscles. Am J Physiol Cell Physiol 280:C1184–C1192

Walker RL, Koh SD, Sergeant GP, Sanders KM, Horowitz B (2002) TRPC4 currents have properties similar to the pacemaker current in interstitial cells of Cajal. Am J Physiol Cell Physiol 283:C1637–C1645

Warnat J, Philipp S, Zimmer S, Flockerzi V, Cavalié A (1999) Phenotype of a recombinant store-operated channel: highly selective permeation of Ca^{2+}. J Physiol 518:631–638

Zeng F, Xu SZ, Jackson PK, McHugh D, Kumar B, Fountain SJ, Beech DJ (2004) Human TRPC5 channel activated by a multiplicity of signals in a single cell. J Physiol 559:739–750

Zhang Z, Tang J, Tikunova S, Johnson JD, Chen Z, Qin N, Dietrich A, Stefani E, Birnbaumer L, Zhu MX (2001) Activation of Trp3 by inositol 1,4,5-trisphosphate receptors through displacement of inhibitory calmodulin from a common binding domain. Proc Natl Acad Sci U S A 98:3168–3173

Zholos AV, Zholos AA, Bolton TB (2004) G-protein-gated TRP-like cationic channel activated by muscarinic receptors: effect of potential on single-channel gating. J Gen Physiol 123:581–598

Zhu MX (2005) Multiple roles of calmodulin and other Ca^{2+}-binding proteins in the functional regulation of TRP channels. Pflügers Arch 451:105–115

Canonical Transient Receptor Potential 5

D. J. Beech

Institute of Membrane and Systems Biology, University of Leeds, Leeds LS2 9JT, UK
d.j.beech@leeds.ac.uk

1	Gene and Protein: Basic Properties and Localisation	110
2	Ion Channel Properties	112
3	Modes of Activation	113
3.1	Receptor Activation	113
3.2	Activation by Store Depletion	114
3.3	Activation by Lanthanides	114
3.4	Activation by Lysophosphatidylcholine	115
3.5	Limited Activation by Intracellular Calcium: A Permissive Role	115
3.6	Multiplicity of Activation	115
4	Vesicular Trafficking	116
5	Na^+-H^+ Exchange Regulatory Factor	116
6	Calcium-Binding Proteins	117
7	Myosin Light Chain Kinase	118
8	Stathmins	118
9	Other Protein Partners	119
10	Desensitisation: Diacylglycerol and Protein Kinase C	119
11	Additional Regulatory Phenomena	119
12	Pharmacology	120
13	Biological Relevance	120
14	Conclusion	121
	References	121

Abstract Canonical transient receptor potential 5 TRPC5 (also TrpC5, trp-5 or trp5) is one of the seven mammalian TRPC proteins. Its known functional property is that of a mixed cationic plasma membrane channel with calcium permeability. It is active alone or as a heteromultimeric assembly with TRPC1; TRPC4 and TRPC3 may also be involved. Multiple activators of TRPC5 are emerging, including various G protein-coupled receptor agonists, lysophospholipids, lanthanide ions and, in some contexts, calcium store depletion. Intracellular calcium has complex impact on TRPC5, including a permissive role for other activators, as well as inhibition at high concentrations. Protein kinase C is inhibitory and mediates desensitisation following receptor activation. Tonic TRPC5 activity is detected and may reflect the presence of constitutive activation signals. The channel has voltage dependence but the biological significance of this is unknown; it is partially due to intracellular magnesium blockade at aspartic acid residue 633. Protein partners include calmodulin, CaBP1, enkurin, Na^+-H^+ exchange regulatory factor (NHERF) and stathmin. TRPC5 is included in local vesicular trafficking regulated by growth factors through phosphatidylinositol (PI)-3-kinase, Rac1 and PIP-5-kinase. Inhibition of myosin light chain kinase suppresses TRPC5, possibly via an effect on trafficking. Biological roles of TRPC5 are emerging but more reports on this aspect are needed. One proposed role is as a mediator of calcium entry and excitation in smooth muscle, another as an inhibitor of neuronal growth cone extension. The latter is intriguing in view of the original cloning of the human *TRPC5* gene from a region of the X chromosome linked to mental retardation. TRPC5 is a broadly expressed calcium channel with capability to act as an integrator of extracellular and intracellular signals at the level of calcium entry.

Keywords Calcium channel · Transient receptor potential · Receptor operated · Store operated · Lanthanide

1
Gene and Protein: Basic Properties and Localisation

The coding region of the canonical transient receptor potential 5 gene (*TRPC5*) was first cloned from rabbit and mouse brain and initially referred to as *CCE2* (Philipp et al. 1998). At a similar time, coding regions of human *TRPC5* were cloned from a region of the X chromosome that follows two genes (*DCX* and *HPAK3*) linked to non-syndromic mental retardation (Sossey-Alaoui et al. 1999). The human *TRPC5* gene (locus *Xq23*) is about 308 kb, the transcript length is 5.84 kb and there are 11 exons (Sossey-Alaoui et al. 1999). No splice variants are reported.

Human versus mouse and human versus rabbit TRPC5 amino acid sequence alignments indicate 96.9% and 98.8% identities, respectively. The protein is 973–975 amino acids in length and has predicted molecular mass of 111.4 kDa. Amino acid sequence analyses and comparisons with related proteins suggest TRPC5 is a membrane protein with six membrane-spanning segments and intracellular N- and C-termini (Fig. 1). Analogy with the structurally related voltage-gated potassium channels further suggests four TRPC5 proteins come together to form a single ion channel. Over-expression of TRPC5 alone leads to functional channels, indicating its capability to operate as a homomeric assembly—i.e. a channel comprising only TRPC5 proteins. Such a homomultimer is suggested to occur natively (Greka et al. 2003). However, TRPC5 also interacts with TRPC1 and TRPC4 (Strubing et al. 2001; Goel et al. 2002; Hofmann et al. 2002; Strubing et al. 2003), and TRPC3 may enter the complex

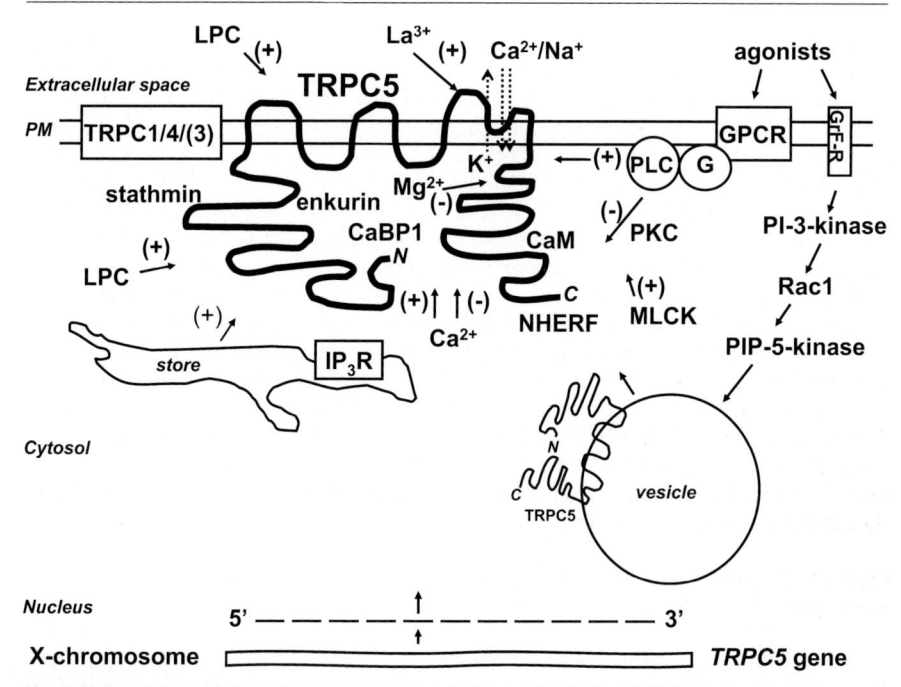

Fig. 1 Schematic summarising current knowledge of TRPC5. TRPC5 is shown in its presumed six membrane-spanning segment orientation; ion permeation is thought to require the co-ordination of four TRPC5, or related, proteins. The schematic is not drawn to scale. To aide clarity, the protein partners FKBP52, MxA and junctate (see text) are not in the schematic. *TRPC*, canonical transient receptor potential; *PM*, plasma membrane; *LPC*, lysophosphatidylcholine; *PLC*, phospholipase C; *GPCR*, G protein-coupled receptor; *G*, G protein; *PKC*, protein kinase C; *IP₃R*, inositol trisphosphate receptor; *GrF-R*, growth factor receptor; *PI-3-kinase*, phosphatidylinositol-3-kinase; *CaM*, calmodulin; *CaBP1*, calcium binding protein-1; *MLCK*, myosin light chain kinase; *NHERF*, Na⁺–H⁺ exchange regulatory factor; Ca^{2+}, calcium; Mg^{2+}, magnesium; Na^+, sodium; K^+, potassium; La^{3+}, lanthanum; *N*, amino terminus; *C*, carboxy terminus

if TRPC1 is present (Strubing et al. 2003). Because these other subunits are widely expressed alongside TRPC5, endogenous TRPC5 may commonly exist as a heterotetrameric arrangement.

In human, the highest levels of TRPC5-encoding messenger RNA (mRNA) are in the brain, with broad but non-uniform expression across different regions (Philipp et al. 1998; Sossey-Alaoui et al. 1999; Riccio et al. 2002). The RNA species is also detected throughout many tissues of the body, albeit at lower levels than the brain (Riccio et al. 2002). TRPC5 protein is reported in neurones (Greka et al. 2003) as well as non-neuronal cells of the murine temporal lobe (von Bohlen et al. 2005), in sperm head (Sutton et al. 2004), and vascular smooth muscle cells (Beech et al. 2004; Yip et al. 2004; Flemming et al. 2005).

2
Ion Channel Properties

TRPC5 is a functional plasma membrane ion channel across a broad range of physiological voltages and is not usually thought of as "voltage-gated". It does not switch off completely at any voltage between −100 and + 100 mV, but activity does have voltage-dependence—that is, voltage can be thought of as a modulator once the channel is activated via some other mechanism (see below). In many recordings, triple rectification is evident (Zeng et al. 2004), such that the current-voltage relationship (I–V) requires three Boltzmann components for satisfactory mathematical description (D.J. Beech, unpublished). One of these components is reminiscent of the behaviour of a voltage-gated potassium channel, showing increasing activation with increasing positive voltage. A key difference, however, is that the sub-threshold activity is not zero. Furthermore, hyperpolarisation does not retain the channel at this constitutive level but enhances opening, conferring inward rectification. With increasing hyperpolarisation the current relaxes (shuts off) to a variable extent. Relaxation at negative potentials is also observed at the single channel level (Yamada et al. 2000). The latter is most variable, in our experience being absent in a significant number of recordings. The combination of the different voltage-dependent modes leads to a "signature I–V" dominated by double-rectification—outward and inward rectification coming together at an inflexion point around 0 mV. TRPC5 is largely devoid of kinetics, except for modest current decay when a square hyperpolarising step is applied. I–Vs constructed by ramps or square step protocols are essentially identical (Xu et al. 2005a). Examples of the TRPC5 I–V can be found, for example, in Schaefer et al. (2000), Strubing et al. (2001), Zhu et al. (2005) and Xu et al. (2005a).

The unitary conductance of the TRPC5 channel is relatively large—for example, 41-pS chord conductance at −60 mV recorded by Jung et al. (2003). These authors also reported a mean open time at −60 mV of 7.5 ms and frequency of opening of 6.2 Hz at 20–25°C. The channel has similar permeability to sodium, caesium and potassium, while lacking permeability to chloride or the large cation N-methyl-D-glucamine (Okada et al. 1998; Schaefer et al. 2000; Strubing et al. 2001; Lee et al. 2003; Obukhov and Nowycky 2004; Xu et al. 2005a). Divalent cations are permeant, including calcium, barium, manganese and strontium (Okada et al. 1998; Schaefer et al. 2000; Venkatachalam et al. 2003). The ratio of calcium to sodium permeability has been estimated to be 9.5 (Okada et al. 1998) or 1.8 (Schaefer et al. 2000). Amino acid residues between the fifth and sixth membrane-spanning segments are highly likely to be involved in forming the ion selectivity filter. Mutation of the conserved LFW sequence (leucine–phenylalanine–tryptophan) in this region (position 575–577 in the murine sequence) to alanine–alanine–alanine creates a dominant-negative TRPC5 mutant (Strubing et al. 2003), presumably because ion flux is damaged.

Intracellular magnesium blocks TRPC5, reducing outward unitary current at +30 mV with an IC_{50} of 457 µM and Hill coefficient of 0.6 (Obukhov and Nowycky 2005). Mutation of aspartic acid residue 633 in the proximal C-terminus reduces the voltage-dependent component of blockade. The blockade accounts partially for rectification in the TRPC5 I–V (see also Schaefer et al. 2000).

Heteromultimeric TRPC5–TRPC1 has a different I–V, with less inflexion and thus greater but not absolute linearity in the physiological range (Strubing et al. 2001; Bezzerides et al. 2004; Obukhov and Nowycky 2005). Outward rectification at positive voltages is retained. Unitary currents are almost ten times smaller compared with TRPC5 alone (Strubing et al. 2001).

3
Modes of Activation

Activation of TRPC5 has been studied by means of over-expression in mammalian cell lines or *Xenopus laevis* oocytes. In these contexts it shows a variable degree of tonic activity. Whether there is tonic activity or not, several stimuli evoke activity or markedly enhance the existing activity, as shown in the following sections.

3.1
Receptor Activation

In various cell types, TRPC5 is activated on application of adenosine 5′-triphosphate, bradykinin, carbachol (or acetylcholine), histamine, IgM (B cell receptor cross-linking), prostaglandin E_2, thrombin or uridine 5′-triphosphate (Schaefer et al. 2000; Tabata et al. 2002; Venkatachalam et al. 2003; Ohta et al. 2004; Zeng et al. 2004). In some of these studies, receptor was concomitantly over-expressed but, nevertheless, the data show clearly that G protein-coupled receptor activation enhances TRPC5 activity at the plasma membrane. It is also apparent that intracellular guanosine triphosphate (GTP)-γ-S (a stable analogue of guanosine triphosphate) stimulates TRPC5, suggesting GTP-bound G proteins are sufficient for TRPC5 activation (Schaefer et al. 2000; Kanki et al. 2001). G proteins of the $G_{q/11}$ type have been shown to be involved (Kanki et al. 2001; Lee et al. 2003), but the signalling mechanism down-stream is unclear. The phospholipase C inhibitor U73122 suppresses TRPC5 activation by muscarinic agonists or epidermal growth factor (EGF) (Schaefer et al. 2000; Kanki et al. 2001; Lee et al. 2003), suggesting signalling involves a type of phospholipase C, or at least a pharmacologically similar protein. Arguments have been put forward for (Kanki et al. 2001) and against (Schaefer et al. 2000; Strubing et al. 2001; Venkatachalam et al. 2003; Xu et al. 2005a) involvement of inositol trisphosphate (IP_3) or its receptor. Diacylglycerol does not stimulate but inhibits TRPC5 and thus cannot be the second messenger underlying receptor activation (Schaefer et al. 2000; Kanki et al. 2001; Zhu et al. 2002). Calcium release is not responsible because prior depletion of calcium stores fails to prevent receptor activation (Schaefer et al. 2000; Zeng et al. 2004). From the latter, however, it should not be presumed that calcium release has no impact on TRPC5 during receptor activation, only that calcium release is not an absolute requirement.

3.2
Activation by Store Depletion

There are differing reports on whether TRPC5 is activated following depletion of intracellular calcium stores, with some investigators observing no association (Schaefer et al. 2000; Strubing et al. 2001; Venkatachalam et al. 2003; Ohta et al. 2004; Zhu et al. 2005; Kinoshita-Kawada et al. 2005) and others observing enhanced TRPC5 activity (Philipp et al. 1998; Kanki et al. 2001; Zeng et al. 2004). Blockade of endogenous store-operated channels is important prior to testing for store-operated function of the over-expressed TRPC5 because of the often large endogenous signals of expression systems (Zeng et al. 2004). It is also apparent that store depletion has modest efficacy as an activator, and thus this type of activation may be weak under some experimental conditions. Other studies now provide evidence for roles of TRPC proteins in some endogenous non-selective cationic store-operated channels (Beech 2005; Rao et al. 2005). Therefore, weak association with store-depletion in the over-expression context may relate to paucity of a necessary endogenous signal or protein partner. The signal that couples stores to TRPC5 is unknown, but may involve binding to the IP_3 receptor or regulation by lysophosphatidylcholine.

3.3
Activation by Lanthanides

An unusual and striking feature of TRPC5 is strong activation by external lanthanides—lanthanum or gadolinium at 1–100 µM (Jung et al. 2003; Zeng et al. 2004; Xu et al. 2005a). This property is shared only by TRPC4; other TRP channels are inhibited or unaffected by lanthanides. Acidic amino acid residues in the outer pore region of TRPC5 are involved—particularly glutamic acid residue 543 at the top end of the fifth membrane-spanning segment—suggesting an external agonist-binding site for these ions (Jung et al. 2003). The biological importance of the lanthanide effect is unknown. Humans contain lanthanides but at relatively low concentrations and the ions are not known to have biological signalling functions. External calcium mimics the lanthanide effect, but only at supra-physiological concentrations (Okada et al. 1998; Jung et al. 2003; Zeng et al. 2004).

3.4
Activation by Lysophosphatidylcholine

Lysophosphatidylcholine is a major component of atherosclerotic plaques and also has important signalling functions in the immune system and other systems. TRPC5 is strongly activated by lysophosphatidylcholine at low micromolar concentrations (Flemming et al. 2005). The effect does not occur through a G protein signalling pathway or generation of reactive oxygen species and is preserved in excised membrane patches, suggesting it is a relatively direct action on the channels—either via a hydrophobic site associated with the channel protein or because of sensitivity of the channel to the structure of the bilayer. TRPC5 is not activated by arachidonic acid (Schaefer et al. 2000; Flemming et al. 2005). The sensing of lysophosphatidylcholine confers on TRPC5 the property of lipid ionotropic receptor, akin to several other TRP channels.

3.5
Limited Activation by Intracellular Calcium: A Permissive Role

Elevation of intracellular calcium to 200 nM stimulates TRPC5 in the absence of an exogenous agonist (Schaefer et al. 2000; Strubing et al. 2001; Zeng et al. 2004). TRPC5 expressed in *Xenopus* oocytes has also been shown to be activated by ionomycin, an ionophore evoking calcium release (Kinoshita-Kawada et al. 2005). However, the activating effect of calcium is relatively small compared with that of lanthanides, carbachol or lysophosphatidylcholine (Zeng et al. 2004; Flemming et al. 2005), and higher micromolar concentrations of calcium are inhibitory and involved in desensitisation (Ordaz et al. 2005; Zhu et al. 2005). It is also striking that buffering of intracellular calcium to low, sub-physiological, levels suppresses activation of TRPC5 by other stimuli, such as ATP or carbachol (Okada et al. 1998; Zhu et al. 2005; Kinoshita-Kawada et al. 2005). Therefore, intracellular calcium at, or just above, the physiological resting concentration would seem to have a permissive role—facilitating responses to other agonists, while not strongly activating the channel on its own. Indeed, the small 'direct' activation of TRPC5 observed on elevating calcium to 200 nM may reflect enhanced tonic activity, which may depend on the presence of low levels of endogenous agonists in the recording medium or within the cells.

3.6
Multiplicity of Activation

Bringing together the above findings leads to the simple conclusion that TRPC5 is vulnerable to a multiplicity of activators, subject to 'permission' by intracellular calcium. This concept is observed within a single cell, showing that TRPC5's different properties do not depend absolutely on discrete cellular contexts but instead reflect intrinsic sensitivity to multiple activators. This may be due to versatility/promiscuity of gating, or different activators may converge on the channel via a common, but, as yet, ill-defined primary activator. Whatever the mechanism, TRPC5 has the capability to act as an integrative sensor with potential to co-ordinate a variety of signals at the level of calcium entry.

4
Vesicular Trafficking

In human embryonic kidney cells, TRPC5 tagged with green fluorescent protein appears as 320-nm punctae that are undefined, non-endocytic vesicles (Schaefer et al. 2000; Bezzerides et al. 2004). In response to EGF, the punctae progress from the sub-plasma membrane to plasma membrane space over a period of 1–2 min (Bezzerides et al. 2004). The outcome is greater surface expression of TRPC5, although not activation (Bezzerides et al. 2004). [Schaefer et al. (2000) observed apparent direct activation by EGF possibly because TRPC5 exhibited significant tonic activity under their experimental conditions]. Co-expression of TRPC1 inhibits the trafficking, resulting in resistance to EGF (Bezzerides et al. 2004).

EGF-evoked trafficking of TRPC5 is prevented by inhibitors of phosphatidylinositol (PI)-3-kinase or a dominant-negative mutant of Rac1, a monomeric G protein (Bezzerides et al. 2004). Furthermore, the constitutively active mutant Rac1 is sufficient to evoke trafficking in the absence of EGF. These observations are relevant because tyrosine kinase receptors (like the EGF receptor) couple to PI-3-kinase, and Rac1 is an effector of PI-3-kinase. In TRPC5 trafficking, the effector down-stream of Rac1 is suggested to be PIP-5-kinase because a dominant-negative mutant of this kinase is also inhibitory. In summary, Bezzerides et al. (2004) suggest the signal cascade: EGF receptor; PI-3-kinase; Rac1; and PIP-5-kinase—which may couple to trafficking via phosphatidylinositol bisphosphate (PIP_2); direct evidence for this last step is lacking, however.

In primary hippocampal neurones, several growth factors stimulate the surface expression of a TRPC5-like protein, including brain-derived neurotrophic factor (BDNF), nerve-growth factor (NGF) and insulin-like growth factor-1 (IGF-1) (Bezzerides et al. 2004). PI-3-kinase inhibitors prevent the NGF effect. Therefore, local vesicular trafficking of TRPC5 regulated by growth factors could be an important event in neuronal development.

G protein-coupled receptor-activation (see Sect. 3.1 above) of TRPC5 is suggested not to involve vesicular trafficking (Ordaz et al. 2005).

5
Na^+–H^+ Exchange Regulatory Factor

Na^+–H^+ exchange regulatory factor (NHERF) is a PDZ protein that interacts with actin cytoskeleton through ezrin/radixin/moesin proteins. It interacts with peptides terminating in TRL sequence (threonine–arginine–leucine), and thus conservation of this sequence at the end of TRPC5 and TRPC4 suggested NHERF as a binding partner. Tang et al. (2000) showed NHERF binds the C-terminus of TRPC5 as well as phospholipase Cβ, indicating a scaffolding function. The role of the interaction with TRPC5 is, however, unclear. Deletion of 'VTTRL' from the end of TRPC5 did not affect localisation to the plasma membrane but increased the punctate appearance (Obukhov and Nowycky 2004). Although deletion of the VTTRL sequence failed to affect activation of TRPC5 by histamine, it suppressed the capacity of over-expressed exogenous NHERF-1 to delay the response time to histamine. Therefore, endogenous NHERF interaction has little role in receptor activation, at least in a TRPC5 over-expression context. The latter may be significant because scaffolding arrangements presumably serve to improve signalling efficiency in the context of endogenous channel expression levels.

6
Calcium-Binding Proteins

Early in studies of TRPC5 the calcium-binding protein calmodulin (CaM) was identified as a protein partner. Binding involves the C-terminus and such binding of CaM is a common feature of TRPC channels, occurring via the CIRB (CaM-IP$_3$ receptor binding) site (Tang et al. 2001). Ordaz et al. (2005) subsequently identified an additional site 95 amino acid residues down-stream of the CIRB site.

Exogenous CaM enhances activation of TRPC5 by thrombin or carbachol (Ordaz et al. 2005; Kim et al. 2005) or has no effect (Kinoshita-Kawada et al. 2005). Kim et al. (2005) observed inhibition of carbachol-activation of TRPC5 by CaM inhibitors or a mutant of CaM lacking calcium affinity, suggesting a role of calcium–CaM in receptor activation, or at least preservation of its integrity at the membrane. Kinoshita-Kawada et al. (2005), however, observed no effect of mutant CaM, and Ordaz et al. (2005) found only a small change in receptor activation of TRPC5 if it carried deletion of the second (down-stream) CaM site. The effect of exogenous CaM was, however, diminished by mutation in this second site (Ordaz et al. 2005). CIRB site mutants were non-functional but localised to the membrane, perhaps suggesting the importance of IP$_3$ receptor interaction for TRPC5 function (Kanki et al. 2001; Ordaz et al. 2005). Calmodulin inhibitors had no effect on the ability of micromolar calcium to inhibit TRPC5 (Ordaz et al. 2005).

Enkurin is a 25-kDa soluble CaM-binding protein identified in a yeast two-hybrid screen using the TRPC2 N-terminus as bait (Sutton et al. 2004). Subsequent results suggested binding also to the N-terminus of TRPC1 or TRPC5. Enkurin is particularly localised to testis, although also elsewhere, including lung, brain and heart. Its role in TRPC5 function is unknown.

Over-expression of calcium-binding protein-1 (CaBP1) strongly inhibits activation of TRPC5 by carbachol or ionomycin without preventing carbachol-evoked calcium release (Kinoshita-Kawada et al. 2005). Mutation of the EF hand of CaBP1 prevents its effect; similarly, CaBP1 binds TRPC5 only in the presence of calcium. Therefore, calcium binding by CaBP1 seems necessary for its binding to TRPC5 and its inhibitory effect on TRPC5. CaBP1 binding to TRPC5 most likely involves tertiary structures, because various N- and C-terminal regions of TRPC5 have the capacity to bind CaBP1 (Kinoshita-Kawada et al. 2005). The CaM sites of the C-terminus fail to bind CaBP1, however, consistent with discrete roles of CaM and CaBP1. CaBP1 has no effect on surface expression of TRPC5, and thus its role may be to inhibit TRPC5's function at the membrane—perhaps in certain cellular localisations (Kinoshita-Kawada et al. 2005).

7
Myosin Light Chain Kinase

Myosin light chain kinase (MLCK) phosphorylates the light chain of myosin— a protein component of various molecular motors—and is activated by calcium–CaM. Chemical inhibitors of CaM or MLCK inhibit receptor activation of TRPC5, leading to the hypothesis that MLCK is involved in this function of TRPC5. Consistently, receptor activation of TRPC5 is suppressed by a dominant-negative mutant of MLCK (Shimizu et al. 2005) or short interfering RNA (siRNA) targeted to messenger RNA (mRNA) encoding MLCK (Kim et al. 2005). It is perplexing, however, that the mutant MLCK also inhibited ATP-evoked calcium release, whereas the chemical inhibitors of MLCK did not (Shimizu et al. 2005). Furthermore, the CaM and MLCK inhibitors ML-7, calmidazolium and W-7 had no effect on TRPC5 activated by intracellular GTP-γ-S (Kim et al. 2005). Therefore, these inhibitors are without direct effect on the function of TRPC5 at the membrane, a finding that is difficult to reconcile with the conclusion that MLCK is involved in the acute trafficking of TRPC5 (Shimizu et al. 2005). Alternatively, the role of MLCK might be in regulation of the receptor-coupling mechanism.

8
Stathmins

Stathmins are phosphoproteins involved in microtubular assembly. A yeast two-hybrid screen using TRPC5 N-terminus as bait pulled out stathmin 3 (Greka et al. 2003). Follow-up immunoprecipitation studies showed other stathmins also bind and that the interaction requires integrity of a predicted coiled-coiled domain of the TRPC5 N-terminus. TRPC5 and stathmin 2 partially co-localise in punctate structures along the processes of growth cones (Greka et al. 2003). Prior studies had shown that stathmin 2-targeting to growth cones requires palmitoylation of stathmin 2. Transfection of neurones with a palmitoylation-defective stathmin-2 mutant revealed growth cones that lack TRPC5, suggesting stathmin 2 is involved in loading of TRPC5 into growth cones (Greka et al. 2003).

9
Other Protein Partners

Other suggested protein partners of TRPC5 are the immunophilin FKBP52 (Sinkins et al. 2004), the dynamin superfamily member MxA (Lussier et al. 2005) and junctate (Stamboulian et al. 2005). See also output from immunoprecipitation proteomic screens (Goel et al. 2005).

10
Desensitisation: Diacylglycerol and Protein Kinase C

Receptor or G protein activation of TRPC5 commonly fades with continuous exposure to the agonist (Zhu et al. 2005). Diacylglycerol and protein kinase C contribute to this desensitisation. Diacylglycerol inhibits the channels via protein kinase C, and inhibitors of protein kinase C suppress the agonist-evoked fade (Venkatachalam et al. 2003; Zhu et al. 2005). Zhu et al. (2005) found that mutation of a threonine residue within the VTTRL sequence of the C-terminus (position 972 in murine TRPC5) slowed the desensitisation process, suggesting the effect may relate to direct phosphorylation of TRPC5. Desensitisation linked to histamine receptor activation was, however, clearly evident in TRPC5 lacking the VTTRL sequence, suggesting threonine 972 is not required (Obukhov and Nowycky 2004). Also in contrast to the findings of Obukhov and Nowycky (2004), Zhu et al. (2005) found that the VTTRL deletion strongly reduced functional expression of TRPC5, and thus Zhu et al. (2005) could not explore effects on the desensitisation process.

11
Additional Regulatory Phenomena

Trans-retinoic acid dramatically up-regulated the expression of mRNA encoding TRPC5 (Chang et al. 1997). Down-regulation of the smooth endoplasmic reticulum Ca^{2+}-ATPase (SERCA-2) in cardiac myocytes led to up-regulation of mRNA encoding TRPC5 (Seth et al. 2004).

12
Pharmacology

TRPC5 is blocked by 2-aminoethoxydiphenyl borate (2APB) with an IC$_{50}$ of 20 µM (Xu et al. 2005a). Blockade occurs strictly via the external and not internal face of the channel. Mild voltage-dependence of block is apparent, which is consistent with the 2APB binding site existing within the electric field, and perhaps within the ion-pore. Intriguingly, however, the voltage-dependence aligns with the intrinsic voltage-dependence of TRPC5, suggesting the alternative possibility that 2APB is a modulator of TRPC5 gating (Xu et al. 2005a).

Okada et al. (1998) observed inhibition in response to 100 µM gadolinium or lanthanum, as did Lee et al. (2003). Jung et al. (2003) also found inhibition, but only at high concentrations (>100 µM) of lanthanum. Inhibitory effects of lanthanides may seem difficult to reconcile with the hallmark agonist action of lanthanides on TRPC5 (see Sect. 3.3), but both effects occur and the inhibitory effect may be more pronounced if the channels already have high activity prior to application of the lanthanide—for example, because of receptor activation. Receptor-activated TRPC5 is also blocked by 25 µM SKF-96365 (Okada et al. 1998), 0.1–10 µM 3,5-bis(trifluoromethyl)pyrazole derivative BTP-2 (He et al. 2005), 100 µM flufenamic acid (Lee et al. 2003), the CaM inhibitors 100 µM W-13 or chlorpromazine (Shimizu et al. 2005), 100 µM W-7 or 5 µM calmidazolium (Kim et al. 2005), and the MLCK inhibitors 3 µM ML-7 or ML-9 (Shimizu et al. 2005; Kim et al. 2005).

Receptor-activated TRPC5 has been found resistant to 10 µM U73343, 30 µM dihydrosphingosine, 10 µM staurosporine, 1 µM bisindolylmaleimide I, 10 µM genistein, 10 µM wortmannin (but see Shimizu et al. 2005), 1 mM sodium orthovanadate, 300 µM indomethacin or 50 µM RHC-80267 (Schaefer et al. 2000), and 1 µM nifedipine, 10 µM methoxyverapamil, 25 µM berberine or 100 µM MRS-1845 (Xu et al. 2005a).

13
Biological Relevance

Despite research on the properties of over-expressed TRPC5, relatively little is yet known of its endogenous biological roles. Nevertheless, Greka et al. (2003) made the striking observation that transfection of hippocampal neurones with dominant-negative mutant TRPC5 leads to growth cone extension. These investigators developed the hypothesis that opening of endogenous TRPC5 channels enables calcium entry that retards outgrowth. Another study showed that sustained lanthanum-evoked calcium entry in arteriolar smooth muscle cells is suppressed by antibody to TRPC5, suggesting functional activity of endogenous TRPC5 channels in these small blood vessels (Xu et al. 2005b). Lastly, a TRPC5-like current is evoked by muscarinic agonists in smooth muscle cells of the stomach; the data suggest TRPC5 or a closely related channel is involved in acetylcholine-evoked depolarisation in visceral smooth muscles (Lee et al. 2003).

14
Conclusion

TRPC5 is an intriguing novel calcium channel protein with emerging and complex biological sensing capabilities. Although there is significant information on the channel, described above and summarised in Fig. 1, the number of published studies is small and consequently understanding of the channel limited. The channel nevertheless shows considerable promise as an integrator of extracellular and intracellular signals at the level of cellular calcium entry through the plasma membrane. There is expectation that, in the near future, investigators will reveal further roles of TRPC5 in physiology and novel roles in disease.

Acknowledgements Supported by the Wellcome Trust and British Heart Foundation.

References

Beech DJ (2005) TRPC1: store-operated channel and more. Pflugers Arch 451:53–60
Beech DJ, Muraki K, Flemming R (2004) Non-selective cationic channels of smooth muscle and the mammalian homologues of Drosophila TRP. J Physiol 559:685–706
Bezzerides VJ, Ramsey IS, Kotecha S, Greka A, Clapham DE (2004) Rapid vesicular translocation and insertion of TRP channels. Nat Cell Biol 6:709–720
Chang AS, Chang SM, Garcia RL, Schilling WP (1997) Concomitant and hormonally regulated expression of trp genes in bovine aortic endothelial cells. FEBS Lett 415:335–340
Flemming PK, Dedman AM, Xu SZ, Li J, Zeng F, Naylor J, Benham CD, Bateson AN, Muraki K, Beech DJ (2005) Sensing of lysophospholipids by TRPC5 calcium channel. J Biol Chem 281:4977–4982
Goel M, Sinkins WG, Schilling WP (2002) Selective association of TRPC channel subunits in rat brain synaptosomes. J Biol Chem 277:48303–48310
Goel M, Sinkins W, Keightley A, Kinter M, Schilling WP (2005) Proteomic analysis of TRPC5- and TRPC6-binding partners reveals interaction with the plasmalemmal Na^+/K^+-ATPase. Pflugers Arch 451:87–98
Greka A, Navarro B, Oancea E, Duggan A, Clapham DE (2003) TRPC5 is a regulator of hippocampal neurite length and growth cone morphology. Nat Neurosci 6:837–845
He LP, Hewavitharana T, Soboloff J, Spassova MA, Gill DL (2005) A functional link between store-operated and TRPC channels revealed by the 3,5-bis(trifluoromethyl)pyrazole derivative, BTP2. J Biol Chem 280:10997–11006
Hofmann T, Schaefer M, Schultz G, Gudermann T (2002) Subunit composition of mammalian transient receptor potential channels in living cells. Proc Natl Acad Sci U S A 99:7461–7466
Jung S, Muhle A, Schaefer M, Strotmann R, Schultz G, Plant TD (2003) Lanthanides potentiate TRPC5 currents by an action at extracellular sites close to the pore mouth. J Biol Chem 278:3562–3571

Kanki H, Kinoshita M, Akaike A, Satoh M, Mori Y, Kaneko S (2001) Activation of inositol 1,4,5-trisphosphate receptor is essential for the opening of mouse TRP5 channels. Mol Pharmacol 60:989–998

Kim MT, Kim BJ, Lee JH, Kwon SC, Yeon DS, Yang DK, So I, Kim KW (2005) Involvement of calmodulin and myosin light chain kinase in the activation of mTRPC5 expressed in HEK cells. Am J Physiol Cell Physiol 290:C1031–C1040

Kinoshita-Kawada M, Tang J, Xiao R, Kaneko S, Foskett JK, Zhu MX (2005) Inhibition of TRPC5 channels by Ca^{2+}-binding protein 1 in Xenopus oocytes. Pflugers Arch 450:345–354

Lee YM, Kim BJ, Kim HJ, Yang DK, Zhu MH, Lee KP, So I, Kim KW (2003) TRPC5 as a candidate for the nonselective cation channel activated by muscarinic stimulation in murine stomach. Am J Physiol Gastrointest Liver Physiol 284:G604–G616

Lussier MP, Cayouette S, Lepage PK, Bernier CL, Francoeur N, St-Hilaire M, Pinard M, Boulay G (2005) MxA, a member of the dynamin superfamily, interacts with the ankyrin-like repeat domain of TRPC. J Biol Chem 280:19393–19400

Obukhov AG, Nowycky MC (2004) TRPC5 activation kinetics are modulated by the scaffolding protein ezrin/radixin/moesin-binding phosphoprotein-50 (EBP50). J Cell Physiol 201:227–235

Obukhov AG, Nowycky MC (2005) A cytosolic residue mediates Mg^{2+} block and regulates inward current amplitude of a transient receptor potential channel. J Neurosci 25:1234–1239

Ohta T, Morishita M, Mori Y, Ito S (2004) Ca^{2+} store-independent augmentation of $[Ca^{2+}]i$ responses to G-protein coupled receptor activation in recombinantly TRPC5-expressed rat pheochromocytoma (PC12) cells. Neurosci Lett 358:161–164

Okada T, Shimizu S, Wakamori M, Maeda A, Kurosaki T, Takada N, Imoto K, Mori Y (1998) Molecular cloning and functional characterization of a novel receptor-activated TRP Ca^{2+} channel from mouse brain. J Biol Chem 273:10279–10287

Ordaz B, Tang J, Xiao R, Salgado A, Sampieri A, Zhu MX, Vaca L (2005) Calmodulin and calcium interplay in the modulation of TRPC5 channel activity. Identification of a novel C-terminal domain for calcium/calmodulin-mediated facilitation. J Biol Chem 280:30788–30796

Philipp S, Hambrecht J, Braslavski L, Schroth G, Freichel M, Murakami M, Cavalie A, Flockerzi V (1998) A novel capacitative calcium entry channel expressed in excitable cells. EMBO J 17:4274–4282

Rao JN, Platoshyn O, Golovina VA, Liu L, Zou T, Marasa BS, Turner DJ, J XJY, Wang JY (2005) TRPC1 functions as a store-operated Ca^{2+} channel in intestinal epithelial cells and regulates early mucosal restitution after wounding. Am J Physiol Gastrointest Liver Physiol 290:G782–G792

Riccio A, Medhurst AD, Mattei C, Kelsell RE, Calver AR, Randall AD, Benham CD, Pangalos MN (2002) mRNA distribution analysis of human TRPC family in CNS and peripheral tissues. Brain Res Mol Brain Res 109:95–104

Schaefer M, Plant TD, Obukhov AG, Hofmann T, Gudermann T, Schultz G (2000) Receptor-mediated regulation of the nonselective cation channels TRPC4 and TRPC5. J Biol Chem 275:17517–17526

Seth M, Sumbilla C, Mullen SP, Lewis D, Klein MG, Hussain A, Soboloff J, Gill DL, Inesi G (2004) Sarco(endo)plasmic reticulum Ca^{2+} ATPase (SERCA) gene silencing and remodeling of the Ca^{2+} signaling mechanism in cardiac myocytes. Proc Natl Acad Sci U S A 101:16683–16688

Shimizu S, Yoshida T, Wakamori M, Ishii M, Okada T, Takahashi M, Seto M, Sakurada K, Kiuchi Y, Mori Y (2005) Ca^{2+}/calmodulin dependent myosin light chain kinase is essential for activation of TRPC5 channels expressed in HEK293 cells. J Physiol 570:219–235

Sinkins WG, Goel M, Estacion M, Schilling WP (2004) Association of immunophilins with mammalian TRPC channels. J Biol Chem 279:34521–34529

Sossey-Alaoui K, Lyon JA, Jones L, Abidi FE, Hartung AJ, Hane B, Schwartz CE, Stevenson RE, Srivastava AK (1999) Molecular cloning and characterization of TRPC5 (HTRP5), the human homologue of a mouse brain receptor-activated capacitative Ca^{2+} entry channel. Genomics 60:330–340

Stamboulian S, Moutin MJ, Treves S, Pochon N, Grunwald D, Zorzato F, De Waard M, Ronjat M, Arnoult C (2005) Junctate, an inositol 1,4,5-triphosphate receptor associated protein, is present in rodent sperm and binds TRPC2 and TRPC5 but not TRPC1 channels. Dev Biol 286:326–337

Strubing C, Krapivinsky G, Krapivinsky L, Clapham DE (2001) TRPC1 and TRPC5 form a novel cation channel in mammalian brain. Neuron 29:645–655

Strubing C, Krapivinsky G, Krapivinsky L, Clapham DE (2003) Formation of novel TRPC channels by complex subunit interactions in embryonic brain. J Biol Chem 278:39014–39019

Sutton KA, Jungnickel MK, Wang Y, Cullen K, Lambert S, Florman HM (2004) Enkurin is a novel calmodulin and TRPC channel binding protein in sperm. Dev Biol 274:426–435

Tabata H, Tanaka S, Sugimoto Y, Kanki H, Kaneko S, Ichikawa A (2002) Possible coupling of prostaglandin E receptor EP(1) to TRP5 expressed in Xenopus laevis oocytes. Biochem Biophys Res Commun 298:398–402

Tang J, Lin Y, Zhang Z, Tikunova S, Birnbaumer L, Zhu MX (2001) Identification of common binding sites for calmodulin and inositol 1,4,5-trisphosphate receptors on the carboxyl termini of trp channels. J Biol Chem 276:21303–21310

Tang Y, Tang J, Chen Z, Trost C, Flockerzi V, Li M, Ramesh V, Zhu MX (2000) Association of mammalian trp4 and phospholipase C isozymes with a PDZ domain-containing protein, NHERF. J Biol Chem 275:37559–37564

Venkatachalam K, Zheng F, Gill DL (2003) Regulation of canonical transient receptor potential (TRPC) channel function by diacylglycerol and protein kinase C. J Biol Chem 278:29031–29040

von Bohlen Und Halbach O, Hinz U, Unsicker K, Egorov AV (2005) Distribution of TRPC1 and TRPC5 in medial temporal lobe structures of mice. Cell Tissue Res 322:201–206

Xu SZ, Zeng F, Boulay G, Grimm C, Harteneck C, Beech DJ (2005a) Block of TRPC5 channels by 2-aminoethoxydiphenyl borate: a differential, extracellular and voltage-dependent effect. Br J Pharmacol 145:405–414

Xu SZ, Zeng F, Lei M, Li J, Gao B, Xiong C, Sivaprasadarao A, Beech DJ (2005b) Generation of functional ion-channel tools by E3 targeting. Nat Biotechnol 23:1289–1293

Yamada H, Wakamori M, Hara Y, Takahashi Y, Konishi K, Imoto K, Mori Y (2000) Spontaneous single-channel activity of neuronal TRP5 channel recombinantly expressed in HEK293 cells. Neurosci Lett 285:111–114

Yip H, Chan WY, Leung PC, Kwan HY, Liu C, Huang Y, Michel V, Yew DT, Yao X (2004) Expression of TRPC homologs in endothelial cells and smooth muscle layers of human arteries. Histochem Cell Biol 122:553–561

Zeng F, Xu SZ, Jackson PK, McHugh D, Kumar B, Fountain SJ, Beech DJ (2004) Human TRPC5 channel activated by a multiplicity of signals in a single cell. J Physiol 559:739–750

Zhu MH, Chae M, Kim HJ, Lee YM, Kim MJ, Jin NG, Yang DK, So I, Kim KW (2005) Desensitization of canonical transient receptor potential channel 5 by protein kinase C. Am J Physiol Cell Physiol 289:C591–600

TRPC6

A. Dietrich (✉) · T. Gudermann

Institut für Pharmakologie u. Toxikologie, FB. Medizin, Philipps-Universität Marburg,
35033 Marburg, Germany
dietrica@staff.uni-marburg.de

1	**Basic Features of the TRPC6 Gene and Protein and the TRPC6 Expression Pattern**	126
1.1	Gene and Protein	126
1.2	Expression Pattern	128
1.3	Ion Channel Properties	129
2	**Regulation of TRPC6 Channels**	129
2.1	SOC Versus ROC?	129
2.2	Regulation by Ca^{2+} and Ca^{2+}/Calmodulin	131
2.3	Regulation by Phosphorylation	131
2.4	Pharmacology of TRPC6	132
3	**Biological Relevance and Emerging/Established Biological Roles for TRPC6**	133
3.1	TRPC6 in Smooth Muscle Cells	133
3.2	TRPC6 in the Kidney	135
3.3	TRPC6 in Immune and Blood Cells	136
3.4	Other Possible Roles of TRPC6	136
References		137

Abstract TRPC6 is a Ca^{2+}-permeable non-selective cation channel expressed in brain, smooth muscle containing tissues and kidney, as well as in immune and blood cells. Channel homomers heterologously expressed have a characteristic doubly rectifying current–voltage relationship and are six times more permeable for Ca^{2+} than for Na^+. In smooth muscle tissues, however, Na^+ influx and activation of voltage-gated calcium channels by membrane depolarization rather than Ca^{2+} elevation by TRPC6 channels is the driving force for contraction. TRPC6 channels are directly activated by the second messenger diacylglycerol (DAG) and regulated by specific tyrosine or serine phosphorylation. Extracellular Ca^{2+} has inhibitory effects, while Ca^{2+}/calmodulin acting from the intracellular side has potentiator effects on channel activity. Given its specific expression, TRPC6 is likely to play a number of physiological roles. Studies with $TRPC6^{-/-}$ mice suggest a role for the channel in the regulation of vascular and pulmonary smooth muscle contraction. TRPC6 was identified as an essential component of the slit diaphragm architecture of kidney podocytes. Other functions in immune and blood cells, as well as in brain and in smooth muscle-containing tissues such as stomach, colon and myometrium, remain elusive.

Keywords Classical transient receptor channel 6 · TRPC6 · Function · Expression · Biological properties

1
Basic Features of the TRPC6 Gene and Protein and the TRPC6 Expression Pattern

1.1
Gene and Protein

Full-length complementary DNA (cDNA) of mouse TRPC6 was isolated from brain (Boulay et al. 1997), while human TRPC6 was cloned from placenta (Hofmann et al. 1999). The gene for human TRPC6 is localized on chromosome 11q21–q22 and has 13 exons (D'Esposito et al. 1998). Its murine homologue with the same number of exons is located on chromosome 9. Full-length human and mouse proteins consist of 931 and 930 amino acids, respectively.

The putative transmembrane structure is similar to that of other transient receptor potential (TRP) channels with intracellular N- and C-termini, six membrane-spanning helices (S1–S6) and a predicted pore forming loop (P) between S5 and S6 (Fig. 1a).

By glycosylation scanning we were able to identify two glycosylation sites (Fig. 1a) in the first and second extracellular loop (Asn^{473}; Asn^{561}), which are important determinants for the tightly receptor-operated behaviour of TRPC6 (Dietrich et al. 2003b). Mutation of Asn^{651} to Gln prevents glycosylation and was sufficient to increase basal activity of TRPC6 (Dietrich et al. 2003b).

The functional characteristics of TRPC6 tetrahomomers over-expressed in different cell lines may not truly represent their physiological properties in vivo, because they form heteromeric channel complexes in native environments. The basic principles of homo- and heteromultimerization of heterologously expressed TRPC6 channel monomers were defined by four different approaches: cellular co-trafficking of TRPC subunits, differential functional suppression by dominant-negative subunits, fluorescence resonance energy transfer (FRET) and co-immunoprecipitation (Hofmann et al. 2002). Experimental approaches employed led to the conclusion that TRPC6 assembles into homo- and heterotetramers within the confines of the TRPC3/6/7 subfamily (see Fig. 1b). These results were confirmed by systematic coimmunoprecipitation of TRPCs from isolated rat brain synaptosomes (Goel et al. 2002). Recently, novel combinations of TRPC1–TRPC4/5 together with TRPC6 were identified in HEK293 cells as well as in embryonic brain but not in adult rat tissues (Strübing et al. 2003; see Fig. 1b). Moreover, native TRPC3/6 heteromers were identified in epithelial cells (Bandyopadhyay et al. 2004).

In general, TRPC channel complexes might be organized in supramolecular signalling complexes with other adaptor proteins (? in Fig. 1b) in native tissues. Such complexes called signalplexes were already identified in photoreceptors of *Drosophila melanogaster* (Montell 2001) but were also observed for TRPC6. A multiprotein complex identified in neuronal PC12 cells was centred on TRPC6 channels and contained protein kinase C (PKC), FK506-binding protein-12 kDa (FKBP12) and calcineurin/calmodulin (Kim and Saffen 2005).

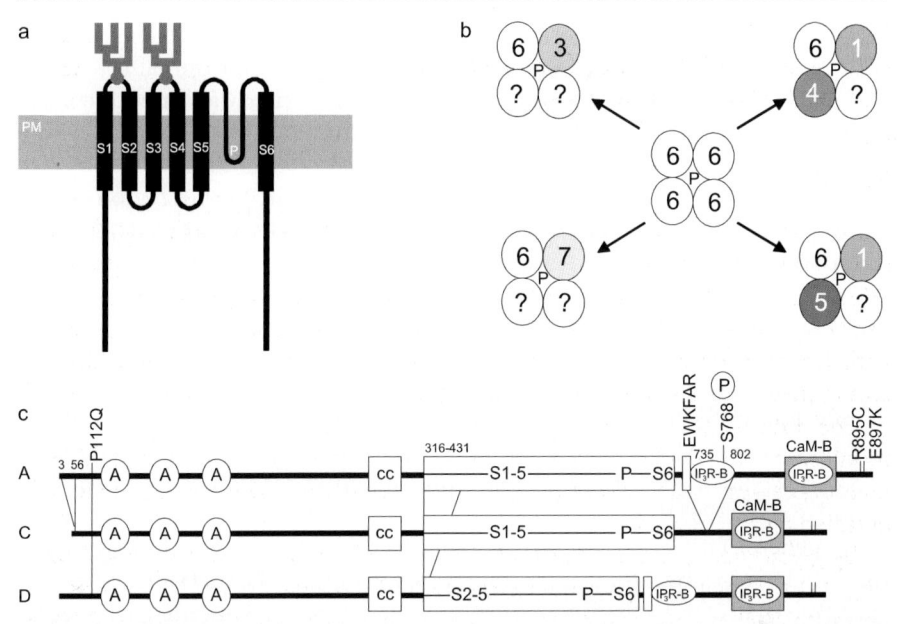

Fig. 1 a–c Structural features of TRPC6. **a** Topology of TRPC6 in the plasma membrane (*PM*) indicating transmembrane regions (*S1–S6*) and the predicted pore domain (*P*). Two glycosylated sites in TRPC6 are indicated by covalently bound carbohydrates (in *grey*). **b** Heteromultimerization potential of TRPC6. TRPC6 can interact with TRPC3 (*3*), TRPC7 (*7*) and with TRPC1 (*1*) only in complexes with TRPC4 (*4*) or TRPC5 (*5*) to form functional tetramers. Unidentified adaptor proteins may stabilize complexes. **c** Domain structure of TRPC6A and its splice variants *C* (Zhang and Saffen 2001) and *D* (Corteling et al. 2004). Deleted regions in the splice variants are indicated by *vertical lines*. IP$_3$R-B, IP$_3$ receptor binding site; *cc*, coiled-coil domain; *CaM*, Ca^{2+}/CaM binding site; *EWKFAR*, conserved TRP-box motif; *S786 P*, PKC phosphorylation site; *P112Q, R895C, E897K*, gain-of-function mutants in FSGS

By a proteomics approach, signalplexes for TRPC5 and TRPC6 containing several cytoskeletal proteins as well as the plasmalemmal Na$^+$/K$^+$-ATPase (NKA) pump were identified from brain lysates. In lysates from rat kidneys and in a heterologous expression system using HEK293 cells, the TRPC6–NKA pump interaction could be confirmed (Goel et al. 2005a).

Three ankyrin domains are found in the amino terminus of TRPC6. The second repeat was recently identified in a yeast two-hybrid screen to interact with MxA, a member of the dynamin superfamily. Moreover MxA was able to enhance TRPC6 activity in a guanosine triphosphate (GTP)-dependent way (Lussier et al. 2005).

The TRPC6 protein also contains an EWKFAR TRP box motif which is conserved in the TRPC family and two inositol 1,4,5-trisphosphate (IP$_3$) receptor binding domains from which the second one overlaps with a calmodulin binding site (Boulay et al. 1999; Birnbaumer et al. 2003; Zhang et al. 2001).

Four additional splice variants with shorter amino and carboxy termini (Zhang and Saffen 2001), as well as shorter transmembrane domains (Corteling et al. 2004), were cloned from rat lung and human airway smooth muscle cells, respectively. TRPC6B and -C lack the aminoterminal aa 3–58, while TRPC6C has an additional deletion in the carboxyl terminus (Δ 735–802; see Fig. 1c). TRPC6D contains an aa 316–431 deletion in the S1 region of the transmembrane domain, while TRPC6E shows a shorter aa 377–431 deletion (Corteling et al. 2004).

1.2
Expression Pattern

While human TRPC6 appears to be expressed ubiquitously with higher expression levels in lung, placenta ovary and spleen (Hofmann et al. 2000), murine TRPC6 was identified in lung and brain using multiple tissue Northern blots (Boulay et al. 1997).

In general TRPC6 is found in numerous tissues enriched with *smooth muscle cells* including lung, stomach, colon, oesophagus and myometrium (reviewed in Beech et al. 2004). Functional roles, however, were only described in vascular and pulmonary arteries (see Sect. 3.1).

Although TRPC6 expression in *brain* is lower than that of other TRPCs, TRPC6 is specifically expressed in the dentate gyrus of the rat hippocampus (Bonaventure et al. 2002). These data were confirmed by in situ hybridization histochemistry, where TRPC6 expression was found to be exclusively localized in the dentate granule cell layer of the adult mouse brain (Otsuka et al. 1998). A TRPC6 messenger RNA (mRNA) expression analysis in adult rat tissues by reverse-transcription polymerase chain reaction (RT-PCR) also revealed an ubiquitous expression with higher expression levels in lung, cerebrum and ovary (Garcia and Schilling 1997).

By positional cloning two research groups independently identified gain-of-function mutations of the TRPC6 channel (see Fig. 1c) in families with focal and segmental glomerulosclerosis (FSGS) in the *kidney* leading to increased calcium influx (P112Q; Winn et al. 2005) and increased current amplitudes (R895C, E897K; Reiser et al. 2005). TRPC6 expression in the kidney was localized to the glomeruli and the tubuli (Winn et al. 2005) as well as the podocytes (Reiser et al. 2005) by immunohistochemistry with a TRPC6-specific antibody. Complexes of TRPC3 and TRPC6 were also found in podocytes and are co-localized with aquaporin-2 (Goel et al. 2005b). In polarized cultures of M1- and IMCD-3-collecting duct cells, however, TRPC3 was localized exclusively to the apical domain, while TRPC6 was found in both the basolateral and apical membranes. Native TRPC3/6 heteromers were identified in apical Madin–Darby canine kidney (MDCK) cells and in rat submandibular glands (Bandyopadhyay et al. 2005).

1.3
Ion Channel Properties

TRPC6 is a Ca^{2+}-permeable non-selective cation channels displaying double rectification with a single-channel conductance of 28–37 pS. The ion permeability ratio P_{Ca}/P_{Na} is approximately 6 (Hofmann et al. 1999; Dietrich et al. 2003; Shi et al. 2004). These features distinguish TRPC6 from the closely related TRPC3 channel, which is significantly less Ca^{2+}-selective with values of 1.1 and has a higher single-channel conductance of 60–66 pS (reviewed in Owsianik et al. 2006). However, Ca^{2+} ions contribute only a small percentage ($\sim$4%) to whole-cell currents of HEK293 cells stably expressing TRPC6 in the presence of extracellular Na^+ (Estacion et al. 2006). Along these lines, application of blockers of voltage-gated calcium channels completely abolishes Ca^{2+} influx after stimulation of TRPC6 in A7r5 smooth muscle cells (Soboloff et al. 2005) favouring a model highlighting Na^+ entry through TRPC6 channels and subsequent membrane depolarization that activates voltage-gated calcium channels, which are responsible for the bulk of Ca^{2+} influx (reviewed in Gudermann et al. 2004).

TRPC6 is a tightly receptor-operated channel and shows little basal activity in contrast to TRPC3, which demonstrates high basal activity, when heterologously expressed in HEK293 cells. As mentioned above, the dual glycosylation pattern is a critical determinant for the quiescence of TRPC6 channel activity (Dietrich et al. 2003b).

In general, one needs to keep in mind that the functional characterization of TRPC channels following heterologous over-expression of its cDNA in different cultured cell lines gave rise to contradictory results regarding intrinsic channel properties, such as activation mechanisms as well as ion selectivity. Yet the analysis of TRPC6 activity by electrophysiological recordings of primary, isolated cells from native tissues is also difficult due to the lack of specific pharmacological blockers of TRPC6.

We recently characterized currents in cells isolated from cerebral arteries of wild-type and TRPC6-deficient mice, but observed increased current densities and more depolarized membrane potentials due to the up-regulation of the closely related, constitutively active TRPC3 channel (Dietrich et al. 2005c).

The identification of cells devoid of compensatory up-regulation in TRPC6-deficient mice will reveal the contribution of TRPC6 to whole-cell currents and physiological processes and will allow for monitoring TRPC6-dependent ion currents in a native environment.

2
Regulation of TRPC6 Channels
2.1
SOC Versus ROC?

TRPs were cloned and identified assuming that they were calcium-selective and activated by emptying of internal Ca^{2+} stores [store-operated channels (SOC)].

After their initial functional characterization it became clear that both assumptions were not entirely true, especially for TRPCs. These ion channels show a moderate Ca^{2+} selectivity (P_{Ca}/P_{Na} from ~0.5 to 9), and at least the TRPC3/6/7 subfamily of TRPC channels can be directly activated by the second messenger diacylglycerol (DAG), which is produced by receptor-activated phospholipase C (PLC) enzymes without the involvement of internal stores [receptor-operated channel (ROC); Hofmann et al. 1999; Okada et al. 1999]. However, store-operated calcium entry (SOCE) occurs when IP_3 or some other signal discharges Ca^{2+} from intracellular stores in the endoplasmic reticulum (ER), and the subsequent fall in the ER Ca^{2+} concentration then signals to the plasma membrane activating SOCs. Experimentally, a similar cellular effect is evoked by thapsigargin or cyclopiazonic acid (CPA), which block sarcoplasmic/endoplasmic reticulum Ca^{2+} (SERCA) pumps, resulting in leakage of calcium ions from internal stores. Neither the exact mechanisms for the SOCE signalling pathway nor the molecular identity of SOC channels is currently known with certainty (reviewed in Dietrich et al. 2005b), but one theory suggests the direct physical coupling of TRPC channels to IP_3 receptors (see Fig. 1c, Boulay et al. 1999; Birnbaumer et al. 2003).

In contrast to other TRPC channels, especially TRPC3, SOC versus ROC regulation was not much of an issue for TRPC6 activation. In the first publication describing TRPC6, it was already demonstrated by single-cell Ca^{2+} imaging experiments that re-addition of external $CaCl_2$ to thapsigargin-treated cells did not result in a significantly higher $[Ca^{2+}]_i$ rise in TRPC6-expressing HEK293 cells compared to control cells, thus allowing the characterization of TRPC6 as a receptor-regulated, but not store-regulated, cation channel (Boulay et al. 1997). Along these lines, thapsigargin treatment only transiently elevates $[Ca^{2+}]_i$ and induces a moderate store depletion-induced Mn^{2+} influx in both TRPC6-expressing and control cells (Hofmann et al. 2000). Moreover, in characterizing TRPC6 by electrophysiological methods, addition of IP_3 did not activate recombinant TRPC6 channels in the whole-cell mode, nor in excised inside-out patches. However, the G protein activator AlF_4^- stimulated and the PLC inhibitor U73122 blocked agonist-induced activation of TRPC6, thus pointing to a G protein- and PLC-dependent activation mechanism (Hofmann et al. 1999). One-oleoyl-1-acetyl-*sn*-glycerol (OAG), a membrane-permeable analogue of DAG, as well as the DAG lipase inhibitor RHC80267, markedly increased TRPC6 activity (Hofmann et al. 1999). TRPC6 was the first ion channel identified that is activated by DAG in a membrane-delimited fashion, independently of PKC. The exact location of a putative binding site for diacylglycerols in the TRPC6 protein is still elusive, because an OAG-insensitive splice variant of TRPC6 [TRPC6B (Δ 3–56 in Fig. 1c), Zhang and Saffen 2001], characterized by means of fluorometry, turned out to be activated by DAG when analysed by electrophysiological methods (Jung et al. 2002).

Recently, the exocytosis of TRPC6 channels to the plasma membrane was analysed. Most interestingly, both mechanisms, receptor activation and store depletion by thapsigargin, induced translocation of TRPC6 channels to the plasma membrane, indicating a store-dependent TRPC6 channel density at the plasma membrane (Cayouette et al. 2004).

The functional significance of two putative IP_3 receptor binding sites characterized in TRPC6 (Fig. 1c) which physically bind two different IP_3 receptor fragments remains elusive (Boulay et al. 1999; Birnbaumer et al. 2003).

2.2
Regulation by Ca^{2+} and Ca^{2+}/Calmodulin

Extracellular Ca^{2+} ($[Ca^{2+}]_o$) has complex effects on TRPC6 activity. A $[Ca^{2+}]_o$ of 2 mM inhibited vasopressin-induced TRPC6 activity in the rat A7r5 smooth muscle cell line. However, reduction of $[Ca^{2+}]_o$ to 50–200 µM facilitated TRPC6 cation currents, whereas complete removal of Ca^{2+} led to a decrease in currents (Jung et al. 2002).

The regulation of TRPC6 by Ca^{2+}/calmodulin was carefully analysed, leading to an overall picture different from that of TRPC3, which is inhibited by Ca^{2+}/calmodulin (Zhang et al. 2001). Calmodulin inhibitors like calmidazolium and trifluoperazine had an inhibitory effect on receptor-operated calcium influx into TRPC6-expressing HEK293 cells, indicating a stimulatory impact of Ca^{2+}/calmodulin on TRPC6 channel activity (Boulay et al. 2002). The latter concept was further extended by the observation that TRPC6 activation and its acceleration by the extracellular calcium concentration ($[Ca^{2+}]_o = 50–100$ µM, see above) most probably involves phosphorylation by calmodulin-dependent kinase II, an effect that was not noted for the closely related TRPC7 protein (Shi et al. 2004). To conclude, TRPC6 is subject to a complex regulation by Ca^{2+} on both sides of the plasma membrane involving calmodulin-dependent and -independent mechanisms (Shi et al. 2004).

2.3
Regulation by Phosphorylation

TRPC6 ion channels are also regulated by protein serine and tyrosine phosphorylation.

The PKC activator phorbol 12-myristate 13-acetate (PMA) has no acute effect on basal TRPC6 activity, but inhibited carbachol-induced TRPC6 activation by more than 90% (Estacion et al. 2004). Along the same line, pharmacological inhibition of PKC by calphostin C resulted in a significantly retarded inactivation time course of TRPC currents (Shi et al. 2004). PKC phosphorylation at an identified TRPC6 phosphorylation site (serine 768), however, is not only responsible for TRPC6 inhibition by PKC, but also enables binding of FKBP12 and calcineurin/calmodulin, resulting in the release of the activated G protein-coupled-receptor from a multiprotein complex centred on TRPC6 channels (Kim and Saffen 2005).

While PKC appears to contribute to channel inactivation (Trebak et al. 2003) Fyn, a member of the Src family of protein tyrosine kinases, increases TRPC6 channel activity. Stimulation of epidermal growth factor (EGF) receptor entails tyrosine phosphorylation of TRPC6, and Fyn and TRPC6 physically interact in mammalian brain as well as after heterologous expression in COS-7 cells (Hisatsune et al. 2004). Recently a tyrosine phosphorylation site in TRPC3 was identified (Y226) and a Y226F mutation resulted in complete TRPC3 inactivation (Kawasaki et al. 2006). The homologous mutation in TRPC6 (Y230F), however, did not impair TRPC6 activation, and TRPC6 was still active in Fyn-, Yes- and Src-deficient cells (Kawasaki et al. 2006). At present the issue remains unresolved whether other Src-family tyrosine kinases are responsible for TRPC6 phosphorylation or whether the discrepancy can be explained by different experimental methods (electrophysiology versus Ca^{2+} imaging) or different activation mechanisms (receptor tyrosine kinase-PLCγ versus G protein-coupled receptor-PLCβ).

2.4
Pharmacology of TRPC6

There are no specific inhibitors for TRPC6. TRPC6 is blocked by La^{3+} ions with an IC_{50} of 4–6 µM (Inoue et al. 2001; Jung et al. 2002) similar to TRPC3 (4 µM; Halaszovich et al. 2000). Electrophysiological whole-cell recordings on TRPC6-expressing HEK293 cells reveal an IC_{50} value of 1.9 µM (Inoue et al. 2001) for a Gd^{3+} block similar to results obtained with the heterologously expressed TRPC3 channel (2.3 µM; Dietrich et al. 2005a).

An arachidonic acid metabolite, 20-hydroxyeicosatetraenoic acid (20-HETE), was identified as a novel activator of TRPC6 (Basora et al. 2003), but its specificity and physiological relevance remain elusive.

The non-specific cation channel blocker flufenamate (FFA; 100 µM), however, turned out to be an activator of heterologously expressed TRPC6 channels in HEK293 cells, while closely related TRPC3 and TRPC7 channel activities were inhibited. Moreover, FFA was able to stimulate the $α_1$-adrenoceptor induced non-selective cation influx in smooth muscle cells from portal vein myocytes (Inoue et al. 2001) and the vasopressin-induced cation currents of the rat smooth muscle cell line A7r5 (Jung et al. 2002), thus identifying native TRPC6 currents. Similar results were obtained with niflumic acid (100 µM; Jung et al. 2002). In the future, these drugs might be helpful to identify native TRPC6 currents in tissues and primary cells (Fig. 2).

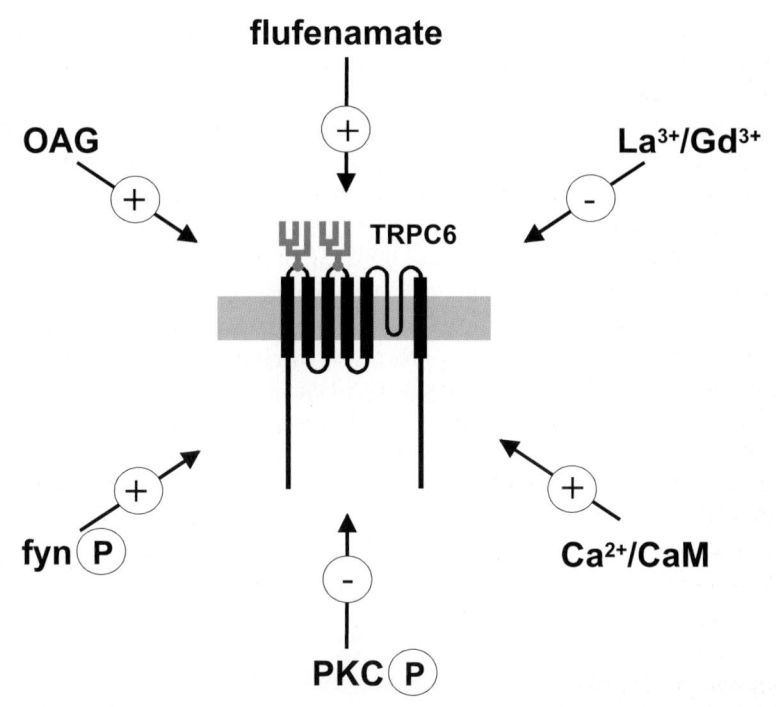

Fig. 2 Pharmacological manipulation of TRPC6. Activating (+) and inhibiting (−) extracellular acting drugs or intracellular protein interactions are depicted. See text for details

3
Biological Relevance and Emerging/Established Biological Roles for TRPC6

The physiological role of TRPC6 is still largely unknown. However, there is growing evidence that TRPC6 is an intrinsic constituent of receptor-operated cation entry involved in numerous physiological processes.

3.1
TRPC6 in Smooth Muscle Cells

It is now clear that TRPC6 plays an important role in vascular and pulmonary smooth muscle cells. By comparative biophysical characterization and gene suppression using antisense oligonucleotides, TRPC6 has been shown to be the molecular correlate of the α_1-adrenoceptor-activated non-selective cation channel in vascular smooth muscle cells (Inoue et al. 2001) and the vasopressin-activated cation channel in an aortic smooth muscle cell line (Jung et al. 2002). In addition, TRPC6 has been proposed to play a critical role in the intravascular pressure-induced depolarization and constriction of small arteries and arterioles (Welsh et al. 2002) known as the Bayliss effect. Myogenic constriction of resistance arteries results from Ca^{2+} influx through voltage-gated Ca^{2+} channels subsequent to membrane depolarization. Apart from TRPC6, the Ca^{2+}-activated cation channel TRPM4 has recently been implicated in myogenic vasoconstriction (Earley et al. 2004), but the precise location of either TRPC6 or TRPM4 in the signalling pathway elicited by elevated intravascular pressure still remains poorly understood.

Recently, expression studies revealed that platelet-derived growth factor (PDGF)-mediated proliferation of pulmonary artery smooth muscle cells (PASMC) is associated with c-jun/STAT3-induced up-regulation of TRPC6 expression (Yu et al. 2003). In this context, it is intriguing to note that excessive PASMC proliferation, a major cause of the elevated pulmonary vascular resistance in patients with idiopathic pulmonary arterial hypertension (IPAH), also correlates with over-expression of TRPC6 and TRPC3 proteins in these tissues. In line with these data, down-regulation of TRPC6 by TRPC6-specific small interfering RNAs resulted in attenuated proliferation of PASMC from IPAH patients (Yu et al. 2004). Moreover, TRPC6 expression is up-regulated in pulmonary arteries of rats kept under chronic hypoxic conditions to induce pulmonary hypertension. As expected, OAG-induced cation entry was significantly increased in hypoxia-treated PASMC as compared to control cells (Lin et al. 2004).

Recently, we were able to present results on the phenotype of mice deficient in TRPC6. Based on the above reviewed data, we predicted that loss of TRPC6 function would lead to diminished airway and vascular smooth muscle tone and attendant reduced airway contractility as well as hypotension. Unexpectedly, we observed airway smooth muscle hyperreactivity after exposure to broncho-constrictors and higher agonist-induced contractility in isolated tracheal rings of *TRPC6$^{-/-}$* mice as compared to control mice (Dietrich et al. 2003b). In line with these findings, aortic rings showed higher smooth muscle contractility. Elevated systemic blood pressure in TRPC6-deficient mice was further increased by inhibition of nitric oxide (NO) synthase (Dietrich et al. 2005c). These effects could be explained by in vivo replacement of TRPC6 by TRPC3-type channels, which are closely related but constitutively active, resulting in enhanced basal and agonist-induced cation entry into smooth muscle cells leading to increased smooth muscle contractility (Dietrich et al. 2005c). Because the expression pattern of TRPC3 and TRPC6 overlaps in most tissues containing smooth muscle cells (reviewed in Beech et al. 2004), a heterotetrameric TRPC3/6 channel complex may be the native molecular correlate of the non-selective cation influx into smooth muscle cells (see Fig. 3). Our findings imply that TRPC3, -6 and -7 are functionally non-redundant and that TRPC6 plays a unique role in the control of airway and vascular smooth muscle contractility.

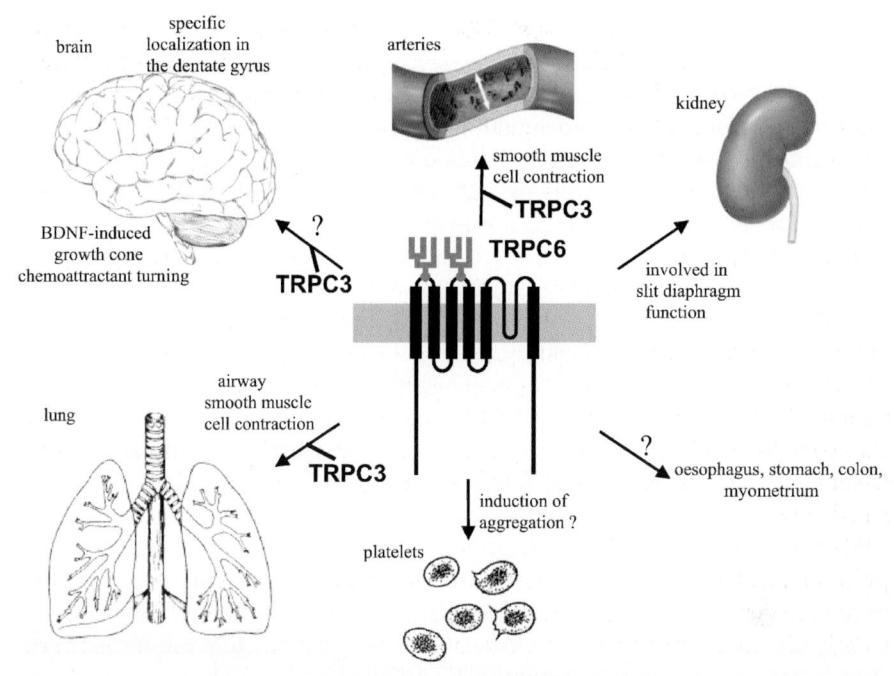

Fig. 3 Functional roles of TRPC6 in different tissues. For some functions TRPC3 expression may compensate for the loss of TRPC6. Not all functional roles of TRPC6 are elucidated so far. See text for details

3.2
TRPC6 in the Kidney

Two recent studies have discovered mutations in the gene encoding for TRPC6 as the molecular cause of focal segmental glomerular sclerosis (FSGS). TRPC6 was identified as an essential component of the glomerular slit diaphragm in the kidney. In a large pedigree from New Zealand, one study identified candidate genes in a locus mapped by linkage analysis. By sequencing candidate genes they identified a missense mutation (P112Q; see Fig. 1b) in TRPC6 that co-segregated with the late onset of FSGS (Winn et al. 2005). By calcium imaging studies they demonstrated increased TRPC6 activity of the P112Q mutant compared to wild-type TRPC6. The latter results were confirmed at least in part by another study which screened 16 additional families with FSGS. Five different TRPC6 mutations (N143S, S270T, R875C and E897K) resulted in amino acid substitutions, while the fifth creates a premature stop codon near the C-terminus.

Consistent with the results of calcium imaging experiments from the previous study, whole-cell patch clamp experiments identified larger current amplitudes of the R895C and E897K mutants than in wild-type TRPC6 (Reiser et al. 2005). Most interestingly, the fact that disruption of the slit diaphragm architecture in nephrin-deficient mice leads to over-expression and mislocalization of TRPC6 in podocytes (Reiser et al. 2005) suggests that TRPC6 is integrated into an organized signalling complex in podocytes with nephrin, podocin and probably the AT1 angiotensin II receptor (reviewed in Gudermann 2005). Thus, mutations in TRPC6 together with other identified mutations (reviewed in Kriz 2005) are responsible for massive proteinuria and finally kidney failure in FSGS.

3.3
TRPC6 in Immune and Blood Cells

Immunohistochemistry revealed TRPC6 expression in neutrophils and lymphocytes of human lung tissues, and most interestingly, preliminary studies indicate the presence of TRPC6 mRNA in a subset of macrophages isolated from the alveolar space and lung tissues of smokers and chronic obstructive pulmonary disease (COPD) patients (Li et al. 2005a).

In a haematopoietic multi-step tumour model, non-tumourigenic PB3c mast cells progress in vivo to interleukin-3 secreting autocrine tumours. One transcript expressed in the precursor, but down-regulated in the tumour cell, was isolated by differential display PCR and identified as TRPC6 mRNA (Buess et al. 1999). At present it remains unclear whether TRPC6 is really involved in tumour suppression or contributes to mast cell function.

In contrast to other in vivo and in vitro studies mentioned above, a TRPC2–TRPC6 complex was reported in primary erythroid cells to modulate calcium influx stimulated by erythropoietin (Chu et al. 2004).

In human platelets, thrombin-activated cation influx is independent of store depletion, consistent with the observation that TRPC6 is highly expressed in these cells. In this model system, TRPC6 does not serve as a substrate of tyrosine kinases, but was phosphorylated in a cyclic adenosine monophosphate (cAMP)-dependent manner (Hassock et al. 2002). It is well documented that phosphoinositide 3 kinase (PI3K) activation resulting in the production of phosphatidylinositol 3,4,5-trisphosphate (PIP$_3$) triggers platelet aggregation by inducing a Ca^{2+} influx (Lu et al. 1998). In accord with this concept, TRPC6 was recently identified as the putative molecular correlate of a PIP$_3$-sensitive calcium entry system in platelets, Jurkat T cells and RBL-2H3 mast cells (Tseng et al. 2004).

3.4
Other Possible Roles of TRPC6

Recently a novel class of microvillous secondary chemosensory cells in the mammalian olfactory system were identified which express TRPC6 co-localized with PLCβ$_2$ and components of the cytoskeleton of microvilli (Elsaesser et al.

2005). But they represent only 5% of all olfactory cells and their function remains unclear.

TRPC6 was also selectively localized to the cell body of rods, while voltage-gated calcium channels were expressed in the synaptic terminal and in the ellipsoid and sub-ellipsoid regions of the retina (Krizaj et al. 2005). However it remains unclear how TRPC6 function might substitute voltage-gated calcium channels in the cell body of rods.

The highly specific expression of TRPC6 in the dentate gyrus may be due to a highly specific function in brain, but has still to be investigated. Together with TRPC3, TRPC6 was identified in cultured cerebellar granule cells as important channel for the brain-derived neurotrophic factor (BDNF)-induced elevation of $[Ca^{2+}]_i$ for growth cone chemo-attractive turning (Li et al. 2005b).

Recently the influence of two familial Alzheimer's disease-linked presenilin 2 variants and a loss-of-function presenilin 2 mutant on TRPC6 activity were analysed in a heterologous expression system. While the Alzheimer's disease-linked mutants abolished agonist-induced TRPC6 activation without affecting direct activation by OAG or interacting with the TRPC6–IP$_3$ receptor interaction, the loss-of-function mutant enhanced agonist- and OAG-induced activation of TRPC6 (Lessard et al. 2005). Although indirect inhibitory effects by an increase in amyloid β-peptides in the medium could be excluded, it remains unclear if presenilin 2 proteins are able to directly interact with TRPC6 proteins (Lessard et al. 2005).

Another interesting signal transduction pathway involving TRPC6 was recently discovered in prostate epithelial cells. Agonist-mediated stimulation of the $α_1$-adrenergic receptor promotes proliferation of primary human prostate cancer epithelial cells by inducing store-independent Ca^{2+} entry mostly relying on TRPC6 and subsequent activation of nuclear factor of activated T cells (NFAT) (Thebault et al. 2006).

To conclude, the exact physiological role of TRPC6 channels is still largely unknown at present. Hopefully, the overt lack of specific pharmacological blockers will be at least partially overcome by the use of small interfering RNAs in cells where no compensatory up-regulation of other TRPC family members occurs.

Future investigations of human pathophysiology involving TRPC6 should be highly rewarding in the years to come.

References

Bandyopadhyay BC, Swaim WD, Liu X, Redman RS, Patterson RL, Ambudkar IS (2005) Apical localization of a functional TRPC3/TRPC6-Ca^{2+}-signaling complex in polarized epithelial cells. Role in apical Ca^{2+} influx. J Biol Chem 280:12908–12916

Basora N, Boulay G, Bilodeau L, Rousseau E, Payet MD (2003) 20-hydroxyeicosatetraenoic acid (20-HETE) activates mouse TRPC6 channels expressed in HEK293 cells. J Biol Chem 278:31709–31716

Beech DJ, Muraki K, Flemming R (2004) Non-selective cationic channels of smooth muscle and the mammalian homologues of Drosophila TRP. J Physiol 559:685–706

Birnbaumer L, Dietrich A, Kawasaki B, Zhu MX (2003) Multiple interactions between TRPCs and IP3Rs: TRPC3-interacting domains of IP3R serve to identify TRPCs as components of CCE channels; yet, IP3R activation does not activate TRPC3 channels. Nova Acta Leopold 89 (336):61–67

Bonaventure P, Guo H, Tian B, Liu X, Bittner A, Roland B, Salunga R, Ma XJ, Kamme F, Meurers B, Bakker M, Jurzak M, Leysen JE, Erlander MG (2002) Nuclei and subnuclei gene expression profiling in mammalian brain. Brain Res 943:38–47

Boulay G, Zhu X, Peyton M, Jiang M, Hurst R, Stefani E, Birnbaumer L (1997) Cloning and expression of a novel mammalian homolog of Drosophila transient receptor potential (Trp) involved in calcium entry secondary to activation of receptors coupled by the Gq class of G protein. J Biol Chem 272:29672–29680

Boulay G, Brown DM, Qin N, Jiang M, Dietrich A, Zhu MX, Chen Z, Birnbaumer M, Mikoshiba K, Birnbaumer L (1999) Modulation of $Ca^{(2+)}$ entry by polypeptides of the inositol 1,4,5-trisphosphate receptor (IP_3R) that bind transient receptor potential (TRP): evidence for roles of TRP and IP_3R in store depletion-activated $Ca^{(2+)}$ entry. Proc Natl Acad Sci U S A 96:14955–14960

Buess M, Engler O, Hirsch HH, Moroni C (1999) Search for oncogenic regulators in an autocrine tumor model using differential display PCR: identification of novel candidate genes including the calcium channel mtrp6. Oncogene 18:1487–1494

Cayouette S, Lussier MP, Mathieu EL, Bousquet SM, Boulay G (2004) Exocytotic insertion of TRPC6 channel into the plasma membrane upon Gq protein-coupled receptor activation. J Biol Chem 279:7241–7246

Chu X, Tong Q, Cheung JY, Wozney J, Conrad K, Mazack V, Zhang W, Stahl R, Barber DL, Miller BA (2004) Interaction of TRPC2 and TRPC6 in erythropoietin modulation of calcium influx. J Biol Chem 279:10514–10522

Corteling RL, Li S, Giddings J, Westwick J, Poll C, Hall IP (2004) Expression of transient receptor potential C6 and related transient receptor potential family members in human airway smooth muscle and lung tissue. Am J Respir Cell Mol Biol 30:145–154

D'Esposito M, Strazzullo M, Cuccurese M, Spalluto C, Rocchi M, D'Urso M, Ciccodicola A (1998) Identification and assignment of the human transient receptor potential channel 6 gene TRPC6 to chromosome 11q21->q22. Cytogenet Cell Genet 83:46–47

Dietrich A, Gollasch M, Chubanov V, Mederos y Schnitzler M, Dubrovska G, Herz U, Renz H, Gudermann T, Birnbaumer L (2003a) Studies on TRPC6 deficient mice reveal its nonredundant role in the regulation of smooth muscle tone. Naunyn Schmiedebergs Arch Pharmacol 367:238

Dietrich A, Mederos y Schnitzler M, Emmel J, Kalwa H, Hofmann T, Gudermann T (2003b) N-linked protein glycosylation is a major determinant for basal TRPC3 and TRPC6 channel activity. J Biol Chem 278:47842–47852

Dietrich A, Mederos y Schnitzler M, Kalwa H, Storch U, Gudermann T (2005a) Functional characterization and physiological relevance of the TRPC3/6/7 subfamily of cation channels. Naunyn Schmiedebergs Arch Pharmacol 371:257–265

Dietrich A, Kalwa H, Rost BR, Gudermann T (2005b) The diacylglycerol-sensitive TRPC3/6/7 subfamily of cation channels: functional characterization and physiological relevance. Pflügers Arch 451:72–80

Dietrich A, Mederos YSM, Gollasch M, Gross V, Storch U, Dubrovska G, Obst M, Yildirim E, Salanova B, Kalwa H, Essin K, Pinkenburg O, Luft FC, Gudermann T, Birnbaumer L (2005c) Increased vascular smooth muscle contractility in TRPC6$^{-/-}$ mice. Mol Cell Biol 25:6980–6989

Earley S, Waldron BJ, Brayden JE (2004) Critical role for transient receptor potential channel TRPM4 in myogenic constriction of cerebral arteries. Circ Res 95:922–929

Elsaesser R, Montani G, Tirindelli R, Paysan J (2005) Phosphatidyl-inositide signalling proteins in a novel class of sensory cells in the mammalian olfactory epithelium. Eur J Neurosci 21:2692–2700

Estacion M, Li S, Sinkins WG, Gosling M, Bahra P, Poll C, Westwick J, Schilling WP (2004) Activation of human TRPC6 channels by receptor stimulation. J Biol Chem 279:22047–22056

Estacion M, Sinkins WG, Jones SW, Applegate MA, Schilling WP (2006) TRPC6 forms non-selective cation channels with limited Ca^{2+} permeability. J Physiol 572:359–377

Garcia RL, Schilling WP (1997) Differential expression of mammalian TRP homologues across tissues and cell lines. Biochem Biophys Res Commun 239:279–283

Goel M, Sinkins WG, Schilling WP (2002) Selective association of TRPC channel subunits in rat brain synaptosomes. J Biol Chem 277:48303–48310

Goel M, Sinkins W, Keightley A, Kinter M, Schilling WP (2005a) Proteomic analysis of TRPC5- and TRPC6-binding partners reveals interaction with the plasmalemmal $Na^{(+)}/K^{(+)}$-ATPase. Pflugers Arch 451:87–98

Goel M, Sinkins WG, Zuo CD, Estacion M, Schilling WP (2005b) Identification and localization of TRPC channels in rat kidney. Am J Physiol Renal Physiol 290:F1241–F1252

Gudermann T (2005) A new TRP to kidney disease. Nat Genet 37:663–664

Gudermann T, Mederos y Schnitzler M, Dietrich A (2004) Receptor-operated cation entry-more than esoteric terminology? Sci STKE 243:pe35

Halaszovich C R, Zitt C, Jüngling E, Lückhoff A (2000) Inhibition of TRP3 channels by lanthanides. Block from the cytosolic side of the plasma membrane. J Biol Chem 275:37423–37428

Hassock SR, Zhu MX, Trost C, Flockerzi V, Authi KS (2002) Expression and role of TRPC proteins in human platelets: evidence that TRPC6 forms the store-independent calcium entry channel. Blood 100:2801–2811

Hisatsune C, Kuroda Y, Nakamura K, Inoue T, Nakamura T, Michikawa T, Mizutani A, Mikoshiba K (2004) Regulation of TRPC6 channel activity by tyrosine phosphorylation. J Biol Chem 279:18887–18894

Hofmann T, Obukhov AG, Schaefer M, Harteneck C, Gudermann T, Schultz G (1999) Direct activation of human TRPC6 and TRPC3 channels by diacylglycerol. Nature 397:259–263

Hofmann T, Schaefer M, Schultz G, Gudermann T (2000) Transient receptor potential channels as molecular substrates of receptor-mediated cation entry. J Mol Med 78:14–25

Hofmann T, Schaefer M, Schultz G, Gudermann T (2002) Subunit composition of mammalian transient receptor potential channels in living cells. Proc Natl Acad Sci U S A 99:7461–7466

Inoue R, Okada T, Onoue H, Hara Y, Shimizu S, Naitoh S, Ito Y, Mori Y (2001) The transient receptor potential protein homologue TRP6 is the essential component of vascular alpha(1)-adrenoceptor-activated $Ca^{(2+)}$-permeable cation channel. Circ Res 88:325–332

Jung S, Strotmann R, Schultz G, Plant TD (2002) TRPC6 is a candidate channel involved in receptor-stimulated cation currents in A7r5 smooth muscle cells. Am J Physiol Cell Physiol 282:C347–359

Kawasaki BT, Liao Y, Birnbaumer L (2006) Role of Src in C3 transient receptor potential channel function and evidence for a heterogeneous makeup of receptor- and store-operated Ca^{2+} entry channels. Proc Natl Acad Sci U S A 103:335–340

Kim JY, Saffen D (2005) Activation of M1 muscarinic acetylcholine receptors stimulates the formation of a multiprotein complex centered on TRPC6 channels. J Biol Chem 280:32035–32047

Kriz W (2005) TRPC6—a new podocyte gene involved in focal segmental glomerulosclerosis. Trends Mol Med 11:527–530

Krizaj D (2005) Compartmentalization of calcium entry pathways in mouse rods. Eur J Neurosci 22:3292–3296

Lessard CB, Lussier MP, Cayouette S, Bourque G, Boulay G (2005) The overexpression of presenilin2 and Alzheimer's-disease-linked presenilin2 variants influences TRPC6-enhanced Ca^{2+} entry into HEK293 cells. Cell Signal 17:437–445

Li S, Gosling M, Poll C (2005a) Determining the functional role of TRPC channels in primary cells. Pflugers Arch 451:43–52

Li Y, Jia YC, Cui K, Li N, Zheng ZY, Wang YZ, Yuan XB (2005b) Essential role of TRPC channels in the guidance of nerve growth cones by brain-derived neurotrophic factor. Nature 434:894–898

Lin MJ, Leung GP, Zhang WM, Yang XR, Yip KP, Tse CM, Sham JS (2004) Chronic hypoxia-induced upregulation of store-operated and receptor-operated Ca^{2+} channels in pulmonary arterial smooth muscle cells: a novel mechanism of hypoxic pulmonary hypertension. Circ Res 95:496–505

Lu PJ, Hsu AL, Wang DS, Chen CS (1998) Phosphatidylinositol 3,4,5-trisphosphate triggers platelet aggregation by activating Ca^{2+} influx. Biochemistry 37:9776–9783

Lussier MP, Cayouette S, Lepage PK, Bernier CL, Francoeur N, St-Hilaire M, Pinard M, Boulay G (2005) MxA, a member of the dynamin superfamily, interacts with the ankyrin-like repeat domain of TRPC. J Biol Chem 280:19393–19400

Montell C (2001) Physiology, phylogeny, and functions of the TRP superfamily of cation channels. Sci STKE 90:RE1

Okada T, Inoue R, Yamazaki K, Maeda A, Kurosaki T, Yamakuni T, Tanaka I, Shimizu S, Ikenaka K, Imoto K, Mori Y (1999) Molecular and functional characterization of a novel mouse transient receptor potential protein homologue TRP7. $Ca^{(2+)}$-permeable cation channel that is constitutively activated and enhanced by stimulation of G protein-coupled receptor. J Biol Chem 274:27359–27370

Otsuka Y, Sakagami H, Owada Y, Kondo H (1998) Differential localization of mRNAs for mammalian trps, presumptive capacitative calcium entry channels, in the adult mouse brain. Tohoku J Exp Med 185:139–146

Owsianik G, Talavera K, Voets T, Nilius B (2005) Permeation and selectivity of TRP channels. Annu Rev Physiol 68:685–717

Reiser J, Polu KR, Moller CC, Kenlan P, Altintas MM, Wei C, Faul C, Herbert S, Villegas I, Avila-Casado C, McGee M, Sugimoto H, Brown D, Kalluri R, Mundel P, Smith PL, Clapham DE, Pollak MR (2005) TRPC6 is a glomerular slit diaphragm-associated channel required for normal renal function. Nat Genet 37:739–744

Shi J, Mori E, Mori Y, Mori M, Li J, Ito Y, Inoue R (2004) Multiple regulation by calcium of murine homologues of transient receptor potential proteins TRPC6 and TRPC7 expressed in HEK293 cells. J Physiol 561:415–432

Soboloff J, Spassova M, Xu W, He LP, Cuesta N, Gill DL (2005) Role of endogenous TRPC6 channels in Ca^{2+} signal generation in A7r5 smooth muscle cells. J Biol Chem 280:39786–39794

Strübing C, Krapivinsky G, Krapivinsky L, Clapham DE (2003) Formation of novel TRPC channels by complex subunit interactions in embryonic brain. J Biol Chem 278:39014–39019

Thebault S, Flourakis M, Vanoverberghe K, Vandermoere F, Roudbaraki M, Lehen'kyi V, Slomianny C, Beck B, Mariot P, Bonnal JL, Mauroy B, Shuba Y, Capiod T, Skyrma R, Prevaraskaya N (2006) Different role of transient receptor potential channels in Ca^{2+} entry and proliferation of prostate cancer epithelial cells. Cancer Res 66:2038–2047

Trebak M, Vazquez G, Bird GS, Putney JW Jr (2003) The TRPC3/6/7 subfamily of cation channels. Cell Calcium 33:451–461

Tseng PH, Lin HP, Hu H, Wang C, Zhu MX, Chen CS (2004) The canonical transient receptor potential 6 channel as a putative phosphatidylinositol 3,4,5-trisphosphate-sensitive calcium entry system. Biochemistry 43:11701–11708

Welsh DG, Morielli AD, Nelson MT, Brayden JE (2002) Transient receptor potential channels regulate myogenic tone of resistance arteries. Circ Res 90:248–250

Winn MP, Conlon PJ, Lynn KL, Farrington MK, Creazzo T, Hawkins AF, Daskalakis N, Kwan SY, Ebersviller S, Burchette JL, Pericak-Vance MA, Howell DN, Vance JM, Rosenberg PB (2005) A mutation in the TRPC6 cation channel causes familial focal segmental glomerulosclerosis. Science 308:1801–1804

Yu Y, Sweeney M, Zhang S, Platoshyn O, Landsberg J, Rothman A, Yuan JX (2003) PDGF stimulates pulmonary vascular smooth muscle cell proliferation by upregulating TRPC6 expression. Am J Physiol Cell Physiol 284:C316–330

Yu Y, Fantozzi I, Remillard CV, Landsberg JW, Kunichika N, Platoshyn O, Tigno DD, Thistlethwaite PA, Rubin LJ, Yuan JX (2004) Enhanced expression of transient receptor potential channels in idiopathic pulmonary arterial hypertension. Proc Natl Acad Sci U S A 101:13861–13866

Zhang L, Saffen D (2001) Muscarinic acetylcholine receptor regulation of TRP6 Ca^{2+} channel isoforms. Molecular structures and functional characterization. J Biol Chem 276:13331–13339

Zhang Z, Tang J, Tikunova S, Johnson JD, Chen Z, Qin N, Dietrich A, Stefani E, Birnbaumer L, Zhu MX (2001) Activation of Trp3 by inositol 1,4,5-trisphosphate receptors through displacement of inhibitory calmodulin from a common binding domain. Proc Natl Acad Sci U S A 98:3168–3173

TRPC7

T. Numaga · M. Wakamori · Y. Mori (✉)

Laboratory of Molecular Biology, Department of Synthetic Chemistry and Biological Chemistry, Graduate School of Engineering, Kyoto University, 615-8510 Kyoto, Japan
mori@sbchem.kyoto-u.ac.jp

1	Introduction	143
2	Basic Features of the Gene, the Cloned cDNAs and Protein	144
2.1	Structural Characteristics of the TRPC7 Gene	144
2.2	Tissue Distribution	144
2.3	Expression Pattern	146
3	Functional Properties	146
3.1	Mode of Activation	146
3.2	Modulation	147
3.3	Ion Permeation Properties	147
3.4	Pharmacology	148
4	Biological Relevance and Roles	148
	References	149

Abstract Canonical transient receptor potential 7 (TRPC7) is the seventh identified member of the mammalian TRPC channel family, comprising nonselective cation channels activated through the phospholipase C (PLC) signaling pathway. TRPC7 is directly activated by diacylglycerol (DAG), one of the PLC products, having high sequence homology with TRPC3 and TRPC6, which are also activated by DAG. TRPC7 shows unique properties of activation, such as constitutive activity and susceptibility to negative regulation by extracellular Ca^{2+}. Although the physiological importance of TRPC7 in the native environment remains elusive, TRPC7 would play important roles in Ca^{2+} signaling pathway through these characteristic features.

Keywords TRPC7 · Diacylglycerol · Constitutive activity · Extracellular · Calcium

1
Introduction

TRPC7 is the seventh mammalian canonical transient receptor potential (TRPC) homolog identified through molecular cloning of complementary DNAs (cDNAs) that are hybridizable with the TRPC3 cDNA probe (Okada et al. 1999). TRPC7 was originally designated TRP7 by Okada et al. (1999); the name was also used for a gene identified by Nagamine et al. (1998). Through the unified nomenclature for the transient receptor potential (TRP) superfamily (Montell et al. 2002), Okada et al. (2002) established the current name for TRP7, TRPC7, as a member of "canonical" TRPC family, while the other TRP7 is now referred to as TRPM2, as a member of the "melastatin" TRPM family. Although mouse TRPC7 (mTRPC7) was found to be very similar to that of TRPC3 in its primary structure (75%), TRPC3 and TRPC7 are distinctly different isoforms, since their genes are mapped at distinct loci of both mouse and human chromosomes. In addition, mouse and human TRPC7 show different functional characteristics and are likely to play different physiological roles compared to other TRPCs including TRPC3. These important aspects characterizing the uniqueness of TRPC7 are discussed below.

2
Basic Features of the Gene, the Cloned cDNAs and Protein

2.1
Structural Characteristics of the TRPC7 Gene

In humans, the TRPC7 gene consists of 12 exons and is mapped to the chromosomal region 5q31.1 (Riccio et al. 2002a). mTRPC7 is mapped to the chromosomal region 13B2 (K. Yamazaki, I. Tanaka, and Y. Mori, unpublished result). Both the human and mouse TRPC7 genes contain an open reading frame of 2,589 bp encoding a protein of 862 amino acids with a predicted molecular mass of 100 kDa (Okada et al. 1999; Riccio et al. 2002a). Two splice variants of mTRPC7 have been reported. Splice variant 1 has a deletion of 2 exons of 345 bp encoding 115 amino acids (amino acid position 261–376) including predicted transmembrane segment S1. Variant 2 retains the upstream exon but has a deletion of the downstream exon. This genetic arrangement deletes 165 bp encoding 55 amino acids (321–376) in the same region of the channel protein (Okada et al. 1999; Walker et al. 2001). TRPC7 shares the common structural feature of the TRP channel with a core of six transmembrane segments and a pore region, with intracellular flanking N- and C-terminal domains. Human TRPC7 (hTRPC7) shares 98% overall identity with mTRPC7 and 81% and 75% overall identity with human TRPC3 and TRPC6, respectively.

2.2
Tissue Distribution

The reported tissue distribution of TRPC7 mRNA and proteins is listed in Table 1. hTRPC7 mRNA is widely expressed in the central nervous system (CNS), with a more restricted pattern of expression in peripheral tissues. In CNS, the highest levels of mRNA are present in the nucleus accumbens, and there is somewhat lower expression in the putamen, striatum, hypothalamus, caudate

Table 1 Tissue distribution of TRPC7

Species	mRNA	Protein
Human	Central nervous system Pituitary gland, kidney, intestine, prostate, cartilage[p] (Riccio et al. 2002) Keratinocytes[p] (Cai et al. 2005) Myometrium[p] (Dalrymple et al. 2004) Endothelial cells[s] (Yip et al. 2004)	Endothelial cells[i] Not present in vascular smooth muscle[i] (Yip et al. 2004) Expression in other cells and tissues not determined
Mouse	Brain, heart, lung, eye[n] (Okada et al. 1999) Smooth muscle[p] (Walker et al. 2001)	Not determined
Rat	Hippocampal neuronal cell[p] (Wu et al. 2004) Myometrium[p] (Babich et al. 2004)	Brain[w] (Goel et al. 2001) Hippocampal neuronal cell[w] (Wu et al. 2004) Carotid body (glomus cell, supportive cell), petrosal ganglion (neuron, satellite cells), carotid sensory nerve (supportive glia)[i] (Buniel et al. 2003) Expression in other cells and tissues not determined

[i] Protein detected by immune staining [n] mRNA detected by Northern blotting [p] mRNA detected by RT-PCR [w] Protein detected by Western blotting with specific antibodies

nucleus, locus coeruleus, and medulla oblongata. Lower levels of mRNA expression are observed in other CNS regions. In peripheral tissues, a high level of hTRP7 expression is detected in pituitary gland and kidney, while lower levels of hTRPC7 are found in intestine, prostate, and cartilage (Riccio et al. 2002a, b). In contrast to the hTRPC7, mTRPC7 mRNA is expressed at the highest level in the heart, lung, and eye and at lower levels in the brain, spleen, and testis. In the brain, cerebellar Purkinje cells are the most prominent expression sites of mTRPC7 RNA expression. mTRPC7 signals were also found intensely in the mitral layer of olfactory bulb and hippocampal neurons and diffusely in other regions including the cerebellar nuclei, pons, and cerebral cortex (Okada et al. 1999). As mentioned above, there are some differences in mRNA distribution between human and mouse orthologues of this channel. It should be noted that mTRPC7 is expressed in smooth muscle cells (Walker et al. 2001), but not hTRPC7 (Yip et al. 2004). These differences suggest that mouse and human TRPC7 channels may have quite distinct physiological roles despite their high sequence homology.

Two recent reports have demonstrated unique expression patterns of TRPC7 compared with other TRPC channels. In the rat carotid body, which is a chemosensory organ, only TRPC7 is expressed in supportive glial cells, while carotid sinus nerve fibers penetrating the carotid body express other TRPC channels

(TRPC1/3/4/5/6; Buniel et al. 2003). In human arteries, TRPC7 is expressed only in endothelial cells but not in smooth muscle cells, while TRPC1/3/4/5/6 were expressed in both cells (Yip et al. 2004).

2.3
Expression Pattern

There are a number of reports indicating the transcriptional regulation of TRPC7 in cell differentiation and development. In hippocampal H19–7 cells, the levels of mRNA and protein for TRPC7 are high in proliferating cells and decline dramatically upon differentiation (Wu et al. 2004). In human gingival keratinocytes, upon induction of keratinocyte differentiation, TRPC7 mRNA expression increases with the initial onset of differentiation and then decreases with increasing differentiation (Cai et al. 2005). TRPC7 mRNA is upregulated in myometrium obtained from pregnant women at term active labor when compared with nonpregnant samples (Dalrymple et al. 2004).

3
Functional Properties

3.1
Mode of Activation

Ionic currents through mTRPC7 are induced by diacylglycerol (DAG) without involvement of protein kinase C (PKC) after receptor-mediated activation of phospholipase C (PLC), but clearly not by store depletion, as observed for TRPC3/7 (Okada et al. 1999). By contrast, the cloned hTRPC7 (Riccio et al. 2002a) transiently expressed in HEK-293 cells behave as a store-operated channel. Recently, Lievremont et al. (2004) have generated the stable transfectant of hTRPC7-expressing HEK-293 cells, and compared the activation mechanism of stably expressed hTRPC7 with that of the transiently expressed hTRPC7 in the HEK-293 cells (Table 2). The transient TRPC7 forms cation channels activated by PLC-stimulating agonists, but not by Ca^{2+} store depletion. However, the stably transfected TRPC7 mediates Ca^{2+} entry in response to thapsigargin, and additionally increases Ca^{2+} entry upon subsequent addition of PLC-stimulating agonists, the latter Ca^{2+} entry being larger than the former. These data suggest that mode of recombinant expression influences the functional behavior of the channel proteins in the same HEK-293 cells. Up to now, the component(s) required for coupling of the TRPC7 to store depletion is unknown. The selection agents in the screening procedure for stable transfectants of TRPC7 may upregulate this component. The mTRPC7 and the hTRPC7 showed constitutive activity in the transient and the stable expression systems (Okada et al. 1999; Lievremont et al. 2004), respectively, although no

Table 2 Activation modes and cation channel blockers

	Activation	Block by trivalent cations
Mouse TRPC7	RACC [ATP (Okada et al. 1999) or methacholine (Lievremont et al. 2004)] Diacylglycerols (but not PKC) ([Ca^{2+}]$_i$ measurement (Okada et al. 1999, Lievremont et al. 2004), patch-clamp (Okada et al. 1999) Constitutive active ([Ca^{2+}]$_i$ measurement (Okada et al. 1999, Lievremont et al. 2004), patch-clamp (Okada et al. 1999)	La^{3+} 100 µM 38% ([Ca^{2+}]$_i$ measurement) (Okada et al. 1999) Gd^{3+} 100 µM 15% ([Ca^{2+}]$_i$ measurement) (Okada et al. 1999)
Human TRPC7	SOC(TG) Dependent on Ca^{2+} store depletion ([Ca^{2+}]$_i$ measurement) (Riccio et al. 2002, Lievremont et al. 2004)	La^{3+} 250 µM 100% ([Ca^{2+}]$_i$ measurement) (Riccio et al. 2002) Gd^{3+} 250 µM 100% ([Ca^{2+}]$_i$ measurement) (Riccio et al. 2002)

such activity has been observed in the transiently expressed hTRPC7 (Riccio et al. 2002a).

In order to investigate the activation mechanisms of the TRPC7 in native preparation, Lievremont et al. (2005) have knocked out TRPC7 in DT40 B cells. The TRPC7 knockout leads to increased size of the Ca^{2+} stores, and reduction of store-operated Ca^{2+} entry and receptor-/DAG-activated Ca^{2+} entry.

3.2
Modulation

Ca^{2+} regulates the mTRPC7 channel negatively through a direct extracellular action, and via Ca^{2+}-calmodulin (CaM)- and PKC-dependent mechanisms (Shi et al. 2004). Inositol 1,4,5-trisphosphate (IP$_3$) enhanced the mTRPC7 channel activity synergistically with its primary activator DAG; however, IP$_3$ failed to enhance the spontaneous mTRPC7 channel activity (Shi et al. 2004; Table 3).

3.3
Ion Permeation Properties

The current–voltage (I–V) relationship of mTRPC7 is almost linear with a slight flattening around the reversal potentials. Reversal potentials measured using

Table 3 Modulation of receptor-activated mTRPC7

	Positive modulator	Negative modulator
Mouse TRPC7	IP$_3$ (patch-clamp) (Shi et al. 2005)	[Ca^{2+}]$_o$, [Ca^{2+}]$_i$ via calmodulin, PKC (patch-clamp) (Shi et al. 2005)

different ionic conditions have suggested that TRPC7 is a nonselective cation channel with the permeability ratios $P_{Cs}:P_{Na}:P_{Ca}:P_{Ba}=1:1.1:5.9:5.0$ (Okada et al. 1999). In TRPC7-deficient DT40 cells, Ba^{2+} entry induced by DAG, receptor stimulation, or store depletion is significantly suppressed (Lievremont et al. 2005b). This is suggestive of a high Ba^{2+} selectivity/conductivity of TRPC7.

3.4
Pharmacology

Broad cation channel blockers, such as 100 µM Gd^{3+} and 100 µM La^{3+}, inhibited the ATP-induced Ca^{2+} influx in mTRPC7-transfected HEK cells by about 15% and 38%, respectively (Okada et al 1999). The trivalent sensitivities of the store-operated hTRPC7 channel are quantitatively different from the receptor-activated mTRPC7 channel. Either Gd^{3+} or La^{3+} at 250 µM completely blocks the Ca^{2+} entry through the hTRPC7 channel (Riccio et al. 2002a). Amiloride is effective to suppress constitutively activated TRPC7 currents (Okada et al. 1999). 2-Aminoethoxydiphenyl borate (2-APB) partially inhibits divalent cation entry via hTRPC7 activated by a muscarinic agonist or by OAG. It has been therefore proposed that 2-APB acts by a direct mechanism but not by simple pore occlusion nor through IP_3 receptors (Lievremont et al. 2005a).

4
Biological Relevance and Roles

The physiological significance of TRPC7 has been deduced from the tissue distribution studies and functional resemblances with native cationic currents (see above). It is unlikely that TRPC7 alone is responsible for specific biological function among TRPCs, since TRPC7 is coexpressed with other TRPCs in most of the tissues except several described already in Sect. 2.3. TRPC7 should rather operate in concert with multiple TRPCs in regulation of cation permeation properties in mammalian cells. Okada et al. (1999) pointed out a unique activation property of TRPC7: TRPC7 gives rise to constitutively activated cationic currents that are susceptible to receptor-induced enhancement.

On the basis of this TRPC7 feature, it has been suggested that TRPC7 is involved in time-independent, spontaneous currents such as background currents in the heart, lung and brain that are abundant in TRPC7 RNA (Okada et al. 1999). TRPC7 current may therefore be involved in the pace-making activity in the sinoatrial node cells, the spontaneous secretion in pace-making endocrine cells, the regulation of resting membrane potential and the frequency and pattern modulation of action potential in smooth muscle cells and neurons. The noradrenaline-evoked cation currents in smooth muscle cells of rabbit ear artery also displays constitutive activity like TRPC7 (Wang et al. 1993).

TRPC7-deficient DT40 cells show a characteristic increase in IP$_3$ receptor-mediated Ca^{2+} release (Lievremont et al. 2005b). This is consistent with the decreased store content or size suggested by the suppression of Ca^{2+} release elicited by recombinant expression of TRPC7 in HEK cells (Okada et al. 1999), but contrasts with the suggested role of TRPC1, whose gene knockout results in decrease of Ca^{2+} release in DT40 cells (Mori et al. 2002). The important role played by TRPC7 in ER Ca^{2+} homeostasis is somewhat puzzling, since the increase of constitutively active Ca^{2+} influx by TRPC7 should result in enhancement of filling the stores with Ca^{2+}. It is possible that TRPC7 interacts with IP$_3$ receptors to suppress their activity. Alternatively, TRPC7 may be also localized in the ER membrane and contribute to passive Ca^{2+} release from stores.

Small interfering RNA has been used to reveal the importance of TRPC7 in native systems (Wu et al. 2004; Zagranichnaya et al. 2005a, b; Cai et al. 2005). TRPC7 participates in formation of heteromeric store-operated and receptor-operated channels in HEK cells, whereas TRPC7 is unlikely to contribute to store-operated channels essential for differentiation of proliferating hippocampal H19-7 cells.

Among TRPC3/6/7, differences exist in Ca^{2+} (mainly extracellular) dependence (Zitt et al. 1997; Okada et al. 1999; Inoue et al. 2001) and in involvement of CaM (Shi et al. 2004). Assuming that this subfamily of TRPCs is activated simultaneously by receptor stimulation via DAG in the same cell, diverse patterns of membrane potential changes and intracellular $[Ca^{2+}]_i$ signaling should be generated, depending upon the relative abundance and geometry of localization of channel complexes. Heteromultimerization of TRPC3/6/7 may further increase the repertoire of the above patterns. In establishing the physiological significance of TRP channels in native tissues, it is interesting to relate characteristic electrical and Ca^{2+} signaling patterns with subcellular distribution of TRPC channels of known molecular compositions.

References

Babich LG, Ku CY, Young HW, Huang H, Blackburn MR, Sanborn BM (2004) Expression of capacitative calcium TrpC proteins in rat myometrium during pregnancy. Biol Reprod 70:919–924

Buniel MC, Schilling WP, Kunze DL (2003) Distribution of transient receptor potential channels in the rat carotid chemosensory pathway. J Comp Neurol 464:404–413

Cai S, Fatherazi S, Presland RB, Belton CM, Izutsu KT (2005) TRPC channel expression during calcium-induced differentiation of human gingival keratinocytes. J Dermatol Sci 40:21–28

Dalrymple A, Slater DM, Poston L, Tribe RM (2004) Physiological induction of transient receptor potential canonical proteins, calcium entry channels, in human myometrium: influence of pregnancy, labor, and interleukin-1 beta. J Clin Endocrinol Metab 89:1291–1300

Goel M, Sinkins WG, Schilling WP (2002) Selective association of TRPC channel subunits in rat brain synaptosomes. J Biol Chem 277:48303–48310

Inoue R, Okada T, Onoue H, Hara Y, Shimizu S, Naitoh S, Ito Y, Mori Y (2001) The transient receptor potential protein homologue TRP6 is the essential component of vascular alpha(1)-adrenoceptor-activated Ca^{2+}-permeable cation channel. Circ Res 88:325–332

Lievremont JP, Bird GS, Putney JW Jr (2004) Canonical transient receptor potential TRPC7 can function as both a receptor- and store-operated channel in HEK-293 cells. Am J Physiol Cell Physiol 287:C1709–C1716

Lievremont JP, Bird GS, Putney JW Jr (2005a) Mechanism of inhibition of TRPC cation channels by 2-aminoethoxydiphenylborane. Mol Pharmacol 68:758–762

Lievremont JP, Numaga T, Vazquez G, Lemonnier L, Hara Y, Mori E, Trebak M, Moss SE, Bird GS, Mori Y, Putney JW Jr (2005b) The role of canonical transient receptor potential 7 in B-cell receptor-activated channels. J Biol Chem 280:35346–35351

Montell C, Birnbaumer L, Flockerzi V, Bindels RJ, Bruford EA, Caterina MJ, Clapham DE, Harteneck C, Heller S, Julius D, Kojima I, Mori Y, Penner R, Prawitt D, Scharenberg AM, Schultz G, Shimizu N, Zhu MX (2002) A unified nomenclature for the superfamily of TRP cation channels. Mol Cell 9:229–231

Nagamine K, Kudoh J, Minoshima S, Kawasaki K, Asakawa S, Ito F, Shimizu N (1998) Molecular cloning of a novel putative Ca^{2+} channel protein (TRPC7) highly expressed in brain. Genomics 54:124–131

Okada T, Inoue R, Yamazaki K, Maeda A, Kurosaki T, Yamakuni T, Tanaka I, Shimizu S, Ikenaka K, Imoto K, Mori Y (1999) Molecular and functional characterization of a novel mouse transient receptor potential protein homologue TRP7. Ca^{2+}-permeable cation channel that is constitutively activated and enhanced by stimulation of G protein-coupled receptor. J Biol Chem 274:27359–27370

Riccio A, Mattei C, Kelsell RE, Medhurst AD, Calver AR, Randall AD, Davis JB, Benham CD, Pangalos MN (2002a) Cloning and functional expression of human short TRP7, a candidate protein for store-operated Ca^{2+} influx. J Biol Chem 277:12302–12309

Riccio A, Medhurst AD, Mattei C, Kelsell RE, Calver AR, Randall AD, Benham CD, Pangalos MN (2002b) mRNA distribution analysis of human TRPC family in CNS and peripheral tissues. Brain Res Mol Brain Res 109:95–104

Shi J, Mori E, Mori Y, Mori M, Li J, Ito Y, Inoue R (2004) Multiple regulation by calcium of murine homologues of transient receptor potential proteins TRPC6 and TRPC7 expressed in HEK293 cells. J Physiol 561:415–432

Walker RL, Hume JR, Horowitz B (2001) Differential expression and alternative splicing of TRP channel genes in smooth muscles. Am J Physiol Cell Physiol 280:C1184–C1192

Wang Q, Hogg RC, Large WA (1993) A monovalent ion-selective cation current activated by noradrenaline in smooth muscle cells of rabbit ear artery. Pflugers Arch 423:28–33

Wu X, Zagranichnaya TK, Gurda GT, Eves EM, Villereal ML (2004) A TRPC1/TRPC3-mediated increase in store-operated calcium entry is required for differentiation of H19-7 hippocampal neuronal cells. J Biol Chem 279:43392–43402

Yang M, Gupta A, Shlykov SG, Corrigan R, Tsujimoto S, Sanborn BM (2002) Multiple Trp isoforms implicated in capacitative calcium entry are expressed in human pregnant myometrium and myometrial cells. Biol Reprod 67:988–994

Yip H, Chan WY, Leung PC, Kwan HY, Liu C, Huang Y, Michel V, Yew DT, Yao X (2004) Expression of TRPC homologs in endothelial cells and smooth muscle layers of human arteries. Histochem Cell Biol 122:553–561

Zagranichnaya TK, Wu X, Villereal ML (2005a) Endogenous TRPC1, TRPC3, and TRPC7 proteins combine to form native store-operated channels in HEK-293 cells. J Biol Chem 280:29559–29569

Zagranichnaya TK, Wu X, Danos AM, Villereal ML (2005b) Gene expression profiles in HEK-293 cells with low or high store-operated calcium entry: can regulatory as well as regulated genes be identified? Physiol Genomics 21:14–33

Zitt C, Obukhov AG, Strubing C, Zobel A, Kalkbrenner F, Luckhoff A, Schultz G (1997) Expression of TRPC3 in Chinese hamster ovary cells results in calcium-activated cation currents not related to store depletion. J Cell Biol 138:1333–1341

Part II
TRPV Channel Subfamily

Capsaicin Receptor: TRPV1
A Promiscuous TRP Channel

S. C. Pingle · J. A. Matta · G. P. Ahern (✉)

Department of Pharmacology, Georgetown University Medical Center,
3900 Reservoir Rd NW, Washington DC, 20007, USA
gpa3@georgetown.edu

1	Introduction	156
2	Tissue Distribution	156
3	Molecular Structure of TRPV1	157
3.1	Splice Variants	157
4	Channel Properties	158
4.1	Permeability	158
4.2	Conductance and Rectification	159
4.3	Desensitization	159
5	Modes of Activation and Regulation	160
5.1	Heat	160
5.2	Voltage	161
5.3	Vanilloids and Lipids	161
5.4	Protons and Cations	162
5.5	Regulation of TRPV1 Channel Activity	162
5.6	Regulation of TRPV1 Expression	163
6	Pharmacology	164
6.1	Agonists	164
6.2	Antagonists	164
7	Biological Relevance	165
7.1	Pain	165
7.2	Urinary Bladder	165
7.3	Gastrointestinal	166
7.4	Vascular	166
7.5	Brain	166
7.6	Temperature Regulation	167
7.7	Ear	167
7.8	Respiratory Tract	167
7.9	Food Intake	168
7.10	Taste	168
7.11	Skin	168
	References	168

Abstract TRPV1, the archetypal member of the vanilloid TRP family, was initially identified as the receptor for capsaicin, the pungent ingredient in hot chili peppers. The receptor has a diverse tissue distribution, with high expression in sensory neurons. TRPV1 is a nonselective cation channel with significant permeability to calcium, protons, and large polyvalent cations. It is the most polymodal TRP channel, being activated by numerous stimuli, including heat, voltage, vanilloids, lipids, and protons/cations. TRPV1 acts as a molecular integrator of physical and chemical stimuli in peripheral nociceptor terminals and plays a critical role in thermal inflammatory hyperalgesia. In addition, TRPV1 may regulate a variety of physiological functions in different organ systems. Various second messenger systems regulate TRPV1 activity, predominantly by serine–threonine phosphorylation. In this review, we provide a concise summary of the information currently available about this channel.

Keywords TRPV1 · Capsaicin · Pain

1
Introduction

TRPV1 was first cloned from rat dorsal root ganglia (DRG) using a functional screening strategy for isolating candidate complementary DNA (cDNA) clones (Caterina et al. 1997). This newly cloned cDNA was named VR1, for vanilloid receptor subtype 1. Later, this receptor was identified to be a member of the transient receptor potential (TRP) family of cation channels and was assigned the nomenclature of TRPV1 to denote this association. To date, TRPV1 has been cloned from human, guinea pig, rabbit, mouse, and porcine tissues. TRPV1 orthologs from different species demonstrate significant differences in their molecular pharmacological profiles, including sensitivity to various agonists and antagonists.

2
Tissue Distribution

TRPV1 has a diverse tissue distribution. High levels of expression are observed in DRG, trigeminal ganglia (TG), and nodose ganglia (NG). TRPV1 is predominantly expressed in small- and medium-diameter peptidergic and nonpeptidergic neurons. Peptidergic neurons are important in the development of neurogenic pain and inflammation while nonpeptidergic neurons play a critical role in mediating chronic pain. TRPV1 is also found in various brain regions including the hypothalamus, cerebellum, cerebral cortex, striatum, midbrain, olfactory bulb, pons, medulla, hippocampus, thalamus, and substantia nigra. In nonneuronal tissues, TRPV1 expression is detected in keratinocytes of the epidermis, bladder urothelium and smooth muscles, glial cells, liver, and polymorphonuclear granulocytes, mast cells, and macrophages (see Tominaga and Tominaga 2005 for a detailed list of references). A reported identification of TRPV1 in dendritic, antigen-presenting cells was later shown to be false (O'Connell et al. 2005).

3
Molecular Structure of TRPV1

Rat TRPV1 cDNA contains an open reading frame of 2,514 nucleotides. TRPV1 is a 95-kDa, 838-amino-acid protein, consisting of six transmembrane (TM) domains, with a short pore-forming region between the fifth and sixth TM domains (Fig. 1). Structurally, TRPV1 consists of a long 400-amino-acid amino-terminus containing three ankyrin-repeat domains and a carboxy-terminus containing a TRP domain close to the sixth TM domain (Fig. 1). Functional TRPV1 channels exist as homo- or heteromultimers. TRPV1 can form functional multimers, with homotetramer being the predominant form of expression (Kedei et al. 2001). In addition to homomers, functional heteromers can be formed between TRPV1 and TRPV3 (Smith et al. 2002) or TRPV1 and TRPV2 (Liapi and Wood 2005; Rutter et al. 2005), which may be responsible, at least in part, for the variable responses to agonists and antagonists. The nature and properties of these channels can be regulated by posttranslational modifications, the presence of TRPV1 splice variants in the multimers, or both factors together.

3.1
Splice Variants

A number of recent studies have focused on alternate splicing and cDNA variants in TRPV1. TRPV1α and TRPV1β are two variants, respectively containing 839 and 829 amino acids. TRPV1β is a dominant-negative regulator of TRPV1 responses, since it is not functional by itself, but inhibits TRPV1α function during coexpression (Wang et al. 2004). TRPV1b, a human RNA splice variant expressed in trigeminal ganglion neurons, is unresponsive to capsaicin or protons, but can be activated by high temperatures (>47°C; Lu et al. 2005). An amino-terminal splice variant of TRPV1 isolated from the supraoptic nucleus (SON) is insensitive to capsaicin, but may be important for the intrinsic osmosensitivity of cells in the SON (Naeini et al. 2006). Rat taste-receptor cells express a TRPV1 splice variant that is constitutively active in the absence of a ligand at 23°C and is not modulated by protons (Lyall et al. 2004). This variant has been proposed to mediate amiloride-insensitive salt taste and function as the nonspecific salt taste receptor. A few other splice variants have been described but their functional significance is not clear.

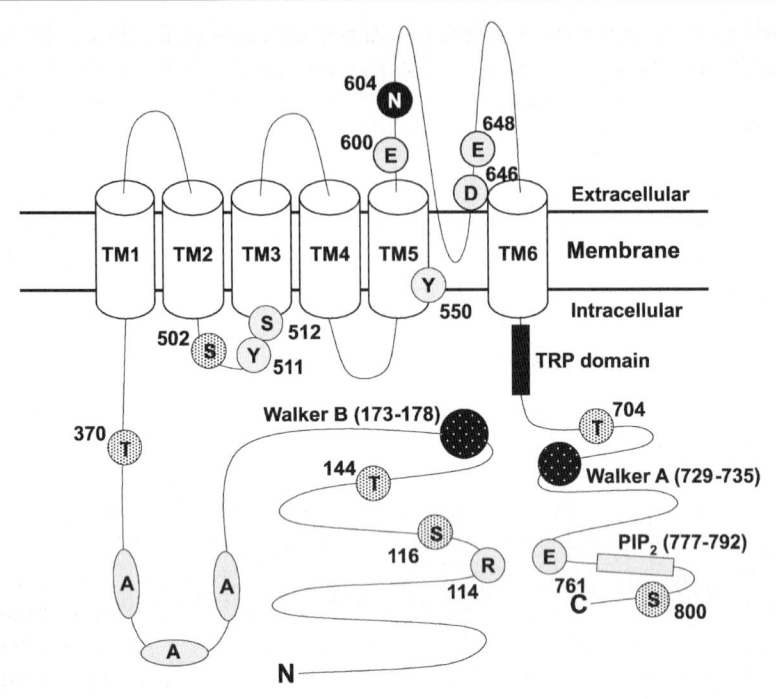

Fig. 1 Model depicting putative membrane topology, domain structure, and regions involved in regulation of TRPV1 function. TRPV1 has six TM domains with the pore-forming region between the 5th and 6th TM domains. Vanilloid-binding sites (Arg-114, Tyr-511, Ser-512, Tyr-550, Glu-761) and residues involved in proton-mediated activation (Glu-648) and sensitization (Glu-600) are depicted as *stippled*. Glu-600, Asp-646, and Glu-648 are involved in cation-induced TRPV1 activation and sensitization. The residue for N-glycosylation (Asn-604) is shown in *black*. Walker B (173–178) and Walker A (729–735) domains are ATP-binding domains at N- and C-termini, respectively. Serine and threonine residues involved in TRPV1 phosphorylation are depicted in *gray* (PKA-mediated phosphorylation: Ser-116, Thr-144, Thr-370, Ser-502; PKC-mediated phosphorylation: Ser-502, Ser-800; and CaMKII-mediated phosphorylation: Ser-502, Thr-704). The putative phosphatidylinositol 4,5-bisphosphate (PIP$_2$)-binding domain at the C-terminus is indicated as *gray*. TRP domain is shown in *black*

4
Channel Properties

4.1
Permeability

TRPV1 is a nonselective cation channel with near equal selectivity for Na$^+$, K$^+$, Li$^+$, Cs$^+$, and Rb$^+$ ions (Caterina et al. 1997) but moderate selectivity for divalent cations. When activated by capsaicin, the permeability of Mg^{2+} and Ca^{2+} relative to Na$^+$ (P_X/P_{Na}) is roughly 5 and 10, respectively (Caterina et al. 1997; Mohapatra et al. 2003; Ahern et al. 2005). Lower P_X/P_{Na} values of 3–4 are reported when the channel is activated by heat (Tominaga et al. 1998). TRPV1 is also highly permeable to protons and large polyvalent cations. The polyamines putrescine, spermidine, and spermine permeate TRPV1 with P_x/P_{Na} values between 3 and 16 (Ahern et al. 2006). Moreover, TRPV1 is permeable to organic cationic dyes (Meyers et al. 2003) and to aminoglycoside antibiotics (Myrdal and Steyger 2005) suggesting the existence of a large pore. Several amino acids in the putative pore-forming region between TM domains 5 and 6 are implicated in cation selectivity. Mutating Glu-648 (E648A) reduces Mg^{2+} permeability and increases Ca^{2+} permeability. Mutation of Asp-646 (D646 N) reduces Mg^{2+} permeability and blockade by the cationic dye, ruthenium red (see Tominaga and Tominaga 2005 for references).

4.2
Conductance and Rectification

The single channel conductance of capsaicin-activated channels is ~90–100 pS at positive potentials (see Fig. 2a). At negative potentials (−60 mV), the conductance is significantly lower, with values of approximately 50 pS (Caterina et al. 1997). Furthermore, divalent cations and protons reduce the single channel conductance. TRPV1 currents exhibit significant outward rectification, due to a combined effect of voltage on both channel conductance and open probability (Nilius et al. 2005; see Sect. 4).

4.3
Desensitization

Upon activation, TRPV1 undergoes desensitization. This phenomenon can occur rapidly during single application of an agonist (fast) or slowly following repeated agonist application (slow). Desensitization is believed to occur predominantly via a Ca^{2+}-dependent process because it is largely abolished by the use of Ca^{2+}-free conditions or Ca^{2+} chelators. It should be noted, however, that some Ca^{2+}-independent desensitization also occurs especially with heat activation. The Ca^{2+}-dependent mechanism arises because TRPV1 has considerable Ca^{2+} permeability, allowing Ca^{2+} influx through the channel to activate an inhibitory feedback signal. Indeed, fast desensitization is removed in a TRPV1 mutant that possesses a markedly reduced Ca^{2+} permeability. Furthermore, desensitization is attenuated by inhibitors of calcineurin, a Ca^{2+}-activated phosphatase, thus linking desensitization to a dephosphorylation event. Accordingly, protein kinase A (PKA)-mediated phosphorylation at Ser-116 in the amino-terminus of TRPV1 reduces desensitization. In addition, Ca^{2+} may signal via calmodulin which interacts with TRPV1 at amino- and carboxyl-terminal regions (positions 189–222 and 767–801). Disruption of this carboxyl-terminal region partially inhibits fast desensitization (for detailed references see Tominaga and Tominaga 2005).

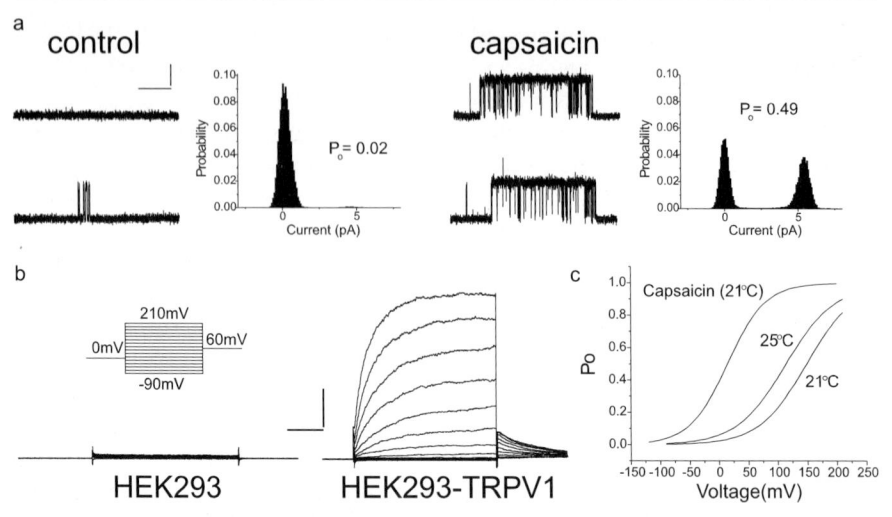

Fig. 2 a–c Agonist and voltage-dependent activation of TRPV1. **a** Single channel currents in an inside-out patch from a sensory dorsal root ganglion neuron (V_h = +60mV, *scale bar*: 4 pA and 1 s). The channel displays a low open probability under control conditions (*left*) but activity is markedly increased in the presence of 1 µM capsaicin (*right*). The single channel conductance is ∼92 pS (with CsCl in the pipette). **b** Voltage-dependent activation of TRPV1. A family of voltage steps in 20 mV increments evoke outward currents at positive potentials in cells expressing TRPV1 (scale bar: 1 nA and 25 ms). **c** Voltage-dependent activation curves at 21°C ($V_{1/2}$ = 149 mV), at 25°C ($V_{1/2}$ = 114 mV), and at 21°C in the presence of 50 nM capsaicin ($V_{1/2}$ = 10.6 mV)

5
Modes of Activation and Regulation

5.1
Heat

TRPV1 is a heat-gated channel with a threshold of approximately 43°C (at resting membrane potential, −60 mV; Caterina et al. 1997) and a steep temperature dependence—the Q_{10} (temperature coefficient over a 10°C range) is greater than 20 (Liu et al. 2003). In addition to direct gating, increases in temperatures in the subthreshold range can synergistically enhance currents produced by TRPV1 ligands, with a Q_{10} of approximately 2–3 (Babes et al. 2002). Consequently, TRPV1 currents are markedly enhanced by a shift from room temperature to physiological temperatures (37°C). These observations suggest that ligands may activate TRPV1 by altering the temperature sensitivity of the channel. The mechanisms underlying heat activation remain unclear; no mutation has been identified to selectively abolish heat activation. Heat activation is preserved in cell-free patches, indicating a membrane-delimited signaling event. Recently it has been proposed that temperature regulates TRPV1 by changing the intrinsic voltage-sensitivity of the channel (Nilius et al. 2005).

5.2
Voltage

TRPV1 has voltage-dependent properties; the channel activates in a time-dependent manner at positive potentials and deactivates at negative potentials. This intrinsic property of the channel contributes (along with single channel conductance) to outward rectification. Furthermore, the sensitivity for voltage-dependent activation and deactivation depends on the temperature and ligand concentration. In the absence of any TRPV1 ligand, large membrane depolarizations are required to activate the channel (Fig. 2), whereas in the presence of ligands, much smaller depolarizations serve to enhance the activation of the channel (Fig. 2; see Nilius et al. 2005 for references). The voltage sensor remains unknown. Unlike voltage-gated channels, TRPV1 and other TRP channels lack an array of charged residues in their TM domains. Thus, the gating currents must be small, as reflected by shallow Boltzmann activation curves (Fig. 2) that reach high positive potentials (the voltage for half-maximal activation, $V_{1/2}$, at 21°C is +150 mV). Although such high potentials necessary for TRPV1 activation seem physiologically irrelevant, small changes in temperature or presence of ligands will shift the activation curve significantly to more negative potentials ($V_{1/2}$ is +114 mV at 25°C, and +10.6 mV at 21°C with 50 nM capsaicin). Thus, the heat or ligand-sensitivity of TRPV1 may reflect a shift in intrinsic voltage-dependence. Consequently, the temperature threshold for TRPV1 activation is not constant, but will fluctuate depending on the membrane potential. Significantly, PKC enhances voltage-dependent activation and this may prime TRPV1 for activation under inflammatory conditions, possibly leading to broader action potentials and increased Ca^{2+} entry (Ahern and Premkumar 2002).

5.3
Vanilloids and Lipids

TRPV1 is activated by capsaicin, the pungent component of hot chili peppers. Capsaicin (a derivative of homovanillic acid and hence termed a "vanilloid"), and related compounds including resiniferatoxin (RTX), and olvanil are highly lipophilic and share structural similarity to several endogenous fatty acids that have been identified as TRPV1 agonists. These include the endocannabinoid anandamide, N-arachidonoyl dopamine (NADA) and oleoyl-dopamine, the lipoxygenase product, 12 hydroperoxyeicosatetraenoic acid (12-HPETE), and 18–20 carbon N-acylethanolamines (Movahed et al. 2005) including oleoylethanolamide (OEA) (Ahern 2003). These lipids may activate TRPV1 by interacting at vanilloid binding sites. In turn, vanilloids interact at intracellular regions of TRPV1; a charged capsaicin analog, which cannot cross the membrane, is only effective when applied to the intracellular surface. Consistent with this observation, several intracellular binding sites have been identified; amino acids residues Arg-114 in the amino-terminus and Glu-761 in the carboxy-terminus play a key role in ligand binding (Fig. 1). In addition, Tyr-511 and Ser-512 located at the transition between the second intracellular loop and third TM domain may play a role in vanilloid binding and channel activation. Tyr-550 is important for vanilloid binding in rat and human TRPV1 (Fig. 1).

A separate form of lipid modulation is mediated by phosphatidylinositol 4,5-bisphosphate (PIP$_2$) which is postulated to tonically bind TRPV1 and hold it in an inhibited state. In turn, hydrolysis of PIP$_2$ by phospholipase C leads to TRPV1 sensitization or activation. A PIP$_2$-binding region has been identified in the C-terminus (Fig. 1; see Tominaga and Tominaga 2005 for detailed references).

5.4
Protons and Cations

Extracellular protons regulate TRPV1 via two distinct mechanisms. At pH 6 to 7, protons sensitize the channel to other stimuli including heat and capsaicin. At higher concentrations (pH<6), protons directly activate the receptor. These effects require application of protons to the extracellular side of the channel. Consistent with this observation, two glutamate residues located near the extracellular pore-forming region appear to be critical for proton regulation (Jordt et al. 2000). Glu-600 mediates sensitization, while Glu-648 is necessary for direct activation (Fig. 1). Cations can regulate TRPV1 in a manner similar to protons, suggesting a generalized activation mechanism based on electrostatic charge. Na$^+$ (additional 50 mM), Mg^{2+}, and Ca^{2+} (1–10 mM range) enhance agonist-evoked currents, while divalent cations (>10 mM) directly gate TRPV1 (at room temperature, −60 mV; Ahern et al. 2005). Polyvalent cations are even more potent regulators; Gd^{3+} (Tousova et al. 2005) and the polyamine spermine (Ahern et al. 2006) sensitize and activate TRPV1 in the micromolar concentration range. However, Gd^{3+} at higher concentrations (>1 mM) blocks the channel. These actions may involve interactions at multiple acidic residues Glu-600, Glu-648, and Asp-646 (Fig. 1). Further, responses to cations are greater at 37°C and after TRPV1 phosphorylation, implicating cation modulation in inflammatory pain signaling (Ahern et al. 2005). Moreover, cation activation of TRPV1 underlies the visceral pain-related behavior induced by injections of Mg^{2+}, and this pain behavior is absent in TRPV1-null animals.

5.5
Regulation of TRPV1 Channel Activity

Various second messengers affect TRPV1 activity, predominantly by phosphorylation of specific residues on the receptor (Fig. 1). TRPV1 activity can be modulated by PKA, PKC, Ca^{2+}/calmodulin-dependent kinase II (CaMKII), or Src kinase. Phosphorylation at Ser-116 in the amino-terminus of TRPV1 is vital in PKA-mediated regulation of TRPV1 desensitization. In addition, Thr-144, Thr-370, and Ser-502 are important in PKA-mediated phosphorylation/sensitization of TRPV1. Phosphorylation of TRPV1 by PKC can induce channel activity at room temperature in a voltage-dependent manner (Premkumar and Ahern 2000). Moreover, PKC-mediated phosphorylation of TRPV1 not only potentiates capsaicin- or proton-evoked responses, but also reduces its temperature threshold such that receptors are active under physiological conditions (37°C). Two serine residues (Ser-502 and Ser-800) on TRPV1 have been recognized to be important in PKC-mediated effects. CaMKII-mediated phosphorylation of TRPV1 at Ser-502 and Thr-704 plays an important role in channel activation in response to capsaicin application. In addition, calcineurin-mediated dephosphorylation at the same sites can produce TRPV1 desensitization. Similarly, the nonreceptor cellular tyrosine kinase c-Src kinase positively regulates TRPV1 channel activity by tyrosine phosphorylation (see Bhave and Gereau 2004 for references).

In addition to phosphorylation, activity of TRPV1 may be regulated by N-glycosylation. Extracellular Asn-604 has been identified as the site for glycosylation of TRPV1 (Fig. 1; Jahnel et al. 2001). ATP may allosterically modulate TRPV1 by directly interacting with the Walker-type nucleotide-binding domains (Fig. 1) and increase vanilloid-induced channel activity. Modulation of the redox state may regulate the physiological activity of TRPV1 (Jin et al. 2004), possibly involving amino acid residue Cys-621, located on the extracellular surface.

5.6
Regulation of TRPV1 Expression

Nerve growth factor (NGF), acting via p38 mitogen-activated protein kinase (MAPK), produces a transcription-independent increase in surface expression of TRPV1 (Ji et al. 2002). One possible mechanism involved in this upregulation may be increased translation, downstream to phosphorylation of the translational factor, eIF4E. NGF can also increase surface expression of TRPV1 by Src kinase-mediated phosphorylation of TRPV1 at Tyr-200, which results in an increased insertion of channels into the surface membrane (Zhang et al. 2005). PKC-mediated potentiation of channel activity may involve a soluble N-ethylmaleimide-sensitive fusion protein (NSF) attachment protein receptor (SNARE)-dependent exocytosis of TRPV1 to the cell surface (Morenilla-Palao et al. 2004). However, transcriptional regulation of TRPV1 has not been well-characterized.

6 Pharmacology

6.1 Agonists

Though a large number of TRPV1 agonists have been identified, the most commonly used agonists are described here. Initially identified in the nineteenth century as the pungent component of peppers of the genus *Capsicum*, capsaicin was later identified as an acylamide derivative of homovanillic acid, 8-methyl-*N*-vanillyl-6-nonenamide. It acts as a potent TRPV1 agonist, evoking increases in intracellular calcium in sensory neurons, TRPV1-expressing HEK293 and CHO cells with EC_{50} values of approximately 270 nM (Acs et al. 1996), 80 nM, and 40 nM (Szallasi et al. 1999), respectively. Electrophysiological recordings give slightly higher values (for example, 500 nM in HEK cells; Gunthorpe et al. 2000). The lower EC_{50} values seen for calcium responses when compared to whole-cell currents could be due either to calcium-induced calcium release associated with calcium influx or to low saturability of calcium responses.

RTX is an ultrapotent vanilloid, which can act at TRPV1 with an EC_{50} of approximately 40 nM for whole-cell currents (Caterina et al. 1997) and approximately 1 nM for calcium response in sensory neurons. Due to its high affinity, the tritiated form of RTX ($[^3H]$-RTX) was developed for use in TRPV1 radioligand-binding assays (Szallasi and Blumberg 1990). RTX binds DRG membranes in a highly cooperative manner with a K_d of roughly 40 pM.

Recently, 2-aminoethoxydiphenyl borate (2-APB) and ethanol were identified as agonists for TRPV1 (Trevisani et al. 2002; Hu et al. 2004).

6.2 Antagonists

Capsazepine is a competitive antagonist for TRPV1, with structural similarity to capsaicin. It competes for the capsaicin-binding site on TRPV1, inhibits capsaicin-mediated channel activation, and can displace RTX from its binding site in radioligand-binding assays. Surprisingly, capsazepine inhibits both heat- and proton-induced channel activation (Tominaga et al. 1998) suggesting a more general disruption of the channel-gating mechanism. However, capsazepine shows only modest potency as an inhibitor of capsaicin-induced responses, with ED_{50} values in the range of 0.2–5 µM (Szallasi et al. 1993). Though RTX is an ultrapotent agonist for TRPV1, the iodinated form of this compound, iodo-resiniferatoxin (iodo-RTX) acts as a potent inhibitor (IC_{50} ~4 nM) of capsaicin-mediated responses (Wahl et al. 2001). However, enthusiasm for its use as a TRPV1 antagonist has been tempered in light of recent evidence showing iodo-RTX activating TRPV1, possibly due to partial agonism or in vivo de-iodination to RTX (Shimizu et al. 2005).

Ruthenium red, an inorganic cationic dye, acts as a TRPV1 blocker (Dray et al. 1990). Interestingly, the sarcoplasmic reticulum ATPase (SERCA) inhibitor, thapsigargin, inhibits [^{3}H]-RTX binding and functional TRPV1 responses (Toth et al. 2002).

In addition, several new compounds are being used as antagonists for TRPV1, including A-425619, IBTU, SB-366791, and AMG 9810. Development of potent and specific TRPV1 antagonists may be useful not only as efficient research tools but also for potential use in clinical conditions that warrant inhibition of TRPV1 channel activity.

7
Biological Relevance

7.1
Pain

TRPV1 acts as a transducer of noxious thermal and chemical stimuli in nociceptive sensory neurons and is vital in mediating enhanced heat sensitivity during inflammation. Inflammatory pain is characterized by allodynia and hyperalgesia, among other factors, due to sensitization of TRPV1. A variety of inflammatory mediators, including NGF, prostaglandins, bradykinin, serotonin, ATP, lipoxygenase products, and adenosine act via second messenger molecules to produce TRPV1 sensitization. In addition, inflammation and ischemia are associated with tissue acidification, further potentiating TRPV1 activity. Taken together, these factors increase TRPV1 responses and lower the temperature threshold for heat activation, such that the channel can be activated at normal body temperatures (∼37°C). Moreover, TRPV1 expression and function is increased in IB4-positive C-fiber nociceptors during inflammation. Modulation of TRPV1 expression and/or activity could form a promising target for pain control (for detailed list of references, refer to Julius and Basbaum 2001). TRPV1 antagonists produce significant attenuation of pain in animal models of bone cancer (Ghilardi et al. 2005). Recently, deletion of TRPV1-expressing primary afferent neurons is being examined as a strategy for pain management (Karai et al. 2004).

7.2
Urinary Bladder

TRPV1 is important in regulating normal lower urinary tract function. TRPV1 knockout animals have greater short-term voluntary urination and abnormal urodynamic responses, with an increase in the frequency of nonvoiding contractions, increased bladder capacity, and inefficient voiding (Birder et al. 2002). Moreover, intravesicular administration of vanilloid compounds is beneficial as a therapeutic strategy for symptomatic treatment of detrusor instability and interstitial cystitis, where vanilloids act by reducing neuronal activity secondary to desensitization of TRPV1 expressed on these neurons (Maggi 1992).

7.3
Gastrointestinal

In the gastrointestinal (GI) tract, TRPV1 mediates afferent and efferent functions. The afferent limb of this pathway transmits information from the GI tract to the spinal cord and brainstem, mainly contributing to GI sensation and pain. On the other hand, the efferent function is reflected in release of calcitonin gene-related peptide (CGRP), substance P (SP), and other mediators from their peripheral fibers. In the proximal part of the GI tract, TRPV1 maintains mucosal homeostasis, and protects against mucosal injury, by increasing blood flow and bicarbonate and mucus secretion. On the other hand, in the pancreas, ileum, and colon, TRPV1 is associated with inflammation and tissue damage (see Holzer 2004 for a detailed list of references). Interestingly, in humans, inflammatory bowel disease is associated with upregulation of TRPV1 in nerve fibers of the colon (Yiangou et al. 2001).

7.4
Vascular

TRPV1-mediated neuropeptide release mediates vasodilatation or vasoconstriction in a tissue-specific manner. Activation of TRPV1 on perivascular sensory nerves produces CGRP-dependent vasodilatation (Zygmunt et al. 1999). On the other hand, following increased intravascular pressure, 20-hydroxyeicosatetraenoic acid (20-HETE) production can activate TRPV1 on C-fiber nerve endings (Scotland et al. 2004). This, in turn, can release vasoactive neuropeptides leading to vasoconstriction. These vascular effects raise the possibility of a novel role for TRPV1 in the pathogenesis of vascular diseases, such as hypertension. Also, neuropeptides released from TRPV1-expressing nerve endings during cardiac ischemia evoke potent coronary vasodilatation and decrease cardiac contractility, thereby facilitating postischemic recovery (Schultz 2003).

7.5
Brain

Though expression of TRPV1 has been demonstrated in a number of brain regions, its exact role has not been characterized. Intranigral injection of capsaicin enhances motor activity in animals (Dawbarn et al. 1981), suggesting a functional role for TRPV1 in the basal ganglia. Activation of TRPV1 in the ventral midbrain stimulates glutamate release onto dopaminergic neurons, without affecting γ-aminobutyric acid (GABA)-ergic outputs (Marinelli et al. 2003). It is possible that activation of TRPV1 by endogenous factors maintains a tonic control of glutamate neurotransmission and likely plays an important role in the functions associated with dopaminergic transmission, including motor activity and reward pathway and motivation.

7.6
Temperature Regulation

TRPV1 contributes to regulation of normal body temperature. Systemic or hypothalamic injection of capsaicin in animals produces a hypothermic response, whereas capsaicin-desensitized animals demonstrate exaggerated fever response and are predisposed to dangerous hyperthermia. Hence it is surprising that TRPV1-knockout mice have an attenuated fever response when compared to normal mice (Iida et al. 2005). To add to the complexity, another study has demonstrated that capsaicin can simultaneously activate independent networks for heat loss and heat production (Kobayashi et al. 1998). Regulation of body temperature by TRPV1 may involve complex mechanisms, both in the brain and at the periphery, and the interplay of multiple factors may decide the ultimate outcome of receptor activation on body temperature.

7.7
Ear

Capsaicin application increases inner ear blood flow, specifically through the basilar and cochlear blood vessels (Vass et al. 1995). TRPV1 modulation may play an important role in the physiology and pathology of the inner ear, especially conditions such as vertigo, tinnitus, hyperacusis, vestibular hypersensitivity, and hearing loss associated with inflammatory conditions, such as Meniere's disease. Recently, TRPV1 expression was demonstrated in the organ of Corti (Zheng et al. 2003), where it may be involved in cochlear nociception.

7.8
Respiratory Tract

Recent studies demonstrate an important role for TRPV1 in mediating airway hypersensitivity. NGF plays an important role in the pathogenesis of asthma. Increased production of NGF during asthma can potentiate airway inflammation, at least in part through TRPV1 sensitization. In addition, lipoxygenase products, such as 15-HPETE, 15-HETE, and leukotriene B_4 (LTB_4) released from epithelial cells in airways can directly activate TRPV1, thus contributing to neuronal hypersensitivity in the airway. In addition to these inflammatory mediators, acidic pH associated with inflammation could further contribute to TRPV1-mediated airway hypersensitivity during asthma (see Jia et al. 2005 for a detailed reference list).

7.9
Food Intake

The fatty acid OEA is a putative, peripheral satiety factor, known to reduce food consumption in both freely feeding and starved rats. OEA can directly activate TRPV1, induce visceral pain-related behavior, and acutely reduce food intake in a TRPV1-dependent manner (Ahern 2003; Wang et al. 2005). Modulation of TRPV1 channel activity may therefore mediate acute anorexigenic effects of OEA.

7.10
Taste

Recently, it was shown that the mammalian nonspecific salt taste receptor is a variant of TRPV1 and mediates amiloride-insensitive salt taste (Lyall et al. 2004). In this regard, it is interesting that TRPV1 activation is regulated by salt (cationic strength; Ahern et al. 2005).

7.11
Skin

TRPV1 in keratinocytes may influence inflammatory processes in the skin and function as a sensor for noxious cutaneous stimulation. TRPV1 was shown to have a protective role in cutaneous contact allergic dermatitis (Banvolgyi et al. 2005). Moreover, TRPV1 activation up-regulates known endogenous hair growth inhibitors and downregulates known hair growth promoters, thus making it a putative target for epithelial growth disorders (Bodo et al. 2005).

References

Acs G, Palkovits M, Blumberg PM (1996) Specific binding of [3H]resiniferatoxin by human and rat preoptic area, locus ceruleus, medial hypothalamus, reticular formation and ventral thalamus membrane preparations. Life Sci 59:1899–1908

Ahern GP (2003) Activation of TRPV1 by the satiety factor oleoylethanolamide. J Biol Chem 278:30429–30434

Ahern GP, Premkumar LS (2002) Voltage-dependent priming of rat vanilloid receptor: effects of agonist and protein kinase C activation. J Physiol 545:441–451

Ahern GP, Brooks IM, Miyares RL, Wang XB (2005) Extracellular cations sensitize and gate capsaicin receptor TRPV1 modulating pain signaling. J Neurosci 25:5109–5116

Ahern GP, Wang X, Miyares RL (2006) Polyamines are potent ligands for the capsaicin receptor TRPV1. J Biol Chem 281:8991–8995

Babes A, Amuzescu B, Krause U, Scholz A, Flonta ML, Reid G (2002) Cooling inhibits capsaicin-induced currents in cultured rat dorsal root ganglion neurones. Neurosci Lett 317:131–134

Banvolgyi A, Palinkas L, Berki T, Clark N, Grant AD, Helyes Z, Pozsgai G, Szolcsanyi J, Brain SD, Pinter E (2005) Evidence for a novel protective role of the vanilloid TRPV1 receptor in a cutaneous contact allergic dermatitis model. J Neuroimmunol 169:86–96

Bhave G, Gereau RW 4th (2004) Posttranslational mechanisms of peripheral sensitization. J Neurobiol 61:88–106

Birder LA, Nakamura Y, Kiss S, Nealen ML, Barrick S, Kanai AJ, Wang E, Ruiz G, De Groat WC, Apodaca G, Watkins S, Caterina MJ (2002) Altered urinary bladder function in mice lacking the vanilloid receptor TRPV1. Nat Neurosci 5:856–860

Bodo E, Biro T, Telek A, Czifra G, Griger Z, Toth BI, Mescalchin A, Ito T, Bettermann A, Kovacs L, Paus R (2005) A hot new twist to hair biology: involvement of vanilloid receptor-1 (VR1/TRPV1) signaling in human hair growth control. Am J Pathol 166:985–998

Caterina MJ, Schumacher MA, Tominaga M, Rosen TA, Levine JD, Julius D (1997) The capsaicin receptor: a heat-activated ion channel in the pain pathway. Nature 389:816–824

Dawbarn D, Harmar AJ, Pycock CJ (1981) Intranigral injection of capsaicin enhances motor activity and depletes nigral 5-hydroxytryptamine but not substance P. Neuropharmacology 20:341–346

Dray A, Forbes CA, Burgess GM (1990) Ruthenium red blocks the capsaicin-induced increase in intracellular calcium and activation of membrane currents in sensory neurones as well as the activation of peripheral nociceptors in vitro. Neurosci Lett 110:52–59

Ghilardi JR, Rohrich H, Lindsay TH, Sevcik MA, Schwei MJ, Kubota K, Halvorson KG, Poblete J, Chaplan SR, Dubin AE, Carruthers NI, Swanson D, Kuskowski M, Flores CM, Julius D, Mantyh PW (2005) Selective blockade of the capsaicin receptor TRPV1 attenuates bone cancer pain. J Neurosci 25:3126–3131

Gunthorpe MJ, Harries MH, Prinjha RK, Davis JB, Randall A (2000) Voltage- and time-dependent properties of the recombinant rat vanilloid receptor (rVR1). J Physiol 525:747–759

Holzer P (2004) TRPV1 and the gut: from a tasty receptor for a painful vanilloid to a key player in hyperalgesia. Eur J Pharmacol 500:231–241

Hu HZ, Gu Q, Wang C, Colton CK, Tang J, Kinoshita-Kawada M, Lee LY, Wood JD, Zhu MX (2004) 2-Aminoethoxydiphenyl borate is a common activator of TRPV1, TRPV2, and TRPV3. J Biol Chem 279:35741–35748

Iida T, Shimizu I, Nealen ML, Campbell A, Caterina M (2005) Attenuated fever response in mice lacking TRPV1. Neurosci Lett 378:28–33

Jahnel R, Dreger M, Gillen C, Bender O, Kurreck J, Hucho F (2001) Biochemical characterization of the vanilloid receptor 1 expressed in a dorsal root ganglia derived cell line. Eur J Biochem 268:5489–5496

Ji RR, Samad TA, Jin SX, Schmoll R, Woolf CJ (2002) p38 MAPK activation by NGF in primary sensory neurons after inflammation increases TRPV1 levels and maintains heat hyperalgesia. Neuron 36:57–68

Jia Y, McLeod RL, Hey JA (2005) TRPV1 receptor: a target for the treatment of pain, cough, airway disease and urinary incontinence. Drug News Perspect 18:165–171

Jin Y, Kim DK, Khil LY, Oh U, Kim J, Kwak J (2004) Thimerosal decreases TRPV1 activity by oxidation of extracellular sulfhydryl residues. Neurosci Lett 369:250–255

Jordt SE, Tominaga M, Julius D (2000) Acid potentiation of the capsaicin receptor determined by a key extracellular site. Proc Natl Acad Sci U S A 97:8134–8139

Julius D, Basbaum AI (2001) Molecular mechanisms of nociception. Nature 413:203–210

Karai L, Brown DC, Mannes AJ, Connelly ST, Brown J, Gandal M, Wellisch OM, Neubert JK, Olah Z, Iadarola MJ (2004) Deletion of vanilloid receptor 1-expressing primary afferent neurons for pain control. J Clin Invest 113:1344–1352

Kedei N, Szabo T, Lile JD, Treanor JJ, Olah Z, Iadarola MJ, Blumberg PM (2001) Analysis of the native quaternary structure of vanilloid receptor 1. J Biol Chem 276:28613–28619

Kobayashi A, Osaka T, Namba Y, Inoue S, Lee TH, Kimura S (1998) Capsaicin activates heat loss and heat production simultaneously and independently in rats. Am J Physiol 275:R92–R98

Liapi A, Wood JN (2005) Extensive co-localization and heteromultimer formation of the vanilloid receptor-like protein TRPV2 and the capsaicin receptor TRPV1 in the adult rat cerebral cortex. Eur J Neurosci 22:825–834

Liu B, Hui K, Qin F (2003) Thermodynamics of heat activation of single capsaicin ion channels VR1. Biophys J 85:2988–3006

Lu G, Henderson D, Liu L, Reinhart PH, Simon SA (2005) TRPV1b, a functional human vanilloid receptor splice variant. Mol Pharmacol 67:1119–1127

Lyall V, Heck GL, Vinnikova AK, Ghosh S, Phan TH, Alam RI, Russell OF, Malik SA, Bigbee JW, DeSimone JA (2004) The mammalian amiloride-insensitive non-specific salt taste receptor is a vanilloid receptor-1 variant. J Physiol 558:147–159

Maggi CA (1992) Therapeutic potential of capsaicin-like molecules: studies in animals and humans. Life Sci 51:1777–1781

Marinelli S, Di Marzo V, Berretta N, Matias I, Maccarrone M, Bernardi G, Mercuri NB (2003) Presynaptic facilitation of glutamatergic synapses to dopaminergic neurons of the rat substantia nigra by endogenous stimulation of vanilloid receptors. J Neurosci 23:3136–3144

Meyers JR, MacDonald RB, Duggan A, Lenzi D, Standaert DG, Corwin JT, Corey DP (2003) Lighting up the senses: FM1-43 loading of sensory cells through nonselective ion channels. J Neurosci 23:4054–4065

Mohapatra DP, Wang SY, Wang GK, Nau C (2003) A tyrosine residue in TM6 of the Vanilloid Receptor TRPV1 involved in desensitization and calcium permeability of capsaicin-activated currents. Mol Cell Neurosci 23:314–324

Morenilla-Palao C, Planells-Cases R, Garcia-Sanz N, Ferrer-Montiel A (2004) Regulated exocytosis contributes to protein kinase C potentiation of vanilloid receptor activity. J Biol Chem 279:25665–25672

Movahed P, Jonsson BA, Birnir B, Wingstrand JA, Jorgensen TD, Ermund A, Sterner O, Zygmunt PM, Hogestatt ED (2005) Endogenous unsaturated C18 N-acylethanolamines are vanilloid receptor (TRPV1) agonists. J Biol Chem 280:38496–38504

Myrdal SE, Steyger PS (2005) TRPV1 regulators mediate gentamicin penetration of cultured kidney cells. Hear Res 204:170–182

Naeini RS, Witty MF, Seguela P, Bourque CW (2006) An N-terminal variant of Trpv1 channel is required for osmosensory transduction. Nat Neurosci 9:93–98

Nilius B, Talavera K, Owsianik G, Prenen J, Droogmans G, Voets T (2005) Gating of TRP channels: a voltage connection? J Physiol 567:35–44

O'Connell PJ, Pingle SC, Ahern GP (2005) Dendritic cells do not transduce inflammatory stimuli via the capsaicin receptor TRPV1. FEBS Lett 579:5135–5139

Premkumar LS, Ahern GP (2000) Induction of vanilloid receptor channel activity by protein kinase C. Nature 408:985–990

Rutter AR, Ma QP, Leveridge M, Bonnert TP (2005) Heteromerization and colocalization of TrpV1 and TrpV2 in mammalian cell lines and rat dorsal root ganglia. Neuroreport 16:1735–1739

Schultz HD (2003) The spice of life is at the root of cardiac pain. J Physiol 551:400

Scotland RS, Chauhan S, Davis C, De Felipe C, Hunt S, Kabir J, Kotsonis P, Oh U, Ahluwalia A (2004) Vanilloid receptor TRPV1, sensory C-fibers, and vascular autoregulation: a novel mechanism involved in myogenic constriction. Circ Res 95:1027–1034

Shimizu I, Iida T, Horiuchi N, Caterina MJ (2005) 5-Iodoresiniferatoxin evokes hypothermia in mice and is a partial transient receptor potential vanilloid 1 agonist in vitro. J Pharmacol Exp Ther 314:1378–1385

Smith GD, Gunthorpe MJ, Kelsell RE, Hayes PD, Reilly P, Facer P, Wright JE, Jerman JC, Walhin JP, Ooi L, Egerton J, Charles KJ, Smart D, Randall AD, Anand P, Davis JB (2002) TRPV3 is a temperature-sensitive vanilloid receptor-like protein. Nature 418:186–190

Szallasi A, Blumberg PM (1990) Specific binding of resiniferatoxin, an ultrapotent capsaicin analog, by dorsal root ganglion membranes. Brain Res 524:106–111

Szallasi A, Goso C, Blumberg PM, Manzini S (1993) Competitive inhibition by capsazepine of [3H]resiniferatoxin binding to central (spinal cord and dorsal root ganglia) and peripheral (urinary bladder and airways) vanilloid (capsaicin) receptors in the rat. J Pharmacol Exp Ther 267:728–733

Szallasi A, Blumberg PM, Annicelli LL, Krause JE, Cortright DN (1999) The cloned rat vanilloid receptor VR1 mediates both R-type binding and C-type calcium response in dorsal root ganglion neurons. Mol Pharmacol 56:581–587

Tominaga M, Tominaga T (2005) Structure and function of TRPV1. Pflugers Arch 451:143–150

Tominaga M, Caterina MJ, Malmberg AB, Rosen TA, Gilbert H, Skinner K, Raumann BE, Basbaum AI, Julius D (1998) The cloned capsaicin receptor integrates multiple pain-producing stimuli. Neuron 21:531–543

Toth A, Kedei N, Szabo T, Wang Y, Blumberg PM (2002) Thapsigargin binds to and inhibits the cloned vanilloid receptor-1. Biochem Biophys Res Commun 293:777–782

Tousova K, Vyklicky L, Susankova K, Benedikt J, Vlachova V (2005) Gadolinium activates and sensitizes the vanilloid receptor TRPV1 through the external protonation sites. Mol Cell Neurosci 30:207–217

Trevisani M, Smart D, Gunthorpe MJ, Tognetto M, Barbieri M, Campi B, Amadesi S, Gray J, Jerman JC, Brough SJ, Owen D, Smith GD, Randall AD, Harrison S, Bianchi A, Davis JB, Geppetti P (2002) Ethanol elicits and potentiates nociceptor responses via the vanilloid receptor-1. Nat Neurosci 5:546–551

Vass Z, Nuttall AL, Coleman JK, Miller JM (1995) Capsaicin-induced release of substance P increases cochlear blood flow in the guinea pig. Hear Res 89:86–92

Wahl P, Foged C, Tullin S, Thomsen C (2001) Iodo-resiniferatoxin, a new potent vanilloid receptor antagonist. Mol Pharmacol 59:9–15

Wang C, Hu HZ, Colton CK, Wood JD, Zhu MX (2004) An alternative splicing product of the murine trpv1 gene dominant negatively modulates the activity of TRPV1 channels. J Biol Chem 279:37423–37430

Wang X, Miyares RL, Ahern GP (2005) Oleoylethanolamide excites vagal sensory neurones, induces visceral pain and reduces short-term food intake in mice via capsaicin receptor TRPV1. J Physiol 564:541–547

Yiangou Y, Facer P, Dyer NH, Chan CL, Knowles C, Williams NS, Anand P (2001) Vanilloid receptor 1 immunoreactivity in inflamed human bowel. Lancet 357:1338–1339

Zhang X, Huang J, McNaughton PA (2005) NGF rapidly increases membrane expression of TRPV1 heat-gated ion channels. EMBO J 24:4211–4223

Zheng J, Dai C, Steyger PS, Kim Y, Vass Z, Ren T, Nuttall AL (2003) Vanilloid receptors in hearing: altered cochlear sensitivity by vanilloids and expression of TRPV1 in the organ of corti. J Neurophysiol 90:444–455

Zygmunt PM, Petersson J, Andersson DA, Chuang H, Sorgard M, Di Marzo V, Julius D, Hogestatt ED (1999) Vanilloid receptors on sensory nerves mediate the vasodilator action of anandamide. Nature 400:452–457

2-Aminoethoxydiphenyl Borate as a Common Activator of TRPV1, TRPV2, and TRPV3 Channels

C. K. Colton · M. X. Zhu (✉)

Department of Neuroscience and Center for Molecular Neurobiology,
The Ohio State University, Columbus OH, 43210, USA
zhu.55@osu.edu

1	Introduction	174
2	2APB Is a Common Activator of TRPV1, V2, and V3	174
2.1	2APB as an Activator of TRPV1	175
2.2	2APB as an Activator of TRPV2	176
2.3	2APB as an Activator of TRPV3	176
2.4	The Stimulatory Effect of 2APB on Native Tissues	177
3	The Effects of 2APB Analogs on TRPV Channels	178
4	Possible Mechanisms of Action of 2APB	181
4.1	Structural Considerations	181
4.2	Site(s) of Action	182
4.3	Effects on Membrane Properties	183
5	Concluding Remarks	184
References		184

Abstract 2-Aminoethoxydiphenyl borate (2APB) had been depicted as a universal blocker of transient receptor potential (TRP) channels. While evidence has accumulated showing that some TRP channels are indeed inhibited by 2APB, especially in heterologous expression systems, there are other TRP channels that are unaffected or affected very little by this compound. More interestingly, the thermosensitive TRPV1, TRPV2, and TRPV3 channels are activated by 2APB. This has been demonstrated both in heterologous systems and in native tissues that express these channels. A number of 2APB analogs have been examined for their effects on native store-operated channels and heterologously expressed TRPV3. These studies revealed a complex mechanism of action for 2APB and its analogs on ion channels. In this review, we have summarized the current results on 2APB-induced activation of TRPV1–3 and discussed the potential mechanisms by which 2APB may regulate TRP channels.

Keywords 2APB · Transient receptor potential · TRPC · Store-operated channel · Thermosensitive channel

1
Introduction

2-Aminoethoxydiphenyl borate (2APB) was first introduced to the biological community in 1997 as an inhibitor of inositol 1,4,5-trisphosphate receptors (IP3Rs) (Maruyama et al. 1997). It was later demonstrated to also block store-operated calcium entry (SOCE) (Dobrydneva and Blackmore 2001; Prakriya and Lewis 2001; Diver et al. 2001; Trebak et al. 2002). Since members of the canonical transient receptor potential (TRPC) family, as well as TRPV6, have been suggested to participate in SOCE (Zhu et al. 1996; Yue et al. 2001), their sensitivity to 2APB has been tested in heterologous expression systems. To date, the inhibition by 2APB has been documented for TRPC1, C3, C5, C6, and C7 (Delmas et al. 2002; Trebak et al. 2002; Hu et al. 2004; Xu et al. 2005; Lievremont et al. 2005). The effect of 2APB on TRPV6 is dependent on the expression level and the host cell type. While the drug slightly increased the constitutive activity of TRPV6 overexpressed in HEK293 and rat basophilic leukemia cells, it indeed blocked the store-operated component acquired by the low expression of TRPV6 in the latter cell type (Voets et al. 2001; Schindl et al. 2002). Furthermore, the inhibition by 2APB of TRPC3, C6, and C7 is dependent on the mode, and perhaps the degree, of activation and is often incomplete (Lievremont et al. 2005). In chicken DT40 cells, the ectopically expressed human TRPC3 was, instead, activated by 2APB (Ma et al. 2003).

Despite the limited number of studies showing consistent inhibition of ectopically expressed TRPC and TRPV6 channels by 2APB, there is ample evidence on 2APB-induced inhibition of endogenous channels presumably composed of various TRPC subunits (Tozzi et al. 2003; Sydorenko et al. 2003; Lucas et al. 2003). In addition, the magnesium-inactivated conductance, which is likely formed by TRPM7, is also inhibited by 2APB (Hermosura et al. 2002; Prakriya and Lewis 2002). In light of these observations, 2APB became recognized as a universal TRP channel blocker (Clapham et al. 2001). However, experimental evidence for this "label" has been scarce. In addition to the data described above, an inhibitory effect of 2APB has been shown for TRPM3, TRPM7, TRPM8, and TRPP2 (Xu et al. 2005; Hanano et al. 2004; Hu et al. 2004; Koulen et al. 2002). However, TRPM2 is unaffected and TRPV5 is only slightly inhibited by 100 µM 2APB (Nilius et al. 2001; Xu et al. 2005). More interestingly, 2APB is able to activate three thermosensitive members of the TRPV family, TRPV1–3 (Hu et al. 2004).

2
2APB Is a Common Activator of TRPV1, V2, and V3

The TRPV family has been studied extensively in recent years due to its involvement in temperature and pain sensation. TRPV1, V2, V3, and V4 are activated by high temperatures from warm to noxious heat with temperature thresholds of 43°C, 52°C, 31°C, and 25°C, respectively. Interestingly, they are expressed not only in the peripheral nervous system where they sense temperature and pain, but also in a wide variety of tissues that are not exposed to significant temperature fluctuation. For example, TRPV1 is expressed in astrocytes and other regions of brain, spinal cord, skin, and tongue; TRPV2 is in brain, vascular smooth muscle cells, intestines, and macrophages; TRPV3 is in keratinocytes, brain, and testis; and TRPV4 is in brain, skin, kidney, liver, trachea, heart, hypothalamus, and airway smooth muscle cells (Patapoutian et al. 2003; Doly et al. 2004; Muraki et al. 2003; Kashiba et al. 2004; Kim et al. 2003; Xu et al. 2002; Peier et al. 2002; Jia et al. 2004). TRP channels are often activated by multiple forms of stimuli. This polymodality, combined with the wide range of tissue distributions, suggests that these channels are involved in many different cellular and physiological functions.

The original drive for testing the effect of 2APB on the TRPV channels was to verify whether 2APB was a universal blocker of all TRP channels. This was done using HEK293 cells that had been transiently transfected with the complementary DNA (cDNA) for all members of the TRPV family (TRPV1–6) in different wells of 96-well plates. The cells were loaded with Fluo4 and assayed for intracellular Ca^{2+} changes using a fluorescence plate reader (FLEXStation, Molecular Devices). To our surprise, 2APB (0.5 mM) evoked a robust increase in Fluo4 fluorescence in cells transfected with TRPV1, V2, and V3. The endogenous response in vector-transfected cells was small and indistinguishable from those in cells that expressed TRPV4, V5, and V6. This initial data suggested that 2APB might be a common activator of TRPV1–3. Concentration-response curves to 2APB obtained from the Ca^{2+} assay at 32°C yielded EC_{50} values of 114±8, 129±13, and 34±12 µM for TRPV1, V2, and V3, respectively. Subsequent experiments were designed to confirm this finding and further characterize the effect of 2APB on TRPV1–3 using electrophysiological methods (Hu et al. 2004).

2.1
2APB as an Activator of TRPV1

In whole-cell recordings, 2APB dose-dependently activated currents in HEK293 cells that expressed mouse TRPV1. Control cells did not show any response to 2APB up to 3 mM. Under similar conditions, TRPC6 and TRPM8 currents were inhibited by 2APB. In order to confirm that the stimulatory effect of 2APB on TRPV1 was not a unique property of HEK293 cells, we also studied this effect in *Xenopus* oocytes injected with cRNA of mouse TRPV1. At 300 µM, 2APB elicited an inward current at −40 mV that was completely blocked by 3 µM ruthenium red (RR) but only partially blocked by 30 µM capsazepine (approx. 30%). Interestingly, 30 µM of capsazepine completely blocked the currents evoked by 1 µM capsaicin in the same cell, suggesting that the site of action for 2APB and capsaicin may be different.

A characteristic feature associated with the polymodality of TRPV1 is that known activators such as heat, protons, and capsaicin act synergistically. This is also true for 2APB in respect to other TRPV1 activators. In HEK293 cells, coapplication of 0.3 µM capsaicin and 100 µM 2APB, or 100 µM 2APB at pH 6.5, greatly increased the TRPV1 current at −100 mV more than 20-fold as compared to the stimulation with capsaicin, 2APB, or the weak acid (pH 6.5) alone. In *Xenopus* oocytes expressing TRPV1, 100 µM 2APB left-shifted the dose-response curve for capsaicin 3.8-fold and the pH dependence 6.6-fold. Conversely, capsaicin (0.3 µM) and weak acid (pH 6.5) also left-shifted the dose-response curve to 2APB 9.3- and 2.0-fold, respectively. Furthermore, about a 9-fold increase in current at −40 mV was obtained when 100 µM 2APB was applied at 40°C as compared to 22°C (Hu et al. 2004).

Chung et al. (2004b, 2005) also examined the effect of 2APB on TRPV channels. Although the initial study only showed a slight activation of TRPV1 by 100 µM 2APB at very positive potentials, subsequent experiments indeed confirmed a robust 2APB-induced intracellular Ca^{2+} increase via rat TRPV1 stably expressed in HEK293 cells at a slightly higher drug concentration of 320 µM.

2.2
2APB as an Activator of TRPV2

2APB-evoked whole-cell currents have been observed in HEK293 cells that expressed mouse TRPV2 (Hu et al. 2004, 2006). At 22°C, this activation was very weak at 1 mM, but became strong at 3 mM 2APB. The currents showed weak double rectification and were blocked by 3 µM RR. Chung et al. (2005) also confirmed the effect of 2APB (320 µM) on eliciting intracellular Ca^{2+} increase at the room temperature in HEK293 cells expressing rat TRPV2. On the other hand, for an endogenous channel encoded by mouse TRPV2 in the F-11 hybridoma derived from rat dorsal root ganglia (DRG) and mouse neuroblastoma, 100 µM 2APB did not significantly change the temperature threshold of current activation at −60 mV (Bender et al. 2005), indicating that there may be other requirement(s) for the activation of TRPV2 by 2APB.

2.3
2APB as an Activator of TRPV3

2APB-evoked TRPV3 currents have been shown in both HEK293 cells and *Xenopus* oocytes (Hu et al. 2004, 2006). In HEK293 cells that expressed mouse TRPV3, 2APB (30–300 µM) invoked dually rectifying currents (stronger in outward direction). RR (3 µM) blocked these currents in the inward direction and potentiated them at potentials higher than 40 mV. Infusion of 1 mM 2APB into the cell through the patch pipette for more than 6 min failed to elicit any current while subsequent application of 2APB in the bath elicited TRPV3 currents, indicating that 2APB acts from the extracellular side. In *Xenopus* oocytes injected with the cRNA for mouse TRPV3, 300 μM 2APB activated an inward current at −40 mV, which was blocked by 3 μM RR but not by 10 μM capsazepine. In addition, although a 40°C temperature challenge did not invoke a significant current, application of 100 μM 2APB at 40°C invoked a current that was 35 ± 6 times in amplitude of that induced by the same concentration at 22°C. These data confirm that TRPV3 is activated by 2APB and 2APB strongly potentiates the thermal response of TRPV3.

Chung et al. (2004b) also showed that in HEK293 cells expressing mouse TRPV3, 32 μM 2APB elicited slowly developing currents that were reversible and sensitized with successive 2APB application. Dose-response curves showed EC_{50} values of 28.3 μM at +80 mV and 41.6 μM at −80 mV, which are within the range (34 ± 12 μM) we obtained from the Ca^{2+} assay. Interestingly and similar to our data, at low 2APB concentrations, the TRPV3 currents exhibited strong outward rectification, with dual rectification gradually increasing at greater than 10 μM 2APB. This change in the current–voltage (IV) relationship indicates a relatively strong and near maximal activation of TRPV3 by the high concentrations of 2APB. The synergy between 2APB and heat was also documented by the 6-fold increase in current amplitude in response to a 37°C heat challenge by 1 μM 2APB, a concentration insufficient to cause TRPV3 activation at 24°C.

In single-channel analysis of inside-out membrane patches excised from TRPV3-expressing HEK293 cells, 1 μM 2APB evoked single-channel openings that were more prolonged than those evoked by heat. The inward and outward slope conductance was 201 and 147 pS, significantly smaller than those elicited by heat at 39°C, which are 337 and 256 pS, respectively. The unitary amplitude of single-channel openings was determined to have a linear relationship with temperature, and extrapolation of the currents observed with 2APB at 24°C and those observed by 39°C alone revealed that the currents resulted from the opening of the same channel (Chung et al. 2004b).

2.4
The Stimulatory Effect of 2APB on Native Tissues

2APB-evoked currents have been demonstrated in neurons from rat DRG and nodose/jugular ganglia (Hu et al. 2004; Gu et al. 2005). In capsaicin-sensitive neurons, 300 μM 2APB directly activated currents that were blocked by 3 μM RR and to a lesser extent by 10 μM capsazepine. At 30 and 100 μM, 2APB also potentiated the response to pH 6.5 and the effect was only partially blocked by 10 μM capsazepine. In addition, 30 μM 2APB strongly potentiated the response to 0.3 μM capsaicin and the current was completely blocked by capsazepine. These data show not only that native TRPV1 channels in rat DRG are activated by 2APB, but that the pharmacology of ectopically expressed TRPV1 is similar to that of native channels.

Similar to DRG neurons, 2APB (30–300 µM) invoked dose-dependent inward current at −70 mV in cultured capsaicin-sensitive rat pulmonary neurons and the current was sensitive to RR and capsazepine (Gu et al. 2005). In addition, intravenous bolus injection of 2APB elicited pulmonary chemoreflex responses, characterized by apnea, bradycardia, and hypotension in anesthetized, spontaneously breathing rats. Although these data cannot distinguish the relative contributions of TRPV1, V3, and perhaps V2, in these responses, similar studies using 2APB and TRPV knockout mice could determine the importance of individual TRPV members in the pulmonary chemoreflex and other sensory responses. In fact, by comparing the heat responses of skin-saphenous nerve preparation and cultured DRG neurons from wild type and $trpv1^{-/-}$ mice, Zimmermann et al. (2005) showed that the 2APB-induced sensitization to thermal stimulation in mouse C-fibers was a TRPV1-facilitated process.

Chung et al. (2004b) have tested the response of cultured mouse keratinocytes to 2APB. Immunostaining revealed that TRPV3 was expressed in most of these cells. However, heat-evoked TRPV3-like sensitizing currents are rarely detectable (5/189 cells; Chung et al. 2004a). Application of 100 µM 2APB at 40°C resulted in outwardly rectifying currents that were sensitized upon repetitive heat challenges in the majority of the wildtype (22/27) and $trpv4^{-/-}$ (23/30) keratinocytes (Chung et al. 2004b). RR (10 µM) inhibited inward currents evoked by 2APB at 42°C in keratinocytes derived from $trpv4^{-/-}$ mice. Together, these data confirm that 2APB can sensitize the response of TRPV3 to heat in mouse keratinocytes independent of TRPV4. More recently, the 2APB-induced activation of native TRPV3 channel and potentiation of its heat response in mouse keratinocytes was confirmed by another group (Moqrich et al. 2005). Unfortunately, whether these responses are missing in $trpv3^{-/-}$ keratinocytes was not reported.

Guatteo et al. (2005) have shown the expression of TRPV3 and V4 in temperature-sensitive dopaminergic neurons of rat substantia nigra pars compacta. Both warming and application of 2APB were found to increase the intracellular Ca^{2+}, suggesting a role for TRPV3 in Ca^{2+} homeostasis near physiological temperatures in these cells.

3
The Effects of 2APB Analogs on TRPV Channels

2APB analogs were first studied in order to identify blockers for Ca^{2+} influx induced by thrombin in human platelets, a process that is believed to involve TRPC1 (Rosado et al. 2002). Dobrydneva and Blackmore (2001) showed that like 2APB, diphenylboronic anhydride (DPBA) and 2,2-diphenyltetrahydrofuran (DPTHF) (see Fig. 1 for structures) could inhibit the thrombin-induced Ca^{2+} signal with a similar affinity as 2APB. This had led Chung et al. (2005) to explore the possibility that these 2APB analogs would activate TRPV3. Using

2APB monomer
(protonated)

2APB monomer
(open chain)

2APB monomer
(ring)

diphenylboronic anhydride
(DPBA)

2APB dimer

dimethyl 2APB

2,2-diphenyltetrahydrofuran
(DPTHF)

diphenhydramine

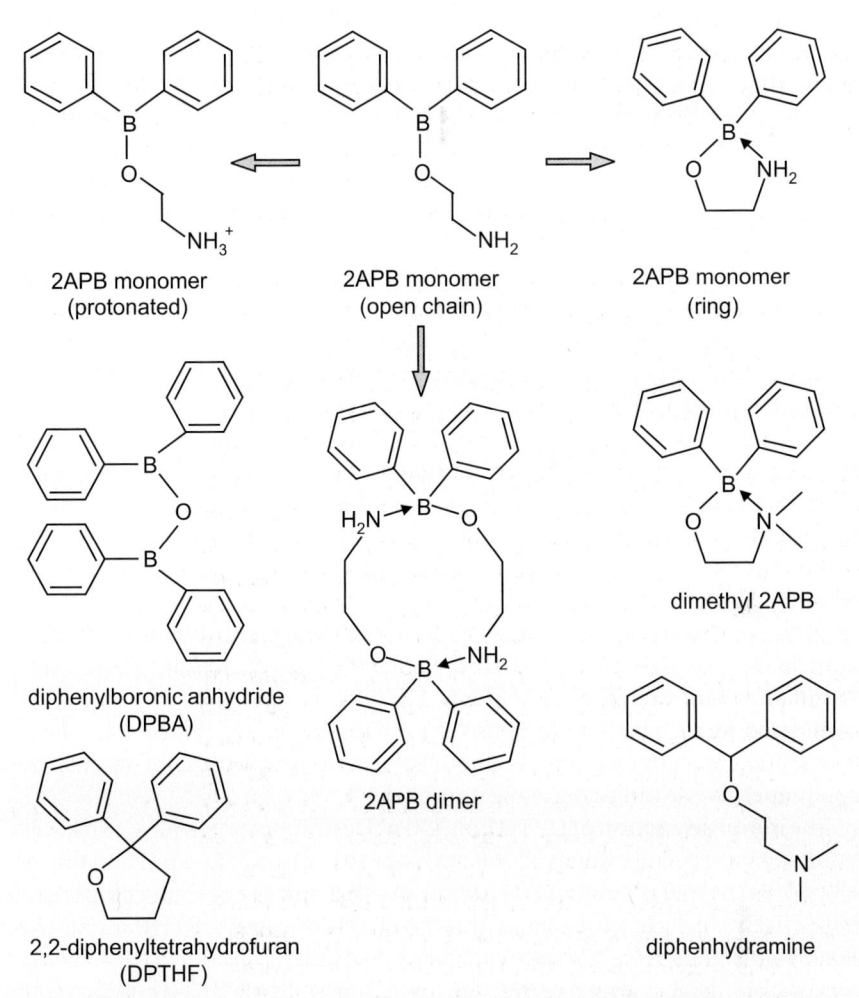

Fig. 1 Various forms of 2APB and several 2APB analogs. The nitrogen of the ethanolamine side chain on the 2APB monomer can become protonated (2APB monomer protonated) or form coordinate bonds with either an internal boron (2APB monomer ring) or the boron on another 2APB molecule (2APB dimer). Most data support the 2APB monomer ring as the predominant form of 2APB. The boron-containing 2APB analog diphenylboronic anhydride (DPBA) cannot be protonated. The nonboron-containing analog 2,2-diphenyltetrahydrofuran (DPTHF) is structurally related to the 2APB monomer ring. Diphenhydramine (Benadryl) is a nonboron-containing antihistamine that is structurally related to the 2APB monomer with the exception that 2APB has a primary amine and diphenhydramine has a tertiary amine. Diphenhydramine is also structurally related to dimethyl 2APB with the exception that dimethyl 2APB should exist predominantly in the ring form, whereas diphenhydramine is unable to form a ring and could be protonated. All of these molecules have a tetrahedral geometry at the equivalent position to the boron of 2APB

Ca^{2+} imaging, they showed that 100 µM DPBA, but not 100 µM DPTHF, caused a rise in intracellular Ca^{2+} in HEK293 cells expressing TRPV1, V2, or V3. Interestingly, 100 µM DPTHF inhibited the response evoked by 100 µM 2APB and 100 µM DPBA by 73.2% and 93.2%, respectively, in TRPV3 cells but not in TRPV1 and TRPV2 cells. Even at 500 µM DPTHF, the inhibition was 25.2% and 33.2% for TRPV1 and V2, respectively. Thus, DPBA activates TRPV1, V2, and V3 in a similar fashion as 2APB, but DPTHF has an opposite action and may be more selective for TRPV3.

In whole-cell patch clamp studies of TRPV3 expressed in HEK293 cells, Chung et al. (2005) demonstrated that 32 µM DPBA evoked outwardly rectifying currents that became dually rectifying with successive application of the drug. In addition, DPBA-evoked currents were blocked by DPTHF (58.9% and 90.8% inhibition at +80 and −80 mV, respectively) or 10 µM RR (99% at −80 mV). A dose–response analysis of DPBA yielded EC$_{50}$ values of 64.1 µM and 85.1 µM at +80 and −80 mV, respectively. The authors also noted an inhibitory effect at high (>100 µM) DPBA and 2APB concentrations, which is characterized by a decline in current amplitude at 1 mM as compared to 0.3 mM DPBA, a desensitization in the continued presence of the drug, and a strong rebound immediately after the washout. The IV relationship during the rebound appeared linear, indicative of a near-maximal activation of TRPV3. The most likely explanation is that DPBA has two sites of action, where one is stimulatory and the other inhibitory. Although a single site of action being modulated by an intrinsic "desensitization" pathway is also possible, the rebound at the washout and the fact that TRPV3 is sensitized but not desensitized upon repetitive stimulation make it unlikely.

The inhibitory action of DPTHF on TRPV3 also appeared to have two kinetic components. The IC$_{50}$ values at −80 mV were 6.0 µM and 151.5 µM and those at +80 mV were 10.0 µM and 226.7 µM for the first and the second components, respectively. In light of the facts that 2APB, DPBA, and DPTHF all blocked SOCE in platelets, and that they each have the ability to inhibit TRPV3 at high concentrations, it is possible that the low-affinity site of DPTHF is shared by 2APB and DPBA at high (>100 µM) concentrations and is inhibitory for all three compounds. This accounts for the rebound during washout. The stimulatory site may also be shared by the three compounds with similar, but nonetheless relatively high, affinities. Hence, they could compete for binding to the same site. However, a structural feature important for activation may be lacking in DPTHF, resulting in inhibition even though it is bound to the "stimulatory site," especially in the presence of other stimulating compounds. Indeed, 100 µM DPTHF was found to potentiate the heat-evoked response of TRPV3 (Chung et al. 2005). Thus, the complex activation/inhibition phenomenon observed with the 2APB analogs could be a result of dual bindings to separate stimulatory and inhibitory sites with different affinities.

4
Possible Mechanisms of Action of 2APB

How 2APB modulates TRP channels is still a mystery. The following sections consider the structural features of the compound, the possible target or binding site(s) on the channel subunits or other protein components associated with the channel complex, and the environment, mainly the lipid bilayers, that surrounds the channels.

4.1
Structural Considerations

Due to the ability of 2APB to form an N→B coordinate bond, this molecule can exist in several different states (Fig. 1). Analyses on 2APB and its analogs by crystallography (Rettig and Trotter 1976), pK_b values in aqueous solution (Dobrydneva and Blackmore 2001), and nuclear magnetic resonance (NMR) (Dobrydneva et al. 2006) support the idea that 2APB exists predominantly in the monomer ring structure as shown in Fig. 1, with the ethanolamine side chain forming a five-membered boroxazolidine heterocyclic ring (Strang et al. 1989; Dobrydneva and Blackmore 2001). The fact that 2APB can block the intracellularly located IP3Rs is consistent with the monomer ring structure. The open chain form would not be expected to pass through the membrane readily because the nitrogen of the ethanolamine side chain is most likely protonated in order to neutralize the free electron pair. 2APB can also form dimers (Nöth 1970; van Rossum et al. 2000). It should be considered that the ability of 2APB to switch between these different forms may also be important for its functional ability to activate or block TRP channels.

The boron on 2APB allows for the formation of coordinate bonds between the electrophilic boron and nucleophiles. 2APB and its boron-containing analogs could form either N→B or O→B coordinate bonds with amino acids that contain amines, imidazoles, and carboxyl groups on TRP channels. Interestingly, even though dimethyl 2APB (Fig. 1) blocked the thrombin-induced SOCE in platelets (Dobrydneva et al. 2006), a nonboron analog with two methyl groups on the secondary amine nitrogen, diphenhydramine, was ineffective in blocking SOCE in platelets (Dobrydneva and Blackmore 2001) and in activating TRPV3 expressed in HEK293 cells (Chung et al. 2004b). It would be interesting to test if dimethyl 2APB activates TRPV3. A positive effect would suggest that boron and/or ring formation is necessary for the stimulatory action of 2APB analogs, since the tertiary carbon and the secondary amine nitrogen of diphenhydramine are unable to make the ring closure like the N→B coordinate bond of the 2APB monomer ring (Fig. 1). The blocking and potentiating effects of DPTHF on TRPV3 (Chung et al. 2005), as well as the ability of several other nonboron analogs of 2APB to block SOCE in platelets (Dobrydneva et al. 2006), suggests that the boron may not be necessary, at least for binding to

TRP channels or a critical auxiliary component(s) of the channel complex. However, without the boron, the compound may not be sufficient to activate the channel because heating appears to be necessary to reveal the stimulatory effect of the nonboron analog, DPTHF, on TRPV3 (Chung et al. 2005).

4.2
Site(s) of Action

Several lines of evidence favor the existence of at least two binding sites or sites of action for 2APB and its analogs, with one being stimulatory and the other inhibitory. First, at low concentrations, 2APB potentiated a native store-operated channel that is normally blocked by higher concentrations (Prakriya and Lewis 2001). Second, at above 100 µM, 2APB- or DPBA-evoked TRPV3 currents tended to reach the maximum and then decrease in mid-response (Chung et al. 2005). This effect became more evident with increasing drug concentrations and led to an apparent decline in maximal current amplitude at 1 mM. Third, even though DPTHF is predominantly inhibitory, it potentiated the heat-induced TRPV3 currents (Chung et al. 2005). Fourth, the inhibition of DPBA-evoked TRPV3 currents by DPTHF extended over several orders of magnitude and had two kinetic components, indicative of two or more sites and/or mechanisms of action. One of these inhibitory actions could result from competition with 2APB or DPBA for binding to the stimulatory site. This two-sites model could explain the concentration-dependent dual actions of the 2APB analogs. If the model holds true, modification of the 2APB structure may generate analogs with greater differences in the affinities to the stimulatory and the inhibitory sites and for different TRP subtypes, allowing for highly specific agonists and/or antagonists for some TRP channels. This exciting possibility warrants an extensive modification of 2APB analogs and evaluation of their effects on multiple TRP channels.

The plasma membrane side of action for 2APB is most likely extracellular. This is supported by the failure of intracellular injection of 2APB to activate any TRPV1 current in *Xenopus* oocytes and intracellular infusion of 2APB and DPBA through patch pipettes to activate TRPV3 expressed in HEK293 cells in whole-cell experiments (Hu et al. 2004; Chung et al. 2005). In HEK293 cells, this same manipulation also failed to inhibit TRPC3 and TRPC5 channels (Trebak et al. 2002; Xu et al. 2005). In all cases, subsequent application of 2APB or DPBA in the bath had elicited either stimulation or inhibitory responses of the TRP channels. In excised inside-out patches, 2APB also failed to inhibit TRPC5 channel activity whereas in outside-out patches, the same concentration of 2APB effectively blocked the channel (Xu et al. 2005). One exception is that TRPV3 is activated by 2APB applied to the intracellular side of the inside-out patches (Chung et al. 2004b). This could be explained by the notion that the membrane permeable 2APB can accumulate at the pipette side (outside) even though it is applied to the exposed side of the membrane patch. Similar accumulation of 2APB at the extracellular side will not occur in the outside-out or whole-cell configurations as the drug will be diluted by the bath solution or washed away by perfusion. However, this does not explain why 2APB failed to inhibit TRPC5 in the inside-out patches.

The available data also suggest that 2APB acts at a different site(s) from those of known TRPV1 agonists. First, TRPV2 and V3 are not activated by capsaicin but they are activated by 2APB. Second, while capsazepine, a competitive antagonist of capsaicin, completely inhibited the capsaicin-induced response, it only partially blocked the 2APB-evoked activation of TRPV1. Third, superimposition of 2APB and capsaicin invoked responses that are more than additive to those elicited by each drug alone. This similar synergistic effect was also observed between 2APB and weak acid, indicating that different mechanisms are involved for the activation of TRPV1 by 2APB, capsaicin, and protons. Most likely, a similar 2APB-binding pocket exists for TRPV1, V2, and V3, but it is very different from the vanilloid-binding pocket, which is mostly intracellular (see Tominaga and Tominaga 2005 for a review on vanilloid binding sites).

4.3
Effects on Membrane Properties

Several observations suggest that membrane properties strongly influence the activities of TRPV channels. First, TRPV1 is activated by a large number of lipophilic molecules, many of which bear no structural similarity (Calixto et al. 2005). Second, increasing the cholesterol content in HEK293 cells shifted the temperature threshold of TRPV1 from 42°C to 46°C (Liu et al. 2003). Third, phosphatidylinositol bisphosphate (PIP_2) has been proposed to hold TRPV1 in an inhibitory state (Prescott and Julius 2003). Fourth, arachidonic acid and other unsaturated fatty acids potentiate the 2APB-induced activation of TRPV3 (Hu et al. 2006). The great variability in the fatty acids used, to include triple bonded analogs may suggest a "loosely" specific activation mechanism that could be accounted for if these molecules cause a change in the membrane biophysical properties that are "sensed" by the channel. Polyunsaturated fatty acids also regulate TRPV channels in *Caenorhabditis elegans* (Kahn-Kirby et al. 2004) and TRPC channels in *Drosophila* (Chyb et al. 1999). Fifth, it has been proposed that mechano- and thermosensitive channels may be modulated by a common mechanism in a membrane-delimited fashion (Kung 2005).

2APB is a lipophilic molecule that possibly could accumulate in the membrane at high concentrations. There are several ways in which a lipophilic molecule such as 2APB could modulate TRP channels. First, when accumulated at high concentrations in the membrane, 2APB and its analogs could disrupt the interaction between various inhibitory phospholipids, such as PIP_2. Second, the observation that 2APB and its analogs affect so many ionic channels and other membrane proteins suggests that 2APB could act in a similar fashion as general anesthetics. A property of the anesthetics is that they usually affect the gating of many different ion channels by altering membrane properties (Antkowiak 2001). It has also been proposed that the best anesthetics accumulate at the membrane-water interface (North and Cafiso 1997). The high degree of lipophilicity along with the polarity of the N→B coordinate bond could result in the accumulation of 2APB in this region. Exactly how 2APB affects different membrane properties remains to be investigated.

5
Concluding Remarks

Numerous studies have documented the effects of 2APB and its analogs on membrane channels. However, the mechanisms by which 2APB regulate ion channels remain a mystery. New evidence suggests that the action of 2APB on TRP channels is not universal. While several TRP channels are inhibited, at least three of them, TRPV1-3, are stimulated by 2APB. Some TRP channels are unaffected by 2APB and many more remain to be tested. The findings that 2APB activates TRPV1-3, while its analog DPTHF shows some selectivity for TRPV3 over TRPV1 and V2, make it promising that specific ligands may be made for TRPV2 and V3 through modification of various 2APB analogs. The identification of specific ligands for TRPV1 (e.g., capsaicin and resiniferatoxin) and TRPV4 (4αPDD) have not only facilitated the identification of physiological processes that these channels are involved in, but also made electrophysiological characterization of these channels more feasible. The recent increase in TRPV3-specific studies is directly related to the identification of 2APB as an agonist for TRPV1 and V3 activation (Gu et al. 2005; Chung et al. 2005; Zimmermann et al. 2005; Guatteo et al. 2005). More specific drugs would certainly accelerate the discovery of the physiological functions and mechanisms of regulation of these amazing channels.

Acknowledgements Supported by US National Institutes of Health grants NS042183 and P30-NS045758. CKC is a recipient of the Meier Schlesinger Graduate Fellowship.

References

Antkowiak B (2001) How do general anaesthetics work? Naturwissenschaften 88:201–213
Bender F, Mederos Y Schnitzler M, Li Y, Ji A, Weihe E, Gudermann T, Schafer M (2005) The temperature-sensitive ion channel TRPV2 is endogenously expressed and functional in the primary sensory cell line F-11. Cell Physiol Biochem 15:183–194
Calixto JB, Kassuya CA, Andre E, Ferreira J (2005) Contribution of natural products to the discovery of the transient receptor potential (TRP) channels family and their functions. Pharmacol Ther 106:179–208

Chung MK, Lee H, Mizuno A, Suzuki M, Caterina MJ (2004a) TRPV3 and TRPV4 mediate warmth-evoked currents in primary mouse keratinocytes. J Biol Chem 279:21569–21575

Chung MK, Lee H, Mizuno A, Suzuki M, Caterina MJ (2004b) 2-Aminoethoxydiphenyl borate activates and sensitizes the heat-gated ion channel TRPV3. J Neurosci 24:5177–5182

Chung MK, Guler AD, Caterina MJ (2005) Biphasic currents evoked by chemical or thermal activation of the heat-gated ion channel, TRPV3. J Biol Chem 280:15928–15941

Chyb S, Raghu P, Hardie RC (1999) Polyunsaturated fatty acids activate the Drosophila light-sensitive channels TRP and TRPL. Nature 397:255–259

Clapham DE, Runnels LW, Strubing C (2001) The TRP ion channel family. Nat Rev Neurosci 2:387–396

Delmas P, Wanaverbecq N, Abogadie FC, Mistry M, Brown DA (2002) Signaling microdomains define the specificity of receptor-mediated InsP(3) pathways in neurons. Neuron 34:209–220

Diver JM, Sage SO, Rosado JA (2001) The inositol trisphosphate receptor antagonist 2-aminoethoxydiphenylborate (2-APB) blocks Ca^{2+} entry channels in human platelets: cautions for its use in studying Ca^{2+} influx. Cell Calcium 30:323–329

Dobrydneva Y, Blackmore P (2001) 2-Aminoethoxydiphenyl borate directly inhibits store-operated calcium entry channels in human platelets. Mol Pharmacol 60:541–552

Dobrydneva Y, Abelt CJ, Dovel B, Thadigiri CM, Williams RL, Blackmore PF (2006) 2-Aminoethoxydiphenyl borate as a prototype drug for a group of structurally related calcium channel blockers in human platelets. Mol Pharmacol 69:247–256

Doly S, Fischer J, Salio C, Conrath M (2004) The vanilloid receptor-1 is expressed in rat spinal dorsal horn astrocytes. Neurosci Lett 357:123–126

Gu Q, Lin RL, Hu HZ, Zhu MX, Lee LY (2005) 2-aminoethoxydiphenyl borate stimulates pulmonary C neurons via the activation of TRPV channels. Am J Physiol Lung Cell Mol Physiol 288:L932–L941

Guatteo E, Chung KK, Bowala TK, Bernardi G, Mercuri NB, Lipski J (2005) Temperature sensitivity of dopaminergic neurons of the substantia nigra pars compacta: involvement of transient receptor potential channels. J Neurophysiol 94:3069–3080

Hanano T, Hara Y, Shi J, Morita H, Umebayashi C, Mori E, Sumimoto H, Ito Y, Mori Y, Inoue R (2004) Involvement of TRPM7 in cell growth as a spontaneously activated Ca^{2+} entry pathway in human retinoblastoma cells. J Pharmacol Sci 95:403–419

Hermosura MC, Monteilh-Zoller MK, Scharenberg AM, Penner R, Fleig A (2002) Dissociation of the store-operated calcium current ICRAC and the Mg-nucleotide-regulated metal ion current MagNuM. J Physiol 539:445–458

Hu HZ, Gu Q, Wang C, Colton CK, Tang J, Kinoshita-Kawada M, Lee LY, Wood JD, Zhu MX (2004) 2-Aminoethoxydiphenyl borate is a common activator of TRPV1, TRPV2, and TRPV3. J Biol Chem 279:35741–35748

Hu HZ, Xiao R, Wang C, Gao N, Colton CK, Wood JD, Zhu MX (2006) Potentiation of TRPV3 channel function by unsaturated fatty acids. J Cell Physiol 208:201–212

Jia Y, Wang X, Varty L, Rizzo CA, Yang R, Correll CC, Phelps PT, Egan RW, Hey JA (2004) Functional TRPV4 channels are expressed in human airway smooth muscle cells. Am J Physiol Lung Cell Mol Physiol 287:L272–L278

Kahn-Kirby AH, Dantzker JL, Apicella AJ, Schafer WR, Browse J, Bargmann CI, Watts JL (2004) Specific polyunsaturated fatty acids drive TRPV-dependent sensory signaling in vivo. Cell 119:889–900

Kashiba H, Uchida Y, Takeda D, Nishigori A, Ueda Y, Kuribayashi K, Ohshima M (2004) TRPV2-immunoreactive intrinsic neurons in the rat intestine. Neurosci Lett 366:193–196

Kim CS, Kawada T, Kim BS, Han IS, Choe SY, Kurata T, Yu R (2003) Capsaicin exhibits anti-inflammatory property by inhibiting IkB-a degradation in LPS-stimulated peritoneal macrophages. Cell Signal 15:299–306

Koulen P, Cai Y, Geng L, Maeda Y, Nishimura S, Witzgall R, Ehrlich BE, Somlo S (2002) Polycystin-2 is an intracellular calcium release channel. Nat Cell Biol 4:191–197

Kung C (2005) A possible unifying principle for mechanosensation. Nature 436:647–654

Lievremont JP, Bird GS, Putney JW Jr (2005) Mechanism of inhibition of TRPC cation channels by 2-aminoethoxydiphenylborane. Mol Pharmacol 68:758–762

Liu B, Hui K, Qin F (2003) Thermodynamics of heat activation of single capsaicin ion channels VR1. Biophys J 85:2988–3006

Lucas P, Ukhanov K, Leinders-Zufall T, Zufall F (2003) A diacylglycerol-gated cation channel in vomeronasal neuron dendrites is impaired in TRPC2 mutant mice: mechanism of pheromone transduction. Neuron 40:551–561

Ma HT, Venkatachalam K, Rys-Sikora KE, He LP, Zheng F, Gill DL (2003) Modification of phospholipase C-gamma-induced Ca^{2+} signal generation by 2-aminoethoxydiphenyl borate. Biochem J 376:667–676

Maruyama T, Kanaji T, Nakade S, Kanno T, Mikoshiba K (1997) 2APB, 2-aminoethoxydiphenyl borate, a membrane-penetrable modulator of Ins(1,4,5)P3-induced Ca^{2+} release. J Biochem (Tokyo) 122:498–505

Moqrich A, Hwang SW, Earley TJ, Petrus MJ, Murray AN, Spencer KS, Andahazy M, Story GM, Patapoutian A (2005) Impaired thermosensation in mice lacking TRPV3, a heat and camphor sensor in the skin. Science 307:1468–1472

Muraki K, Iwata Y, Katanosaka Y, Ito T, Ohya S, Shigekawa M, Imaizumi Y (2003) TRPV2 is a component of osmotically sensitive cation channels in murine aortic myocytes. Circ Res 93:829–838

Nilius B, Prenen J, Vennekens R, Hoenderop JG, Bindels RJ, Droogmans G (2001) Pharmacological modulation of monovalent cation currents through the epithelial Ca^{2+} channel ECaC1. Br J Pharmacol 134:453–462

Nöth H (1970) Some recent developments in boron-nitrogen chemistry. In: Brotherton RJ, Steinberg H (eds) Progress in boron chemistry. Pergamon Press, New York, pp 211–311

North C, Cafiso DS (1997) Contrasting membrane localization and behavior of halogenated cyclobutanes that follow or violate the Meyer-Overton hypothesis of general anesthetic potency. Biophys J 72:1754–1761

Patapoutian A, Peier AM, Story GM, Viswanath V (2003) ThermoTRP channels and beyond: mechanisms of temperature sensation. Nat Rev Neurosci 4:529–539

Peier AM, Reeve AJ, Andersson DA, Moqrich A, Earley TJ, Hergarden AC, Story GM, Colley S, Hogenesch JB, McIntyre P, Bevan S, Patapoutian A (2002) A heat-sensitive TRP channel expressed in keratinocytes. Science 296:2046–2049

Prakriya M, Lewis RS (2001) Potentiation and inhibition of Ca^{2+} release-activated Ca^{2+} channels by 2-aminoethyldiphenyl borate (2-APB) occurs independently of IP3 receptors. J Physiol 536:3–19

Prakriya M, Lewis RS (2002) Separation and characterization of currents through store-operated CRAC channels and Mg^{2+}-inhibited cation (MIC) channels. J Gen Physiol 119:487–507

Prescott ED, Julius D (2003) A modular PIP_2 binding site as a determinant of capsaicin receptor sensitivity. Science 300:1284–1288

Rettig SJ, Trotter J (1976) Crystal and molecular structure of B, B-bis(p-tolyl)boroxazolidine and the orthorhombic form of B,B-diphenylboroxazolidine. Can J Chem 54:3130–3141

Rosado JA, Brownlow SL, Sage SO (2002) Endogenously expressed Trp1 is involved in store-mediated Ca^{2+} entry by conformational coupling in human platelets. J Biol Chem 277:42157–42163

Schindl R, Kahr H, Graz I, Groschner K, Romanin C (2002) Store depletion-activated CaT1 currents in rat basophilic leukemia mast cells are inhibited by 2-aminoethoxydiphenyl borate. Evidence for a regulatory component that controls activation of both CaT1 and CRAC (Ca^{2+} release-activated Ca^{2+} channel) channels. J Biol Chem 277:26950–26958

Strang CJ, Henson E, Okamoto Y, Paz MA, Gallop PM (1989) Separation and determination of alpha -amino acids by boroxazolidone formation. Anal Biochem 178:278–286

Sydorenko V, Shuba Y, Thebault S, Roudbaraki M, Lepage G, Prevarskaya N, Skryma R (2003) Receptor-coupled, DAG-gated Ca^{2+}-permeable cationic channels in LNCaP human prostate cancer epithelial cells. J Physiol 548:823–836

Tominaga M, Tominaga T (2005) Structure and function of TRPV1. Pflugers Arch 451:143–150

Tozzi A, Bengtson CP, Longone P, Carignani C, Fusco FR, Bernardi G, Mercuri NB (2003) Involvement of transient receptor potential-like channels in responses to mGluR-I activation in midbrain dopamine neurons. Eur J Neurosci 18:2133–2145

Trebak M, Bird GS, McKay RR, Putney JW Jr (2002) Comparison of human TRPC3 channels in receptor-activated and store-operated modes. Differential sensitivity to channel blockers suggests fundamental differences in channel composition. J Biol Chem 277:21617–21623

van Rossum DB, Patterson RL, Ma HT, Gill DL (2000) Ca^{2+} entry mediated by store depletion, S-nitrosylation, and TRP3 channels. Comparison of coupling and function. J Biol Chem 275:28562–28568

Voets T, Prenen J, Fleig A, Vennekens R, Watanabe H, Hoenderop JG, Bindels RJ, Droogmans G, Penner R, Nilius B (2001) CaT1 and the calcium release-activated calcium channel manifest distinct pore properties. J Biol Chem 276:47767–47770

Xu H, Ramsey IS, Kotecha SA, Moran MM, Chong JA, Lawson D, Ge P, Lilly J, Silos-Santiago I, Xie Y, DiStefano PS, Curtis R, Clapham DE (2002) TRPV3 is a calcium-permeable temperature-sensitive cation channel. Nature 418:181–186

Xu SZ, Zeng F, Boulay G, Grimm C, Harteneck C, Beech DJ (2005) Block of TRPC5 channels by 2-aminoethoxydiphenyl borate: a differential, extracellular and voltage-dependent effect. Br J Pharmacol 145:405–414

Yue L, Peng JB, Hediger MA, Clapham DE (2001) CaT1 manifests the pore properties of the calcium-release-activated calcium channel. Nature 410:705–709

Zhu X, Jiang M, Peyton M, Boulay G, Hurst R, Stefani E, Birnbaumer L (1996) trp, a novel mammalian gene family essential for agonist-activated capacitative Ca^{2+} entry. Cell 85:661–671

TRPV4

T. D. Plant[1] (✉) · R. Strotmann[2]

[1]Institut für Pharmakologie u. Toxikologie, FB-Medizin, Philipps-Universität Marburg, 35032 Marburg, Germany
plant@staff.uni-marburg.de

[2]Institut für Biochemie, Abteilung Molekulare Biochemie, Medizinische Fakultät, Universität Leipzig, 04103 Leipzig, Germany

1	Basic Features of the TRPV4 Gene and Protein and the TRPV4 Expression Pattern	190
1.1	Gene and Protein	190
1.2	Expression Pattern	191
2	Ion Channel Properties	192
2.1	Current–Voltage Relation and Conductance	192
2.2	Ion Selectivity	194
3	Modes of TRPV4 Activation	194
3.1	Spontaneous TRPV4 Activity	195
3.2	Osmosensitivity of TRPV4-Expressing Cells	195
3.3	Sensitivity of TRPV4 to Lipid Messengers	196
3.4	Mechanosensitivity of TRPV4	196
3.5	Activation of TRPV4 by 4α-Phorbol Ester Derivatives	197
3.6	Temperature Sensitivity of TRPV4	197
3.7	Phorbol Esters and Temperature Use a Common Activation Mechanism	197
3.8	Regulation of TRPV4 by Ca^{2+}	198
3.9	Activation of TRPV4 by H^+ and Citrate	199
4	Pharmacology of TRPV4	199
5	Biological Relevance and Emerging/Established Biological Roles for TRPV4	200
5.1	TRPV4 in Systemic Osmoregulation	200
5.2	TRPV4 in Mechano-/Osmosensation	200
5.3	TRPV4 as a Thermosensor	201
5.4	TRPV4 in Vascular Regulation	201
5.5	Other Possible Roles of TRPV4	202
References		203

Abstract TRPV4 is a non-selective cation channel subunit expressed in a wide variety of tissues. TRP channels are formed by a tetrameric complex of channel subunits. The available evidence suggests that TRPV4 cannot form heteromultimers with other TRPV isoforms, and that TRPV4-containing channels are homotetramers. These channels have a characteristic outwardly rectifying current–voltage relation, and are 5–10 times more permeable for Ca^{2+} than for Na^+. TRPV4 can be activated by a wide range of stimuli including physical (cell swelling, heat, mechanical stimulation) and chemical stimuli (endocannabinoids, arachidonic acid, and, surprisingly, 4α-phorbol esters). Activation by swelling and endocannabinoids involves cytochrome P450 epoxygenase-dependent arachidonic acid metabolism to the epoxyeicosatrienoic acids (EETs). Heat and 4α-phorbol esters also seem to share a common mechanism of activation, but the endogenous messenger involved in the response to heat has not yet been identified. Ca^{2+} acting from the intracellular side can have both potentiating and inhibitory effects on channel activity and is involved in channel activation and inactivation. Given its wide expression and the variety of activatory stimuli, TRPV4 is likely to play a number of physiological roles. Studies with $TRPV4^{-/-}$ mice suggest a role for the channel in the regulation of body osmolarity, mechanosensation, temperature sensing, vascular regulation and, possibly, hearing.

Keywords TRP channel · TRPV channel · TRPV4 · Non-selective cation channel · Calcium entry · Messenger-gated · Mechanosensation · Thermosensation · Osmoregulation

1
Basic Features of the TRPV4 Gene and Protein and the TRPV4 Expression Pattern

1.1
Gene and Protein

TRPV4 was found by screening expressed sequence tag databases for sequences with similarity to TRPV1, TRPV2 and the *Caenorhabditis elegans* TRPV isoform Osm-9. TRPV4 was cloned from the kidney, hypothalamus and auditory epithelium and given a number of names: Osm-9-like TRP channel 4 (OTRPC4, Strotmann et al. 2000), vanilloid receptor-related osmotically activated channel (VR-OAC, Liedtke et al. 2000), TRP12 (Wissenbach et al. 2000) and vanilloid receptor-like channel 2 (VRL-2, Delany et al. 2001). The gene for human TRPV4 is localized on chromosome 12q23–q24.1 and has 15 exons (Liedtke et al. 2000; Wissenbach et al. 2000; Delany et al. 2001). These exons code for a full-length protein with 871 amino acids (aa). The putative transmembrane structure is similar to that of other transient receptor potential (TRP) channels with intracellular N- and C-termini, six membrane-spanning helices (S1–S6), and a pore-forming loop between S5 and S6. TRPV4 has at least three ankyrin domains in its cytosolic N terminus (Fig. 1). From the evidence currently available, TRPV4 is likely to form homotetramers (Hellwig et al. 2005).

Five splice variants of TRPV4 have been described (Arniges et al. 2006). The full-length form of TRPV4, which has been studied extensively, has been denoted TRPV4A. Compared to this, TRPV4B lacks exon 6 (Δ384–444 aa), TRPV4C lacks exon 5 (Δ237–284 aa), TRPV4D has a small deletion inside exon 2 (Δ27–61 aa), and TRPV4E lacks both exons 5 and 7 (Δ237–284 and Δ384–444 aa). Only two of these splice variants, TRPV4A and TRPV4D, traffic

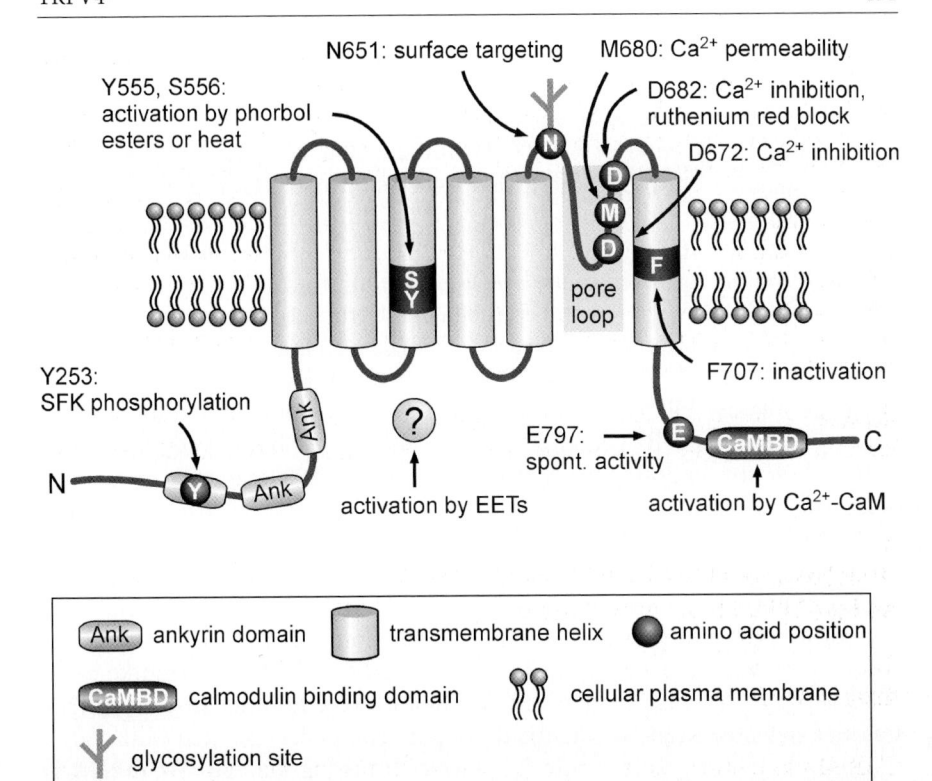

[Figure: Structural features of TRPV4 showing transmembrane topology with labeled residues:
- Y555, S556: activation by phorbol esters or heat
- N651: surface targeting
- M680: Ca²⁺ permeability
- D682: Ca²⁺ inhibition, ruthenium red block
- D672: Ca²⁺ inhibition
- Y253: SFK phosphorylation
- F707: inactivation
- E797: spont. activity
- activation by EETs
- activation by Ca²⁺-CaM
Legend: Ank = ankyrin domain, CaMBD = calmodulin binding domain, transmembrane helix, amino acid position, cellular plasma membrane, glycosylation site]

Fig. 1 Structural features of TRPV4. Transmembrane topology of TRPV4 indicating characteristic regions of the protein and amino acids described to be involved in channel regulation or determination of channel properties

correctly to the cell membrane. The others are retained in the endoplasmic reticulum, probably because splicing leads to the loss of parts of the ankyrin domains and these are important in trafficking (Arniges et al. 2006). Recently, a glycosylation site, Asn^{651}, close to S5 in the pore-forming loop between S5 and S6 (Fig. 1) has been identified that influences trafficking of TRPV4 (Xu et al. 2005). Mutation of Asn^{651} to Gln prevented glycosylation and increased both cell surface expression of TRPV4 and functional responses mediated by TRPV4.

1.2
Expression Pattern

TRPV4 has been found in many tissues which have various physiological functions. In multiple tissue Northern blots, TRPV4 messenger RNA (mRNA) has been detected in heart, endothelium, brain, liver, placenta, lung, trachea and salivary gland (Liedtke et al. 2000; Strotmann et al. 2000; Wissenbach et al. 2000; Delany et al. 2001).

Strong expression is detected in the epithelia of the kidney, particularly in the distal tubule (Liedtke et al. 2000; Strotmann et al. 2000; Delany et al. 2001). A more detailed study indicated that expression of TRPV4 is restricted to nephron segments with a constitutively or conditionally [vasopressin (AVP)-dependent] low water permeability (Tian et al. 2004). According to this study, TRPV4 is mainly localized to the basolateral membrane of the renal epithelial cells. TRPV4 is also expressed in epithelia of the trachea, lung, oviduct and the stria vascularis of the cochlea (Liedtke et al. 2000; Delany et al. 2001; Andrade et al. 2005).

In the airway, TRPV4 is not only expressed in the epithelium, but also in smooth muscle cells (Jia et al. 2004). Similarly, in blood vessels, TRPV4 is found in the endothelium (Wissenbach et al. 2000) and in some vascular smooth muscle cells, such as those of the cerebral arteries (Earley et al. 2005).

In the brain, in situ hybridization shows expression of TRPV4 mRNA in neurones of the circumventricular nuclei of the hypothalamus and in ependymal cells of the choroid plexus of the lateral and fourth, but not third, ventricles, and in scattered neurones in other regions of the brain (Liedtke et al. 2000; Güler et al. 2002; Liedtke and Friedman 2003). In other studies, mRNA was also detected in the substantia nigra pars compacta (Guatteo et al. 2005). Consistent with a possible role in sensory transduction, TRPV4 mRNA is present in large sensory neurones of the trigeminal ganglion and dorsal root ganglia (Liedtke et al. 2000; Delany et al. 2001; Alessandri-Haber et al. 2003), and in the inner ear, in inner and outer hair cells of the organ of Corti, and in hair cells of the semicircular canals and utricles (Liedtke et al. 2000; Takumida et al. 2005). TRPV4 is also expressed in keratinocytes where it may play a role in sensory transduction (Güler et al. 2002; Chung et al. 2003, 2004). TRPV4 protein was found in sympathetic ganglia, and in sympathetic and parasympathetic nerve fibres in a number of tissues (Delany et al. 2001).

2
Ion Channel Properties

2.1
Current–Voltage Relation and Conductance

Heterologously expressed and native TRPV4 forms Ca^{2+}-permeable, non-selective cation channels. Consistent with these permeability properties, with physiological cation concentrations, channel currents reverse at potentials just positive to 0 mV. The current–voltage (IV) relation displays outward rectification (Fig. 2) and has a similar shape to that of TRPV3, with a higher proportion of inward current to outward current than e.g. TRPV1 or TRPV2, which rectify more strongly. Currents through TRPV4 also do not remain low at negative membrane potentials, but increase with stronger membrane hyperpolarization. The outwardly rectifying shape of the IV relation in physiological solutions results from a block by extracellular Ca^{2+} ions. Reducing $[Ca^{2+}]_o$ leads to a progressive loss of rectification, until the IV relation becomes linear in Ca^{2+}-free media (Voets et al. 2002). Raising $[Ca^{2+}]_o$ above physiolog-

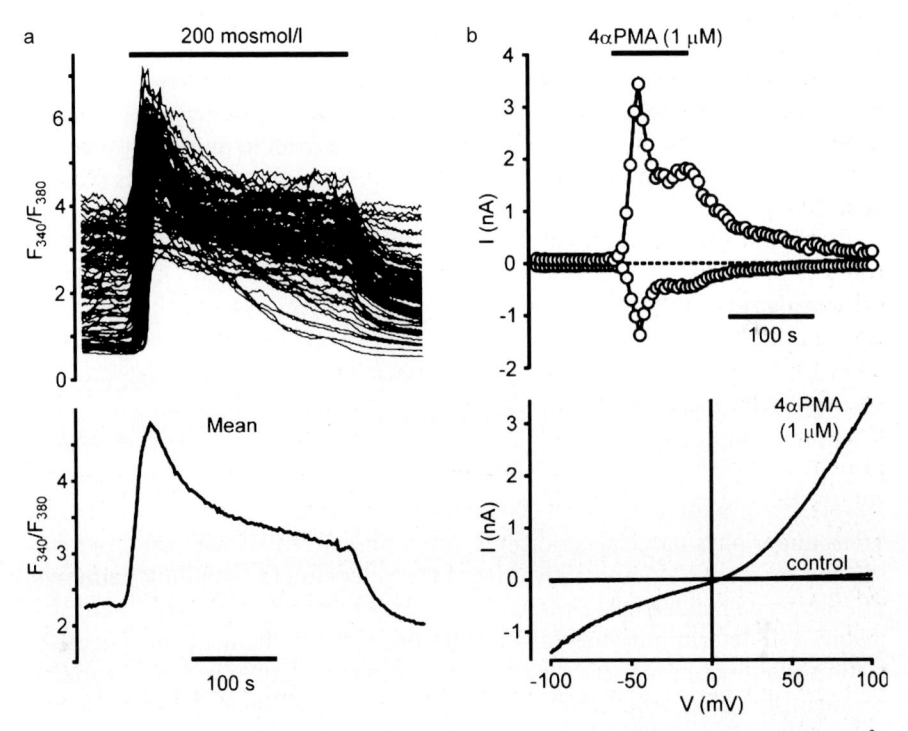

Fig. 2a,b Basic functional properties of heterologously expressed TRPV4. **a** Intracellular Ca^{2+} concentration in TRPV4-expressing cells. The *upper panel* shows traces from individual cells, the *lower panel* the mean response. The data illustrate that many, but not all, TRPV4-expressing cells show an elevated Ca^{2+} level before stimulation, resulting in an elevated mean basal $[Ca^{2+}]_i$. A number of cells do not display elevated Ca^{2+} levels (fluorescence ratios of around 0.8 to 0.9). Application of a hypotonic solution (200 mosmol/l) reversibly increases $[Ca^{2+}]_i$ in all cells. The data are fluorescence ratios (F_{340}/F_{380}) measured from TRPV4-expressing cells loaded with fura-2AM. **b** Characteristics of membrane currents through TRPV4. The *upper panel* shows the time course of currents at −100 (negative current values) and +100 mV (positive current values), the *lower panel* current–voltage (IV) relationships of currents before (control) and close to the maximum of channel activation by the 4α-phorbol ester derivative 4α-phorbol myristate acetate (4αPMA). The response to 4αPMA is rapid and transient. The IV relationship in the presence of 4αPMA shows the outwardly rectifying shape that is characteristic of TRPV4 and the reversal potential just positive to 0 mV

ical values increases rectification. The block by Ca^{2+} results from binding to negatively charged amino acids (Asp^{672} and Asp^{682}) in the pore loop (Fig. 1; Voets et al. 2002). Since these effects are mediated by Ca^{2+} binding to the channel pore, they also influence the single channel properties of TRPV4. In cells heterologously expressing TRPV4 as well as in native endothelial cells, the single channel conductance, which was consistent in different patch configurations with different activating stimuli, was larger for outward currents (88–105 pS) than for inward currents (30–61 pS; Strotmann et al. 2000; Watanabe et al. 2002b). The higher values of the conductance for inward currents were obtained with nominally Ca^{2+}-free solutions bathing the extracellular face of the patch (Watanabe et al. 2002b), which is consistent with the block by Ca^{2+} described above. In another study on heterologously expressed TRPV4, a much higher value of around 310 pS for outward currents was obtained (Liedtke et al. 2000), and in cell-attached patches from keratinocytes, values of 140 and 150 pS compare with a value of 90 pS in inside-out patches (Chung et al. 2003).

2.2
Ion Selectivity

TRPV4 is cation-selective but does not discriminate well between cations. The relative permeabilities for monovalent cations are $P_K:P_{Cs}:P_{Na}:P_{Li}$=1.3–2:1.3:1:0.9 (Nilius et al. 2001; Voets et al. 2002; Watanabe et al. 2002a). For divalent cations, the relative permeabilities are $P_{Ca}:P_{Sr}:P_{Ba}:P_{Mg}:P_{Na}$=6–10:9:0.7–7:2–3:1 (Strotmann et al. 2000, 2003; Voets et al. 2002; Watanabe et al. 2002a). Thus, under physiological conditions, activation of TRPV4 will result in a significant influx of Ca^{2+} and activation of Ca^{2+}-dependent signalling pathways. The aspartate residues that influence Ca^{2+} block (Asp^{672} and Asp^{682}) are also involved in determining the Ca^{2+} permeability of the channel, since neutralization of either one of these aspartates reduced the relative Ca^{2+} permeability, and neutralization of both resulted in a stronger reduction in Ca^{2+} permeability and a modification of the permeability to monovalent cations (Voets et al. 2002). Replacement of Met^{680}, located between the two aspartates (Fig. 1), by a negatively charged aspartate strongly reduced the divalent cation permeability (Voets et al. 2002).

3
Modes of TRPV4 Activation

TRPV4 can be activated by a wide range of signals that include physical stimuli like heat, extracellular osmolarity and mechanical stimulation, and chemical stimuli like lipid agonists.

3.1
Spontaneous TRPV4 Activity

Cells heterologously expressing TRPV4 can be divided into two groups; those that show spontaneous channel activity and those that show no basal activity (Fig. 2a; Strotmann et al. 2000). Like the spontaneously active cells, the latter do, however, respond to channel activators. Spontaneous activity, which is evidenced by an elevated $[Ca^{2+}]_i$ (Fig. 2a) and spontaneous ionic currents, is dependent on the extracellular $[Ca^{2+}]$, and the spontaneous currents are strongly reduced in nominally Ca^{2+}-free solutions (Strotmann et al. 2003). There are discrepancies between studies from different laboratories regarding the presence of spontaneous TRPV4 activity, but why there have been differences is unclear.

3.2
Osmosensitivity of TRPV4-Expressing Cells

The first modulator of TRPV4 activity that was found was changes in the extracellular osmolarity. Increases in osmolarity from 300 mosmol/l reduced TRPV4 activity (Strotmann et al. 2000; Nilius et al. 2001), whereas reductions (hypotonic solutions) led to an increase in activity (Liedtke et al. 2000; Strotmann et al. 2000; Wissenbach et al. 2000; Nilius et al. 2001). Thus, the channel is spontaneously active at physiological osmolarities and can respond to changes in osmolarity in both directions. The sensitivity of TRPV4 is high, and cells expressing the channel can respond to 1% changes in osmolarity (Liedtke et al. 2000). The range of responsiveness stretches from osmolarities of around 350 mosmol/l to values under 200 mosmol/l (Liedtke et al. 2000; Strotmann et al. 2000).

The mechanism of sensitivity to changes in osmolarity was initially unclear, but did not seem to involve membrane stretch (Strotmann et al. 2000) or changes in intracellular ionic strength (Nilius et al. 2001), the latter being the link between swelling and signalling pathways activating volume-sensitive anion channels. Tyrosine phosphorylation by members of the src family of tyrosine kinases was suggested to be involved in the response to hypotonic solutions, and mutation of Tyr[253] (Fig. 1) was found in one study to abolish responses to hypotonic solutions (Xu et al. 2003b). However, Vriens et al. (2004) could not confirm this result and we also observed normal Ca^{2+} responses to hypotonic solutions with the same mutant (R. Strotmann, T.D. Plant, unpublished). Since the work on tyrosine phosphorylation, it has been shown that TRPV4 can be activated by products of arachidonic acid breakdown (Watanabe et al. 2003b), notably by epoxyeicosatrienoic acids (EETs), the products of arachidonic acid metabolism by the cytochrome P450 (CYP) epoxygenase activity. Arachidonic acid is released from membrane phospholipids by the action of phospholipase A_2 (PLA_2), which is known to be activated by cell swelling (see e.g. Hoffmann 2000 for review); indeed, inhibition of PLA_2 prevents the activation of TRPV4 by cell swelling (Watanabe et al. 2003b). Thus, the signalling pathway involved in the activation of TRPV4 by decreased osmolarity is likely to be:

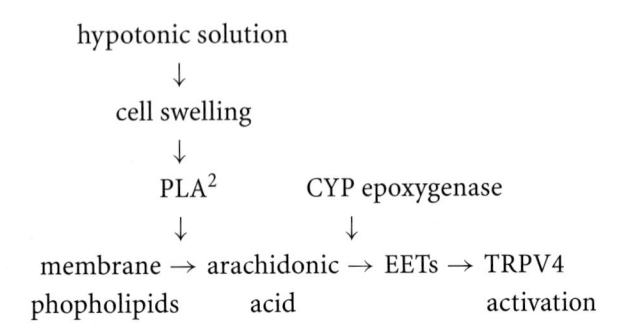

$$\text{hypotonic solution} \downarrow$$
$$\text{cell swelling} \downarrow$$
$$PLA^2 \quad\quad \text{CYP epoxygenase}$$
$$\downarrow \quad\quad\quad\quad \downarrow$$
$$\text{membrane} \to \text{arachidonic} \to \text{EETs} \to \text{TRPV4}$$
$$\text{phopholipids} \quad\quad \text{acid} \quad\quad\quad\quad\quad \text{activation}$$

It is not yet clear how swelling activates PLA_2, although this might be mediated by swelling-activated kinases, nor is it clear whether EETs directly activate the channel or whether other steps are involved. The PLA_2 pathway does not seem to be involved in the maintenance of spontaneous TRPV4 activity because this was not influenced by inhibition of PLA_2 (Vriens et al. 2004).

3.3
Sensitivity of TRPV4 to Lipid Messengers

The same pathway responsible for the response to hypotonic stimulation mediates activation of TRPV4 by endocannabinoids like anandamide (arachidonylethanolamide, AEA) (Watanabe et al. 2003b). AEA is hydrolysed by the fatty acid amidohydrolase to arachidonic acid, which then activates the channel as described above. Thus, the same signalling pathway mediates the response of TRPV4 to an endogenous chemical messenger and a physical stimulus.

3.4
Mechanosensitivity of TRPV4

The localization of TRPV4 in tissues known to express mechanosensitive channels and the response to swelling led to the suggestion that TRPV4 may be a mechanosensitive channel. One of the first studies on TRPV4 found no response of the channel to mechanical stimulation (Strotmann et al. 2000). However, another study has found responses of TRPV4 to mechanical stimuli like cell inflation (Suzuki et al. 2003). In addition, mammalian TRPV4 expressed in *C. elegans* neurones can restore mechanosensitivity to mutants lacking the worm osmo- and chemosensitive TRPV isoform Osm-9 (Liedtke et al. 2003).

3.5
Activation of TRPV4 by 4α-Phorbol Ester Derivatives

Exogenous agonists which strongly activate TRPV4, and for which endogenous counterparts have yet to be found, are the 4α-phorbol ester derivatives. Watanabe et al. showed that 4α-phorbol didecanoate (4αPDD) activated TRPV4 with an EC$_{50}$ of 0.2 µM (Watanabe et al. 2002a). We could show similar activation with 4α-phorbol myristate acetate (4αPMA; Fig. 2b; Strotmann et al. 2003). The 4α-phorbol ester derivatives are usually used as negative controls for the 4β-derivatives, and unlike the latter do not activate protein kinase C (PKC). The 4α-phorbol ester derivatives have few other biological effects: they only very weakly activate TRPV1 (Bhave et al. 2003), and inhibit some other ion channels. Thus, as relatively specific activators of TRPV4, they are likely to be important tools for the identification of TRPV4-mediated effects in native tissues. Not only the 4α-phorbol esters activate TRPV4, but also the 4β-derivatives. At room temperature, the latter are less effective activators than the 4α-derivatives (Watanabe et al. 2002a), but at more physiological temperatures their effects are stronger, and mediated to a large extent by a PKC-dependent mechanism (Gao et al. 2003; Xu et al. 2003a).

3.6
Temperature Sensitivity of TRPV4

TRPV1–4, but not TRPV5 and -6, are activated by heat with quite different temperature sensitivities. TRPV4 is sensitive to warm temperatures. The activity of the channel is strongly increased upon raising the temperature above approximately 25°C. For cells heterologously expressing TRPV4, different studies have found temperature thresholds that lay between 25°C and 34°C (Güler et al. 2002; Watanabe et al. 2002b). Keratinocytes display a TRPV4-mediated current with a similar threshold (Chung et al. 2003, 2004) and endothelial cells show similar responses (Watanabe et al. 2002b). Responses show no saturation at temperatures up to 43°C, but prolonged warming leads to desensitization of the channel. However, desensitization is not complete, and a constitutively active current component remains, which is significant also at physiological temperatures. Repeated heating decreases the responsiveness to temperature and shifts the threshold for channel activation to higher temperatures (Güler et al. 2002; Watanabe et al. 2002b).

3.7
Phorbol Esters and Temperature Use a Common Activation Mechanism

Temperature and phorbol esters seem to use a common activation mechanism, independent of that used by hypotonic solutions and of substances metabolized to EETs via arachidonic acid. Knowing that binding of the lipophilic TRPV1 agonist capsaicin to TRPV1 involves a tyrosine–serine motif in the intracellular loop between the second and third transmembrane segments (S2 and S3) (Jordt and Julius 2002), Vriens et al. looked for a similar motif in TRPV4 (Vriens et al. 2004). They found one (Tyr^{555} and Ser^{556}) at the intracellular N terminal end of S3 (Fig. 1). Mutation of the tyrosine led to a loss of responsiveness to 4αPDD and to heat, but not to hypotonic solutions or arachidonic acid. This result suggests that 4α-phorbol ester derivatives may activate TRPV4 in a similar way to that by which capsaicin activates TRPV1, i.e. by binding to a region of the channel involving the intracellular part of S3. Furthermore, activation by heat may involve an as-yet-unidentified endogenous lipid mediator that binds to the same region. The involvement of a diffusible messenger in the response to heat is supported by the loss of the response in excised membrane patches (Güler et al. 2002; Watanabe et al. 2002b).

3.8
Regulation of TRPV4 by Ca^{2+}

Like many other Ca^{2+}-permeable ion channels, the activity of TRPV4 is strongly regulated by Ca^{2+}. In the case of TRPV4, Ca^{2+} regulates the channel in both directions, i.e. it controls both the activation and inactivation of TRPV4. Spontaneous TRPV4 activity is strongly reduced in the absence of extracellular Ca^{2+}, or by the replacement of extracellular Ca^{2+} by Sr^{2+} or Ba^{2+} (Strotmann et al. 2003). In addition, activation of TRPV4 by 4α-phorbol esters or by hypotonic solutions is slower in the absence of Ca^{2+} or in Ba^{2+}. Re-addition of Ca^{2+} during the activation leads to an acceleration of activation and a larger increase in current (Strotmann et al. 2003). Watanabe et al. also found a dependence of current activation on extracellular Ca^{2+} with currents activating more rapidly but also being smaller and turning off earlier the higher the [Ca^{2+}]$_o$ (Watanabe et al. 2003a). In the continuous presence of an activating stimulus, currents through TRPV4 are transient and decay to varying degrees following a current maximum (Fig. 2b). This decay is also Ca^{2+}-dependent (Watanabe et al. 2003a). Furthermore, raising [Ca^{2+}]$_i$ via the patch pipette led to a reduction in the current activated by 4αPDD with an IC$_{50}$ of between 0.4 and 0.6 µM (Watanabe et al. 2002a, 2003a). The regions of the channel involved in Ca^{2+}-dependent channel inactivation are unknown, but seem to be different from those in the S2–S3 linker and C terminus that are involved in the Ca^{2+}-dependent inactivation of TRPV6 (Niemeyer et al. 2001; Lambers et al. 2004). An amino acid in S6, Phe^{707} (Fig. 1), is involved in inactivation. Mutation to alanine increases currents and decreases current decay, but does not change channel sensitivity to [Ca^{2+}]$_i$ (Watanabe et al. 2003a). The processes involved in channel activation/potentiation by Ca^{2+} are more clearly understood. This process is mediated by a helical domain, VGRLRRDRWSSVVPRV, starting at position 814 in the C terminus of TRPV4 (Fig. 1), which is similar to one involved in inactivation of TRPV6 (Niemeyer et al. 2001; Lambers et al. 2004) that likewise binds Ca^{2+}-calmodulin (CaM) (Strotmann et al. 2003). Mutations of the CaM-binding site that prevent Ca^{2+}-CaM binding decrease the rate and extent of channel activation, and prevent the current potentiation that occurs on switching from Ca^{2+}-free solutions to those containing Ca^{2+} (Strotmann et al. 2003). Mutation of an amino acid just proximal to this domain, Glu^{797} (Fig. 1), leads to increased spontaneous TRPV4 activity (Watanabe et al. 2003a).

3.9
Activation of TRPV4 by H⁺ and Citrate

There is one report that TRPV4 heterologously expressed in CHO cells can be activated by lowering the pH to values less than 6, and activation reached a maximum at around pH 4 (Suzuki et al. 2003). Responses to low pH were larger than those to cell inflation, but in this study TRPV4 was, surprisingly, not responsive to low osmolarity or to heat. Other groups have not confirmed the effect of low pH, nor have the amino acids involved in activation by H^+ been identified. TRPV4 lacks a Glu at the equivalent position in the pore loop to Glu^{600} that is responsible for potentiation of TRPV1 by H^+ (Jordt et al. 2000), but TRPV4 does have a Glu at the equivalent position to Glu^{648} involved in activation of TRPV1 by H^+ (Jordt et al. 2000). TRPV4 was also reported to be activated by citrate, but not by other organic acids at neutral pH (Suzuki et al. 2003).

4
Pharmacology of TRPV4

There are no specific inhibitors of TRPV4. TRPV4 is inhibited by micromolar concentrations of ruthenium red. However, in a similar concentration range, ruthenium red also inhibits other channels of the TRPV family, ryanodine receptors, voltage-gated calcium channels and mitochondrial function. Inhibition by extracellularly applied ruthenium red is voltage-dependent, with a stronger block of inward than of outward currents (Voets et al. 2002). Block by the polycation involves an aspartate residue (Asp^{682}) in the extracellular mouth of the channel pore (Fig. 1), since neutralization of this amino acid leads to a slowing of block and a strong (80 mV) shift in the voltage dependence of the block, but not in abolition of the block (Voets et al. 2002). Other substances shown to inhibit TRPV4 include SKF96365 (Liedtke et al. 2000) and trivalent cations like Gd^{3+} (Liedtke et al. 2000; Strotmann et al. 2000), but these also inhibit a wide range of other channels. Similarly, Mn^{2+} (200 µM) reduced $[Ca^{2+}]_i$ and currents in TRPV4-expressing cells (Strotmann et al. 2000). Activators of TRPV4 are discussed in detail in Sect. 3. The 4α-phorbol esters are potent and relatively specific activators of the channel (Watanabe et al. 2002a), which may interact directly with the channel (Vriens et al. 2004).

Too little is known about the biological effects and pathophysiological role of TRPV4 to speculate on the channel as a therapeutic target. However, given the wide expression and varied functions of the channel, it may be difficult to target a specific function.

5
Biological Relevance and Emerging/Established Biological Roles for TRPV4

In line with its expression in a wide variety of tissues, and the ability of cells expressing the channel to respond to a range of stimuli, TRPV4 has been implicated to play a role in a number of physiological responses.

5.1
TRPV4 in Systemic Osmoregulation

Expression of TRPV4 in appropriate regions of the brain, the circumventricular organs (the organum vasculosum of the lamina terminalis, and the subfornical organ), together with the response of TRPV4-expressing cells to changes in osmolarity suggest a role for the channel in systemic osmoregulation. $TRPV4^{-/-}$ mice displayed defects in osmoregulation evidenced by diminished drinking, an elevated systemic osmotic pressure, and reduced AVP synthesis in response to systemic hypertonicity (Liedtke and Friedman 2003). These effects were associated with a decreased expression of the immediate–early response gene, c-FOS, in osmotically responsive cells of the organum vasculosum. In contrast to these data, another study described an increase in AVP secretion in response to salt ingestion or hypertonicity in $TRPV4^{-/-}$ mice (Mizuno et al. 2003).

5.2
TRPV4 in Mechano-/Osmosensation

TRPV4 has been suggested to be involved in the nociceptive response of primary sensory neurones to hypotonic stimulation (Alessandri-Haber et al. 2003). Furthermore, taxol-induced hyperalgesia to osmotic and mechanical stimuli is decreased following a reduction in TRPV4 expression (Alessandri-Haber et al. 2004). A role of TRPV4 in responses to noxious mechanical stimuli has also been suggested from a study of $TRPV4^{-/-}$ mice (Liedtke and Friedman 2003; Suzuki et al. 2003). Evidence for the ability of TRPV4 to respond to osmotic and mechanical stimuli was also provided by a behavioural study of the worm *C. elegans*. TRPV4 targeted to the sensory neurones restored responses to osmotic (hypertonic solutions) and mechanical (nose touch) stimuli, but not those to chemical stimuli (odorants) (Liedtke et al. 2003) that are impaired in mutant worms lacking the TRPV isoform Osm-9 (Colbert et al. 1997). Surprisingly, TRPV4 restores responses to hypertonic solutions, whereas in mammalian cells it is activated by changes in osmolarity in the opposite direction.

TRPV4 was initially a major candidate for a mechanosensitive channel in hair cells in the cochlea involved in hearing. However, one report described no differences in the response of $TRPV4^{-/-}$ mice to acoustic startle (Liedtke and Friedman 2003), and another reported a delayed onset hearing loss and an increased susceptibility to acoustic injury in $TRPV4^{-/-}$ mice (Tabuchi et al. 2005). The precise role of TRPV4 in hearing remains unclear, but the channel does not appear to be the mechanosensitive channel in hair cells. Since this study, TRPA1 has been proposed to contribute to the mechanosensitive channel in hair cells and has properties similar to those of the native channel (Corey et al. 2004; Nagata et al. 2005).

5.3
TRPV4 as a Thermosensor

The responsiveness of TRPV4 to warm temperatures and its expression in sensory neurones, keratinocytes and in the hypothalamus point to a role for TRPV4 in thermosensation and thermoregulation. $TRPV4^{-/-}$ mice show decreased frequencies of sensory nerve discharge in response to thermal stimulation and a decrease in the number of responsive fibres (Todaka et al. 2004). In a model of thermal hyperalgesia, $TRPV4^{-/-}$ mice displayed longer latencies to escape from thermal stimuli (Todaka et al. 2004), but latencies were unaffected in the absence of hyperalgesia (Liedtke and Friedman 2003; Suzuki et al. 2003; Todaka et al. 2004). In contrast, another study found increases in tail withdrawal latency to moderately hot temperatures in untreated $TRPV4^{-/-}$ mice (Lee et al. 2005). This study also showed that $TRPV4^{-/-}$ mice chose warmer floor temperatures on a thermal gradient than wild-type mice, and that, in contrast to wild-type mice, which did not discriminate between 30°C and 34°C, $TRPV4^{-/-}$ mice preferred a floor temperature of 34°C. The combination of TRPV4 expression in the hypothalamus together with responsiveness to temperatures around body temperature lends support for a possible role in thermoregulation. However, Liedtke and Friedman (2003) found no difference in body temperature, nor differences in the temperature response to cold stress in $TRPV4^{-/-}$ mice (Liedtke and Friedman 2003).

5.4
TRPV4 in Vascular Regulation

TRPV4 in endothelial cells could have a number of possible roles (Nilius et al. 2003). Most likely, as a Ca^{2+} permeable channel, it could, on activation, increase $[Ca^{2+}]_i$ resulting in the Ca^{2+}-induced release of the vasodilator NO. Stimuli reported to activate TRPV4 that also produce vasodilation include mechanical stimuli (e.g. shear stress), temperature (warming) or local messengers (e.g. anandamide or EETs). TRPV4 has not yet been shown to be mechanosensitive in the endothelial cell, but it is possible that mechanical stimuli like shear stress via stimulation of PLA$_2$ could activate TRPV4. From the temperature sensitivity of TRPV4, the channel is likely to be constitutively active at body temperature (Watanabe et al. 2002b). It could respond to changes in temperature in both directions, by closure on cooling and opening on heating leading to vasoconstriction and vasodilation respectively. TRPV4 may be the target of local messengers which activate the channel. Anandamide is known to regulate vascular tone (see Nilius et al. 2003), and EETs have anti-hypertensive and anti-inflammatory properties and regulate renal vascular function (see Vriens et al. 2005 for references).

Endothelial cells produce an endothelium-derived hyperpolarizing factor (EDHF), a local mediator that hyperpolarizes vascular smooth muscle cells. In some vessels, EDHF has shown to be EETs. It was recently shown that EET-induced hyperpolarization of cerebral artery smooth muscle cells involves local, ryanodine receptor-mediated Ca^{2+} release from intracellular stores (sparks) and activation of large Ca^{2+}-activated K$^+$ channels (BK$_{Ca}$; Earley et al. 2005). TRPV4 was found to be expressed in these cells, and EET-induced, Ca^{2+} entry-dependent Ca^{2+} release was prevented by TRPV4 antisense oligonucleotides. Thus, TRPV4, in a signalling complex with ryanodine receptors and BK$_{Ca}$ channels, is likely to be involved in the response of vascular smooth muscle cells to EDHF.

5.5
Other Possible Roles of TRPV4

Among the physiological roles proposed for TRPV4 is its function as an osmosensitive channel in airway smooth muscle cells which might be involved in bronchoconstriction in response to inhalation of hypotonic aerosols (Jia et al. 2004). TRPV4 is strongly expressed in the kidney and has been proposed to be an osmotically or flow-sensitive channel (Tian et al. 2004; O'Neil and Heller 2005), but there are, to date, few functional data to support this role. Flow and low osmolarity are stimuli to which the apical membrane of tubular cells is exposed, but not the basolateral membrane in which TRPV4 expression has been reported to predominate (Tian et al. 2004). TRPV4 has been reported to be involved in the regulation of ciliary beating in response to changes in fluid viscosity in the oviduct (Andrade et al. 2005). After initial swelling, many cell types respond to hypotonic solutions with a regulatory volume decrease (RVD), a Ca^{2+}-dependent loss of electrolytes and water to restore the cell volume. Keratinocytes, which strongly express TRPV4, undergo Ca^{2+}-dependent, Gd^{3+}-sensitive RVD. CHO cells which lack TRPV4 show no RVD, but those cells heterologously expressing TRPV4 do, suggesting that TRPV4 may play an important role in RVD in some cells (Becker et al. 2005).

References

Alessandri-Haber N, Yeh JJ, Boyd AE, Parada CA, Chen X, Reichling DB, Levine JD (2003) Hypotonicity induces TRPV4-mediated nociception in rat. Neuron 39:497–511

Alessandri-Haber N, Dina OA, Yeh JJ, Parada CA, Reichling DB, Levine JD (2004) Transient receptor potential vanilloid 4 is essential in chemotherapy-induced neuropathic pain in the rat. J Neurosci 24:4444–4452

Andrade YN, Fernandes J, Vazquez E, Fernandez-Fernandez JM, Arniges M, Sanchez TM, Villalon M, Valverde MA (2005) TRPV4 channel is involved in the coupling of fluid viscosity changes to epithelial ciliary activity. J Cell Biol 168:869–874

Arniges M, Fernandez-Fernandez JM, Albrecht N, Schaefer M, Valverde MA (2006) Human TRPV4 channel splice variants revealed a key role of ankyrin domains in multimerization and trafficking. J Biol Chem 281:1580–1586

Becker D, Blase C, Bereiter-Hahn J, Jendrach M (2005) TRPV4 exhibits a functional role in cell-volume regulation. J Cell Sci 118:2435–2440

Bhave G, Hu HJ, Glauner KS, Zhu W, Wang H, Brasier DJ, Oxford GS, Gereau RW (2003) Protein kinase C phosphorylation sensitizes but does not activate the capsaicin receptor transient receptor potential vanilloid 1 (TRPV1). Proc Natl Acad Sci USA 100:12480–12485

Chung MK, Lee H, Caterina MJ (2003) Warm temperatures activate TRPV4 in mouse 308 keratinocytes. J Biol Chem 278:32037–32046

Chung MK, Lee H, Mizuno A, Suzuki M, Caterina MJ (2004) TRPV3 and TRPV4 mediate warmth-evoked currents in primary mouse keratinocytes. J Biol Chem 279:21569–21575

Colbert HA, Smith TL, Bargmann CI (1997) Osm-9, a novel protein with structural similarity to channels, is required for olfaction, mechanosensation and olfactory adaptation in Caenorhabditis elegans. J Neurosci 17:8259–8269

Corey DP, Garcia-Anoveros J, Holt JR, Kwan KY, Lin SY, Vollrath MA, Amalfitano A, Cheung EL, Derfler BH, Duggan A, Geleoc GS, Gray PA, Hoffman MP, Rehm HL, Tamasauskas D, Zhang DS (2004) TRPA1 is a candidate for the mechanosensitive transduction channel of vertebrate hair cells. Nature 432:723–730

Delany NS, Hurle M, Facer P, Alnadaf T, Plumpton C, Kinghorn I, See CG, Costigan M, Anand P, Woolf CJ, Crowther D, Sanseau P, Tate SN (2001) Identification and characterization of a novel human vanilloid receptor-like protein, VRL-2. Physiol Genomics 4:165–174

Earley S, Heppner TJ, Nelson MT, Brayden JE (2005) TRPV4 forms a novel Ca^{2+} signaling complex with ryanodine receptors and BKCa channels. Circ Res 97:1270–1279

Gao X, Wu L, O'Neil RG (2003) Temperature-modulated diversity of TRPV4 channel gating: activation by physical stresses and phorbol ester derivatives through protein kinase C-dependent and -independent pathways. J Biol Chem 278:27129–27137

Guatteo E, Chung KK, Bowala TK, Bernardi G, Mercuri NB, Lipski J (2005) Temperature sensitivity of dopaminergic neurons of the substantia nigra pars compacta: involvement of TRP channels. J Neurophysiol 94:3069–3080

Güler AD, Lee H, Iida T, Shimizu I, Tominaga M, Caterina M (2002) Heat-evoked activation of the ion channel, TRPV4. J Neurosci 22:6408–6414

Hellwig N, Albrecht N, Harteneck C, Schultz G, Schaefer M (2005) Homo- and heteromeric assembly of TRPV channel subunits. J Cell Sci 118:917–928

Hoffmann EK (2000) Intracellular signalling involved in volume regulatory decrease. Cell Physiol Biochem 10:273–288

Jia Y, Wang X, Varty L, Rizzo CA, Yang R, Correll CC, Phelps PT, Egan RW, Hey JA (2004) Functional TRPV4 channels are expressed in human airway smooth muscle cells. Am J Physiol Lung Cell Mol Physiol 287:L272–278

Jordt SE, Julius D (2002) Molecular basis for species-specific sensitivity to "hot" chili peppers. Cell 108:421–430

Jordt SE, Tominaga M, Julius D (2000) Acid potentiation of the capsaicin receptor determined by a key extracellular site. Proc Natl Acad Sci USA 97:8134–8139

Lambers TT, Weidema AF, Nilius B, Hoenderop JG, Bindels RJ (2004) Regulation of the mouse epithelial Ca^{2+} channel TRPV6 by the Ca^{2+}-sensor calmodulin. J Biol Chem 279:28855–28861

Lee H, Iida T, Mizuno A, Suzuki M, Caterina MJ (2005) Altered thermal selection behavior in mice lacking transient receptor potential vanilloid 4. J Neurosci 25:1304–1310

Liedtke W, Friedman JM (2003) Abnormal osmotic regulation in trpv4$^{-/-}$ mice. Proc Natl Acad Sci USA 100:13698–13703

Liedtke W, Choe Y, Marti-Renom MA, Bell AM, Denis CS, Sali A, Hudspeth AJ, Friedman JM, Heller S (2000) Vanilloid receptor-related osmotically activated channel (VR-OAC), a candidate vertebrate osmoreceptor. Cell 103:525–535

Liedtke W, Tobin DM, Bargmann CI, Friedman JM (2003) Mammalian TRPV4 (VR-OAC) directs behavioral responses to osmotic and mechanical stimuli in Caenorhabditis elegans. Proc Natl Acad Sci USA 100 Suppl 2:14531–14536

Mizuno A, Matsumoto N, Imai M, Suzuki M (2003) Impaired osmotic sensation in mice lacking TRPV4. Am J Physiol Cell Physiol 285:C96–101

Nagata K, Duggan A, Kumar G, Garcia-Anoveros J (2005) Nociceptor and hair cell transducer properties of TRPA1, a channel for pain and hearing. J Neurosci 25:4052–4061

Niemeyer BA, Bergs C, Wissenbach U, Flockerzi V, Trost C (2001) Competitive regulation of CaT-like-mediated Ca^{2+} entry by protein kinase C and calmodulin. Proc Natl Acad Sci USA 98:3600–3605

Nilius B, Prenen J, Wissenbach U, Bödding M, Droogmans G (2001) Differential activation of the volume-sensitive cation channel TRP12 (OTRPC4) and volume-regulated anion currents in HEK-293 cells. Pflügers Arch 443:227–233

Nilius B, Droogmans G, Wondergem R (2003) Transient receptor potential channels in endothelium: solving the calcium entry puzzle? Endothelium 10:5–15

O'Neil RG, Heller S (2005) The mechanosensitive nature of TRPV channels. Pflügers Arch 451:193–203

Strotmann R, Harteneck C, Nunnenmacher K, Schultz G, Plant TD (2000) OTRPC4, a nonselective cation channel that confers sensitivity to extracellular osmolarity. Nat Cell Biol 2:695–702

Strotmann R, Schultz G, Plant TD (2003) Ca^{2+}-dependent potentiation of the nonselective cation channel TRPV4 is mediated by a C-terminal calmodulin binding site. J Biol Chem 278:26541–26549

Suzuki M, Mizuno A, Kodaira K, Imai M (2003) Impaired pressure sensation in mice lacking TRPV4. J Biol Chem 278:22664–22668

Tabuchi K, Suzuki M, Mizuno A, Hara A (2005) Hearing impairment in TRPV4 knockout mice. Neurosci Lett 382:304–308

Takumida M, Kubo N, Ohtani M, Suzuka Y, Anniko M (2005) Transient receptor potential channels in the inner ear: presence of transient receptor potential channel subfamily 1 and 4 in the guinea pig inner ear. Acta Otolaryngol 125:929–934

Tian W, Salanova M, Xu H, Lindsley JN, Oyama TT, Anderson S, Bachmann S, Cohen DM (2004) Renal expression of osmotically responsive cation channel TRPV4 is restricted to water-impermeant nephron segments. Am J Physiol Renal Physiol 287:F17–24

Todaka H, Taniguchi J, Satoh J, Mizuno A, Suzuki M (2004) Warm temperature-sensitive transient receptor potential vanilloid 4 (TRPV4) plays an essential role in thermal hyperalgesia. J Biol Chem 279:35133–35138

Voets T, Prenen J, Vriens J, Watanabe H, Janssens A, Wissenbach U, Bödding M, Droogmans G, Nilius B (2002) Molecular determinants of permeation through the cation channel TRPV4. J Biol Chem 277:33704–33710

Vriens J, Watanabe H, Janssens A, Droogmans G, Voets T, Nilius B (2004) Cell swelling, heat, and chemical agonists use distinct pathways for the activation of the cation channel TRPV4. Proc Natl Acad Sci USA 101:396–401

Vriens J, Owsianik G, Fisslthaler B, Suzuki M, Janssens A, Voets T, Morisseau C, Hammock BD, Fleming I, Busse R, Nilius B (2005) Modulation of the Ca^{2+} permeable cation channel TRPV4 by cytochrome P450 epoxygenases in vascular endothelium. Circ Res 97:908–915

Watanabe H, Davis JB, Smart D, Jerman JC, Smith GD, Hayes P, Vriens J, Cairns W, Wissenbach U, Prenen J, Flockerzi V, Droogmans G, Benham CD, Nilius B (2002a) Activation of TRPV4 channels (hVRL-2/mTRP12) by phorbol derivatives. J Biol Chem 277:13569–13577

Watanabe H, Vriens J, Suh SH, Benham CD, Droogmans G, Nilius B (2002b) Heat-evoked activation of TRPV4 channels in an HEK293 cell expression system and in native mouse aorta endothelial cells. J Biol Chem 277:47044–47051

Watanabe H, Vriens J, Janssens A, Wondergem R, Droogmans G, Nilius B (2003a) Modulation of TRPV4 gating by intra- and extracellular Ca^{2+}. Cell Calcium 33:489–495

Watanabe H, Vriens J, Prenen J, Droogmans G, Voets T, Nilius B (2003b) Anandamide and arachidonic acid use epoxyeicosatrienoic acids to activate TRPV4 channels. Nature 424:434–438

Wissenbach U, Bödding M, Freichel M, Flockerzi V (2000) Trp12, a novel Trp related protein from kidney. FEBS Lett 485:127–134

Xu F, Satoh E, Iijima T (2003a) Protein kinase C-mediated Ca^{2+} entry in HEK 293 cells transiently expressing human TRPV4. Br J Pharmacol 140:413–421

Xu H, Zhao H, Tian W, Yoshida K, Roullet JB, Cohen DM (2003b) Regulation of a transient receptor potential (TRP) channel by tyrosine phosphorylation. SRC family kinase-dependent tyrosine phosphorylation of TRPV4 on TYR-253 mediates its response to hypotonic stress. J Biol Chem 278:11520–11527

Xu H, Fu Y, Tian W, Cohen DM (2005) Glycosylation of the osmoresponsive transient receptor potential channel TRPV4 on Asn-651 influences membrane trafficking. Am J Physiol Renal Physiol 290:F1103–F1109

ion
TRPV5, the Gateway to Ca^{2+} Homeostasis

A. R. Mensenkamp · J. G. J. Hoenderop · R. J. M. Bindels (✉)

Department of Physiology, Radboud University Nijmegen Medical Center,
286 Cell Physiology, PO Box 9101, 6500 HB Nijmegen, The Netherlands
r.bindels@ncmls.ru.nl

1	Introduction	208
2	Molecular Properties of TRPV5	210
3	Electrophysiology	210
4	Modes of Activation	211
4.1	Hormonal Control	211
4.1.1	Vitamin D$_3$	211
4.1.2	Parathyroid Hormone	212
4.1.3	Klotho	212
4.2	Binding Partners of TRPV5	213
4.2.1	S100A10/Annexin 2	213
4.2.2	80K-H	213
4.2.3	Rab11a	214
4.2.4	BSPRY	214
4.2.5	Mg^{2+} and PIP$_2$	215
5	Biological Function	215
6	Pharmacology	216
References		217

Abstract Ca^{2+} homeostasis in the body is tightly controlled, and is a balance between absorption in the intestine, excretion via the urine, and exchange from bone. Recently, the epithelial Ca^{2+} channel (TRPV5) has been identified as the gene responsible for the Ca^{2+} influx in epithelial cells of the renal distal convoluted tubule. TRPV5 is unique within the family of transient receptor potential (TRP) channels due to its high Ca^{2+} selectivity. Ca^{2+} flux through TRPV5 is controlled in three ways. First, TRPV5 gene expression is regulated by calciotropic hormones such as vitamin D$_3$ and parathyroid hormone. Second, Ca^{2+} transport through TRPV5 is controlled by modulating channel activity. Intracellular Ca^{2+}, for example, regulates channel activity by feedback inhibition. Third, TRPV5 is controlled by mobilization of the channel through trafficking toward the plasma membrane. The newly identified anti-aging hormone Klotho regulates TRPV5 by cleaving off sugar residues from the extracellular domain of the protein, resulting in a prolonged expression of TRPV5 at the plasma membrane. Inactivation of TRPV5 in mice leads to severe hypercalciuria, which is compensated by increased intestinal Ca^{2+} absorption due to augmented vitamin D$_3$ levels. Furthermore, TRPV5 deficiency in mice is associated with polyuria, urine acidification, and reduced bone thickness. Some pharmaceutical compounds, such as the immunosuppressant FK506, affect the Ca^{2+} balance by modulating TRPV5 gene expression. This underlines the importance of elucidating the role of TRPV5 in Ca^{2+}-related disorders, thereby enhancing the possibilities for pharmacological intervention. This chapter describes a unique TRP channel and highlights its regulation and function in renal Ca^{2+} reabsorption and overall Ca^{2+} homeostasis.

Keywords Ca^{2+} · TRP · Distal convoluted tubule · Kidney · Regulation

1
Introduction

The Ca^{2+} ion is an essential ion in all organisms, where it plays a crucial role in processes ranging from the formation and maintenance of the skeleton to the temporal and spatial regulation of neuronal excitability and muscle contraction. Therefore, the extracellular Ca^{2+} concentration is tightly controlled by the concerted action of the intestine, kidney, and bone. Ca^{2+} uptake occurs in the intestine, where it is transported through the epithelium into the blood, and excreted by the kidney via the urine. The main storage of Ca^{2+} takes place in the bone. The net Ca^{2+} balance across the bone is zero in healthy adults, when bone formation is in equilibrium with bone resorption (van Os 1987).

Only 2%–3% of Ca^{2+} that is filtered by the kidney is eventually excreted. Ca^{2+} is predominantly reabsorbed along the proximal tubule (PT) and the thick ascending limb of Henle's loop (TAL). This passive reabsorption of Ca^{2+} is a paracellular process via the tight junctions of the epithelial cells and is primarily driven by the electrochemical gradient generated by Na^+ (PT), K^+ (TAL), and water reabsorption (Hoenderop et al. 2002b). The last 15% of Ca^{2+} reabsorption is based on active transcellular Ca^{2+} transport across the epithelium of the distal convoluted tubule (DCT) and the connecting tubule (CNT). This process can be divided in three discrete steps. The first step requires Ca^{2+} transport across the apical membrane. Transient receptor potential channel vanilloid 5 (TRPV5) has been identified as the channel responsible for this process and is the main topic of this chapter. The second step is the diffusion of Ca^{2+} through the cytosol. Calbindin-D_{28k} binds intracellular Ca^{2+} trans-

Fig. 1 Mechanism of active epithelial Ca^{2+} transport. Renal Ca^{2+} transport is divided into three steps. The first step is the transport of lumenal Ca^{2+} into the cell by TRPV5. Cytosolic diffusion is facilitated by binding of intracellular Ca^{2+} to calbindin-D_{28k} and the subsequent shuttling to the basolateral membrane (step 2). Ca^{2+} is extruded into the blood compartment via NCX1 and PMCA1b (step 3). In this way, net Ca^{2+} reabsorption occurs from the luminal space to the extracellular compartment. These steps are coordinately regulated by 1,25-$(OH)_2D_3$ and parathyroid hormone (PTH)

TRPV5, the Gateway to Ca^{2+} Homeostasis

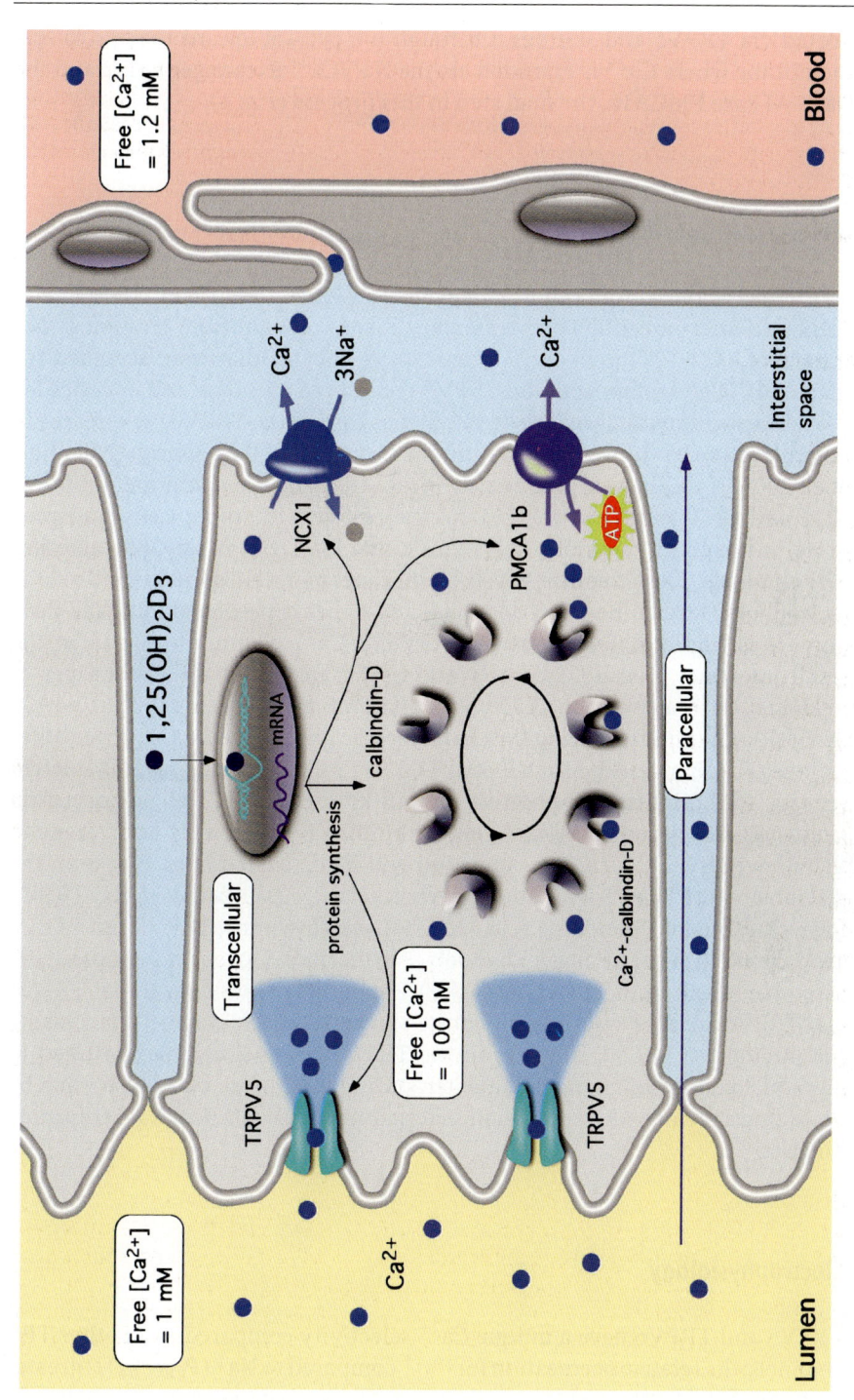

ported via TRPV5 and shuttles it through the cytosol toward the basolateral membrane where Ca^{2+} is extruded via the Na^+/Ca^{2+} exchanger NCX1 and the Ca^{2+}-ATPase PMCA1b, the final step in this process (Fig. 1).

2
Molecular Properties of TRPV5

TRPV5 was cloned from rabbit kidney in 1999 (Hoenderop et al. 1999). It was identified after injecting complementary RNAs (cRNAs) from a rabbit cDNA library of DCT/CNT cells in *Xenopus laevis* oocytes that were screened for maximal $^{45}Ca^{2+}$ influx activity. TRPV5 consists of 15 exons and encodes for 730 amino acids with a predicted protein mass of 83 kDa. TRPV5 is expressed in placenta, bone, and, most importantly, the distal part of the nephron (Nijenhuis et al. 2003). In kidney, it is primarily localized along the apical membrane of DCT and CNT epithelia, where it colocalizes with the other Ca^{2+} transport proteins including calbindin-D_{28k}, NCX1, and PMCA1b. In line with all other TRP channels, TRPV5 consists of six membrane-spanning domains and a short hydrophobic stretch between domain 5 and 6 that is predicted to be the Ca^{2+} pore. It is highly homologous to TRPV6, a Ca^{2+} channel localized in the small intestine, prostate, stomach, and brain, and likely the gate-keeper of intestinal Ca^{2+} absorption. TRPV5 and TRPV6 form a distinct class within the TRP superfamily, sharing the characteristic prevalence for Ca^{2+} over other cations (Clapham 2003; Montell et al. 2002). TRPV5 contains several ankyrin repeats, PDZ motifs, and putative protein kinase C (PKC) phosphorylation sites, suggesting that many intracellular binding partners may exist. Some of them have already been identified, and will be discussed later. Based on the resemblance of TRPV5 with other ion channels, it was anticipated that TRPV5 forms a tetramer. Hoenderop et al. (2003b) have shown that TRPV5 and TRPV6 are expressed as tetrameric channels. Furthermore, by using concatamers assembled from both TRPV5 and TRPV6 subunits, it was shown that TRPV5 and TRPV6 are able to form both homo- and heteromeric channels. Increasing the number of TRPV6 subunits in the TRPV5 channel complex resulted in a gradual change of channel characteristics toward those of TRPV6, thereby providing a pleiotropic set of functional channels with different Ca^{2+} transport kinetics.

3
Electrophysiology

TRPV5 and TRPV6 have a unique Ca^{2+} selectivity compared to all other TRP channels; the relative permeation for Ca^{2+} compared to Na^+ (P_{Ca}/P_{Na}) is greater than 100 for TRPV5/TRPV6, while other TRPV channels have a P_{Ca}/P_{Na} in the range of 1–5. This preference for Ca^{2+} is due to a single aspartate at position 542 in the pore of TRPV5. Mutation into a neutral amino acid or even in a negatively charged glutamate abolished Ca^{2+} permeation, while the permeation by monovalent ions was unaffected (Nilius et al. 2001b).

TRPV5 is constitutively active at physiological Ca^{2+} concentrations and membrane potentials, and shows a strong inward-rectifying current/voltage (I–V) relationship. TRPV5 is only permeable for monovalent cations at very low Ca^{2+} concentrations (~1 nM). At increasing extracellular Ca^{2+} concentrations, these currents quickly disappeared due to Ca^{2+}-mediated feedback inhibition of the channel. Currents were only stimulated again at extracellular Ca^{2+} concentrations above 1 mM (Nilius et al. 2001a; Vennekens et al. 2000). This Ca^{2+} dependence shows the typical anomalous mole-fraction behavior, which has been described for highly Ca^{2+}-selective channels (Dang and McCleskey 1998). Repeated exposure to Ca^{2+} resulted in a run-down of channel activity. This effect was also observed with repeated exposure to Mg^{2+}, but could be reversed by application of phosphatidylinositol 4,5-bisphosphate (PIP_2) (Nilius et al. 2000; Yeh et al. 2005). This effect is described in Sect. 4.2.5. TRPV5 currents are also dependent on pH since both intra- and extracellular acidification reduced TRPV5 currents. Inhibition by intracellular acidification is likely caused by clockwise rotation of the pore helix, leading to narrowing of the selectivity filter gate (Vennekens et al. 2001; Yeh et al. 2005).

4
Modes of Activation

TRPV5 expression and activity is controlled by calciotropic hormones and by interactions with other proteins. In this following, the various forms of (in)activation are discussed.

4.1
Hormonal Control

4.1.1
Vitamin D₃

It is well established that vitamin D_3 plays a key role in Ca^{2+} homeostasis. Vitamin D_3 enters the body via the diet and is synthesized in the skin by sunlight (Neer 1975). Activation of vitamin D_3 is under control of 25-hydroxy-1α-hydroxylase (1α-OHase), the enzyme responsible for hydroxylation of vitamin D_3 into 1,25-dihydroxyvitamin D_3 [1,25-$(OH)_2D_3$] in the PT (Fraser and Kodicek 1970). Although the function of 1,25-$(OH)_2D_3$ in the intestinal absorption of Ca^{2+} has been known for some time (Harrison and Harrison 1960), its role in renal DCT has evolved more recently (Bindels et al. 1991).

TRPV5 gene expression is strongly under the control of 1,25-$(OH)_2D_3$. Hoenderop et al. (2001) showed that treatment of vitamin D_3-depleted rats with 1,25-$(OH)_2D_3$ results in increased levels of TRPV5 messenger RNA (mRNA) in the kidney. This is supported by the fact that the putative promoter region of TRPV5 contains potential vitamin D response elements (VDREs) (Muller et al. 2000; Weber et al. 2001). The role of 1,25-$(OH)_2D_3$ in Ca^{2+} transport was further substantiated in mice deficient for 1α-OHase (Dardenne et al. 2001; Panda et al. 2001). These mice had undetectable levels of 1,25-$(OH)_2D_3$, were severely hypocalcemic, and had reduced mRNA levels of renal TRPV5, calbindin-D_{28k}, and NCX1, which could be normalized by administration of 1,25-$(OH)_2D_3$ (Hoenderop et al. 2002a).

4.1.2
Parathyroid Hormone

Parathyroid hormone (PTH) is an essential component of Ca^{2+} homeostasis. High plasma Ca^{2+} concentrations are sensed by the parathyroid Ca^{2+}-sensing receptor, resulting in a decrease in PTH secretion from the parathyroid glands (Brown et al. 1993). Van Abel et al. (2005) have shown that parathyroidectomized rats exhibit reduced levels of renal TRPV5, calbindin-D_{28k}, and NCX1. This was correlated with a decrease in Ca^{2+} reabsorption and hypocalcemia. Supplementation with PTH restored protein expression, Ca^{2+} transport, and plasma Ca^{2+} levels, convincingly showing that PTH not only controls Ca^{2+} resorption from bone but also regulates the key players in renal transcellular Ca^{2+} transport.

4.1.3
Klotho

Klotho was identified in a mouse strain showing features resembling human aging (Kuro-o et al. 1997). Mice homozygous for the mutated allele featured ectopic calcification, skin and muscle atrophy, atherosclerosis, and osteoporosis. Klotho exhibits β-glucuronidase activity (Tohyama et al. 2004) and is activated by cleavage of the N-terminal extracellular tail. This part is then released into the urine, serum, and cerebrospinal fluid (Chang et al. 2005; Imura et al. 2004). Klotho is linked to Ca^{2+} metabolism in many ways. Klotho-deficient mice did not only develop ectopic calcification and osteoporosis, but 1,25-$(OH)_2D_3$ levels were also upregulated due to increased levels of 1α-OHase expression. In addition, application of 1,25-$(OH)_2D_3$ increased *Klotho* gene expression in wildtype mice. It was therefore suggested that klotho may play a role in feedback inhibition of vitamin D_3 (Tsujikawa et al. 2003). Microarray data showed that the expression of *klotho* is downregulated in *Trpv5*$^{-/-}$ mice. Therefore, Chang et al. (2005) studied the functional relation between klotho and TRPV5 showing that incubation of TRPV5-expressing HEK293 cells with culture medium containing klotho resulted in a significantly increased Ca^{2+} uptake. These effects could be mimicked by a purified β-glucuronidase, indicating that the enzymatic activity of klotho is responsible for the increased TRPV5 activity. Mutation of the predicted N-glycosylation site of TRPV5 (N358Q) abolished both klotho- and β-glucuronidase-mediated activation of TRPV5. Cell surface biotinylation indicated a strong increase in plasma membrane localization of TRPV5 after klotho or β-glucuronidase stimulation. These data revealed a new concept of channel regulation showing that partial deglycosylation of TRPV5 by extracellular klotho controls its membrane expression and thereby biological activity.

4.2
Binding Partners of TRPV5

4.2.1
S100A10/Annexin 2

One of the first binding partners being identified for both TRPV5 and TRPV6 was S100A10 (van de Graaf et al. 2003). S100A10 is a distinct member of the Ca^{2+}-binding EF-hand-containing S100 protein family. S100A10 forms a tetramer with annexin 2 by binding of two homodimers (Gerke and Moss 2002) and belongs to a family of Ca^{2+} and phospholipid-binding proteins. Annexin 2/S100A10 colocalized with TRPV5 in kidney along the apical membrane of DCT and CNT. Annexin 2 was also identified as a binding partner of TRPV5, but only in the presence of S100A10, showing that S100A10 forms the bridge between TRPV5 and annexin 2. Suppressing annexin 2 gene expression by short interfering RNA (siRNA) resulted in decreased channel activity in HEK293 cells heterologously expressing TRPV5, confirming the essential role of the annexin 2/S100A10 complex in TRPV5 channel activity.

4.2.2
80K-H

80K-H was identified by microarray analysis in 1α-OHase-deficient mice receiving supplemental 1,25-$(OH)_2D_3$ (Gkika et al. 2004). These mice are not able to synthesize 1,25-$(OH)_2D_3$, resulting in severe hypocalcemia (Dardenne et al. 2001; Panda et al. 2001). The gene for 80K-H was significantly downregulated in the absence of 1,25-$(OH)_2D_3$, while dietary Ca^{2+} supplementation resulted in a significant upregulation. 80K-H was previously identified as a PKC substrate, but its biological function remained elusive for a long time (Hirai and Shimizu 1990). Recently, Drenth et al. (2004) showed that mutations in the gene for 80K-H cause autosomal-dominant polycystic liver disease (PCLD), which is clinically characterized by the presence of multiple liver cysts. Association of 80K-H with TRPV5 was determined by coimmunoprecipitation and pull-down experiments (Gkika et al. 2004). The binding site consists of a short region in the carboxy-terminus between position 598 and 608. Disruption of the two putative Ca^{2+}-binding EF-hand motifs did not prevent binding to TRPV5. Coexpression of wildtype 80K-H with TRPV5 did not have an effect on Ca^{2+} and Na^+ currents of TRPV5. Co-expression of TRPV5 with the EF-hand mutant, however, resulted in a significant decrease of the currents and reduced the amount of intracellular Ca^{2+} needed for the feedback inhibition of the channel. 80K-H could, therefore, play a role in the Ca^{2+}-sensing of TRPV5.

4.2.3
Rab11a

Rab11a was identified as a binding partner of TRPV5 and TRPV6 (van de Graaf et al. 2006a) by yeast two-hybrid assay using the carboxy-terminus of TRPV6 as a bait. Rab11a belongs to the family of Rab GTPases and is involved in the trafficking of vesicles from Golgi toward the plasma membrane. Its feature to act as a molecular switch that cycles between GDP- and GTP-bound states underlies the functionality of this protein. To differentiate between these two states, mutants were used that were locked in either the GDP- ($Rab11^{S25N}$) or the GTP-bound ($Rab11^{S20V}$) form. The carboxy-termini of TRPV5 and TRPV6 had a strong binding preference for $Rab11^{S25N}$ in pull-down experiments. Ca^{2+} uptake was decreased in $X.$ $laevis$ oocytes coexpressing $Rab11^{S25N}$ and TRPV5, correlating with a defect in trafficking of TRPV5 toward the plasma membrane. Likewise, introduction of $Rab11^{S25N}$ in primary kidney cells from CNT and cortical collecting ducts (CCD) displaying endogenous TRPV5-mediated Ca^{2+} transport resulted in diminished transcellular Ca^{2+} transport. All forms of Rab11a colocalized with TRPV5 in vesicular structures, both in heterologously expressed HeLa cells and in primary kidney cells. Site-directed mutagenesis studies identified a stretch of five amino acids (MLERK) in a helical stretch at the carboxy terminus as the binding site, the same region as is required for 80K-H interaction (Gkika et al. 2004). TRPV5 is one of the first cargo proteins that physically interact with a Rab protein. It is postulated that cytosolic GDP-bound Rab11a interacts with TRPV5 in a cytosolic compartment. Subsequent exchange of GDP to GTP initiates other effectors to bind Rab11a in this compartment while binding to TRPV5 is released. Rab11a-GTP-containing vesicles will translocate toward the apical plasma membrane, where the cargo, including TRPV5, is released to the cell surface, thus allowing channel activity.

4.2.4
BSPRY

Another binding partner of TRPV5 that was identified by a yeast two-hybrid screen was BSPRY (B-box and SPRY-domain containing protein) (van de Graaf et al. 2006b). The function of this protein is thus far unknown, although it was previously postulated to play a role in protein–protein interactions (Ponting et al. 1997). BSPRY was also identified as a binding partner of 14-3-3ζ (Birkenfeld et al. 2003). 14-3-3 Proteins represent a family of conserved eukaryotic proteins participating in a wide range of cellular processes, including exocytosis (Fu et al. 2000). Van de Graaf et al. (2006) showed that BSPRY completely colocalizes with TRPV5 in the Ca^{2+}-transporting DCT and CNT segments of the kidney. Coexpression of TRPV5 and BSPRY in polarized confluent monolayers of Madin–Darby canine kidney (MDCK) cells resulted in a significant decrease of Ca^{2+} influx, while membrane-associated TRPV5 protein expression remained the same. The latter finding suggests that BSPRY is not involved in the trafficking of TRPV5, making a role for BSPRY in controlling the signaling cascades of channel activity more likely.

4.2.5
Mg^{2+} and PIP_2

Mg^{2+} affects TRPV5 currents in three ways (Lee et al. 2005; Nilius et al. 2000). First, application of 1 mM Mg^{2+} results in a reversible inhibition over tens of seconds in inside-out patch membranes containing TRPV5. Second, Mg^{2+} also causes a faster reversible voltage-dependent block. This effect can be distinguished from the previous by its different kinetics, as a shift of the reversal potential, a steeper I–V curve near the reversal potential and a block of the outward current. Neutralization of aspartate-542 in the pore selectivity filter abolished both effects. Third, a progressive irreversible run-down of current occurs after repetitive application of Mg^{2+}. Supplementation with exogenous PIP_2 reactivated these currents, indicating that Mg^{2+} exerts its run-down effect by removing PIP_2 from the plasma membrane (Lee et al. 2005). The authors supported these observations by showing that Mg-ATP, a putative activator of lipid kinases, could reverse Mg^{2+}-induced current rundown, which in its turn could be blocked by an inhibitor of phosphoinositide 4-kinase, wortmannin. Application of PIP_2 also reduced the TRPV5 sensitivity to the slow reversible inhibition by increasing the Mg^{2+} concentration that is necessary for complete inhibition of the currents. Conversely, hydrolysis of PIP_2 by receptor activation of phospholipase C (PLC) increased the sensitivity to Mg^{2+}-induced inhibition. PLC-activating hormones may regulate TRPV5 channel activity via this mechanism.

5
Biological Function

To investigate the physiological function of TRPV5 and to reveal diseases associated with TRPV5 dysfunction, Hoenderop et al. (2003a) generated TRPV5-null ($Trpv5^{-/-}$) mice. These mice were characterized by a robust hypercalciuria, polyuria, and urinary acidification. Ca^{2+} loss via the urine was compensated by intestinal Ca^{2+} hyperabsorption due to the high levels of 1,25-$(OH)_2D_3$ (Renkema et al. 2005). This was associated with increased levels of intestinal TRPV6 and calbindin-D_{9k} expression in small intestine. Ca^{2+} reabsorption was affected only in DCT/CNT, where TRPV5 is localized. The absence of renal stone formation is likely due to the acidification of urine and polyuria (Baumann 1998). A surprising finding was the downregulation of renal calbindin-D_{28k} and NCX1, despite the increased 1,25-$(OH)_2D_3$ levels, which suggests a regulatory role of TRPV5. This was confirmed by van Abel et al. (2005) by showing that ruthenium red-induced downregulation of TRPV5 resulted in decreased mRNA levels for calbindin-D_{28k} and NCX1 in kidney. $Trpv5^{-/-}$ mice also displayed reduced bone thickness in the femoral head (Hoenderop et al. 2003a). The role of TRPV5 in bone formation, however, is still not clear. TRPV5 is only present in osteoclasts, the cells responsible for bone resorption, but not in bone-forming osteoblasts. As osteoclasts derived from $Trpv5^{-/-}$ bone marrow were not able to form resorption pits in vitro, the cause of the reduced bone formation in $Trpv5^{-/-}$ mice is still unanswered (van der Eerden et al. 2005). Although plasma Ca^{2+} levels were not affected in $Trpv5^{-/-}$ mice, it is likely that the maintenance of Ca^{2+} balance in these mice is fragile due to Ca^{2+} wasting in the urine. This may affect bone formation, but additional studies are needed to unravel the origin of bone defects in these mice.

6
Pharmacology

Some pharmacological agents used in the clinic affect Ca^{2+} reabsorption in the kidney. It is widely known that the thiazide diuretic, a blocker of the Na^+/Cl^- cotransporter (NCC), leads to hypocalciuria (Ellison 2000). It was anticipated that this is due to increased Ca^{2+} absorption in DCT, where TRPV5 and NCC reside (Costanzo and Windhager 1978). However, it has recently been shown that thiazide-induced hypocalciuria is due to an increased passive absorption of Ca^{2+} in PT rather than in the distal part of the nephron. This effect is likely due to contraction of the extracellular volume after the NaCl and water loss during thiazide treatment. The contraction of the extracellular volume is compensated by increased expression of the renal proximal Na^+/H^+ exchanger, thereby enhancing the electrochemical gradient and ultimately increasing paracellular Ca^{2+} absorption. This was confirmed by the observation that thiazide-induced hypocalciuria also occurs in $Trpv5^{-/-}$ mice, excluding a role for this Ca^{2+} channel (Nijenhuis et al. 2005).

The immunosuppressant FK506 (or tacrolimus) has major side effects on mineral homeostasis, including hypercalciuria, ultimately leading to osteoporosis (Stempfle et al. 2002). Treatment of rats with FK506 leads to decreased mRNA and protein levels of renal TRPV5 and calbindin-D_{28k}, explaining the hypercalciuria in these animals (Nijenhuis et al. 2004). The mechanism responsible for this effect has yet to be clarified, although a role for an FK-binding protein (FKBP) is postulated, as this protein family is known to be involved in ion channel regulation. Recently, Gkika et al. (2005) have shown that submicromolar concentrations of FK506 stimulate Ca^{2+} transport in primary CNT/CCD cells via suppression of the FK-binding protein FKBP52. Only at higher doses could a decrease in Ca^{2+} transport be observed, supporting the in vivo effect of FK506.

This chapter has provided insight in the mechanism and regulation of epithelial Ca^{2+} transport in the kidney. Cloning of TRPV5 and the characterization of the $Trpv5^{-/-}$ mouse model has been a major step forward in elucidating the molecular aspects of Ca^{2+} homeostasis. Dysfunction of this channel may well be associated with several disorders. $Trpv5^{-/-}$ mice, tissue-specific genetic ablation of this gene, and genomic analyses of patients resembling the $Trpv5^{-/-}$ phenotype will help to reveal the diseases associated with epithelial Ca^{2+} transport.

Acknowledgements This work was supported by the Human Frontiers Science Program (RGP32/2004), the Dutch Organization of Scientific Research (Zon-Mw 016.006.001, Zon-Mw 902.18.298, NWO-ALW 810.38.004, NWO-ALW 805-09.042, NWO-ALW 814-02.001, NWO 812-08.002), the Stomach, Liver, Intestine Foundation (MWO 03–19), and the Dutch Kidney Foundation (C10.1881 and C03.6017).

References

Baumann JM (1998) Stone prevention: why so little progress? Urol Res 26:77–81
Bindels RJ, Hartog A, Timmermans J, Van Os CH (1991) Active Ca^{2+} transport in primary cultures of rabbit kidney CCD: stimulation by 1,25-dihydroxyvitamin D3 and PTH. Am J Physiol 261:F799–807
Birkenfeld J, Kartmann B, Anliker B, Ono K, Schlotcke B, Betz H, Roth D (2003) Characterization of zetin 1/rBSPRY, a novel binding partner of 14-3-3 proteins. Biochem Biophys Res Commun 302:526–533
Brown EM, Gamba G, Riccardi D, Lombardi M, Butters R, Kifor O, Sun A, Hediger MA, Lytton J, Hebert SC (1993) Cloning and characterization of an extracellular Ca^{2+}-sensing receptor from bovine parathyroid. Nature 366:575–580
Chang Q, Hoefs S, van der Kemp AW, Topala CN, Bindels RJ, Hoenderop JG (2005) The beta-glucuronidase klotho hydrolyzes and activates the TRPV5 channel. Science 310:490–493
Clapham DE (2003) TRP channels as cellular sensors. Nature 426:517–524
Costanzo LS, Windhager EE (1978) Calcium and sodium transport by the distal convoluted tubule of the rat. Am J Physiol Renal Physiol 235:F492–506
Dang TX, McCleskey EW (1998) Ion channel selectivity through stepwise changes in binding affinity. J Gen Physiol 111:185–193
Dardenne O, Prud'homme J, Arabian A, Glorieux FH, St-Arnaud R (2001) Targeted inactivation of the 25-hydroxyvitamin D(3)-1(alpha)-hydroxylase gene (CYP27B1) creates an animal model of pseudovitamin D-deficiency rickets. Endocrinology 142:3135–3141

Drenth JP, Tahvanainen E, te Morsche RH, Tahvanainen P, Kaariainen H, Hockerstedt K, van de Kamp JM, Breuning MH, Jansen JB (2004) Abnormal hepatocystin caused by truncating PRKCSH mutations leads to autosomal dominant polycystic liver disease. Hepatology 39:924–931

Ellison DH (2000) Divalent cation transport by the distal nephron: insights from Bartter's and Gitelman's syndromes. Am J Physiol Renal Physiol 279:F616–625

Fraser DR, Kodicek E (1970) Unique biosynthesis by kidney of a biological active vitamin D metabolite. Nature 228:764–766

Fu H, Subramanian RR, Masters SC (2000) 14-3-3 proteins: structure, function, and regulation. Annu Rev Pharmacol Toxicol 40:617–647

Gerke V, Moss SE (2002) Annexins: from structure to function. Physiol Rev 82:331–371

Gkika D, Mahieu F, Nilius B, Hoenderop JG, Bindels RJ (2004) 80K-H as a new Ca^{2+} sensor regulating the activity of the epithelial Ca^{2+} channel transient receptor potential cation channel V5 (TRPV5). J Biol Chem 279:26351–26357

Gkika D, Topala CN, Hoenderop JGJ, Bindels RJM (2005) The immunophilin FKBP52 inhibits the activity of the epithelial Ca^{2+} channel TRPV5. Am J Physiol Renal Physiol 290:F1253–F1259

Harrison HE, Harrison HC (1960) Transfer of Ca45 across intestinal wall in vitro in relation to action of vitamin D and cortisol. Am J Physiol 199:265–271

Hirai M, Shimizu N (1990) Purification of two distinct proteins of approximate Mr 80,000 from human epithelial cells and identification as proper substrates for protein kinase C. Biochem J 270:583–589

Hoenderop JG, van der Kemp AW, Hartog A, van de Graaf SF, van Os CH, Willems PH, Bindels RJ (1999) Molecular identification of the apical Ca^{2+} channel in 1, 25-dihydroxyvitamin D3-responsive epithelia. J Biol Chem 274:8375–8378

Hoenderop JG, Muller D, Van Der Kemp AW, Hartog A, Suzuki M, Ishibashi K, Imai M, Sweep F, Willems PH, Van Os CH, Bindels RJ (2001) Calcitriol controls the epithelial calcium channel in kidney. J Am Soc Nephrol 12:1342–1349

Hoenderop JG, Dardenne O, Van Abel M, Van Der Kemp AW, Van Os CH, St -Arnaud R, Bindels RJ (2002a) Modulation of renal Ca^{2+} transport protein genes by dietary Ca^{2+} and 1,25-dihydroxyvitamin D3 in 25-hydroxyvitamin D3–1alpha-hydroxylase knockout mice. FASEB J 16:1398–1406

Hoenderop JG, Nilius B, Bindels RJ (2002b) Molecular mechanism of active Ca^{2+} reabsorption in the distal nephron. Annu Rev Physiol 64:529–549

Hoenderop JG, van Leeuwen JP, van der Eerden BC, Kersten FF, van der Kemp AW, Merillat AM, Waarsing JH, Rossier BC, Vallon V, Hummler E, Bindels RJ (2003a) Renal Ca^{2+} wasting, hyperabsorption, and reduced bone thickness in mice lacking TRPV5. J Clin Invest 112:1906–1914

Hoenderop JG, Voets T, Hoefs S, Weidema F, Prenen J, Nilius B, Bindels RJ (2003b) Homo- and heterotetrameric architecture of the epithelial Ca^{2+} channels TRPV5 and TRPV6. EMBO J 22:776–785

Imura A, Iwano A, Tohyama O, Tsuji Y, Nozaki K, Hashimoto N, Fujimori T, Nabeshima Y (2004) Secreted Klotho protein in sera and CSF: implication for post-translational cleavage in release of Klotho protein from cell membrane. FEBS Lett 565:143–147

Kuro-o M, Matsumura Y, Aizawa H, Kawaguchi H, Suga T, Utsugi T, Ohyama Y, Kurabayashi M, Kaname T, Kume E, Iwasaki H, Iida A, Shiraki-Iida T, Nishikawa S, Nagai R, Nabeshima YI (1997) Mutation of the mouse klotho gene leads to a syndrome resembling ageing. Nature 390:45–51

Lee J, Cha SK, Sun TJ, Huang CL (2005) PIP2 activates TRPV5 and releases its inhibition by intracellular Mg^{2+}. J Gen Physiol 126:439–451

Montell C, Birnbaumer L, Flockerzi V (2002) The TRP channels, a remarkably functional family. Cell 108:595–598
Muller D, Hoenderop JGJ, Merkx GFM, van Os CH, Bindels RJM (2000) Gene structure and chromosomal mapping of human epithelial calcium channel. Biochem Biophys Res Commun 275:47
Neer RM (1975) The evolutionary significance of vitamin D, skin pigment, and ultraviolet light. Am J Phys Anthropol 43:409–416
Nijenhuis T, Hoenderop JG, Nilius B, Bindels RJ (2003) (Patho)physiological implications of the novel epithelial Ca^{2+} channels TRPV5 and TRPV6. Pflugers Arch 446:401–409
Nijenhuis T, Hoenderop JG, Bindels RJ (2004) Downregulation of Ca^{2+} and Mg^{2+} transport proteins in the kidney explains tacrolimus (FK506)-induced hypercalciuria and hypomagnesemia. J Am Soc Nephrol 15:549–557
Nijenhuis T, Vallon V, van der Kemp AW, Loffing J, Hoenderop JG, Bindels RJ (2005) Enhanced passive Ca^{2+} reabsorption and reduced Mg^{2+} channel abundance explains thiazide-induced hypocalciuria and hypomagnesemia. J Clin Invest 115:1651–1658
Nilius B, Vennekens R, Prenen J, Hoenderop JG, Bindels RJ, Droogmans G (2000) Whole-cell and single channel monovalent cation currents through the novel rabbit epithelial Ca^{2+} channel ECaC. J Physiol 527:239–248
Nilius B, Prenen J, Vennekens R, Hoenderop JG, Bindels RJ, Droogmans G (2001a) Modulation of the epithelial calcium channel, ECaC, by intracellular Ca^{2+}. Cell Calcium 29:417–428
Nilius B, Vennekens R, Prenen J, Hoenderop JG, Droogmans G, Bindels RJ (2001b) The single pore residue Asp542 determines Ca^{2+} permeation and Mg^{2+} block of the epithelial Ca^{2+} channel. J Biol Chem 276:1020–1025
Panda DK, Miao D, Tremblay ML, Sirois J, Farookhi R, Hendy GN, Goltzman D (2001) Targeted ablation of the 25-hydroxyvitamin D 1alpha -hydroxylase enzyme: evidence for skeletal, reproductive, and immune dysfunction. Proc Natl Acad Sci U S A 98:7498–7503
Ponting C, Schultz J, Bork P (1997) SPRY domains in ryanodine receptors (Ca^{2+}-release channels). Trends Biochem Sci 22:193–194
Renkema KY, Nijenhuis T, van der Eerden BC, van der Kemp AW, Weinans H, van Leeuwen JP, Bindels RJ, Hoenderop JG (2005) Hypervitaminosis D mediates compensatory Ca^{2+} hyperabsorption in TRPV5 knockout mice. J Am Soc Nephrol 16:3188–3195
Stempfle HU, Werner C, Siebert U, Assum T, Wehr U, Rambeck WA, Meiser B, Theisen K, Gartner R (2002) The role of tacrolimus (FK506)-based immunosuppression on bone mineral density and bone turnover after cardiac transplantation: a prospective, longitudinal, randomized, double-blind trial with calcitriol. Transplantation 73:547–552
Tohyama O, Imura A, Iwano A, Freund JN, Henrissat B, Fujimori T, Nabeshima Y (2004) Klotho is a novel beta-glucuronidase capable of hydrolyzing steroid beta-glucuronides. J Biol Chem 279:9777–9784
Tsujikawa H, Kurotaki Y, Fujimori T, Fukuda K, Nabeshima Y (2003) Klotho, a gene related to a syndrome resembling human premature aging, functions in a negative regulatory circuit of vitamin D endocrine system. Mol Endocrinol 17:2393–2403
van Abel M, Hoenderop JG, van der Kemp AW, Friedlaender MM, van Leeuwen JP, Bindels RJ (2005) Coordinated control of renal Ca^{2+} transport proteins by parathyroid hormone. Kidney Int 68:1708–1721
van de Graaf SF, Hoenderop JG, Gkika D, Lamers D, Prenen J, Rescher U, Gerke V, Staub O, Nilius B, Bindels RJ (2003) Functional expression of the epithelial Ca^{2+} channels (TRPV5 and TRPV6) requires association of the S100A10-annexin 2 complex. EMBO J 22:1478–1487

van de Graaf SF, van der Kemp AW, van den Berg D, van Oorschot M, Hoenderop JG, Bindels RJ (2006b) Identification of BSPRY as a novel auxiliary protein inhibiting TRPV5 activity. J Am Soc Nephrol 17:26–30

van de Graaf SFJ, Chang Q, Mensenkamp AR, Hoenderop JGJ, Bindels RJM (2006a) Direct interaction with Rab11a targets the epithelial Ca^{2+} channels TRPV5 and TRPV6 to the plasma membrane. Mol Cell Biol 26:303–312

van der Eerden BC, Hoenderop JG, de Vries TJ, Schoenmaker T, Buurman CJ, Uitterlinden AG, Pols HA, Bindels RJ, van Leeuwen JP (2005) The epithelial Ca^{2+} channel TRPV5 is essential for proper osteoclastic bone resorption. Proc Natl Acad Sci U S A 102:17507–17512

van Os CH (1987) Transcellular calcium transport in intestinal and renal epithelial cells. Biochim Biophys Acta 906:195–222

Vennekens R, Hoenderop JG, Prenen J, Stuiver M, Willems PH, Droogmans G, Nilius B, Bindels RJ (2000) Permeation and gating properties of the novel epithelial Ca^{2+} channel. J Biol Chem 275:3963–3969

Vennekens R, Prenen J, Hoenderop JG, Bindels RJ, Droogmans G, Nilius B (2001) Modulation of the epithelial Ca^{2+} channel ECaC by extracellular pH. Pflugers Arch 442:237–242

Weber K, Erben RG, Rump A, Adamski J (2001) Gene structure and regulation of the murine epithelial calcium channels ECaC1 and 2. Biochem Biophys Res Commun 289:1287

Yeh BI, Kim YK, Jabbar W, Huang CL (2005) Conformational changes of pore helix coupled to gating TRPV5 by protons. EMBO J 24:3224–3234

TRPV6

U. Wissenbach (✉) · B. A. Niemeyer

Experimentelle und Klinische Pharmakologie und Toxikologie, Medizinische Fakultät, Universität des Saarlandes, 66421 Homburg, Germany
ulrich.wissenbach@uniklinikum-saarland.de

1	Basic Features	222
1.1	Nomenclature	222
1.2	Chromosomal Localization	222
1.3	Expression Pattern	223
2	Ion Channel Properties	224
3	Modes of Activation and Inactivation	227
4	Pharmacology	228
5	Biological Relevance	229
5.1	Involvement of TRPV6 in Prostate Cancer	230
References		230

Abstract The ion channel TRPV6 is likely to function as an epithelial calcium channel in organs with high calcium transport requirements such as the intestine, kidney, and placenta. Transcriptional regulation of TRPV6 messenger RNA (mRNA) is controlled by 1,25-dihydroxyvitamin D, which is the active hormonal form of vitamin D_3, and by additional calcium-dependent and vitamin D_3-independent mechanisms. Under physiological conditions, the conductance of the channel itself is highly calcium-selective and underlies complex inactivation mechanisms triggered by intracellular calcium and magnesium ions. There is growing evidence that transcriptional regulation of TRPV6 in certain tissues undergoing malignant transformation, such as prostate cancer, is linked to cancer progression.

Keywords TRPV6 · Prostate cancer · Endometrial cancer · Vitamin D_3 · Calcium channel

1
Basic Features

1.1
Nomenclature

At the beginning of the 1970s, O'Donnell and Smith demonstrated calcium and magnesium uptake in epithelial cells of the rat duodenum (O'Donnell and Smith 1973), and a few years later it was shown that lanthanum inhibited calcium uptake into these cells (Pento et al. 1978). To identify the molecule responsible for the calcium uptake, Peng and coworkers identified a protein through functional cloning from rat duodenum and named this protein CaT1 (calcium transport protein). CaT1 constitutes an ion channel that was shown to be sensitive to the trivalent ions lanthanum and gadolinium (Peng et al. 1999; Nilius et al. 2001). In an independent, in silico cloning approach, a human gene was isolated from placenta which was termed CaT-like (CaT-L) (Wissenbach et al. 2001). Because the overall identity between CaT1 and CaT-L (90% identical amino acids) is remarkably low in comparison to other TRP-orthologous proteins, and the expression pattern appeared to show some species-dependent differences in humans and rat, it was initially believed that CaT1 and CaT-L might be the products of individual genes. However, completion of the human genome sequencing project revealed that CaT1 and CaT-L are orthologous proteins with similarity to the family of transient receptor potential channels. Due to its homology to the epithelial calcium channel ECaC, a third nomenclature (ECaC2) was used by the group of Rene Bindels (Hoenderop et al. 2001). In 2002 a unified nomenclature was developed for proteins of the transient receptor potential (TRP) family and CaT1, ECaC2, and CaT-L are now named TRPV6 and belong to the TRPV subfamily consisting of the six members, TRPV1 to TRPV6 (Montell et al. 2002).

1.2
Chromosomal Localization

The human TRPV6 gene is located on chromosome 7q33–q34 in close proximity to its closest relative, TRPV5 on 7q35, and exhibits approximately 75% identical amino acids to TRPV5 (Peng et al. 2000; Müller et al. 2000; Peng et al. 2001; Hoenderop et al. 1999). The human chromosomal region corresponds to the murine chromosome 6 and to chromosome 4 of rat (Hirnet et al. 2003). It is likely that TRPV5 and TRPV6 arose by gene duplication from an ancestral gene; the pufferfish *Takifugu rubripes*, for example, has only one gene which is slightly more similar to TRPV6 than to TRPV5 (Qiu and Hogstrand 2004). The chromosomal organization of TRPV6 is conserved among several species. In the mouse genome, TRPV6 spans 15 exons and extends over a region of approximately 15.7 kb. Depending on the species, the deduced amino acid sequence results in 719, 725, and 727 amino acids in pufferfish, humans, and rodents, respectively. The calculated molecular weight of TRPV6 in humans is 83,21 kDa, with a pI of 7.56. Due to posttranslational glycosylation, heterologously expressed TRPV6 protein can be detected at molecular weights of around 75 to 85–100 kDa in a denaturing sodium dodecyl sulfate (SDS)-polyacrylamide gel (see below).

1.3
Expression Pattern

The TRPV6 gene was originally cloned from rat duodenum. Expression, as determined by Northern blot analysis, occurs predominantly in the small intestine (duodenum, proximal jejunum, cecum) and colon of rat (Peng et al. 1999). Interestingly, in human and murine tissues, no detectable level of TRPV6 was found in the small intestine and colon by Wissenbach et al. (2001) and Hirnet et al. (2003). However, duodenal TRPV6 transcripts were detected in variable amounts in an analysis of several human patients (Barley et al. 2001). Peng et al. demonstrated TRPV6 expression in isolated messenger RNA (mRNA) from human duodenum whereas no expression was detectable in mRNA samples of total small intestine as well as in colon (Peng et al. 2000). These discrepancies most likely arise from differences in the relative amount of TRPV6 mRNA in total mRNA samples from intestine versus duodenum. However, further differences seem to exist between rat and human (Table 1). In the rat, but not in human, TRPV6 transcripts were detectable in cecum and colon, whereas in stomach, transcripts were detected in human but not in rat (Peng et al. 1999, 2000; Wissenbach et al. 2001). Expression analysis is complicated because mRNA expression is differentially regulated depending on the hormonal status, calcium diet, and age (see Sect. 5 below).

Predominant TRPV6 expression is detected in placenta and pancreas of human and murine origin (Wissenbach et al. 2001; Peng et al. 2000; Hirnet et al. 2003). The expression pattern was confirmed by in situ hybridization, which showed expression in placental trophoblasts and syncytiotrophoblasts and in pancreatic acinar cells. In addition, expression was found in human

Table 1 Expression of TRPV6 transcripts in human and rat tissues as detected by Northern blot analysis. (See text for references)

Tissue	Small intestine	Duodenum	Ileum	Jejunum	Cecum	Colon	Placenta	Pancreas	Kidney	Stomach
Human	−	+	−	−	−	−	++	++	−/+[a]	+
Rat	++	++	−	−	++	+	nd	nd	−	−

nd, not determined [a]Different results reported

myoepithelial cells of the salivary gland (Wissenbach et al. 2001). Similar to rat, mice also show detectable duodenal TRPV6 mRNA levels by RT-PCR and real-time PCR (Hoenderop et al. 2001; Bouillon et al. 2003; van Abel et al. 2003; Nijenhuis et al. 2003).

However, contradictory results were obtained for TRPV6 expression in kidney. TRPV6 antibodies positively stain the apical domain of the distal convoluted tubules, connecting tubules, and cortical and medullary collecting ducts of the murine kidney, although comparatively low levels of mRNA were detectable by RT-PCR (Nijenhuis et al. 2003; Hoenderop et al. 2003). Additional data indicate that TRPV6 shows little to no expression in adult murine kidney (Hirnet et al. 2003; Song et al. 2003). In rat as well as in human kidney, TRPV6 transcripts were not detectable by Northern blot (Peng et al. 1999; Brown et al. 2005; Wissenbach et al. 2001; but see very weak signal in human kidney by Peng et al. 2000) or PCR analysis (Hoenderop et al. 2001). Similar to duodenal expression, regulation by age, salt levels, and hormonal status are likely to confound detection in the kidney (Lee et al. 2004; see review by Hoenderop et al. 2005). For reliable expression data on the TRPV6 protein, comparisons between TRPV6 wildtype and knockout mice will be necessary. In summary, by Northern blot analysis TRPV6 expression is clearly apparent in human and mouse placenta and pancreas, and in the small intestine of rat.

2
Ion Channel Properties

Hydropathy prediction algorithms suggest that TRPV6, like most of the other TRP proteins, contains six membrane-spanning regions with intracellularly located amino- and carboxytermini. The extracellular loop between the first and the second transmembrane domain (S1/S2 loop) contains an N-glycosylation site that is the only site glycosylated in TRPV6 (Hoenderop et al. 2003; Hirnet et al. 2003; Chang et al. 2005). In analogy to the voltage-gated potassium channels, it is likely that native TRPV6 channels are composed of tetramers with the ionic pore located between S5 and S6. The properties and architecture of the pore region for TRPV5 and TRPV6 have been studied in great detail. The most salient features of the pore is its high calcium selectivity ($P_{Ca}/P_{Na} > 100$) with a divalent permeation sequence of $Ca^{2+} > Ba^{2+} > Sr^{2+} > Mn^{2+}$ (Peng et al. 2000; Vennekens 2002). Under physiological conditions, TRPV6 thus conducts calcium exclusively, and only upon switching to a divalent-free extracellular solution does the channel become permeable to monovalent ions; upon reexposure to physiological solution, inward currents are initially reduced and then increase, indicating an anomalous mole fraction behavior (Wissenbach et al. 2001; Vennekens 2002). In the absence of divalent cations, the pore shows a monovalent permeation sequence of $Rb^+ \cong K^+ > Na^+ > Li^+$ for rat TRPV6 (Eisenman V or VI, Peng et al. 1999) but a permeation sequence of $Na^+ \cong Li^+ > K^+ > Cs^+$ for mouse TRPV6 (Eisenman sequence X or XI, Hoenderop et al. 2001). Single channel conductance measurements have only been performed for monovalent ions and range between 40 and 70 pS (Yueet al. 2001; Vassilevet al. 2001a) and have been used as one argument why TRPV6 might constitute a part of the calcium release-activated calcium current (CRAC) channel (Hoth and Penner 1992). However, a number of distinct properties including relative permeabilities for monovalent ions, block by 2-aminoethoxydiphenyl borate (2-APB), and open pore blockage by intracellular Mg^{2+}, among others, argue against a contributory role of TRPV6 in CRAC (Voets et al. 2001; for extensive discussion see review by Hoenderop et al. 2005).

The selectivity filter of TRPV6/TRPV5 is formed by the acidic side chains of aspartate residues within the narrowest point (5.4 Å) of the pore. A single aspartate pore residue thus determines calcium selectivity, inward rectification, and pore size and is also the binding site for Mg^{2+}-dependent block (Voets et al. 2003, 2004; for an extensive review on TRPV5/TRPV6 pore properties see Owsianiket al. 2006). Three-dimensional pore architecture has been modeled based on similarities to the bacterial potassium channel KcsA (Voets et al. 2004). Tetramer formation has been shown for TRPV5 and TRPV6, and it has furthermore been demonstrated that TRPV6 can also form functional heteromeric channels with TRPV5 (Hoenderop et al. 2003a; Erler et al. 2004). Primary sequence analysis of TRPV6 furthermore suggests that the hydrophilic N-terminal region of TRPV6 contains at least five, possibly six, ankyrin (ANK) repeats (Wissenbach et al. 2001). ANK repeats are 33-amino-acid-long motifs whose name stems from finding 22 of these motifs in the ankyrin protein (Lux et al. 1990). ANK repeats are rather variable in primary sequence, but show a high degree of conservation regarding their secondary and tertiary structure. Mosavi et al. analyzed approximately 4,000 ANK repeat-containing proteins and used the consensus information to design an idealized ANK repeat and to define amino acids necessary for structure versus more variable surface residues that may confer specificity upon the interaction with other proteins (Mosavi et al. 2002; Yuan et al. 2004; for reviews see Sedgwick and Smerdon 1999; Mosavi et al. 2004).

For TRPV6 it has been shown that for dominant-negative effects on TRPV6-mediated currents, the first 154 amino acids containing the first three ANK repeats are sufficient, whereas the N-terminal region of TRPC3 with four ANK repeats does not exert any dominant-negative effect (Kahret al. 2004). To decipher the codes for intersubunit assembly sites within TRPV6 proteins, Erler et al. used both a bacterial two-hybrid interaction screen as well as coimmunoprecipitation of tagged protein fragments with full-length TRPV6 from a stable cell line to narrow down interacting domains. Two N-terminal interactions sites were found that both reside in the ANK-repeat domains. The primary site, which included the third ANK repeat, appeared to be the dominant site, as deletions or point mutations within this domain abolished interaction with the full-length channel, destroyed the ability to form tetramers as detected on sucrose density gradients, and could furthermore relieve the dominant-negative effect of TRPV6 subunits with a mutated ionic pore (Erler et al. 2004). These studies suggest a model where tetramerization initiates via a zippering of the ANK repeats. Chang et al. identified interaction motives both within the hydrophilic C-terminus as well as an N-terminal region around the first ANK repeat of the TRPV5 protein (Chang et al. 2004) that are important for channels assembly. Because both channels are able to form functional heteromultimers, one would expect conserved sites for subunit assembly (for a review see also Niemeyer 2005). In a very recent study, the structure of the N-terminal Ankyrin repeat domain (ARD) of the TRPV2 ion channel has been solved and there has been no indication that TRPV2-ARD alone can form homo-oligomers in solution (Jin et al. 2006). As there is significant homology in the ARD region of all TRPV channels, it appears that the final mechanism of assembly still awaits clarification.

Shortly after the last transmembrane domain (S6), members of the TRPC, TRPV, and TRPM subfamilies contain a highly conserved Trp box that is followed by a less well conserved Trp domain (both together encompass 25 amino acids). Conserved positively charged amino acids in this region can confer binding of phosphatidylinositol-4,5 bisphosphate [PI(4,5)P$_2$] to the channel (Rohacs et al. 2005). Binding of PI(4,5)P$_2$ activates TRPV5 and TRPV6 and releases channel inhibition by intracellular Mg^{2+} (shown for TRPV5 by Lee et al. 2005).

The TRP domain within TRPV5 and TRPV6 has also been implicated in binding to a variety of different binding partners that also regulate channel activity. The VATTV region within this domain has been shown to be involved in the binding of S100A10-annexin 2, an interaction that is important for surface expression of the channel (van de Graaf 2003).

Interestingly, the site immediately adjacent to the VATTV motif, namely the MLERK motif, has been shown to bind the GDP bound form of rab11a (van de Graaf 2006), and this interaction is also critical for surface expression of TRPV5/6. The same amino acids are also implicated in binding of 80K-H, a calcium-binding protein (Gkika et al. 2004), and in the self-assembly of

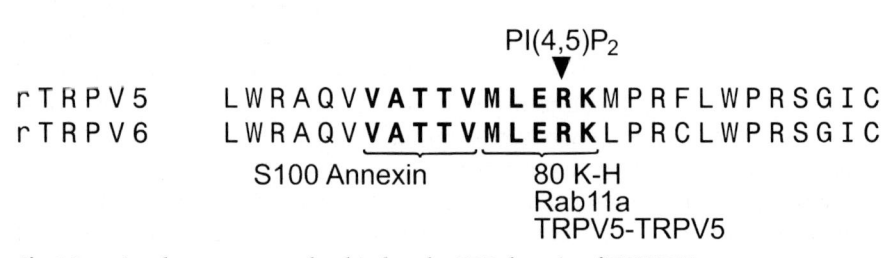

PI(4,5)P$_2$
▼
rTRPV5   LWRAQV**VATTV****MLERK**MPRFLWPRSGIC
rTRPV6   LWRAQV**VATTV****MLERK**LPRCLWPRSGIC
              S100 Annexin   80 K-H
                                  Rab11a
                                  TRPV5-TRPV5

Fig. 1 Proteins that are reported to bind to the TRP domain of TRPV5/6

TRPV5 multimers (Chang et al. 2004). Given the fact that the arginine of this motif is also critical for binding of PI(4,5)P$_2$ (Rohacs et al. 2005), it will be necessary to separate the individual contributions of all interacting molecules on the activity of TRPV5/6 and place them in a temporal and local native context (see Fig. 1).

3
Modes of Activation and Inactivation

Heterologously expressed, TRPV6 forms constitutively active ion channels at low intracellular calcium concentrations and negative voltage. In physiological solutions, the selectivity filter of the pore is occupied by Ca^{2+} and blocks movement of monovalent ions. Ca^{2+} permeates the pore when hyperpolarization increases its driving force and intracellular calcium is low to relieve calcium dependent inactivation. Outward currents are very small, indicating that the channel is almost entirely inward rectifying. In the absence of divalent cations, TRPV6 channels become permeable for monovalent ions (Peng et al. 1999; Wissenbach et al. 2001; Hoenderop et al. 2001; Bödding 2005).

Several mechanisms controlling the activity of TRPV6 have been proposed. Due to its constitutive activity, regulation of plasma membrane localization is one way to alter total currents. Evidence includes regulation of surface expression by binding of S100 annexin A2 and rab11a (see also above). Small annexin 2-specific interfering RNA experiments indicated reduced TRPV5 activity as a result from reduced trafficking to the plasma membrane, and a similar mechanism was postulated to regulate TRPV6 transport to the membrane (van de Graaf et al. 2003). Coexpression of TRPV5/6 with a GDP-locked form of rab11a also reduces surface expression and thereby current density (van de Graaf 2006).

Trapping TRPV5/6 in the plasma membrane can be achieved by involvement of the β-glucuronidase klotho. Klotho hydrolyzes extracellular sugar residues on TRPV5, enhancing currents by locking the channel in the plasma membrane (Chang et al. 2005). This effect can be blocked by D-saccharic acid 1,4-lactone, an inhibitor of klotho activity. TRPV6 channels with an identical glycosylation site (Hirnet et al. 2003) are also influenced by the action of klotho. Sternfeld demonstrated that Src-dependent tyrosine phosphorylation enhanced TRPV6 activity in an overexpression system. This activation can be prevented by dephosphorylation by the tyrosine phosphatase PTP1B (Sternfeld et al. 2005), but it is unclear whether or not surface expression is affected.

Binding of PI(4,5)P$_2$ can activate TRPV5 and TRPV6 and furthermore decreases sensitivity of the channel to Mg^{2+}-induced slow inhibition (Rohacs et al. 2005; Lee et al. 2005).

One characteristic feature of TRPV6 is the initially rapid and subsequently slower decay of its calcium currents during prolonged stimulation by hyperpolarizing potentials. The calcium-dependent components become apparent when Ca^{2+} is substituted with Ba^{2+}. When applying a hyperpolarizing pulse in the presence of Ca^{2+}, TRPV6 initially show very rapid inactivation kinetics followed by a multiphasic slower phase (Niemeyer et al. 2001). The initial rapid inactivation is not as prominent in TRPV5, and the region responsible for this difference was mapped to the intracellular loop between transmembrane regions S3 and S4 (Nilius et al. 2002). The subsequent slower phase of inactivation is governed by calcium-dependent binding of calmodulin (CAM) to a very C-terminal binding site. Calmodulin binding occurs when calcium concentrations rise significantly above resting levels, thus contributing to inactivation. In human TRPV6, a protein kinase C (PKC) phosphorylation site within the calmodulin-binding region exists, and phosphorylation by PKC prevents binding of calmodulin, resulting in a dual regulation of the channel's activity by PKC and calmodulin (Niemeyer et al. 2001). More recently, a different group investigated CAM-binding sites within mouse TRPV6 and found additional binding sites within the N-terminal domain, the transmembrane domain, and a different C-terminal domain (Lambers et al. 2004).

4
Pharmacology

Extracellular addition of trivalent ions such as La^{3+} and Gd^{3+}—as well as by Cu^{2+}, Pb^{2+}, and Cd^{2+} (at 100 µM each)—inhibits $^{45}Ca^{2+}$ uptake mediated by TRPV6 by more than 50% (see Peng et al. 1999, 2000). Extracellular Mg^{2+} blocks TRPV6 with an IC_{50} value of approximately 62 µM in the presence of calcium and an IC_{50} of approximately 328 µM in the absence of extracellular Ca^{2+} (Vennekens 2002). The characteristic voltage-dependent gating and inward rectification depend on the voltage-dependent binding of magnesium ions within the channel pore. Neutralization of a single aspartate residue (D542A) in the pore region thus results in a channel with monovalent currents that are insensitive to the inhibition by magnesium (Nilius et al. 2001b; Voets et al. 2003). This aspartate residue is critical for the calcium selectivity of TRPV5/6 channels and is not conserved among other members of the nonselective TRPV channels. Ruthenium red blocks TRPV6 with an IC_{50} of 9 µM, demonstrating that these channels are much less sensitive to ruthenium red than TRPV5 (IC_{50} 121 nM; Hoenderop 2001). The IC_{50} value for block of TRPV5 by econazole is 1.3 µM (Nilius et al. 2001a) and 0.6 µM econazole has been used to reduce TRPV6-mediated cell proliferation (Schwarz et al. 2006). The noncompetitive IP_3 receptor antagonist xestospongin C seems to block single channel currents of a fraction of TRPV6-expressing oocytes (Vassilev et al. 2001b); however, further validation is necessary. Econazole and ruthenium red are rather unspecific channel blockers (Franzius et al. 1994; Tapia and Velasco 1997); a specific TRPV6 blocker is currently not known.

5
Biological Relevance

The gastrointestinal tract, bone, and kidney are involved in the regulation of calcium balance. Normally, secretion of calcium by the kidney is compensated for through absorption of calcium in the small intestine. Calcium flux is controlled by the synergistic action of 1,25-dihydroxyvitamin D_3 and the parathyroid hormone, involving the bones as calcium stores. Under conditions of limiting Ca^{2+} concentration, synthesis of 1,25-dihydroxyvitamin D_3 is stimulated by the parathyroid hormone, resulting in accelerated calcium uptake by the small intestine, reabsorption by the kidney, and resorption of calcium from existing bone.

Initially, Peng et al. demonstrated by Northern blot analysis that TRPV6 transcripts are not upregulated in duodenal epithelial cells of rats after treatment with vitamin D_3 [1,25(OH)$_2$ D_3] (Peng et al. 1999); however, these experiments were conducted in animals that were not depleted of vitamin D_3.

A mouse knockout model characterized by hyperparathyroidism, rickets, hypocalcemia, and undetectable levels of 1,25(OH)$_2$ D_3 (Dardenne et al. 2001) was used to test the vitamin D_3-dependence of TRPV6 expression. In the 25-hydroxyvitamin D_3-1α-hydroxylase-deficient mouse model, van Abel et al. demonstrated upregulation of duodenal TRPV6 transcripts by quantitative real time PCR after supplementation with 1,25(OH)$_2$ D_3 (van Abel et al. 2003). In addition, a cyclic time-dependent regulation of TRPV6 mRNA levels after 1,25(OH)$_2$ D_3 or 1(OH) D_2 supplementation in the same mouse model has also been shown (Hoenderop et al. 2004). In agreement with these results, in vitamin D_3-depleted rats, duodenal TRPV6 transcripts returned to normal levels after supplementation with vitamin D_3. Furthermore, duodenal TRPV6 expression was shown to be age-dependent (Brown et al. 2005). In TRPV5 knockout mice, which exhibit a compensatory enhanced vitamin D_3 level, upregulation of intestinal TRPV6 accompanied by intestinal hyperabsorption of Ca^{2+} was demonstrated (Hoenderop et al. 2003b). This hyperabsorption was abolished in a TRPV5/25-hydroxyvitamin D_3-1α-hydroxylase double knockout (Renkema et al. 2005). Together these data indicate that TRPV6 is involved in vitamin D_3, depending intestinal calcium adsorption. Given its strong expression in placenta, it has been postulated that TRPV6 might take part of the transplacental calcium transport in trophoblasts (Wissenbach et al. 2001) Expression of TRPV6 and TRPV5 in cultured human syncytiotrophoblasts has also been demonstrated by Northern blot analysis as well as by RT-PCR, and a dose-dependent inhibition of Ca^{2+} uptake into these cells by ruthenium red and magnesium has been demonstrated (Moreau et al. 2002). Duodenal upregulation of TRPV6 transcripts was demonstrated in lactating as well as in estrogen-treated mice, indicating that changes in hormonal status are involved in TRPV6 regulation (van Cromphaut et al. 2001).

5.1
Involvement of TRPV6 in Prostate Cancer

Surprisingly, while a commercially available Northern blot showed a strong expression of TRPV6 transcripts in prostate, a Northern blot containing only benign prostate tissue collected from ten individuals did not exhibit any detectable levels of TRPV6 transcripts (Wissenbach et al. 2001, 2004b; Peng et al. 2001). However, the commercially available Northern blot contained pooled tissues from humans age 14 to 70, with unidentified prostate pathology. In situ hybridization of prostate samples from 140 individuals with a TRPV6-specific probe showed a strict correlation between the histopathological tumor staging and degree of TRPV6 expression, while TRPV6 was absent in benign or healthy tissue (Fixemer et al. 2003). Small tumors of stage pT1 (which are restricted to the prostate) do not express detectable amounts of TRPV6 transcripts, whereas expression was detectable in 20% of all cases with tumor stage pT2 (restricted to the prostate), 79% in the stage pT3a (breakout of the prostate capsule) and in 90% of the stage pT3b (metastasis in the seminal vesicles). Therefore, TRPV6 expression correlates with the malignancy of prostate cancer and might be useful for differential diagnosis and a promising target for treatment of prostate cancer. Similar to its expression in prostate cancer, TRPV6 transcripts are detected in endometrial cancer but not in the corresponding healthy endometrium (Wissenbach et al. 2004b). Although all endometrial cancers tested were TRPV6-positive, more statistical data are necessary. Immunohistochemical analysis with polyclonal TRPV6 antibodies indicated that TRPV6 protein expression is also upregulated in breast, thyroid, ovary, and colon cancer, but statistical data are lacking (Zhuang et al. 2002).

Evidence for a physiological link stems from experiments by Schwarz and coworkers who demonstrate enhanced proliferation rates of TRPV6-expressing cells. This enhanced proliferation is due to the increased intracellular calcium levels in TRPV6-expressing cells and can be blocked by econazole (Schwarz et al. 2006).

References

Barley NF, Howard A, O'Callaghan D, Legon S, Walters JR (2001) Epithelial calcium transporter expression in human duodenum. Am J Physiol Gastrointest Liver Physiol 280: G285–G290

Bödding M (2005) Voltage-dependent changes of TRPV6-mediated Ca^{2+} currents. J Biol Chem 280:7022–7029

Bouillon R, van Cromphaut S, Carmeliet G (2003) Intestinal calcium absorption: molecular vitamin D mediated mechanisms. J Cell Biochem 88:332–339

Brown AJ, Krits I, Armbrecht HJ (2005) Effect of age, vitamin D, and calcium on the regulation of rat intestinal epithelial calcium channels. Arch Biochem Biophys 437:51–58

Chang Q, Gyftogianni E, van de Graaf SF, Hoefs S, Weidema FA, Bindels RJ, Hoenderop JG (2004) Molecular determinants in TRPV5 channel assembly. J Biol Chem 279:54304–54311

Chang Q, Hoefs S, van der Kemp AW, Topala CN, Bindels RJ, Hoenderop JG (2005) The beta-glucuronidase klotho hydrolyzes and activates the TRPV5 channel. Science 310:490–493

Dardenne O, Prud'homme J, Arabian A, Glorieux FH, St-Arnaud R (2001) Targeted inactivation of the 25-hydroxyvitamin D(3)-1(alpha)-hydroxylase gene (CYP27B1) creates an animal model of pseudovitamin D-deficiency rickets. Endocrinology 142:3135–3141

Erler I, Hirnet D, Wissenbach U, Flockerzi V, Niemeyer BA (2004) Ca^{2+}-selective transient receptor potential V channel architecture and function require a specific ankyrin repeat. J Biol Chem 279:34456–34463

Fixemer T, Wissenbach U, Flockerzi V, Bonkhoff H (2003) Expression of the Ca^{2+}-selective cation channel TRPV6 in human prostate cancer: a novel prognostic marker for tumor progression. Oncogene 22:7858–7861

Franzius D, Hoth M, Penner R (1994) Non-specific effects of calcium entry antagonists in mast cells. Pflugers Arch 428:433–438

Gkika D, Mahieu F, Nilius B, Hoenderop JG, Bindels RJ (2004) 80K-H as a new Ca^{2+} sensor regulating the activity of the epithelial Ca^{2+} channel transient receptor potential cation channel V5 (TRPV5). J Biol Chem 279:26351–26357

Hirnet D, Olausson J, Fecher-Trost C, Bödding M, Nastainczyk W, Wissenbach U, Flockerzi V, Freichel M (2003) The TRPV6 gene, cDNA and protein. Cell Calcium 33:509–518

Hoenderop JG, van der Kemp AW, Hartog A, van de Graaf SF, van Os CH, Willems PH, Bindels RJ (1999) Molecular identification of the apical Ca^{2+} channel in 1, 25-dihydroxyvitamin D3-responsive epithelia. J Biol Chem 274:8375–8378

Hoenderop JG, Vennekens R, Muller D, Prenen J, Droogmans G, Bindels RJ, Nilius B (2001) Function and expression of the epithelial Ca^{2+} channel family: comparison of mammalian ECaC1 and 2. J Physiol 537:747–761

Hoenderop JG, Voets T, Hoefs S, Weidema F, Prenen J, Nilius B, Bindels RJ (2003a) Homo- and heterotetrameric architecture of the epithelial Ca^{2+} channels TRPV5 and TRPV6. EMBO J 22:776–785

Hoenderop JG, van Leeuwen JP, van der Eerden BC, Kersten FF, van der Kemp AW, Merillat AM, Waarsing JH, Rossier BC, Vallon V, Hummler E, Bindels RJ (2003b) Renal Ca^{2+} wasting, hyperabsorption, and reduced bone thickness in mice lacking TRPV5. J Clin Invest 112:1906–1914

Hoenderop JG, van der Kemp AW, Urben CM, Strugnell SA, Bindels RJ (2004) Effects of vitamin D compounds on renal and intestinal Ca^{2+} transport proteins in 25-hydroxyvitamin D3-1alpha-hydroxylase knockout mice. Kidney Int 66:1082–1089

Hoenderop JG, Nilius B, Bindels RJ (2005) Calcium absorption across epithelia. Physiol Rev 85:373–422

Hoth M, Penner R (1992) Depletion of intracellular calcium stores activates a calcium current in mast cells. Nature 355:353–356

Jin X, Touhey J, Gaudet R (2006) Structure of the N-terminal Ankyrin repat domain of the TRPV2 ion channel. J Biol Chem Jun 29 [Epub ahead of print]

Kahr H, Schindl R, Fritsch R, Heinze B, Hofbauer M, Hack ME, Mortelmaier MA, Groschner K, Peng JB, Takanaga H, Hediger MA, Romanin C (2004) CaT1 knock-down strategies fail to affect CRAC channels in mucosal-type mast cells. J Physiol 557:121–132

Lambers TT, Weidema AF, Nilius B, Hoenderop JG, Bindels RJ (2004) Regulation of the mouse epithelial $Ca^{2(+)}$ channel TRPV6 by the $Ca^{(2+)}$-sensor calmodulin. J Biol Chem 279:28855–28861

Lee CT, Shang S, Lai LW, Yong KC, Lien YH (2004) Effect of thiazide on renal gene expression of apical calcium channels and calbindins. Am J Physiol Renal Physiol 287:F1164–F1170

Lee J, Cha SK, Sun TJ, Huang CL (2005) PIP$_2$ activates TRPV5 and releases its inhibition by intracellular Mg^{2+}. J Gen Physiol 126:439–451

Lux SE, John KM, Bennett V (1990) Analysis of cDNA for human erythrocyte ankyrin indicates a repeated structure with homology to tissue-differentiation and cell-cycle control proteins. Nature 344:36–42

Montell C, Birnbaumer L, Flockerzi V, Bindels RJ, Bruford EA, Caterina MJ, Clapham DE, Harteneck C, Heller S, Julius D, Kojima I, Mori Y, Penner R, Prawitt D, Scharenberg AM, Schultz G, Shimizu N, Zhu MX (2002) A unified nomenclature for the superfamily of TRP cation channels. Mol Cell 9:229–231

Moreau R, Daoud G, Bernatchez R, Simoneau L, Masse A, Lafond J (2002) Calcium uptake and calcium transporter expression by trophoblast cells from human term placenta. Biochim Biophys Acta 1564:325–332

Mosavi LK, Minor DL Jr, Peng ZY (2002) Consensus-derived structural determinants of the ankyrin repeat motif. Proc Natl Acad Sci U S A 99:16029–16034

Mosavi LK, Cammett TJ, Desrosiers DC, Peng ZY (2004) The ankyrin repeat as molecular architecture for protein recognition. Protein Sci 13:1435–1448

Müller D, Hoenderop JG, Merkx GF, van Os CH, Bindels RJ (2000) Gene structure and chromosomal mapping of human epithelial calcium channel. Biochem Biophys Res Commun 275:47–52

Niemeyer BA (2005) Structure-function analysis of TRPV channels. Naunyn Schmiedebergs Arch Pharmacol 371:285–294

Niemeyer BA, Bergs C, Wissenbach U, Flockerzi V, Trost C (2001) Competitive regulation of CaT-like-mediated Ca^{2+} entry by protein kinase C and calmodulin. Proc Natl Acad Sci U S A 98:3600–3605

Nijenhuis T, Hoenderop JG, van der Kemp AW, Bindels RJ (2003) Localization and regulation of the epithelial Ca^{2+} channel TRPV6 in the kidney. J Am Soc Nephrol 14:2731–2740

Nilius B, Prenen J, Vennekens R, Hoenderop JG, Bindels RJ, Droogmans G (2001a) Pharmacological modulation of monovalent cation currents through the epithelial Ca^{2+} channel ECaC1. Br J Pharmacol 134:453–462

Nilius B, Vennekens R, Prenen J, Hoenderop JG, Droogmans G, Bindels RJ (2001b) The single pore residue Asp542 determines Ca^{2+} permeation and Mg^{2+} block of the epithelial Ca^{2+} channel. J Biol Chem 276:1020–1025

Nilius B, Prenen J, Hoenderop JG, Vennekens R, Hoefs S, Weidema AF, Droogmans G, Bindels RJ (2002) Fast and slow inactivation kinetics of the Ca^{2+} channels ECaC1 and ECaC2 (TRPV5 and TRPV6) Role of the intracellular loop located between transmembrane segments 2 and 3. J Biol Chem 277:30852–30858

O'Donnell JM, Smith MW (1973) Uptake of calcium and magnesium by rat duodenal mucosa analysed by means of competing metals. J Physiol 229:733–749

Owsianik G, Talavera K, Voets T, Nilius B (2006) Permeation and selectivity of TRP channels. Annu Rev Physiol 68:685–717

Peng JB, Chen XZ, Berger UV, Vassilev PM, Tsukaguchi H, Brown EM, Hediger MA (1999) Molecular cloning and characterization of a channel-like transporter mediating intestinal calcium absorption. J Biol Chem 274:22739–22746

Peng JB, Chen XZ, Berger UV, Weremowicz S, Morton CC, Vassilev PM, Brown EM, Hediger MA (2000) Human calcium transport protein CaT1. Biochem Biophys Res Commun 278:326–332

Peng JB, Brown EM, Hediger MA (2001) Structural conservation of the genes encoding CaT1, CaT2, and related cation channels. Genomics 76:99–109

Pento JT (1978) Influence of lanthanum on calcium transport and retention in the rat duodenum. Nutr Metab 22:362–367

Qiu A, Hogstrand C (2004) Functional characterisation and genomic analysis of an epithelial calcium channel (ECaC) from pufferfish, *Fugu rubripes*. Gene 342:113–123

Renkema KY, Nijenhuis T, van der Eerden BC, van der Kemp AW, Weinans H, van Leeuwen JP, Bindels RJ, Hoenderop JG (2005) Hypervitaminosis D mediates compensatory Ca^{2+} hyperabsorption in TRPV5 knockout mice. J Am Soc Nephrol 16:3188–3195

Rohacs T, Lopes CM, Michailidis I, Logothetis DE (2005) PI(4,5)P2 regulates the activation and desensitization of TRPM8 channels through the TRP domain. Nat Neurosci 8:626–634

Schwarz EC, Wissenbach U, Niemeyer BA, Strauss B, Philipp SE, Flockerzi V, Hoth M (2006) TRPV6 potentiates calcium-dependent cell proliferation. Cell Calcium 39:163–173

Sedgwick SG, Smerdon SJ (1999) The ankyrin repeat: a diversity of interactions on a common structural framework. Trends Biochem Sci 24:311–316

Sternfeld L, Krause E, Schmid A, Anderie I, Latas A, Al-Shaldi H, Kohl A, Evers K, Hofer HW, Schulz I (2005) Tyrosine phosphatase PTP1B interacts with TRPV6 in vivo and plays a role in TRPV6-mediated calcium influx in HEK293 cells. Cell Signal 17:951–960

Tapia R, Velasco I (1997) Ruthenium red as a tool to study calcium channels, neuronal death and the function of neural pathways. Neurochem Int 30:137–147

van Abel M, Hoenderop JG, van der Kemp AW, van Leeuwen JP, Bindels RJ (2003) Regulation of the epithelial Ca^{2+} channels in small intestine as studied by quantitative mRNA detection. Am J Physiol Gastrointest Liver Physiol 285:G78–85

van Cromphaut SJ, Rummens K, Stockmans I, van Herck E, Dijcks FA, Ederveen AG, Carmeliet P, Verhaeghe J, Bouillon R, Carmeliet G (2003) Intestinal calcium transporter genes are upregulated by estrogens and the reproductive cycle through vitamin D receptor-independent mechanisms. J Bone Miner Res 18:1725–1736

van de Graaf SF, Hoenderop JG, Gkika D, Lamers D, Prenen J, Rescher U, Gerke V, Staub O, Nilius B, Bindels RJ (2003) Functional expression of the epithelial $Ca^{(2+)}$ channels (TRPV5 and TRPV6) requires association of the S100A10-annexin 2 complex. EMBO J 22:1478–1487

van de Graaf SF, Chang Q, Mensenkamp AR, Hoenderop JG, Bindels RJ (2006) Direct interaction with Rab11a targets the epithelial Ca^{2+} channels TRPV5 and TRPV6 to the plasma membrane. Mol Cell Biol 26:303–312

Vassilev PM, Peng JB, Hediger MA, Brown EM (2001a) Single-channel activities of the human epithelial Ca^{2+} transport proteins CaT1 and CaT2. J Membr Biol 184:113–120

Vassilev PM, Peng JB, Johnson J, Hediger MA, Brown EM (2001b) Inhibition of CaT1 channel activity by a noncompetitive IP_3 antagonist. Biochem Biophys Res Commun 280:145–150

Vennekens R (2002) Molecular and biophysical analysis of the epithelial calcium ECAC1 and ECAC2. Leuven University Press, Leuven

Vennekens R, Hoenderop JG, Prenen J, Stuiver M, Willems PH, Droogmans G, Nilius B, Bindels RJ (2000) Permeation and gating properties of the novel epithelial $Ca^{(2+)}$ channel. J Biol Chem 275:3963–3969

Voets T, Prenen J, Fleig A, Vennekens R, Watanabe H, Hoenderop JG, Bindels RJ, Droogmans G, Penner R, Nilius B (2001) CaT1 and the calcium release-activated calcium channel manifest distinct pore properties. J Biol Chem 276:47767–47770

Voets T, Janssens A, Prenen J, Droogmans G, Nilius B (2003) Mg^{2+}-dependent gating and strong inward rectification of the cation channel TRPV6. J Gen Physiol 121:245–260

Voets T, Janssens A, Droogmans G, Nilius B (2004) Outer pore architecture of a Ca^{2+}-selective TRP channel. J Biol Chem 279:15223–15230

Wissenbach U, Niemeyer BA, Fixemer T, Schneidewind A, Trost C, Cavalie A, Reus K, Meese E, Bonkhoff H, Flockerzi V (2001) Expression of CaT-like, a novel calcium-selective channel, correlates with the malignancy of prostate cancer. J Biol Chem 276:19461–19468

Wissenbach U, Bonkhoff H, Flockerzi V (2004a) TRP TRPV6 and cancer. Nova Acta Leopold 336:69–74

Wissenbach U, Niemeyer B, Himmerkus N, Fixemer T, Bonkhoff H, Flockerzi V (2004b) TRPV6 and prostate cancer: cancer growth beyond the prostate correlates with increased TRPV6 Ca^{2+} channel expression. Biochem Biophys Res Commun 322:1359–1363

Yuan C, Li J, Mahajan A, Poi MJ, Byeon IJ, Tsai MD (2004) Solution structure of the human oncogenic protein gankyrin containing seven ankyrin repeats and analysis of its structure-function relationship. Biochemistry 43:12152–12161

Yue L, Peng JB, Hediger MA, Clapham DE (2001) CaT1 manifests the pore properties of the calcium-release-activated calcium channel. Nature 410:705–709

Zhuang L, Peng JB, Tou L, Takanaga H, Adam RM, Hediger MA, Freeman MR (2002) Calcium-selective ion channel, CaT1, is apically localized in gastrointestinal tract epithelia and is aberrantly expressed in human malignancies. Lab Invest 82:1755–1764

Part III
TRPM Channel Subfamily

TRPM2

J. Eisfeld · A. Lückhoff (✉)

Institut für Physiologie, Medizinische Fakultät, RWTH Aachen,
Pauwelsstr. 30, 52057 Aachen, Germany
luckhoff@physiology.rwth-aachen.de

1	Features of Genes and Proteins, Expression Pattern	238
2	Ion Channel Properties	239
3	Modes of Activation	240
3.1	ADPR	240
3.2	NAD	241
3.3	H_2O_2 and Oxidative Stress	241
3.4	Alloxan	242
3.5	Co-operation of Intracellular Ca^{2+} with ADPR	242
3.6	cADPR	243
3.7	Other Regulators of TRPM2 Channels	243
4	Pharmacology	244
4.1	Channel Blockers	244
4.1.1	Fenamates	244
4.1.2	The Anti-fungal Agents Clotrimazole and Econazole	244
4.1.3	Lanthanide and Heavy Metal Ions	245
4.2	Drugs Interfering with TRPM2 Activation	245
4.2.1	Anti-oxidants	245
4.2.2	PARP Inhibitors	246
4.2.3	Glycohydrolase Inhibitors	246
4.2.4	Inhibition of Receptor-Mediated Activation of CD38	247
5	Biological Relevance	247
5.1	Pancreatic β-Cells and Diabetes Mellitus	248
5.2	Cell Death	248
5.3	White Blood Cells	248
5.4	Outlook	249
	References	249

Abstract TRPM2 is a cation channel enabling influx of Na^+ and Ca^{2+}, leading to depolarization and increases in the cytosolic Ca^{2+} concentration ($[Ca^{2+}]_i$). It is widely expressed, e.g. in many neurons, blood cells and the endocrine pancreas. Channel gating is induced by ADP-ribose (ADPR) that binds to a Nudix box motif in the cytosolic C-terminus of the channel. Endogenous ADPR concentrations in leucocytes are sufficiently high to activate TRPM2 in the presence of an increased $[Ca^{2+}]_i$ but probably not at resting $[Ca^{2+}]_i$. Another channel activator is oxidative stress, especially hydrogen peroxide (H_2O_2) that may act through ADPR after ADPR polymers have been formed by poly(ADP-ribose) polymerases (PARPs) and hydolysed by glycohydrolases. H_2O_2-stimulated TRPM2 channels essentially contribute to insulin secretion in pancreatic β-cells and alloxan-induced diabetes mellitus. Inhibition of TRPM2 channels may be achieved by channel blockers such as flufenamic acid or the anti-fungal agents clotrimazole or econazole. Selective blockers of TRPM2 are not yet available; those would be valuable for a characterization of biological roles of TRPM2 in various tissues and as potential drugs directed against oxidative cell damage, reperfusion injury or leucocyte activation. Activation of TRPM2 may be prevented by anti-oxidants, PARP inhibitors and glycohydrolase inhibitors. In future, binding of ADPR to the Nudix box may be targeted. In light of the wide-spread expression and growing list of cellular functions of TRPM2, useful therapeutic applications are expected for future drugs that block TRPM2 channels or inhibit their activation.

Keywords Ca^{2+} entry · Granulocytes · Pancreatic beta cells · ADP ribose · Oxidative stress

1
Features of Genes and Proteins, Expression Pattern

TRPM2 is a member of the M-family of transient receptor potential (TRP) channels named after the first identified member of the family, the tumour-suppressor melastatin, now TRPM1 (Duncan et al. 1998). This subfamily shares a TRPM-homology region of about 700 amino acids in the N-terminus. The general structure, with six membrane-spanning α-helices, a pore between S5 and S6, and two cytosolic tails is common within the TRP family and the cation superfamily. The chromosomal location of the TRPM2 gene is 21q22.3; it consists of 33 exons spanning about 90 kb and 1,503 amino acids (Nagamine et al. 1998; Uemura et al. 2005). Its closest relative is TRPM8 (Tsavaler et al. 2001), a cation channel acting as a sensor for cold temperatures and cool-tasting agents such as menthol. TRPM2 and TRPM8 are 42% identical (Peier et al. 2002). In spite of the close relationship, TRPM2 and TRPM8 are quite different in their biological functions. The predominant feature of TRPM2 is the so-called Nudix box, a consensus region for pyrophosphatases. The Nudix box is localized in the cytoplasmatic C-terminal tail of the channel protein. It conveys a unique mode of activation to TRPM2: the channel is gated by cytosolic adenosine 5′-diphosphoribose (ADPR), a novel messenger for channel activation (Perraud et al. 2001).

The Nudix box motif consists of 22 amino acids highly conserved within many pyrophosphatases (Bessman et al. 1996, 2001; Dunn et al. 1999). In a more general sense, it is part of a larger domain of some 300 amino acids conserved also within pyrophosphatases, although to a lesser degree. A very close homology exists between this part of TRPM2 and the "human nucleoside diphosphate-linked moiety X-type motif 9" (NUDT9), a mitochondrial ADPR hydrolase that degrades ADPR to AMP and ribose 5-phosphate (Shen et al. 2003). Within the NUDT9-H domain, the Nudix box is the catalytic side. N-terminally of the Nudix box, a region can be defined that is critical for binding of ADPR. Interestingly, there are two amino acids in the catalytic domain that differ between TRPM2 and the NUDT9 pyrophosphatase (E1405I/F1406L). When the two amino acids in the NUDT9 pyrophosphatase are substituted with those of TRPM2, the enzymatic activity of the pyrophosphatase is dramatically reduced or even abolished (Perraud et al. 2003). Conversely, after exchange of the two amino acids in TRPM2, any channel activity of TRPM2 is vanished (Kühn and Lückhoff 2004). Therefore, it seems that no enzymatic activity is required for the cation channel TRPM2. This leaves the likely assumption that the NUDT9-H domain governs long-lasting binding of ADPR that is responsible for channel gating.

This assumption has been backed by the discovery of a splice variant of TRPM2, TRPM2-ΔC, that lacks a stretch of 34 amino acids within the binding region and is not activated by ADPR (Wehage et al. 2002). Further studies using artificial mutants have revealed the importance of one particular amino acid (Asn-1326) within the binding region for ADPR-gating of TRPM2 (Kühn and Lückhoff 2004).

Other isoforms of TRPM2 include a variant in which 20 amino acids in the N-terminal cytosolic tail are lacking (TRPM2-ΔN; Wehage et al. 2002), a quite incomplete variant lacking the pore, where an alternatively spliced stop codon terminates the protein after the second transmembrane domain (TRPM2-S; Zhang et al. 2003), and an N-terminally truncated variant (SSF-TRPM2) in the striatum, where 214 amino acid residues are truncated (Uemura et al. 2005).

Neither TRPM2ΔN nor, as expected, TRPM2-S functions as an ion channel. But co-expression of TRPM2-S with TRPM2 reduces functional expression of TRPM2. Thus, the variant may have an inhibitory role in TRPM2-expressing cells.

The expression pattern of TRPM2 is widespread (for details see Harteneck 2005; McNulty and Fonfria 2005). TRPM2 is found in various brain regions such as (among others) cerebellum, cortex, medulla and hippocampus. Outside the central nervous system, TRPM2 is also present in many tissues, notably in pancreatic β-cells, various immune cells and the intestines. Several cell lines exist in which TRPM2 is a predominant cation channel, in particular the insulinoma cell line CRI-G1 (Inamura et al. 2003) and other cells derived from the endogenous part of the pancreas.

2
Ion Channel Properties

TRPM2 is a cation-selective channel. Within various cations there exists little selectivity. Ca^{2+} is weakly permeant; from shifts in the reversal potential, the permeability ratio $P_{Ca}:P_{Na}$ has been determined as about 0.6 to 0.7 (Kraft et al. 2004; Sano et al. 2001). Functionally, the Ca^{2+} permeability is important, as evidenced by sizeable increases in the cytosolic Ca^{2+} concentration ($[Ca^{2+}]_i$) brought about after stimulation of TRPM2. As has been demonstrated for other channels, e.g. CNG channels, the permeability of Ca^{2+} found in the presence of unphysiologically high extracellular Ca^{2+} concentrations may lead to an underestimation of Ca^{2+} fluxes in physiological concentrations (Dzeja et al. 1999; Frings et al. 1995). Current–voltage relations of TRPM2 currents obtained by voltage ramps show a reversal potential close to zero millivolts and an outward rectification. With symmetric Na^+ concentrations on both sides of the patch membrane, single channel openings were shorter at negative potentials (Perraud et al. 2001). These findings demonstrate the non-selectivity of the cation channel and a mild voltage-dependence. In comparison to TRPM8, the outward rectification is much weaker. Thus, TRPM2 produces cation inward currents mostly carried by Na^+ but also by Ca^{2+} in physiological conditions.

The single channel properties of TRPM2 are unique because the channel displays extremely long open times. When one single channel opens, it may remain open without apparent flickering for many seconds with a single-channel conductance of about 60 to 80 pS (Heiner et al. 2003; Perraud et al. 2001; Sano et al. 2001). Such long open times prevent an appropriate analysis of the mean open time that would require hundreds of channel openings and closings; those do not happen in a reasonable time in the case of TRPM2. Instead, TRPM2 channels remain open most of the time after stimulation.

Remarkably, the properties of TRPM2 are virtually identical after heterologous expression of TRPM2 constructs as compared with channels in native cells expressing TRPM2 endogenously (Inamura et al. 2003; Kraft et al. 2004; Perraud et al. 2001). This includes not only the activation in most cases (see Sect. 3), e.g. by ADPR and the ion selectivity, but also the long open times and the single-channel conductance. Therefore, it is likely that TRPM2 does not require accessory or regulatory subunits. In contrast to many other members of the TRP family, in particular of the TRPC subfamily, TRPM2 appears to form homomultimers.

TRPM2 has been dubbed a "chanzyme" (channel-enzyme) because of its relation to pyrophosphatases and its Nudix box. A similar dual function as channel and enzyme has been attributed to TRPM7, which possesses kinase activity and may display autophosphorylation (Cahalan 2001; Levitan and Cibulsky 2001; Runnels et al. 2001; Scharenberg 2005). However, in the case of TRPM2, the enzymatic activity is probably not important; on the contrary, as mentioned above, the enzymatic activity of TRPM2's Nudix box seems to have to be extremely low or the channel function of TRPM2 is abrogated.

3
Modes of Activation

3.1
ADPR

The first discovered activator of TRPM2 has been ADPR. ADPR has not previously been on the list of substances thought to be involved in ion channel regulation. The discovery happened because the substrates accepted by pyrophosphatases possessing the Nudix box homology region may include ADPR. In patch-clamp experiments on TRPM2-transfected HEK-293 cells or monocytic U937 cells endogenously expressing TRPM2, infusion of ADPR into the cells through the patch pipette consistently induced large characteristic cation currents. Moreover, application of ADPR to excised inside-out patches from TRPM2-expressing cells led to single-channel activity. Minimally required concentrations were in the range from 10 µM to 100 µM, dependent on the Ca^{2+} concentration facing the cytosolic side of the channels (Perraud et al. 2001).

3.2
NAD

Other compounds related to ADPR such as nicotinamide adenine dinucleotide (NAD), cyclic ADPR (cADPR), ATP, ADP-glucose or the ADPR breakdown products AMP and ribose-5-phosphate were originally described as inactive on TRPM2 (Perraud et al. 2001). This view has since been revised in parts. In particular, NAD has been described as an activator of TRPM2 similar to ADPR (Hara et al. 2002; Heiner et al. 2003; Inamura et al. 2003; Sano et al. 2001), while cADPR and AMP may have modulatory functions at least in some experimental conditions (see Sects. 3.6 and 3.7). NAD was effective on TRPM2 even in excised inside-out patches (Inamura et al. 2003; Hara et al. 2002; Sano et al. 2001). Consequently, the authors have proposed an effect of NAD on its own, without requirement for conversion to a metabolite such as ADPR. Moreover, it was proposed that NAD may bind to the Nudix box as ADPR (Hara et al. 2002). In our own hands, NAD is ineffective in TRPM2-transfected HEK-293 cells (Wehage et al. 2002). In TRPM2-transfected CHO cells, however, NAD leads to an activation of TRPM2 currents when infused into the cell through the patch-clamp pipette in a concentration of 1 mM (Kühn and Lückhoff 2004). Moreover, NAD induces characteristic TRPM2 currents and channel activity in neutrophil granulocytes, in the same way as ADPR (Heiner et al. 2003). Interestingly, mutants of TRPM2 that are not activated by ADPR are not activated by NAD either (Kühn and Lückhoff 2004). These results do not yet completely clarify the role of NAD, since they would be compatible with a direct effect of NAD in a cell-specific manner as well as with a required conversion of NAD to ADPR.

3.3
H₂O₂ and Oxidative Stress

The second described activator of TRPM2 is oxidative stress for which application of hydrogen peroxide (H_2O_2) is an experimental paradigm. H_2O_2 induces TRPM2 currents as well as increases in $[Ca^{2+}]_i$ in various cell types transfected with TRPM2 (Hara et al. 2002; Wehage et al. 2002). Moreover, cation channels in pancreatic β-cells stimulated by H_2O_2 have been identified as TRPM2 (Inamura et al. 2003). H_2O_2 is so far the only stimulus of TRPM2 that may be applied extracellularly. The mode of action of H_2O_2 has been debated. Oxidative stress is well known to lead to the formation of ADPR polymers by poly(ADP-ribose) polymerases (PARPs). These polymers can be degraded by poly(ADP-ribose) glycohydrolases, resulting in the formation of ADPR (Ame et al. 2004). Thus, in many experimental conditions, the TRPM2 activation by H_2O_2 is well explained by formation of ADPR. Additionally, there may be a direct effect of H_2O_2 on TRPM2 not necessitating ADPR. In particular, the TRPM2 variant TRPM2ΔC was insensitive to ADPR but responded to H_2O_2 (Wehage et al. 2002), although this was observed only in a fairly small fraction of experiments and was dependent on the cell model used for expression.

3.4
Alloxan

By an mechanism related to that used by H_2O_2, the pro-diabetic drug alloxan (Dunn et al. 1943) leads to TRPM2 activation. Alloxan has been used for years to induce diabetes mellitus in animals because it destroys insulin-secreting pancreatic β-cells. It has been shown that it acts through activation of a cation channel and induction of massive influx of cations including Ca^{2+} that mediates cell death (Herson and Ashford 1997). Convincing evidence has been provided that this channel is TRPM2 (Inamura et al. 2003). Alloxan is taken up into β-cells by a cell-specific glucose transporter (GLUT2) physiologically involved in the glucose-regulated insulin secretion and leads intracellularly to the production of free radicals and H_2O_2 (Elsner et al. 2002; im Walde et al. 2002). The strictly intracellular action depends on the availability of the reducing agent glutathione. Other cell types lacking an alloxan uptake mechanism are not affected by alloxan. Thus, in pancreatic β-cells, the H_2O_2-induced cation influx through TRPM2 is of high relevance, at least in this standard pharmacological model.

3.5
Co-operation of Intracellular Ca^{2+} with ADPR

A very important role for the activation of TRPM has to be attributed to intracellular Ca^{2+} (McHugh et al. 2003). Ca^{2+} does not lead to channel activation by itself. However, it dramatically shifts the concentration–response curve to ADPR to the left (Perraud et al. 2001). In the presence of $[Ca^{2+}]_i$ corresponding to those found in many cells after receptor-mediated stimulation (i.e. 1 µM), small ADPR concentrations induce TRPM2 currents, whereas the same ADPR concentrations fail to induce channel activation in the presence of resting $[Ca^{2+}]_i$ (Heiner et al. 2006). Recently, ADPR concentrations have been determined in Jurkat cells (Gasser and Guse 2005). The concentrations are in the range that enables full activation of TRPM2 in the presence of elevated $[Ca^{2+}]_i$ levels, whereas the measured ADPR concentrations are probably ineffective on TRPM2 at resting $[Ca^{2+}]_i$. These findings open the question as to whether ADPR acts a second messenger on TRPM2. Indeed, stimulation of Jurkat cells with concavalin A increased their ADPR content (Gasser et al. 2006). On the other side, these increases may be considered too small for a second-messenger role of ADPR unless the co-effect with $[Ca^{2+}]_i$ (considerably increased by the same stimuli) is taken into account. Our own data on ADPR levels in neutrophil granulocytes support the view that endogenous ADPR enables a regulation of TRPM2 by $[Ca^{2+}]_i$, whereas no changes of ADPR levels occur that would be sufficient for ADPR-mediated regulation of TRPM2.

It remains unresolved by which metabolic pathways (except PARP activation) endogenous ADPR is formed and possibly regulated. Enzymes that convert NAD to ADPR and cADPR are named ADPR cyclases or NAD glycohydrolases. Both reactions may be achieved by the same enzyme. An example is CD38 expressed in several white blood cell types (Mehta et al. 1996). However, CD38 is an ectoenzyme in these cells (Howard et al. 1993), raising the questions of how its products may enter the cytosol to reach the Nudix box of TRPM2 strictly localized intracellularly. Moreover, there is no regulatory pathway known that would link receptor stimulation with glycohydrolase activity changes.

3.6
cADPR

In a recent study, cADPR has been described as a regulatory factor of TRPM2. In concentrations of 10 µM, cADPR enhances the Ca^{2+}-supported activation by ADPR by shifting the EC_{50} for ADPR from 50 µM to 90 nM (Kolisek et al. 2005). However, these concentrations are probably far above physiological levels, in contrast to the endogenous levels of ADPR that are in the range active on TRPM2. Therefore, it remains to be established whether cADPR has a functional role in the regulation of TRPM2, possibly in co-operation with other modulatory factors.

3.7
Other Regulators of TRPM2 Channels

An inhibition of TRPM2 has been demonstrated for intracellular AMP that may affect binding of ADPR to the Nudix box. Importantly, the required concentrations (IC_{50} 70 µM; Kolisek et al. 2005) are in the range expected to be present in many cells. Again, it is not yet known whether this mechanism is used for a regulation of TRPM2 in any cells.

Further reported activators of TRPM2 include arachidonic acid (Hara et al. 2002). The molecular mechanisms and their relevance have not yet been elucidated.

4
Pharmacology

As is the case for the TRPM family as a whole, the pharmacology of TRPM2 has not yet reached a satisfactory state, although TRPM2 seems to have attracted major interest in the pharmaceutical industry. Therefore, it can be hoped that specific modulators of TRPM2 functions may become available in the not-too-distant future. Pharmacological tools would make research on the biological function of TRPM2 much easier and more convincing.

Inhibition of TRPM2 activity may be achieved, in principle, by blockers that act on the channel pore or by drugs that interfere with the activation of TRPM2. Unfortunately, blockers with the desirable specificity are not yet available. Some compounds with established effects on other targets have been reported to exhibit effects on TRPM2 as well.

4.1
Channel Blockers

4.1.1
Fenamates

Fenamates such as flufenamic acid (FFA) are non-steroidal anti-inflammatory agents that are capable of producing anti-inflammatory effects in the central nervous system (Chen et al. 1998), probably by inhibiting a wide (as yet poorly defined) spectrum of cation influx pathways. Examples are Ca^{2+}-activated chloride currents (Kim et al. 2003) and voltage gated Na^{+}- and K^{+}-channels (Lee et al. 2003a; Lee and Wang 1999). In TRPM2-expressing HEK-293 cells, flufenamic acid evoked a pH-dependent inhibition of ADPR- or H_2O_2-induced cation currents (Hill et al. 2004a). Concentrations of 50 µM or above produced a complete inhibition of TRPM2-mediated currents. Within the TRP family, similar effects of FFA have been reported on TRPC5 (Lee et al. 2003c), TRPM4 and TRPM5 (Ullrich et al. 2005; Lee et al. 2003c), whereas TRPC6 was even stimulated by FFA (Inoue et al. 2001). Thus, FFA is hardly a suitable tool that would allow us to clarify differential roles of various cation channels in native cells. A further drawback of FFA is the fact that the inhibition of TRPM2 is irreversible within a time frame of several minutes. Only extremely short applications of FFA could be made reversible. Therefore, structural alterations of the TRPM2 protein may occur, rather than a simple binding to the pore. In contrast, the inhibitory effect of FFA on TRPM2-like currents in CRI-G1 cells was completely reversible with a time constant of 5 s (Hill et al. 2004a).

4.1.2
The Anti-fungal Agents Clotrimazole and Econazole

Further non-specific blockers of TRPM2 are the anti-fungals clotrimazole and econazole. They exert an open-channel block of TRPM2 channels activated by ADPR in HEK-293 cells (Hill et al. 2004b). The IC$_{50}$ was estimated to be in the sub-micromolar range. For comparison, the IC$_{50}$ for block of Ca^{2+}-activated K$^+$ channels (hIK) was 153 nM (Jensen et al. 1998). The block of TRPM2 occurred with a half-time kinetic that was dependent on the inhibitor concentration, for example with a $t_{1/2}$ of 310 s at 1 µM clotrimazole. Again, the inhibition was irreversible in HEK-293 cells, even after very short exposure times. Interestingly, again there was no such irreversibility in CRI-G1 cells where currents returned some minutes after clotrimazole or econazole had been removed.

4.1.3
Lanthanide and Heavy Metal Ions

Most TRP channels are blocked by inorganic blockers such as the lanthanides La^{3+} or Gd^{3+} in the micromolar or millimolar range (Halaszovich et al. 2000; Lee et al. 2003b; Tousova et al. 2005; Becker et al. 2005; Jensen et al. 1998). TRPM2 seems to be an exception (Kraft et al. 2004). No block by any lanthanides has been reported.

4.2
Drugs Interfering with TRPM2 Activation

With our growing understanding of the mechanisms that induce activation of TRPM2 channels in various cells, pharmacological possibilities arise that allow researchers to manipulate the channel activation. So far, effects of substances have best been studied that inhibit the H$_2$O$_2$-initiated pathways, in particular those that result in the generation of ADPR. Other pharmacological principles use the receptor-mediated ADPR formation, even though the underlying mechanisms of this formation have not yet been elucidated in detail. Finally, binding of ADPR to TRPM2 will probably be a pharmacological target in the near future.

4.2.1
Anti-oxidants

It might be expected that various anti-oxidants attenuate TRPM2 activation by H$_2$O$_2$. Surprisingly, few studies have analysed this in detail. Catalase prevented the stimulation of cation channels (now known to be TRPM2) by alloxan in CRI-G1 cells (Herson and Ashford 1997). Similar data were obtained with dimethylthiourea (DMTU; Smith et al 2003). Mannitol can be used as an intracellular radical scavenger that prevents TRPM2 currents in response to extracellular H$_2$O$_2$ (Wehage et al. 2002).

4.2.2
PARP Inhibitors

PARP inhibitors have been used as tools to demonstrate the involvement of PARPs in H_2O_2-mediated TRPM2 activation (Miller 2004; Fonfria et al. 2004). On the other hand, the development of further PARP inhibitors offers pharmaceutical prospects for various conditions characterized by DNA damage and ADPR polymers.

There are about 18 proteins of the PARP family (Ame et al. 2004); most efforts have been directed on PARP-1, which is the classical enzyme catalysing polymerization in response to oxidative stress. The following PARP inhibitors have been demonstrated to attenuate H_2O_2-induced increases in $[Ca^{2+}]_i$ in TRPM2-expressing cell models: PJ34, SB750139-B and DPQ (Fonfria et al. 2004; Yang et al. 2005). The required concentrations vary greatly, reflecting differential potency of the compounds. In most cases, the concentrations effective on TRPM2 activation are several orders of magnitude above those that achieve PARP inhibition in vitro. This fact has been used to criticize the specificity of the inhibitors (Scharenberg 2005). On the other hand, little is known on the permeability, uptake and metabolism of the compounds, and a far lower potency in vivo than in vitro is not an unexpected feature of inhibitors. At any rate, the development of more specific and potent drugs is certainly under way even though they might not immediately be made available to the scientific community.

As a specific test that PARP inhibitors do not interfere with the TRPM2 channel directly or with the interaction of ADPR and TRPM2, stimulation with intracellular ADPR has been performed in patch-clamp experiments. Indeed, ADPR activated TRPM2 in the presence as well as in the absence of the PARP inhibitors (Fonfria et al. 2004). Nevertheless, it should be kept in mind that PARP inhibitors are thought to act by preventing NAD binding to PARP (Ruf et al. 1998; Scharenberg 2005). Since NAD may bind to TRPM2 as well and since NAD may use the same binding site on TRPM2 as ADPR (see Sect. 3.2), interference of the tested PARP inhibitors with the binding of TRPM2 to its activating ligands remains to be considered. This is not necessarily a weakness of PARP inhibitors; in the contrary, major progress in the field would be achieved if drugs should come available that would more specifically target TRPM2-ADPR interactions. Such drugs might allow us to pinpoint TRPM2-mediated effects of oxidative stress, whereas PARP inhibitors do not differentiate within the rather broad spectrum of consequences of ADPR polymerization.

4.2.3
Glycohydrolase Inhibitors

ADPR formation in cells may be brought about by enzymes called glycohydrolases, a class of enzymes not well defined molecularly, with the exception of CD38 (Guse 2000). As mentioned, the CD38 glycohydrolase is a multifunctional enzyme accepting NAD and nicotinamide adenine dinucleotide phosphate (NADP) as substrates and forming several products, including ADPR, cADPR, and nicotinic acid adenine dinucleotide phosphate (NAADP) (Schuber and Lund 2004). There are no good inhibitors of glycohydrolases. Cibacron blue (synonym: 3GA) is a compound classified as a NAD glycohydrolase inhibitor (Li et al. 2002; Kim et al. 1993; Yost and Anderson 1981). It is unlikely that this is its only effect. However, 3GA has been an extremely important tool because it prevents concavalin A-dependent increases in intracellular ADPR concentrations without affecting basal concentrations. 3GA has been used in a cell death assay with Jurkat cells (Gasser et al. 2006). Remarkably, it completely abrogated the effects of concavalin A. This was taken as evidence for a role of ADPR and TRPM2 in this cellular function.

4.2.4
Inhibition of Receptor-Mediated Activation of CD38

In neutrophil granulocytes, CD38 is highly expressed and is likely to play a major role during activation of the cells. Granulocytes from CD38 knockout mice display a diminished chemoattraction in response to the formylated oligopeptide N-formyl-L-methionyl-L-leucyl-L-phenylalanine (fMLP) as well as attenuated increases in $[Ca^{2+}]_i$ (Partida-Sanchez et al. 2001). Originally, this has been attributed to the lack of CD38-derived cADPR. However, the main (>90%) product of CD38 is ADPR (Pfister et al. 2001). Therefore, it may well be speculated that CD38 signals to TRPM2 through ADPR and evokes its effects in part by Ca^{2+} influx through TRPM2. Unfortunately, there is no signalling pathway known that would link the fMLP receptor (or receptors for other chemoattractants) to CD38. fMLP receptor antagonists would be valuable tools to provide initial evidence for such a link. In combination with TRPM2 channel blockers, they might be used to probe the relative contribution of ADPR and TRPM2 to the activation process in granulocytes.

5
Biological Relevance

Research on biologically important roles of TRPM2 in various cells and tissues is currently performed in many laboratories, but as yet our understanding is far from complete—in part due to the fact that no good pharmacological tools are available.

Because of its activation by H_2O_2, TRPM2 is a channel that enables Ca^{2+} influx in response to oxidative stress. At the same time, TRPM2 creates a sizeable permeability for Na^+ and evokes depolarization that may be an important regulatory factor in some cells (especially electrically excitable cells) and conditions. Thus, TRPM2 may be a key player in pathological situations when oxidative damage to cells and tissues occurs, such as in reperfusion injury. Here, it may be worthwhile to explore the therapeutic usefulness of TRPM2 inhibitors. We will include (see Sects. 5.1 and 5.2) some notes on TRPM2-induced cell damage in pancreatic β-cells and neurons. Moreover, the activation of TRPM2 by ADPR is likely to be of relevance in activation processes in white blood cells (granulocytes and T lymphocytes). Again, pharmacological intervention may prove useful in future, with the aim of a modulation of immune processes.

5.1
Pancreatic β-Cells and Diabetes Mellitus

TRPM2 is abundantly expressed in pancreatic β-cells and cell lines derived from tumours of these cells, particularly the CRI-G1 and RIN-5F cell lines (Hara et al. 2002; Inamura et al. 2003; Qian et al. 2002; Herson et al. 1999). As mentioned, TRPM2 is responsible for the cell death that is specifically induced by alloxan. Most recently, evidence has been presented for a role of TRPM2 in the process of insulin secretion by means of Ca^{2+} entry and depolarization (Togashi et al. 2006). Moreover, it may be speculated that oxidative damage is a confounding factor in the development of juvenile insulin-dependent diabetes mellitus.

5.2
Cell Death

In cultured rat striatal cells expressing TRPM2 endogenously, cell death can be induced by the amyloid β-protein and by H_2O_2 (Fonfria et al. 2005). Remarkably, cell death was prevented by transfection of the cells with TRPM2-S, the shortened variant without the pore region demonstrated to inhibit the functional expression of TRPM2 channels (Zhang et al. 2003). Thus, in this experimental model of Alzheimer's disease, TRPM2 clearly has a decisive role and may be pharmacologically targeted. With some caution, it may be furthermore concluded that PARP inhibitors that also preclude cell death in this model act by preventing TRPM2 activation. In other models, TRPM2 proved to be relevant for apoptosis (Zhang et al. 2003, 2006). TRPM2 channels may be involved in reperfusion injury (Smith et al. 2003), probably related to free radicals, superoxide anions and H_2O_2. Moreover, cell death involving TRPM2 activation is not confined to neurons but has been found also in haematopoietic cells (Zhang et al. 2006).

5.3
White Blood Cells

In neutrophil granulocytes, TRPM2 seems to constitute a predominant Ca^{2+} influx pathway (Heiner et al. 2003) that critically contributes to the activating effects of chemoattractant peptides, such as the oxidative burst and chemotaxis (Heiner et al. 2005; Krause et al. 1990). ADPR is an essential factor in the gating of TRPM2. In T lymphocytes, ADPR levels are regulated and exert regulatory functions on TRPM2 and thereby on cellular responses to extracellular receptor-dependent stimuli (Gasser et al. 2006). Therefore, pathways involving TRPM2 offer several targets for therapeutic drugs that modulate immune responses.

5.4
Outlook

Modulation TRPM2 channels and their activation may become an important pharmacological principle in future that is expected to be valuable in a wide range of pathological situations. As-yet-unexpected conditions may be associated with TRPM2. For example, in a genetic linkage study on patients with bipolar disorder, a correlation between a mutation in the TRPM2 gene and the disease has been found (McQuillin et al. 2006; Xu et al. 2006). At the same time, drugs affecting TRPM2 will be indispensable pharmacological tools to analyse the biological significance of TRPM2 in far greater detail than is possible at present.

References

Ame JC, Spenlehauer C, de Murcia G (2004) The PARP superfamily. Bioessays 26:882–893
Becker D, Blase C, Bereiter-Hahn J, et al (2005) TRPV4 exhibits a functional role in cell-volume regulation. J Cell Sci 118:2435–2440
Bessman MJ, Frick DN, O'Handley SF (1996) The MutT proteins or "Nudix" hydrolases, a family of versatile, widely distributed, "housecleaning" enzymes. J Biol Chem 271: 25059–25062
Bessman MJ, Walsh JD, Dunn CA, et al (2001) The gene ygdP, associated with the invasiveness of Escherichia coli K1, designates a Nudix hydrolase, Orf176, active on adenosine (5')-pentaphospho-(5')-adenosine (Ap5A). J Biol Chem 276:37834–37838
Cahalan MD (2001) Cell biology. Channels as enzymes. Nature 411:542–543
Chen Q, Olney JW, Lukasiewicz PD, et al (1998) Fenamates protect neurons against ischemic and excitotoxic injury in chick embryo retina. Neurosci Lett 242:163–166
Duncan LM, Deeds J, Hunter J, et al (1998) Down-regulation of the novel gene melastatin correlates with potential for melanoma metastasis. Cancer Res 58:1515–1520
Dunn CA, O'Handley SF, Frick DN, et al (1999) Studies on the ADP-ribose pyrophosphatase subfamily of the nudix hydrolases and tentative identification of trgB, a gene associated with tellurite resistance. J Biol Chem 274:32318–32324
Dunn JS, Sheehan HL, McLetchie NGB (1943) Necrosis of islets of Langerhans. Lancet 1:484–487
Dzeja C, Hagen V, Kaupp UB, et al (1999) Ca^{2+} permeation in cyclic nucleotide-gated channels. EMBO J 18:131–144
Elsner M, Tiedge M, Guldbakke B, et al (2002) Importance of the GLUT2 glucose transporter for pancreatic beta cell toxicity of alloxan. Diabetologia 45:1542–1549
Fonfria E, Marshall IC, Benham CD, et al (2004) TRPM2 channel opening in response to oxidative stress is dependent on activation of poly(ADP-ribose) polymerase. Br J Pharmacol 143:186–192

Fonfria E, Marshall IC, Boyfield I, et al (2005) Amyloid beta-peptide(1-42) and hydrogen peroxide-induced toxicity are mediated by TRPM2 in rat primary striatal cultures. J Neurochem 95:715-723

Frings S, Seifert R, Godde M, et al (1995) Profoundly different calcium permeation and blockage determine the specific function of distinct cyclic nucleotide-gated channels. Neuron 15:169-179

Gasser A, Guse AH (2005) Determination of intracellular concentrations of the TRPM2 agonist ADP-ribose by reversed-phase HPLC. J Chromatogr B Analyt Technol Biomed Life Sci 821:181-187

Gasser A, Glassmeier G, Fliegert R, et al (2006) Activation of T cell calcium influx by the second messenger ADP-ribose. J Biol Chem 281:2489-2496

Guse HA (2000) Cyclic ADP-ribose. J Mol Med 78:26-35

Halaszovich CR, Zitt C, Jüngling E, et al (2000) Inhibition of TRP3 channels by lanthanides. Block from the cytosolic side of the plasma membrane. J Biol Chem 275:37423-37428

Hara Y, Wakamori M, Ishii M, et al (2002) LTRPC2 Ca^{2+}-permeable channel activated by changes in redox status confers susceptibility to cell death. Mol Cell 9:163-173

Harteneck C (2005) Function and pharmacology of TRPM cation channels. Naunyn Schmiedebergs Arch Pharmacol 371:307-314

Heiner I, Eisfeld J, Halaszovich CR, et al (2003) Expression profile of the transient receptor potential (TRP) family in neutrophil granulocytes: evidence for currents through long TRP channel 2 induced by ADP-ribose and NAD. Biochem J 371:1045-1053

Heiner I, Radukina N, Eisfeld J, et al (2005) Regulation of TRPM2 channels in neutrophil granulocytes by ADP-ribose: a promising pharmacological target. Naunyn Schmiedebergs Arch Pharmacol 371:325-333

Heiner I, Eisfeld J, Warnstedt M, et al (2006) Endogenous ADP-ribose enables calcium-regulated cation currents through TRPM2 channels in neutrophil granulocytes. Biochem J 398:225-232

Herson PS, Ashford ML (1997) Activation of a novel non-selective cation channel by alloxan and H2O2 in the rat insulin-secreting cell line CRI-G1. J Physiol 501:59-66

Herson PS, Lee K, Pinnock RD, et al (1999) Hydrogen peroxide induces intracellular calcium overload by activation of a non-selective cation channel in an insulin-secreting cell line. J Biol Chem 274:833-841

Hill K, Benham CD, McNulty S, et al (2004a) Flufenamic acid is a pH-dependent antagonist of TRPM2 channels. Neuropharmacology 47:450-460

Hill K, McNulty S, Randall AD (2004b) Inhibition of TRPM2 channels by the antifungal agents clotrimazole and econazole. Naunyn Schmiedebergs Arch Pharmacol 370:227-237

Howard M, Grimaldi JC, Bazan JF, et al (1993) Formation and hydrolysis of cyclic ADP-ribose catalyzed by lymphocyte antigen CD38. Science 262:1056-1059

im Walde SS, Dohle C, Schott-Ohly P, et al (2002) Molecular target structures in alloxan-induced diabetes in mice. Life Sci 71:1681-1694

Inamura K, Sano Y, Mochizuki S, et al (2003) Response to ADP-ribose by activation of TRPM2 in the CRI-G1 insulinoma cell line. J Membr Biol 191:201-207

Inoue R, Okada T, Onoue H, et al (2001) The transient receptor potential protein homologue TRP6 is the essential component of vascular alpha(1)-adrenoceptor-activated $Ca^{(2+)}$-permeable cation channel. Circ Res 88:325-332

Jensen BS, Strobaek D, Christophersen P, et al (1998) Characterization of the cloned human intermediate-conductance Ca^{2+}-activated K^+ channel. Am J Physiol 275:C848-C856

Kim SJ, Shin SY, Lee JE, et al (2003) Ca^{2+}-activated Cl^- channel currents in rat ventral prostate epithelial cells. Prostate 55:118-127

Kim UH, Kim MK, Kim JS, et al (1993) Purification and characterization of NAD glycohydrolase from rabbit erythrocytes. Arch Biochem Biophys 305:147–152

Kolisek M, Beck A, Fleig A, et al (2005) Cyclic ADP-ribose and hydrogen peroxide synergize with ADP-ribose in the activation of TRPM2 channels. Mol Cell 18:61–69

Kraft R, Grimm C, Grosse K, et al (2004) Hydrogen peroxide and ADP-ribose induce TRPM2-mediated calcium influx and cation currents in microglia. Am J Physiol Cell Physiol 286:C129–C137

Krause KH, Campbell KP, Welsh MJ, et al (1990) The calcium signal and neutrophil activation. Clin Biochem 23:159–166

Kühn FJ, Lückhoff A (2004) Sites of the NUDT9-H domain critical for ADP-ribose activation of the cation channel TRPM2. J Biol Chem 279:46431–46437

Lee HM, Kim HI, Shin YK, et al (2003a) Diclofenac inhibition of sodium currents in rat dorsal root ganglion neurons. Brain Res 992:120–127

Lee N, Chen J, Sun L, et al (2003b) Expression and characterization of human transient receptor potential melastatin 3 (hTRPM3). J Biol Chem 278:20890–20897

Lee YM, Kim BJ, Kim HJ, et al (2003c) TRPC5 as a candidate for the nonselective cation channel activated by muscarinic stimulation in murine stomach. Am J Physiol Gastrointest Liver Physiol 284:G604–G616

Lee YT, Wang Q (1999) Inhibition of hKv2.1, a major human neuronal voltage-gated K$^+$ channel, by meclofenamic acid. Eur J Pharmacol 378:349–356

Levitan IB, Cibulsky SM (2001) Biochemistry. TRP ion channels—two proteins in one. Science 293:1270–1271

Li PL, Zhang DX, Ge ZD, et al (2002) Role of ADP-ribose in 11,12-EET-induced activation of KCa channels in coronary arterial smooth muscle cells. Am J Physiol Heart Circ Physiol 282:H1229–H1236

McHugh D, Flemming R, Xu SZ, et al (2003) Critical intracellular Ca^{2+} dependence of transient receptor potential melastatin 2 (TRPM2) cation channel activation. J Biol Chem 278:11002–11006

McNulty S, Fonfria E (2005) The role of TRPM channels in cell death. Pflugers Arch 451:235–242

McQuillin A, Bass NJ, Kalsi G, et al (2006) Fine mapping of a susceptibility locus for bipolar and genetically related unipolar affective disorders, to a region containing the C21ORF29 and TRPM2 genes on chromosome 21q22.3. Mol Psychiatry 11:134–142

Mehta K, Shahid U, Malavasi F (1996) Human CD38, a cell-surface protein with multiple functions. FASEB J 10:1408–1417

Miller BA (2004) Inhibition of TRPM2 function by PARP inhibitors protects cells from oxidative stress-induced death. Br J Pharmacol 143:515–516

Nagamine K, Kudoh J, Minoshima S, et al (1998) Molecular cloning of a novel putative Ca^{2+} channel protein (TRPC7) highly expressed in brain. Genomics 54:124–131

Partida-Sanchez S, Cockayne DA, Monard S, et al (2001) Cyclic ADP-ribose production by CD38 regulates intracellular calcium release, extracellular calcium influx and chemotaxis in neutrophils and is required for bacterial clearance in vivo. Nat Med 7:1209–1216

Peier AM, Moqrich A, Hergarden AC, et al (2002) A TRP channel that senses cold stimuli and menthol. Cell 108:705–715

Perraud AL, Fleig A, Dunn CA, et al (2001) ADP-ribose gating of the calcium-permeable LTRPC2 channel revealed by Nudix motif homology. Nature 411:595–599

Perraud AL, Shen B, Dunn CA, et al (2003) NUDT9, a member of the Nudix hydrolase family, is an evolutionarily conserved mitochondrial ADP-ribose pyrophosphatase. J Biol Chem 278:1794–1801

Pfister M, Ogilvie A, da Silva CP, et al (2001) NAD degradation and regulation of CD38 expression by human monocytes/macrophages. Eur J Biochem 268:5601–5608

Qian F, Huang P, Ma L, et al (2002) TRP genes: candidates for nonselective cation channels and store-operated channels in insulin-secreting cells. Diabetes 51 Suppl 1:S183–S189

Ruf A, de Murcia G, Schulz GE (1998) Inhibitor and NAD^+ binding to poly(ADP-ribose) polymerase as derived from crystal structures and homology modeling. Biochemistry 37:3893–3900

Runnels LW, Yue L, Clapham DE (2001) TRP-PLIK, a bifunctional protein with kinase and ion channel activities. Science 291:1043–1047

Sano Y, Inamura K, Miyake A, et al (2001) Immunocyte Ca^{2+} influx system mediated by LTRPC2. Science 293:1327–1330

Scharenberg AM (2005) TRPM2 and TRPM7: channel/enzyme fusions to generate novel intracellular sensors. Pflugers Arch 451:220–227

Schuber F, Lund FE (2004) Structure and enzymology of ADP-ribosyl cyclases: conserved enzymes that produce multiple calcium mobilizing metabolites. Curr Mol Med 4:249–261

Shen BW, Perraud AL, Scharenberg A, et al (2003) The crystal structure and mutational analysis of human NUDT9. J Mol Biol 332:385–398

Smith MA, Herson PS, Lee K, et al (2003) Hydrogen-peroxide-induced toxicity of rat striatal neurones involves activation of non-selective cation channel. J Physiol (Lond) 547:417–425

Togashi K, Hara Y, Tominaga T, Higashi T, Konishi Y, Mori Y, Tominaga M (2006) TRPM2 activation by cyclic ADP-ribose at body temperature is involved in insulin secretion. EMBO J 25:1804–1815

Tousova K, Vyklicky L, Susankova K, et al (2005) Gadolinium activates and sensitizes the vanilloid receptor TRPV1 through the external protonation sites. Mol Cell Neurosci 30:207–217

Tsavaler L, Shapero MH, Morkowski S, et al (2001) Trp-p8, a novel prostate-specific gene, is up-regulated in prostate cancer and other malignancies and shares high homology with transient receptor potential calcium channel proteins. Cancer Res 61:3760–3769

Uemura T, Kudoh J, Noda S, et al (2005) Characterization of human and mouse TRPM2 genes: identification of a novel N-terminal truncated protein specifically expressed in human striatum. Biochem Biophys Res Commun 328:1232–1243

Ullrich ND, Voets T, Prenen J, et al (2005) Comparison of functional properties of the Ca^{2+}-activated cation channels TRPM4 and TRPM5 from mice. Cell Calcium 37:267–278

Wehage E, Eisfeld J, Heiner I, et al (2002) Activation of the cation channel long transient receptor potential channel 2 (LTRPC2) by hydrogen peroxide. A splice variant reveals a mode of activation independent of ADP-ribose. J Biol Chem 277:23150–23156

Xu C, Macciardi F, Li PP, et al (2006) Association of the putative susceptibility gene, transient receptor potential protein melastatin type 2, with bipolar disorder. Am J Med Genet B Neuropsychiatr Genet 141:36–43

Yang KT, Chang WL, Yang PC, et al (2006) Activation of the transient receptor potential M2 channel and poly(ADP-ribose) polymerase is involved in oxidative stress-induced cardiomyocyte death. Cell Death Differ 13:1815–1826

Yost DA, Anderson BM (1981) Purification and properties of the soluble NAD glycohydrolase from Bungarus fasciatus venom. J Biol Chem 256:3647–3653

Zhang W, Chu X, Tong Q, et al (2003) A novel TRPM2 isoform inhibits calcium influx and susceptibility to cell death. J Biol Chem 278:16222–16229

Zhang W, Hirschler-Laszkiewicz I, Tong Q, et al (2006) TRPM2 is an ion channel which modulates hematopoietic cell death through activation of caspases and PARP cleavage. Am J Physiol Cell Physiol 290:C1146–C1159

TRPM3

J. Oberwinkler (✉) · S. E. Philipp

Institut für klinische und experimentelle Pharmakologie und Toxikologie der Universität des Saarlandes, 66421 Homburg, Germany
johannes.oberwinkler@uniklinikum-saarland.de
stephan.philipp@uniklinikum-saarland.de

1	Basic Features of the TRPM3 Gene and the Encoded Proteins	254
1.1	The TRPM3 Gene Encodes Many Different Variants	254
1.2	The TRPM3 Gene Hosts a MicroRNA Gene	256
1.3	Basic Features of TRPM3 Proteins	257
1.4	Expression Pattern of TRPM3	258
2	Ion Channel Properties	259
2.1	TRPM3 Proteins Build an Ion Conducting Pore	259
2.2	TRPM3 Channels Show Constitutive Activity	259
2.3	TRPM3 Channels Are Outwardly Rectifying and Voltage Dependent	260
2.4	Permeation Through TRPM3 Channels	260
2.5	TRPM3 Channels Are Inhibited by Intracellular Mg^{2+}	261
2.6	Block by Extracellular Cations	262
2.7	Single Channel Conductance	262
3	Modes of Activation	263
3.1	Store Depletion	263
3.2	Hypotonic Stimulation	263
3.3	Activation by D-*erythro*-Sphingosine	264
4	Pharmacology	265
5	Biological Relevance of TRPM3 Channels	265
References		266

Abstract TRPM3 is the last identified member of the TRPM subfamily and is most closely related to TRPM1. Due to alternative splicing, the TRPM3 gene encodes a large number of different variants. One splice event, affecting the pore-forming region of the channel, changes its selectivity for divalent cations. In this review, we give an overview of the identified TRPM3 variants and compare their functional properties.

Keywords TRPM3 variants · Alternative splicing · Channel pore

1
Basic Features of the TRPM3 Gene and the Encoded Proteins

1.1
The TRPM3 Gene Encodes Many Different Variants

The TRPM3 genes of mouse, rat and human are large and show a highly conserved organization. In the mouse, the gene spans more than 850 kb on chromosome 19b and contains 28 exons (Fig. 1a, Oberwinkler et al. 2005). The exons 1, 2 and 3 are separated by huge intronic sequences of approximately 309 and 249 kb, respectively. The rat gene is located on chromosome 1q51 and the human gene is placed on chromosome 9q21.11. Compared to other members of the transient receptor potential (TRP) gene family, the TRPM3 gene encodes the largest number of variants. These variants mainly arise by alternative splicing of their primary transcripts (Fig. 1b). Five different complementary DNAs (cDNAs) called mTRPM3α1 to mTRPM3α5 have been cloned from mouse brain (Oberwinkler et al. 2005). Their reading frames are flanked by stop codons establishing entire protein coding sequences of 1,699 to 1,721 amino acid residues (Fig. 1c). Their first exon (exon 1) encodes 61 amino acids which are fused to the following sequence determined by exon 3. However, none of these variants contain sequence information of a predicted exon 2, which is the first exon of a human variant described by Grimm et al. (2005). This variant has been predicted from the identification of three overlapping partial cDNA clones but has not yet been confirmed by cloning of a complete transcript. It contains 1,325 amino acid residues only and differs from the α-variants also at

Fig. 1 a Genomic organization of the mouse TRPM3 gene on chromosome 19b. Exons are indicated by *numbers*. The localization of a microRNA sequence (*miR-204*) is shown. **b** Structure of identified TRPM3 transcripts. Transcripts are shown as *grey bars* and scaled to their relative size (*upper bar*). Exons 1–28 are numbered. Start codons (*ATG*) present in exons 1, 2 and 4 and stop codons (*stop*) in exon 28 are indicated. **c** Schematic presentation of TRPM3 variants (*black bars*) encoded by the transcripts shown in **b**. A linear representation of the subdomain organization of TRPM3 proteins in general is shown (*upper bar*). Putative calmodulin-binding sites (*CamBS*), the TRPM homology region, the transmembrane region including the channel pore (*P*), the TRP motif (*TRP*) and a coiled-coil region (*cc*) are indicated. The variants (*black bars*) are scaled to their relative size and the numbers of amino acid residues are indicated in *brackets*. Internal protein domains removed due to alternative splicing are given as dashed line. **d** Pore regions of TRPM3α1 and TRPM3α2 compared to the corresponding sequences of TRPM1 variants (Lis et al. 2005). Identical residues are *boxed* in *black*, conserved in *grey*

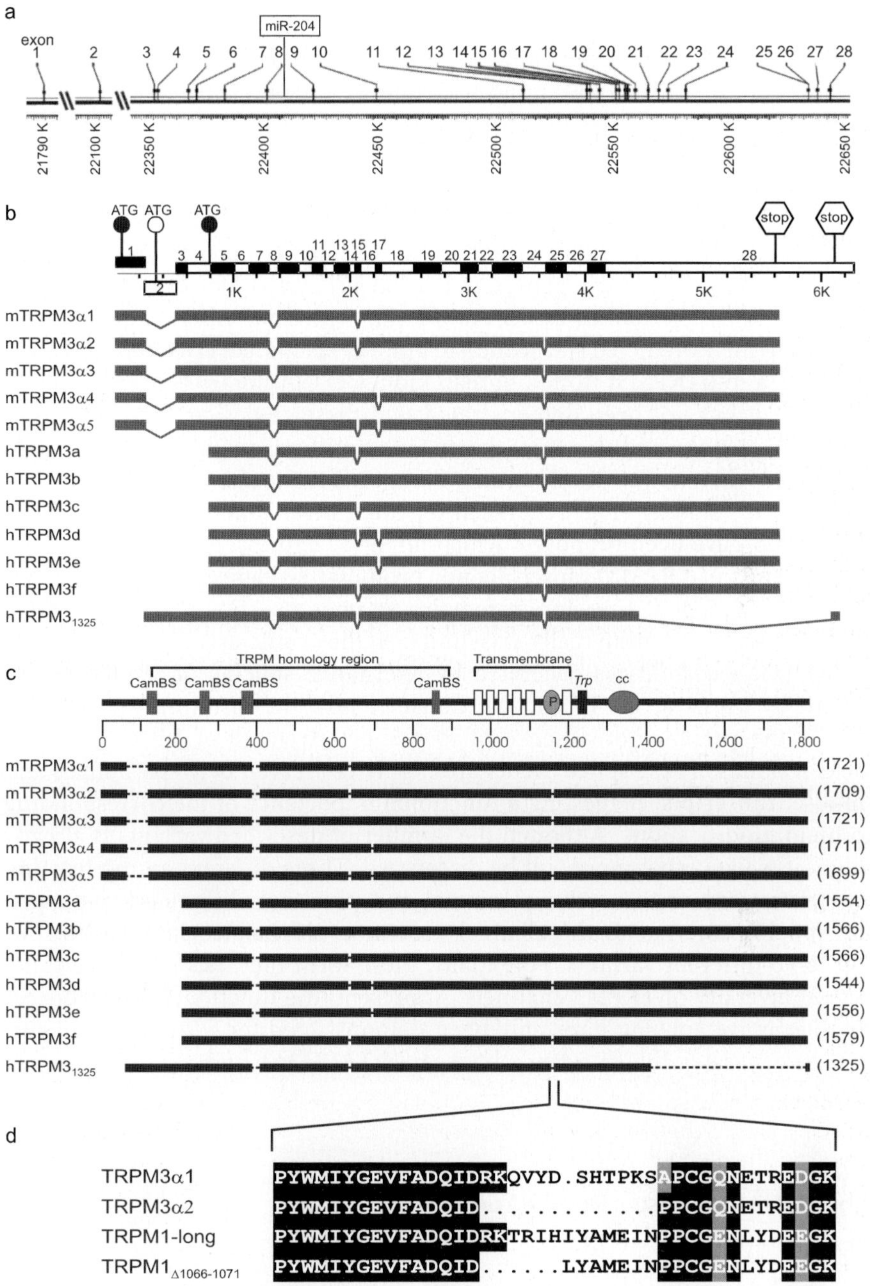
a
miR-204
exon
1 2 3 4 5 6 7 8 9 10 11 12 13 14 15 16 17 18 19 20 21 22 23 24 25 26 27 28
21790 K
22100 K
22350 K
22400 K
22450 K
22500 K
22550 K
22600 K
22650 K
b
ATG ATG ATG
stop stop
1 2 3 4 5 6 7 8 9 10 11 12 13 14 15 16 17 18 19 20 21 22 23 24 25 26 27 28
1K 2K 3K 4K 5K 6K
mTRPM3α1
mTRPM3α2
mTRPM3α3
mTRPM3α4
mTRPM3α5
hTRPM3a
hTRPM3b
hTRPM3c
hTRPM3d
hTRPM3e
hTRPM3f
hTRPM3₁₃₂₅
c
TRPM homology region
Transmembrane
CamBS CamBS CamBS CamBS
P
Trp
cc
0 200 400 600 800 1,000 1,200 1,400 1,600 1,800
mTRPM3α1 (1721)
mTRPM3α2 (1709)
mTRPM3α3 (1721)
mTRPM3α4 (1711)
mTRPM3α5 (1699)
hTRPM3a (1554)
hTRPM3b (1566)
hTRPM3c (1566)
hTRPM3d (1544)
hTRPM3e (1556)
hTRPM3f (1579)
hTRPM3₁₃₂₅ (1325)
d
TRPM3α1 PYWMIYGEVFADQIDRKQVYD.SHTPKSAPCGQNETREDGK
TRPM3α2 PYWMIYGEVFADQID.............PPCGQNETREDGK
TRPM1-long PYWMIYGEVFADQIDRKTRIHIYAMEINPPCGENLYDEEGK
TRPM1Δ1066-1071 PYWMIYGEVFADQID......LYAMEINPPCGENLYDEEGK

its carboxyterminus, which is considerably shorter. Throughout this review we will refer to this variant as hTRPM3$_{1325}$. Splicing within the last exon (exon 28) removes approximately 350 amino acid residues and part of the 3′-untranslated region (Fig. 1b). Splicing also introduces a frameshift, and the consequence is that hTRPM3$_{1325}$ contains seven alternative amino acid residues at the very C-terminal end.

By computational analysis of genomic sequences using the TRPM1 cDNA as a template, Lee et al. (2003) predicted the amino-terminal end of TRPM3 to be encoded by the last triplet present in exon 4 (Fig. 1b). However, this prediction obviously overlooked exons 1 and 2 (Fig. 1a). They used primers exactly matching their predicted start codon to amplify six human cDNA clones called hTRPM3a to hTRPM3f from a human kidney cDNA library (Fig. 1b). Unfortunately, it has not been demonstrated that these clones contain full-length cDNA sequences, and therefore the encoded proteins might lack amino acid residues at their amino-terminal end. Compared to the α-variants, they are 155 amino acids shorter (Fig. 1c). Taken together, three different amino termini of TRPM3 have been proposed, which might be expressed alternatively. Neither 5′-ends of TRPM3 transcripts nor promoter sequences of TRPM3 genes have been mapped. Therefore it remains an open question whether all three different amino-termini really exist and how they could arise.

Further variability is generated by alternative splicing of sequences encoded by exons 8, 15, 17 and 24. Consequently, the resulting variants differ by the presence or absence of four short stretches of 10 to 25 amino acid residues. These splice events are highly conserved between human and mouse transcripts, suggesting a functional importance of the corresponding protein modifications. Although the number of described variants is already high, it is likely that more will be discovered. Three different amino termini, two variations of the carboxyterminal end plus four differences generated by splicing internal to the channel protein could in principle account for up to 96 different variants. Potentially, these form the basis of a large functional diversity of TRPM3 channels. At present, the functional consequences of alternative splicing have only been demonstrated for a single splice site, which generates the difference between mTRPM3α1 and mTRPM3α2 (see Sect. 2).

1.2
The TRPM3 Gene Hosts a MicroRNA Gene

MicroRNAs (miRNAs) are endogenous, approximately 22-nt RNAs that can play an important regulatory role by targeting mRNAs for cleavage or translational repression (Bartel 2004). At present, 326 and 249 microRNA genes are known within the human and mouse genomes, respectively. A microRNA designated miR-204 has been cloned from mouse eye (Lagos-Quintana et al. 2003), and expression of its orthologue has also been detected in zebrafish (Chen et al. 2005). The sequence of miR-204 has been attributed to the TRPM3 gene where it resides in intron 8 (Fig. 1a; Rodriguez et al. 2004; Weber 2005). It might therefore share the regulatory elements and primary transcript with the TRPM3 pre-mRNA. The miR-204 sequence seems to be highly conserved throughout phylogeny since orthologous sequences have been detected in the genomes of pufferfish, chicken, rat, pig and a variety of different primates including humans (Berezikov et al. 2005). A highly similar microRNA (miR-211) is also present in the TRPM1 gene (Rodriguez et al. 2004; Weber 2005) whereas no other TRPM gene contains microRNA sequence information (Weber 2005). Thus, within the TRPM family this feature is unique to TRPM1 and TRPM3 and may add an additional functional property to these genes. Above and beyond encoding large numbers of ion channel proteins, the TRPM3 gene might regulate the expression of a variety of target genes on the post-transcriptional level. In contrast to small interfering RNAs (siRNAs) that typically cause the silencing of the same locus from which they originate, miRNAs induce down-regulation of genes very different from their host genes (Bartel 2004). Thus miR-204 and miR-211 likely do not provide a mechanism to control the expression TRPM3 or TRPM1 in a direct autoregulatory manner, and there are also no other TRP genes among the predicted targets of miR-204 (http://microrna.sanger.ac.uk; Griffiths-Jones 2004).

1.3
Basic Features of TRPM3 Proteins

All TRPM3 variants described so far show the typical features of a TRP protein with six putative membrane-spanning domains, a conserved TRP motif and a coiled-coil region in its C-terminus (Fig. 1c; Lee et al. 2003; Oberwinkler et al. 2005). Ankyrin repeats, which are present in TRPC and TRPV proteins, are lacking at the amino terminus of TRPM3 channels. As is the case for all TRPM proteins, a large, roughly 700-amino-acid-long, N-terminal "TRPM homology region" is present instead, which bears no obvious resemblance to other sequences outside the TRPM subfamily. This region starts with a sequence motif [consensus sequence $(W/F)IX_3-(F/L/I)CK(R/K)EC(V/I/S)X_{12-24}CXCG$; Grimm et al. 2003], which in TRPM3 is encoded by exon 3. This motif is present in all TRPM3α variants as well as in hTRPM3$_{1325}$, but is absent in the amino-terminally truncated variants hTRPM3a–f. A comparison of the amino acid sequence with sequences of calmodulin-binding proteins (http://calcium.uhnres.utoronto.ca/ctdb/flash.htm; Yap et al. 2000) indicates the presence of four putative calmodulin-binding sites (CamBS, Fig. 1c) within the amino terminus of TRPM3. This suggests a Ca^{2+}/Calmodulin-dependent regulation of TRPM3 channel activity, which, however, has not yet been substantiated experimentally. Interestingly, the CamBS 1 is encoded partly by exon 2. It is therefore predicted only for hTRPM3$_{1325}$ and not for the other TRPM3 variants. Thus, hTRPM3$_{1325}$ may form ion channels that have a different dependence on intracellular Ca^{2+}.

1.4
Expression Pattern of TRPM3

In humans, the most prominent TRPM3 expression seems to be in the kidney, where TRPM3 transcripts could be detected by semi-quantitative PCR, Northern blot and, in renal tubules, in situ hybridization (Lee et al. 2003; Grimm et al. 2003). Correspondingly, TRPM3 could also be detected in membrane fractions of human and bovine kidney in Western blots using a polyclonal anti-TRPM3 antibody (Grimm et al. 2003). Surprisingly, TRPM3 seems to be absent in mouse kidney as revealed by Northern and Western blots (Grimm et al. 2003; Oberwinkler et al. 2005). This indicates a species-specific expression of TRPM3 in this tissue.

Preliminary data for TRPM3 expression in other tissues were obtained with RT-PCR experiments: partial TRPM3 transcripts were amplified from cells derived from the human pulmonary artery endothelium (Fantozzi et al. 2003) and a neuroblastoma cell line (Bollimuntha et al. 2005). Furthermore, RT-PCR suggested the presence of TRPM3 in a subset of primary cultured trigeminal neurons (Nealen et al. 2003) and in a variety of tissues (liver, pancreas, ovary, spinal cord, testis and brain; Grimm et al. 2003; Lee et al. 2003). Because of the high sensitivity of the method and the amplification of only parts of the transcripts, these experiments did not prove the presence of TRPM3 channels in a given tissue or cell. However, expression in the brain has been confirmed by Northern blot analysis (Lee et al. 2003; Grimm et al. 2003; Oberwinkler et al. 2005) and by in situ hybridization studies (Oberwinkler et al. 2005). In mouse brain, transcripts could be detected in several regions including the dentate gyrus, lateral septal nuclei, indusium griseum and tenia tecta. The most prominent signals within the brain were obtained from epithelial cells of the choroid plexus.

Expressed sequence tag analysis of adult human iris and lens indicated TRPM3 expression in pigmented and non-pigmented cells of the eye (Wistow et al. 2002a, b). In addition, analysis of differentially expressed genes in retinal pigment epithelium cells by suppression subtractive hybridization revealed that TRPM3 belongs to the ten most abundantly expressed genes in these cells (Schulz et al. 2004). Expression of the TRPM3 gene in the eye could also be confirmed by Northern blot hybridization (Oberwinkler et al. 2005) and indirectly by the identification of miR-204 from this tissue (Lagos-Quintana et al. 2003).

2
Ion Channel Properties

2.1
TRPM3 Proteins Build an Ion Conducting Pore

As outlined above, 11 different TRPM3 splice variants have been reported, but only 4 of them have been investigated with functional methods (Fig. 1): mTRPM3α1 (Oberwinkler et al. 2005), mTRPM3α2 (Oberwinkler et al. 2005), hTRPM3a (Lee et al. 2003) and hTRPM3$_{1325}$ (Grimm et al. 2003, 2005; Xu et al. 2005). The splice variants mTRPM3α2, hTRPM3a and hTRPM3$_{1325}$ share the same shorter pore region in contrast to TRPM3α1, which contains a longer pore region (Fig. 1c, d). The functional properties of these proteins have only been examined after heterologous expression of cloned genes. This methodology, however, can have severe pitfalls. For example, expression of a given protein may induce or up-regulate the expression of other genes intrinsic to the expression system that may also encode for ion channels (e.g. Zhang et al. 2001). It is therefore crucial to establish that the ionic currents observed after heterologous expression of a gene are mediated by the encoded recombinant protein. An accepted way to accomplish this is the demonstration that specific alterations of the primary sequence of the recombinant protein change the permeation properties of the resulting channels (Voets and Nilius 2003). The existence of splice variants that vary only in the presumed pore region between the predicted transmembrane helices 5 and 6 of TRPM3 (Fig. 1c) directly allowed testing this. It turned out that the ratio of monovalent to divalent cation permeability was significantly different between the mTRPM3α1 and mTRPM3α2 variants (Oberwinkler et al. 2005), constituting strong evidence that TRPM3 proteins indeed form bona fide ion-conducting channels in the plasma membrane of transfected HEK 293 cells.

2.2
TRPM3 Channels Show Constitutive Activity

All available studies agree that TRPM3 channels show constitutive activity (Lee et al. 2003; Grimm et al. 2003; Oberwinkler et al. 2005). Fluorometric measurements demonstrated that the cytosolic Ca^{2+} concentration is higher in cells that express TRPM3 channels with the short pore region compared to controls cells, even in the absence of any stimulus (Grimm et al. 2003; Oberwinkler et al. 2005). In experiments where Ca^{2+} was added back to Ca^{2+}-free extracellular solutions, TRPM3-expressing cells showed a significantly larger Ca^{2+} increase compared to control cells (Lee et al. 2003; Grimm et al. 2003; Oberwinkler et al. 2005). Also, Mn^{2+} influx, as assessed by measuring Fura-2 quench, was larger in hTRPM3$_{1325}$-expressing cells than in control cells (Grimm et al. 2003; Xu et al. 2005). The constitutive activity of short-pore TRPM3 channels was also observed in electrophysiological experiments (Grimm et al. 2003; Oberwinkler et al. 2005). Similarly, electrophysiological experiments showed that channels encoded by the long-pore splice variant, TRPM3α1, are constitutively active (Oberwinkler et al. 2005).

2.3
TRPM3 Channels Are Outwardly Rectifying and Voltage Dependent

All TRPM3 splice variants that have been investigated electrophysiologically were found to have outwardly rectifying current–voltage relationships, but the reported degree of outward rectification varied considerably (Grimm et al. 2003; Oberwinkler et al. 2005). Interestingly, the outwardly rectifying current–voltage relationship of TRPM3 channels does not seem to depend on divalent cations, as it persisted in the absence of free intra- and extracellular divalent cations (Oberwinkler et al. 2005). This feature of TRPM3 channels contrasts with the behaviour of the closely related channels TRPM6 and TRPM7, whose current–voltage relationship is nearly linear under divalent-free conditions (Nadler et al. 2001; Kozak and Cahalan 2003; Voets et al. 2004). When using very high extracellular Ca^{2+} concentrations (100–120 mM), the TRPM3α2 and hTRPM3$_{1325}$ channels display S-shaped (inwardly and outwardly rectifying) current–voltage relationships (Grimm et al. 2005; Oberwinkler et al. 2005).

All investigated splice variants seem to be voltage-dependent, as depolarizing the membrane potential to values more positive than +40 mV leads to a time-dependent increase of the outward current (Grimm et al. 2005; Oberwinkler et al. 2005). Although the observed voltage dependence was only weak, it may contribute to the outward rectification of the current–voltage relationship. Quite diverging results were obtained for the time constants of the voltage dependence. In TRPM3α1 and TRPM3α2 channels, we reported that the time-dependent current relaxations were essentially complete after 20–40 ms (Oberwinkler et al. 2005), whereas agonist-stimulated hTRPM3$_{1325}$ channels took hundreds of milliseconds to reach a steady-state (Grimm et al. 2005). The cause of this difference is not known, but the structural differences between the different splice variants could conceivably have influenced the temporal behaviour of the channels. Grimm et al. (2005) also determined the voltage of half-maximal activation, but did not find it to be well-behaved, since it varied by more than 150 mV between individual cells. This could indicate that the voltage dependence of TRPM3 is regulated in complex ways that we still do not understand.

2.4
Permeation Through TRPM3 Channels

Depending on the splice event in the linker region between the putative transmembrane helices 5 and 6, the permeation characteristics of TRPM3-encoded channels differ markedly. Splice variants with the shorter pore region

(TRPM3α2 and hTRPM3$_{1325}$) clearly show significant inward currents when Ca^{2+} or Mg^{2+} are the only cations present extracellularly. They are therefore permeable for these divalent cations (Grimm et al. 2003; Grimm et al. 2005; Oberwinkler et al. 2005). Permeability of these channels to Ca^{2+} and Mn^{2+} was also inferred from measuring the fluorescence intensity from Ca^{2+} indicators (Lee et al. 2003; Grimm et al. 2003; Oberwinkler et al. 2005). Measuring the reversal potentials under bi-ionic conditions confirmed that TRPM3α2 and hTRPM3$_{1325}$ are permeable for divalent cations. TRPM3α1-encoded channels, on the other hand, had significantly lower reversal potentials under these conditions (Oberwinkler et al. 2005). In addition, no significant inward currents through TRPM3α1 could be observed even with very high extracellular divalent concentrations (Oberwinkler et al. 2005). These data suggest that TRPM3α1 channels are less well permeated by divalent cations than their counterparts with the shorter pore region. This is important for at least three reasons. (1) As pointed out before, it provides very strong evidence that the recombinant TRPM3 channels participate in the formation of the ion-conducting pore. (2) It provides equally strong evidence that the location of the pore-lining amino acids of TRPM3 is between the putative transmembrane helices 5 and 6. This location of the ion-conducting pore was suspected due to the topological similarity between TRP family members and voltage-gated cation channels. However, experimental confirmation for this hypothesis in the TRPM subfamily has been obtained only for TRPM3 and TRPM4 (Nilius et al. 2005). (3) TRPM3 is the first ion channel described to regulate its ionic selectivity by alternative splicing. The relative Ca^{2+} permeability is strongly affected, which possibly might have important functional consequences for intracellular signalling events induced by Ca^{2+}. The closest relative of TRPM3, TRPM1, might be similar in this regard, since the TRPM1 gene also encodes splice variants that differ in the corresponding region (Fig. 1d; Lis et al. 2005).

2.5
TRPM3 Channels Are Inhibited by Intracellular Mg^{2+}

The constitutive activity of TRPM3 channels was shown to be diminished by millimolar concentrations of intracellular Mg^{2+} (Oberwinkler et al. 2005). Since both TRPM3α2 and TRPM3α1 channels were similarly inhibited, the precise properties of the selectivity filter, which is different between the two splice variants (Fig. 1), seems unimportant for the effects of intracellular Mg^{2+}. In TRPM3α1-expressing cells, the intracellular Mg^{2+} concentration is in a range that does not fully inhibit TRPM3 channels, as constitutively activated currents could be measured immediately after rupturing the cell membrane (when the intracellular Mg^{2+} concentration is still unaffected by the composition of the pipette solution). Interestingly, the closely related channels TRPM6 and TRPM7 are also inhibited by intracellular Mg^{2+} in the millimolar range (Voets et al. 2004; Nadler et al. 2001; Kozak and Cahalan 2003), suggesting that inhibition by intracellular Mg^{2+} is a general feature of this subgroup of TRPM channels.

2.6
Block by Extracellular Cations

Divalent cations on the extracellular side block the outward currents through TRPM3, regardless of the length of the pore region (Oberwinkler et al. 2005). Monovalent cations on the extracellular side, however, reduce the outward currents through TRPM3α2 (short pore), but not through TRPM3α1 (long pore). Na^+ had a stronger inhibitory effect than K^+ (Oberwinkler et al. 2005). This finding again emphasizes the importance of the splice event in the pore region for the functional properties of the resulting channels. Block by extracellular Na^+ is a rather unusual feature of ion channels and has only been described for inward rectifier and human ether-a-go-go-related gene (HERG) potassium channels (Wischmeyer et al. 1995; Numaguchi et al. 2000). Because it is observed rarely, this feature might be useful in future studies for identifying native TRPM3 channels, especially, because it allows a functional discrimination between splice variants.

2.7
Single Channel Conductance

Only the $hTRPM3_{1325}$ splice variant has been investigated with respect to its single channel conductance. For spontaneously active channels, Grimm et al. (2003) found in cell-attached experiments that the single channel conductance did depend only very little on the membrane potential ($\leq 20\%$ variation), but was dependent on the extracellular cation species. The reported values at -60 mV were 133 pS with 140 mM extracellular Cs^+, 83 pS with 140 mM extracellular Na^+ and 65 pS with 100 mM extracellular Ca^{2+}. Additionally, the same group reported values for the single channel conductance of agonist-stimulated TRPM3 channels measured in the whole cell configuration. Using an extracellular solution with nearly physiological cation concentrations, they reported a value of 73 pS at negative membrane potentials (Grimm et al. 2005). All these values are very high and rivalled in the TRPC, TRPV and TRPM families only by TRPV3 (Xu et al. 2002). If correct, they indicate that TRPM3 channels have a very large pore diameter, which could be consistent with a nonselective ion channel. It should, however, be kept in mind that the number of observations is still very small (1–3 per data point; Grimm et al. 2003, 2005). In order to unambiguously identify the investigated channels as encoded by TRPM3, it will be instrumental to show that the ensemble average of the observed single channels events correlates with the biophysical properties of the corresponding whole cell currents.

3
Modes of Activation

As mentioned in the previous section, all studies that have functionally investigated TRPM3 channels agree that these channels are constitutively active when heterologously expressed in HEK 293 cells (Lee et al. 2003; Grimm et al. 2003; Oberwinkler et al. 2005). Below we summarize the studies that sought to identify stimuli that enhance TRPM3 channel activity.

3.1
Store Depletion

Lee et al. (2003) added Ca^{2+} at various concentrations to cells expressing hTRPM3a after a number of different pre-treatments designed to affect the Ca^{2+} content of intracellular stores (block of SERCA-ATPase with thapsigargin, G_q-coupled receptor stimulation with carbachol, passive store depletion in extracellular Ca^{2+}-free solution). They invariably found that the increase in intracellular Ca^{2+} concentration was larger in hTRPM3a-expressing cells compared to control cells (Lee et al. 2003). This result was not affected by the nature of the pre-treatment. These data are compatible with the notion that hTRPM3a channels are constitutively active. Under some, but not all, conditions Lee et al. (2003) reported an increase of Ca^{2+} influx through hTRPM3a channels after store depletion. When observed, this increase was very small and it is thus unlikely that Ca^{2+} store depletion is a major factor regulating TRPM3 activity. Using the hTRPM3$_{1325}$ variant, Grimm et al. (2003) could not detect any enhancement of Mn^{2+} entry after thapsigargin treatment or activation of G_q-coupled receptors. Also, Grimm et al. (2005) showed that D-*erythro*-sphingosine-induced Ca^{2+} entry was not influenced by thapsigargin. In conclusion, neither store depletion nor activation of G_q-coupled receptors has been convincingly shown to activate or to substantially regulate TRPM3 channels.

3.2
Hypotonic Stimulation

Application of hypotonic extracellular solution with an osmolarity of 200 mosmol/l induces strong cell swelling in HEK 293 cells. Under such conditions, cells transfected with the cDNA of hTRPM3$_{1325}$ show a much stronger increase in intracellular Ca^{2+} concentration than non-transfected controls (Grimm et al. 2003). Conversely, hypertonic extracellular solution (400 mosmol/l) led to a decrease of the intracellular Ca^{2+} concentration in hTRPM3$_{1325}$-expressing cells. The increase in intracellular Ca^{2+} in hypotonic conditions was not sensitive to 1 µM ruthenium red (Grimm et al. 2003), a blocker that strongly inhibits TRPV4 channels, which are also activated by hypotonic extracellular solutions (Strotmann et al. 2000; Liedtke et al. 2000).

Grimm et al. (2003) also reported currents induced by hypotonicity from TRPM3-transfected HEK 293 cells. These currents could only be detected in the perforated patch configuration, but not in the ruptured patch configuration. The hypotonicity-induced currents were insensitive to the Cl^--channel blocker 5-nitro-2-(3-phenylpropylamino)benzoic acid (NPPB), making it unlikely that they were caused by volume-regulated anion currents. However, less than 50% of the investigated cells responded in that way. The reason for this cell-to-cell variability is not clear. Together, the data suggest that TRPM3 channels are not directly activated by membrane stretch, but rather by a signal that is produced in HEK 293 cells in response to cell swelling. However, the nature of this cellular signal is unknown. Consequently, we do not know whether cells that endogenously express TRPM3 are capable of producing this signal and whether endogenous TRPM3 channels activate in response to hypotonic stimulation.

3.3
Activation by D-*erythro*-Sphingosine

Screening a variety of lipophilic substances, Grimm et al. (2005) found that D-*erythro*-sphingosine (SPH) can induce an increase in intracellular Ca^{2+} concentration in cells that express $hTRPM3_{1325}$. Untransfected cells and cells that expressed other TRP channels (from the TRPC, TRPV and TRPM subfamilies) reacted significantly less to stimulation with SPH. The effect of SPH on the cytosolic Ca^{2+} concentration took place 20–30 s after extracellular application of the substance and was dose-dependent with an estimated EC_{50} value of 12 µM. SPH appeared to act in a specific way, since a number of other lipophilic substances did not activate TRPM3. The only other compounds that were shown to increase intracellular Ca^{2+} were the SPH analogues dihydro-D-*erythro*-sphingosine and *N,N*-dimethyl-D-*erythro*-sphingosine. The effect of SPH was found to be independent of the function of protein kinase C (PKC) since several substances known not inhibit PKC function did not block the SPH-induced Ca^{2+} influx in TRPM3-expressing cells. Equally, 5 µM thapsigargin and 1 µM Xestospongin C were not effective as inhibitors. From these data, Grimm et al. (2005) concluded that neither intracellular Ca^{2+} stores, nor inositol tris-phosphate receptors are involved in the effects of SPH on TRPM3-expressing cells.

Grimm et al. (2005) also reported that SPH induces currents in TRPM3-expressing cells. The observed currents were weakly voltage dependent, weakly outwardly rectifying and non-selectively permeable for cations (see Sect. 2 for details). The current amplitudes were small ($\leq$250 pA at −80 mV), especially compared to the large single channel conductance of 73 pS measured under the same conditions. Given that TRPM3 channels are constitutively active, it is unfortunate that it has not been specified by how much SPH enhances currents through TRPM3. Throughout these experiments, SPH was only applied to the outside of the cells, and it is unclear if the extracellular SPH concentration in vivo can reach values sufficient for activation of native TRPM3 channels. Alternatively, SPH may also be effective from the intracellular side of the membrane, but this has not been shown. These issues need to be addressed before it can be concluded that SPH is a physiologically relevant activator of TRPM3 channels. To summarize, there is evidence that hypotonicity and SPH activate heterologously expressed TRPM3 channels. The available data, however, do not allow concluding that these stimuli are physiologically relevant modulators of endogenous TRPM3 channels.

4
Pharmacology

The substances that are capable of enhancing TRPM3 activity are reviewed in the previous section. Only a few compounds were tested for inhibitory action on TRPM3 channels. Lee et al. (2003) reported that Gd^{3+} at a concentration of 100 μM partially blocks TRPM3-dependent Ca^{2+} influx in HEK 293 cells. On the other hand, Grimm et al. (2003) found a complete block of constitutively active TRPM3 channels by 100 μM Gd^{3+} and 100 μM La^{3+}. SK&F-96365, a substance known to block calcium release-activated calcium currents (I_{crac}; Franzius et al. 1994) and some TRP channels (e.g. TRPC3; Zhu et al. 1998), did not affect spontaneously active TRPM3 channels. It is not known if these substances also block hypotonicity or SPH-induced TRPM3 activity, but 1 μM ruthenium red was found to be ineffective on hypotonicity-induced Ca^{2+} influx in TRPM3-expressing cells (Grimm et al. 2003). In a recent study, 2-aminoethoxydiphenyl borate (2-APB) at a concentration of 100 μM was found to inhibit spontaneous activity of recombinant TRPM3 channels (Xu et al. 2005). None of these substances is a specific pharmacological tool (see e.g. Bootman et al. 2002). It is therefore unlikely that they will be useful for identifying endogenous TRPM3 channels or to unravel the function of TRPM3 channels in their native environment.

5
Biological Relevance of TRPM3 Channels

Based on TRPM3 channels' presence in renal tubules, permeability to divalent cations and sensitivity to changes in extracellular osmolarity observed in heterologous expression systems, it has been proposed that they play a role in osmoregulation and renal Ca^{2+} homeostasis (Lee et al. 2003; Grimm et al. 2003). Similarly, the high expression of TRPM3 in epithelial cells of the choroid plexus and its regulation by mono- and divalent cations has led us to suggest that TRPM3 channels might be involved in the cation homeostasis of the cerebrospinal fluid (Oberwinkler et al. 2005). While these speculations are interesting and helpful in formulating testable working hypotheses, for now they rest on an insufficient experimental foundation. At present, there are no established biological roles for TRPM3 channels.

References

Bartel DP (2004) MicroRNAs: genomics, biogenesis, mechanism, and function. Cell 116:281–297

Berezikov E, Guryev V, van de Belt J, Wienholds E, Plasterk RH, Cuppen E (2005) Phylogenetic shadowing and computational identification of human microRNA genes. Cell 120:21–24

Bollimuntha S, Singh BB, Shavali S, Sharma SK, Ebadi M (2005) TRPC1-mediated inhibition of 1-methyl-4-phenylpyridinium ion neurotoxicity in human SH-SY5Y neuroblastoma cells. J Biol Chem 280:2132–2140

Bootman MD, Collins TJ, MacKenzie L, Roderick HL, Berridge MJ, Peppiatt CM (2002) 2-Aminoethoxydiphenyl borate (2-APB) is a reliable blocker of store-operated Ca^{2+} entry but an inconsistent inhibitor of InsP3-induced Ca^{2+} release. FASEB J 16:1145–1150

Chen PY, Manninga H, Slanchev K, Chien M, Russo JJ, Ju J, Sheridan R, John B, Marks DS, Gaidatzis D, Sander C, Zavolan M, Tuschl T (2005) The developmental miRNA profiles of zebrafish as determined by small RNA cloning. Genes Dev 19:1288–1293

Fantozzi I, Zhang S, Platoshyn O, Remillard CV, Cowling RT, Yuan JX (2003) Hypoxia increases AP-1 binding activity by enhancing capacitative Ca^{2+} entry in human pulmonary artery endothelial cells. Am J Physiol Lung Cell Mol Physiol 285:L1233–L1245

Franzius D, Hoth M, Penner R (1994) Non-specific effects of calcium entry antagonists in mast cells. Pflugers Arch 428:433–438

Griffiths-Jones S (2004) The microRNA Registry. Nucleic Acids Res 32:D109–D111

Grimm C, Kraft R, Sauerbruch S, Schultz G, Harteneck C (2003) Molecular and functional characterization of the melastatin-related cation channel TRPM3. J Biol Chem 278:21493–21501

Grimm C, Kraft R, Schultz G, Harteneck C (2005) Activation of the melastatin-related cation channel TRPM3 by D-*erythro*-sphingosine. Mol Pharmacol 67:798–805

Kozak JA, Cahalan MD (2003) MIC channels are inhibited by internal divalent cations but not ATP. Biophys J 84:922–927

Lagos-Quintana M, Rauhut R, Meyer J, Borkhardt A, Tuschl T (2003) New microRNAs from mouse and human. RNA 9:175–179

Lee N, Chen J, Wu S, Sun L, Huang M, Levesque PC, Rich A, Feder JN, Gray KR, Lin JH, Janovitz EB, Blanar MA (2003) Expression and characterization of human transient receptor potential melastatin 3 (hTRPM3). J Biol Chem 278:20890–20897

Liedtke W, Choe Y, Marti-Renom MA, Bell AM, Denis CS, Sali A, Hudspeth AJ, Friedman JM, Heller S (2000) Vanilloid receptor-related osmotically activated channel (VR-OAC), a candidate vertebrate osmoreceptor. Cell 103:525–535

Lis A, Wissenbach U, Philipp SE (2005) Transcriptional regulation and processing increase the functional variability of TRPM channels. Naunyn Schmiedebergs Arch Pharmacol 371:315–324

Nadler MJ, Hermosura MC, Inabe K, Perraud AL, Zhu Q, Stokes AJ, Kurosaki T, Kinet JP, Penner R, Scharenberg AM, Fleig A (2001) LTRPC7 is a MgATP-regulated divalent cation channel required for cell viability. Nature 411:590–595

Nealen ML, Gold MS, Thut PD, Caterina MJ (2003) TRPM8 mRNA is expressed in a subset of cold-responsive trigeminal neurons from rat. J Neurophysiol 90:515–520

Nilius B, Prenen J, Janssens A, Owsianik G, Wang C, Zhu MX, Voets T (2005) The selectivity filter of the cation channel TRPM4. J Biol Chem 280:22899–22906

Numaguchi H, Johnson JP Jr, Petersen CI, Balser JR (2000) A sensitive mechanism for cation modulation of potassium current. Nat Neurosci 3:429–430

Oberwinkler J, Lis A, Giehl KM, Flockerzi V, Philipp SE (2005) Alternative splicing switches the divalent cation selectivity of TRPM3 channels. J Biol Chem 280:22540–22548

Rodriguez A, Griffiths-Jones S, Ashurst JL, Bradley A (2004) Identification of mammalian microRNA host genes and transcription units. Genome Res 14:1902–1910

Schulz HL, Rahman FA, Fadl El Moula FM, Stojic J, Gehrig A, Weber BH (2004) Identifying differentially expressed genes in the mammalian retina and the retinal pigment epithelium by suppression subtractive hybridization. Cytogenet Genome Res 106:74–81

Strotmann R, Harteneck C, Nunnenmacher K, Schultz G, Plant TD (2000) OTRPC4, a nonselective cation channel that confers sensitivity to extracellular osmolarity. Nat Cell Biol 2:695–702

Voets T, Nilius B (2003) The pore of TRP channels: trivial or neglected? Cell Calcium 33:299–302

Voets T, Nilius B, Hoefs S, van der Kemp AW, Droogmans G, Bindels RJ, Hoenderop JG (2004) TRPM6 forms the Mg^{2+} influx channel involved in intestinal and renal Mg^{2+} absorption. J Biol Chem 279:19–25

Weber MJ (2005) New human and mouse microRNA genes found by homology search. FEBS J 272:59–73

Wischmeyer E, Lentes KU, Karschin A (1995) Physiological and molecular characterization of an IRK-type inward rectifier K^+ channel in a tumour mast cell line. Pflugers Arch 429:809–819

Wistow G, Bernstein SL, Ray S, Wyatt MK, Behal A, Touchman JW, Bouffard G, Smith D, Peterson K (2002a) Expressed sequence tag analysis of adult human iris for the NEIBank Project: steroid-response factors and similarities with retinal pigment epithelium. Mol Vis 8:185–195

Wistow G, Bernstein SL, Wyatt MK, Behal A, Touchman JW, Bouffard G, Smith D, Peterson K (2002b) Expressed sequence tag analysis of adult human lens for the NEIBank Project: over 2000 non-redundant transcripts, novel genes and splice variants. Mol Vis 8:171–184

Xu H, Ramsey IS, Kotecha SA, Moran MM, Chong JA, Lawson D, Ge P, Lilly J, Silos-Santiago I, Xie Y, DiStefano PS, Curtis R, Clapham DE (2002) TRPV3 is a calcium-permeable temperature-sensitive cation channel. Nature 418:181–186

Xu SZ, Zeng F, Boulay G, Grimm C, Harteneck C, Beech DJ (2005) Block of TRPC5 channels by 2-aminoethoxydiphenyl borate: a differential, extracellular and voltage-dependent effect. Br J Pharmacol 145:405–414

Yap KL, Kim J, Truong K, Sherman M, Yuan T, Ikura M (2000) Calmodulin target database. J Struct Funct Genomics 1:8–14

Zhang Z, Tang Y, Zhu MX (2001) Increased inwardly rectifying potassium currents in HEK-293 cells expressing murine transient receptor potential 4. Biochem J 354:717–725

Zhu X, Jiang M, Birnbaumer L (1998) Receptor-activated Ca^{2+} influx via human Trp3 stably expressed in human embryonic kidney (HEK)293 cells. Evidence for a non-capacitative Ca^{2+} entry. J Biol Chem 273:133–142

Insights into TRPM4 Function, Regulation and Physiological Role

R. Vennekens (✉) · B. Nilius

Laboratory of Physiology, Katholieke Universiteit Leuven, Campus Gasthuisberg O/N1, Herestraat 49-Bus 802, 3000 Leuven, Belgium
rudi.vennekens@med.kuleuven.be

1	Introduction	270
2	Cloning, Expression and Gene Structure	270
3	Calcium- and Voltage-Dependent Activation	272
4	The Selectivity Filter	276
5	Modulation	277
6	Pharmacology	279
7	Physiological Role	279
8	Endogenous TRPM4-Like Currents	280
References		283

Abstract In the current review we will summarise data from the recent literature describing molecular and functional properties of TRPM4. Together with TRPM5, these channels are up till now the only molecular candidates for a class of non-selective, Ca^{2+}-impermeable cation channels which are activated by elevated Ca^{2+} levels in the cytosol. Apart from intracellular Ca^{2+}, TRPM4 activation is also dependent on membrane potential. Additionally, channel activity is modulated by ATP, phosphatidylinositol bisphosphate (PiP$_2$), protein kinase C (PKC) phosphorylation and heat. The molecular determinants for channel activation, permeation and modulation are increasingly being clarified, and will be discussed here in detail. The physiological role of Ca^{2+}-activated non-selective cation channels is unclear, especially in the absence of gene-specific knock-out mice, but evidence indicates a role as a regulator of membrane potential, and thus the driving force for Ca^{2+} entry from the extracellular medium.

Keywords TRP channels · TRPM4 · Ca^{2+}-activated non-selective cation channel

1
Introduction

TRPM4 is a non-selective cation channel which is activated by a high intracellular Ca^{2+} concentration (a so-called CAN channel, for Ca^{2+}-activated non-selective). This class of ion channels has long remained an enigma. They were first described in cultured rat cardiac myocytes (Colquhoun et al. 1981) and have been found since then in multiple tissues ranging from cardiac muscle, a variety of neuron types, in exocrine tissues (including pancreatic acini), renal tubules, intestine and endothelium (Partridge and Swandulla 1988; Siemen 1993; Teulon 2000). Experimental findings have suggested that CAN channels could mediate and maintain cell depolarisation and support cellular functions such as neuronal bursting activity, kidney cell osmotic regulation, pancreatic acinar fluid secretion and cardiac rhythmicity (Partridge and Swandulla 1988; Siemen 1993; Teulon 2000), but in the absence of CAN channel knock-out mice and selective pharmacological blockers this issue remains largely speculation. The cloning of TRPM4 thus serves as a major breakthrough in the field. In the current review, we will summarise data on the molecular and functional characterisation of this exciting new channel.

2
Cloning, Expression and Gene Structure

The first report on TRPM4, published on 11 September 2001 in *The Proceedings of the National Academy of Sciences*, described cloning of a 4.0-kb complementary DNA (cDNA) after screening of human brain, placenta and testis cDNA libraries with an expressed sequence tags (EST) clone (Xu et al. 2001). Subsequent efforts (described in Launay et al. 2002; Nilius et al. 2003) revealed that this clone is in fact a short splice variant of full-length TRPM4, lacking 174 aa in the N-terminus. Accordingly, the full-length human clone was designated as TRPM4b [accession number (acc. nr.) AF497623] and the short variant as TRPM4a (acc. nr. AY046396). In Nilius et al. (2003), a further human splice variant lacking 537 N-terminal amino acids, TRPM4c (acc. nr. AX443225), was described. Through analysis of the ENSEMBL mouse genome database, the mouse orthologue of TRPM4 was identified. Subsequent RT-PCR cloning of the cDNA from a mouse heart poly-A RNA sample revealed the full-length TRPM4b clone (acc. nr. AJ575814) and two additional splice variants, distinct from TRPM4a and c. Additionally, Murakami et al. (2003) reported two more short variants isolated from mouse brain (acc. nr. AB112658 and AB112657). Thus, several TRPM4 variants have been identified, but the functional significance of these splice variants remains in question. The majority of the functional characterisation described below has been performed with mouse and human TRPM4b.

TRPM4 is a member of the transient receptor potential (TRP) superfamily of channel proteins (Fig. 1). This large group of ion channel proteins is subdivided into roughly three subfamilies, TRPC (6 members), TRPM (8 members) and TRPV (6 members), and a group of more distantly related proteins, TRPML, TRPP, TRPA and TRPN (Ramsey et al. 2005). The TRPM4 gene is located on human chromosome 19, and in mouse on chromosome 7. It consists of 25 exons, spanning 54 kb in the human and 31 kb in the mouse genome. Transmembrane (TM) regions are coded from exons 15 to 20. As for other members of the TRP family, TRPM4 is likely to have six TM domains with a pore region between TM regions 5 and 6. It is plausible that four subunits are required to form a functional channel. Notably in this regard is the fact that TRPM4 subunits can homo-associate (Murakami et al. 2003). Formation of heteromers of TRPM4 and another TRP member has not been reported to date. Within the TRPM subfamily, TRPM4 is most closely related to TRPM5, sharing approximately 50% homology. Unlike other members of the TRP superfamily of membrane proteins, apparently no ankyrin repeats are present in the N-terminus of TRPM4. In the N-terminal region, however, four stretches of moderate sequence homology to other members of the TRPM family are found. Several protein domains were identified in the TRPM4b protein sequence, including putative calmodulin binding sites in the N- and C-terminus, as well as phosphorylation sites for protein kinase (PK)A and PKC, four Walker B motifs, a phosphatidylinositol bisphosphate (PiP_2) binding site with homology to a pleckstrin homology domain (PH) and comprising a decavanadate binding site and two ABC transporter-like signature motifs (Nilius et al. 2005b). A coiled-coil domain is predicted in the C-terminus. The functional significance of these domains is becoming increasingly clarified and will be discussed in detail below.

Northern blot analysis revealed human TRPM4b expression in numerous tissues including placenta, skeletal muscle, heart, kidney, liver, pancreas, thymus, spleen, prostate, small intestine, colon and lung (Launay et al. 2002; Nilius et al. 2003; Xu et al. 2001). Remarkably, in contrast to Launay et al. (2002), Nilius et al. (2003) could not detect TRPM4 messenger RNA (mRNA) in spleen and thymus. In mouse tissue, TRPM4 expression was shown in stomach, intestine, placenta, oesophagus, aortic endothelium, kidney, heart, pancreas and placenta (Nilius et al. 2003). No transcripts were found in leucocytes. Both in human and mouse tissues no indications were found for TRPM4 expression in whole brain. Using serial analysis of gene expression in the kidney glomerulus and seven different nephron segments, Chabardes-Garonne et al. (2003) showed expression of TRPM4 in the proximal convoluted tubule, proximal straight tubule, medullary thick ascending limb, distal convoluted tubule, cortical collecting duct and outer medullary collecting duct.

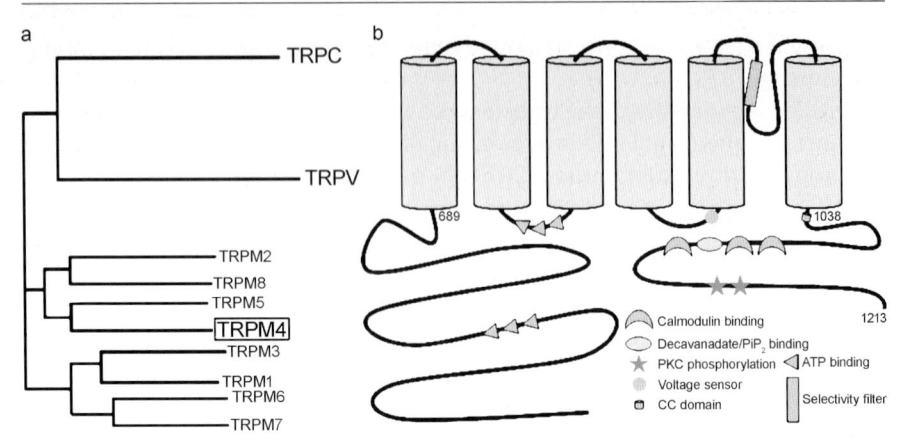

Fig. 1 a Phylogenetic tree of the TRP family of membrane proteins. For clarity TRP-P, TRPA, TRP-ML and TRPN were omitted from the scheme. Only the TRPM subfamily is shown with all its members. TRPM4 is most closely related to TRPM5. **b** Schematic representation of the TRPM4 protein. Except for the coiled-coil domain (*CC*) in the C-terminus, the functional significance of all indicated domains was determined experimentally through site-directed mutagenesis. See text for more details

3
Calcium- and Voltage-Dependent Activation

Thus far, the majority of the functional data available on TRPM4 is gathered through work with both human and mouse, full-length, TRPM4b cDNA over-expressed in HEK293 cells. Extensive electrophysiological analysis, using the patch clamp technique, showed that this protein functions as a Ca^{2+}-activated cation channel (Nadler et al. 2001; Nilius et al. 2003, 2004a, 2005a, b; Ullrich et al. 2005; Fig. 2), which is a finding with significant impact, since this is the first channel of this class to have been identified on the molecular level. In Xu et al. (2001) it was shown, using Ca^{2+} imaging, that over-expression of the shorter variant, TRPM4a, in HEK293 cells promotes Ca^{2+} accumulation upon Ca^{2+} re-addition, apparently without a clear activating stimulus. This finding was, however, not followed by electrophysiological characterisation and was not reproduced by other groups (B. Nilius, J. Prenen). Recent work claims that TRPM4a actually functions as a dominant-negative construct for TRPM4b (Launay et al. 2004).

Upon loading HEK293 cells with high intracellular Ca^{2+} concentrations through the pipette solution, large cation currents can be measured in the whole-cell mode in TRPM4b over-expressing cells. The steady-state current–voltage relationship was reported to be either quasi-linear (Launay et al. 2002) or strongly outwardly rectifying (Nilius et al. 2003, 2004a, 2005a, b; Ullrich et al. 2005). In Launay et al. (2002), currents were stable for at least 1 min when intracellular Ca^{2+} is kept constantly elevated, while Nilius et al. (2003) reported strong inactivation of currents within 30 to 120 s. This inactivation could be partly reversed or retarded in excised, cell-free patches, indicating that in the whole-cell experiments a crucial cytoplasmic factor for channel function is washed out of the cell. The single channel conductance of TRPM4b amounts approximately 23 pS (Launay et al. 2002; Nilius et al. 2003). The sensitivity of TRPM4b to intracellular calcium ($[Ca^{2+}]_i$) as determined by different research groups varies greatly. When Ca^{2+} was loaded through the patch pipette in whole-cell experiments, Launay et al. (2002) reported an IC_{50} value of channel activation between 320 and 520 nM, with a Hill coefficient between 6 and 4. In Nilius et al. (2003), on the other hand, significant current activation was only seen at 1 µM $[Ca^{2+}]_i$ and higher. Subsequently, Ullrich et al. (2005) described an IC_{50} value of 20 µM with a Hill coefficient of 1.6. In Ca^{2+} uncaging experiments, a threshold for current activation of 5 µM was found (Ullrich et al. 2005). In excised inside-out patches, on the other hand, an EC_{50} of 374 µM was reported (Nilius et al. 2004a), which is at least 20 times greater compared to whole-cell experiments, indicating again that a cytoplasmic factor is important for Ca^{2+} sensitivity of TRPM4b. When the Ca^{2+} sensitivity of TRPM4b was determined in relation to the inactivation state in inside-out patches, an EC_{50} of 4.4 µM was found immediately after current activation, compared to 140 µM after reaching a steady-state current level. These apparently severe discrepancies will be discussed further below.

Additional to Ca^{2+}-dependent activation, TRPM4 currents are also strongly voltage-dependent (Nilius et al. 2003). Voltage steps to positive potentials induce slowly activating currents, whereas steps to negative potentials induce deactivating currents. This reflects channel activation at positive potentials and channel closure at negative potentials. When a classic Boltzmann analysis of voltage-dependent open probability was applied, a $V_{1/2}$ for activation in the range between −20 and +60 mV was determined, depending on a variety of factors such as $[Ca^{2+}]_i$, presence of calmodulin, phosphorylation, temperature, phosphatidylinositol bisphosphate (PIP_2) content, etc. (Nilius et al. 2005b, c). The apparent gating charge obtained from Boltzmann fits is in the range between 0.6 and 0.8. As shown for voltage-activated Ca^{2+} or K^+ channels, voltage dependence involves the movement of a charged voltage sensor, characterised by an effective gating charge of valence z moving from the inner membrane surface to the outer. If this gating charge is large, as shown for voltage-dependent K^+ channels such as the Shaker K^+ channel, activation occurs in a very narrow voltage range. In case of TRPM4b the calculated gating charge is small, resulting in a relatively shallow activation curve and activation in a comparatively broad voltage range. In addition, because $V_{1/2} = (\Delta H - T \times \Delta S)/zF$, small changes in enthalpy (ΔH) or entropy (ΔS) will induce larges shifts in the voltage range of activation (Nilius et al. 2005c; Talavera et al. 2005). It should be clear, however, that Ca^{2+} is a necessary requirement to open the channel, and that voltage by itself is insufficient to activate the channel. In fact, it was shown that higher cytoplasmic Ca^{2+} concentrations actually induce a slight leftward shift of the voltage-dependent activation curve (towards more physiological values) and faster time-constants for current activation at positive potentials (Nilius et al. 2004a), thus providing a rationale for channel opening at physiological membrane potentials when $[Ca^{2+}]_i$ is elevated. To elaborate these findings further, a minimal kinetic model was designed to describe Ca^{2+} and voltage-dependent gating of TRPM4b. A sequential model was conceived in which binding of Ca^{2+} to the channel precedes voltage-dependent activation of the channel. Thus, two closed states are considered, one Ca^{2+} unbound and one Ca^{2+} bound, and channel opening is achieved when a depolarising voltage step is applied. This scheme is compatible with the observed desensitisation of the channel after patch excision and the fact that the open probability of the channel reaches unity at high Ca^{2+} concentrations and at very positive membrane potentials (Nilius et al. 2004a).

As mentioned, upon activation the TRPM4b channel will completely inactivate within 30–120 s. In whole-cell measurements, inactivation is faster and more complete compared to experiments in the excised patch mode where a variable Ca^{2+}-dependent rest activity is left over, indicating that a cellular process plays a role in shutting down the channel completely. Partly this inactivation can be explained by a decrease in the sensitivity for Ca^{2+}-dependent current activation. Indeed, depending on whether the EC_{50} in inside-out patches is determined soon after patch break or after current decay, a 30-fold decrease in Ca^{2+} sensitivity is found. Concomitantly, current decay is slower when higher $[Ca^{2+}]_i$ are applied to activate the channel. The rate of activation at positive potentials and the rate of deactivation at negative potentials also decline dur-

Fig. 2 a Schematic view of factors regulating TRPM4b activity as assessed from data gathered in over-expression experiments in HEK293 cells. See text for more details. **b** Whole-cell currents from TRPM4b over-expressing HEK293 cells. Cells were loaded with 1 µM Ca^{2+} through the pipette solution. Time-course of an experiment at −80 mV and at +80 mV. Current traces are shown at the peak current level (*1*), and after rundown (*2*). Note the strong outward rectification of the current–voltage (I–V) curve at peak current levels, and the transient nature of the current despite constant $[Ca^{2+}]_i$. **c** Single channel recordings from inside-out patches of TRPM4b over-expressing HEK293 cells in response to a voltage step to −100 mV. The *dotted line* represents the zero current level. Patches were exposed to 300 µM Ca^{2+} at the cytosolic side. **d** Representative current analysis in excised inside-out patches after application of 300 µM Ca^{2+} to the cytosolic side of the channel. *a* Current traces in response to a voltage protocol depicted in the above panel. *b* I–V curves derived by taking data points at the indicated times from currents in panel *a*. Note that the instantaneous I–V curve is linear (*1*), the steady state I–V curve shows strong outward rectification (*2*). Trace 3 shows the relation between the amplitude of tail currents in relation to the voltage of the pre-step. *c* Voltage dependence of the open probability of TRPM4b obtained from tail current measurements as in panel *a*. Solid line represents a fit of the data with the Boltzmann equation, $V_{1/2} = -7.8$ mV and a slope of 38.7 mV which gives an estimate of a gating charge of $z \sim 0.62$

Insights into TRPM4 Function, Regulation and Physiological Role

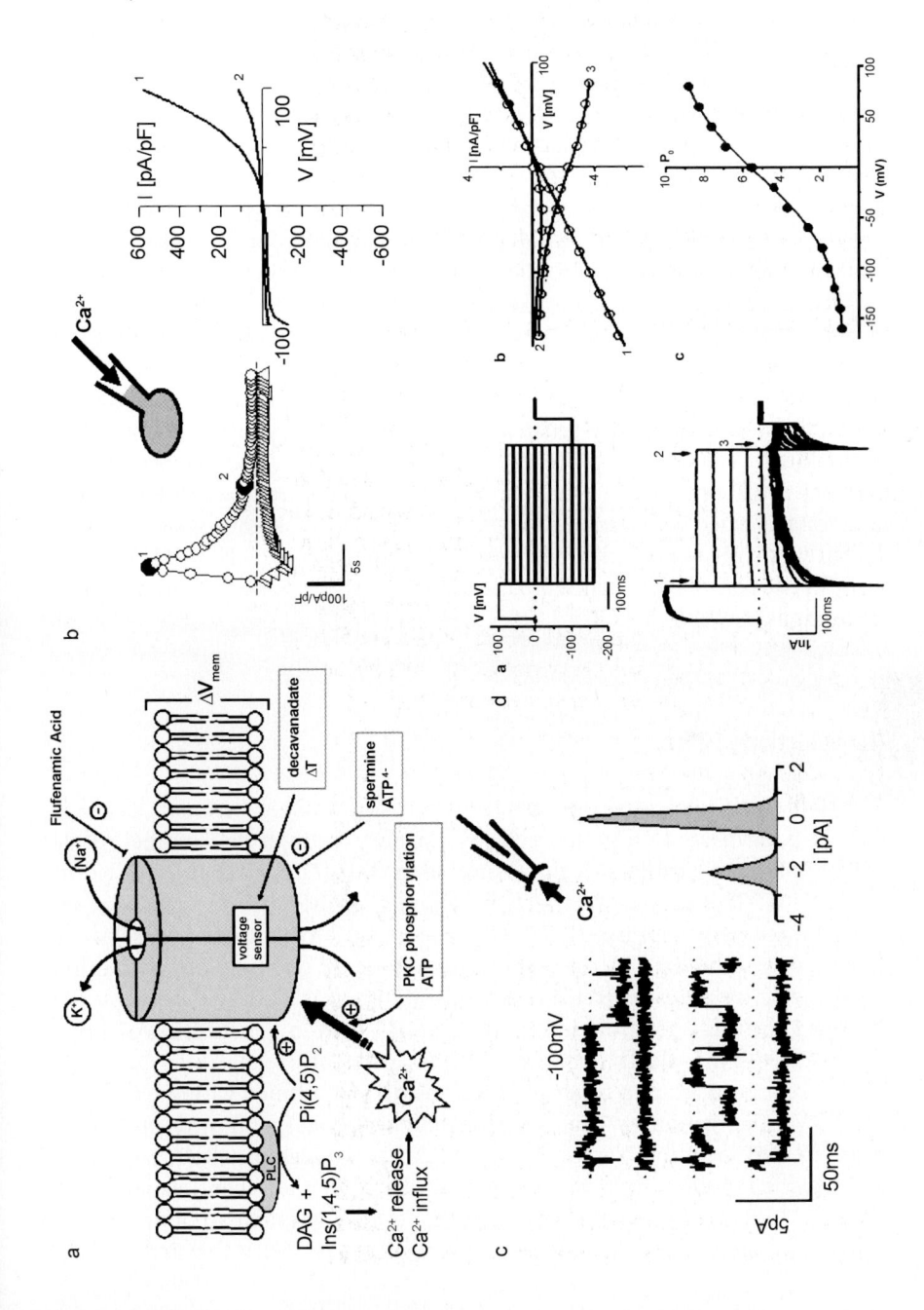

ing this desensitisation phase. Obviously, desensitisation of TRPM4b activity might also be the source for the above-mentioned scattering of EC_{50} values for Ca^{2+}-dependent current activation reported in the literature. Furthermore, because by definition EC_{50} values result from an estimation of the ratio between the open probability at a certain $[Ca^{2+}]_i$ and the maximal open probability, the measured EC_{50} value depends both on the 'real' K_d for Ca^{2+} binding and voltage, since the open probability of the channel is also voltage-dependent. Changes in voltage dependence will lead inevitably to changes in the apparent EC_{50} value (Nilius et al. 2004a). As discussed below, several cellular factors can modulate the voltage sensitivity of TRPM4b, such as ATP, PiP$_2$ and temperature.

The molecular determinants for Ca^{2+} and voltage dependence of TRPM4b are close to being elucidated. Over-expression of a calmodulin mutant unable to bind Ca^{2+} dramatically reduced TRPM4b activation by activation by Ca^{2+}. Concomitantly, mutation of any of the three putative calmodulin binding sites in the C-terminus of TRPM4b strongly impaired current activation by reducing the Ca^{2+} sensitivity of TRPM4b and shifting the voltage dependence of activation to very positive potentials. This indicates a crucial role of calmodulin in inferring Ca^{2+} sensitivity to TRPM4b. However, since Ca^{2+} sensitivity is never completely lost in TRPM4b mutants unable to bind calmodulin, it is conceivable that another mechanism also plays a role (Nilius et al. 2005b). Other members of the TRP family, including the cold- and menthol-activated TRPM8, also exhibit voltage-sensitivity. Recent work has shown that mutations in the fourth TM region of TRPM8 significantly influence voltage sensitivity, pointing to an analogous mechanism of voltage sensing between TRP channels and voltage-gated K$^+$ channels (Voets et al. 2004). Analogously, neutralising a positive charge in the linker between TM domains 4 and 5 of TRPM4b significantly reduces voltage sensitivity and shifts the activation curve dramatically to the right (Nilius et al. 2005c).

4
The Selectivity Filter

TRPM4b constitutes a cation-selective pore. Monovalent cations (Na$^+$, K$^+$, Cs$^+$, Li$^+$) permeate in a poorly selective fashion through the channel, while the TRPM4b pore is virtually impermeable to Ca^{2+} (Launay et al. 2002; Nilius et al. 2003). This is a unique feature within the TRP superfamily, since all other functionally expressed TRPs either form Ca^{2+}-permeable non-selective cation pores or highly Ca^{2+}-selective channels. Based on homology with other cation-selective pores, a stretch of 6 aa, EDMDVA, was identified between TM regions 5 and 6 as a potential selectivity filter of TRPM4b. Substitution of this 6-aa stretch with the selectivity filter of TRPV6, a distantly related member of the TRP family, resulted in a functional channel that combined gating hallmarks of TRPM4 (activation by $[Ca^{2+}]_i$, voltage dependence) with pore properties from TRPV6 including sensitivity to block by extracellular Ca^{2+} and Mg^{2+} and, strikingly, Ca^{2+} permeation. Neutralisation of the second aspartate in the EDMDVA stretch resulted in a non-functional channel with a dominant-negative phenotype when co-expressed with wild-type TRPM4b. Furthermore, selected point mutations in this region altered the inactivation properties and monovalent permeability profile of TRPM4b. Thus, the TRPM4 selectivity filter could be effectively delineated. Furthermore, this study actually provides the first insights into molecular determinants for monovalent cation selectivity ion channels (Nilius et al. 2005a).

5
Modulation

TRPM4b activity is modulated by PKC activity, temperature and binding of intracellular ATP, PiP$_2$ and decavanadate to the channel. When TRPM4b overexpressing HEK293 cells were pre-incubated with phorbol 12-myristate 13-acetate (PMA), the EC$_{50}$ value for channel activation by Ca^{2+} decreased fourfold. This effect was abolished when either of the two C-terminal Ser-residues predicted to have the highest score for PKC phosphorylation was mutated. The same mutations also substantially decrease desensitisation.

ATP, on the other hand, helps to restore the Ca^{2+} sensitivity of TRPM4b. It was shown that TRPM4b recovered from desensitisation when the cytoplasmic side of the membrane in inside-out patches was exposed to a Ca^{2+}-free solution containing MgATP. To elaborate the mechanism (direct binding of ATP to the protein or indirect action through activation of an ATP consuming enzyme), mutations were generated in putative ATP binding sites in the TRPM4b protein. Multiple ATP binding sites can be predicted from the amino acid sequence of TRPM4b, including two Walker B motifs in the N-terminus and two more in cytoplasmic loop between TM3 and 4. When either of these motifs was mutated, the ATP-induced recovery was strongly reduced in all mutants. Moreover, these mutations drastically accelerated the channel desensitisation to Ca^{2+}. Thus, these findings indicate that ATP plays a crucial role in maintaining Ca^{2+} sensitivity of TRPM4b through direct binding to the channel protein (Nilius et al. 2005b). Surprisingly, decavanadate, a compound known to interfere with ATP binding in ATP-dependent transporters, does not have opposite effects on TRPM4b function compared with ATP. Instead, decavanadate is a strong modulator of voltage-dependent gating of the TRPM4b. In the presence of decavanadate on the cytosolic side of excised inside-out patches, TRPM4b currents are sustained, not desensitising, and linear over a voltage range from −180 to +140 mV. Again, the binding site for decavanadate to the TRPM4b channel was identified and located to the C-terminal tail of TRPM4b (Nilius et al. 2004a).

Another TRPM4b modulator is phosphatidylinositol (4,5) bisphosphate [Pi(4,5)P$_2$] (Nilius et al. 2006; Zhang et al. 2005). Besides being the substrate for phospholipase C (PLC), generating second messengers as inositol 1,4,5-trisphosphate and diacylglycerol, Pi(4,5)P$_2$ has emerged as an important regulator of many ion channels and transporters, including voltage-gated K$^+$

and Ca^{2+} channels and a growing number of TRP channels. In the TRP family, the effect of PiP_2 on channel activity can be either stimulatory (as for TRPM5, TRPM7, TRPM8 and TRPV5) or inhibitory (as for TRPV1 and TRPL). PiP_2 is unable to gate TRPM4b directly when Ca^{2+} is buffered at low levels. Instead, PiP_2 acts as a modulator of the channels sensitivity to both Ca^{2+} and voltage: increasing PiP_2 levels causes a 100-fold increase in Ca^{2+} sensitivity and a dramatic shift to more negative potentials of the voltage dependence of activation, thereby strongly increasing the open probability of the channel at physiological membrane potentials. To show this, several tools were used, including depletion of PiP_2 from the cell via receptor stimulation, incubation of cells with wortmannin, an inhibitor of PI-4-kinase which delays PiP_2 replenishment, application of the PiP_2 scavenging agent poly-L-lysine and over-expression of a PiP_2-consuming enzyme 5ptase IV, all leading to a reduction of current amplitudes and fast inactivating of currents. These effects could be reversed when PiP_2 was reapplied to the cytosolic side of excised patches. On the other hand, when PLC activity was inhibited using U73122 (and PiP_2 levels are likely increased), TRPM4b current desensitisation was strongly attenuated. Both application of PiP_2 and U73122 led to an almost complete loss of time dependence of TRPM4b activation at positive potentials, a dramatic slowing of current deactivation at negative potentials, significant steady-state inward currents and a dramatic shift of the steady-state open probability towards more negative potentials. Interestingly, PiP_2 also reduced the slope of voltage dependence of open probability of TRPM4, suggesting that PiP_2 reduces the effective gating charge of TRPM4. Two putative PiP_2-binding pleckstrin homology domains were identified in the C-terminus of TRPM4b. Only the first one—closest to TM6 and also the site of interaction with the channel for the highly negatively charged decavanadate—could be implicated in TRPM4b modulation. Neutralisation of all four positively charged amino acids in this stretch resulted in a channel exhibiting very rapid desensitisation and highly reduced sensitivity to PiP_2.

Very recent data have shown that TRPM4b is also a heat-activated channel. All ion channels, as all other types of enzymes, show some temperature dependence, quantified with the 10-degree temperature coefficient Q_{10} value, indicating defined as rate(T+10)/rate(T) (Hille 2001). Ion channels regarded as temperature independent display Q_{10} values in the range of 1–4. Analysing current amplitude at +25 mV showed a Q_{10} of 8.5±0.6 between 15°C and 25°C, indicating strong temperature dependence of the channel activity. Heating shifted the activation curve for voltage-dependent opening of the channel towards negative, more physiological potentials and increased the rate of current relaxation at every potential between −100 and +180 mV. On the other hand, temperature had little effect on the Ca^{2+} dependence of channel activation. Thus, the heat dependence is not due to modulation of the Ca^{2+} sensitivity of the channel, but likely through shifting the voltage-dependent activation curve (Talavera et al. 2005).

6
Pharmacology

Concerning the pharmacological block of TRPM4, not much is known. Sensitive blockers include intracellular spermine and flufenamic acid and clotrimazole applied from the extracellular side, both with IC$_{50}$ values in the range of 1–10 µM (Nilius et al. 2004b; B. Nilius, personal communication). These compounds are, however, poorly selective among other ion channels and thus provide not much of a pharmacological basis for current dissection in primary cells. TRPM4b is inhibited by intracellular adenine nucleotides, including ATP, ADP, AMP and AMP-PNP with an IC$_{50}$ value between 2 and 19 µM. Adenosine also blocked TRPM4 at 630 µM. GTP, UTP and CTP do not exert any effect at concentrations up to 1 mM. The most sensitive compound, the ionic form of ATP, ATP^{4-}, when applied to the cytosolic side of the channel, inhibits currents with an IC$_{50}$ value of 1.3 µM (Nilius et al. 2004b). This block is voltage independent (both inward and outward currents are reduced) and, surprisingly, not affected in a negative fashion by the presence of decavanadate (see Sect. 5). In fact, sensitivity for ATP^{4-} block is augmented tenfold in the presence of decavanadate. Thus, ATP can both block the channel and facilitate its activation. It is currently unclear, however, whether the inhibitory site and the facilitory ATP binding site on the TRPM4 protein are identical (Nilius et al. 2004a).

7
Physiological Role

At the time of writing no data are available from TRPM4 knock-out mice. Data concerning TRPM4's physiological role are gathered solely through gene knock-down studies, using RNA interference (RNAi) or expression of a dominant-negative TRPM4 splice variant (Earley et al. 2004; Launay et al. 2004). In Launay et al. (2002), it was already suggested that TRPM4 can control Ca^{2+} influx after receptor stimulation through depolarising the membrane potential and thus limiting the driving force for Ca^{2+} entry into TRPM4 overexpressing HEK293 cells. This idea was further elaborated in Jurkat T cells (Launay et al. 2004). Here endogenous TRPM4 expression was reduced using RNAi. Additionally, TRPM4 was functionally inhibited through expression of TRPM4a, which seems to function as a dominant-negative variant. It was shown that endogenous CAN currents could be inhibited to 25% of control values with both methods. To assess the functional role of this current in Ca^{2+} signalling, Jurkat T cells were stimulated using phytohaemagglutinins (PHA), while monitoring Ca^{2+} levels in the cell. Upon stimulation of control cells, a pattern characterised by oscillations is apparent in untreated cells. In RNAi-treated cells and TRPM4a-expressing cells, these Ca^{2+} signals were transformed in a prolonged, sustained Ca^{2+} increase, amounting to significantly higher values compared to control cells. Concomitantly, interleukin (IL)-2 production in TRPM4-down-regulated cells was significantly increased. Thus, it was hypothesised that TRPM4 functions through limiting the driving force for Ca^{2+} entry upon activation, since cation influx will depolarise the membrane potential. In such a system TRPM4 would work together with Ca^{2+}-activated K^+ channels, depolarising and hyperpolarising the cell membrane in a cyclic manner. It is important, however, to make some technical remarks concerning this paper. First, using the RNAi approach, TRPM4 protein levels in Jurkat T cells were hardly reduced, although current levels were knocked down significantly. This could indicate off-target effects of siRNA molecules in this study, which is a serious concern in all transient gene-knock-down experiments. Second, it is questionable how Jurkat T cells, a cell line used for many years in Ca^{2+}-signalling research, relate to primary T cells in functional properties.

In a second study, TRPM4 expression was knocked down in vascular smooth muscle cells from cerebral artery (Earley et al. 2004). The authors observed a 24-pS Ca^{2+}-sensitive cation current, of which the activity in excised patches was upregulated when cells were pre-treated with PMA, a non-selective PKC activator. The Ca^{2+} sensitivity of channel activity is comparatively low, with an apparent IC_{50} value around 100 μM $[Ca^{2+}]_i$. Since TRPM5 is not expressed in these cells, these features could point to TRPM4 as the ion-carrying protein, but further characterisation of the current was not provided. In TRPM4 antisense-treated cells, the occurrence of these TRPM4-like currents was reduced to 10% of cell patches, compared to 53% in untreated cells. When whole cerebral arteries were treated with TRPM4 antisense oligonucleotides and compared to untreated controls, it was found that pressure-induced depolarisation was lost in treated tissue, indicating that TRPM4 functions as a mechanosensitive channel in this system. When myogenic constriction (a.k.a. the Bayliss effect, or the phenomenon that vessel diameter is reduced when pressure is raised, due to constriction of the vessel) was studied, it was shown that pressure-induced constriction of vessels was impaired in antisense-treated vessels. Again, as in the previous study, actual knock-down of TRPM4 expression was not complete and hard to quantify from the presented data. No effect on the protein level was shown, and off-target effects of the antisense strategy cannot be excluded. Likely, the only conclusive data on the physiological role of TRPM4 will be provided from analysing the knock-out mouse.

8
Endogenous TRPM4-Like Currents

As mentioned in the introduction, over the years several Ca^{2+}-activated non-selective cation currents have been reported in a variety of tissues and cell lines. In Table 1, a selection from the literature is summarised. These refer-

Table 1 Endogenous TRPM4b-like currents

Tissue	Gs	Permeant ions	Ca^{2+} permeation	Ca^{2+} sensitivity	Voltage dependence	Remarks	Reference
Mouse and rat type II pneumocytes	26–29 pS	Na, K	No	>100 nM	Yes	Block by intracell. nucleotides	Mair et al. 2004
Guinea-pig cochlear hair cells	21–29 pS	Na, K	No	>100 nM	Yes	Block by intracell. nucleotides	Van den Abbeele et al. 1994
Rat neonatal atrial myocytes	26 pS	Na, K, Cs, Li	No	≥ 10 µM	Yes		Zhainazarov 2003
Hamster VNO neurons	22 pS	Na, K	No	K_d=500 µM	Yes	Block by intracell. nucleotides	Liman 2003
Rat reactive astrocytes	35 pS	Na, K, Li	No	K_d=300 nM	Yes	ATP sensitive	Chen and Simard 2001
Rat brown adipocytes	30 pS	Na, K, NH$_4$	No	>100 µM	Yes	Block by intracell. nucleotides	Halonen and Nedergaard 2002
Rat brain capillary endothelium	30 pS	Na, K	No	K_d=20 µM	Yes	Block by intracell. nucleotides	Csanady and Adam-Vizi 2003
Mouse neuroblastoma	22 pS	Na, K, Li, Cs	No	K_d=1 µM	n.d.	-	Yellen 1982
Mouse collecting duct cells	23 pS	Na, K, Li, Cs	No	≥ 1 µM	n.d.	Block by intracell. nucleotides	Korbmacher et al. 1995
Rat collecting duct cells	28 pS	Na, K, Li, Cs	No	K_d=5 µM	n.d.	Block by intracell. nucleotides	Nonaka et al. 1995
Rabbit smooth muscle cells	28 pS	Na	No	≥ 100 nM	n.d.	-	Wang et al. 1993
Chick dorsal root ganglion	38 pS	Na, K	No	K_d=400 nM	n.d.	-	Razani-Boroujerdi and Partridge 1993
Human umbilical vein endothelium	25 pS	Na, Cs,	No	K_d=400 nM	n.d.	Block by intracell. nucleotides	Kamouchi et al. 1999

Table 1 continued

Tissue	Gs	Permeant ions	Ca^{2+} permeation	Ca^{2+} sensitivity	Voltage dependence	Remarks	Reference
Human macrovascular endothelium	25 pS	Na, Cs	No	K_d=420 nM	n.d.	Sensitive to intracellular ATP and NO	Suh et al. 2002
Human atrial myocytes	19 pS	Na, K	Yes	K_d=21 μM	Yes	Block by intracell. nucleotides	Guinamard et al. 2004
Rat dorsal root ganglion neurons	35 pS	Na	Yes	n.d.	Yes	Heat-sensitive	Reichling and Levine 1997
Human red blood cells	21 pS	Na, K	Yes	n.d.	Yes	-	Kaestner et al. 1999; Rodighiero et al. 2004
Mouse kidney, TAL	27 pS	Na, K	n.d.	≥1 μM	Yes	Block by intracell. nucleotides	Teulon et al. 1987
Human, mouse, rat beta cells	25–30 pS	Na, Cs, Li	n.d.	>100 nM	n.d.	MTX sensitive	Leech and Habener 1998
Rat cardiac myocytes	30 pS	Na, K	n.d.	≥1 μM	n.d.	-	Colquhoun et al. 1981
Guinea pig cardiac myocytes	15 pS	Na, K, Li, Cs	n.d.	K_d=1.2 μM	n.d.	-	Ehara et al. 1988
Mouse pancreatic acinar cells	30 pS	Na, K	n.d.	≥1 μM	n.d.	-	Maruyama and Petersen 1982

Gs, single channel conductance; Intracell., intracellular; n.d, not determinded

ences were chosen because the currents described show striking similarities with the properties of TRPM4b over-expressed in HEK293 cells, especially concerning permeation, voltage dependence and block by intracellular adenosine nucleotides. The single channel conductance reported for CAN channels ranges between 15 and 38 pS. Voltage dependence was not determined always, but almost all reported CANs have a higher open probability at positive potentials. The sensitivity for intracellular Ca^{2+} is also a variable feature, with the activation threshold reported between 10^{-7} M and 10^{-4} M. Permeability for Ca^{2+} seems to be a distinguishing feature between different classes of Ca^{2+}-activated cation channels. In endothelial cells, hepatocytes and neutrophils, Ca^{2+}-permeable Ca^{2+}-activated channels have been reported, but others are not Ca^{2+}-permeable at all, or only to a very small extent. It is clear that in the absence of TRPM4 knock-out mice or highly selective pharmacological blockers no current can be unequivocally assigned to TRPM4. Also, since it cannot be ruled out that functional TRPM4 channels in vivo are heteromers with other partners (TRPM5 and maybe other TRPs), it is not unexpected that several of the listed currents show some functional similarities with TRPM4 but deviate when it comes to, for instance, Ca^{2+} sensitivity and Ca^{2+} permeation.

References

Chabardes-Garonne D, Mejean A, Aude JC, Cheval L, Di Stefano A, Gaillard MC, Imbert-Teboul M, Wittner M, Balian C, Anthouard V, Robert C, Segurens B, Wincker P, Weissenbach J, Doucet A, Elalouf JM (2003) A panoramic view of gene expression in the human kidney. Proc Natl Acad Sci U S A 100:13710–13715

Chen M, Simard JM (2001) Cell swelling and a nonselective cation channel regulated by internal Ca^{2+} and ATP in native reactive astrocytes from adult rat brain. J Neurosci 21:6512–6521

Colquhoun D, Neher E, Reuter H, Stevens CF (1981) Inward current channels activated by intracellular Ca in cultured cardiac cells. Nature 294:752–754

Csanady L, Adam-Vizi V (2003) Ca^{2+}- and voltage-dependent gating of Ca^{2+}- and ATP-sensitive cationic channels in brain capillary endothelium. Biophys J 85:313–327

Earley S, Waldron BJ, Brayden JE (2004) Critical role for transient receptor potential channel TRPM4 in myogenic constriction of cerebral arteries. Circ Res 95:922–929

Ehara T, Noma A, Ono K (1988) Calcium-activated non-selective cation channel in ventricular cells isolated from adult guinea-pig hearts. J Physiol 403:117–133

Guinamard R, Chatelier A, Lenfant J, Bois P (2004) Activation of the $Ca^{(2+)}$-activated nonselective cation channel by diacylglycerol analogues in rat cardiomyocytes. J Cardiovasc Electrophysiol 15:342–348

Halonen J, Nedergaard J (2002) Adenosine 5′-monophosphate is a selective inhibitor of the brown adipocyte nonselective cation channel. J Membr Biol 188:183–197

Hille B (2001) Ionic channels of excitable membranes, third edn. Sinauer Associates, Sunderland

Kaestner L, Bollensdorff C, Bernhardt I (1999) Non-selective voltage-activated cation channel in the human red blood cell membrane. Biochim Biophys Acta 1417:9–15

Kamouchi M, Philipp S, Flockerzi V, Wissenbach U, Mamin A, Raeymaekers L, Eggermont J, Droogmans G, Nilius B (1999) Properties of heterologously expressed hTRP3 channels in bovine pulmonary artery endothelial cells. J Physiol (Lond) 518:345–358

Korbmacher C, Volk T, Segal AS, Boulpaep EL, Fromter E (1995) A calcium-activated and nucleotide-sensitive nonselective cation channel in M-1 mouse cortical collecting duct cells. J Membr Biol 146:29–45

Launay P, Fleig A, Perraud AL, Scharenberg AM, Penner R, Kinet JP (2002) TRPM4 is a Ca^{2+}-activated nonselective cation channel mediating cell membrane depolarization. Cell 109:397–407

Launay P, Cheng H, Srivatsan S, Penner R, Fleig A, Kinet JP (2004) TRPM4 regulates calcium oscillations after T cell activation. Science 306:1374–1377

Leech CA, Habener JF (1998) A role for Ca^{2+}-sensitive nonselective cation channels in regulating the membrane potential of pancreatic beta-cells. Diabetes 47:1066–1073

Liman ER (2003) Regulation by voltage and adenine nucleotides of a Ca^{2+}-activated cation channel from hamster vomeronasal sensory neurons. J Physiol (Lond) 548:777–787

Mair N, Frick M, Bertocchi C, Haller T, Amberger A, Weiss H, Margreiter R, Streif W, Dietl P (2004) Inhibition by cytoplasmic nucleotides of a new cation channel in freshly isolated human and rat type II pneumocytes. Am J Physiol Lung Cell Mol Physiol 287:L1284–L1292

Maruyama Y, Petersen OH (1982) Cholecystokinin activation of single-channel currents is mediated by internal messenger in pancreatic acinar cells. Nature 300:61–63

Murakami M, Xu F, Miyoshi I, Sato E, Ono K, Iijima T (2003) Identification and characterization of the murine TRPM4 channel. Biochem Biophys Res Commun 307:522–528

Nadler MJ, Hermosura MC, Inabe K, Perraud AL, Zhu Q, Stokes AJ, Kurosaki T, Kinet JP, Penner R, Scharenberg AM, Fleig A (2001) LTRPC7 is a Mg.ATP-regulated divalent cation channel required for cell viability. Nature 411:590–595

Nilius B, Prenen J, Droogmans G, Voets T, Vennekens R, Freichel M, Wissenbach U, Flockerzi V (2003) Voltage dependence of the Ca^{2+}-activated cation channel TRPM4. J Biol Chem 278:30813–30820

Nilius B, Prenen J, Janssens A, Voets T, Droogmans G (2004a) Decavanadate modulates gating of TRPM4 cation channels. J Physiol 560:753–765

Nilius B, Prenen J, Voets T, Droogmans G (2004b) Intracellular nucleotides and polyamines inhibit the $Ca^{(2+)}$-activated cation channel TRPM4b. Pflugers Arch 448:70–75

Nilius B, Prenen J, Janssens A, Owsianik G, Wang C, Zhu MX, Voets T (2005a) The selectivity filter of the cation channel TRPM4. J Biol Chem 280:22899–22906

Nilius B, Prenen J, Tang J, Wang C, Owsianik G, Janssens A, Voets T, Zhu MX (2005b) Regulation of the Ca^{2+} sensitivity of the nonselective cation channel TRPM4. J Biol Chem 280:6423–6433

Nilius B, Talavera K, Owsianik G, Prenen J, Droogmans G, Voets T (2005c) Gating of TRP channels: a voltage connection? J Physiol 567:35–44

Nilius B, Mahieu F, Prenen J, Janssens A, Owsianik G, Vennekens R, Voets T (2006) The Ca^{2+}-activated cation channel TRPM4 is regulated by phosphatidylinositol 4,5-biphosphate. EMBO J 25:467–478

Nonaka T, Matsuzaki K, Kawahara K, Suzuki K, Hoshino M (1995) Monovalent cation selective channel in the apical membrane of rat inner medullary collecting duct cells in primary culture. Biochim Biophys Acta 1233:163–174

Partridge LD, Swandulla D (1988) Calcium-activated non-specific cation channels. Trends Neurosci 11:69–72

Ramsey IS, Delling M, Clapham DE (2005) An introduction to TRP channels. Annu Rev Physiol 68:619–647

Razani-Boroujerdi S, Partridge LD (1993) Activation and modulation of calcium-activated non-selective cation channels from embryonic chick sensory neurons. Brain Res 623:195–200

Reichling DB, Levine JD (1997) Heat transduction in rat sensory neurons by calcium-dependent activation of a cation channel. Proc Natl Acad Sci U S A 94:7006–7011

Rodighiero S, De Simoni A, Formenti A (2004) The voltage-dependent nonselective cation current in human red blood cells studied by means of whole-cell and nystatin-perforated patch-clamp techniques. Biochim Biophys Acta 1660:164–170

Siemen D (1993) Nonselective cation channels. EXS 66:3–25

Suh SH, Watanabe H, Droogmans G, Nilius B (2002) ATP and nitric oxide modulate a Ca^{2+}-activated non-selective cation current in macrovascular endothelial cells. Pflugers Arch 444:438–445

Talavera K, Yasumatsu K, Voets T, Droogmans G, Shigemura N, Ninomiya Y, Margolskee RF, Nilius B (2005) Heat activation of TRPM5 underlies thermal sensitivity of sweet taste. Nature 438:1022–1025

Teulon J (2000) Ca^{2+}-activated nonselective cation channels. In: Endo M, Kurachi Y, Mishina M (eds) Pharmacology of ionic channel function: activators and inhibitors. Springer-Verlag, Heidelberg, Berlin, New York, pp 625–649

Teulon J, Paulais M, Bouthier M (1987) A Ca2-activated cation-selective channel in the basolateral membrane of the cortical thick ascending limb of Henle's loop of the mouse. Biochim Biophys Acta 905:125–132

Ullrich ND, Voets T, Prenen J, Vennekens R, Talavera K, Droogmans G, Nilius B (2005) Comparison of functional properties of the Ca^{2+}-activated cation channels TRPM4 and TRPM5 from mice. Cell Calcium 37:267–278

Van den Abbeele T, Tran Ba Huy P, Teulon J (1994) A calcium-activated nonselective cationic channel in the basolateral membrane of outer hair cells of the guinea-pig cochlea. Pflugers Arch 427:56–63

Voets T, Droogmans G, Wissenbach U, Janssens A, Flockerzi V, Nilius B (2004) The principle of temperature-dependent gating in cold- and heat-sensitive TRP channels. Nature 430:748–754

Wang Q, Hogg RC, Large WA (1993) A monovalent ion-selective cation current activated by noradrenaline in smooth muscle cells of rabbit ear artery. Pflugers Arch 423:28–33

Xu XZ, Moebius F, Gill DL, Montell C (2001) Regulation of melastatin, a TRP-related protein, through interaction with a cytoplasmic isoform. Proc Natl Acad Sci U S A 98:10692–10697

Yellen G (1982) Single Ca^{2+}-activated nonselective cation channels in neuroblastoma. Nature 296:357–359

Zhainazarov AB (2003) Ca^{2+}-activated nonselective cation channels in rat neonatal atrial myocytes. J Membr Biol 193:91–98

Zhang Z, Okawa H, Wang Y, Liman ER (2005) Phosphatidylinositol 4,5-bisphosphate rescues TRPM4 channels from desensitization. J Biol Chem 280:39185–39192

TRPM5 and Taste Transduction

E. R. Liman

University of Southern California, 3641 Watt Way, Los Angeles CA, 90089, USA
liman@usc.edu

1	Gene and Protein Structure of TRPM5	288
2	Expression Pattern and Biological Function of TRPM5	288
3	Ion Channel Properties	289
3.1	Activation by Ca^{2+}	289
3.1.1	Sensitivity to Activation by Intracellular Ca^{2+}	290
3.2	Ion Selectivity	292
3.3	Unitary Properties	293
4	Modulation of TRPM5 Function	293
4.1	Voltage-Dependent Activation	293
4.2	Desensitization and Regulation by $PI(4,5)P_2$	293
4.3	Temperature Modulation of TRPM5	294
5	Pharmacology: Block by Acid pH	295
6	A Bitter-Sweet Conclusion	295
References		296

Abstract TRPM5 is a cation channel that it is essential for transduction of bitter, sweet and umami tastes. Signaling of these tastes involves the activation of G protein-coupled receptors that stimulate phospholipase C (PLC) β2, leading to the breakdown of phosphatidylinositol bisphosphate (PIP_2) into diacylglycerol (DAG) and inositol trisphosphate (IP_3), and release of Ca^{2+} from intracellular stores. TRPM5 forms a nonselective cation channel that is directly activated by Ca^{2+} and it is likely to be the downstream target of this signaling cascade. Therefore, study of TRPM5 promises to provide insight into fundamental mechanisms of taste transduction. This review highlights recent work on the mechanisms of activation of the TRPM5 channel. The mouse TRPM5 gene encodes a protein of 1,158 amino acids that is proposed to have six transmembrane domains and to function as a tetramer. TRPM5 is structurally most closely related to the Ca^{2+}-activated channel TRPM4 and it is more distantly related to the cold-activated channel TRPM8. In patch clamp recordings, TRPM5 channels are activated by micromolar concentrations of Ca^{2+} and are permeable to monovalent but not divalent cations. TRPM5 channel activity is strongly regulated by voltage, phosphoinositides and temperature, and is blocked by acid pH. Study of TRPM4 and TRPM8, which show similar modes of regulation, has yielded insights into possible structural domains of TRPM5. Understanding the structural basis for TRPM5 function will ultimately allow the design of pharmaceuticals to enhance or interfere with taste sensations.

Keywords Transient receptor potential · $PI(4,5)P_2$ · Taste · Bitter · Sweet

1
Gene and Protein Structure of TRPM5

TRPM5 was first identified in an effort to find genes associated with the tumor-producing condition known as Beckwith–Wiedemann syndrome (BWS) and, although failing to link TRPM5 with BWS, these initial studies defined the basic structure of the gene (Enklaar et al. 2000; Prawitt et al. 2000). The human TRPM5 gene comprises 24 exons on chromosome 11 and it contains an open reading frame of 3,495 bp, which predicts a protein of 1,165 amino acids (Prawitt et al. 2000). The orthologous mouse gene is located on the syntenic distal end of chromosome 7, and it contains an open reading frame that predicts a protein of 1,158 amino acids (Enklaar et al. 2000). TRPM5 shows highest homology to TRPM4 (40% identity at the amino acid level) and it is more distantly related to other TRPM channels, such as the cold and menthol receptor TRPM8 (McKemy et al. 2002; Peier et al. 2002). Like other transient receptor potential (TRP) channels, TRPM5 is thought to contain six transmembrane domains and to assemble as a tetramer (Montell et al. 2002; Clapham 2003).

2
Expression Pattern and Biological Function of TRPM5

A major advance in understanding the physiological significance of TRPM5 came with the discovery that its expression is largely restricted to taste receptor cells (Perez et al. 2002; Zhang et al. 2003). There are five modalities of taste of which three—bitter, sweet and umami—are mediated by G protein-coupled receptors that bind their respective tastant (Lindemann 2001; Margolskee 2002). These receptors activate the G protein gustducin and phospholipase C (PLC) β2 (Lindemann 2001; Margolskee 2002), thereby initiating an intracellular signaling cascade that leads to an electrical response, the nature of which is not well understood (Medler and Kinnamon 2004). The ion channel TRPM5 may be the ultimate target of this cascade, transducing the biochemical changes into an electrical signal. This is supported by the observation that TRPM5 is coexpressed with receptors for all three modalities and with gustducin and PLCβ2 (Perez et al. 2002; Zhang et al. 2003; Fig. 1a). Moreover, in a striking series of experiments, the Zuker and Ryba labs showed that both PLCβ2 and TRPM5 are essential for normal bitter, sweet and umami taste in mice (Zhang et al. 2003). Both behavioral studies and nerve recording show that mice lacking either gene are dramatically less sensitive to these three types of tastes, but retain their ability to detect sour and salty. TRPM5 is also expressed in the small intestine and stomach, where it may play a role in postingestive chemosensation (Perez et al. 2002).

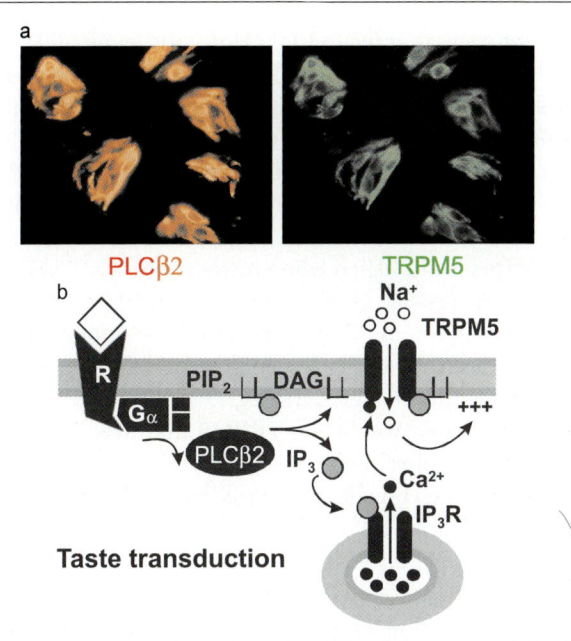

Fig. 1 a,b TRPM5 is a component of taste transduction. **a** Immunoreactivity for taste signaling components in taste buds of the mouse circumvallate papillae, showing colocalization of TRPM5 with PLCβ2 (from Perez et al. 2002, with permission). **b** A model for taste transduction. Binding of taste stimuli to G protein-coupled taste receptors (R) leads to dissociation of the heterotrimeric G protein. βγ subunits of the G protein activate PLCβ2, which in turn hydrolyzes phosphatidylinositol bisphosphate (PIP$_2$) into diacylglycerol (DAG) and inositol trisphosphate (IP$_3$). IP$_3$ activates IP$_3$ receptors, which release Ca^{2+} from intracellular stores. Intracellular Ca^{2+} opens TRPM5 channels, leading to an influx of Na$^+$ and depolarization of the cell. Note that TRPM5 is not permeable to Ca^{2+} and, therefore, there is no positive feedback loop. (From Liu and Liman 2003, with permission)

3
Ion Channel Properties

3.1
Activation by Ca^{2+}

A key to understanding the contribution of TRPM5 to taste and other physiological processes is to identify the mechanisms by which TRPM5 channels are activated. Fortunately the TRPM5 protein is well-trafficked to the plasma membrane when expressed in heterologous cell types (Liu et al. 2005), making it possible to study its functional properties with patch-clamp recording. The expression pattern of TRPM5 suggests that the channel is activated downstream of a PLC-mediated signaling cascade. Consistent with this interpretation, TRPM5 currents can be gated in heterologous cell types by stimulation of

G_q-coupled receptors that activate PLC (Hofmann et al. 2003; Liu and Liman 2003; Prawitt et al. 2003; Zhang et al. 2003). PLC hydrolyzes phosphatidylinositol 4,5-bisphosphate $PI(4,5)P_2$ into diacylglycerol (DAG) and inositol trisphosphate (IP_3), and IP_3 causes release of Ca^{2+} from intracellular stores and presumably one or more of these small molecules activates TRPM5. Of these, only Ca^{2+} is able to directly gate TRPM5 channels (Hofmann et al. 2003; Liu and Liman 2003; Prawitt et al. 2003; see Fig. 2a). Moreover, activation of TRPM5 by G_q-coupled receptors is abolished when intracellular Ca^{2+} is strongly buffered (Liu and Liman 2003; Prawitt et al. 2003) or when IP_3 receptors are inhibited with heparin (Hofmann et al. 2003). Overall, these data support a model for activation of TRPM5 shown in Fig. 1b. In this model, taste receptors (or other G protein-coupled receptors) signal through PLCβ2 to release Ca^{2+} from intracellular stores, which rapidly activates TRPM5 channels (Liu and Liman 2003). This is consistent with physiological data from taste cells and results from targeted deletion of taste transduction molecules (Akabas et al. 1988; Hwang et al. 1990; Bernhardt et al. 1996; Wong et al. 1996; Ogura et al. 1997; Huang et al. 1999; Ogura et al. 2002; Zhang et al. 2003).

Ca^{2+} signals can be generated from a number of different sources and they can vary in magnitude and temporal properties (Hille 2001). For example, release of Ca^{2+} through ryanodine receptors generates a rapid elevation of local Ca^{2+} ("spark") that can reach levels as high as 20–30 µM, a concentration that is able to activate closely opposed plasma membrane Ca^{2+}-activated K^+ channels (Wellman and Nelson 2003). On the other hand, global changes in Ca^{2+} concentration rarely exceed one micromolar and can last for many seconds (Hille 2001). In understanding how TRPM5 channels are gated under physiological conditions, two questions must be answered: (1) Are the channels localized in close proximity to a Ca^{2+} source? (2) How sensitive to Ca^{2+} is the gating of the channels? The second question will be dealt with in Sect. 3.1.1. In answer to the first question, we know that TRPM5 channels are distributed across the entire plasma membrane of taste cells (Perez et al. 2002; Fig. 1a). This is in striking contrast to the distribution of the pheromone-transduction channel TRPC2, which is localized to sensory microvilli of vomeronasal sensory neurons (Liman et al. 1999). The IP_3 receptor and PLCβ2 show a similarly diffuse expression pattern in taste cells (Clapp et al. 2001) and therefore it is conceivable that the three molecules are localized in a signaling complex, like that which organizes signaling components of fly phototransduction (Montell et al. 2002).

3.1.1
Sensitivity to Activation by Intracellular Ca^{2+}

Determination of the Ca^{2+} sensitivity of TRPM5, and of the related channel TRPM4, has been more difficult than might be expected and there is a great deal of variation in the values for half-activation of the channels by Ca^{2+}

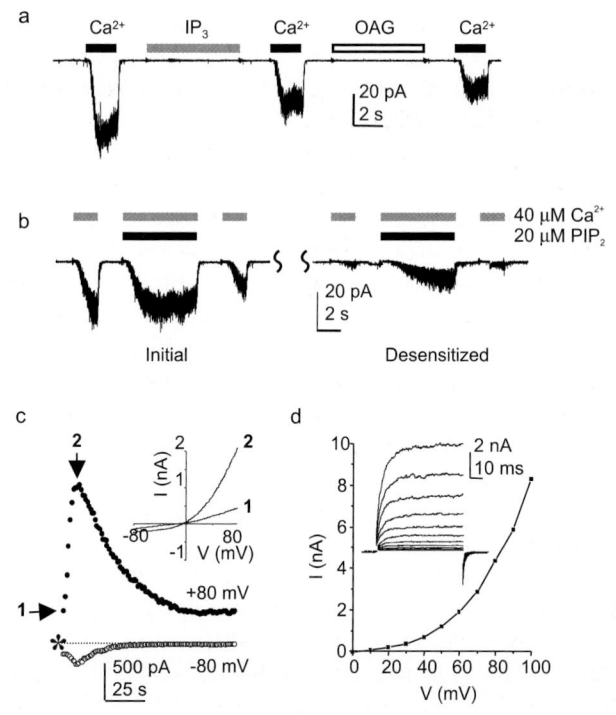

Fig. 2 a–d Functional properties of TRPM5. **a** Activation by 40 μM Ca^{2+} of an inward current in a patch excised from a TRPM5-transfected CHO-K1 cell ($V_m = -80$ mV). Neither 10 μM IP_3 nor 100 μM 1-oleoyl-2 acetyl-*sn* glycerol (OAG) elicited a current in the same patch. **b** PIP_2 partially restores TRPM5 channel activity following desensitization. The illustration shows responses to 40 μM Ca^{2+} in the presence and absence of 20 μM PIP_2 before and after desensitization. Desensitization was induced by a 30-s exposure to 40 μM Ca^{2+} ($V_m = -80$ mV). **c** Electrophysiological properties of TRPM5 expressed in HEK-293 M1 cells. Whole-cell recording mode, 40 μM Ca^{2+} in the pipette, elicited a large rectifying current. Recording began shortly after break in to the whole-cell mode. *Inset* shows the current in response to a ramp depolarization (1 V/s). **d** Currents in response to a family of step depolarizations and the resulting I–V relationship for the peak current at each voltage. Steps are to 0–100 mV from a holding potential of −80 mV with repolarization to −50 mV. Note the prominent relaxation of the current upon depolarization, consistent with voltage-dependent gating of the channels. (From Liu and Liman 2003, with permission)

reported by different groups. This may be in part due to the fact that the Ca^{2+} sensitivity of these channels is subject to modulatory influences that are not completely understood. Perhaps the most robust and reproducible way to measure intracellular Ca^{2+} sensitivity is in inside-out patches (Fig. 2a). In this mode, immediately after patch excision TRPM5 channels are activated by intracellular Ca^{2+} with an EC_{50} of 20–30 μM (Liu and Liman 2003; Ullrich et al. 2005). This value increases over time, possibly as a result of loss of $PI(4,5)P_2$ from the channels, to 80 μM (Liu and Liman 2003). Under the same conditions, the structurally related channel TRPM4 is five times less sensitive to activation by intracellular Ca^{2+} (Ullrich et al. 2005; Zhang et al. 2005). The low sensitivity of TRPM5 channels in this recording mode argues that to be activated by physiological stimuli the channels are most likely localized in close proximity to a Ca^{2+} source.

Somewhat mysteriously, the sensitivity of TRPM5 channels to activation by Ca^{2+} in whole-cell recording mode is several orders of magnitude higher than it is in excised inside-out patches. While dose–response data are more difficult to obtain in this mode due to rundown of the current (Fig. 2c) and the need to use population data, there is nonetheless general consensus that TRPM5 can be near-maximally activated by intracellular dialysis of 1 µM Ca^{2+} (Prawitt et al. 2003; Ullrich et al. 2005; but see also Hofmann et al. 2003). This might reflect the loss of a factor that enhances the sensitivity of the channels to Ca^{2+}, as will be discussed in Sect. 4.2. In addition, it is possible that perfusion of the cells with relatively low concentrations of Ca^{2+} elicits Ca^{2+} release in the vicinity of the TRPM5 channels, which further augments their activation. The high sensitivity to Ca^{2+} of TRPM5 channels in whole-cell recoding mode could be used to argue that the channels detect global changes in Ca^{2+} (Prawitt et al. 2003), a conclusion at odds with that based on data from excised inside-out patch recording. Clearly we need to understand more about how these channels are regulated to resolve this discrepancy.

3.2
Ion Selectivity

The ease of activation of TRPM5 in heterologous expression systems has allowed careful investigation of the channel's selectivity and gating. These experiments have shown that TRPM5 channels show little discrimination among the monovalent cations Na^+, K^+, and Cs^+ and do not conduct divalent cations (Hofmann et al. 2003; Liu and Liman 2003; Prawitt et al. 2003). It is probably not a coincidence that the two Ca^{2+}-activated TRP channels are also the only two that are impermeable to Ca^{2+}. The structural basis for the differential Ca^{2+} permeability of TRP channels is not known. In a detailed set of experiments, residues in the putative pore of TRPM4 were changed to the corresponding residues in a Ca^{2+}-permeable TRP channel, TRPV6, and this conferred moderate Ca^{2+} permeability to the chimeric channel (Nilius et al. 2005a). However, the fact that these authors were not able to identify a mutant that could confer more substantial Ca^{2+} permeability suggests that multiple residues or regions of the channel contribute to this process. Nonetheless, these experiments have provided experimental evidence that the region between the fifth and sixth transmembrane domain of TRPM4, and by homology TRPM5, contains the pore of the channel (Owsianik et al. 2005).

3.3
Unitary Properties

Single TRPM5 channels show a conductance of approximately 16–25 pS (Hofmann et al. 2003; Liu and Liman 2003; Prawitt et al. 2003). Channel openings are short-lived and flickery (Liu and Liman 2003), which has precluded a detailed analysis of gating properties. This is in contrast to the long-lived openings of TRPM4 channels (bursts can last several seconds) (Launay et al. 2002; Zhang et al. 2005), and may serve as a defining feature in categorizing native channels.

4
Modulation of TRPM5 Function

4.1
Voltage-Dependent Activation

Although activation of TRPM5 requires elevated Ca^{2+} levels, gating of the channel is also strongly affected by voltage (Hofmann et al. 2003; Liu and Liman 2003; Talavera et al. 2005). This is apparent in the outward rectification of TRPM5 currents in response to a voltage ramps, despite a linear current–voltage relationship (I–V) for the single channel conductance, and in the time-dependent relaxation of the current following a voltage step (Hofmann et al. 2003; Liu and Liman 2003; Prawitt et al. 2003; Talavera et al. 2005; Fig. 2c,d). Voltage-dependent activation has also been reported for TRPM8 and TRPM4 and thus may be a common feature of TRPM channels (Hofmann et al. 2003; Nilius et al. 2003; Nilius et al. 2005b; Rohacs et al. 2005). Why should channels whose primary role is to transduce sensory signals be voltage-dependent? While the answer to this question is not known, it has been hypothesized that the weak voltage dependence of these channels allows their gating to be easily modulated (Nilius et al. 2005b), a hypothesis that is supported by work on cold regulation of TRPM8, TRPV1, and TRPM5 (Voets et al. 2004; Talavera et al. 2005), decavanadate modulation of TRPM4 (Nilius et al. 2004), and $PI(4,5)P_2$ regulation of TRPM4 and TRPM8 (Rohacs et al. 2005; Zhang et al. 2005; see Sect. 4.2). At present the structural mechanism of voltage sensing of any of the TRP channels is not known. The fourth transmembrane of these channels contain several charged residues that might act as the voltage sensor, by analogy to voltage-activation of K^+ channels (Jiang et al. 2003; Nilius et al. 2005b).

4.2
Desensitization and Regulation by $PI(4,5)P_2$

A consistent observation is that TRPM5 currents rapidly desensitize after activation (Hofmann et al. 2003; Liu and Liman 2003; Prawitt et al. 2003), a process that may play a role in sensory adaptation of taste cells. In whole-cell recording mode, rundown is observed following dialysis of intracellular Ca^{2+} (Fig. 2c) and similar rundown is observed in perforated patch recording following activation by bath applied Ca^{2+} ionophore, arguing that rundown is not due to washout of signaling components (Liu and Liman 2003). In excised inside-out patches, rundown of TRPM5 currents is accompanied by both a change in the Ca^{2+} sensitivity and in the maximal magnitude of the current (Liu and Liman 2003).

Recently the phosphoinositide $PI(4,5)P_2$ has emerged as an important cofactor in the activation of ion channels, and hydrolysis of $PI(4,5)P_2$ has been proposed to underlie rundown of many types of $PI(4,5)P_2$-sensitive ion channels (Suh and Hille 2005). $PI(4,5)P_2$ is likewise a cofactor for activation of TRPM5 (Liu and Liman 2003). Exogenous $PI(4,5)P_2$ enhances both the Ca^{2+} sensitivity and magnitude of TRPM5 currents following rundown, but is ineffective prior to rundown (see Fig. 2b), suggesting that loss of this signaling molecule underlies desensitization (Liu and Liman 2003). $PI(4,5)P_2$ is expected to be hydrolyzed by ubiquitous membrane bound Ca^{2+}-dependent PLCs in response to the elevated Ca^{2+} levels used to evoke TRPM5 currents in excised patches or whole-cell recording (Varnai and Balla 1998). Consistent with this possibility, desensitization of TRPM5 is Ca^{2+}-dependent (Liu and Liman 2003; Ullrich et al. 2005). Similar desensitization and recovery by $PI(4,5)P_2$ is also observed for TRPM8 and TRPM4, and for both these channels additional data support the conclusion that hydrolysis of $PI(4,5)P_2$ mediates desensitization (Liu and Qin 2005; Rohacs et al. 2005; Zhang et al. 2005; Nilius et al. 2006). A structural determinant for $PI(4,5)P_2$ modulation of TRPM8 has been identified in positively charged residues among the 25 amino acids proximal to the sixth transmembrane domain (Rohacs et al. 2005). It is not known yet whether this region also contains key structural determinants for regulation of TRPM5 by $PI(4,5)P_2$.

4.3
Temperature Modulation of TRPM5

TRPM5 is structurally related to the cold-activated channel TRPM8 and more distantly related to heat-activated TRPV channels, suggesting the possibility that its activity is also thermal-sensitive. Indeed, warm temperatures promote activation of TRPM5, similar to the effects of heat on the TRPV channels (Talavera et al. 2005). An elegant theoretical framework has been developed to explain heat and cold activation of TRP channels, which postulates that their extreme thermal sensitivity derives from their small voltage dependence (Nilius et al. 2005b). Heat acts by shifting the midpoint for voltage-dependent activation to negative voltages for TRPV1 and TRPM5 and positive voltages for TRPM8, leading to opposing thermal sensitivities of the channels (Voets et al. 2004; Talavera et al. 2005). However, unlike TRPV1, heat is not sufficient to activate TRPM5, which even at warm temperatures requires elevated Ca^{2+} (Talavera et al. 2005). The thermal sensitivity of TRPM5 suggests that sensation of bitter, sweet, and umami tastes might be reduced at cold temperatures. Electrophysiological recordings from mice indeed show that sweet taste is highly sensitive to temperature, although bitter and umami are unaffected (Talavera et al. 2005). Thus, while these data support a role for TRPM5 in the thermal sensitivity of sweet taste, there are likely other factors that contribute to the thermal sensitivity of this process.

5
Pharmacology: Block by Acid pH

Blockers of TRPM5 have the potential to alter taste sensation, and therefore identification of these molecules is of great interest. Although no specific blockers have yet been reported, acid pH has been found to be a very potent and relatively specific blocker of the channel (Liu et al. 2005). TRPM5 is sensitive to pH levels below 7.0 and is completely blocked by pH 5.9. In comparison, TRPM4 is insensitive to pH levels as low as 5.4. By comparing sequence differences between TRPM4 and TRPM5, two residues were identified that account for most of the pH sensitivity of TRPM5—a Glu residue in the S3–S4 linker and a His residue in the pore region (S5–S6 linker; Liu et al. 2005). Acid pH also enhances the rate of inactivation of TRPM5 channels through its effects on these residues. It is tempting to speculate that acid block of TRPM5 may play a functional role in taste sensation, possibly decreasing responses of taste cells to activation by bitter, sweet, or umami when consumed at acid pH.

6
A Bitter-Sweet Conclusion

TRPM5 plays an essential role in the detection of bitter, sweet, and umami tastes and therefore a better understanding of its regulation will lead to insights into taste sensory transduction. The elegant molecular work of the last decade has shown that bitter, sweet, and umami tastes are mediated by distinct G protein-coupled receptors, and that these receptors activate a common downstream signaling cascade of which PLCβ2 and TRPM5 are critical components. Currently it is well established that Ca^{2+} is the primary stimulus for activating TRPM5 in heterologous cells types. Thus, the simplest model for taste sensation envisions that receptor activation promotes hydrolysis of $PI(4,5)P_2$ by PLCβ2 leading to the generation of IP_3 and release of Ca^{2+} from intracellular stores. Ca^{2+} then activates TRPM5, which conducts an inward Na^+ current and depolarizes the cell. Whether, indeed, activation of TRPM5 by Ca^{2+} underlies the electrical response of taste cells to sensory stimuli remains to be established.

The identification of a robust mechanism for activation of TRPM5 in heterologous cells has facilitated the discovery of basic features of the channel and novel regulatory mechanisms that are likely to be of physiological significance. PI(4,5)P$_2$ hydrolysis has been proposed to play an important role in desensitization of TRPM5 and may mediate sensory adaptation of taste. Thermal sensitivity of TRPM5 has been shown to contribute to the temperature dependence of sweet sensation, and acid inhibition of TRPM5 may also modulate sensory responses to taste. Finally, in the future we can look forward to structural information that will allow the design of rational chemicals to block or enhance TRPM5 function and thereby remove some of the bitterness or enhance some of the sweetness of pharmaceuticals and foods we consume.

Acknowledgements Support was provided by the National Institutes of Health, R01DC004564 and R01DC005000.

References

Akabas MH, Dodd J, Al-Awqati Q (1988) A bitter substance induces a rise in intracellular calcium in a subpopulation of rat taste cells. Science 242:1047–1050

Bernhardt SJ, Naim M, Zehavi U, Lindemann B (1996) Changes in IP$_3$ and cytosolic Ca^{2+} in response to sugars and non-sugar sweeteners in transduction of sweet taste in the rat. J Physiol 490:325–336

Clapham DE (2003) TRP channels as cellular sensors. Nature 426:517–524

Clapp TR, Stone LM, Margolskee RF, Kinnamon SC (2001) Immunocytochemical evidence for co-expression of type III IP$_3$ receptor with signaling components of bitter taste transduction. BMC Neurosci 2:6

Enklaar T, Esswein M, Oswald M, Hilbert K, Winterpacht A, Higgins M, Zabel B, Prawitt D (2000) Mtr1, a novel biallelically expressed gene in the center of the mouse distal chromosome 7 imprinting cluster, is a member of the Trp gene family. Genomics 67:179–187

Hille B (2001) Ionic channels of excitable membranes. Sinauer Associates, Sunderland

Hofmann T, Chubanov V, Gudermann T, Montell C (2003) TRPM5 is a voltage-modulated and Ca$^{(2+)}$-activated monovalent selective cation channel. Curr Biol 13:1153–1158

Huang L, Shanker YG, Dubauskaite J, Zheng JZ, Yan W, Rosenzweig S, Spielman AI, Max M, Margolskee RF (1999) Ggamma13 colocalizes with gustducin in taste receptor cells and mediates IP$_3$ responses to bitter denatonium. Nat Neurosci 2:1055–1062

Hwang PM, Verma A, Bredt DS, Snyder SH (1990) Localization of phosphatidylinositol signaling components in rat taste cells: role in bitter taste transduction. Proc Natl Acad Sci U S A 87:7395–7399

Jiang Y, Ruta V, Chen J, Lee A, MacKinnon R (2003) The principle of gating charge movement in a voltage-dependent K+ channel. Nature 423:42–48

Launay P, Fleig A, Perraud AL, Scharenberg AM, Penner R, Kinet JP (2002) TRPM4 is a Ca^{2+}-activated nonselective cation channel mediating cell membrane depolarization. Cell 109:397–407

Liman ER, Corey DP, Dulac C (1999) TRP2: a candidate transduction channel for mammalian pheromone sensory signaling. Proc Natl Acad Sci U S A 96:5791–5796

Lindemann B (2001) Receptors and transduction in taste. Nature 413:219–225

Liu B, Qin F (2005) Functional control of cold- and menthol-sensitive TRPM8 ion channels by phosphatidylinositol 4,5-bisphosphate. J Neurosci 25:1674–1681

Liu D, Liman ER (2003) Intracellular Ca^{2+} and the phospholipid PIP_2 regulate the taste transduction ion channel TRPM5. Proc Natl Acad Sci U S A 100:15160–15165

Liu D, Zhang Z, Liman ER (2005) Extracellular acid block and acid-enhanced inactivation of the Ca^{2+}-activated cation channel TRPM5 involve residues in the S3-S4 and S5-S6 extracellular domains. J Biol Chem 280:20691–20699

Margolskee RF (2002) Molecular mechanisms of bitter and sweet taste transduction. J Biol Chem 277:1–4

McKemy DD, Neuhausser WM, Julius D (2002) Identification of a cold receptor reveals a general role for TRP channels in thermosensation. Nature 416:52–58

Medler K, Kinnamon S (2004) Transduction mechanisms in taste cells. In: Frings S, Bradley J (eds) Transduction channels in sensory cells. Wiley-VCH, Weinheim, pp 153–174

Montell C, Birnbaumer L, Flockerzi V (2002) The TRP channels, a remarkably functional family. Cell 108:595–598

Nilius B, Prenen J, Droogmans G, Voets T, Vennekens R, Freichel M, Wissenbach U, Flockerzi V (2003) Voltage dependence of the Ca^{2+}-activated cation channel TRPM4. J Biol Chem 278:30813–30820

Nilius B, Prenen J, Janssens A, Voets T, Droogmans G (2004) Decavanadate modulates gating of TRPM4 cation channels. J Physiol 560:753–765

Nilius B, Prenen J, Janssens A, Owsianik G, Wang C, Zhu MX, Voets T (2005a) The selectivity filter of the cation channel TRPM4. J Biol Chem 280:22899–22906

Nilius B, Talavera K, Owsianik G, Prenen J, Droogmans G, Voets T (2005b) Gating of TRP channels: a voltage connection? J Physiol 567:35–44

Nilius B, Mahieu F, Prenen J, Janssens A, Owsianik G, Vennekens R, Voets T (2006) The Ca^{2+}-activated cation channel TRPM4 is regulated by phosphatidylinositol 4,5-biphosphate. EMBO J 25:467–478

Ogura T, Mackay-Sim A, Kinnamon SC (1997) Bitter taste transduction of denatonium in the mudpuppy Necturus maculosus. J Neurosci 17:3580–3587

Ogura T, Margolskee RF, Kinnamon SC (2002) Taste receptor cell responses to the bitter stimulus denatonium involve Ca^{2+} influx via store-operated channels. J Neurophysiol 87:3152–3155

Owsianik G, Talavera K, Voets T, Nilius B (2005) Permeation and selectivity of TRP channels. Annu Rev Physiol 68:685–717

Peier AM, Moqrich A, Hergarden AC, Reeve AJ, Andersson DA, Story GM, Earley TJ, Dragoni I, McIntyre P, Bevan S, Patapoutian A (2002) A TRP channel that senses cold stimuli and menthol. Cell 108:705–715

Perez CA, Huang L, Rong M, Kozak JA, Preuss AK, Zhang H, Max M, Margolskee RF (2002) A transient receptor potential channel expressed in taste receptor cells. Nat Neurosci 5:1169–1176

Prawitt D, Enklaar T, Klemm G, Gartner B, Spangenberg C, Winterpacht A, Higgins M, Pelletier J, Zabel B (2000) Identification and characterization of MTR1, a novel gene with homology to melastatin (MLSN1) and the trp gene family located in the BWS-WT2 critical region on chromosome 11p15.5 and showing allele-specific expression. Hum Mol Genet 9:203–216

Prawitt D, Monteilh-Zoller MK, Brixel L, Spangenberg C, Zabel B, Fleig A, Penner R (2003) TRPM5 is a transient Ca^{2+}-activated cation channel responding to rapid changes in $[Ca^{2+}]i$. Proc Natl Acad Sci U S A 100:15166–15171

Rohacs T, Lopes CM, Michailidis I, Logothetis DE (2005) PI(4,5)P(2) regulates the activation and desensitization of TRPM8 channels through the TRP domain. Nat Neurosci 8:626–634

Suh BC, Hille B (2005) Regulation of ion channels by phosphatidylinositol 4,5-bisphosphate. Curr Opin Neurobiol 15:370–378

Talavera K, Yasumatsu K, Voets T, Droogmans G, Shigemura N, Ninomiya Y, Margolskee RF, Nilius B (2005) Heat activation of TRPM5 underlies thermal sensitivity of sweet taste. Nature 438:1022–1025

Ullrich ND, Voets T, Prenen J, Vennekens R, Talavera K, Droogmans G, Nilius B (2005) Comparison of functional properties of the Ca^{2+}-activated cation channels TRPM4 and TRPM5 from mice. Cell Calcium 37:267–278

Varnai P, Balla T (1998) Visualization of phosphoinositides that bind pleckstrin homology domains: calcium- and agonist-induced dynamic changes and relationship to myo-[3H]inositol-labeled phosphoinositide pools. J Cell Biol 143:501–510

Voets T, Droogmans G, Wissenbach U, Janssens A, Flockerzi V, Nilius B (2004) The principle of temperature-dependent gating in cold- and heat-sensitive TRP channels. Nature 430:748–754

Wellman GC, Nelson MT (2003) Signaling between SR and plasmalemma in smooth muscle: sparks and the activation of Ca^{2+}-sensitive ion channels. Cell Calcium 34:211–229

Wong GT, Gannon KS, Margolskee RF (1996) Transduction of bitter and sweet taste by gustducin. Nature 381:796–800

Zhang Y, Hoon MA, Chandrashekar J, Mueller KL, Cook B, Wu D, Zuker CS, Ryba NJ (2003) Coding of sweet, bitter, and umami tastes: different receptor cells sharing similar signaling pathways. Cell 112:293–301

Zhang Z, Okawa H, Wang Y, Liman ER (2005) Phosphatidylinositol 4,5-bisphosphate rescues TRPM4 channels from desensitization. J Biol Chem 280:39185–39192

TRPM6: A Janus-Like Protein

M. Bödding

Experimentelle und Klinische Pharmakologie und Toxikologie,
Universität des Saarlandes, 66421 Homburg, Germany
matthias.boedding@uniklinik-saarland.de

1	Introduction	300
2	Tissue Distribution	300
3	Molecular Structure	301
3.1	Gene and Protein	301
3.2	Splice Variants	301
3.3	Protein kinase	301
3.4	Homo/Heteromultimerisation	302
4	Channel Properties	302
4.1	Biophysical Properties of TRPM6-Mediated Currents in Over-expression Systems	302
4.1.1	Activation Mechanism	302
4.1.2	Current–Voltage Relationship	303
4.1.3	Permeability	303
4.1.4	TRPM7 and Store-Operated Ca^{2+} Currents	303
4.2	Biophysical Properties of TRPM6-Like Currents in Native Cells	304
5	Kinase Properties	304
6	Pharmacology	305
6.1	Ruthenium Red	305
6.2	2-Aminoethoxydiphenyl Borate	305
7	Physiology	305
7.1	Mg^{2+} Re/Absorption	305
7.2	Regulation of TRPM6 Expression Levels	306
7.2.1	Acid Base Status	306
7.2.2	Dietary Mg^{2+} Intake	306
7.2.3	FK506 (Tacrolimus) and Glucocorticoids	307
8	Pathophysiology	307
8.1	Hypomagnesaemia with Secondary Hypocalcaemia	307
8.2	Diabetes Mellitus	308
References		308

Abstract TRPM6 and TRPM7 proteins share similar molecular structures and biophysical properties. Both proteins are Mg^{2+}- and Ca^{2+}-permeable cation channels with the typical topology of six transmembrane domains. In addition, TRPM6 and TRPM7 function as serine/threonine kinases with kinase domains at their C-terminal tails. At present, the role of the association of kinase and channel domains in TRPM6 and TRPM7 remains elusive. TRPM6 is mainly expressed in kidney and intestine, where it might be responsible for epithelial Mg^{2+} re/absorption. This hypothesis is strengthened by the identification of TRPM6 mutants in patients with a rare but severe hereditary disease called hypomagnesaemia with secondary hypocalcaemia. The aim of this review is to provide a brief but concise overview of the information currently available about TRPM6.

Keywords TRPM7 · Transient receptor potential (TRP) channel · Melastatin · TRPM6 · Magnesium absorption · Hypomagnesaemia with secondary hypocalcaemia (HSH)

1
Introduction

The TRPM6 gene was first cloned by Riazanova et al. in 2001 by screening for homologues of elongation factor-2 kinase (Riazanova et al. 2001). Just 1 year later two groups published that spontaneous mutations in TRPM6 result in severe electrolyte disturbances called hypomagnesaemia with secondary hypocalcaemia (Schlingmann et al. 2002; Walder et al. 2002). Indeed, electrophysiological experiments showed that Mg^{2+} and Ca^{2+} ions can pass the TRPM6 channel pore (Voets et al. 2004) which is also characteristic for the structurally nearest relative TRPM7. Thus, TRPM6 is a remarkable protein, since it is a Mg^{2+}-permeable channel with kinase activity and mutations can lead to severed electrolyte disturbances.

2
Tissue Distribution

TRPM6 is predominantly expressed in kidney and colon (Schlingmann et al. 2002; Walder et al. 2002), which is in contrast to the ubiquitous expression of TRPM7 (Nadler et al. 2001; Runnels et al. 2001). The TRPM6 messenger RNA (mRNA) was detected by Northern blot analysis in human colon and kidney (Walder et al. 2002). Similar results were reported by in situ hybridisation in colon, duodenum, jejunum, ileum and distal renal tubule cells (Schlingmann et al. 2002). Immunohistochemical experiments revealed expression of TRPM6 along the brush-border membrane of the small intestine and the apical membrane of the renal distal convoluted tubule in mice (Voets et al. 2004). Both epithelia are particularly associated with active Mg^{2+} re/absorption.

3
Molecular Structure

3.1
Gene and Protein

TRPM6 is closely related to TRPM7 and the founding member melastatin (TRPM1). They share several identical amino acids predominantly within their N-terminal sequences, the putative transmembrane domains and pore region. TRPM6 is a protein of 2,022 amino acids encoded by 39 exons. It is the longest member of the TRP channel family. The hydrophobicity plot indicates six putative transmembrane domains with a pore loop between the fifth and sixth transmembrane domains. In the pore-forming region the sequence identity between TRPM6 and TRPM7 increases to more than 80%.

3.2
Splice Variants

Alternative splicing gives rise to at least seven TRPM6 variants with partially different expression patterns (Chubanov et al. 2004). Three of these splice variants lack the hexahelical transmembrane region, indicating that the function of the kinase domain is not restricted to ion channel modulation (Chubanov et al. 2004). The three splice variants differ in their N-terminal region and expression pattern. They derive from alternative use of the exons 1a, 1b and 1c, indicating that the TRPM6 gene has a promoter with alternative transcription start sites (Chubanov et al. 2004).

Finally, splicing in exon 36 leads to a premature stop. The C-terminal truncation of this mRNA is specifically expressed in testis (Chubanov et al. 2004).

A detailed functional characterisation of these splice variants is lacking, and so far most experiments have been performed on the full-length versions of TRPM6 (NM_017662).

3.3
Protein kinase

Both channels, TRPM6 and TRPM7, are unique since they contain an atypical serine/threonine protein kinase domain. It is located at the C-terminal region and has similarities with members of the α-kinase family (Riazanova et al. 2001; Runnels et al. 2001; Ryazanova et al. 2004). Despite containing a zinc-finger domain (Yamaguchi et al. 2001; Ryazanova et al. 2004), TRPM6 and TRPM7 offer little sequence identity of their kinase domains to other eukaryotic protein kinases.

3.4
Homo/Heteromultimerisation

Fluorescence resonance energy transfer (FRET) analysis indicates that TRPM6 specifically interacted with itself and its closest homologue TRPM7 (Chubanov et al. 2004), similar to what has been described for other members of the TRP family, with a high degree of homology such as TRPV5/6 (Hoenderop et al. 2003) and TRPC1/5 (Strübing et al. 2001). Based on experiments with fusion proteins in over-expressing *Xenopus* oocytes and human embryonic kidney (HEK)-293 cells, it has been suggested that the interaction between TRPM6 and TRPM7 is required for trafficking TRPM6 to the cell surface (Chubanov et al. 2004). However, TRPM6 does not increase TRPM7 expression (Chubanov et al. 2004).

4
Channel Properties

4.1
Biophysical Properties of TRPM6-Mediated Currents in Over-expression Systems

4.1.1
Activation Mechanism

TRPM6, like TRPM7, is a Mg^{2+}- and Ca^{2+}-permeable channels which is activated by a decrease in the $[Mg^{2+}]_i$ (Voets et al. 2004). It is, however, not clear whether TRPM6 is a constitutively active channel (Chubanov et al. 2004; Voets et al. 2004) as has been suggested for TRPM7 (Nadler et al. 2001).

The injection of TRPM6 complementary RNA (cRNA) into oocytes did not entail significant ion currents compared to water-injected control oocytes, and co-expression of TRPM6 and TRPM7 produces larger currents than TRPM7 alone (Chubanov et al. 2004). These findings indicate that TRPM6 is unable to form functional channels when expressed alone, whereas TRPM7 is able to usher TRPM6 to the cell membrane, presumably by forming mixed multimers.

In contrast, HEK-293 cells transfected with TRPM6 cDNA exhibited the characteristic current immediately upon establishment of the whole-cell configuration (Voets et al. 2004). When interpreting these data, one has to consider that HEK-293 cells endogenously express both TRPM6 and TRPM7 as revealed by RT-PCR (Chubanov et al. 2004). Whether oocytes express TRPM7 is not known.

TRPM6-induced currents increased over minutes when over-expressing HEK-293 cells were dialysed with an intracellular solution containing high concentrations of the unspecific chelator ethylenediaminetetraacetate (EDTA) (Voets et al. 2004). Under these conditions, metal ions such as Mg^{2+} are rapidly bound and TRPM6 is relieved from intracellular Mg^{2+} block. Flash experiments with the photolabile Mg^{2+} chelator 1-(2-nitro-4,5-dimethoxyphenyl)-N,N,N′,N′-tetrakis[(oxycarbonyl)methyl]-1,2-ethanediamine (DM-nitrophen) support the hypothesis that TRPM6 is inhibited by intracellular Mg^{2+} (Voets et al. 2004).

4.1.2
Current–Voltage Relationship

The current–voltage (IV) relationship of TRPM6 (Voets et al. 2004) is indistinguishable from that of TRPM7 (Nadler et al. 2001; Runnels et al. 2001). In both cases, currents are outwardly rectifying and reverse close to 0 mV. The outward flux of monovalent cations helps identify this current. However, the small inward flux of divalent cations takes place within the typical range of the membrane potential and is therefore of major physiological importance.

The IV curve is linear in the absence of extracellular divalent cations. This is in contrast to voltage-modulated TRP proteins such as TRPM8, TRPM4, TRPM5 and TRPV1 (reviewed by Nilius et al. 2005b; see also Voets et al. 2002; Nilius et al. 2003, 2005a; Talavera et al. 2005). Therefore, the voltage dependence of TRPM6 might arise from ionic interaction within the channel. Such a voltage-dependent permeation block by divalent cations has also been proposed for TRPM7. For both channels, TRPM6 and TRPM7, a voltage-dependent gating behaviour is less likely.

4.1.3
Permeability

The permeation rank order of the inward current is $Ba^{2+} \geq Ni^{2+} > Mg^{2+} > Ca^{2+}$ (Voets et al. 2004). Permeability ratios based on the reversal potential yielded P_X/P_{Na} values of 9 for Mg^{2+} and 7 for Ca^{2+} (Voets et al. 2004). However, these data should be interpreted with caution since the reversal potential is difficult to determine due to the shallow slope of the IV curve around the reversal potential, and the calculation of P_X/P_{Na} from the reversal potential is based on the Goldmann–Hodgkin–Katz equation, which is based on wrong assumptions in the case of TRPM6 as discussed (Voets et al. 2004). In addition, changes in the reversal potential from non-transfected HEK cells were not considered even though it has been shown that endogenous TRPM7 channels are constitutively open in these cells and become further activated by the Mg^{2+}-free internal solution (Nadler et al. 2001).

4.1.4
TRPM7 and Store-Operated Ca^{2+} Currents

In summary, TRPM6 and TRPM7 channel properties are essentially indistinguishable yet. Their IV relationships are similar, both channels are selective for Ca^{2+} and Mg^{2+}, and currents are inhibited by intracellular Mg^{2+} (Voets et al. 2004). TRPM6 is far less well characterised by electrophysiological means than TRPM7. Several open questions remain: For instance, no single channel data have been reported for TRPM6 to determine the unitary conductance, and it is not known whether receptor-mediated depletion of phosphatidylinositol 4,5-bisphosphate (PIP_2) inhibits TRPM6 as suggested for TRPM7 (Runnels et al. 2002). Thus, future studies could lead to functional differences between TRPM6 and TRPM7.

Another aspect needs attention concerning store-operated Ca^{2+} currents: When measuring them by using high concentrations of intracellular chelators, one needs to keep in mind that TRPM6 and TRPM7 could be activated due to the decrease of the intracellular free Mg^{2+} concentration (Hermosura et al. 2002; Kozak et al. 2002; Prakriya and Lewis 2002).

4.2
Biophysical Properties of TRPM6-Like Currents in Native Cells

Very similar currents to those measured in over-expression systems were detected in several primary tissues and cell lines (Nadler et al. 2001; Hermosura et al. 2002; Prakriya and Lewis 2002; Kozak and Cahalan 2003; Gwanyanya et al. 2004). It is likely that TRPM6 or TRPM7 channels (or both) represent the molecular correlate for these currents, but this awaits further support by experiments on knock-out mice or cells where expression levels of these channels are markedly reduced e.g. by small inhibitory RNA (siRNA) or cDNA in antisense orientation.

5
Kinase Properties

TRPM6, like TRPM7, is a protein which is both an ion channel and a protein kinase. The zinc finger-containing α-kinase domain is functional (Runnels et al. 2001). Both channels are unique since they show protein kinase activity. The enzymatically active C-terminal kinase domain has little sequence identity to other protein kinases despite the zinc-finger domain (Yamaguchi et al. 2001; Ryazanova et al. 2004). In contrast to TRPM7, no data have been published concerning the functional role of the kinase domain of TRPM6. Thus, it is not known whether the kinase activity of TRPM6 is required for channel function, as has been discussed for TRPM7 (Nadler et al. 2001; Runnels et al. 2001; Schmitz et al. 2003). However, it has been shown that truncation of TRPM6 prior to the kinase domain leads to a loss of function as observed in hypomagnesaemia with secondary hypocalcaemia (Schlingmann et al. 2002; Walder et al. 2002).

No substrate has been identified yet for the TRPM6 kinase, while the TRPM7 kinase phosphorylates annexin I at a serine residue (Ser^5) located within the

N-terminal amphipathic alpha-helix (Dorovkov and Ryazanov 2004). As both proteins, TRPM7 and annexin I, play a role in the regulation of cell death and proliferation (Aarts et al. 2003; Hanano et al. 2004; Perretti and Solito 2004), it is speculated that the phosphorylation of annexin I by TRPM7 kinase modulates these processes (Dorovkov and Ryazanov 2004).

6
Pharmacology

6.1
Ruthenium Red

Ruthenium red (10 µM) causes a strong inhibition of inward monovalent currents while leaving outward currents unaltered (Voets et al. 2004). This voltage-dependent block can be interpreted as a direct block of the TRPM6 channel pore within the transmembrane electrical field. The action of ruthenium red on TRPM6 is unspecific since it also inhibits other TRP channels (Caterina et al. 1997; Nilius et al. 2001) as well as Ca^{2+}-activated K^+ channels (Wu et al. 1999). This is not unexpected because ruthenium red is a synthetic inorganic polycationic dye originally used in electron microscopy for cell staining (Luft 1971) and is widely used to inhibit Ca^{2+} release channels from sarcoplasmic reticulum (reviewed by Ehrlich et al. 1994).

6.2
2-Aminoethoxydiphenyl Borate

It has not been reported whether 2-aminoethoxydiphenyl borate (2APB) (Maruyama et al. 1997) modulates TRPM6-mediated currents as has been shown for other TRP channels (see for example Hu et al. 2004; Xu et al. 2005).

7
Physiology

7.1
Mg^{2+} Re/Absorption

TRPM6, probably together with TRPM7, might be responsible for transcellular Mg^{2+} transport in colon and kidney. The ubiquitous expression of TRPM7 suggests that this channel is necessary for cellular Mg^{2+} transport in virtually every cell.

Mg^{2+} re/absorption in the intestine and kidney takes place by two transport mechanisms: a saturable, transcellular transport and a passive, paracellular transport. At physiological concentrations of intraluminal Mg^{2+} the transcellular pathway is more important than the paracellular pathway. The active transcellular Mg^{2+} re/absorption consists of an apical entry via TRPM6, TRPM7 or both into the epithelial cell and basolateral extrusion, presumably by a Na^+-coupled exchanger.

TRPM6 and TRPM7 are not only permeable for Mg^{2+} but also for Ca^{2+}. One might, therefore, speculate whether TRPM6 and TRPM7 are also involved in epithelial Ca^{2+} re/absorption. This cannot be excluded, although TRPV5 seems to be the gatekeeper for Ca^{2+} in the kidney (Hoenderop et al. 1999) and TRPV6 in the intestine (Peng et al. 1999).

7.2
Regulation of TRPM6 Expression Levels

7.2.1
Acid Base Status

Acid base status influences renal Mg^{2+} and Ca^{2+} excretion, and it was questioned whether the epithelial transport proteins TRPM6 and TRPV5 are responsible for these adaptations. Chronic metabolic acidosis results in renal Mg^{2+} and Ca^{2+} losses, whereas the reverse effects take place during metabolic acidosis. Mice with metabolic acidosis (induced by NH_4Cl loading or administration of the carbonic anhydrase inhibitor acetazolamide for 6 days) had decreased renal TRPM6 expression levels. Up-regulation of TRPM6 was shown in mice treated with $NaHCO_3$ for 6 days to induce chronic metabolic alkalosis (Nijenhuis et al. 2006). Changes in TRPV5 expression levels were also detected by RT-PCR and semiquantitative immunohistochemistry (Nijenhuis et al. 2006). Unfortunately, the expression levels of the close relatives TRPV6 and TRPM7 were not evaluated.

7.2.2
Dietary Mg^{2+} Intake

TRPM6 does not only regulate transepithelial Mg^{2+} transport but is also regulated by Mg^{2+} intake. Dietary Mg^{2+} restriction resulted in hypomagnesaemia and renal Mg^{2+} and Ca^{2+} conservation, whereas a Mg^{2+}-enriched diet has the opposite effect. Accordingly, Mg^{2+} restriction up-regulates renal TRPM6 mRNA levels in mice, and a Mg^{2+}-enriched diet increased TRPM6 mRNA expression in the colon as determined by RT-PCR (Groenestege et al. 2006). No effect was observed for TRPM7 mRNA in both tissues (Groenestege et al. 2006). This is in line with the hypothesis of TRPM6 being the gatekeeper in transepithelial Mg^{2+} transport, whereas TRPM7 is responsible for cellular Mg^{2+} homeostasis.

Confusingly, TRPM6 mRNA was initially shown to be expressed in the small intestine of mice (Voets et al. 2004), while later it was no longer detected in duodenum, jejunum and ileum but in cecum and colon (Groenestege et al. 2006).

This point needs further clarification since if TRPM6 is the gatekeeper for Mg^{2+} then it should be expressed in that region of the intestine where Mg^{2+} is actually absorbed (Kayne and Lee 1993; Schweigel and Martens 2000).

7.2.3
FK506 (Tacrolimus) and Glucocorticoids

The immunosuppressants FK506 (tacrolimus) and dexamethasone induce hypomagnesaemia and hypercalciuria as side-effects. Down-regulation of renal Mg^{2+} and Ca^{2+} transport proteins has been proposed as the underlying molecular mechanisms.

Administration of FK506 and dexamethasone to rats changed the expression levels of TRPM6 and TRPV5 in the kidney as determined by real-time quantitative PCR and immunohistochemical analysis. Down-regulation of TRPM6 and TRPV6 may provide a molecular mechanism for FK506-induced hypercalciuria and hypomagnesaemia. However, dexamethasone increases TRPM6 and TRPV6 expression levels, suggesting a different mechanism leading to renal Ca^{2+} and Mg^{2+} wasting (Nijenhuis et al. 2004).

8
Pathophysiology

8.1
Hypomagnesaemia with Secondary Hypocalcaemia

Patients suffering from hypomagnesaemia with secondary hypocalcaemia have mutations in TRPM6 (Schlingmann et al. 2002; Walder et al. 2002). Stop mutations, splice site mutations, frame shift mutations, deletions of exons and single point mutations (S141L, P1017R) were identified in these patients, and all mutations led to non-functional proteins (Schlingmann et al. 2002; Walder et al. 2002; Chubanov et al. 2006).

The familial hypomagnesaemia with secondary hypocalcaemia was first described by Paunier et al. in 1968. It is a rare inherited disorder (Chery et al. 1994; Walder et al. 1997) that is characterised by very low serum Mg^{2+} and low calcium levels (reviewed by Konrad and Weber 2003). Hypomagnesaemia with secondary hypocalcaemia has been previously reported to be X-linked, since there is a preponderance of male patients (Yamamoto et al. 1985; Abdulrazzaq et al. 1989; Pronicka and Gruszczynska 1991; Chery et al. 1994). However, the gene underlying this disorder was identified on chromosome 9q in three large inbred Bedouin kindreds from Israel, and within these families the disease segregates with an autosomal recessive mode of inheritance (Walder et al. 1997). The clinical features appear shortly after birth (Meyer and Boettger 2001). As a consequence of the hypocalcaemia the affected individuals suffer from restlessness, tremor, tetany and overt seizures (Abdulrazzaq et al. 1989; Pronicka and Gruszczynska 1991; Challa et al. 1995). If left undiagnosed or untreated, the disorder may result in neurological damage or may be fatal (Abdulrazzaq et al. 1989; Pronicka and Gruszczynska 1991). It is, therefore, of the utmost importance that a correct and prompt diagnosis be made, as therapy is simple and effective. High-dose Mg^{2+} supplementation is required lifelong (Visudhiphan et al. 2005). The correction of the Mg^{2+} deficiency with high oral Mg^{2+} intake supports the idea of at least two independent pathways for intestinal Mg^{2+} absorption. One of them is TRPM6, which is non-functional due to spontaneous mutations in individuals suffering from hypomagnesaemia with secondary hypocalcaemia. Whether TRPM7 is up-regulated in the kidney or intestine of these patients is not known. The other Mg^{2+} transport mechanism is the passive paracellular re/absorption, which gains importance at high intraluminal Mg^{2+} concentrations.

8.2
Diabetes Mellitus

Diabetes is associated with renal Mg^{2+} and Ca^{2+} wasting, but the molecular mechanisms of these defects are unknown. An increased abundance of Mg^{2+} and Ca^{2+} transporters in the kidney was detected in streptozotocin-induced diabetic rats by RT-PCR, immunofluorescence and Western blotting. The up-regulation of TRPM6, TRPV5 and TRPV6 may represent a compensatory adaptation to avoid Mg^{2+} and Ca^{2+} losses in the urine. Interestingly, these changes in gene expression can be reversed by insulin treatment (Lee et al. 2006).

References

Aarts M, Iihara K, Wei WL, Xiong ZG, Arundine M, Cerwinski W, MacDonald JF, Tymianski M (2003) A key role for TRPM7 channels in anoxic neuronal death. Cell 115:863–877

Abdulrazzaq YM, Smigura FC, Wettrell G (1989) Primary infantile hypomagnesaemia; report of two cases and review of literature. Eur J Pediatr 148:459–461

Caterina MJ, Schumacher MA, Tominaga M, Rosen TA, Levine JD, Julius D (1997) The capsaicin receptor: a heat-activated ion channel in the pain pathway. Nature 389:816–824

Challa A, Papaefstathiou I, Lapatsanis D, Tsolas O (1995) Primary idiopathic hypomagnesemia in two female siblings. Acta Paediatr 84:1075–1078

Chery M, Biancalana V, Philippe C, Malpuech G, Carla H, Gilgenkrantz S, Mandel JL, Hanauer A (1994) Hypomagnesemia with secondary hypocalcemia in a female with balanced X;9 translocation: mapping of the Xp22 chromosome breakpoint. Hum Genet 93:587–591

Chubanov V, Waldegger S, Mederos y Schnitzler M, Vitzthum H, Sassen MC, Seyberth HW, Konrad M, Gudermann T (2004) Disruption of TRPM6/TRPM7 complex formation by a mutation in the TRPM6 gene causes hypomagnesemia with secondary hypocalcemia. Proc Natl Acad Sci U S A 101:2894–2899

Chubanov V, Schlingmann KP, Waring J, Mederos y Schnitzler M, Waldegger S, Gudermann T (2006) Dominat-negativer Effekt einer neuen Missensmutation im menschlichen TRPM6-Gen führt zu Hypomagnesiämie mit sekundären Hypokalzämie. Naunyn Schmiedebergs Arch Pharmacol 372:60

Dorovkov MV, Ryazanov AG (2004) Phosphorylation of annexin I by TRPM7 channel-kinase. J Biol Chem 279:50643–50646

Ehrlich BE, Kaftan E, Bezprozvannaya S, Bezprozvanny I (1994) The pharmacology of intracellular Ca^{2+}-release channels. Trends Pharmacol Sci 15:145–149

Groenestege WM, Hoenderop JG, van den Heuvel L, Knoers N, Bindels RJ (2006) The epithelial Mg^{2+} channel transient receptor potential melastatin 6 is regulated by dietary Mg^{2+} content and estrogens. J Am Soc Nephrol 17:1035–1043

Gwanyanya A, Amuzescu B, Zakharov SI, Macianskiene R, Sipido KR, Bolotina VM, Vereecke J, Mubagwa K (2004) Magnesium-inhibited, TRPM6/7-like channel in cardiac myocytes: permeation of divalent cations and pH-mediated regulation. J Physiol 559:761–776

Hanano T, Hara Y, Shi J, Morita H, Umebayashi C, Mori E, Sumimoto H, Ito Y, Mori Y, Inoue R (2004) Involvement of TRPM7 in cell growth as a spontaneously activated Ca^{2+} entry pathway in human retinoblastoma cells. J Pharmacol Sci 95:403–419

Hermosura MC, Monteilh-Zoller MK, Scharenberg AM, Penner R, Fleig A (2002) Dissociation of the store-operated calcium current I(CRAC) and the Mg-nucleotide-regulated metal ion current MagNuM. J Physiol 539:445–458

Hoenderop JG, van der Kemp AW, Hartog A, van de Graaf SF, van Os CH, Willems PH, Bindels RJ (1999) Molecular identification of the apical Ca^{2+} channel in 1, 25-dihydroxyvitamin D3-responsive epithelia. J Biol Chem 274:8375–8378

Hoenderop JG, Voets T, Hoefs S, Weidema F, Prenen J, Nilius B, Bindels RJ (2003) Homo- and heterotetrameric architecture of the epithelial Ca^{2+} channels TRPV5 and TRPV6. EMBO J 22:776–785

Hu HZ, Gu Q, Wang C, Colton CK, Tang J, Kinoshita-Kawada M, Lee LY, Wood JD, Zhu MX (2004) 2-Aminoethoxydiphenyl borate is a common activator of TRPV1, TRPV2, and TRPV3. J Biol Chem 279:35741–35748

Kayne LH, Lee DB (1993) Intestinal magnesium absorption. Miner Electrolyte Metab 19:210–217

Konrad M, Weber S (2003) Recent advances in molecular genetics of hereditary magnesium-losing disorders. J Am Soc Nephrol 14:249–260

Kozak JA, Cahalan MD (2003) MIC channels are inhibited by internal divalent cations but not ATP. Biophys J 84:922–927

Kozak JA, Kerschbaum HH, Cahalan MD (2002) Distinct properties of CRAC and MIC channels in RBL cells. J Gen Physiol 120:221–235

Lee CT, Lien YH, Lai LW, Chen JB, Lin CR, Chen HC (2006) Increased renal calcium and magnesium transporter abundance in streptozotocin-induced diabetes mellitus. Kidney Int 69:1786–1791

Luft JH (1971) Ruthenium red and violet. II. Fine structural localization in animal tissues. Anat Rec 171:369–415

Maruyama T, Kanaji T, Nakade S, Kanno T, Mikoshiba K (1997) 2APB, 2-aminoethoxydiphenyl borate, a membrane-penetrable modulator of Ins(1,4,5)P3-induced Ca^{2+} release. J Biochem (Tokyo) 122:498–505

Meyer P, Boettger MB (2001) Familial hypomagnesaemia with secondary hypocalcaemia: a new case that indicates autosomal recessive inheritance. J Inherit Metab Dis 24:875–876

Nadler MJ, Hermosura MC, Inabe K, Perraud AL, Zhu Q, Stokes AJ, Kurosaki T, Kinet JP, Penner R, Scharenberg AM, Fleig A (2001) LTRPC7 is a Mg.ATP-regulated divalent cation channel required for cell viability. Nature 411:590–595

Nijenhuis T, Hoenderop JG, Bindels RJ (2004) Downregulation of Ca^{2+} and Mg^{2+} transport proteins in the kidney explains tacrolimus (FK506)-induced hypercalciuria and hypomagnesemia. J Am Soc Nephrol 15:549–557

Nijenhuis T, Renkema KY, Hoenderop JG, Bindels RJ (2006) Acid-base status determines the renal expression of Ca^{2+} and Mg^{2+} transport proteins. J Am Soc Nephrol 17:617–626

Nilius B, Prenen J, Vennekens R, Hoenderop JG, Bindels RJ, Droogmans G (2001) Pharmacological modulation of monovalent cation currents through the epithelial Ca^{2+} channel ECaC1. Br J Pharmacol 134:453–462

Nilius B, Prenen J, Droogmans G, Voets T, Vennekens R, Freichel M, Wissenbach U, Flockerzi V (2003) Voltage dependence of the Ca^{2+}-activated cation channel TRPM4. J Biol Chem 278:30813–30820

Nilius B, Talavera K, Owsianik G, Prenen J, Droogmans G, Voets T (2005a) Gating of TRP channels: a voltage connection? J Physiol 567:35–44

Nilius B, Voets T, Peters J (2005b) TRP channels in disease. Sci STKE 2005:eg7

Paunier L, Radde IC, Kooh SW, Conen PE, Fraser D (1968) Primary hypomagnesemia with secondary hypocalcemia in an infant. Pediatrics 41:385–402

Peng JB, Chen XZ, Berger UV, Vassilev PM, Tsukaguchi H, Brown EM, Hediger MA (1999) Molecular cloning and characterization of a channel-like transporter mediating intestinal calcium absorption. J Biol Chem 274:22739–22746

Perretti M, Solito E (2004) Annexin 1 and neutrophil apoptosis. Biochem Soc Trans 32:507–510

Prakriya M, Lewis RS (2002) Separation and characterization of currents through store-operated CRAC channels and Mg^{2+}-inhibited cation (MIC) channels. J Gen Physiol 119:487–507

Pronicka E, Gruszczynska B (1991) Familial hypomagnesaemia with secondary hypocalcaemia–autosomal or X-linked inheritance? J Inherit Metab Dis 14:397–399

Riazanova LV, Pavur KS, Petrov AN, Dorovkov MV, Riazanov AG (2001) Novel type of signaling molecules: protein kinases covalently linked to ion channels (in Russian). Mol Biol (Mosk) 35:321–332

Runnels LW, Yue L, Clapham DE (2001) TRP-PLIK, a bifunctional protein with kinase and ion channel activities. Science 291:1043–1047

Runnels LW, Yue L, Clapham DE (2002) The TRPM7 channel is inactivated by PIP(2) hydrolysis. Nat Cell Biol 4:329–336

Ryazanova LV, Dorovkov MV, Ansari A, Ryazanov AG (2004) Characterization of the protein kinase activity of TRPM7/ChaK1, a protein kinase fused to the transient receptor potential ion channel. J Biol Chem 279:3708–3716

Schlingmann KP, Weber S, Peters M, Niemann Nejsum L, Vitzthum H, Klingel K, Kratz M, Haddad E, Ristoff E, Dinour D, Syrrou M, Nielsen S, Sassen M, Waldegger S, Seyberth HW, Konrad M (2002) Hypomagnesemia with secondary hypocalcemia is caused by mutations in TRPM6, a new member of the TRPM gene family. Nat Genet 31:166–170

Schmitz C, Perraud AL, Johnson CO, Inabe K, Smith MK, Penner R, Kurosaki T, Fleig A, Scharenberg AM (2003) Regulation of vertebrate cellular Mg^{2+} homeostasis by TRPM7. Cell 114:191–200

Schweigel M, Martens H (2000) Magnesium transport in the gastrointestinal tract. Front Biosci 5:D666–D677

Strübing C, Krapivinsky G, Krapivinsky L, Clapham DE (2001) TRPC1 and TRPC5 form a novel cation channel in mammalian brain. Neuron 29:645–655

Talavera K, Yasumatsu K, Voets T, Droogmans G, Shigemura N, Ninomiya Y, Margolskee RF, Nilius B (2005) Heat activation of TRPM5 underlies thermal sensitivity of sweet taste. Nature 438:1022–1025

Visudhiphan P, Visudtibhan A, Chiemchanya S, Khongkhatithum C (2005) Neonatal seizures and familial hypomagnesemia with secondary hypocalcemia. Pediatr Neurol 33:202–205

Voets T, Prenen J, Vriens J, Watanabe H, Janssens A, Wissenbach U, Bödding M, Droogmans G, Nilius B (2002) Molecular determinants of permeation through the cation channel TRPV4. J Biol Chem 277:33704–33710

Voets T, Nilius B, Hoefs S, van der Kemp AW, Droogmans G, Bindels RJ, Hoenderop JG (2004) TRPM6 forms the Mg^{2+} influx channel involved in intestinal and renal Mg^{2+} absorption. J Biol Chem 279:19–25

Walder RY, Shalev H, Brennan TM, Carmi R, Elbedour K, Scott DA, Hanauer A, Mark AL, Patil S, Stone EM, Sheffield VC (1997) Familial hypomagnesemia maps to chromosome 9q, not to the X chromosome: genetic linkage mapping and analysis of a balanced translocation breakpoint. Hum Mol Genet 6:1491–1497

Walder RY, Landau D, Meyer P, Shalev H, Tsolia M, Borochowitz Z, Boettger MB, Beck GE, Englehardt RK, Carmi R, Sheffield VC (2002) Mutation of TRPM6 causes familial hypomagnesemia with secondary hypocalcemia. Nat Genet 31:171–174

Wu SN, Jan CR, Li HF (1999) Ruthenium red-mediated inhibition of large-conductance Ca^{2+}-activated K^+ channels in rat pituitary GH3 cells. J Pharmacol Exp Ther 290:998–1005

Xu SZ, Zeng F, Boulay G, Grimm C, Harteneck C, Beech DJ (2005) Block of TRPC5 channels by 2-aminoethoxydiphenyl borate: a differential, extracellular and voltage-dependent effect. Br J Pharmacol 145:405–414

Yamaguchi H, Matsushita M, Nairn AC, Kuriyan J (2001) Crystal structure of the atypical protein kinase domain of a TRP channel with phosphotransferase activity. Mol Cell 7:1047–1057

Yamamoto T, Kabata H, Yagi R, Takashima M, Itokawa Y (1985) Primary hypomagnesemia with secondary hypocalcemia. Report of a case and review of the world literature. Magnesium 4:153–164

1
Introduction

TRPM7 belongs to the melastatin-related subfamily of TRP-related ion channels, which comprises eight members with homologous architecture, but distinctive biophysical properties and activation mechanisms. TRPM7 is a bifunctional protein that contains both ion channel and protein kinase domains (Nadler et al. 2001; Runnels et al. 2001; Ryazanova et al. 2001) and provides a pathway for the transport of Ca^{2+} and Mg^{2+} as well trace metal ions (Nadler et al. 2001; Monteilh-Zoller et al. 2003). The protein was originally discovered in three independent studies that used very different approaches to clone it: Ryazanova et al. identified TRPM7 by screening data bases for homologs of human eukaryotic elongation factor 2 kinase (Ryazanova et al. 2001), Runnels et al. found TRPM7 in a yeast two-hybrid screen using the C2 domain of phospholipase C as a bait (Runnels et al. 2001), and Nadler et al. obtained TRPM7 using a bioinformatics approach aimed at identifying novel ion channels expressed in immune cells (Nadler et al. 2001). The latter two studies overexpressed TRPM7 in a heterologous expression system and provided the first electrophysiological characterization of the channel; however, they arrived at very different conclusions about the selectivity, activation mechanism, and the role of the channel's endogenous kinase domain. Since then, TRPM7 has been the subject of numerous studies and at the center of lively and controversial discussions.

2
Channel Properties

The full-length TRPM7 protein is composed of 1,863 amino acid residues (Nadler et al. 2001; Runnels et al. 2001; Ryazanova et al. 2001). Its currents are characterized by a reversal potential of approximately 0 mV and very strong outward rectification, with little inward current in the physiological rage of −70 mV to 0 mV and large outward currents above +50 mV (Fig. 1a, b). Endogenous TRPM7 currents in various cell types are small and rarely exceed 20 pA/pF (Fig. 1c, d; Nadler et al. 2001; Hermosura et al. 2002; Kozak et al. 2002; Prakriya and Lewis 2002).

At physiological, negative membrane potentials TRPM7 conducts very little inward current and exclusively transports divalent cations such as Ca^{2+} and Mg^{2+} from the extracellular space into the cytosol down their concentration gradient (Nadler et al. 2001). However, at positive membrane potentials, where divalent ions do not experience sufficient driving force to enter the cell, the outward transport rates of intracellular cations such as K^+ or Cs^+ increase, and at potentials beyond +50 mV the monovalent cation fluxes become quite prominent and shape the characteristic outwardly rectifying current–voltage

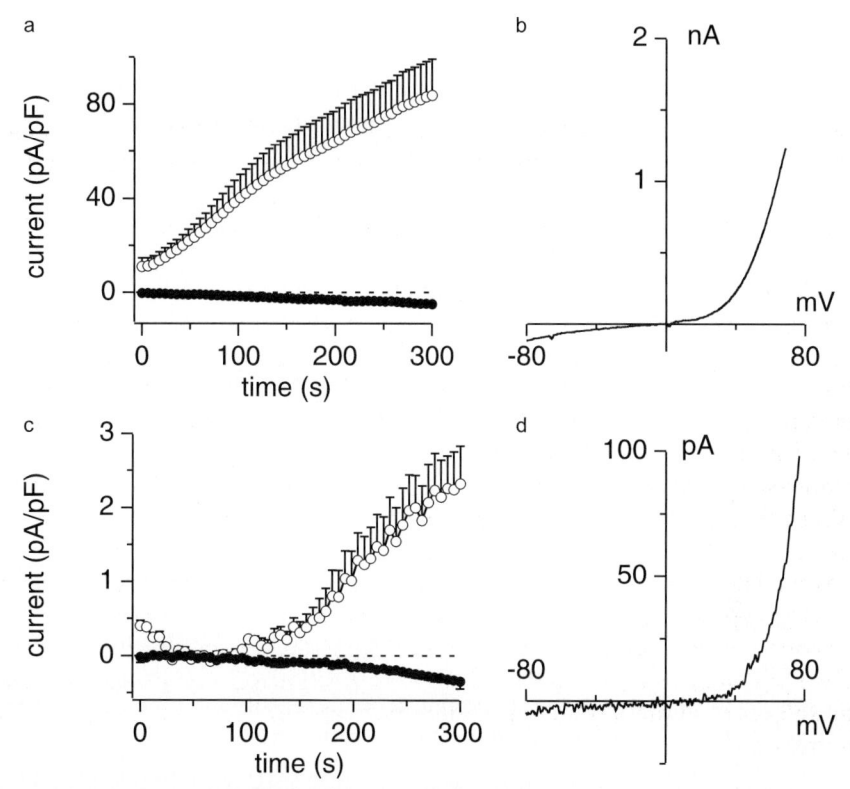

Fig. 1 a–d Development of heterologous and endogenous TRPM7 currents. **a** Representative development of heterologous TRPM7 currents in overexpressing HEK-293 cells where free [Mg^{2+}]$_i$ was kept at 800 µM and MgATP was omitted ($n = 6$). Data points correspond to average and normalized current amplitudes measured at −80 mV (*closed circles*) and +80 mV (*open circles*) and plotted as a function of time. Note that TRPM7 develops immediately. **b** Current–voltage (I–V) relationship of heterologous TRPM7 extracted from an example cell at 200 s. **c** Representative development of endogenous TRPM7-like MagNuM currents measured in RBL-2H3 cells. Average inward (*closed circles*) and outward (*open circles*) currents at −80 mV and +80 mV, respectively ($n = 6$), recorded under conditions that suppress I$_{CRAC}$ and support MagNuM activation (omission of MgATP, free [Mg^{2+}]$_i$ = 760 µM, free [Ca^{2+}]$_i$ = 100 nM). Note that MagNuM starts to develop around 80 s into the experiment. **d** Representative I–V relationship of endogenous TRPM7-like MagNuM extracted at 300 s after whole-cell establishment

relationship of TRPM7. Only when completely removing extracellular divalent cations, TRPM7 will transport monovalent cations inwardly and this will linearize the current–voltage relationship, revealing both TRPM7's exquisite specificity for divalent cation transport and the lack of any significant voltage dependence (Nadler et al. 2001).

Since the outwardly rectifying current–voltage relationship of TRPM7 macroscopic currents is based on the channel's permeation properties, the microscopic behavior of TRPM7 in single-channel recordings also exhibits nonlinear outward rectification when divalent cations are present. Under these conditions, single-channel events are not detectable at negative voltages, and single-channel outward currents gradually increase in a nonlinear fashion at positive voltages (Runnels et al. 2001). It is therefore not possible to determine an accurate slope conductance estimate. Because of this, the slope conductance of 105 pS obtained by linear regression by Runnels et al. (2001) is likely an overestimate. Under similar experimental conditions, Nadler et al. (2001) arrived at a single-channel conductance of approximately 40 pS at +60 mV, and this is in good agreement with the single-channel conductance of approximately 40 pS obtained by Kerschbaum et al. (1999) under symmetric divalent-free conditions at negative membrane voltages (Kerschbaum and Cahalan 1999). The latter study attributed these 40-pS single channels to store-operated calcium release-activated calcium (CRAC) channels; however, in retrospect, these channels have been identified as TRPM7 (Nadler et al. 2001; Hermosura et al. 2002; Kozak et al. 2002; Prakriya and Lewis 2002). The asymmetrical permeation properties of TRPM7, with exclusive divalent ion influx at negative voltages and exclusive efflux of monovalent cations at high positive voltages, as well as the pronounced block of monovalent cation permeation by divalent cations, renders classical analyses of selectivity based on constant field equations inaccurate, since these rely on the assumption that the permeating ion species move through the channel independently of each other. Therefore, TRPM7 cannot be classified as a nonselective cation channel, and the proposed permeability of cations relative to that of Cs^+ of 1.1, 0.97, and 0.34 for K^+, Na^+, and Ca^{2+}, respectively (Runnels et al. 2001), does not reflect the channel's true selectivity (Monteilh-Zoller et al. 2003).

The unusual permeation profile of TRPM7 is further evident from ion substitution experiments in which other divalent cations were tested for permeation relative to Ca^{2+} with the sequence $Zn^{2+} \approx Ni^{2+} \gg Ba^{2+} > Co^{2+} > Mg^{2+} \geq Mn^{2+} \geq$ $\geq Sr^{2+} \geq Cd^{2+} \geq Ca^{2+}$ (Monteilh-Zoller et al. 2003). Interestingly, it has recently been proposed that the selectivity of TRPM7 may be regulated by extracellular pH (Jiang et al. 2005).

3
Channel Activation and Regulation

TRPM7 channels appear to be constitutively active in resting cells, albeit at very low levels of less than 10% of maximal activity (Nadler et al. 2001), and this is true for both heterologously expressed and native channels. TRPM7 channel activity can be either up- or downregulated, and a number of positive and negative modulators have been identified, although some of the proposed mechanisms are controversial.

3.1
Regulation by Free Mg^{2+} and Mg·ATP

The two most significant factors that control TRPM7 channel activity are free Mg^{2+} ions and Mg-complexed nucleotides, which tonically inhibit the channels and represent the primary reason for the low activity observed in resting cells (Nadler et al. 2001). Nadler et al. (2001) observed that perfusing cells with intracellular solutions in which MgATP is omitted but free Mg^{2+} is present at physiological levels of approximate 700–900 µM can elicit large TRPM7-mediated currents in whole-cell patch-clamp experiments. Increasing MgATP concentrations in the pipette solution while keeping free Mg^{2+} levels constant gradually decreased the TRPM7 activation in a dose-dependent manner so that at physiological levels of 3–4 mM MgATP there was no increase in current, and higher concentrations even inhibited the resting TRPM7 current. The Mg-nucleotide-mediated regulation of TRPM7 was observed both in human embryonic kidney (HEK)-293 cells overexpressing TRPM7 as well as in cells such as Jurkat T cells and RBL-2H3, where MgATP-free pipette solutions activate native currents with identical properties as TRPM7 that are completely suppressed by millimolar concentrations of MgATP. A similar nucleotide-mediated regulation of TRPM7 has also been observed in human retinoblastoma cells, where MgATP appears to be very effective in inhibiting TRPM7 whole-cell currents (Hanano et al. 2004). The MgATP-dependent inhibition of TRPM7 is not due to ATP hydrolysis, since MgATPγS is also effective in suppressing the current (Nadler et al. 2001). Interestingly, the suppression of TRPM7 is not limited to adenine nucleotides, since GTP was found to be similarly effective (Nadler et al. 2001; Demeuse et al. 2006).

The nucleotide-dependent regulation of TRPM7 is not mediated by the free nucleotide, since perfusing cells with NaATP fails to suppress TRPM7 currents and in fact causes a massive activation of TRPM7 (Nadler et al. 2001; Demeuse et al. 2006). The strong activation of TRPM7 by Na-ATP was also observed by Clapham and colleagues and was interpreted in favor of a kinase-mediated activation mechanism (Runnels et al. 2001). However, this interpretation turned out to be incorrect, since Nadler et al. (2001) demonstrated that free Mg^{2+} is another important factor in regulating TRPM2 activity. Since the phosphate groups of nucleotides can coordinate Mg^{2+} ions, they represent very strong Mg^{2+} chelators, so that NaATP will chelate free cytosolic Mg^{2+} to very low levels and thereby cause a strong activation of TRPM7. Therefore, the physiological, Mg-complexed form of ATP inhibits the channel, whereas the free nucleotide activates the channel by chelating free Mg^{2+}. Since both Mg-nucleotides and free Mg^{2+} can regulate TRPM7 activity, omitting both and additionally chelating residual Mg^{2+} by strong buffers such as N-hydroxyethylenediaminetriacetic acid (HEDTA) or NaATP yields maximal activation of TRPM7 (Demeuse et al. 2006).

It should be noted that Kozak and Cahalan have questioned the regulation by Mg-nucleotides (Kozak and Cahalan 2003). While they confirmed the original observation that MgATP in the presence of ethyleneglycoltetraacetic acid (EGTA) can inhibit TRPM7, they also reported that this inhibition is lost when using a strong Mg^{2+} chelator, HEDTA, leading them to conclude that free Mg^{2+} alone can explain the effect. The study compared experiments in which free Mg^{2+} was buffered to 270 µM with either 12 mM EGTA or by 3 mM EGTA+2.5 mM HEDTA+5 mM MgATP, and its conclusion regarding the absence of MgATP-dependent regulation of TRPM7 rests entirely on the accuracy of these two data sets. Both solutions reportedly caused a very similar inhibition of TRPM7 by approximate 75%. The strong suppression of TRPM7 by the reference solution containing 270 µM certainly is surprising, as it would suggest an IC_{50} of free Mg^{2+} that is well below the typical IC_{50} values of approximately 700 µM observed by several laboratories (Nadler et al. 2001; Kozak et al. 2002; Prakriya and Lewis 2002; Schmitz et al. 2003; Demeuse et al. 2006). A recent study by Demeuse et al. (2006) has reevaluated the experimental conditions employed by Kozak and Cahalan and confirmed that the MgATP-containing solution indeed causes approximately 80% inhibition of TRPM7 currents (Demeuse et al. 2006). However, the reference solution containing only EGTA and 270 µM free Mg^{2+} did not produce any significant inhibition compared to completely Mg^{2+}-free conditions (Demeuse et al. 2006). The study further demonstrated that MgATP can inhibit TRPM7 in the complete absence of any exogenous chelator, suggesting that MgATP can also regulate TRPM7 channel activity in physiologically more relevant circumstances.

Demeuse et al. (2006) also provided a comprehensive assessment of the dual regulation of TRPM7 by free Mg^{2+} and Mg-nucleotides and found that the two factors act in synergy to regulate TRPM7 (Demeuse et al. 2006). They established dose-response curves of TRPM7 current inhibition by increasing the MgATP concentration at fixed low (~200 µM), physiological (~800 µM), or high (~1600 µM) free Mg^{2+} concentrations. There was only a modest shift in the IC_{50} for MgATP-mediated inhibition of TRPM7 from 3.3 mM (at 200 µM free Mg^{2+}) to 2 mM (at 800 µM free Mg^{2+}) and 1.6 mM (at 1.6 mM free Mg^{2+}). Although this amounts to a twofold shift in the IC_{50} for MgATP, this shift occurs at relatively low concentrations of free Mg^{2+} and is essentially complete at 800 µM free Mg^{2+}, suggesting that slight variations in free Mg^{2+} around physiological levels may not have a major impact on the nucleotide sensitivity of TRPM7. Conversely, the efficacy of free Mg^{2+} is shifted by about sixfold when increasing MgATP from 0 to 6 mM. The IC_{50} of free Mg^{2+} in the absence of any added ATP is very close to physiological levels in resting cells (720 µM), whereas the IC_{50} for free Mg^{2+} in the presence of 6 mM MgATP is 130 µM. For physiologically relevant MgATP concentrations between 1 and 4 mM, the shift in IC_{50} for Mg^{2+} is fourfold. Therefore, it appears that free Mg^{2+} and MgATP are acting in synergy (or cooperatively) to inhibit TRPM7. However, free Mg^{2+} is not the only divalent cation that can inhibit TRPM7 intracellularly, since most divalent cations will inhibit TRPM7 with various degrees of potency from the cytosol (Kozak and Cahalan 2003) and this effect is also seen with various polyamines and even protons (Kerschbaum et al. 2003; Kozak et al. 2005). Kozak and Cahalan (2005) have proposed that this inhibitory effect of charged cations is due to screening the head group phosphates on membrane phospholipids that would otherwise promote TRPM7 activity (Kozak et al. 2005). If this were the case, then the Mg^{2+}-dependent inhibition of TRPM7 would be extraneous to the protein itself and would be difficult to reconcile with the significant shifts of Mg^{2+} sensitivity of point mutations and kinase deletion mutants of TRPM7 (Schmitz et al. 2003; Demeuse et al. 2006). These studies presented an alternative model that takes this into account and favors a separate Mg^{2+} binding site within the TRPM7 protein itself (Schmitz et al. 2003; Demeuse et al. 2006).

In addition to MgATP, Demeuse et al. (2006) also found other purine and pyrimidine nucleotide triphosphates to inhibit TRPM7 currents. Except for 5′ triphosphate (ITP), all nucleotides synergize with free Mg^{2+}, with ATP being the most potent. The sequence of potency was determined as follows: ATP>TTP>CTP$\geq$GTP$\geq$UTP. Not only did the triphosphate nucleotide prove effective in suppressing TRPM7, but to a lesser degree also the dinucleotides, whereas nucleotide monophosphates, which cannot coordinate Mg^{2+} into a complex, were completely ineffective in modulating TRPM7 currents.

Since free Mg^{2+} and MgATP differentially affect each other's inhibitory efficacy, the nucleotide-dependent regulation of TRPM7 appears to be mediated by two distinct binding sites. Demeuse et al. (2006) proposed that a binding site within the endogenous kinase domain mediates the nucleotide effect, since two point mutations that affect nucleotide binding and phosphotransferase activity of the kinase domain reduce both free Mg^{2+} and Mg-nucleotide-mediated inhibition of channel activity (Schmitz et al. 2003; Demeuse et al. 2006). Additional characterizations of these mutants demonstrate the loss of MgATP and MgGTP block in the phosphotransferase-deficient mutant K1648R. The lysine residue mutated is implicated in nucleotide binding and catalysis and is conserved in classical kinases (Yamaguchi et al. 2001). In the second mutant, G1799D, a site that might coordinate the Mg-phosphate complex, some regulation remains, even though phosphotransferase activity is lost. Interestingly, a truncation mutant of TRPM7 that lacks the entire kinase domain exhibits supersensitivity to both Mg^{2+} and MgATP (Schmitz et al. 2003), but exhibits no specificity toward nucleotide species (Demeuse et al. 2006). A hypothesis that explains this seemingly paradoxical behavior proposes that the truncation of the C-terminal kinase domain exposes the Mg^{2+} binding site normally masked by the kinase domain and converts it into a high-affinity site that can be blocked by very low levels of free Mg^{2+}. In addition, this site now becomes available for binding Mg-nucleotides. However, since the kinase deletion mutant does not discriminate between nucleotide species, it might simply accept the Mg-phosphate moiety as a binding partner.

3.2
Regulation by the Endogenous Kinase Domain

Since TRPM7 contains both an ion channel and a kinase domain, an important question is whether these two functional domains interact and regulate each other. Conflicting accounts of the kinase's importance for ion channel activity have been presented, ranging from one extreme that the kinase domain is essential for gating the channel (Runnels et al. 2001)—others support a modulatory role of the kinase in shaping Mg^{2+} and Mg-nucleotide sensitivity (Schmitz et al. 2003; Demeuse et al. 2006)—to the opposite extreme of complete dissociation of kinase and channel activities (Matsushita et al. 2005).

Clapham and colleagues (Runnels et al. 2001) suggested that the kinase domain is absolutely required for channel activation based on the effects of ATP and the absence of TRPM7 currents in two mouse TRPM7 mutants that were designed to inactivate phosphotransferase activity [G1796D and the double mutant C1809A/C1812A; although based on structural considerations, the latter mutant may not represent a phosphotransferase-deficient construct (Yamaguchi et al. 2001)]. This interpretation is probably incorrect, since the ATP effect is attributable to Mg^{2+} chelation and the lack of TRPM7 currents when expressing the mutant channels was likely due to failed expression of TRPM7 mutant proteins, since studies from two other laboratories have demonstrated that mutated human TRPM7 channels without phosphotransferase activity still exhibit channel activation comparable to wildtype TRPM7 (Schmitz et al. 2003; Matsushita et al. 2005). Both studies are in general agreement that kinase activity is not essential to activate TRPM7. They mutated a lysine residue to aspartate (K1648R), which is implicated in catalysis and is conserved in classical kinases, and found that this mutant is activated normally when reducing cytosolic Mg^{2+} levels. Matsushita et al. (2005) observed that this mutant channel exhibited the same Mg^{2+} sensitivity as wildtype TRPM7 (Matsushita et al. 2005), leading them to conclude that the kinase and channel activity are completely dissociated. However, this interpretation may be overstated, as it is based on just two relatively high concentrations of Mg^{2+} (4 and 6 mM) and the data set for 4 mM actually hints at a reduced efficacy of Mg^{2+}-induced inhibition in the mutant compared to wildtype channels. Schmitz et al. (2003) had previously provided a comprehensive dose-response analysis of phosphotransferase-deficient mutants of human TRPM7 (K1648R and G1799D). Based on their analysis, the differences in the inhibition by Mg^{2+} at high concentrations are indeed relatively small; however, at intermediate concentrations around the IC_{50}, there is in fact a reduced sensitivity of these mutants to inhibition by Mg^{2+}. In a more recent study, this reduced sensitivity to Mg^{2+} has been confirmed and extended, suggesting that the kinase domain provides the site for Mg-nucleotide binding, which interacts with a separate Mg^{2+} binding site to regulate TRPM7 channel activity (Demeuse et al. 2006).

An important question relates to whether the nucleotide binding alone is sufficient to regulate TRPM7 channel activity or whether kinase activity per se also plays a role, since this domain can autophosphorylate the channel. The primary sites of autophosphorylation are a matter of controversy, since some groups mainly observe phosphorylation of serine (Ryazanova et al. 2004; Matsushita et al. 2005), while another study proposes threonine as a primary target (Hermosura et al. 2005). Matsushita et al. (2005) mutated two serine residues at positions 1511 and 1567 to alanine, and this resulted in a large decrease in autophosphorylation compared with wildtype proteins; however, this mutant did not show any obvious changes in its regulation of Mg^{2+}. However, as discussed above, this analysis was performed at relatively high concentrations of Mg^{2+}, so that subtle changes at intermediate or low Mg^{2+} concentrations or possible effects on MgATP-mediated regulation may have gone unnoticed. At variance with this report, Hermosura et al. (2005) identified threonine 1428 as a significant locus for TRPM7 autophosphorylation, since a T1428I point mutation not only reduced the phosphate incorporation by 30%, but also changed the sensitivity of the channel toward inhibition by free Mg^{2+}. The authors found that 1 mM of added free Mg^{2+} reduced the currents of the T1428I mutant by approximately 50% compared to Mg^{2+}-free solutions. This value on its own would not suggest an overly sensitive channel, since it is in fact very similar to the degree of Mg^{2+}-induced inhibition observed in numerous studies on heterologous or native TRPM7 (Nadler et al. 2001; Kozak et al. 2002; Prakriya and Lewis 2002; Schmitz et al. 2003; Demeuse et al. 2006). However, in the context of that study, the same concentration of Mg^{2+} produced a surprisingly small inhibition of wildtype TRPM7 (less than 20%), which is at variance with values obtained by others. Therefore it remains doubtful that the phosphorylation state of that particular threonine residue is important for shaping Mg^{2+} and/or Mg-nucleotide sensitivity. It should be noted that a sensitization would be rather unexpected and in apparent conflict with the reduced Mg^{2+} and MgATP sensitivity reported for phosphotransferase-deficient mutants (Schmitz et al. 2003), which would likely also result in a dephosphorylated protein.

In summary, it would appear that the kinase domain, while not essential for gating TRPM7 channels, might modulate channel activity by shaping its sensitivity toward free Mg^{2+} and Mg-nucleotides. Whether or not the autophosphorylated channel contributes to this modulation awaits additional experimental efforts.

3.3
Regulation by Receptor Stimulation

While concentrations of cellular free Mg^{2+} and Mg-nucleotide levels provide an important "passive" regulatory mechanism that tonically reduces TRPM7 activity to approximately 10% of maximum, the channels can also be "actively" modulated following receptor stimulation. Two conflicting reports have proposed regulatory mechanisms of TRPM7 by phosphatidylinositol bisphosphate (PIP$_2$) (Runnels et al. 2002) and by cyclic AMP (cAMP) signaling (Takezawa et al. 2004). Clapham and colleagues (2002) suggested that receptor-mediated PIP$_2$ is required to maintain TRPM7 activity, and its breakdown by receptor-mediated stimulation of phospholipase C (PLC) is responsible for the inhibition TRPM7 they observed when stimulating cells with agonists that couple to PLC. Although several ion channels, including members of the TRPM family [TRPM4 (Zhang et al. 2005; Nilius et al. 2006), TRPM5 (Liu and Liman 2003), and TRPM8 (Liu and Qin 2005; Rohacs et al. 2005)] are subject to PIP$_2$-mediated regulation, the case for TRPM7 has not been made without ambiguity. Most of the experiments presented in support of PIP$_2$ regulation were performed in cells that overexpress TRPM7 or with receptor agonists that can couple to PLC (or both). Such overexpression is prone to significant pitfalls. Thus, the overexpression of TRPM7 leads to a complete suppression of PLC-mediated signaling and likely abolishes any PIP$_2$ breakdown, without affecting the receptor-mediated regulation of TRPM7 by muscarinic agonists (Takezawa et al. 2004). This is presumably due to the fact that TRPM7 binds to and inhibits PLC (Runnels et al. 2001). The overexpression of Gq-coupled receptors faces the problem of G protein promiscuity (Xiao 2000; Milligan and Kostenis 2006), where receptors may couple to other G proteins and activate signaling pathways unrelated to PLC. Another potential problem is the fact that the cells under investigation can have endogenous receptors that are activated in parallel with the overexpressed receptors, which is the case for muscarinic receptors, as some of them couple to PLC (M1, M3, M5) whereas others couple to Gi (M2, M4).

Takezawa et al. (2004) demonstrated that the muscarinic receptor-mediated inhibition of TRPM7 in HEK-293 cells by endogenous muscarinic receptors likely occurs via Gi-coupled pathways, since the effect was suppressed by pertussis toxin. They further demonstrated that TRPM7 currents are upregulated by isoproterenol (a beta receptor agonist that couples to Gs) or by perfusing cells with elevated cAMP. Furthermore, these modulations were inhibited by protein kinase A inhibitors but not by the PLC antagonist U73122, suggesting that the receptor signaling cascade that regulates TRPM7 indeed involves Gi and/or Gs-coupled receptors that elevate or decrease cAMP levels, which in turn regulate the activity of PKA. Interestingly, this PKA-dependent regulation additionally requires TRPM7's functional endogenous kinase, since phosphotransferase-deficient mutants lose their ability to be regulated via this mechanism (Takezawa et al. 2004). It remains to be established whether the cAMP-dependent and the MgATP-mediated regulation of TRPM7 is caused by two separate and unrelated mechanisms or the two are somehow linked. For example, it is conceivable that the cAMP-induced activation of PKA could induce a shift in MgATP sensitivity of TRPM7 or conversely, changes in MgATP could determine the activity of PKA and TRPM7's endogenous kinase activity.

4
Physiological Functions

TRPM7 is found in virtually every mammalian cell, although it appears to be expressed in variable amounts in different cells. Nadler et al. (2001) and Hermosura et al. (2002) characterized native TRPM7-like currents in Jurkat T, RBL-2H3, and HEK-293 cells and named these endogenous currents Mag-NuM (for magnesium-nucleotide-regulated metal ion currents), because they were inhibited by high concentrations of MgATP and transport divalent metal ions. Unfortunately, two later publications (Kozak et al. 2002; Prakriya and Lewis 2002) did not follow the original nomenclature and introduced the new term MIC (for Mg^{2+}-inhibited cation current), which is based on the observation that Mg^{2+} also inhibits TRPM7 as described by the same report that originally designated the current as MagNuM (Nadler et al. 2001). Because MagNuM activates under the same experimental conditions typically used to study the store-operated Ca^{2+} current I_{CRAC} (Hermosura et al. 2002; Kozak et al. 2002; Prakriya and Lewis 2002), and both I_{CRAC} and MagNuM will carry large monovalent cation currents when using divalent-free extracellular solutions (Hermosura et al. 2002), endogenous TRPM7 channels had in fact been recorded before (Kerschbaum and Cahalan 1998, 1999; Fomina et al. 2000), but were erroneously interpreted to represent CRAC channels (Kerschbaum and Cahalan 1999). Monovalent currents that are unmasked by removal of divalent cations from the extracellular solution and are possibly due to TRPM7 have also been observed previously in myocytes (Mubagwa et al. 1997), aortic smooth muscle cells (Zakharov et al. 1999), and hippocampal neurons (Xiong et al. 1997). Based on the whole-cell current densities of approximately 2–10 pA/pF at +80 mV that one typically sees in a range of different cell types, and assuming a single-channel conductance of approximately 40 pS and open probabilities of approximately 90%, one would estimate about 10–100 channels per cell. Fomina et al. (2000) estimated the number of TRPM7 channels in resting lymphocytes to be approximately 15, but this number is upregulated to approximately 140 channels in activated cells. In smooth muscle cells, the number of TRPM7 channels can be doubled by quick recruitment from a cytosolic pool in response to laminar flow (Oancea et al. 2006).

So far, most of the information available on the physiological functions of TRPM7 in mammalian systems has been derived from cellular assays. Unfortunately, a conventional TRPM7 knockout in the mouse has proved to be lethal at early embryonic stages (A. Ryazanov, personal communication) and characterization of TRPM7 at the rodent animal level will have to await an inducible knockout. TRPM7 orthologs have been identified in *Danio rerio* (zebrafish) and *Caenorhabditis elegans*, although it remains to be determined whether the physiological context in these organisms translates into mammalian systems. The zebrafish mutant touchtone/nutria is caused by mutations in TRPM7 and exhibits severe growth retardation and gross alterations in skeletal development in addition to embryonic melanophore and touch-response defects (Elizondo et al. 2005). In *C. elegans* a Mg^{2+}-sensitive current was found in intestinal cells with electrophysiological properties that are very similar to TRPM7/MagNuM (Estevez et al. 2003; Estevez and Strange 2005), and Teramoto et al. (2005) have recently identified GTL-1 and GON-2 as possible candidates underlying this current (Teramoto et al. 2005).

The divalent cation transport function of TRPM7 appears to be vital to cell survival, since cellular knockout of TRPM7 in avian B cells induces death of the cells over a 48–72 h time frame (Schmitz et al. 2003). The crucial role of Mg^{2+} transport, at least in the case of avian B cells, is evident from ion substitution experiments, which demonstrate that growing TRPM7-deficient cells in medium with high Mg^{2+} concentrations, but not high Ca^{2+}, can rescue the lethal phenotype and restore normal cell growth and proliferation (Schmitz et al. 2003). On the other hand, in retinoblastoma cells, the role of Ca^{2+} influx through TRPM7 may be more significant, since molecular suppression of TRPM7 in these cells by small interfering RNA (siRNA) silencing markedly reduces TRPM7 currents and the magnitude of resting Ca^{2+} influx, which is paralleled by decreased TRPM7 immunoreactivity, decelerated cell proliferation, and retarded G1/S cell cycle progression (Hanano et al. 2004). In addition to Ca^{2+} and Mg^{2+}, TRPM7 is also capable of transporting other divalent cations, exhibiting a strong preference for the essential trace metals Zn^{2+}, Mn^{2+}, and Co^{2+} (Monteilh-Zoller et al. 2003). While undeniably crucial for cellular function, if they accumulate above trace levels, these ions are highly toxic. This is particularly true for zinc, which is one of the most abundant transition metals in the brain. Excessive vesicular release of Zn^{2+} is thought to play a key role in neuronal cell death following ischemia (Lee et al. 2000). In addition to its acute neurotoxicity, zinc may contribute to the pathogenesis of chronic neurodegenerative conditions such as Alzheimer's disease, where it is found to accumulate in amyloid plaques (Maynard et al. 2005). Furthermore, TRPM7 also allows significant entry of toxic metals such as Ni^{2+} and Cd^{2+} (Monteilh-Zoller et al. 2003). This permeation profile coupled with the constitutive activity of TRPM7 indicates that it may represent a ubiquitous basal influx pathway for metal ions with the potential of significant physiological and toxicological repercussions that remain to be studied in more detail.

While the knock down or complete knock out of TRPM7 compromises cell growth and proliferation, overexpression of TRPM7 in heterologous cell systems such as HEK-293 cells also has harmful consequences for cell viability. Nadler et al. (2001) observed that these normally adherent cells detach after about 24–48 h and eventually die (Nadler et al. 2001), presumably because of excessive TRPM7 channel activity. Two recent studies have investigated this phenomenon in more detail, and although they arrive at different conclusions regarding the underlying mechanisms that may be responsible for it, both studies implicate TRPM7 in affecting cell adhesion. Su et al. (2006) found that overexpression of TRPM7 in cells causes loss of cell adhesion. This was determined to be due to the Ca^{2+}-dependent protease m-calpain requiring TRPM7 activity by presumably creating a high Ca^{2+} environment that is localized to adhesion points.

Clark et al. (2006), studying NIE-E115 neuroblastoma cells, also found TRPM7-mediated changes in cell adhesion. This study suggested that both Ca^{2+}- and kinase-dependent mechanisms were involved. It was found that mild overexpression of TRPM7 in NIE-E115 cells causes increases intracellular Ca^{2+} levels accompanied by kinase-independent cell spreading and the formation of focal adhesions. Kinase-dependent effects of TRPM7 were seen after bradykinin stimulation, which caused increased Ca^{2+} entry into these cells that the authors attribute to TRPM7 activation. This was accompanied by the formation of podosomes in TRPM7 overexpressing cells, but not in wildtype cells, and the effect involves the TRPM7 kinase-mediated phosphorylation of myosin IIA heavy chain (Clark et al. 2006). The interaction between TRPM7 and myosin IIA is strictly Ca^{2+}-dependent, but additionally requires an active kinase domain, as the kinase-dead mutant does not support this interaction. It remains to be determined whether the discrepancies between these two studies reflect differences in TRPM7 expression levels or cell type-dependent mechanisms.

A further role of TRPM7 in neuronal cells has been suggested by Aarts et al. (2003), who subjected cultured cortical neurons to prolonged oxygen and glucose deprivation (OGD), an experimental model for ischemia that causes anoxic cell death. OGD caused significant influx of Ca^{2+} that correlated with the amount of cell death. The underlying membrane currents were attributed to TRPM7, since gene silencing by siRNA targeted against TRPM7 inhibits the currents and protects neurons from cell death. One complicating factor in these studies was that molecular silencing of TRPM7 was accompanied by a parallel reduction in TRPM2, which represents another candidate channel for mediating neuronal cell death, as it is activated by reactive oxygen species (Hara et al. 2002). It is therefore possible that both channels may have contributed to some of the effects on cell death.

Acknowledgements This work was supported by NIH grants R01-GM065360 to A.F. and R01-NS040927 to R.P.

References

Aarts M, Iihara K, Wei WL, Xiong ZG, Arundine M, Cerwinski W, MacDonald JF, Tymianski M (2003) A key role for TRPM7 channels in anoxic neuronal death. Cell 115:863–877

Clark K, Langeslag M, van Leeuwen B, Ran L, Ryazanov AG, Figdor CG, Moolenaar WH, Jalink K, van Leeuwen FN (2006) TRPM7, a novel regulator of actomyosin contractility and cell adhesion. EMBO J 25:290–301

Demeuse P, Penner R, Fleig A (2006) TRPM7 channel is regulated by magnesium nucleotides via its kinase domain. J Gen Physiol 127:421–434

Elizondo MR, Arduini BL, Paulsen J, MacDonald EL, Sabel JL, Henion PD, Cornell RA, Parichy DM (2005) Defective skeletogenesis with kidney stone formation in dwarf zebrafish mutant for trpm7. Curr Biol 15:667–671

Estevez AY, Strange K (2005) Calcium feedback mechanisms regulate oscillatory activity of a TRP-like Ca^{2+} conductance in C. elegans intestinal cells. J Physiol 567:239–251

Estevez AY, Roberts RK, Strange K (2003) Identification of store-independent and store-operated Ca^{2+} conductances in Caenorhabditis elegans intestinal epithelial cells. J Gen Physiol 122:207–223

Fomina AF, Fanger CM, Kozak JA, Cahalan MD (2000) Single channel properties and regulated expression of Ca^{2+} release-activated Ca^{2+} (CRAC) channels in human T cells. J Cell Biol 150:1435–1444

Hanano T, Hara Y, Shi J, Morita H, Umebayashi C, Mori E, Sumimoto H, Ito Y, Mori Y, Inoue R (2004) Involvement of TRPM7 in cell growth as a spontaneously activated Ca^{2+} entry pathway in human retinoblastoma cells. J Pharmacol Sci 95:403–419

Hara Y, Wakamori M, Ishii M, Maeno E, Nishida M, Yoshida T, Yamada H, Shimizu S, Mori E, Kudoh J, Shimizu N, Kurose H, Okada Y, Imoto K, Mori Y (2002) LTRPC2 Ca^{2+}-permeable channel activated by changes in redox status confers susceptibility to cell death. Mol Cell 9:163–173

Hermosura MC, Monteilh-Zoller MK, Scharenberg AM, Penner R, Fleig A (2002) Dissociation of the store-operated calcium current ICRAC and the Mg-nucleotide-regulated metal ion current MagNuM. J Physiol 539:445–458

Hermosura MC, Nayakanti H, Dorovkov MV, Calderon FR, Ryazanov AG, Haymer DS, Garruto RM (2005) A TRPM7 variant shows altered sensitivity to magnesium that may contribute to the pathogenesis of two Guamanian neurodegenerative disorders. Proc Natl Acad Sci U S A 102:11510–11515

Jiang J, Li M, Yue L (2005) Potentiation of TRPM7 inward currents by protons. J Gen Physiol 126:137–150

Kerschbaum HH, Cahalan MD (1998) Monovalent permeability, rectification, and ionic block of store-operated calcium channels in Jurkat T lymphocytes. J Gen Physiol 111:521–537

Kerschbaum HH, Cahalan MD (1999) Single-channel recording of a store-operated Ca^{2+} channel in Jurkat T lymphocytes. Science 283:836–839

Kerschbaum HH, Kozak JA, Cahalan MD (2003) Polyvalent cations as permeant probes of MIC and TRPM7 pores. Biophys J 84:2293–2305

Kozak JA, Cahalan MD (2003) MIC channels are inhibited by internal divalent cations but not ATP. Biophys J 84:922–927

Kozak JA, Kerschbaum HH, Cahalan MD (2002) Distinct properties of CRAC and MIC channels in RBL cells. J Gen Physiol 120:221–235

Kozak JA, Matsushita M, Nairn AC, Cahalan MD (2005) Charge screening by internal pH and polyvalent cations as a mechanism for activation, inhibition, and rundown of TRPM7/MIC channels. J Gen Physiol 126:499–514

Lee JM, Grabb MC, Zipfel GJ, Choi DW (2000) Brain tissue responses to ischemia. J Clin Invest 106:723–731

Liu B, Qin F (2005) Functional control of cold- and menthol-sensitive TRPM8 ion channels by phosphatidylinositol 4,5-bisphosphate. J Neurosci 25:1674–1681

Liu D, Liman ER (2003) Intracellular Ca^{2+} and the phospholipid PIP_2 regulate the taste transduction ion channel TRPM5. Proc Natl Acad Sci U S A 100:15160–15165

Matsushita M, Kozak JA, Shimizu Y, McLachlin DT, Yamaguchi H, Wei FY, Tomizawa K, Matsui H, Chait BT, Cahalan MD, Nairn AC (2005) Channel function is dissociated from the intrinsic kinase activity and autophosphorylation of TRPM7/ChaK1. J Biol Chem 280:20793–20803

Maynard CJ, Bush AI, Masters CL, Cappai R, Li QX (2005) Metals and amyloid-beta in Alzheimer's disease. Int J Exp Pathol 86:147–159

Milligan G, Kostenis E (2006) Heterotrimeric G-proteins: a short history. Br J Pharmacol 147 Suppl 1:S46-55

Monteilh-Zoller MK, Hermosura MC, Nadler MJ, Scharenberg AM, Penner R, Fleig A (2003) TRPM7 provides an ion channel mechanism for cellular entry of trace metal ions. J Gen Physiol 121:49–60

Mubagwa K, Stengl M, Flameng W (1997) Extracellular divalent cations block a cation non-selective conductance unrelated to calcium channels in rat cardiac muscle. J Physiol 502:235–247

Nadler MJ, Hermosura MC, Inabe K, Perraud AL, Zhu Q, Stokes AJ, Kurosaki T, Kinet JP, Penner R, Scharenberg AM, Fleig A (2001) LTRPC7 is a MgATP-regulated divalent cation channel required for cell viability. Nature 411:590–595

Nilius B, Mahieu F, Prenen J, Janssens A, Owsianik G, Vennekens R, Voets T (2006) The Ca$^{(2+)}$-activated cation channel TRPM4 is regulated by phosphatidylinositol 4,5-biphosphate. EMBO J 25:467–478

Oancea E, Wolfe JT, Clapham DE (2006) Functional TRPM7 channels accumulate at the plasma membrane in response to fluid flow. Circ Res 98:245–253

Prakriya M, Lewis RS (2002) Separation and characterization of currents through store-operated CRAC channels and Mg^{2+}-inhibited cation (MIC) channels. J Gen Physiol 119:487–507

Rohacs T, Lopes CM, Michailidis I, Logothetis DE (2005) PI(4,5)P2 regulates the activation and desensitization of TRPM8 channels through the TRP domain. Nat Neurosci 8:626–634

Runnels LW, Yue L, Clapham DE (2001) TRP-PLIK, a bifunctional protein with kinase and ion channel activities. Science 291:1043–1047

Runnels LW, Yue L, Clapham DE (2002) The TRPM7 channel is inactivated by PIP$_2$ hydrolysis. Nat Cell Biol 4:329–336

Ryazanova LV, Pavur KS, Petrov AN, Dorovkov MV, Ryazanov AG (2001) Novel type of signaling molecules: protein kinases covalently linked with ion channels. Mol Biol 35:271–283

Ryazanova LV, Dorovkov MV, Ansari A, Ryazanov AG (2004) Characterization of the protein kinase activity of TRPM7/ChaK1, a protein kinase fused to the transient receptor potential ion channel. J Biol Chem 279:3708–3716

Schmitz C, Perraud AL, Johnson CO, Inabe K, Smith MK, Penner R, Kurosaki T, Fleig A, Scharenberg AM (2003) Regulation of vertebrate cellular Mg^{2+} homeostasis by TRPM7. Cell 114:191–200

Su LT, Agapito MA, Li M, W TNS, Huttenlocher A, Habas R, Yue L, Runnels LW (2006) Trpm7 regulates cell adhesion by controlling the calcium dependent protease calpain. J Biol Chem 281:11260–11270

Takezawa R, Schmitz C, Demeuse P, Scharenberg AM, Penner R, Fleig A (2004) Receptor-mediated regulation of the TRPM7 channel through its endogenous protein kinase domain. Proc Natl Acad Sci U S A 101:6009–6014

Teramoto T, Lambie EJ, Iwasaki K (2005) Differential regulation of TRPM channels governs electrolyte homeostasis in the C. elegans intestine. Cell Metab 1:343–354

Xiao RP (2000) Cell logic for dual coupling of a single class of receptors to G(s) and G(i) proteins. Circ Res 87:635–637

Xiong Z, Lu W, MacDonald JF (1997) Extracellular calcium sensed by a novel cation channel in hippocampal neurons. Proc Natl Acad Sci U S A 94:7012–7017

Yamaguchi H, Matsushita M, Nairn AC, Kuriyan J (2001) Crystal structure of the atypical protein kinase domain of a TRP channel with phosphotransferase activity. Mol Cell 7:1047–1057

Zakharov SI, Mongayt DA, Cohen RA, Bolotina VM (1999) Monovalent cation and L-type Ca^{2+} channels participate in calcium paradox-like phenomenon in rabbit aortic smooth muscle cells. J Physiol 514:71–81

Zhang Z, Okawa H, Wang Y, Liman ER (2005) Phosphatidylinositol 4,5-bisphosphate rescues TRPM4 channels from desensitization. J Biol Chem 280:39185–39192

TRPM8

T. Voets (✉) · G. Owsianik · B. Nilius

Laboratory of Physiology, KU Leuven, Campus Gasthuisberg O&N, 3000 Leuven, Belgium
Thomas.voets@med.kuleuven.be

1	Cloning	329
2	Gene Structure and Expression	330
3	Pore Properties	330
4	Gating Mechanisms	331
4.1	Cold	331
4.2	Agonists	332
4.3	Antagonists	336
4.4	Desensitisation: PIP$_2$ and PKC	337
5	(Patho)physiology	338
5.1	Cold Sensing	338
5.2	Tumor Growth?	339
5.3	Calcium Release from Intracellular Stores?	340
6	Perspectives	341
References		341

Abstract Originally cloned as a prostate-specific protein, TRPM8 is now best known as a cold- and menthol-activated channel implicated in thermosensation. In this chapter we provide a brief review of current knowledge concerning the biophysical properties, gating mechanisms, pharmacology and (patho)physiology of this TRP channel.

Keywords Temperature sensing · Voltage-dependent gating

1
Cloning

In 2001, Tsavaler and colleagues screened a human prostate-specific subtracted complementary DNA (cDNA) library to identify novel prostate-specific genes (Tsavaler et al. 2001). They identified a novel gene whose expression was mainly restricted to prostate and up-regulated in prostate cancer and other malignancies. The gene coded for a protein with an estimated molecular weight of 130 kDa and significant homology to known TRP channels, which was baptised TRP-p8. Tsavaler and colleagues proposed that TRP-p8 might be a Ca^{2+}-permeable channel with oncogene or tumour promoter potential (Tsavaler et al. 2001). However, functional expression data were not provided in this initial study. Following the unified nomenclature for TRP channels, TRP-p8 was subsequently classified into the TRPM subfamily and re-baptised TRPM8 (Montell et al. 2002; Montell 2005).

One year later, the Julius and Patapoutian laboratories (re)discovered TRPM8 as a cold-activated 'thermoTRP' channel (McKemy et al. 2002; Peier et al. 2002). McKemy and colleagues used an expression-cloning strategy using a cDNA library from rat trigeminal neurons to identify a receptor for menthol, the refreshing component of the mint plant (McKemy et al. 2002). They identified a cDNA clone that encoded a Ca^{2+}-permeable cation channel activated by cold, and by cooling agents such as menthol, eucalyptol or icilin. This channel, which they termed CMR1 (for cold and menthol receptor 1), turned out to be the rat orthologue of TRP-p8 (TRPM8) (McKemy et al. 2002). Peier and colleagues used a bioinformatics approach to search for thermosensitive channels related to the heat-activated vanilloid receptor TRPV1. They isolated TRPM8 from dorsal root ganglia (DRG) and subsequently heterologously expressed it as a cold- and menthol sensitive cation channel (Peier et al. 2002).

2
Gene Structure and Expression

In humans, the TRPM8 gene is located in chromosome region 2q37.1. The gene spans 102.12 kb and encompasses 25 exons. It transcribes into a messenger RNA (mRNA) coding for a protein of 1,104 amino acids. Analysis of published genome sequences indicates the presence of TRPM8 orthologues in all investigated mammals as well as in chicken and toad.

The precise expression pattern and tissue distribution of TRPM8 is still unclear. In the first study by Tsavaler and colleagues, TRPM8 was proposed to be expressed almost exclusively in the prostate and in a number of non-prostatic primary tumours of breast, colon, lung and skin origin (Tsavaler et al. 2001). However, later studies detected TRPM8 mRNA or protein (or both) in a subset of sensory neurons from DRG and trigeminal ganglia (McKemy et al. 2002; Peier et al. 2002; Nealen et al. 2003), in nodose ganglion cells innervating the upper gut (Zhang et al. 2004), gastric fundus (Mustafa and Oriowo 2005), vascular smooth muscle (Yang et al. 2006), liver (Henshall et al. 2003), and in bladder urothelium and different tissues of the male genital tract (Stein et al. 2004).

3
Pore Properties

Like most other TRP channels, TRPM8 is a non-selective cation channel permeable to both monovalent and divalent cations (McKemy et al. 2002; Peier et al. 2002). The permeability sequence for monovalent cations was reported to be $Cs^+>K^+>Na^+$, which corresponds to Eisenman sequence I or II (weak-field-strength site), but with very little discrimination between these cations (P_{Cs}/P_{Na} ~1.4) (McKemy et al. 2002; Peier et al. 2002). Reported values for the relative permeability for Ca^{2+} versus Na^+ (P_{Ca}/P_{Na}) range from 0.97 to 3.3 (McKemy et al. 2002; Peier et al. 2002). Similar to most other TRP channels, unitary TRPM8 channel events indicate an intermediate single-channel conductance of 40–83 pS (McKemy et al. 2002; Brauchi et al. 2004; Hui et al. 2005).

At present, a detailed structure-function analysis of the TRPM8 pore is not available, but in analogy with related TRP family members it is generally assumed that TRPM8 functions as a tetramer (Hoenderop et al. 2003; Owsianik et al. 2006) and that the loop between transmembrane (TM)5 and TM6 constitutes the outer pore region and selectivity filter (Voets et al. 2004a; Nilius et al. 2005b; Owsianik et al. 2006).

4
Gating Mechanisms

Figure 1 illustrates some of the basic properties of the whole-cell TRPM8 current, and its modulation by temperature and menthol. Below we briefly review recent insights into the gating mechanisms of TRPM8 in response to thermal and chemical stimuli.

4.1
Cold

In studies that describe the temperature sensitivity of thermoTRPs, including TRPM8, very often a thermal threshold for channel activation is reported. The

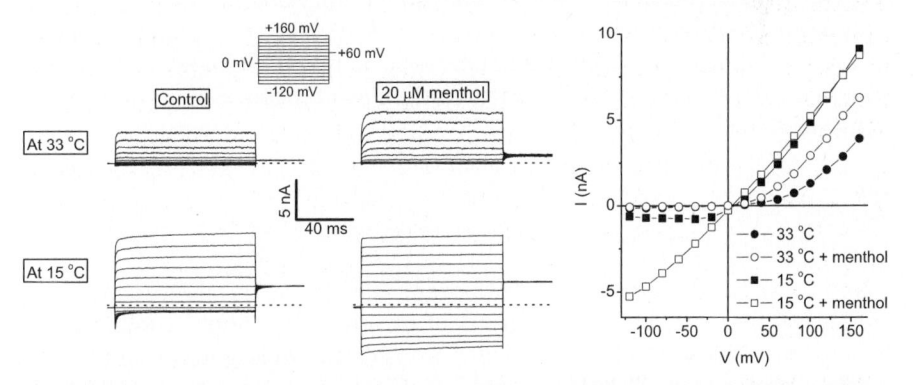

Fig. 1 Effects of temperature and menthol on whole-cell TRPM8 currents. (*Left*) Current traces in response to the indicated voltage protocol at two different temperatures and in the absence and presence of menthol. (*Right*) Steady-state current-voltage relations obtained from these current traces

concept of thermal thresholds originates from studies on native thermoreceptor cells, in which they can be determined unambiguously as the temperature at which the sensory nerve starts firing action potentials. At voltages that correspond to typical resting membrane potentials of sensory neurons (∼−70 mV), TRPM8 starts to carry significant inward current upon cooling below ∼25°C (McKemy et al. 2002), and similar apparent thermal thresholds were obtained in Ca^{2+} imaging experiments (Peier et al. 2002).

However, the temperature sensitivity of TRPM8 is strongly dependent on the transmembrane voltage: at depolarised potentials, TRPM8 activated at much higher temperatures than at more physiological, negative potentials (Fig. 1; see also Brauchi et al. 2004; Voets et al. 2004b). We found that TRPM8 is a voltage-gated channel activated upon membrane depolarisation, comparable to classical voltage-gated K^+, Na^+ and Ca^{2+}, but with a much weaker voltage sensitivity (Voets et al. 2004b). Cooling causes a robust but graded shift of the voltage dependence of activation from strongly depolarised potentials towards the physiological potential range (Voets et al. 2004b). Thus, cold activation of TRPM8 (and other thermoTRPs) should not be regarded as a threshold phenomenon, but rather as a gradual shift of voltage sensitivity.

We found that the temperature dependence of many thermoTRPs, including TRPM8, can be faithfully reproduced using a simple two-state gating model for a voltage-dependent channel (Voets et al. 2004b; Talavera et al. 2005). In the case of TRPM8, the temperature dependence of channel opening is much less steep than that of channel closing, which leads to channel activation upon cooling (Voets et al. 2004b). The opposite is true for heat-activated thermoTRPs such as TRPV1, TRPM4 and TRPM5 (Voets et al. 2004b; Talavera et al. 2005). Figure 2 illustrates the model, its parameters and model predictions, which match very well to the published data (McKemy et al. 2002; Peier et al. 2002; Brauchi et al. 2004; Voets et al. 2004b). Note that a leftward shift of the voltage-dependent activation curve upon cooling implies that channel opening leads to a decrease in entropy, which in the case of TRPM8 amounts to approximately 550 kJ mol^{-1}. In contrast, opening of the heat-activated TRP channels TRPV1, TRPM4 and TRPM5 is associated with a significant increase in entropy in the range of 300–600 kJ mol^{-1}.

4.2
Agonists

Table 1 gives a list of known TRPM8 agonists, their chemical structures and potencies. Best studied are the effects of menthol, the natural cooling compound from the mint plant, and icilin, a synthetic cooling agent and up till now the most potent TRPM8 activator (McKemy et al. 2002; Peier et al. 2002). Several synthetic menthol-derivatives such as Frescolat ML, Coolact P, Cooling agent 10 and WS-3 are also TRPM8 agonists, with EC_{50} values in the micromolar range, comparable to the potency of menthol (Behrendt et al. 2004).

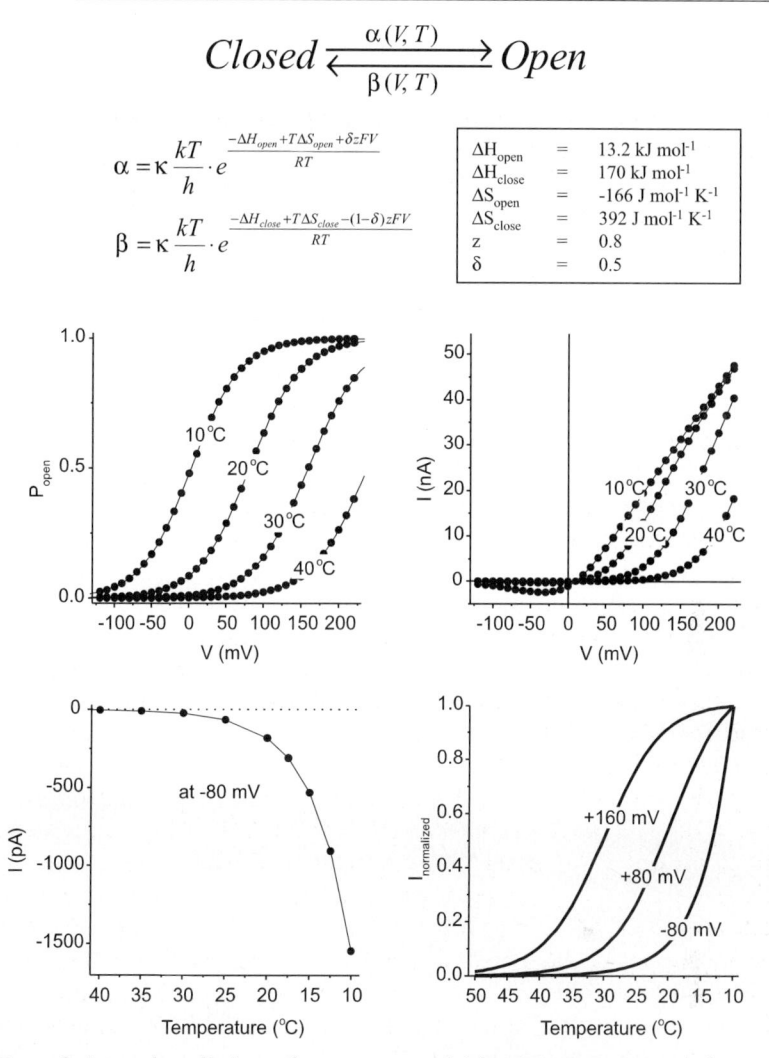

$$Closed \underset{\beta(V,T)}{\overset{\alpha(V,T)}{\rightleftarrows}} Open$$

$$\alpha = \kappa \frac{kT}{h} \cdot e^{\frac{-\Delta H_{open} + T\Delta S_{open} + \delta zFV}{RT}}$$

$$\beta = \kappa \frac{kT}{h} \cdot e^{\frac{-\Delta H_{close} + T\Delta S_{close} - (1-\delta)zFV}{RT}}$$

ΔH_{open}	=	13.2 kJ mol^{-1}
ΔH_{close}	=	170 kJ mol^{-1}
ΔS_{open}	=	-166 J mol^{-1} K^{-1}
ΔS_{close}	=	392 J mol^{-1} K^{-1}
z	=	0.8
δ	=	0.5

Fig. 2 Formulation and predictions of a two-state model for TRPM8. For more information, see Voets et al. (2004b)

Other structurally unrelated, naturally occurring cooling substances such as eucalyptol, linalool and geraniol are less potent TRPM8 agonists, with EC$_{50}$ values in the millimolar range (McKemy et al. 2002; Behrendt et al. 2004). Many of these compounds are used in the food, pharmaceutical and cosmetic industries, not only because of their cooling effect, but in many case also for their pleasant scents, which range from minty (menthol and derivatives), to camphor-like (eucalyptol) and flowery (linalool, hydroxy-citronellal).

We found that menthol activates TRPM8 in a membrane-delimited manner by shifting the voltage dependence of activation towards physiological voltages,

Table 1 Chemical agonists and antagonists of TRPM8

Compounds	Structure	Concentration range[a]
Agonists		
Icilin		Nanomolar
Frescolat ML		Micromolar
WS-3		Micromolar
Menthol		Micromolar
Frescolat MGA		Micromolar
Cooling agent 10		Micromolar
PMD-38		Micromolar
WS-23		Micromolar
Coolact P		Micromolar

Table 1 continued

Compounds	Structure	Concentration range[a]
PIP$_2$		Micromolar
Geraniol		Millimolar
Linalool		Millimolar
Eucalyptol		Millimolar
Hydroxy-citronellal		Millimolar
Antagonists		
BCTC		Micromolar
Thio-BCTC		Micromolar
2-APB		Micromolar

Table 1 continued

Compounds	Structure	Concentration range[a]
Capsazepine		Micromolar

APB, 2-aminoethoxydiphenyl borate; BCTC, N-(4-tertiarybutylphenyl)-4-(3-chloropyridin-2-yl) tetrahydropyrazine-1(2H)-carboxamide; PMD-38, p-menthane-3,8-diol [a]Agonists and antagonists are sorted following estimated potency, with the most potent (ant)agonists first. As EC_{50} and IC_{50} values strongly depend on the used techniques (Ca^{2+}-imaging, patch-clamp) and the experimental conditions (temperature, voltage) we only provide an indicative concentration range

comparable to the effect of cooling (Voets et al. 2004b). Moreover, the effects of cold and menthol on the voltage-dependent activation curve were additive (Voets et al. 2004b), which explains why low doses of menthol only increase the cold sensitivity of inward TRPM8 currents, whereas higher concentrations are able to induce inward currents even at 37°C (McKemy et al. 2002; Peier et al. 2002). Thus, strictly speaking, menthol should be considered as a gating modifier rather than an activator of TRPM8. Whether other chemical agonists of TRPM8 have a similar effect on the channel's voltage dependence is likely but not proven.

Chuang and colleagues (2004) have studied the mode of action of icilin on TRPM8 in more detail. They found that icilin, unlike cold or menthol, can only activate TRPM8 when intracellular Ca^{2+} is elevated. As such, TRPM8 can act as a coincidence detector, responding only when a chemical stimulus (i.e. icilin) and a stimulus that increases intracellular Ca^{2+} (e.g. receptor activation) are paired. Moreover, they found the chicken TRPM8 orthologue to be cold- and menthol-sensitive but icilin-insensitive. Using rat-chicken TRPM8 chimeras and point mutations they were able to find critical residues for icilin activation in TM3 and the TM2–TM3 linker (Chuang et al. 2004). Interestingly, the same region is also involved in the activation of TRPV1 by vanilloid compounds (Jordt and Julius 2002; Gavva et al. 2004) and of TRPV4 by 4α-PDD (Vriens et al. 2004), suggesting a conserved mechanism for gating of TRP channels by chemical agonists.

4.3
Antagonists

Currently, the most potent inhibitors of the channel are N-(4-tertiarybutylphenyl)-4-(3-chloropyridin-2-yl) tetrahydropyrazine-1(2H)-carboxamide (BCTC) and its derivative thio-BCTC, which inhibit menthol-induced Ca^{2+} responses with IC_{50} values of 800 nM and 3.5 µM, respectively (Behrendt et al. 2004). However, BCTC and thio-BCTC are also known to inhibit TRPV1 at significantly lower concentrations (IC_{50} values of 35 nM and 55 nM, respectively) (Valenzano et al. 2003; Behrendt et al. 2004). Likewise, the competitive TRPV1-antagonist capsazepine inhibits menthol-induced Ca^{2+} responses in TRPM8-expressing cells with an IC_{50} of 18 µM (Behrendt et al. 2004), a potency that is one order of magnitude lower than what has been reported for TRPV1 (Jerman et al. 2000).

Another TRPM8 antagonist is 2-aminoethoxydiphenyl borate (2-APB), which rapidly and reversibly blocks menthol-induced TRPM8 currents with an IC_{50} of approximately 10 µM (Hu et al. 2004). However, 2-APB is notoriously unspecific. Initially, it was put forward as a membrane-permeable inhibitor of inositol 1,4,5-trisphosphate receptors (Maruyama et al. 1997), but later studies have shown that it can also directly block native store-operated channels (Iwasaki et al. 2001), sarco/endoplasmic reticulum Ca^{2+}-ATPase pumps (Bilmen et al. 2002), the mitochondrial permeability transition pore (Missiaen et al. 2001) and a few other ion channels including several TRP channels of the TRPC subfamily (Hu et al. 2004). Moreover, 2-APB was shown to activate and/or sensitise TRPV1, TRPV2, TRPV3 and TRPV6 (Voets et al. 2001; Chung et al. 2004; Hu et al. 2004).

In addition to these non-specific drugs, TRPM8 activity is also inhibited by acidic pH, although there are some conflicting data concerning the mechanism. One study indicated that protons inhibit cold-induced and icilin-induced TRPM8 activation, whereas the responses to menthol were pH insensitive (Andersson et al. 2004). However, in our hands acidic pH also inhibits the TRPM8 response to menthol, in line with the results from Behrendt and colleagues (Behrendt et al. 2004).

Clearly, highly specific antagonists for TRPM8, which would be highly instrumental in dissecting the physiological role of TRPM8, remain to be discovered. One possibility to obtain a specific TRPM8 inhibitor might be to raise antibodies against the N-terminal part of the extracellular loop between TM5 and TM6. Such antibodies have recently been shown to potently inhibit other channels with six transmembrane domains and to exhibit high specificity between closely homologous channels (Xu et al. 2005).

4.4
Desensitisation: PIP$_2$ and PKC

Activation of TRPM8 by cold, menthol or icilin is followed by significant channel desensitisation (McKemy et al. 2002; Peier et al. 2002), and this desensitisation depends on the presence of Ca^{2+} in the extracellular medium (McKemy et al. 2002). Recently, Rohacs and colleagues (2005) provided evidence for an involvement of phosphatidylinositol 4,5-bisphosphate (PIP_2) in this process. PIP_2 acts as a positive modulator of the channel's sensitivity for cold or menthol and prevent s channel rundown in cell-free patches (Liu and Qin 2005; Rohacs et al.

2005), most likely by shifting the voltage-sensitivity of activation towards physiological voltages (Rohacs et al. 2005). Rohacs and colleagues proposed that Ca^{2+} influx through TRPM8 leads to activation of Ca^{2+}-dependent phospholipase C (e.g. $PLC_{\delta 1}$), leading to depletion of cellular PIP_2 levels and channel desensitisation (Rohacs et al. 2005). They also provided evidence that positively charged residues in the TRP domain, which is located C-terminal of TM6, participate in the interaction of the channel with membrane-bound PIP_2 (Rohacs et al. 2005).

An alternative or complementary pathway for TRPM8 desensitisation occurs via activation of Ca^{2+}-dependent protein kinase C (PKC) (Abe et al. 2005; Premkumar et al. 2005). Activation of PKC indirectly causes dephosphorylation of TRPM8, leading to down-regulation of TRPM8 activity (Premkumar et al. 2005).

The sensitivity of TRPM8 to PKC and PIP_2 suggest that in vivo TRPM8 is highly sensitive to stimulation of PLC-coupled receptors. Indeed, activation of PLC would have a dual inhibitory effect on the channel, via a reduction of cellular PIP_2 levels and via a diacyl glycerol-induced activation of PKC. Interestingly, the effects of PIP_2 and PKC activation on TRPM8 are opposite of what has been described for the heat-activated TRPV1, where both PIP_2 depletion and PKC activation sensitise the channel to heat and ligands (Premkumar and Ahern 2000; Chuang et al. 2001; Prescott and Julius 2003). Several inflammatory mediators lead to PIP_2 depletion and PKC activation, which has been shown to cause heat hyperalgesia by sensitising TRPV1 (Caterina et al. 2000; Davis et al. 2000; Julius and Basbaum 2001). Heat hypersensitivity may, however, be further aggravated by down-regulation of TRPM8, as its activation may provide a cool and soothing sensation.

5
(Patho)physiology

5.1
Cold Sensing

The most obvious role attributed to TRPM8 is that of a cold sensor in the somatosensory system. In mammals, cold sensation is thought to be mediated by a subset of Aδ and C afferent sensory fibres (Julius and Basbaum 2001; Patapoutian et al. 2003). Several non-mutually exclusive mechanisms have been proposed to explain cold-induced depolarisation of these cold-sensitive neurons, including inhibition of a Na^+/K^+-ATPase (Reid and Flonta 2001b), blockade of background K^+ conductances (Reid and Flonta 2001b; Viana et al. 2002) or activation of a cold-dependent depolarising current (Reid and Flonta 2001a). Reid and Flonta (2001a) were the first to directly demonstrate cold-induced activation of a non-selective cation channel in a subset of DRG neurons. This current was activated between 23°C and 34°C, potentiated by menthol and carried by a mixture of cations. These authors proposed that this cold current is the principal determinant of cold-receptor activity, and that inhibition of background K^+ currents or of the Na^+/K^+-ATPase are of secondary importance (Reid and Flonta 2001a).

When reporting the expression-cloning of TRPM8 from trigeminal neurons, McKemy and colleagues (2002) also included a comparison between the current in cells heterologously expressing TRPM8 and the endogenous cold-activated current in trigeminal neurons and found that both currents were identical with respect to their ionic selectivity, cold- and menthol-sensitivity, and outwardly rectifying current-voltage relationship. Based on these results, it seems safe to predict that TRPM8 is the molecular correlate of the cold-induced current in the cold thermosensory cells, although analysis of a TRPM8 knock-out mouse is required to further substantiate this notion.

Whereas a role for TRPM8 as a cold-activated depolarising channel in cold-sensitive neurons is highly probable, cumulating evidence points at the existence of additional cold-activated channels in these cells. Cold-sensitive DRG neurons can be subdivided into menthol-sensitive and menthol-insensitive groups, with the latter group requiring stronger cooling for activation (Nealen et al. 2003; Babes et al. 2004). These data suggest the existence of a menthol-insensitive cold-activated channel with a sensitivity that is more in the noxious cold range. These properties are reminiscent of heterologously expressed TRPA1, which forms a menthol-insensitive channel activated by temperatures below 18°C (Story et al. 2003). However, menthol-insensitive neurons also did not respond to known chemical TRPA1 activators such as icilin or mustard oil, which may point to the existence of additional cold sensing (TRP) channels and/or of heteromultimeric TRP channels with different thermal and ligand sensitivities than the homomultimeric channels (Babes et al. 2004). Finally, it should be noted that the behaviour and sensitivity of an intact cold receptor depends not only on cold-sensitive conductances but also on the other ion channels and transporters, which can amplify or dampen the effect of the thermosensitive channel (Viana et al. 2002; Reid 2005).

A recent report provides intriguing evidence for a role of TRPM8 in bladder function (Tsukimi et al. 2005). It was shown that filling of the bladder with cold water induces a detrusor reflex and voiding in patients with supraspinal lesions, whereas no such reflex was observed in patients with peripheral lesions or in normal subjects. The cold reflex depends on stimulation of sensory C fibres in the bladder, most likely via activation of TRPM8. In line with a role for TRPM8 in bladder function, the cold-induced detrusor reflex was shown to be significantly sensitised by menthol (Tsukimi et al. 2005).

5.2
Tumor Growth?

TRPM8 is highly expressed in several types of tumour cells (Tsavaler et al. 2001). In particular, TRPM8 expression is up-regulated in prostate cancer, but the expression is almost lost in the transition to androgen independence (Henshall et al. 2003; Bidaux et al. 2005). As such, TRPM8 represents a novel biomarker in prostate cancer (Tsavaler et al. 2001; Fuessel et al. 2003). Moreover, the correlation between TRPM8 expression and the disease stage suggests that the channel may be involved in the survival and/or growth of tumour cells (Henshall et al. 2003). Zhang and Barritt (2004) presented some evidence for such a pathophysiological role using the androgen-responsive prostate cancer cell line LNCaP. They found that both stimulation (using menthol) and inhibition (using capsazepine or siRNA) of TRPM8 reduced the viability of these cells. These results suggest that some TRPM8 activity is beneficial for survival of these cancer cells, but at the same time, overstimulation of the channel can become toxic for the cells, probably due to calcium-overload (Zhang and Barritt 2004). Such a dual role in cell growth and cell survival is reminiscent to what has been described for the closely related channel TRPM7. Indeed, inhibition of TRPM7 expression blocks proliferation and leads to cell death, which has been attributed to intracellular Mg^{2+}-deficiency, but over-expression of TRPM7 is also lethal to cells (Nadler et al. 2001; Schmitz et al. 2003).

5.3
Calcium Release from Intracellular Stores?

Two recent reports have presented results suggesting that TRPM8 may also act as an intracellular Ca^{2+} release channel. In the first study, Zhang and Barritt (2004) used immunofluorescence to demonstrate that about half of TRPM8 protein is present in the plasma membrane and the other half in the endoplasmic reticulum of LNCaP cells. Moreover, they found that, even in the absence of extracellular Ca^{2+}, cooling and high concentrations of menthol (500 µM) caused a small but significant increase in intracellular Ca^{2+} in LNCaP cells (Zhang and Barritt 2004). In the second study, Thebault and colleagues (2005) reported that in the same LNCaP cell line, TRPM8 is fully absent from the plasma membrane, and entirely restricted to intracellular membranes. These authors also reported that menthol induces an inward current in LNCaP cells, which is not carried by TRPM8 but rather by store-operated channels activated due to menthol-induced store depletion (Thebault et al. 2005).

Although these results are certainly intriguing, more evidence is needed to firmly establish the role of TRPM8 as an intracellular Ca^{2+} release channel, and some discrepancies need to be resolved. For example, the effects of menthol on intracellular Ca^{2+} release are quite modest, relatively slow and require higher menthol concentrations than for activation of plasma membrane TRPM8. Moreover, the two studies are in disagreement with respect to the subcellular localisation of TRPM8 in LNCaP cells, which raises some doubts about the specificity of the used antibodies. Finally, the mechanisms determining whether TRPM8 is located in the plasma membrane or in intracellular membranes are completely unknown.

6
Perspectives

At present, there are strong arguments supporting a major role for TRPM8 in the sensation of cold and cooling substances. The functions of TRPM8 outside the sensory system are much less obvious. Clearly, a better understanding of its role in the pathophysiology of (prostate) cancer may reveal whether TRPM8 is a potential new target for anti-cancer treatments (Nilius et al. 2005a). Better tools such as potent and specific TRPM8 antagonists are mandatory for such progress. In addition, the normal physiological role of TRPM8 in healthy tissues that do not normally experience considerable fluctuations in temperature is still puzzling. It will be highly interesting to learn whether TRPM8-deficient mice exhibit defects unrelated to thermosensation.

On a more fundamental level, important questions in future research relate to the nature and structure of the voltage sensor and ligand binding sites. Detailed structure-function analyses and hopefully detailed crystallographic information are expected to provide novel insight into the modus operandi of this fascinating biological thermometer.

References

Abe J, Hosokawa H, Sawada Y, Matsumura K, Kobayashi S (2005) Ca(2+)-dependent PKC activation mediates menthol-induced desensitization of transient receptor potential M8. Neurosci Lett 397:140–144

Andersson DA, Chase HW, Bevan S (2004) TRPM8 activation by menthol, icilin, and cold is differentially modulated by intracellular pH. J Neurosci 24:5364–5369

Babes A, Zorzon D, Reid G (2004) Two populations of cold-sensitive neurons in rat dorsal root ganglia and their modulation by nerve growth factor. Eur J Neurosci 20:2276–2282

Behrendt HJ, Germann T, Gillen C, Hatt H, Jostock R (2004) Characterization of the mouse cold-menthol receptor TRPM8 and vanilloid receptor type-1 VR1 using a fluorometric imaging plate reader (FLIPR) assay. Br J Pharmacol 141:737–745

Bidaux G, Roudbaraki M, Merle C, Crepin A, Delcourt P, Slomianny C, Thebault S, Bonnal JL, Benahmed M, Cabon F, Mauroy B, Prevarskaya N (2005) Evidence for specific TRPM8 expression in human prostate secretory epithelial cells: functional androgen receptor requirement. Endocr Relat Cancer 12:367–382

Bilmen JG, Wootton LL, Godfrey RE, Smart OS, Michelangeli F (2002) Inhibition of SERCA Ca^{2+} pumps by 2-aminoethoxydiphenyl borate (2-APB). 2-APB reduces both Ca^{2+} binding and phosphoryl transfer from ATP, by interfering with the pathway leading to the Ca^{2+}-binding sites. Eur J Biochem 269:3678–3687

Brauchi S, Orio P, Latorre R (2004) Clues to understanding cold sensation: thermodynamics and electrophysiological analysis of the cold receptor TRPM8. Proc Natl Acad Sci U S A 101:15494–15499

Caterina MJ, Leffler A, Malmberg AB, Martin WJ, Trafton J, Petersen-Zeitz KR, Koltzenburg M, Basbaum AI, Julius D (2000) Impaired nociception and pain sensation in mice lacking the capsaicin receptor. Science 288:306–313

Chuang HH, Prescott ED, Kong H, Shields S, Jordt SE, Basbaum AI, Chao MV, Julius D (2001) Bradykinin and nerve growth factor release the capsaicin receptor from PtdIns(4,5)P2-mediated inhibition. Nature 411:957–962

Chuang HH, Neuhausser WM, Julius D (2004) The super-cooling agent icilin reveals a mechanism of coincidence detection by a temperature-sensitive TRP channel. Neuron 43:859–869

Chung MK, Lee H, Mizuno A, Suzuki M, Caterina MJ (2004) 2-Aminoethoxydiphenyl borate activates and sensitizes the heat-gated ion channel TRPV3. J Neurosci 24:5177–5182

Davis JB, Gray J, Gunthorpe MJ, Hatcher JP, Davey PT, Overend P, Harries MH, Latcham J, Clapham C, Atkinson K, Hughes SA, Rance K, Grau E, Harper AJ, Pugh PL, Rogers DC, Bingham S, Randall A, Sheardown SA (2000) Vanilloid receptor-1 is essential for inflammatory thermal hyperalgesia. Nature 405:183–187

Fuessel S, Sickert D, Meye A, Klenk U, Schmidt U, Schmitz M, Rost AK, Weigle B, Kiessling A, Wirth MP (2003) Multiple tumor marker analyses (PSA, hK2, PSCA, trp-p8) in primary prostate cancers using quantitative RT-PCR. Int J Oncol 23:221–228

Gavva NR, Klionsky L, Qu Y, Shi L, Tamir R, Edenson S, Zhang TJ, Viswanadhan VN, Toth A, Pearce LV, Vanderah TW, Porreca F, Blumberg PM, Lile J, Sun Y, Wild K, Louis JC, Treanor JJ (2004) Molecular determinants of vanilloid sensitivity in TRPV1. J Biol Chem 279:20283–20295

Henshall SM, Afar DE, Hiller J, Horvath LG, Quinn DI, Rasiah KK, Gish K, Willhite D, Kench JG, Gardiner-Garden M, Stricker PD, Scher HI, Grygiel JJ, Agus DB, Mack DH, Sutherland RL (2003) Survival analysis of genome-wide gene expression profiles of prostate cancers identifies new prognostic targets of disease relapse. Cancer Res 63:4196–4203

Hoenderop JGJ, Voets T, Hoefs S, Weidema F, Prenen J, Nilius B, Bindels RJM (2003) Homo- and heterotetrameric architecture of the epithelial Ca^{2+} channels, TRPV5 and TRPV6. EMBO J 22:776–785

Hu HZ, Gu Q, Wang C, Colton CK, Tang J, Kinoshita-Kawada M, Lee LY, Wood JD, Zhu MX (2004) 2-Aminoethoxydiphenyl borate is a common activator of TRPV1, TRPV2, and TRPV3. J Biol Chem 279:35741–35748

Hui K, Guo Y, Feng ZP (2005) Biophysical properties of menthol-activated cold receptor TRPM8 channels. Biochem Biophys Res Commun 333:374–382

Iwasaki H, Mori Y, Hara Y, Uchida K, Zhou H, Mikoshiba K (2001) 2-Aminoethoxydiphenyl borate (2-APB) inhibits capacitative calcium entry independently of the function of inositol 1,4,5-trisphosphate receptors. Receptors Channels 7:429–439

Jerman JC, Brough SJ, Prinjha R, Harries MH, Davis JB, Smart D (2000) Characterization using FLIPR of rat vanilloid receptor (rVR1) pharmacology. Br J Pharmacol 130:916–922

Jordt SE, Julius D (2002) Molecular basis for species-specific sensitivity to "hot" chili peppers. Cell 108:421–430

Julius D, Basbaum AI (2001) Molecular mechanisms of nociception. Nature 413:203–210

Liu B, Qin F (2005) Functional control of cold- and menthol-sensitive TRPM8 ion channels by phosphatidylinositol 4,5-bisphosphate. J Neurosci 25:1674–1681

Maruyama T, Kanaji T, Nakade S, Kanno T, Mikoshiba K (1997) 2APB, 2-aminoethoxydiphenyl borate, a membrane-penetrable modulator of Ins(1,4,5)P3-induced Ca^{2+} release. J Biochem (Tokyo) 122:498–505

McKemy DD, Neuhausser WM, Julius D (2002) Identification of a cold receptor reveals a general role for TRP channels in thermosensation. Nature 416:52–58

Missiaen L, Callewaert G, De Smedt H, Parys JB (2001) 2-Aminoethoxydiphenyl borate affects the inositol 1,4,5-trisphosphate receptor, the intracellular Ca^{2+} pump and the non-specific Ca^{2+} leak from the non-mitochondrial Ca^{2+} stores in permeabilized A7r5 cells. Cell Calcium 29:111–116

Montell C (2005) The TRP superfamily of cation channels. Sci STKE 2005:re3

Montell C, Birnbaumer L, Flockerzi V, Bindels RJ, Bruford EA, Caterina MJ, Clapham DE, Harteneck C, Heller S, Julius D, Kojima I, Mori Y, Penner R, Prawitt D, Scharenberg AM, Schultz G, Shimizu N, Zhu MX (2002) A unified nomenclature for the superfamily of TRP cation channels. Mol Cell 9:229–231

Mustafa S, Oriowo M (2005) Cooling-induced contraction of the rat gastric fundus: mediation via transient receptor potential (TRP) cation channel TRPM8 receptor and Rho-kinase activation. Clin Exp Pharmacol Physiol 32:832–838

Nadler MJ, Hermosura MC, Inabe K, Perraud AL, Zhu Q, Stokes AJ, Kurosaki T, Kinet JP, Penner R, Scharenberg AM, Fleig A (2001) LTRPC7 is a Mg.ATP-regulated divalent cation channel required for cell viability. Nature 411:590–595

Nealen ML, Gold MS, Thut PD, Caterina MJ (2003) TRPM8 mRNA is expressed in a subset of cold-responsive trigeminal neurons from rat. J Neurophysiol 90:515–520

Nilius B, Voets T, Peters J (2005a) TRP channels in disease. Sci STKE 2005:re8

Nilius B, Prenen J, Janssens A, Owsianik G, Wang C, Zhu MX, Voets T (2005b) The selectivity filter of the cation channel TRPM4. J Biol Chem 280:22899–22906

Owsianik G, Talavera K, Voets T, Nilius B (2006) Permeation and selectivity of TRP channels. Annu Rev Physiol 68:685–717

Patapoutian A, Peier AM, Story GM, Viswanath V (2003) ThermoTRP channels and beyond: mechanisms of temperature sensation. Nat Rev Neurosci 4:529–539

Peier AM, Moqrich A, Hergarden AC, Reeve AJ, Andersson DA, Story GM, Earley TJ, Dragoni I, McIntyre P, Bevan S, Patapoutian A (2002) A TRP channel that senses cold stimuli and menthol. Cell 108:705–715

Premkumar LS, Ahern GP (2000) Induction of vanilloid receptor channel activity by protein kinase C. Nature 408:985–990

Premkumar LS, Raisinghani M, Pingle SC, Long C, Pimentel F (2005) Downregulation of transient receptor potential melastatin 8 by protein kinase C-mediated dephosphorylation. J Neurosci 25:11322–11329

Prescott ED, Julius D (2003) A modular PIP2 binding site as a determinant of capsaicin receptor sensitivity. Science 300:1284–1288

Reid G (2005) ThermoTRP channels and cold sensing: what are they really up to? Pflugers Arch 451:250–263

Reid G, Flonta M (2001b) Cold transduction by inhibition of a background potassium conductance in rat primary sensory neurones. Neurosci Lett 297:171–174

Reid G, Flonta ML (2001a) Physiology. Cold current in thermoreceptive neurons. Nature 413:480

Rohacs T, Lopes CM, Michailidis I, Logothetis DE (2005) PI(4,5)P2 regulates the activation and desensitization of TRPM8 channels through the TRP domain. Nat Neurosci 8:626–634

Schmitz C, Perraud AL, Johnson CO, Inabe K, Smith MK, Penner R, Kurosaki T, Fleig A, Scharenberg AM (2003) Regulation of vertebrate cellular Mg^{2+} homeostasis by TRPM7. Cell 114:191–200

Stein RJ, Santos S, Nagatomi J, Hayashi Y, Minnery BS, Xavier M, Patel AS, Nelson JB, Futrell WJ, Yoshimura N, Chancellor MB, De Miguel F (2004) Cool (TRPM8) and hot (TRPV1) receptors in the bladder and male genital tract. J Urol 172:1175–1178

Story GM, Peier AM, Reeve AJ, Eid SR, Mosbacher J, Hricik TR, Earley TJ, Hergarden AC, Andersson DA, Hwang SW, McIntyre P, Jegla T, Bevan S, Patapoutian A (2003) ANKTM1, a TRP-like channel expressed in nociceptive neurons, is activated by cold temperatures. Cell 112:819–829

Talavera K, Yasumatsu K, Voets T, Droogmans G, Shigemura N, Ninomiya Y, Margolskee RF, Nilius B (2005) Heat activation of TRPM5 underlies thermal sensitivity of sweet taste. Nature 438:1022–1025

Thebault S, Lemonnier L, Bidaux G, Flourakis M, Bavencoffe A, Gordienko D, Roudbaraki M, Delcourt P, Panchin Y, Shuba Y, Skryma R, Prevarskaya N (2005) Novel role of cold/menthol-sensitive transient receptor potential melastatine family member 8 (TRPM8) in the activation of store-operated channels in LNCaP human prostate cancer epithelial cells. J Biol Chem 280:39423–39435

Tsavaler L, Shapero MH, Morkowski S, Laus R (2001) Trp-p8, a novel prostate-specific gene, is up-regulated in prostate cancer and other malignancies and shares high homology with transient receptor potential calcium channel proteins. Cancer Res 61:3760–3769

Tsukimi Y, Mizuyachi K, Yamasaki T, Niki T, Hayashi F (2005) Cold response of the bladder in guinea pig: involvement of transient receptor potential channel, TRPM8. Urology 65:406–410

Valenzano KJ, Grant ER, Wu G, Hachicha M, Schmid L, Tafesse L, Sun Q, Rotshteyn Y, Francis J, Limberis J, Malik S, Whittemore ER, Hodges D (2003) N-(4-tertiarybutylphenyl)-4-(3-chloropyridin-2-yl)tetrahydropyrazine-1(2H)-carbox-amide (BCTC), a novel, orally effective vanilloid receptor 1 antagonist with analgesic properties. I. In vitro characterization and pharmacokinetic properties. J Pharmacol Exp Ther 306:377–386

Viana F, de la Pena E, Belmonte C (2002) Specificity of cold thermotransduction is determined by differential ionic channel expression. Nat Neurosci 5:254–260

Voets T, Prenen J, Fleig A, Vennekens R, Watanabe H, Hoenderop JG, Bindels RJ, Droogmans G, Penner R, Nilius B (2001) CaT1 and the calcium release-activated calcium channel manifest distinct pore properties. J Biol Chem 276:47767–47770

Voets T, Janssens A, Droogmans G, Nilius B (2004a) Outer pore architecture of a Ca^{2+}-selective TRP channel. J Biol Chem 279:15223–15230

Voets T, Droogmans G, Wissenbach U, Janssens A, Flockerzi V, Nilius B (2004b) The principle of temperature-dependent gating in cold- and heat-sensitive TRP channels. Nature 430:748–754

Vriens J, Watanabe H, Janssens A, Droogmans G, Voets T, Nilius B (2004) Cell swelling, heat, and chemical agonists use distinct pathways for the activation of the cation channel TRPV4. Proc Natl Acad Sci U S A 101:396–401

Xu SZ, Zeng F, Lei M, Li J, Gao B, Xiong C, Sivaprasadarao A, Beech DJ (2005) Generation of functional ion-channel tools by E3 targeting. Nat Biotechnol 23:1289–1293

Yang XR, Lin MJ, McIntosh LS, Sham JS (2006) Functional expression of transient receptor potential melastatin- (TRPM) and vanilloid-related (TRPV) channels in pulmonary arterial and aortic smooth muscle. Am J Physiol Lung Cell Mol Physiol 290:L1267–L1276

Zhang L, Barritt GJ (2004) Evidence that TRPM8 is an androgen-dependent Ca^{2+} channel required for the survival of prostate cancer cells. Cancer Res 64:8365–8373

Zhang L, Jones S, Brody K, Costa M, Brookes SJ (2004) Thermosensitive transient receptor potential channels in vagal afferent neurons of the mouse. Am J Physiol Gastrointest Liver Physiol 286:G983–991

Part IV
Other TRP Channels

TRPA1

J. García-Añoveros[1] (✉) · K. Nagata[2]

[1] Departments of Anesthesiology, Physiology, and Neurology, Northwestern University Institute for Neuroscience, Feinberg School of Medicine,
Ward 10-070, 303 E Chicago Ave., Chicago IL, 60611, USA
anoveros@northwestern.edu

[2] Department of Anesthesiology, Northwestern University Feinberg School of Medicine,
Ward 10-070, 303 E Chicago Ave., Chicago IL, 60611, USA

1	Basic Features of the Gene, the Cloned cDNAs and the Protein	348
2	Expression Pattern	348
3	Activation of TRPA1 Channels	350
3.1	Pungent Compounds and Environmental Irritants	350
3.2	Bradykinin	353
3.3	Noxious Cold	353
3.4	Mechanical Force	354
4	Pharmacology	354
5	Ion Channel Properties	355
6	Biological Relevance and Emerging/Established Biological Roles for TRPA1	357
References		361

Abstract The TRPA1 protein has up to 18 N-terminal and presumed cytoplasmic ankyrin repeats followed by the six membrane spanning and single pore-loop domains characteristic of all TRPs. In mice, TRPA1 is almost exclusively expressed in nociceptive neurons of peripheral ganglia and in all the mechanosensory epithelia of inner ear. In nociceptive neurons, TRPA1 mediates the response to the proalgesic bradykinin as well as the response to pungent irritants found in mustards and garlic, and probably also to those found in cinnamon and tear gas. The channel properties of TRPA1 are discussed and compared to those of sensory transducers. TRPA1 is well conserved across the animal kingdom, with likely orthologs from human to nematode, which suggest an ancestral role for this channel, probably in sensation.

Keywords Nociceptor · Hyperalgesia · Mechanotransduction · Inner ear · Support cell · Hair cell · Channel · Sensory transduction · Pain · Hearing · Mustard oil · Allyl isothiocyanate · AITC · Icilin · A-G-35 · Gadolinium · Amiloride · Ruthenium red · Gentamicin · Calcium · Channel potentiation · Inactivation · Desensitization · Adaptation

1
Basic Features of the Gene, the Cloned cDNAs and the Protein

TRPA1, previously called ANKTM1 and P120, is a large protein of 1,119 amino acids in human (1,125 in rat and mouse, 1,115 and 1,120 in zebrafish, 1,197 in fly, 1,193 in worm) with a calculated molecular weight of 127.4 kDa, although in Western blots it often runs with an altered mobility, appearing larger. Like all other TRP proteins and many other channels, TRPA1 has six predicted membrane-spanning domains (S1 to S6) and presumed pore loop between S5 and S6. In the four short loops predicted to be extracellular there are two potential sites for N-linked glycosylation. Because there is no apparent signal peptide at the N-terminus, both the N- and C-termini are predicted to be cytoplasmic. A distinguishing feature of TRPA1 is its long N-terminal region with up to 18 predicted ankyrin repeats. Most other TRPs have zero to eight ankyrin repeats, and the only other TRP protein with more than that is the mechanosensory channel TRPN1 (29 ankyrin repeats) of flies (Walker et al. 2000), nematodes (Walker et al. 2000; Li et al. 2006), and fish (Sidi et al. 2003), but not found in mammalian genomes. Ankyrin repeats, each 33 amino acids long, have long been implicated in protein–protein interactions, although more recently they were proposed (cited as V. Bennet, personal communication in Corey et al. 2004; Howard and Bechstedt 2004; Sotomayor et al. 2005), and then found (Lee et al. 2006), to provide elasticity and make molecular springs.

The genes encoding TRPA1 in mammals are large, around 50 kb, and have 27 exons. The TRPA1 gene is in human chromosome 8q13 (Jaquemar et al. 1999) and the syntenic region of mouse chromosome 1. Nearby in the human chromosome, at 13.3, there is a region with very high homology to the TRPA1 gene called LOC392232. It is probably a human-specific pseudogene, as it has numerous mutations, and a nonsense stop at codon 74, in the second exon.

Phylogenetic comparisons of TRP channels from different species reveals the existence of one potential ortholog (two in zebrafish) in species as diverse as *Caenorhabditis elegans* (Duggan et al. 2000). All these TRPA1 proteins bear the same arrangement of protein domains (up to 18 predicted ankyrin repeats followed by six transmembranes and a p-loop between the last two). This remarkable conservation in TRPA1 proteins suggests that their function has been conserved for over 600 million years of evolution.

2
Expression Pattern

By Northern blot, TRPA1 messenger RNA (mRNA) was detected at very low levels in human fibroblasts and found to be lost upon oncogenic transformation. However, its expression in most tissues may be very low or negligible, as it was not detected with Northern blots from several adult or fetal human organs (Jaquemar et al. 1999). In addition, by in situ hybridization TRPA1 mRNA was not detected in most organs of late embryonic or adult mice: brain, heart, liver, kidney, skeletal muscle, lung, spleen, and testis, as well as whisker pad skin and superior cervical ganglia (Nagata et al. 2005). Because five complementary DNA (cDNA) clones of mouse TRPA1 were generated from mesencephalon (corpora quadrigemina; see the UniGene cluster Mm.186329), there may be restricted or low level expression of TRPA1 in brain.

So far, however, the only organs where TRPA1 mRNA and protein expression has been reliably and reproducibly detected are two specialized in sensation: the peripheral ganglia that contain nociceptive neurons and the mechanosensory epithelia of the inner ear.

In dorsal root, trigeminal, and nodose ganglia, TRPA1 mRNA is detected by in situ hybridization (ISH) in a large percentage of the small-diameter neurons, which are generally regarded as nociceptors (Table 1). Although the precise estimates on the proportion of TRPA1-expressing neurons per ganglion varies from study to study, this may be largely attributed to differences in species, strain, sex, type of ganglia, sensitivity of experimental method, and perhaps also to inherent variations due to the nociceptive history of the ganglion, if we consider that pain exposure may affect the proportion of neurons that express TRPA1 (Obata et al. 2005).

TRPA1-expressing peripheral neurons do not coexpress neurofilament 200 nor the growth factor receptors TrkC and TrkB, but do express the nerve growth factor (NGF) receptor TrkA (or no Trk receptor at all), and most express other markers of nociceptive neurons such as the capsaicin receptor TRPV1, calcitonin gene related peptide (CGRP), substance P (SP), or peripherin (Bautista et al. 2005; Kobayashi et al. 2005; Nagata et al. 2005). All these experiments confirm that TRPA1 is expressed by C-fiber nociceptors and not by A-fiber

Table 1 Distribution of TRPA1 mRNA in neurons of peripheral ganglia, estimated from in situ hybridization studies

Ganglion	Type	Species	% of all neurons	Reference
Dorsal root	Cervical	Mouse	56.5	Nagata et al. 2005
Dorsal root	L5 and 6	Rat	39.5	Kobayashi et al. 2005
Dorsal root		Mouse	3.6	Story et al. 2003
Trigeminal		Mouse	36.5	Nagata et al. 2005
Trigeminal		Rat	36.7	Kobayashi et al. 2005
Trigeminal	Cultured neonatal	Rat	20	Jordt et al. 2004
Nodose		Mouse	28.4	Nagata et al. 2005

Table 2 continued

Chemical	Structure	EC$_{50}$, or single activating dose	Examples of origin (not limited)
Eugenol		600 µM[a] (Bandell et al. 2004)	Clove oil
Gingerol		600 µM[a] (Bandell et al. 2004)	Ginger
Δ-9-Tetrahydro-cannabinol		12±2 µM (Jordt et al. 2004)	*Cannabis sativa*
Acrolein (2-propenal)		5±1 µM (Bautista et al. 2006)	Tear gas, vehicle exhaust, and smoke from burning vegetation. Byproduct of chemotherapy
2-Pentenal		10 µM[a] (Bautista et al. 2006)	Structurally related compound of acrolein

[a] Single activating dose

in tear gas), methyl salicylate (found in mouthwashes such as Listerine), gingerol (found in ginger), and eugenol (found in clove). These last three, however, have only been tested at an extremely high concentration (600 µM). Finally, the initial chemical found to activate TRPA1, the synthetic compound icilin or A-G-35, produces sensations of cold in addition to prickling. Icilin also activates the cold-sensitive channel TRPM8. But although some of these agonists activate other channels (Bandell et al. 2004), some others may exert their nociceptive effects primarily or exclusively through TRPA1.

Mice with a deletion of part of TRPA1 have been generated and found to be insensitive to the TRPA1 agonists mustard oil and allicin (Bautista et al. 2006). Dissociated trigeminal ganglion neurons from mutants did not respond (with Ca^{2+} influx) to these agonists, and mutant animals did not react adversely to their acute application and did not develop neurogenic inflammation and hyperalgesia, which wildtype animals do. This suggests that TRPA1 is the primary and perhaps only receptor for these pungent compounds in nociceptive neurons.

How all these different compounds activate TRPA1 is not clear. Because α,β-unsaturated double bonds are common to several of these agonists (all isothiocyanates, allicin, acrolein, and 2-pentenal), this feature may relate to their ability to activate TRPA1 (Bautista et al. 2006). In addition, the low hydrophilicity and extremely slow activation (seconds to minutes) of TRPA1 agonists, plus the fact that they can activate a cell-attached patch by exposure to the outside of the rest of the cell membrane (K. Nagata and J. García-Añoveros, unpublished observation), imply a mechanism of action that might require prior partition of these chemicals into the lipid membrane to then, directly or indirectly, activate TRPA1.

3.2
Bradykinin

BK is another agent that activates TRPA1. It does so not as a direct agonist but via the BK receptor, which activates phospholipase C (PLC) second messenger signaling pathways (Bandell et al. 2004; Jordt et al. 2004). Unlike the exogenous pungent agonists described above, BK is an endogenously produced proalgesic and proinflammatory agent. Indeed, mice with a mutation in TRPA1 did not develop hyperalgesia after exposure to BK (Bautista et al. 2006). Interestingly, many other proalgesic and proinflammatory agents also stimulate the PLC pathway, suggesting that they may exert their effects via TRPA1 (Jordt et al. 2004). However, trigeminal neurons also require TRPV1 for responding to BK. Because TRPV1 channels also gate in response to acute painful stimulation like heat and acidosis, we wonder if there may also be an endogenous noxious stimulus that activates TRPA1 directly.

3.3
Noxious Cold

Whether cold activates TRPA1 channels is unresolved (Reid 2005). Some studies report activation of heterologously expressed TRPA1 by temperatures below 17°C (Story et al. 2003; Bandell et al. 2004) and some report no activation (Jordt et al. 2004; Nagata et al. 2005). It has been suggested that the occasionally observed activation of TRPA1 upon cooling may have to do with nonspecific changes in $[Ca^{2+}]i$, which could affect TRPA1 gating. Clearly the cold-producing agent icilin activates TRPA1 (Story et al. 2003; Nagata et al. 2005), but none of the many other TRPA1 agonists produce a cooling sensation in humans, and icilin also activates the cold-sensing channel TRPM8 (McKemy et al. 2002). In addition, sensory ganglia of TRPA1 mutant mice have a normal distribution of menthol sensitive and insensitive cold-responsive neurons. Further, the mutant mice themselves respond to cold with the same aversion as wildtype mice (Bautista et al. 2006).

3.4
Mechanical Force

There are no reports of mechanical activation of heterologously expressed TRPA1 although, as described below, there are reasons to suspect that this might occur in the appropriate cellular context. Mice with a deletion of the pore domain of TRPA1 have no overt vestibular or auditory defect, as determined by behavioral observation, auditory brainstem response thresholds, and distortion product otoacoustic emissions (Bautista et al. 2006). It should be noted, however, that these mice might still make a truncated TRPA1 protein bearing the N-terminal ankyrin repeats and transmembrane domains S1 to S5. Such a protein would not form a pore, but it could conceivably still function as part of a multimeric complex. By contrast, acute inhibition of TRPA1 mRNA with viral-mediated siRNA in mouse utricular hair cells, or with morpholinos in zebrafish hair cells, reduced but did not eliminate their mechanotransducing currents (Corey et al. 2004). These apparently contradictory results could be reconciled if TRPA1 participates in, but is not essential for, hair cell mechanotransduction complexes.

4
Pharmacology

In addition to the numerous pain-producing chemicals described above as TRPA1 agonists (Table 2), four antagonists have been studied to date: gentamicin, ruthenium red, gadolinium, and amiloride (Table 3). Although each of these antagonists also blocks other types of channels, block by all four is characteristic of mechanosensory channels from various cell types, including hair cells (Hamill and McBride 1996). The similar Hill coefficients with which each of these antagonists blocks heterologously expressed TRPA1 and the hair cell mechanotransducer suggests a shared mechanism of action. The IC$_{50}$ are also indistinguishable for two of these antagonists, gentamycin and ruthenium red (both simple pore blockers that plug into the pore), but differs by a factor of 10 for amiloride and by a factor of 100 for gadolinium, which is a more potent blocker of heterologously expressed TRPA1 than of any other channel.

Table 3 Blockers of heterologous TRPA1 channels activated by AITC and of the endogenous transduction channels of hair cells activated mechanically. Note that—while blocking effects on TRPA1 were tested in the absence of external calcium to avoid calcium-induced potentiation and inactivation—the blocking effects on hair cells were tested in the presence of external calcium, since in its absence the tip links disassemble and mechanotransduction is abolished. Data for TRPA1 show mean±confidential interval

Blocker	TRPA1 in HEKs (Nagata et al. 2005)	Hair cell transducers	Reference for hair cell transducers
Gentamicin—IC50	6.7 ± 0.7 µM	7.6 µM	Kroese et al. 1989
Gentamicin—Hill coef.	1.2	1.2	Kroese et al. 1989
Ruthenium red—IC50	3.4 ± 0.1 µM	3.6 ± 0.3 µM	Farris et al. 2004
Ruthenium red—Hill coef.	1.1 ± 0.03	1.4 ± 0.2	Farris et al. 2004
Gd3+−IC50	0.1 ± 0.02 µM	10.1 µM	Kimitsuki et al. 1996
Gd3+−Hill coef.	1.2 ± 0.25	1.1	Kimitsuki et al. 1996
Amiloride—IC50	511 ± 12 µM	50 µM	Jorgensen and Ohmori 1988; Rüsch et al. 1994
Amiloride—Hill coef.	2.4 ± 0.1	2.2 ± 0.1	Ricci 2002

5 Ion Channel Properties

Heterologously expressed TRPA1 channels, primarily studied by activation with AITC and other agonists, are permeable to both monovalent and divalent cations, although the ionic selectivity has not been calculated in detail. The estimated conductance of a single TRPA1 channel in the negative voltage range was 100 pS, but traces with estimated values as low as 38 pS were observed. In fact, channel amplitude varied widely, even during the course of a single channel recording, suggesting that TRPA1 channels do not have a few discrete conductance levels but a range of them. Addition of external calcium gradually decreased the amplitude of single channel inward current to 54% %(Nagata et al. 2005).

Calcium exerts pronounced effects on TRPA1 gating (Nagata et al. 2005). Whole-cell recordings of human embryonic kidney (HEK)-293 cells clamped at −80 mV, transiently expressing TRPA1 and activated with AITC in the presence of a physiological external solution with Ca^{2+}, reveals a multiphasic inward current that starts slowly, and then gains in amplitude (potentiation) before subsiding (inactivation), after which the channels remain inactivated for at least tens of seconds (Fig. 1a). These phenomena depend on extracellular Ca^{2+}. In the absence of external Ca^{2+}, the slowly developing inward currents do not potentiate or inactivate (Fig. 1b). However, if Ca^{2+} is suddenly added to the bath solution when the channels are already opened by AITC, the currents potentiate and then inactivate (Figs. 1d, e). These phenomena are also voltage-dependent. At more depolarized potentials (−20 mV in Fig. 1c), inactivation of channels (activated in the continuous presence of 2 mM external Ca^{2+}) does not occur (it does at −80 mV). At +80 mV, the potentiation and inactivation induced by sudden exposure to external Ca^{2+} at −80 mV are absent or much reduced (arrowhead in Fig. 1f). In addition, channels inactivated at −80 mV will reverse from this inactivation (although in about 30 s) and reopen if held at +80 mV (asterisk in Fig. 1f). It appears that Ca^{2+} must be entering the cell to exert its effects on TRPA1 channels. However, these effects occurred whether the cytosol of the cells were loaded with the Ca^{2+} chelators EGTA or BAPTA, or by contrast with 2 or 3 mM $[Ca^{2+}]_i$, and therefore it is unlikely that Ca^{2+} is exerting these effects inside the cell. Our working model proposes that permeating Ca^{2+} binds to a site within the channel, either at the pore or at any other site not accessible to cytosolic Ca^{2+}, to cause potentiation followed by closure and inactivation (Fig. 2c). The delay in recovery from inactivation by depolarization cannot be explained if inactivation occurred by simple Ca^{2+} block, and suggests that other independent events must take place for the channels to open.

Single channel recordings from outside-out patches reveal the effects of Ca^{2+} on channel conductance and open probability (P_o). In the absence of external Ca^{2+}, AITC induces the TRPA1 channel to adopt a flickery, high conductance (5.7±0.4 pA current amplitude) but low P_o, state (II of Fig. 2a). Upon external Ca^{2+} addition, the channel reduces its conductance (to 3.1±0.4 pA current amplitude) but increases its P_o by reducing closed time and increasing burst duration (potentiation; III in Fig. 2a) for several seconds before reducing it to zero (inactivation; IV on Fig. 2a; Table 4).

In addition to these effects of external Ca^{2+} on TRPA1 gating, it has also been reported that thapsigargin treatment, which would increase intracellular Ca^{2+} levels due to release from intracellular stores, activates TRPA1 channels at the plasma membrane (Jordt et al. 2004). The mechanisms for this effect, and whether it relates to the above-described potentiation, is unknown.

Fig. 1 a Current induced by 10 μM AITC in normal external solution with 2 mM Ca^{2+}. **b** In the absence of external Ca^{2+}, AITC elicits only slow activation. **c** In the presence of external Ca^{2+}, inactivation is complete at hyperpolarized (−80 mV) holding potentials but not at depolarized (−20 mV) potentials. **d** and **e** Addition of 2 mM Ca^{2+} to the external medium of slowly activated TRPA1 channels elicits fast potentiation followed by inactivation regardless of $[Ca^{2+}]_i$ levels. **f** Ca^{2+}-induced inactivation, which is reduced at positive holding potentials (+80 mV) and maximal at negative (−80 mV) holding potentials, is reversed by positive holding potentials (*asterisk*).

6
Biological Relevance and Emerging/Established Biological Roles for TRPA1

The highly restricted expression pattern of TRPA1 to nociceptive neurons of peripheral ganglia and to the mechanosensory epithelia of the inner ear is suggestive of sensory roles, although not necessarily mechanical transduction. Studies of heterologously expressed TRPA1 are well suited to determine which

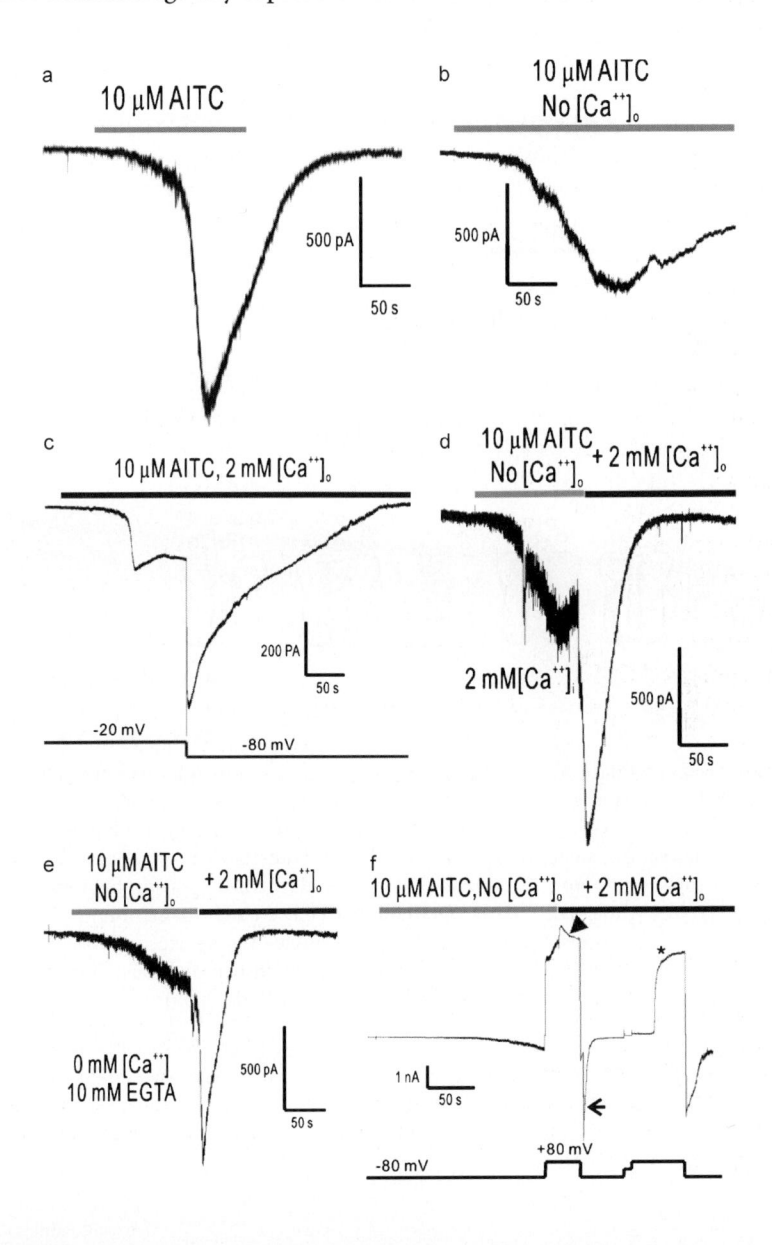

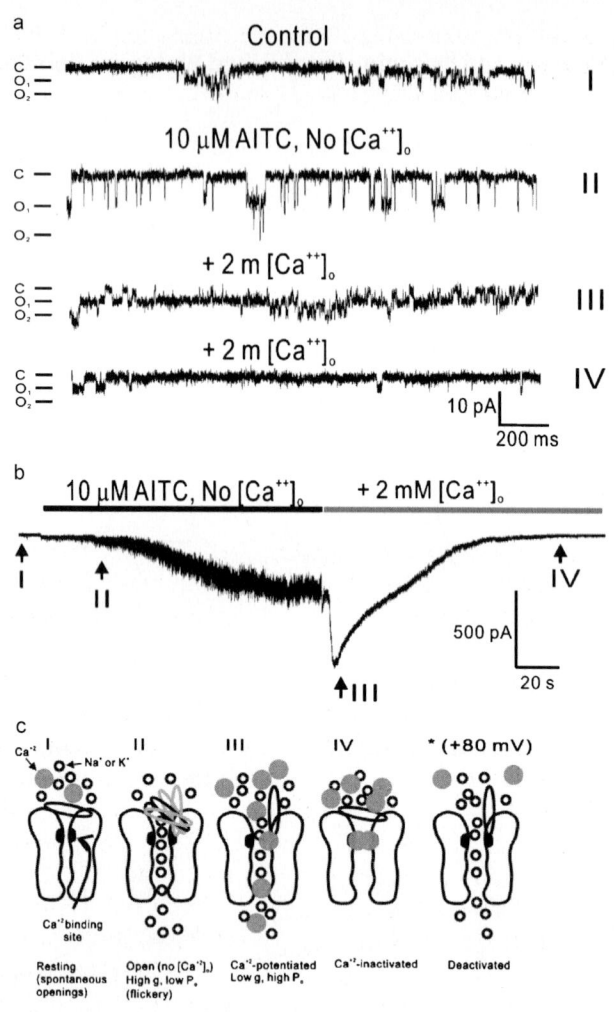

Fig. 2 a–c a Recapitulation of TRPA1 slow activation, calcium-induced potentiation, and subsequent inactivation at the single-channel level in outside-out patches (at a holding potential of −80 mV). AITC, in the absence of calcium, activates a large conductance, flickery channel (*II*, activation) that, on exposure to calcium, transitions to a low conductance but high open-probability state (*III*, potentiation) followed by a low open-probability state (*IV*, inactivation). **b** A whole-cell record, obtained just before patch formation, indicating (in *Roman numerals*) the times that correspond to the single-channel traces above. **c** A model for how Ca^{2+} and voltage affects TRPA1 channels, which can also describe the similar behaviors of hair cell transducers. Modified from Nagata et al. (2005)

Table 4 Single-channel parameters of AITC-induced activation (in the absence of external Ca^{2+}), Ca^{2+}-induced potentiation, and Ca^{2+}-induced inactivation. Data show mean±SD estimated from 1,117 events pooled from four separate experiments. From Nagata et al. (2005)

Phase	P_{open}	Burst duration (ms)	Open time (ms)	Close time (ms)	Amplitude (pA)
Ca^{2+}-free activation	0.17 ± 0.11	11.7 ± 24.4	4.4 ± 5.9	78.5 ± 177	5.7 ± 0.4
Ca^{2+}-induced potentiation	0.51 ± 0.30	94.2 ± 133	6.1 ± 8.9	7.4 ± 25.6	3.1 ± 0.4
Ca^{2+}-induced inactivation	0.02 ± 0.01	4.1 ± 1.3	ND	ND	3.1 ± 0.4

ND, not determined

stimuli can directly activate these channels. They are also useful to compare TRPA1 with endogenous channels. While data from heterologous expression may suggest biological roles for TRPA1, additional evidence must come from gene targeting studies.

In nociceptors, TRPA1 clearly acts as a receptor channel for pain-producing chemicals, mediating both their acute noxious effects as well as the sensitization that these chemicals produce. Not only does heterologously expressed TRPA1 open in response to these pungent compounds, but mice bearing a deletion in part of TRPA1 lack the nocifensive behaviors elicited by at least two of them: AITC (mustard oil) and allicin (garlic). These compounds, however, are not endogenous, and are rarely in contact with most TRPA1-expressing nociceptive neurons. Although it is possible that there are similar substances endogenously produced by tissue damage or inflammation, it is also possible that these compounds are agonists that bear no structural relationship with the nociceptive signals that under normal physiological conditions activate TRPA1, such as BK.

In nociceptors, TRPA1 is also clearly a mediator of BK's proalgesic effects. Not only are TRPA1 channels opened if coexpressed with the BK receptor and exposed to BK, but TRPA1 mutant mice did not develop hyperalgesia when exposed to BK. As the BK receptor activates PLC, and this enzyme also mediates the effects of other proalgesic agents such as ATP, monoamines, and neurotrophins, TRPA1 channels may also mediate their effects. Surprisingly, induction of hyperalgesia by BK needed both TRPA1 and TRPV1, which appear to act in concert for this function. Therefore, in the same way that TRPV1 plays other apparently unrelated roles in nociceptors, such as heat and acid sensation, so might TRPA1 play some yet undetermined role as a primary transducer.

The proposed role of TRPA1 as a painful cold receptor is controversial because: (1) heterologous activation of TRPA1 by cold, although observed several times, is often not observed; and (2) although cold hyperalgesia was blocked by intrathecal injection of antisense oligonucleotides targeting TRPA1 mRNA (Obata et al. 2005), mice with a deletion in the TRPA1 gene displayed normal nocifensive responses to cold temperatures (Bautista et al. 2006). What is surprising is that a *Drosophila* ortholog of TRPA1, when heterologously expressed, makes channels that are opened by warm temperatures (Viswanath et al. 2003). Larvae in which TRPA1 mRNA was targeted with RNA interference (RNAi) displayed defective thermotaxis (Rosenzweig et al. 2005), suggesting that in insects TRPA1 is a thermosensitive channel. Another *Drosophila* TRPA homolog, *Painless*, appears to mediate thermal and mechanical nociception in the fly as mutant larvae fail to respond aversively to prodding with a heated object (Tracey et al. 2003). In mammals, however, there is no definitive evidence that TRPA1 mediates thermal sensation.

The similar pharmacology and pore properties between heterologously expressed TRPA1 and mechanosensory channels, the wide distribution of TRPA1 mRNA among nociceptors, many of which are mechanosensory, and the structural or phylogenetic similarity of TRPA1 with two mechanosensory channels from *Drosophila*, TRPN1 (Walker et al. 2000) and Painless (Tracey et al. 2003), all suggests a potential role of TRPA1 in mechanonociception (Nagata et al. 2005). Although this function has not been extensively tested in the TRPA1 mutant mice, they had normal paw withdrawal thresholds to mechanical stimulation with von Frey hairs. But, given that the touch and mechanonociceptive sensory systems may be largely redundant (as evidenced by the paucity of naturally occurring mutations that impair these senses), detailed tests must be performed.

Regardless of the precise function of TRPA1 in nociceptors, the voltage sensitivity of its Ca^{2+}-dependent inactivation render it well suited for a role in pain (Nagata et al. 2005). In a sensory terminal exposed to low threshold (i.e., innocuous) stimulation, and thus insufficiently depolarized, TRPA1 channels would close. Therefore, in the presence of persistent but innocuous stimulation, TRPA1 would not activate a nociceptor. But with stimulation above threshold (i.e., noxious), enough TRPA1 channels would open and depolarize the terminal. Under these conditions the TRPA1 channels fail to inactivate, and remain open as long as noxious stimulation persists.

Perhaps the biggest unanswered question regarding the function of TRPA1 is what it may be doing in the sensory epithelia of the inner ear. It should be noted that TRPA1 is prominently expressed in supporting cells, where its function or even its distribution have not been addressed (García-Añoveros and Duggan 2006). But there is evidence that TRPA1 is also expressed in the mechanosensory hair cells. One reason to consider that TRPA1—or a similar channel protein—contributes to hair cell transducers is the many similarities in the behavior of the pores of heterologously expressed TRPA1 and endogenous hair cell mechanotransducers (Ca^{2+}-effects on conductance and gating, and similar pharmacological profile) (Nagata et al. 2005). Gene and mRNA targeting methods reveal that intact TRPA1 is not absolutely required for hair cell transduction. But the mechanotransducers of hair cells may be very heterogeneous, as suggested by the large number of conductance levels found tonotopically distributed in hair cells along the cochlea. We have speculated that hair cell transducers may be the product of several ion channel subunits, differentially heteromultimerized in each hair cell (García-Añoveros and Duggan 2006). In this context, lacking one of these subunits may not necessarily produce a severe or even noticeable defect in transduction. Clearly, further tests are required to determine whether TRPA1, perhaps in combination with other TRP channels that we have found in hair cells, makes transduction channels, or whether it plays other roles, particularly in support cells.

References

Bandell M, Story GM, Hwang SW, Viswanath V, Eid SR, Petrus MJ, Earley TJ, Patapoutian A (2004) Noxious cold ion channel TRPA1 is activated by pungent compounds and bradykinin. Neuron 41:849–857

Bautista DM, Movahed P, Hinman A, Axelsson HE, Sterner O, Hèogestèatt ED, Julius D, Jordt SE, Zygmunt PM (2005) Pungent products from garlic activate the sensory ion channel TRPA1. Proc Natl Acad Sci U S A 102:12248–12252

Bautista DM, Jordt SE, Nikai T, Tsuruda PR, Read AJ, Poblete J, Yamoah EN, Basbaum AI, Julius D (2006) TRPA1 mediates the inflammatory actions of environmental irritants and proalgesic agents. Cell 124:1269–1282

Corey DP, García-Añoveros J, Holt JR, Kwan KY, Lin SY, Vollrath MA, Amalfitano A, Cheung EL, Derfler BH, Duggan A, Géléoc GS, Gray PA, Hoffman MP, Rehm HL, Tamasauskas D, Zhang DS (2004) TRPA1 is a candidate for the mechanosensitive transduction channel of vertebrate hair cells. Nature 432:723–730

Duggan A, García-Añoveros J, Corey DP (2000) Insect mechanoreception: what a long, strange TRP it's been. Curr Biol 10:R384–R387

Farris HE, LeBlanc CL, Goswami J, Ricci AJ (2004) Probing the pore of the auditory hair cell mechanotransducer channel in turtle. J Physiol 558:769–792

García-Añoveros J, Duggan A (2006) TRPA1 in auditory and nociceptive organs. In: Liedtke W, Heller S (eds) TRP ion channels in transduction of sensory stimuli and cellular signaling cascades. CRC Publishing/Taylor and Francis, Boca Raton

Hamill OP, McBride DW Jr (1996) The pharmacology of mechanogated membrane ion channels. Pharmacol Rev 48:231–252

Howard J, Bechstedt S (2004) Hypothesis: a helix of ankyrin repeats of the NOMPC-TRP ion channel is the gating spring of mechanoreceptors. Curr Biol 14:R224–R226

Jaquemar D, Schenker T, Trueb B (1999) An ankyrin-like protein with transmembrane domains is specifically lost after oncogenic transformation of human fibroblasts. J Biol Chem 274:7325–7333

Jordt SE, Bautista DM, Chuang HH, McKemy DD, Zygmunt PM, Hogestatt ED, Meng ID, Julius D (2004) Mustard oils and cannabinoids excite sensory nerve fibres through the TRP channel ANKTM1. Nature 427:260–265

Jorgensen F, Ohmori H (1988) Amiloride blocks the mechano-electrical transduction channel of hair cells of the chick. J Physiol 403:577–588

Kimitsuki T, Nakagawa T, Hisashi K, Komune S, Komiyama S (1996) Gadolinium blocks mechano-electric transducer current in chick cochlear hair cells. Hear Res 101:75–80

Kobayashi K, Fukuoka T, Obata K, Yamanaka H, Dai Y, Tokunaga A, Noguchi K (2005) Distinct expression of TRPM8, TRPA1, and TRPV1 mRNAs in rat primary afferent neurons with adelta/c-fibers and colocalization with trk receptors. J Comp Neurol 493:596–606

Kroese AB, Das A, Hudspeth AJ (1989) Blockage of the transduction channels of hair cells in the bullfrog's sacculus by aminoglycoside antibiotics. Hear Res 37:203–217

Lee G, Abdi K, Jiang Y, Michaely P, Bennett V, Marszalek PE (2006) Nanospring behaviour of ankyrin repeats. Nature 440:246–249

Li W, Feng Z, Sternberg PW, Shawn Xu XZ (2006) A C. elegans stretch receptor neuron revealed by a mechanosensitive TRP channel homologue. Nature 440:684–687

McKemy DD, Neuhausser WM, Julius D (2002) Identification of a cold receptor reveals a general role for TRP channels in thermosensation. Nature 416:52–58

Nagata K, Duggan A, Kumar G, García-Añoveros J (2005) Nociceptor and hair cell transducer properties of TRPA1, a channel for pain and hearing. J Neurosci 25:4052–4061

Obata K, Katsura H, Mizushima T, Yamanaka H, Kobayashi K, Dai Y, Fukuoka T, Tokunaga A, Tominaga M, Noguchi K (2005) TRPA1 induced in sensory neurons contributes to cold hyperalgesia after inflammation and nerve injury. J Clin Invest 115:2393–2401

Reid G (2005) ThermoTRP channels and cold sensing: what are they really up to? Pflugers Arch 451:250–263

Ricci A (2002) Differences in mechano-transducer channel kinetics underlie tonotopic distribution of fast adaptation in auditory hair cells. J Neurophysiol 87:1738–1748

Rosenzweig M, Brennan KM, Tayler TD, Phelps PO, Patapoutian A, Garrity PA (2005) The Drosophila ortholog of vertebrate TRPA1 regulates thermotaxis. Genes Dev 19:419–424

Rüsch A, Kros CJ, Richardson GP (1994) Block by amiloride and its derivatives of mechano-electrical transduction in outer hair cells of mouse cochlear cultures. J Physiol 474:75–86

Sidi S, Friedrich RW, Nicolson T (2003) NompC TRP channel required for vertebrate sensory hair cell mechanotransduction. Science 301:96–99

Sotomayor M, Corey DP, Schulten K (2005) In search of the hair-cell gating spring elastic properties of ankyrin and cadherin repeats. Structure 13:669–682

Stepanyan R, Boger ET, Friedman TB, Frolenkov GL (2006) TRPA1, a hair cell channel with unknown function? (abstract No. 1213). Abstr Midwinter Res Meet Assoc Res Otolaryngol, pp 211–212

Story GM, Peier AM, Reeve AJ, Eid SR, Mosbacher J, Hricik TR, Earley TJ, Hergarden AC, Andersson DA, Hwang SW, McIntyre P, Jegla T, Bevan S, Patapoutian A (2003) ANKTM1, a TRP-like channel expressed in nociceptive neurons, is activated by cold temperatures. Cell 112:819–829

Tracey WD Jr, Wilson RI, Laurent G, Benzer S (2003) painless, a Drosophila gene essential for nociception. Cell 113:261–273

Viswanath V, Story GM, Peier AM, Petrus MJ, Lee VM, Hwang SW, Patapoutian A, Jegla T (2003) Opposite thermosensor in fruitfly and mouse. Nature 423:822–823

Walker RG, Willingham AT, Zuker CS (2000) A Drosophila mechanosensory transduction channel. Science 287:2229–2234

TRPP2 Channel Regulation

R. Witzgall

Institute for Molecular and Cellular Anatomy, University of Regensburg,
Universitätsstrasse 31, 93053 Regensburg, Germany
ralph.witzgall@vkl.uni-regensburg.de

1	A Brief Introduction to Cystic Kidney Diseases	363
2	Organization and Expression Pattern of the *PKD2* Gene	364
3	Domain Structure and Intracellular Distribution of Polycystin-2	365
4	Ion Channel Properties and Pharmacology of Polycystin-2	366
5	Modes of Activation of Polycystin-2	369
6	Biological Relevance, Emerging and Established Roles for Polycystin-2	370
7	Perspective	371
	References	371

Abstract Polycystin-2, or TRPP2 according to the TRP nomenclature, is encoded by *PKD2*, a gene mutated in patients with autosomal-dominant polycystic kidney disease. Its precise subcellular location and its intracellular trafficking are a matter of intense debate, although a consensus has emerged that it is located in primary cilia, a long-neglected organelle possibly involved in sensory functions. Polycystin-2 has a calculated molecular mass of 110 kDa, and according to structural predictions it contains six membrane-spanning domains and a pore-forming region between the 5th and 6th membrane-spanning domain. This section first introduces the reader to the field of cystic kidney diseases and to the *PKD2* gene, before the ion channel properties of polycystin-2 are discussed in great detail.

Keywords Polycystic kidney disease · Cation channel · Mechanosensation · Chemosensation · Primary cilia

1
A Brief Introduction to Cystic Kidney Diseases

Cystic kidneys are a cardinal symptom of not only one, but a whole class of diseases comprising the autosomal-dominant and autosomal-recessive forms of polycystic kidney disease, medullary cystic kidney disease, nephronophthisis, Bardet–Biedl syndrome, von Hippel–Lindau syndrome, oro-facial digital syndrome type I and tuberous sclerosis (Witzgall 2005a). Although the age of

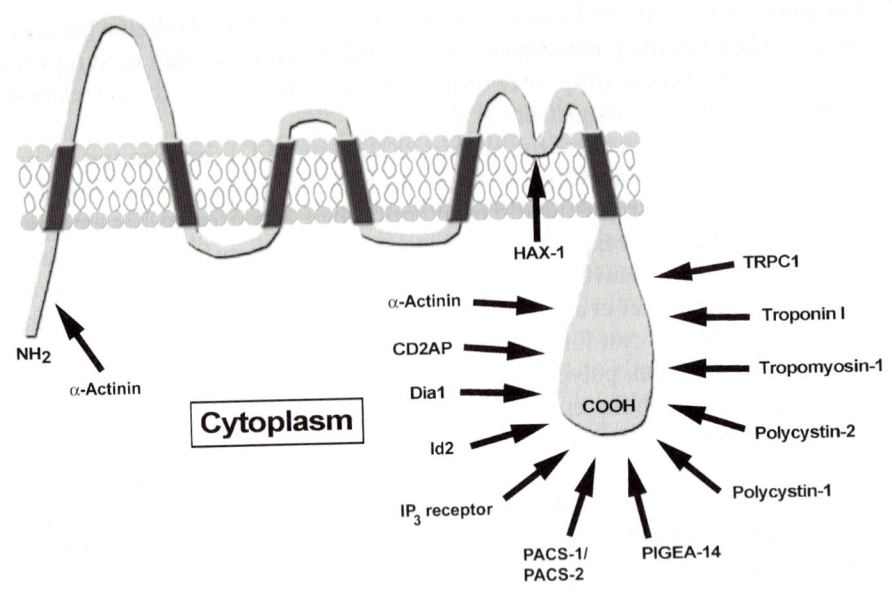

Fig. 1 Putative topology of polycystin-2. Polycystin-2 is predicted to span the membrane six times, with a supposed pore-forming region between the 5th and 6th membrane-spanning domains. The NH$_2$- and COOH-terminus are believed to extend into the cytoplasm, although these assumptions have not been corroborated experimentally. As can be easily appreciated, the COOH-terminus has received the most attention. It has been shown to interact with polycystin-2 itself (Qian et al. 1997; Tsiokas et al. 1997), with α-actinin (Li et al. 2005a), CD2AP (Lehtonen et al. 2000), Dia1 (Rundle et al. 2004), Id2 (Li et al. 2005b), the IP$_3$ receptor (Li et al. 2005c), PACS-1 and PACS-2 (Köttgen et al. 2005), PIGEA-14 (Hidaka et al. 2004), polycystin-1 (Qian et al. 1997; Tsiokas et al. 1997), tropomyosin-1 (Li et al. 2003a), troponin I (Li et al. 2003b) and TRPC1 (Tsiokas et al. 1999). The actin cytoskeleton-associated protein α-actinin has also been demonstrated to interact with the NH$_2$-terminus of polycystin-2, and HAX-1 was shown to associate with the putative pore-forming region of polycystin-2 (Gallagher et al. 2000)

Witzgall 2005b). The intracellular distribution of polycystin-2 has become even more complicated by the fact that polycystin-2 was detected in primary cilia (Pazour et al. 2002; Yoder et al. 2002).

4
Ion Channel Properties and Pharmacology of Polycystin-2

Due to the sequence and structural similarity of polycystin-2 to α-subunits of voltage-activated calcium channels (Mochizuki et al. 1996) and to the interaction of polycystin-2 with TRPC1 (Tsiokas et al. 1999), it was speculated already early on that polycystin-2 acts as a cation channel. The finding that the closely related protein polycystin-2L acts as a non-selective cation channel with a large unitary conductance further supported this assumption (Chen et al. 1999). Since then, a large number of publications have reported on the ion channel properties of polycystin-2; they all agree that polycystin-2 represents a non-selective ion channel with a large conductivity of approximately 100 pS. A more detailed presentation of the additional, sometimes contradictory findings is presented in the rest of this section.

The first report on the channel properties of polycystin-2 was based on the expression of the human PKD2 complementary DNA (cDNA) in transiently transfected CHO cells (Hanaoka et al. 2000). Measurements of whole-cell currents showed novel cation currents with slightly outwardly rectifying characteristics. La^{3+} ions, which inhibit non-selective cation channels, and niflumic acid, which inhibits non-selective cation and Cl^- channels, blocked the current in a dose-dependent manner. Consistent with these observations was the finding that polycystin-2 conducted Ca^{2+}, Na^+ and Cs^+ (with permeability ratios of 1.12:1:0.57, $P_{Ca2+}:P_{Na+}:P_{Cs+}$), but not Cl^- ions. In this publication the remarkable observation was made that polycystin-2 synthesized in CHO cells in the absence of polycystin-1 remained in the endoplasmic reticulum, whereas in its presence it reached the plasma membrane, which correlated with the fact that novel currents were only seen when polycystin-2 was synthesized together with polycystin-1 but not without it. A truncation mutant of polycystin-2 which lacked a large portion of the COOH-terminus including the putative EF-hand (R742X mutant) generated no currents (Hanaoka et al. 2000).

The main observations of this first report were confirmed by another group which expressed the murine Pkd2 cDNA in *Xenopus* oocytes, although there were some differences. Most of the full-length polycystin-2 protein was retained in intracellular membranes, but chemical chaperones and proteasome inhibitors induced its trafficking to the plasma membrane (Vassilev et al. 2001). Whole-cell conductance of *Xenopus* oocytes did not change after injection of the murine Pkd2 cDNA, consistent with the predominant intracellular location of full-length polycystin-2, although the uptake of radioactive Ca^{2+} ions increased fourfold (Chen et al. 2001). Using cell-attached or excised membrane patches, again a novel non-selective cation channel was detected which conducted both monovalent (Na^+, K^+, Li^+, Rb^+, NH_4^+) and divalent (Ca^{2+}, Sr^{2+}, Ba^{2+}) ions and was inhibited by La^{3+} but not nifedipine, a blocker of voltage-gated Ca^{2+} channels and ryanodine and inositol 1,4,5-trisphosphate (IP_3) receptors (Vassilev et al. 2001). This time, however, the channel was slightly inwardly rectifying and no single channel recordings were observed in the presence of Cs^+ (permeability ratios were 0.14:1:0.73, $P_{Na+}:P_{K+}:P_{NH4+}$, and 0.21:1, $P_{Ca2+}:P_{K+}$) (Vassilev et al. 2001). The addition of 1 µM Ca^{2+} to the bath solution transiently increased the open probability of polycystin-2, but millimolar concentrations of Ca^{2+} inhibited currents. The R742X mutant displayed a slightly lower conductivity than the wild-type protein, was (almost) impermeable to Ba^{2+} and Sr^{2+} (Chen et al. 2001) and did not respond to the addition of Ca^{2+} (Vassilev et al. 2001). The findings in *Xenopus* oocytes were extended to non-transfected and transfected inner medullary collecting duct (IMCD) cells in which very similar conductiviti es and permeability ratios for different cations were obtained as in the oocytes (Luo et al. 2003). Finally, synthesis of the full-length polycystin-2 protein in sympathetic neurons together with polycystin-1 led to the accumulation of polycystin-2 in the plasma membrane, where it acted as a non-selective cation channel whose activity could be inhibited by La^{3+} ions and amiloride (Delmas et al. 2004). In contrast to the results obtained with oocytes, however, the $P_{Na+}:P_{K+}$ permeability ratio was 0.98 and not 0.14. The R742X polycystin-2 mutant protein again displayed a somewhat lower conductivity than the full-length protein (90 pS vs 110 ps), but its open probability was markedly higher (Delmas et al. 2004).

A different approach was used by a third group which first investigated the channel properties of human syncytiotrophoblast cells (González-Perrett et al. 2001). For this, membrane vesicles (apical enriched) of human syncytiotrophoblast cells were reconstituted onto planar lipid bilayers. An endogenous high-conductance, non-selective, voltage-dependent cation channel was detected with a $P_{Na+}:P_{K+}$ permeability ratio of 1.1. Again the channel was inhibited by La^{3+} and high concentrations (90 mM) of Ca^{2+} ions, and also by Gd^{3+} ions and amiloride. The expression of the endogenous PKD2 gene was confirmed on the messenger RNA (mRNA) and protein level, and an anti-polycystin-2 antibody inhibited this particular channel. Subsequent synthesis of the human polycystin-2 protein in Sf9 insect cells and use of in vitro-translated protein confirmed the data obtained with the human syncytiotrophoblast cells and further demonstrated a $P_{Cs+}:P_{Ba2+}$ permeability ratio of 0.65 (González-Perrett et al. 2001). This group was also able to confirm that the R742X mutant of polycystin-2 acted as a channel, although with a lower conductivity and open probability than the full-length protein (Xu et al. 2003).

A comparable strategy from yet another group, i.e. expression of the human PKD2 cDNA in stably transfected cells (this time the porcine kidney epithelial cell line LLC-PK$_1$), isolation of endoplasmic reticulum-enriched membrane vesicle and reconstitution onto lipid bilayers again demonstrated that polycystin-2 represented a high-conductance, non-selective, voltage-dependent cation channel (Koulen et al. 2002). Using various organic cations of different sizes, it was estimated that the pore diameter of polycystin-2 was at least 1.1 nm (Anyatonwu and Ehrlich 2005). Another truncated polycystin-2 protein, the L703X mutant, showed a lower conductivity (28 pS vs 114 pS for full-length polycystin-2 in the presence of Ba^{2+}) and required a larger negative membrane potential for activation. Confirming previous results, [Ca^{2+}] up to 1,260 µM increased the open probability of full-length polycystin-2, whereas higher concentrations were inhibitory; no such modulatory activity of Ca^{2+} was observed for the L703X mutant protein (Koulen et al. 2002). Phosphorylation of the serine residue at position 812, possibly by casein kinase 2, appears to increase the calcium sensitivity of polycystin-2 (Cai et al. 2004).

5
Modes of Activation of Polycystin-2

So far only a very limited amount of data are available on how polycystin-2 is activated. In particular it is not known whether there is a natural ligand for this protein. Probably the most exciting data indicate that polycystin-2 (together with polycystin-1) plays an essential role for mechanosensation of kidney epithelial cells by primary cilia (Nauli et al. 2003). There are several open questions as to whether the diameter of renal tubules can be determined by a flow-based mechanism (Witzgall 2005a), but the flow model currently is the most widely discussed one regarding the pathomechanism of cyst formation. Consistent with the biochemical interaction between polycystin-2 and polycystin-1 (Qian et al. 1997; Tsiokas et al. 1997), polycystin-1 was able to reactivate the channel activity of polycystin-2 in lipid bilayer membranes (Xu et al. 2003). The interaction of polycystin-2 with cytoskeletal components such as Hax-1 (Gallagher et al. 2000), troponin I (Li et al. 2003b), tropomyosin-1 (Li et al. 2003a) and α-actinin (Li et al. 2005a) could also be very important in such a scenario, and indeed it was shown for α-actinin (Li et al. 2005a) that it drastically increases the channel activity of polycystin-2. Employing membrane preparations from human syncytiotrophoblast cells, it was demonstrated that the activity of a cation channel, possibly polycystin-2, was modulated by gelsolin and the polymerization status of the actin cytoskeleton (Montalbetti et al. 2005b).

Other factors directly or indirectly modulating the activity of polycystin-2 are the pH value, hydro-osmotic pressure, vasopressin, epidermal growth factor (EGF) and ATP. The channel activity of full-length polycystin-2 (González-Perrett et al. 2002) and its R742X truncation mutant (Chen et al. 2001) was inhibited by an increased $[H^+]$. Using again apical membrane preparations from human syncytiotrophoblast cells, it was also shown that polycystin-2 channel activity can be modulated by hydro-osmotic pressure (Montalbetti et al. 2005a). LLC-PK_1 cells stably transfected with the human PKD2 cDNA were used to demonstrate that polycystin-2 can be activated by vasopressin (Koulen et al. 2002), EGF (apparently through releasing the inhibition by phosphatidylinositol-4,5-bisphosphate) (Ma et al. 2005) and ATP (Gallagher et al. 2006). In addition to acting as a non-selective cation channel itself, polycystin-2 also interacts with the IP_3 receptor and may thereby influence $[Ca^{2+}]_i$. The latter findings deserve note because vasopressin (Torres et al. 2004), EGF (Sweeney et al. 2000; Torres et al. 2003) and ATP (Hooper et al. 2003; Schwiebert et al. 2002) are felt to promote cyst formation. Since most of the mutations in the *PKD2* gene are believed to be inactivating (Deltas 2001), the cyst-promoting activity of vasopressin, EGF and ATP should be prevented as long as functional polycystin-2 is present; only if polycystin-2 is lost by mutations should vasopressin, EGF and ATP be able to induce cystogenesis.

6
Biological Relevance, Emerging and Established Roles for Polycystin-2

As already elaborated in the introductory remarks and in the preceding section, polycystin-2 is believed to be responsible for the flow-mediated rise in $[Ca^{2+}]_i$ (Nauli et al. 2003) and thereby possibly for determining the correct width of renal tubules. At this point it is not clear how the mechanical stimulus is transmitted through polycystin-2—whether it is a direct action on polycystin-2 or whether other proteins sense the mechanical stress and subsequently act on polycystin-2. Experiments performed with the worm *Caenorhabditis elegans* have provided evidence for a chemosensory role of cilia (Barr et al. 2001; Barr and Sternberg 1999; Haycraft et al. 2001) and it therefore cannot be ruled out that the orthologue of polycystin-2 is activated by chemical stimuli in the worm. The activation of polycystin-2, possibly by phosphorylation, in turn can inhibit the cell cycle through its interaction with the helix-loop-helix protein Id2 (Li et al. 2005b). Such a scenario is consistent with the finding that cyst wall epithelial cells proliferate at a higher rate than regular tubular epithelial cells (e.g. Lanoix et al. 1996; Ramasubbu et al. 1998).

Another surprising finding in the *Pkd2* knock-out mice was the development of situs inversus (Pennekamp et al. 2002), thus indicating that polycystin-2 also plays an important function in left-right axis differentiation. A phenotypic connection between polycystic kidney disease and situs inversus, for which primary cilia may be the key, had already been made earlier for other proteins (Mochizuki et al. 1998; Murcia et al. 2000). The correct placement of the internal organs, e.g. of the heart on the left and the liver on the right, may already be determined very early on (Levin et al. 2002), although the best investigated stage of development represents that of the primitive node. The primitive node is a temporary structure during embryonic development which serves as an essential signalling centre for left–right axis development. Cells in the primitive node elaborate primary cilia, but surprisingly some of these cilia are motile—they are able to rotate, and this rotational movement has been proposed to break the left–right axis (Cartwright et al. 2004). There is another population of cells in the primitive node which are immotile but which (like primary cilia of motile cells) contain polycystin-2. According to the current model, the motile cilia are responsible for generating a gradient for an extracellular signalling molecule which in turn is transduced into an intracellular calcium signal by polycystin-2 (McGrath et al. 2003). It is noteworthy that *Pkd1* knock-out mice do not present with situs inversus, which is consistent with the absence of polycystin-1 in primary cilia of the node (Karcher et al. 2005).

7
Perspective

The emergence of the polycystins and their connection to primary cilia have opened up a whole new arena in the field of cell biology and renal biology. Despite a lot of exciting new evidence, we are still in the early stages of the investigations and many questions remain to be answered. A host of proteins has been described to interact with the COOH-terminus of polycystin-2, and it appears very difficult conceptually to see them all interacting at the same time in this rather short region of the protein. It will be a challenge to sort out the respective interacting domains and to determine whether there are cell type-specific interactions. What about proteins interacting with the rest of polycystin-2, are they involved in directing polycystin-2 to primary cilia? What is the natural ligand for polycystin-2? It remains to be seen whether the stimulation of cells with vasopressin, EGF and ATP is the physiologically relevant stimulus for polycystin-2, how mechanical stimuli are transmitted to and by polycystin-2, and whether polycystin-2 is also involved in chemosensation. We certainly also have much to discover about the signalling events downstream of polycystin-2, and this may lead us to target genes activated by polycystin-2 and ultimately to the "Holy Grail" of the mechanism through which the width of the renal tubules is determined. Obviously the field of cystic kidney diseases, polycystin-2 and primary cilia will keep us busy for quite some time to come.

Acknowledgements   Figure 1 has been skilfully arranged by Antje Zenker and Ton Maurer. The author gratefully acknowledges financial support from the German Research Council (SFB 699).

References

Andrews PM, Porter KR (1974) A scanning electron microscopic study of the nephron. Am J Anat 140:81–115
Anyatonwu GI, Ehrlich BE (2005) Organic cation permeation through the channel formed by polycystin-2. J Biol Chem 280:29488–29493
Barr MM, Sternberg PW (1999) A polycystic kidney-disease gene homologue required for male mating behaviour in C. elegans. Nature 401:386–389
Barr MM, DeModena J, Braun D, Nguyen CQ, Hall DH, Sternberg PW (2001) The Caenorhabditis elegans autosomal dominant polycystic kidney disease gene homologs lov-1 and pkd-2 act in the same pathway. Curr Biol 11:1341–1346
Cai Y, Maeda Y, Cedzich A, Torres VE, Wu G, Hayashi T, Mochizuki T, Park JH, Witzgall R, Somlo S (1999) Identification and characterization of polycystin-2, the PKD2 gene product. J Biol Chem 274:28557–28565
Cai Y, Anyatonwu G, Okuhara D, Lee KB, Yu Z, Onoe T, Mei CL, Qian Q, Geng L, Witzgall R, Ehrlich BE, Somlo S (2004) Calcium dependence of polycystin-2 channel activity is modulated by phosphorylation at Ser812. J Biol Chem 279:19987–19995

Cartwright JHE, Piro O, Tuval I (2004) Fluid-dynamical basis of the embryonic development of left-right asymmetry in vertebrates. Proc Natl Acad Sci USA 101:7234–7239

Chauvet V, Qian F, Boute N, Cai Y, Phakdeekitacharoen B, Onuchic LF, Attié-Bitach T, Guicharnaud L, Devuyst O, Germino GG, Gubler MC (2002) Expression of PKD1 and PKD2 transcripts and proteins in human embryo and during normal kidney development. Am J Pathol 160:973–983

Chen XZ, Vassilev PM, Basora N, Peng JB, Nomura H, Segal Y, Brown EM, Reeders ST, Hediger MA, Zhou J (1999) Polycystin-L is a calcium regulated cation channel permeable to calcium ions. Nature 401:383–386

Chen XZ, Segal Y, Basora N, Guo L, Peng JB, Babakhanlou H, Vassilev PM, Brown EM, Hediger MA, Zhou J (2001) Transport function of the naturally occurring pathogenic polycystin-2 mutant, R742X. Biochem Biophys Res Commun 282:12511256

Delmas P, Nauli SM, Li X, Coste B, Osorio N, Crest M, Brown DA, Zhou J (2004) Gating of the polycystin ion channel signaling complex in neurons and kidney cells. FASEB J 18:740–742

Deltas CC (2001) Mutations of the human polycystic kidney disease 2 (PKD2) gene. Hum Mutat 18:13–24

European Polycystic Kidney Disease Consortium (1994) The polycystic kidney disease 1 gene encodes a 14 kb transcript and lies within a duplicated region on chromosome 16. Cell 77:881–894

Foggensteiner L, Bevan AP, Thomas R, Coleman N, Boulter C, Bradley J, Ibraghimov-Beskrovnaya O, Klinger K, Sandford R (2000) Cellular and subcellular distribution of polycystin-2, the protein product of the PKD2 gene. J Am Soc Nephrol 11:814–827

Gallagher AR, Cedzich A, Gretz N, Somlo S, Witzgall R (2000) The polycystic kidney disease protein PKD2 interacts with Hax-1, a protein associated with the actin cytoskeleton. Proc Natl Acad Sci USA 97:4017–4022

Gallagher AR, Hidaka S, Gretz N, Witzgall R (2002) Molecular basis of autosomal-dominant polycystic kidney disease. Cell Mol Life Sci 59:682–693

Gallagher AR, Hoffmann S, Brown N, Cedzich A, Meruvu S, Podlich D, Feng Y, Könecke V, de Vries U, Hammes HP, Gretz N, Witzgall R (2006) A truncated polycystin-2 protein causes polycystic kidney disease and retinal degeneration in transgenic rats. J Am Soc Nephrol 17:2719–2730

González-Perrett S, Kim K, Ibarra C, Damiano AE, Zotta E, Batelli M, Harris PC, Reisin IL, Arnaout MA, Cantiello HF (2001) Polycystin-2, the protein mutated in autosomal dominant polycystic kidney disease (ADPKD), is a Ca^{2+}-permeable nonselective cation channel. Proc Natl Acad Sci USA 98:1182–1187

González-Perrett S, Batelli M, Kim K, Essafi M, Timpanaro G, Moltabetti N, Reisin IL, Arnaout MA, Cantiello HF (2002) Voltage dependence and pH regulation of human polycystin-2-mediated cation channel activity. J Biol Chem 277:24959–24966

Hanaoka K, Qian F, Boletta A, Bhunia AK, Piontek K, Tsiokas L, Sukhatme VP, Guggino WB, Germino GG (2000) Co-assembly of polycystin-1 and -2 produces unique cation-permeable currents. Nature 408:990–994

Hateboer N, Veldhuisen B, Peters D, Breuning MH, San-Millán JL, Bogdanova N, Coto E, van Dijk MA, Afzal AR, Jeffery S, Saggar-Malik AK, Torra R, Dimitrakov D, Martinez I, de Castro SS, Krawczak M, Ravine D (2000) Location of mutations within the PKD2 gene influences clinical outcome. Kidney Int 57:1444–1451

Hayashi T, Mochizuki T, Reynolds DM, Wu G, Cai Y, Somlo S (1997) Characterization of the exon structure of the polycystic kidney disease 2 gene (PKD2). Genomics 44:131–136

Haycraft CJ, Swoboda P, Taulman PD, Thomas JH, Yoder BK (2001) The C. elegans homolog of the murine cystic kidney disease gene Tg737 functions in a ciliogenic pathway and is disrupted in osm-5 mutant worms. Development 128:1493–1505

Hidaka S, Könecke V, Osten L, Witzgall R (2004) PIGEA-14, a novel coiled-coil protein affecting the intracellular distribution of polycystin-2. J Biol Chem 279:35009–35016

Hooper KM, Unwin RJ, Sutters M (2003) The isolated C-terminus of polycystin-1 promotes increased ATP-stimulated chloride secretion in a collecting duct cell line. Clin Sci 104:217–221

Karcher C, Fischer A, Schweickert A, Bitzer E, Horie S, Witzgall R, Blum M (2005) Lack of a laterality phenotype in Pkd1 knock-out embryos correlates with absence of polycystin-1 in nodal cilia. Differentiation 73:425–432

Köttgen M, Benzing T, Simmen T, Tauber R, Buchholz B, Feliciangeli S, Huber TB, Schermer B, Kramer-Zucker A, Höpker K, Simmen KC, Tschucke CC, Sandford R, Kim E, Thomas g, Walz G (2005) Trafficking of TRPP2 by PACS proteins represents a novel mechanism of ion channel regulation. EMBO J 24:705–716

Koulen P, Cai Y, Geng L, Maeda Y, Nishimura S, Witzgall R, Ehrlich BE, Somlo S (2002) Polycystin-2 is an intracellular calcium release channel. Nat Cell Biol 4:191–197

Lanoix J, D'Agati V, Szabolcs M, Trudel M (1996) Dysregulation of cellular proliferation and apoptosis mediates human autosomal dominant polycystic kidney disease (ADPKD). Oncogene 13:1153–1160

Latta H, Maunsbach AB, Madden SC (1961) Cilia in different segments of the rat nephron. J Biophys Biochem Cytol 11:248–252

Lehtonen S, Ora A, Olkkonen VM, Geng L, Zerial M, Somlo S, Lehtonen E (2000) In vivo interaction of the adapter protein CD2-associated protein with the type 2 polycystic kidney disease protein, polycystin-2. J Biol Chem 275:32888–32893

Levin M, Thorlin T, Robinson KR, Nogi T, Mercola M (2002) Asymmetries in H^+/K^+-ATPase and cell membrane potentials comprise a very early step in left-right patterning. Cell 111:77–89

Li Q, Dai Y, Guo L, Liu Y, Hao C, Wu G, Basora N, Michalak M, Chen XZ (2003a) Polycystin-2 associates with tropomyosin-1, an actin microfilament component. J Mol Biol 325:949–962

Li Q, Shen PY, Wu G, Chen XZ (2003b) Polycystin-2 interacts with troponin I, an angiogenesis inhibitor. Biochemistry 42:450–457

Li Q, Montalbetti N, Shen PY, Dai XQ, Cheeseman CI, Karpinski E, Wu G, Cantiello HF, Chen XZ (2005a) Alpha-actinin associates with polycystin-2 and regulates its channel activity. Hum Mol Genet 14:1587–1603

Li X, Luo Y, Starremans PG, McNamara CA, Pai Y, Zhou J (2005b) Polycystin-1 and polycystin-2 regulate the cell cycle through the helix-loop-helix inhibitor Id2. Nat Cell Biol 7:1202–1212

Li Y, Wright JM, Qian F, Germino GG, Guggino WB (2005c) Polycystin 2 interacts with type I inositol 1,4,5-trisphosphate receptor to modulate intracellular Ca^{2+} signaling. J Biol Chem 280:41298–41306

Luo Y, Vassilev PM, Li X, Kawanabe Y, Zhou J (2003) Native polycystin 2 functions as a plasma membrane Ca^{2+}-permeable cation channel in renal epithelia. Mol Cell Biol 23:2600–2607

Ma R, Li WP, Rundle D, Kong J, Akbarali HI, Tsiokas L (2005) PKD2 functions as an epidermal growth factor-activated plasma membrane channel. Mol Cell Biol 25:8285–8298

Magistroni R, He N, Wang K, Andrew R, Johnson A, Gabow P, Dicks E, Parfrey P, Torra R, San-Millan JL, Coto E, van Dijk M, Breuning M, Peters D, Bogdanova N, Ligabue G, Albertazzi A, Hateboer N, Demetriou K, Pierides A, Deltas C, St George-Hyslop P, Ravine D, Pei Y (2003) Genotype-renal function correlation in type 2 autosomal dominant polycystic kidney disease. J Am Soc Nephrol 14:1164–1174

Markowitz GS, Cai Y, Li L, Wu G, Ward LC, Somlo S, D'Agati VD (1999) Polycystin-2 expression is developmentally regulated. Am J Physiol 277:F17–F25

McGrath J, Somlo S, Makova S, Tian X, Brueckner M (2003) Two populations of node monocilia initiate left-right asymmetry in the mouse. Cell 114:61–73

Mochizuki T, Wu G, Hayashi T, Xenophontos SL, Veldhuisen B, Saris JJ, Reynolds DM, Cai Y, Gabow PA, Pierides A, Kimberling WJ, Breuning MH, Deltas CC, Peters DJM, Somlo S (1996) PKD2, a gene for polycystic kidney disease that encodes an integral membrane protein. Science 272:1339–1342

Mochizuki T, Saijoh Y, Tsuchiya K, Shirayoshi Y, Takai S, Taya C, Yonekawa H, Yamada K, Nihei H, Nakatsuji N, Overbeek PA, Hamada H, Yokoyama T (1998) Cloning of inv, a gene that controls left/right asymmetry and kidney development. Nature 395:177–181

Montalbetti N, Li Q, González-Perrett S, Semprine J, Chen XZ, Cantiello HF (2005a) Effect of hydro-osmotic pressure on polycystin-2 channel function in the human syncytiotrophoblast. Pflügers Arch 451:294–303

Montalbetti N, Li Q, Timpanaro GA, González-Perrett S, Dai XQ, Chen XZ, Cantiello HF (2005b) Cytoskeletal regulation of calcium-permeable cation channels in the human syncytiotrophoblast: role of gelsolin. J Physiol 566:309–325

Murcia NS, Richards WG, Yoder BK, Mucenski ML, Dunlap JR, Woychik RP (2000) The Oak Ridge Polycystic Kidney (orpk) disease gene is required for left-right axis determination. Development 127:2347–2355

Nauli SM, Alenghat FJ, Luo Y, Williams E, Vassilev P, Li X, Elia AEH, Lu W, Brown EM, Quinn SJ, Ingber DE, Zhou J (2003) Polycystins 1 and 2 mediate mechanosensation in the primary cilium of kidney cells. Nat Genet 33:129–137

Obermüller N, Gallagher AR, Cai Y, Gassler N, Gretz N, Somlo S, Witzgall R (1999) The rat Pkd2 protein assumes distinct subcellular distributions in different organs. Am J Physiol 277:F914–F925

Park JH, Li L, Cai Y, Hayashi T, Dong F, Maeda Y, Rubin C, Somlo S, Wu G (2000) Cloning and characterization of the murine Pkd2 promoter. Genomics 66:305–312

Pazour GJ, San Agustin JT, Follit JA, Rosenbaum JL, Witman GB (2002) Polycystin-2 localizes to kidney cilia and the ciliary level is elevated in orpk mice with polycystic kidney disease. Curr Biol 12:R378–R380

Pennekamp P, Bogdanova N, Wilda M, Markoff A, Hameister H, Horst J, Dworniczak B (1998) Characterization of the murine polycystic kidney disease (Pkd2) gene. Mamm Genome 9:749–752

Pennekamp P, Karcher C, Fischer A, Schweickert A, Skryabin B, Horst J, Blum M, Dworniczak B (2002) The ion channel protein polycystin-2 is required for left-right axis determination in mice. Curr Biol 12:938–943

Peters DJM, Sandkuijl LA (1992) Genetic heterogeneity of polycystic kidney disease in Europe. Contrib Nephrol 97:128–139

Praetorius HA, Spring KR (2001) Bending the MDCK cell primary cilium increases intracellular calcium. J Membr Biol 184:71–79

Qian F, Germino FJ, Cai Y, Zhang X, Somlo S, Germino GG (1997) PKD1 interacts with PKD2 through a probable coiled-coil domain. Nat Genet 16:179–183

Ramasubbu K, Gretz N, Bachmann S (1998) Increased epithelial cell proliferation and abnormal extracellular matrix in rat polycystic kidney disease. J Am Soc Nephrol 9:937–945

Roscoe JM, Brissenden JE, Williams EA, Chery AL, Silverman M (1993) Autosomal dominant polycystic kidney disease in Toronto. Kidney Int 44:1101–1108

Rundle DR, Gorbsky G, Tsiokas L (2004) PKD2 interacts and co-localizes with mDia1 to mitotic spindles of dividing cells. Role of mDia1 in PKD2 localization to mitotic spindles. J Biol Chem 279:29728–29739

Schwiebert EM, Wallace DP, Braunstein GM, King SR, Peti-Peterdi J, Hanaoka K, Guggino WB, Guay-Woodford LM, Bell PD, Sullivan LP, Grantham JJ, Taylor AL (2002) Autocrine extracellular purinergic signaling in epithelial cells derived from polycystic kidneys. Am J Physiol Renal Physiol 282:F763–F775

Sweeney WE Jr, Chen Y, Nakanishi K, Frost P, Avner ED (2000) Treatment of polycystic kidney disease with a novel tyrosine kinase inhibitor. Kidney Int 57:33–40

Torra R, Badenas C, Darnell A, Nicolau C, Volpini V, Revert L, Estivill X (1996) Linkage, clinical features, and prognosis of autosomal dominant polycystic kidney disease types 1 and 2. J Am Soc Nephrol 7:2142–2151

Torres VE, Sweeney WE Jr, Wang X, Qian Q, Harris PC, Frost P, Avner ED (2003) EGF receptor tyrosine kinase inhibition attenuates the development of PKD in Han:SPRD rats. Kidney Int 64:1573–1579

Torres VE, Wang X, Qian Q, Somlo S, Harris PC, Gattone VH II (2004) Effective treatment of an orthologous model of autosomal dominant polycystic kidney disease. Nat Med 10:363–364

Tsiokas L, Kim E, Arnould T, Sukhatme VP, Walz G (1997) Homo- and heterodimeric interactions between the gene products of PKD1 and PKD2. Proc Natl Acad Sci USA 94:6965–6970

Tsiokas L, Arnould T, Zhu C, Kim E, Walz G, Sukhatme VP (1999) Specific association of the gene product of PKD2 with the TRPC1 channel. Proc Natl Acad Sci USA 96:3934–3939

Vassilev PM, Guo L, Chen XZ, Segal Y, Peng JB, Basora N, Babakhanlou H, Cruger G, Kanazirska M, Ye Cp, Brown EM, Hediger MA, Zhou J (2001) Polycystin-2 is a novel cation channel implicated in defective intracellular Ca^{2+} homeostasis in polycystic kidney disease. Biochem Biophys Res Commun 282:341–350

Wheatley DN (1995) Primary cilia in normal and pathological tissues. Pathobiology 63:222–238

Wheatley DN, Wang AM, Strugnell GE (1996) Expression of primary cilia in mammalian cells. Cell Biol Int 20:73–81

Witzgall R (2005a) New developments in the field of cystic kidney diseases. Curr Mol Med 5:455–465

Witzgall R (2005b) Polycystin-2—an intracellular or plasma membrane channel? Naunyn Schmiedebergs Arch Pharmacol 371:342–347

Wright AF, Teague PW, Pound SE, Pignatelli PM, Macnicol AM, Carothers AD, de Mey RJ, Allan PL, Watson ML (1993) A study of genetic linkage heterogeneity in 35 adult-onset polycystic kidney disease families. Hum Genet 90:569–571

Wu G, D'Agati V ,Cai Y, Markowitz G, Park JH, Reynolds DM, Maeda Y, Le TC, Hou H Jr, Kucherlapati R, Edelmann W, Somlo S (1998) Somatic inactivation of Pkd2 results in polycystic kidney disease. Cell 93:177–188

Xu GM, González-Perrett S, Essafi M, Timpanaro GA, Montalbetti N, Arnaout MA, Cantiello HF (2003) Polycystin-1 activates and stabilizes the polycystin-2 channel. J Biol Chem 278:1457–1462

Yoder BK, Hou X, Guay-Woodford LM (2002) The polycystic kidney disease proteins, polycystin-1, polycystin-2, polaris, and cystin, are co-localized in renal cilia. J Am Soc Nephrol 13:2508–2516

Part V
TRP Proteins and Specific Cellular Functions

Know Thy Neighbor: A Survey of Diseases and Complex Syndromes that Map to Chromosomal Regions Encoding TRP Channels

J. Abramowitz · L. Birnbaumer (✉)

Transmembrane Signaling Group, Laboratory of Signal Transduction,
Division of Intramural Research, National Institutes of Environmental Health Sciences,
National Institutes of Health, Department of Health and Human Services,
Building 101, Room A214,
111 T.W. Alexander Drive, Research Triangle Park NC, 27709, USA
birnbau1@niehs.nih.gov

1	Introduction	379
2	On the Naming of TRPs and Their Genes	383
3	Chromosomal Distribution of TRP Channel Genes	383
4	Disease Associations	397
	References	398

Abstract On the basis of their ever-expanding roles, not only in sensory signaling but also in a plethora of other, often Ca^{2+}-mediated actions in cell and whole body homeostasis, it is suggested that mutations in TRP channel genes not only cause disease states but also contribute in more subtle ways to simple and complex diseases. A survey is therefore presented of diseases and syndromes that map to one or multiple chromosomal loci containing TRP channel genes. A visual map of the chromosomal locations of TRP channel genes in man and mouse is also presented.

Keywords TRP Channels · Genetic diseases

1
Introduction

Knowledge about the existence of the transient receptor potential (TRP) superfamily of cation channels has accumulated thanks to seven or more independent research lines. Each of these research lines had evolved with the purpose of elucidating the molecular basis of a different set of biological or clinical problems. The TRPC channels were discovered in response to hypothesizing that an ortholog of the *Drosophila* TRP visual transduction channel might be responsible for mediating Ca^{2+} entry into mammalian cells after activation of the Gq signal transduction pathway and attendant store depletion, and that this or a similar channel might also mediate Ca^{2+} entry triggered by store depletion without previous activation of the Gq signaling pathway. The TRPM subfamily emerged from studying differences in gene expression in melanomas as compared to wildtype cells. The TRPV subfamily was uncovered when four laboratories independently cloned and then compared the complementary DNAs (cDNAs) of the capsaicin receptor, the renal and intestinal epithelial Ca^{2+} transporters, and a growth factor-activated Ca^{2+} entry channel. The TRPP (also polycystin or PKD subfamily) emerged from research into the genetic basis of polycystic kidney disease, while the TRPML (also mucolipin) subfamily emerged from cloning the gene responsible for type 4 mucolipidosis. The lone and final member of the TRP superfamily is TRPA1, identified as a force transduction channel that operates in inner ear hair cell stereocilia (for further details on these historical aspects of the discovery of TRP channel members see E. Yildirim and L. Birnbaumer, this volume).

From the foregoing, as well as from the elucidation of the broad spectrum of the roles that TRP channels play in a wide variety of physiological functions—leading to the collection of chapters in this book—it follows that failure to function properly due to either loss of function (*lof*) or gain of function (*gof*) mutations is likely to result in simple or complex pathological consequences. Some consequences of mutations in TRP channel encoding loci are obvious, e.g., that *lof* at the TRPM7 locus is incompatible with cellular viability (TRPM7; Nadler et al. 2001); that *lof* at the PKD2 locus (encoding TRPP2) is responsible for one form of autosomal dominant familial polycystic kidney disease (PKD2; Mochizuki et al. 1996); and that a *gof* mutation in TRPC6 is responsible for familial focal segmental glomerulosclerosis (FSGS) (Winn et al. 2005). Consequences of other mutations are more subtle, such as the contribution to complex conditions like the development of a melanoma (Hunter et al. 1998). Knockout studies—such as that of TRPC4, showing diminished increases in vascular endothelial fluid permeability (Tirrupathi et al. 2002) and NO production (Freichel et al. 2001), increased vascular smooth muscle contractility in TRPC6-deficient mice (Dietrich et al. 2005), or a severely blunted allergic inflammatory response in mice lacking TRPC1 (E. Yildirim, M.A. Carey, J.W. Card, A. Dietrich, J.A. Bradbury, Y. Rebolloso, G. Flake, D. Morgan, L. Birnbaumer, D. Zeldin et al., unpublished)—predict that *lof* and *gof* mutations in human TRPC genes ought to have occurred, and that there ought to be not just a few but a plethora of diseases and syndromes that trace to mutations in TRP genes. Below we present a survey of diseases and syndromes that map to chromosomal loci that encode TRP genes.

Table 1 TRP nomenclature and chromosomal location of TRP channel genes in the human and mouse genomes. The locations of human TRP genes are denoted on the basis of cytological banding. Those of mouse TRP genes are in genetic recombination units (cM) or on the basis of cytological banding

TRP channel	Chromosomal locus	
	Human	Mouse
TRPC1	3q22-q24	9 51.0 cM
TRPC2	11p15.3-p15.4; 11q13.2	7 50.0 cM
TRPC3	4q27	3 18.2 cM
TRPC4	13q13.1-q13.2	3 28.6 cM
TRPC5	Xq23	X 63.0 cM
TRPC6	11q21-q22	9 1.0 cM
TRPC7	5q31.1	13 B2
TRPM1	15q13-q14	7 27.0 cM
TRPM2	21q22.3	10 41.7 cM
TRPM3	9q21.11	19 B
TRPM4	19q13.32	7 B3
TRPM5	11p15.5	7 E5
TRPM6	9q21.13	19 B
TRPM7	15q21	2 F2
TRPM8	2q37.1	1 D
TRPV1	17p13.3	11 44.1 cM
TRPV2	17p11.2	11 B2
TRPV3	17p13.3	11 44.1 cM
TRPV4	12q23-q24.1	5 F
TRPV5	7q35	6 B2
TRPV6	7q33-q34	6 B2
TRPP1	16p13.3	17 10.4 cM
TRPP2	4q21-23	5 55.0 cM
TRPP3	10q24	19 44.0 cM
TRPP4	22q13.31	15 E2
TRPP5	5q31	18 C
TRPML1	19p13.3-p13.2	8 1.1 cM
TRPML2	1p22	3 74.8 cM
TRPML3	1p22.3	3 74.8 cM
TRPA1	8q13	1 A3

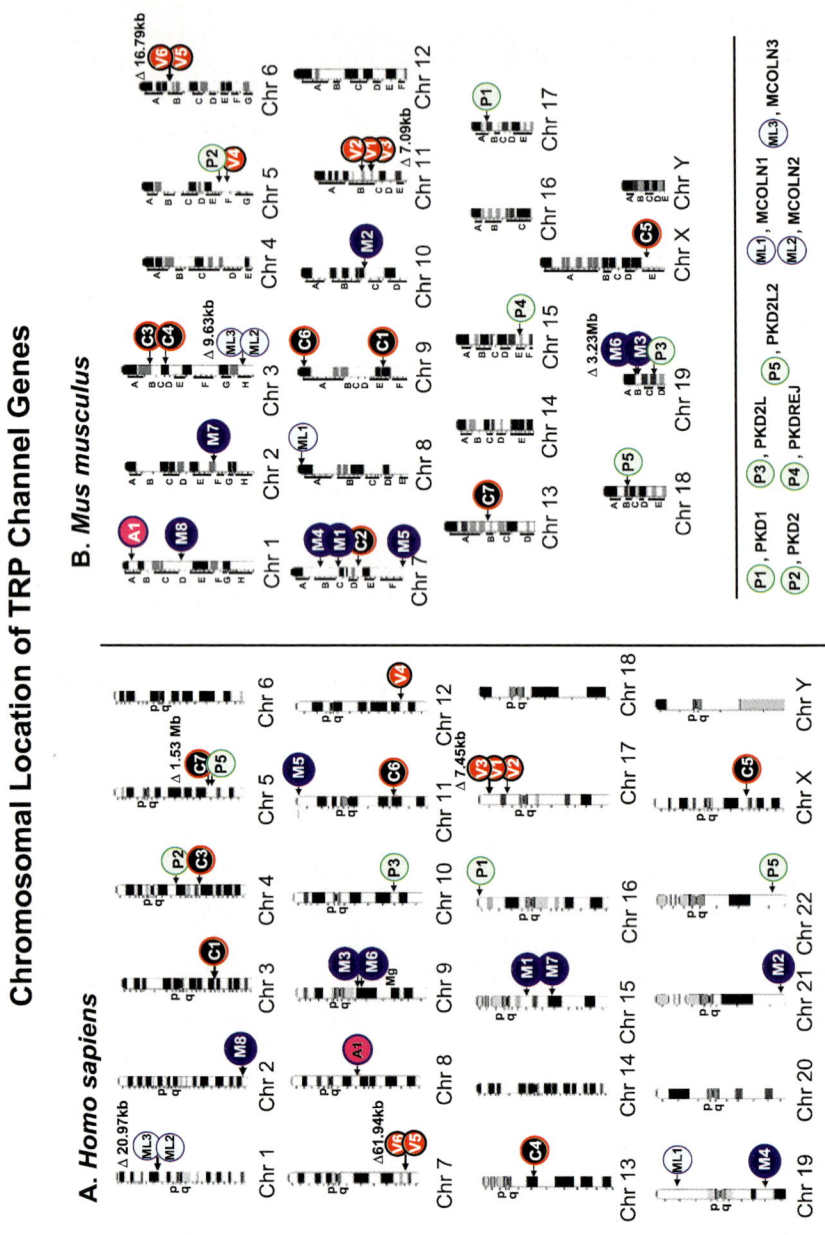

Fig. 1 Chromosomal location of human and mouse TRP channel genes. Ideograms of chromosomes with banding patterns are from http://www.ncbi.nlm.nih.gov/mapview/maps.cgi. *Numbers* next to closely located genes are chromosomal coordinates in millions of nucleotides. For the human genome, all chromosomes are scaled to the same size; for the mouse genome, the chromosomes have been scaled to their number of nucleotides. The TRPP and TRPML coding is presented at the *bottom of the right panel*

2
On the Naming of TRPs and Their Genes

Table 1 summarizes the nomenclature used in this chapter and lists the chromosomal locations of the 30 TRP channel genes analyzed for this chapter. Although accepted by the International Union of Pharmacology (IUPHAR) (Clapham et al. 2005), searches of the National Center for Biotechnical Information (NCBI) genomic BLAST site do not respond to the terms TRPP or TRPML, instead requiring use of the disease-derived PKD- and mucolipin (MCOLN)-based names. We will be using the TRPP and TRPML nomenclature throughout this chapter.

3
Chromosomal Distribution of TRP Channel Genes

A visual image of the chromosomal locations of TRP channel genes and their distribution throughout the human and mouse genomes is presented in Fig. 1.

Table 2 Diseases and/or syndromes caused by mutations or associated with altered expression

Disease	TRP channel	Reference(s)
A. *Mutants*		
Aromatase excess syndrome	TRPM7	Tiulpakov et al. 2005
Deafness	TRPML3	Di Palma et al. 2002
Focal segmental glomerulosclerosis	TRPC6[a]	Winn et al. 2005; Reiser et al. 2005
Polycystic kidney disease	TRPP1	The European Polycystic Kidney Disease Consortium 1994
	TRPP2	Mochizuki et al. 1996
Hypomagnesemia with secondary hypocalcemia	TRPM6	Schlingmann et al. 2002; Walder et al. 2002
Mucolipidosis type IV	TRPML1	Sun et al. 2000; Bargal et al. 2000
B. *Altered expression*		
Melanoma	TRPM1	Hunter et al. 1998
Muscular dystrophy	TRPV2	Iwata et al. 2003
Prostate cancer	TRPM8	Tsavaler et al. 2001
	TRPV6	Wissenbach et al. 2001

[a] Activating mutation

Table 3 Diseases and/or complex syndromes that show associations with (map to) chromosomal regions that encode TRP channel genes

Disease	TRP channel loci[a]	Reference(s)
Map to five or more TRP loci		
Multiple sclerosis	TRPC1	Ban et al. 2002
	TRPC3	Ban et al. 2002
	TRPC5	Ban et al. 2002
	TRPM3/M6	Ban et al. 2002
	TRPM4	Haines et al. 2002; Jonasdottir et al. 2003
	TRPM5	Coraddu et al. 2001
	TRPV1/V3	Ban et al. 2002
	TRPV4	Haines et al. 2002
	TRPV5	Xu et al. 2001a; Ban et al. 2002
	TRPV6	Ban et al. 2002
	TRPP1	Haines et al. 2002
	TRPP4	GAMES, Transatlantic Multiple Sclerosis Genetics Cooperative 2003
	TRPML	Sawcer et al. 2005
Schizophrenia	TRPC1	Camp et al. 2001; Bulayeva et al. 2005
	TRPC3	Gurling et al. 2001
	TRPC4	Camp et al. 2001
	TRPC6	Gurling et al. 2001
	TRPC7/P5	Schwab et al. 1997
	TRPM1	Liu et al. 2001
	TRPM2	Demirhan and Tastemir 2003; Lerer et al. 2003
	TRPM3/M6	Hovatta et al. 1999
	TRPM8	Lerer et al. 2003
	TRPV2	Bulayeva et al. 2005
	TRPP2	Gurling et al. 2001
	TRPP3	Lerer et al. 2003
	TRPP4	Gross et al. 2001; Devaney et al. 2002; Demirhan and Tastemir 2003
Bipolar disorder	TRPC7/P5	Hong et al. 2004
	TRPM2	Straub et al. 1994
	TRPM5	Zandi et al. 2003
	TRPV4	Berrettini 2000
	TRPV6	Liu et al. 2003
	TRPA1	Liu et al. 2003

Table 3 continued

Disease	TRP channel loci[a]	Reference(s)
	TRPP1	Cheng et al. 2006
	TRPP3	Liu et al. 2003
	TRPP4	Berrettini 2000; Liang et al. 2002
	TRPML	Cheng et al. 2006
Charcot–Marie–Tooth neuropathy	TRPC1	Gemignani and Marbini 2001
	TRPC7/P5	Guilbot et al. 1999
	TRPM1	Gemignani and Marbini 2001
	TRPM4	Berghoff et al. 2004
	TRPM5	Senderek et al. 2003
	TRPV2	Thomas et al. 1997
	TRPV4	Klein et al. 2003
	TRPA1	Ben Othmane et al. 1998
	TRPP3	Verhoeven et al. 2001
Prostate cancer	TRPC6	Xu et al. 2005
	TRPC7/P5	Nwosu et al. 2001
	TRPM4	Wiklund et al. 2003
	TRPM8	Tsavaler et al. 2001
	TRPV2	Nwosu et al. 2001
	TRPV6	Wissenbach et al. 2001
	TRPA1	Xu et al. 2005
	TRPP1	Xu et al. 2005
	TRPP3	Matsuyama et al. 2003
Asthma	TRPC4	Xu et al. 2001b
	TRPC6	Xu et al. 2001b
	TRPC7/P5	Shek et al. 2001; Xu et al. 2001b
	TRPM1	Xu et al. 2001b
	TRPM4	Venanzi et al. 2001
Cervical carcinoma	TRPC1	Rao et al. 2004
	TRPC6	Rao et al. 2004
	TRPV2	Rao et al. 2004
	TRPP3	Rao et al. 2004
	TRPML	Rao et al. 2004
Obesity	TRPC4	Dong et al. 2005
	TRPM8	Saar et al. 2003
	TRPV6	Bouchard et al. 1996
	TRPP2	Saar et al. 2003
	TRPML	Saar et al. 2003

Table 3 continued

Disease	TRP channel loci[a]	Reference(s)
Retinitis pigmentosa	TRPC1	Wang et al. 2001
	TRPC7/P5	Wang et al. 2001
	TRPM4	Wang et al. 2001
	TRPM8	Wang et al. 2001
	TRPV1/V3	Kojis et at. 1996
Systemic sclerosis	TRPC5	Zhou et al. 2003
	TRPM4	Zhou et al. 2003
	TRPV5	Zhou et al. 2003
	TRPP4	Zhou et al. 2003
	TRPML	Zhou et al. 2003
Type 2 diabetes	TRPC1	Pratley et al. 1998
	TRPM1	Mori et al. 2002
	TRPM3/M6	Pratley et al. 1998
	TRPM7	Mori et al. 2002
	TRPP4	Pratley et al. 1998
Map to four TRP loci		
Breast cancer	TRPV1/V3	Hirano et al. 2001; Nagahata et al. 2002
	TRPP4	Hirano et al. 2001; Nagahata et al. 2002
	TRPML2/3	Nagahata et al. 2002
	TRPC6	Launonen et al. 1999
Lupus erythematosus	TRPM5	Quintero-Del-Rio et al. 2002
	TRPM8	Kelly et al. 2002
	TRPV1/V3	Kelly et al. 2002
	TRPP1	Kelly et al. 2002
Myeloid leukemia	TRPC1	Stark et al. 1994
	TRPC5	Wong et al. 2000
	TRPM5	Stark et al. 1994
	TRPV6	Koike et al. 1999
Map to three TRP loci		
Alzheimer's	TRPM2	Blacker et al. 2003
	TRPM4	Blacker et al. 2003
	TRPP3	Blacker et al. 2003
Atopic dermatitis	TRPC1	Beyer et al. 2001
	TRPC7/P5	Beyer et al. 2001
	TRPV4	Beyer et al. 2001
Barrett's esophagus (cancer)	TRPM3/M6	Riegman et al. 2001
	TRPV1/V3	Riegman et al. 2001
	TRPV6	Riegman et al. 2001

Table 3 continued

Disease	TRP channel loci[a]	Reference(s)
Lipid abnormalities	TRPM4	Malhotra et al. 2005
	TRPM5	Malhotra et al. 2005
	TRPML1	Malhotra et al. 2005
Melanoma	TRPM1	Hunter et al. 1998
	TRPM3/M6	Jonsson et al. 2005
	TRPML2/3	Gillanders et al. 2003
Metabolic syndrome	TRPM3/M6	Loos et al. 2003
	TRPM4	Loos et al. 2003
	TRPP1	Francke et al. 2001
Otosclerosis (hearing loss)	TRPC1	Van Den Bogaert et al. 2004
	TRPV5	Van Den Bogaert et al. 2001, 2004
	TRPV6	Van Den Bogaert et al. 2001, 2004
Spinocerebellar ataxia	TRPP3	Nikali et al. 1997
	TRPP4	Matsuura et al. 1999
	TRPML1	Yu et al. 2005
Usher syndrome	TRPC1	Petit 2001
	TRPM2	Petit 2001
	TRPM5	Ouyang et al. 2005
Wilms tumor	TRPM4	Ruteshouser and Huff 2004
	TRPM8	Viot-Szoboszlai et al. 1998
	TRPV1/V3	Ruteshouser and Huff 2004
Map to two TRP loci		
Adrenocortical tumors	TRPM5	Gicquel et al. 2001
	TRPV1/V3	Gicquel et al. 2001
Atypical cerebral asymmetry	TRPV2	Smalley et al. 2005
	TRPP1	Smalley et al. 2005
Behcet's disease	TRPV1/V3	Karasneh et al. 2005
	TRPP3	Karasneh et al. 2005
Celiac disease	TRPC7/P5	Liu et al. 2002
	TRPM3/M6	Liu et al. 2002
Cleft lip	TRPV1/V3	Wyszynski et al. 2003
	TRPV6	Wyszynski et al. 2003
Deafness	TRPM4	Mirghomizadeh et al. 2002
	TRPML2/3	Di Palma et al. 2002
Drusen formation	TRPM4	Jun et al. 2005
	TRPP1	Jun et al. 2005
End-stage renal disease	TRPC4	Freedman et al. 2004
	TRPP3	Freedman et al. 2002

Table 3 continued

Disease	TRP channel loci[a]	Reference(s)
Inflammatory bowel disease	TRPM4	Paavola-Sakki et al. 2003
	TRPV4	Paavola-Sakki et al. 2003
Lymphocytic leukemia	TRPC6	Cuneo et al. 2002
	TRPV1/V3	Cuneo et al. 2002
Nonsyndromic deafness	TRPM2	Berry et al. 2000
	TRPV4	Blanton et al. 2002
Parkinson's disease	TRPP2	Nussbaum and Polymeropoulos 1997
	TRPP4	Wilhelmsen et al. 1997
Polycystic kidney disease	TRPP1	European Polycystic Kidney Disease Consortium 1994
	TRPP2	Mochizuki et al. 1996
Sarcoidosis	TRPM5	Iannuzzi et al. 2005
	TRPML2/3	Iannuzzi et al. 2005
Vitiligo	TRPV1/V3	Zhang et al. 2005
	TRPP2	Zhang et al. 2005
Map to one TRP Locus		
Albright's hereditary osteodystrophy	TRPM8	Phelan et al. 1995; Sakaguchi et al. 1998
Aromatase excess syndrome	TRPM7	Tiulpakov et al. 2005
Autosomal dominant medullary cystic Kidney disease	TRPC1	Scolari et al. 2003
Beckwith–Wiedemann syndrome with renal abnormalities-cysts	TRPM5	Prawitt et al. 2000; Goldman et al. 2002
Birt–Hogg–Dube syndrome	TRPV2	Schmidt et al. 2001
Bone mineral density	TRPP1	Ralston et al. 2005
Clark–Baraitser syndrome	TRPP4	Tabolacci et al. 2005
Congenital bile duct dilatation	TRPA1	Parada et al. 1999
Darier's disease	TRPV4	Jacobsen et al. 2001
Dementia	TRPML1	Hahs et al. 2006
Diabetes insipidus	TRPP4	Barakat et al. 2004
Febrile convulsions	TRPA1	Wallace et al. 1996

Table 3 continued

Disease	TRP channel loci[a]	Reference(s)
Focal segmental glomerulosclerosis	TRPC6	Reiser et al. 2005; Winn et al. 2005
Hereditary neuropathy with liability to pressure palsies	TRPV2	Hong et al. 2003
Hereditary spastic paraplegia	TRPP3	Meijer et al. 2004
Hypomagnesemia with secondary hypocalcemia	TRPM3/M6	Schlingmann et al. 2002; Walder et al. 2002
Idiopathic generalized epilepsy	TRPM2	Rees et al. 1994
Juvenile myoclonic epilepsy	TRPM1	Taske et al. 2002
Limb girdle muscular dystrophy	TRPM7	Fougerousse et al. 1994
Long QT syndrome	TRPM5	Schulz-Bahr et al. 1999
Mental retardation	TRPV1/V3	Walter et al. 2004
Miller–Dieker epilepsy	TRPV1/V3	Schinzel et al. 2001
Mucolipidosis type IV	TRPML1	Sun et al. 2000; Bargal et al. 2000
Muscular dystrophy	TRPV2	Iwata et al. 2003
Nephronophthisis	TRPC1	Omran et al. 2000, 2001
Nephrotic syndrome	TRPM4	Vats et al. 2000
Non-myotonic dystrophy	TRPM7	Le Ber et al. 2004
Oguchi disease-night blindness	TRPM8	Maw et al. 1995
Partial epilepsy	TRPP3	Nobile et al. 1999
Photoparoxysmal response	TRPP1	Pinto et al. 2005
Pigment dispersion syndrome	TRPV5	Andersen et al. 1997
Polycystic ovary syndrome	TRPML1	Urbanek et al. 2005
Progressive external ophthalmoplegia	TRPP3	Li et al. 1999
Progressive myoclonus epilepsy	TRPM2	Delgado-Escueta et al. 2001
Rolandic epilepsy	TRPM1	Neubauer et al. 1998

Table 3 continued

Disease	TRP channel loci[a]	Reference(s)
Schizoaffective disorder	TRPML1	Hamshere et al. 2005
Seckel syndrome	TRPC1	Goodship et al. 2000
Stargardt disease	TRPML2/3	Hoyng et al. 1996
Type 1 diabetes	TRPM5	Cox et al. 2001
Weill–Marchesani syndrome	TRPM7	Wirtz et al. 1996
X-linked mental retardation	TRPC5	Gregg et al. 1996

[a]Channels located at the same locus are listed together

With the exception of the Y chromosomes, human chromosomes (Chr)6, 14, and 20, and mouse Chr4, 14 and 16, all other chromosomes encode one or more TRP channel genes. The chromosomes that carry the largest number of TRP channel genes are Chr3 and Chr7 in the mouse, with four TRPs each, and human Chr17 with three genes. The very close location of the TRPV1/TRPV3, TRPV5/TRPV6, and TRLML2/TRPML3 genes, with intergenic distances of less than 100 kb in both the human and the mouse genomes, is noteworthy. The intergenic distances between human (mouse) *TRPV1* and *TRPV3* are just 7.45 kb (7.09 kb) with no known gene between them. Likewise, the distances between human (mouse) *TRPV5* and *TRPV6*, and *TRPML2* and *TRPML3* are (V5/6) 11.94 kb (16.79 kb) and (ML2/3) 20.79 kb (9.63 kb), also with no known intervening gene. In the human genome, the next closest pair of TRP genes is that of *TRPC7* and *TRPP5* (*PKD2L2*). They are encoded in opposite (tail-to-tail) directions separated by 1.5 Mb that code for five known genes. In the mouse genome the next closest pair is that of *TRPM3* and *TRPM6*. They are encoded by the same strand, separated by 3.2 Mb encoding 16 identifiable loci.

In both humans and mice, the TRPV2 gene maps very close to the TRPV1/TRPV3 gene doublets, with the TRPV3 gene being the "middle" gene in both genomes. An amino acid sequence alignment shows the TRPV1 to be more closely related to the more distant TRPV2 gene than the more closely positioned TRPV3 gene. It would appear that the three genes are the result of rather recent and independent gene duplication events, the first duplication generating the TRPV3 gene and the second duplicating the ancestor of the TRPV1 and TRPV3 genes.

TRP Channels and Disease

Table 4 Genetic loci encoding TRP channels that contribute or may contribute to one or more diseases and/or syndromes

TRP channel and locus[a]	Disease	Reference(s)
TRPC		
TRPC1 3q22–24	Atopic dermatitis	Beyer et al. 2001
	Autosomal dominant medullary cystic kidney disease	Scolari et al. 2003
	Cervical carcinoma	Rao et al. 2004
	Charcot–Marie–Tooth neuropathy	Gemignani and Marbini 2001
	Multiple sclerosis	Ban et al. 2002
	Myeloid leukemia	Stark et al. 1994
	Nephronophthisis	Omran et al. 2000, 2001
	Otosclerosis (hearing loss)	Van Den Bogaert et al. 2004
	Retinitis pigmentosa	Wang et al. 2001
	Schizophrenia	Camp et al. 2001; Bulayeva et al. 2005
	Seckel syndrome	Goodship et al. 2000
	Type 2 diabetes	Pratley et al. 1998
	Usher syndrome	Petit 2001
TRPC3 4q27	Multiple sclerosis	Ban et al. 2002
	Schizophrenia	Gurling et al. 2001
TRPC4 13q13.1–q13.2	Asthma	Xu et al. 2001b
	End-stage renal disease	Freedman et al. 2004
	Obesity	Dong et al. 2005
	Schizophrenia	Camp et al. 2001
TRPC5 Xq23	Multiple sclerosis	Ban et al. 2002
	Myeloid leukemia	Wong et al. 2000
	Systemic sclerosis	Zhou et al. 2003
	X-linked mental retardation	Gregg et al. 1996
TRPC6 11q21–q22	Asthma	Xu et al. 2001b
	Breast cancer	Launonen et al. 1999
	Cervical carcinoma	Rao et al. 2004
	Focal segmental glomerulosclerosis	Reiser et al. 2005; Winn et al. 2005
	Lymphocytic leukemia	Cuneo et al. 2002
	Schizophrenia	Gurling et al. 2001
	Prostate cancer	Xu et al. 2005

Table 4 continued

TRP channel and locus[a]	Disease	Reference(s)
TRPC7 5q31.1/ TRPP5 5q31	Asthma	Shek et al. 2001; Xu et al. 2001b
	Atopic dermatitis	Beyer et al. 2001
	Bipolar disorder	Hong et al. 2004
	Celiac disease	Liu et al. 2002
	Charcot–Marie–Tooth neuropathy	Guilbot et al. 1999
	Prostate cancer	Nwosu et al. 2001
	Retinitis pigmentosa	Wang et al. 2001
	Schizophrenia	Schwab et al. 1997
TRPM		
TRPM1 15q13–q14	Asthma	Xu et al. 2001b
	Charcot–Marie–Tooth neuropathy	Gemignani and Marbini 2001
	Juvenile myoclonic epilepsy	Taske et al. 2002
	Melanoma	Hunter et al. 1998
	Rolandic epilepsy	Neubauer et al. 1998
	Schizophrenia	Liu et al. 2001
	Type 2 diabetes	Mori et al. 2002
TRPM2 21q22.3	Alzheimer's	Blacker et al. 2003
	Bipolar disorder	Straub et al. 1994
	Idiopathic generalized epilepsy	Rees et al. 1994
	Nonsyndromic deafness	Berry et al. 2000
	Progressive myoclonus epilepsy	Delgado-Escueta et al. 2001
	Schizophrenia	Demirhan and Tastemir 2003; Lerer et al. 2003
	Usher syndrome	Petit 2001
TRPM3 9q21.11/ TRPM6 9q21.13	Barrett's esophagus (cancer)	Riegman et al. 2001
	Celiac disease	Liu et al. 2002
	Hypomagnesemia with secondary hypocalcemia	Schlingmann et al. 2002; Walder et al. 2002
	Melanoma	Jonsson et al. 2005
	Metabolic syndrome	Loos et al. 2003
	Multiple sclerosis	Ban et al. 2002
	Schizophrenia	Hovatta et al. 1999
	Type 2 diabetes	Pratley et al. 1998

Table 4 continued

TRP channel and locus[a]	Disease	Reference(s)
TRPM4 19q13.32	Alzheimer's	Blacker et al. 2003
	Asthma	Venanzi et al. 2001
	Charcot–Marie–Tooth neuropathy	Berghoff et al. 2004
	Deafness	Mirghomizadeh et al. 2002
	Drusen formation	Jun et al. 2005
	Inflammatory bowel disease	Paavola-Sakki et al. 2003
	Lipid abnormalities	Malhotra et al. 2005
	Metabolic syndrome	Loos et al. 2003
	Multiple sclerosis	Haines et al. 2002; Jonasdottir et al. 2003
	Nephrotic syndrome	Vats et al. 2000
	Prostate cancer	Wiklund et al. 2003
	Retinitis pigmentosa	Wang et al. 2001
	Systemic sclerosis	Zhou et al. 2003
	Wilms tumor	Ruteshouser and Huff 2004
TRPM5 11p15.5	Adrenocortical tumors	Gicquel et al. 2001
	Beckwith–Wiedemann syndrome with renal abnormalities–cysts	Prawitt et al. 2000; Goldman et al. 2002
	Bipolar disorder	Zandi et al. 2003
	Charcot–Marie–Tooth neuropathy	Senderek et al. 2003
	Lipid abnormalities	Malhotra et al. 2005
	Long QT syndrome	Schulz-Bahr et al. 1999
	Lupus erythematosus	Quintero-Del-Rio et al. 2002
	Multiple sclerosis	Coraddu et al. 2001
	Myeloid leukemia	Stark et al. 1994
	Sarcoidosis	Iannuzzi et al. 2005
	Type 1 diabetes	Cox et al. 2001
	Usher syndrome	Ouyang et al. 2005
TRPM6 see TRPM3		
TRPM7 15q21	Aromatase excess syndrome	Tiulpakov et al. 2005
	Limb girdle muscular dystrophy	Fougerousse et al. 1994
	Non-myotonic dystrophy	Le Ber et al. 2004
	Type 2 diabetes	Mori et al. 2002
	Weill–Marchesani syndrome	Wirtz et al. 1996

Table 4 continued

TRP channel and locus[a]	Disease	Reference(s)
TRPM8 2q37.1	Albright's hereditary osteodystrophy	Phelan et al. 1995; Sakaguchi et al. 1998
	Lupus erythematosus	Kelly et al. 2002
	Obesity	Saar et al. 2003
	Oguchi disease-night blindness	Maw et al. 1995
	Prostate cancer	Tsavaler et al. 2001
	Retinitis pigmentosa	Wang et al. 2001
	Schizophrenia	Lerer et al. 2003
	Wilms tumor	Viot-Szoboszlai et al. 1998
TRPV		
TRPV1 17p13.3/ TRPV3 17p13.3	Adrenocortical tumors	Gicquel et al. 2001
	Barrett's esophagus (cancer)	Riegman et al. 2001
	Behcet's disease	Karasneh et al. 2005
	Breast cancer	Hirano et al. 2001; Nagahata et al. 2002
	Cleft lip	Wyszynski et al. 2003
	Lupus erythematosus	Kelly et al. 2002
	Lymphocytic leukemia	Cuneo et al. 2002
	Mental retardation	Walter et al. 2004
	Miller–Dieker epilepsy	Schinzel et al. 2001
	Multiple sclerosis	Ban et al. 2002
	Retinitis pigmentosa	Kojis et at. 1996
	Vitiligo	Zhang et al. 2005
	Wilms tumor	Ruteshouser and Huff 2004
TRPV2 17p11.2	Atypical cerebral asymmetry	Smalley et al. 2005
	Birt–Hogg–Dube syndrome	Schmidt et al. 2001
	Cervical carcinoma	Rao et al. 2004
	Charcot–Marie–Tooth neuropathy	Thomas et al. 1997
	Hereditary neuropathy with liability to pressure palsies	Hong et al. 2003
	Muscular dystrophy	Iwata et al. 2003
	Prostate cancer	Nwosu et al. 2001
	Schizophrenia	Bulayeva et al. 2005
TRPV3 see TRPV1		

Table 4 continued

TRP channel and locus[a]	Disease	Reference(s)
TRPV4 12q23–q24.1	Atopic dermatitis	Beyer et al. 2001
	Bipolar disorder	Berrettini 2000
	Charcot-Marie-Tooth Neuropathy	Klein et al. 2003
	Darier's disease	Jacobsen et al. 2001
	Inflammatory bowel disease	Paavola-Sakki et al. 2003
	Multiple sclerosis	Haines et al. 2002
	Nonsyndromic deafness	Blanton et al. 2002
TRPV5 7q35	Otosclerosis (hearing loss)	Van Den Bogaert et al. 2001, 2004
	Multiple sclerosis	Xu et al. 2001a; Ban et al. 2002
	Pigment dispersion syndrome	Andersen et al. 1997
	Systemic sclerosis	Zhou et al. 2003
TRPV6 7q33–q34	Barrett's esophagus (cancer)	Riegman et al. 2001
	Bipolar disorder	Liu et al. 2003
	Cleft lip	Wyszynski et al. 2003
	Multiple sclerosis	Ban et al. 2002
	Myelogenous leukemia	Koike et al. 1999
	Obesity	Bouchard et al. 1996
	Otosclerosis (hearing loss)	Van Den Bogaert et al. 2001, 2004
	Prostate cancer	Wissenbach et al. 2001
TRPA		
TRPA1 8q13	Bipolar disorder	Liu et al. 2003
	Charcot–Marie–Tooth neuropathy	Ben Othmane et al. 1998
	Congenital bile duct dilatation	Parada et al. 1999
	Febrile convulsions	Wallace et al. 1996
	Prostate cancer	Xu et al. 2005
TRPP		
TRPP1 16p13.3	Atypical cerebral asymmetry	Smalley et al. 2005
	Bipolar disorder	Cheng et al. 2006
	Bone mineral density	Ralston et al. 2005
	Drusen formation	Jun et al. 2005
	Lupus erythematosus	Kelly et al. 2002
	Metabolic syndrome	Francke et al. 2001
	Multiple sclerosis	Haines et al. 2002
	Photoparoxysmal response	Pinto et al. 2005

Table 4 continued

TRP channel and locus[a]	Disease	Reference(s)
TRPP2 4q21–q23	Polycystic kidney disease	European Polycystic Kidney Disease Consortium 1994
	Prostate cancer	Xu et al. 2005
	Obesity	Saar et al. 2003
	Parkinson's disease	Nussbaum and Polymeropoulos 1997
TRPP3 10q24	Polycystic kidney disease	Mochizuki et al. 1996
	Schizophrenia	Gurling et al. 2001
	Vitiligo	Zhang et al. 2005
	Alzheimer's	Blacker et al. 2003
	Behcet's disease	Karasneh et al. 2005
	Bipolar disorder	Liu et al. 2003
	Cervical carcinoma	Rao et al. 2004
	Charcot–Marie–Tooth neuropathy	Verhoeven et al. 2001
	End-stage renal disease	Freedman et al. 2002
	Hereditary spastic paraplegia	Meijer et al. 2004
	Partial epilepsy	Nobile et al. 1999
	Progressive external ophthalmoplegia	Li et al. 1999
	Prostate cancer	Matsuyama et al. 2003
	Schizophrenia	Lerer et al. 2003
	Spinocerebellar ataxia	Nikali et al. 1997
TRPP4 22q13.31	Bipolar disorder	Berrettini 2000; Liang et al. 2002
	Breast cancer	Hirano et al. 2001; Nagahata et al. 2002
	Clark–Baraitser syndrome	Tabolacci et al. 2005
	Diabetes insipidus	Barakat et al. 2004
	Spinocerebellar ataxia with epilepsy	Matsuura et al. 1999
	Multiple sclerosis	GAMES, Transatlantic Multiple Sclerosis Genetics Cooperative 2003
	Parkinson's disease	Wilhelmsen et al. 1997

Table 4 continued

TRP channel and locus[a]	Disease	Reference(s)
	Schizophrenia	Gross et al. 2001; Devaney et al. 2002; Demirhan and Tastemir 2003
	Systemic sclerosis	Zhou et al. 2003
	Type 2 diabetes	Pratley et al. 1998
TRPP5 see TRPC7		
TRPML		
TRPML1	Bipolar disorder	Cheng et al. 2006
19p13.3–p13.2	Cervical carcinoma	Rao et al. 2004
	Dementia	Hahs et al. 2006
	Lipid abnormalities	Malhotra et al. 2005
	Mucolipidosis type IV	Sun et al. 2000; Bargal et al. 2000
	Multiple sclerosis	Sawcer et al. 2005
	Obesity	Saar et al. 2003
	Polycystic ovary syndrome	Urbanek et al. 2005
	Schizoaffective disorder	Hamshere et al. 2005
	Spinocerebellar ataxia	Yu et al. 2005
	Systemic sclerosis	Zhou et al. 2003
TRPML2 1p22/	Breast cancer	Nagahata et al. 2002
TRPML3 1p22.3	Deafness	Di Palma et al. 2002
	Melanoma	Gillanders et al. 2003
	Sarcoidosis	Iannuzzi et al. 2005
	Stargardt disease	Hoyng et al. 1996
TRPML3 see TRPML2		

[a]Channels located at the same locus are listed together

4
Disease Associations

Confirmed associations of mutations in the coding regions of TRP channel genes with human diseases are listed in Table 2. Except one, all are *lof* mutations. The exception is FSGS, which was found to be due to a *gof* mutation in TRPC6 resulting in a proline to glutamine change at position 112 that confers constitutive activity to the TRPC6 channel. Glomerulosclerosis in the affected individuals is likely due to excess Ca^{2+} influx.

Table 3 lists TRP-disease associations that emerge if one asks which diseases or complex syndromes map to chromosomal regions encoding TRP channels. It should be noted that these are very broad associations and that in each case much work is needed to provide even an initial estimate of significance. Even so, it is interesting that for many diseases the association is not only with a single but multiple TRPs, indeed up to ten channels distributed over a large portion of the genome. It is tempting to suggest that in case of a disease pointing to multiple loci, this might be indicative of an alteration of a codon underlying function, in this case altered ionic influxes, or altered excitability if it is a neuronal syndrome. Table 4 presents the same data shown in Table 3, sorted according to TRP channels.

Acknowledgements Supported by the Intramural Research Division of the NIH, NIEHS.

References

Andersen JS, Pralea AM, DelBono EA, Haines JL, Gorin MB, Schuman JS, Mattox CG, Wiggs JL (1997) A gene responsible for the pigment dispersion syndrome maps to chromosome 7q35-q36. Arch Ophthalmol 115:384–388

Ban M, Stewart GJ, Bennetts BH, Heard R, Simmons R, Maranian M, Compston A, Sawcer SJ (2002) A genome screen for linkage in Australian sibling-pairs with multiple sclerosis. Genes Immun 3:464–469

Barakat AJ, Pearl PL, Acosta MT, Runkle BP (2004) 22q13 deletion syndrome with central diabetes insipidus: a previously unreported association. Clin Dysmorphol 13:191–194

Bargal R, Avidan N, Ben-Asher E, Olender Z, Zeigler M, Frumkin A, Raas-Rothschild A, Glusman G, Lancet D, Bach G (2000) Identification of the gene causing mucolipidosis type IV. Nat Genet 26:118–123

Ben Othmane K, Rochelle JM, Ben Hamida M, Slotterbeck B, Rao N, Hentati F, Pericak-Vance MA, Vance JM (1998) Fine localization of the CMT4A locus using a PAC contig and haplotype analysis. Neurogenetics 2:18–23

Berghoff C, Berghoff M, Leal A, Morera B, Barrantes R, Reis A, Neundorfer B, Rautenstrauss B, Del Valle G, Heuss D (2004) Clinical and electrophysiological characteristics of autosomal recessive axonal Charcot-Marie-Tooth disease (ARCMT2B) that maps to chromosome 19q13.3. Neuromuscul Disord 14:301–306

Berrettini WH (2000) Susceptibility loci for bipolar disorder: overlap with inherited vulnerability to schizophrenia. Biol Psychiatry 47:245–251

Berry A, Scott HS, Kudoh J, Talior I, Korostishevsky M, Wattenhofer M, Guipponi M, Barras C, Rossier C, Shibuya K, Wang J, Kawasaki K, Asakawa S, Minoshima S, Shimizu N, Antonarakis S, Bonne-Tamir B (2000) Refined localization of autosomal recessive nonsyndromic deafness DFNB10 locus using 34 novel microsatellite markers, genomic structure, and exclusion of six known genes in the region. Genomics 68:22–29

Beyer K, Nickel R, Freidhoff L, Bjorksten B, Huang SK, Barnes KC, MacDonald S, Forster J, Zepp F, Wahn V, Beaty TH, Marsh DG, Wahn U (2000) Association and linkage of atopic dermatitis with chromosome 13q12–14 and 5q31–33 markers. J Invest Dermatol 115:906–908

Blacker D, Bertram L, Saunders AJ, Moscarillo TJ, Albert MS, Wiener H, Perry RT, Collins JS, Harrell LE, Go RC, Mahoney A, Beaty T, Fallin MD, Avramopoulos D, Chase GA, Folstein MF, McInnis MG, Bassett SS, Doheny KJ, Pugh EW, Tanzi RE; NIMH Genetics Initiative Alzheimer's Disease Study Group (2003) Results of a high-resolution genome screen of 437 Alzheimer's disease families. Hum Mol Genet 12:23–32

Blanton SH, Liang CY, Cai MW, Pandya A, Du LL, Landa B, Mummalanni S, Li KS, Chen ZY, Qin XN, Liu YF, Balkany T, Nance WE, Liu XZ (2002) A novel locus for autosomal dominant non-syndromic deafness (DFNA41) maps to chromosome 12q24-qter. J Med Genet 39:567–570

Bouchard C, Perusse L (1996) Current status of the human obesity gene map. Obes Res 4:81–90

Bulayeva KB, Leal SM, Pavlova TA, Kurbanov RM, Glatt SJ, Bulayev OA, Tsuang MT (2005) Mapping genes of complex psychiatric diseases in Daghestan genetic isolates. Am J Med Genet B Neuropsychiatr Genet 132:76–84

Camp NJ, Neuhausen SL, Tiobech J, Polloi A, Coon H, Myles-Worsley M (2001) Genomewide multipoint linkage analysis of seven extended Palauan pedigrees with schizophrenia, by a Markov-chain Monte Carlo method. Am J Hum Genet 69:1278–1289

Cheng R, Juo SH, Loth JE, Nee J, Iossifov I, Blumenthal R, Sharpe L, Kanyas K, Lerer B, Lilliston B, Smith M, Trautman K, Gilliam TC, Endicott J, Baron M (2006) Genome-wide linkage scan in a large bipolar disorder sample from the National Institute of Mental Health genetics initiative suggests putative loci for bipolar disorder, psychosis, suicide, and panic disorder. Mol Psychiatry 11:252–260

Clapham DE, Julius D, Montell C, Schultz G (2005) International Union of Pharmacology. XLIX. Nomenclature and structure–function relationships of transient receptor potential channels. Pharmacol Rev 57:427–450

Coraddu F, Sawcer S, D'Alfonso S, Lai M, Hensiek A, Solla E, Broadley S, Mancosu C, Pugliatti M, Marrosu MG, Compston A (2001) A genome screen for multiple sclerosis in Sardinian multiplex families. Eur J Hum Genet 9:621–626

Cox NJ, Wapelhorst B, Morrison VA, Johnson L, Pinchuk L, Spielman RS, Todd JA, Concannon P (2001) Seven regions of the genome show evidence of linkage to type 1 diabetes in a consensus analysis of 767 multiplex families. Am J Hum Genet 69:820–830

Cuneo A, Bigoni R, Rigolin GM, Roberti MG, Bardi A, Cavazzini F, Milani R, Minotto C, Tieghi A, Della Porta M, Agostini P, Tammiso E, Negrini M, Castoldi G (2002) Late appearance of the 11q22.3-23.1 deletion involving the ATM locus in B-cell chronic lymphocytic leukemia and related disorders. Clinico-biological significance. Haematologica 87:44–51

Delgado-Escueta AV, Ganesh S, Yamakawa K (2001) Advances in the genetics of progressive myoclonus epilepsy. Am J Med Genet 106:129–138

Demirhan O, Tastemir D (2003) Chromosome aberrations in a schizophrenia population. Schizophr Res 65:1–7

Devaney JM, Donarum EA, Brown KM, Meyer J, Stober G, Lesch KP, Nestadt G, Stephan DA, Pulver AE (2002) No missense mutation of WKL1 in a subgroup of probands with schizophrenia. Mol Psychiatry 7:419–423

Di Palma F, Belyantseva IA, Kim HJ, Vogt TF, Kachar B, Noben-Trauth K (2002) Mutations in Mcoln3 associated with deafness and pigmentation defects in varitint-waddler (Va) mice. Proc Natl Acad Sci U S A 99:14994–14999

Dong C, Li WD, Li D, Price RA (2005) Interaction between obesity-susceptibility loci in chromosome regions 2p25-p24 and 13q13-q21. Eur J Hum Genet 13:102–108

European Polycystic Kidney Disease Consortium (1994) The polycystic kidney disease 1 gene encodes a 14 kb transcript and lies within a duplicated region on chromosome 16. Cell 77:881–894

Fougerousse F, Broux O, Richard I, Allamand V, de Souza AP, Bourg N, Brenguier L, Devaud C, Pasturaud P, Roudaut C, et al (1994) Mapping of a chromosome 15 region involved in limb girdle muscular dystrophy. Hum Mol Genet 3:285–293

Francke S, Manraj M, Lacquemant C, Lecoeur C, Lepretre F, Passa P, Hebe A, Corset L, Yan SL, Lahmidi S, Jankee S, Gunness TK, Ramjuttun US, Balgobin V, Dina C, Froguel P (2001) A genome-wide scan for coronary heart disease suggests in Indo-Mauritians a susceptibility locus on chromosome 16p13 and replicates linkage with the metabolic syndrome on 3q27. Hum Mol Genet 10:2751–2765

Freedman BI, Rich SS, Yu H, Roh BH, Bowden DW (2002) Linkage heterogeneity of end-stage renal disease on human chromosome 10. Kidney Int 62:770–774

Freedman BI, Langefeld CD, Rich SS, Valis CJ, Sale MM, Williams AH, Brown WM, Beck SR, Hicks PJ, Bowden DW (2004) A genome scan for ESRD in black families enriched for nondiabetic nephropathy. J Am Soc Nephrol 15:2719–2727

Freichel M, Suh SH, Pfeifer A, Schweig U, Trost CP, Weissgerber P, Biel M, Philipp S, Freise D, Droogmans G, Hofmann F, Flockerzi V, Nilius B (2001) Lack of an endothelial store-operated Ca^{2+} current impairs agonist dependent vasorelaxation in $Trp4^{-/-}$ mice. Nat Cell Biol 3:121–127

GAMES; Transatlantic Multiple Sclerosis Genetics Cooperative (2003) A meta-analysis of whole genome linkage screens in multiple sclerosis. J Neuroimmunol 143:39–46

Gemignani F, Marbini A (2001) Charcot-Marie-Tooth disease (CMT): distinctive phenotypic and genotypic features in CMT type 2. J Neurol Sci 184:1–9

Gicquel C, Bertagna X, Gaston V, Coste J, Louvel A, Baudin E, Bertherat J, Chapuis Y, Duclos JM, Schlumberger M, Plouin PF, Luton JP, Le Bouc Y (2001) Molecular markers and long-term recurrences in a large cohort of patients with sporadic adrenocortical tumors. Cancer Res 61:6762–6767

Gillanders E, Juo SH, Holland EA, et al (2003) Localization of a novel melanoma susceptibility locus to 1p22. Am J Hum Genet 73:301–313

Goldman M, Smith A, Shuman C, Caluseriu O, Wei C, Steele L, Ray P, Sadowski P, Squire J, Weksberg R, Rosenblum ND (2002) Renal abnormalities in Beckwith-Wiedemann syndrome are associated with 11p15.5 uniparental disomy. J Am Soc Nephrol 13:2077–2084

Goodship J, Gill H, Carter J, Jackson A, Splitt M, Wright M (2000) Autozygosity mapping of a Seckel syndrome locus to chromosome 3q22. 1-q24. Am J Hum Genet 67:498–503

Gregg RG, Palmer C, Kirkpatrick S, Simantel A (1996) Localization of a non-specific X-linked mental retardation gene, MRX23, to Xq23-q24. Hum Mol Genet 5:411–414

Gross J, Grimm O, Ortega G, Teuber I, Lesch KP, Meyer J (2001) Mutational analysis of the neuronal cadherin gene CELSR1 and exclusion as a candidate for catatonic schizophrenia in a large family. Psychiatr Genet 11:197–200

Guilbot A, Ravise N, Bouhouche A, Coullin P, Birouk N, Maisonobe T, Kuntzer T, Vial C, Grid D, Brice A, LeGuern E (1999) Genetic, cytogenetic and physical refinement of the autosomal recessive CMT linked to 5q31-q33: exclusion of candidate genes including EGR1. Eur J Hum Genet 7:849–859

Gurling HM, Kalsi G, Brynjolfson J, Sigmundsson T, Sherrington R, Mankoo BS, Read T, Murphy P, Blaveri E, McQuillin A, Petursson H, Curtis D (2001) Genomewide genetic linkage analysis confirms the presence of susceptibility loci for schizophrenia, on chromosomes 1q32.2, 5q33.2, and 8p21-22 and provides support for linkage to schizophrenia, on chromosomes 11q23.3-24 and 20q12.1-11.23. Am J Hum Genet 68:661-673

Hahs DW, McCauley JL, Crunk AE, McFarland LL, Gaskell PC, Jiang L, Slifer SH, Vance JM, Scott WK, Welsh-Bohmer KA, Johnson SR, Jackson CE, Pericak-Vance MA, Haines JL (2006) A genome-wide linkage analysis of dementia in the Amish. Am J Med Genet B Neuropsychiatr Genet 141:160-166

Hamshere ML, Bennett P, Williams N, Segurado R, Cardno A, Norton N, Lambert D, Williams H, Kirov G, Corvin A, Holmans P, Jones L, Jones I, Gill M, O'Donovan MC, Owen MJ, Craddock N (2005) Genomewide linkage scan in schizoaffective disorder: significant evidence for linkage at 1q42 close to DISC1, and suggestive evidence at 22q11 and 19p13. Arch Gen Psychiatry 62:1081-1088

Hirano A, Emi M, Tsuneizumi M, Utada Y, Yoshimoto M, Kasumi F, Akiyama F, Sakamoto G, Haga S, Kajiwara T, Nakamura Y (2001) Allelic losses of loci at 3p25.1, 8p22, 13q12, 17p13.3, and 22q13 correlate with postoperative recurrence in breast cancer. Clin Cancer Res 7:876-882

Hong KS, McInnes LA, Service SK, Song T, Lucas J, Silva S, Fournier E, Leon P, Molina J, Reus VI, Sandkuijl LA, Freimer NB (2004) Genetic mapping using haplotype and model-free linkage analysis supports previous evidence for a locus predisposing to severe bipolar disorder at 5q31-33. Am J Med Genet B Neuropsychiatr Genet 125:83-86

Hong YH, Kim M, Kim HJ, Sung JJ, Kim SH, Lee KW (2003) Clinical and electrophysiologic features of HNPP patients with 17p11.2 deletion. Acta Neurol Scand 108:352-358

Hovatta I, Varilo T, Suvisaari J, Terwilliger JD, Ollikainen V, Arajarvi R, Juvonen H, Kokko-Sahin ML, Vaisanen L, Mannila H, Lonnqvist J, Peltonen L (1999) A genomewide screen for schizophrenia genes in an isolated Finnish subpopulation, suggesting multiple susceptibility loci. Am J Hum Genet 65:1114-1124

Hoyng CB, Poppelaars F, van de Pol TJ, Kremer H, Pinckers AJ, Deutman AF, Cremers FP (1996) Genetic fine mapping of the gene for recessive Stargardt disease. Hum Genet 98:500-504

Hunter JJ, Shao J, Smutko JS, Dussault BJ, Nagle DL, Woolf EA, Holmgren LM, Moore KJ, Shyjan AW (1998) Chromosomal localization and genomic characterization of the mouse melastatin gene (Mlsn1). Genomics 54:116-123

Iannuzzi MC, Iyengar SK, Gray-McGuire C, Elston RC, Baughman RP, Donohue JF, Hirst K, Judson MA, Kavuru MS, Maliarik MJ, Moller DR, Newman LS, Rabin DL, Rose CS, Rossman MD, Teirstein AS, Rybicki BA (2005) Genome-wide search for sarcoidosis susceptibility genes in African Americans. Genes Immun 6:509-518

Iwata Y, Katanosaka Y, Arai Y, Komamura K, Miyatake K, Shigekawa M (2003) A novel mechanism of myocyte degeneration involving the Ca^{2+}-permeable growth factor-regulated channel. J Cell Biol 161:957-967

Jacobsen NJ, Franks EK, Elvidge G, Jones I, McCandless F, O'Donovan MC, Owen MJ, Craddock N (2001) Exclusion of the Darier's disease gene, ATP2A2, as a common susceptibility gene for bipolar disorder. Mol Psychiatry 6:92-97

Jonasdottir A, Thorlacius T, Fossdal R, Jonasdottir A, Benediktsson K, Benedikz J, Jonsson HH, Sainz J, Einarsdottir H, Sigurdardottir S, Kristjansdottir G, Sawcer S, Compston A, Stefansson K, Gulcher J (2003) A whole genome association study in Icelandic multiple sclerosis patients with 4804 markers. J Neuroimmunol 143:88-92

Jonsson G, Bendahl PO, Sandberg T, Kurbasic A, Staaf J, Sunde L, Cruger DG, Ingvar C, Olsson H, Borg A (2005) Mapping of a novel ocular and cutaneous malignant melanoma susceptibility locus to chromosome 9q21.32. J Natl Cancer Inst 97:1377–1382

Jun G, Klein BE, Klein R, Fox K, Millard C, Capriotti J, Russo K, Lee KE, Elston RC, Iyengar SK (2005) Genome-wide analyses demonstrate novel loci that predispose to drusen formation. Invest Ophthalmol Vis Sci 46:3081–3088

Karasneh J, Gul A, Ollier WE, Silman AJ, Worthington J (2005) Whole-genome screening for susceptibility genes in multicase families with Behcet's disease. Arthritis Rheum 52:1836–1842

Kelly JA, Moser KL, Harley JB (2002) The genetics of systemic lupus erythematosus: putting the pieces together. Genes Immun 3 Suppl 1:S71–S85

Klein CJ, Cunningham JM, Atkinson EJ, Schaid DJ, Hebbring SJ, Anderson SA, Klein DM, Dyck PJ, Litchy WJ, Thibodeau SN, Dyck PJ (2003) The gene for HMSN2C maps to 12q23–24: a region of neuromuscular disorders. Neurology 60:1151–1156

Koike M, Tasaka T, Spira S, Tsuruoka N, Koeffler HP (1999) Allelotyping of acute myelogenous leukemia: loss of heterozygosity at 7q31.1 (D7S486) and q33–34 (D7S498, D7S505). Leuk Res 23:307–310

Kojis TL, Heinzmann C, Flodman P, Ngo JT, Sparkes RS, Spence MA, Bateman JB, Heckenlively JR (1996) Map refinement of locus RP13 to human chromosome 17p13.3 in a second family with autosomal dominant retinitis pigmentosa. Am J Hum Genet 58:347–355

Launonen V, Laake K, Huusko P, Niederacher D, Beckmann MW, Barkardottir RB, Geirsdottir EK, Gudmundsson J, Rio P, Bignon YJ, Seitz S, Scherneck S, Bieche I, Champeme MH, Birnbaum D, White G, Varley J, Sztan M, Olah E, Osorio A, Benitez J, Spurr N, Velikonja N, Peterlin B, Winqvist R, et al (1999) European multicenter study on LOH of APOC3 at 11q23 in 766 breast cancer patients: relation to clinical variables. Breast Cancer Somatic Genetics Consortium. Br J Cancer 80:879–882

Le Ber I, Martinez M, Campion D, Laquerriere A, Betard C, Bassez G, Girard C, Saugier-Veber P, Raux G, Sergeant N, Magnier P, Maisonobe T, Eymard B, Duyckaerts C, Delacourte A, Frebourg T, Hannequin D (2004) A non-DM 1, non-DM 2 multisystem myotonic disorder with frontotemporal dementia: phenotype and suggestive mapping of the DM 3 locus to chromosome 15q21–24. Brain 127:1979–1992

Lerer B, Segman RH, Hamdan A, Kanyas K, Karni O, Kohn Y, Korner M, Lanktree M, Kaadan M, Turetsky N, Yakir A, Kerem B, Macciardi F (2003) Genome scan of Arab Israeli families maps a schizophrenia susceptibility gene to chromosome 6q23 and supports a locus at chromosome 10q24. Mol Psychiatry 8:488–498

Li FY, Tariq M, Croxen R, Morten K, Squier W, Newsom-Davis J, Beeson D, Larsson C (1999) Mapping of autosomal dominant progressive external ophthalmoplegia to a 7-cM critical region on 10q24. Neurology 53:1265–1271

Liang SG, Sadovnick AD, Remick RA, Keck PE, McElroy SL, Kelsoe JR (2002) A linkage disequilibrium study of bipolar disorder and microsatellite markers on 22q13. Psychiatr Genet 12:231–235

Liu CM, Hwu HG, Lin MW, Ou-Yang WC, Lee SF, Fann CS, Wong SH, Hsieh SH (2001) Suggestive evidence for linkage of schizophrenia to markers at chromosome 15q13–14 in Taiwanese families. Am J Med Genet 105:658–661

Liu J, Juo SH, Holopainen P, Terwilliger J, Tong X, Grunn A, Brito M, Green P, Mustalahti K, Maki M, Gilliam TC, Partanen J (2002) Genomewide linkage analysis of celiac disease in Finnish families. Am J Hum Genet 70:51–59

Liu J, Juo SH, Dewan A, Grunn A, Tong X, Brito M, Park N, Loth JE, Kanyas K, Lerer B, Endicott J, Penchaszadeh G, Knowles JA, Ott J, Gilliam TC, Baron M (2003) Evidence for a putative bipolar disorder locus on 2p13-16 and other potential loci on 4q31, 7q34, 8q13, 9q31, 10q21-24, 13q32, 14q21 and 17q11-12. Mol Psychiatry 8:333-342

Loos RJ, Katzmarzyk PT, Rao DC, Rice T, Leon AS, Skinner JS, Wilmore JH, Rankinen T, Bouchard C; HERITAGE Family Study (2003) Genome-wide linkage scan for the metabolic syndrome in the HERITAGE Family Study. J Clin Endocrinol Metab 88:5935-5943

Malhotra A, Wolford JK; American Diabetes Association GENNID Study Group (2005) Analysis of quantitative lipid traits in the genetics of NIDDM (GENNID) study. Diabetes 54:3007-3014

Matsuura T, Achari M, Khajavi M, Bachinski LL, Zoghbi HY, Ashizawa T (1999) Mapping of the gene for a novel spinocerebellar ataxia with pure cerebellar signs and epilepsy. Ann Neurol 45:407-411

Matsuyama H, Pan Y, Yoshihiro S, Kudren D, Naito K, Bergerheim US, Ekman P (2003) Clinical significance of chromosome 8p, 10q, and 16q deletions in prostate cancer. Prostate 54:103-111

Maw MA, John S, Jablonka S, Muller B, Kumaramanickavel G, Oehlmann R, Denton MJ, Gal A (1995) Oguchi disease: suggestion of linkage to markers on chromosome 2q. J Med Genet 32:396-398

Meijer IA, Cossette P, Roussel J, Benard M, Toupin S, Rouleau GA (2004) A novel locus for pure recessive hereditary spastic paraplegia maps to 10q22.1-10q24.1. Ann Neurol 56:579-582

Mirghomizadeh F, Bardtke B, Devoto M, Pfister M, Oeken J, Konig E, Vitale E, Riccio A, De Rienzo A, Zenner HP, Blin N (2002) Second family with hearing impairment linked to 19q13 and refined DFNA4 localisation. Eur J Hum Genet 10:95-99

Mochizuki T, Wu G, Hayashi T, Xenophontos SL, Veldhuisen B, Saris JJ, Reynolds DM, Cai Y, Gabow PA, Pierides A, Kimberling WJ, Breuning MH, Deltas CC, Peters DJ, Somlo S (1996) PKD2, a gene for polycystic kidney disease that encodes an integral membrane protein. Science 272:1339-1342

Mori Y, Otabe S, Dina C, Yasuda K, Populaire C, Lecoeur C, Vatin V, Durand E, Hara K, Okada T, Tobe K, Boutin P, Kadowaki T, Froguel P (2002) Genome-wide search for type 2 diabetes in Japanese affected sib-pairs confirms susceptibility genes on 3q, 15q, and 20q and identifies two new candidate Loci on 7p and 11p. Diabetes 51:1247-1255

Nadler MSJ, Hermosura MC, Inabe K, Perraud AL, Zhu Q, Stoles AJ, Kinet JP, Penner R, Scharenberg AM, Feilg A (2001) LTRPC7 (TRP-PLIK) is a Mg.ATP-regulated divalent cation channel required for cell viability. Nature 441:590-595

Nagahata T, Hirano A, Utada Y, Tsuchiya S, Takahashi K, Tada T, Makita M, Kasumi F, Akiyama F, Sakamoto G, Nakamura Y, Emi M (2002) Correlation of allelic losses and clinicopathological factors in 504 primary breast cancers. Breast Cancer 9:208-215

Neubauer BA, Fiedler B, Himmelein B, Kampfer F, Lassker U, Schwabe G, Spanier I, Tams D, Bretscher C, Moldenhauer K, Kurlemann G, Weise S, Tedroff K, Eeg-Olofsson O, Wadelius C, Stephani U (1998) Centrotemporal spikes in families with rolandic epilepsy: linkage to chromosome 15q14. Neurology 51:1608-1612

Nikali K, Isosomppi J, Lonnqvist T, Mao JI, Suomalainen A, Peltonen L (1997) Toward cloning of a novel ataxia gene: refined assignment and physical map of the IOSCA locus (SCA8) on 10q24. Genomics 39:185-191

Nobile C, Pitzalis S (1999) Expression analysis of 21 transcripts physically anchored within the chromosomal region 10q24. Genomics 62:86-89

Nussbaum RL, Polymeropoulos MH (1997) Genetics of Parkinson's disease. Hum Mol Genet 6:1687-1691

Nwosu V, Carpten J, Trent JM, Sheridan R (2001) Heterogeneity of genetic alterations in prostate cancer: evidence of the complex nature of the disease. Hum Mol Genet 10:2313-2318

Omran H, Fernandez C, Jung M, Haffner K, Fargier B, Villaquiran A, Waldherr R, Gretz N, Brandis M, Ruschendorf F, Reis A, Hildebrandt F (2000) Identification of a new gene locus for adolescent nephronophthisis, on chromosome 3q22 in a large Venezuelan pedigree. Am J Hum Genet 66:118-127

Omran H, Haffner K, Burth S, Fernandez C, Fargier B, Villaquiran A, Nothwang HG, Schnittger S, Lehrach H, Woo D, Brandis M, Sudbrak R, Hildebrandt F (2001) Human adolescent nephronophthisis: gene locus synteny with polycystic kidney disease in pcy mice. J Am Soc Nephrol 12:107-113

Ouyang XM, Yan D, Du LL, Hejtmancik JF, Jacobson SG, Nance WE, Li AR, Angeli S, Kaiser M, Newton V, Brown SD, Balkany T, Liu XZ (2005) Characterization of Usher syndrome type I gene mutations in an Usher syndrome patient population. Hum Genet 116:292-299

Paavola-Sakki P, Ollikainen V, Helio T, Halme L, Turunen U, Lahermo P, Lappalainen M, Farkkila M, Kontula K (2003) Genome-wide search in Finnish families with inflammatory bowel disease provides evidence for novel susceptibility loci. Eur J Hum Genet 11:112-120

Parada LA, Hallen M, Hagerstrand I, Tranberg KG, Johansson B (1999) Clonal chromosomal abnormalities in congenital bile duct dilatation (Caroli's disease). Gut 45:780-782

Petit C (2001) Usher syndrome: from genetics to pathogenesis. Annu Rev Genomics Hum Genet 2:271-297

Phelan MC, Rogers RC, Clarkson KB, Bowyer FP, Levine MA, Estabrooks LL, Severson MC, Dobyns WB (1995) Albright hereditary osteodystrophy and del(2) (q37.3) in four unrelated individuals. Am J Med Genet 58:1-7

Pinto D, Westland B, de Haan GJ, Rudolf G, da Silva BM, Hirsch E, Lindhout D, Trenite DG, Koeleman BP (2005) Genome-wide linkage scan of epilepsy-related photoparoxysmal electroencephalographic response: evidence for linkage on chromosomes 7q32 and 16p13. Hum Mol Genet 14:171-178

Pratley RE, Thompson DB, Prochazka M, Baier L, Mott D, Ravussin E, Sakul H, Ehm MG, Burns DK, Foroud T, Garvey WT, Hanson RL, Knowler WC, Bennett PH, Bogardus C (1998) An autosomal genomic scan for loci linked to prediabetic phenotypes in Pima Indians. J Clin Invest 101:1757-1764

Prawitt D, Enklaar T, Klemm G, Gartner B, Spangenberg C, Winterpacht A, Higgins M, Pelletier J, Zabel B (2000) Identification and characterization of MTR1, a novel gene with homology to melastatin (MLSN1) and the trp gene family located in the BWS-WT2 critical region on chromosome 11p15.5 and showing allele-specific expression. Hum Mol Genet 9:203-216

Quintero-Del-Rio AI, Kelly JA, Kilpatrick J, James JA, Harley JB (2002) The genetics of systemic lupus erythematosus stratified by renal disease: linkage at 10q22.3 (SLEN1), 2q34-35 (SLEN2), and 11p15.6 (SLEN3). Genes Immun 3 Suppl 1:S57-S62

Ralston SH, Galwey N, MacKay I, Albagha OM, Cardon L, Compston JE, Cooper C, Duncan E, Keen R, Langdahl B, McLellan A, O'Riordan J, Pols HA, Reid DM, Uitterlinden AG, Wass J, Bennett ST (2005) Loci for regulation of bone mineral density in men and women identified by genome wide linkage scan: the FAMOS study. Hum Mol Genet 14:943–951

Rao PH, Arias-Pulido H, Lu XY, Harris CP, Vargas H, Zhang FF, Narayan G, Schneider A, Terry MB, Murty VV (2004) Chromosomal amplifications, 3q gain and deletions of 2q33-q37 are the frequent genetic changes in cervical carcinoma. BMC Cancer 4:5

Rees M, Curtis D, Parker K, Sundqvist A, Baralle D, Bespalova IN, Burmeister M, Chung E, Gardiner RM, Whitehouse WP (1994) Linkage analysis of idiopathic generalised epilepsy in families of probands with juvenile myoclonic epilepsy and marker loci in the region of EPM 1 on chromosome 21 q: Unverricht-Lundborg disease and JME are not allelic variants. Neuropediatrics 25:20–25

Reiser J, Polu KR, Moller CC, Kenlan P, Altintas MM, Wei C, Faul C, Herbert S, Villegas I, Avila-Casado C, McGee M, Sugimoto H, Brown D, Kalluri R, Mundel P, Smith PL, Clapham DE, Pollak MR (2005) TRPC6 is a glomerular slit diaphragm-associated channel required for normal renal function. Nat Genet 37:739–744

Riegman PH, Vissers KJ, Alers JC, Geelen E, Hop WC, Tilanus HW, van Dekken H (2001) Genomic alterations in malignant transformation of Barrett's esophagus. Cancer Res 61:3164–3170

Ruteshouser EC, Huff V (2004) Familial Wilms tumor. Am J Med Genet C Semin Med Genet 129:29–34

Saar K, Geller F, Ruschendorf F, Reis A, Friedel S, Schauble N, Nurnberg P, Siegfried W, Goldschmidt HP, Schafer H, Ziegler A, Remschmidt H, Hinney A, Hebebrand J (2003) Genome scan for childhood and adolescent obesity in German families. Pediatrics 111:321–327

Sakaguchi H, Sanke T, Ohagi S, Iiri T, Nanjo K (1998) A case of Albright's hereditary osteodystrophy-like syndrome complicated by several endocrinopathies: normal Gs alpha gene and chromosome 2q37. J Clin Endocrinol Metab 83:1563–1565

Sawcer S, Ban M, Maranian M, et al (2005) A high-density screen for linkage in multiple sclerosis. Am J Hum Genet 77:454–467

Schinzel A, Niedrist D (2001) Chromosome imbalances associated with epilepsy. Am J Med Genet 106:119–124

Schlingmann KP, Weber S, Peters M, Niemann Nejsum L, Vitzthum H, Klingel K, Kratz M, Haddad E, Ristoff E, Dinour D, Syrrou M, Nielsen S, Sassen M, Waldegger S, Seyberth HW, Konrad M (2002) Hypomagnesemia with secondary hypocalcemia is caused by mutations in TRPM6, a new member of the TRPM gene family. Nat Genet 31:166–170

Schmidt LS, Warren MB, Nickerson ML, Weirich G, Matrosova V, Toro JR, Turner ML, Duray P, Merino M, Hewitt S, Pavlovich CP, Glenn G, Greenberg CR, Linehan WM, Zbar B (2001) Birt-Hogg-Dube syndrome, a genodermatosis associated with spontaneous pneumothorax and kidney neoplasia, maps to chromosome 17p11.2. Am J Hum Genet 69:876–882

Schulze-Bahr E, Wedekind H, Haverkamp W, Borggrefe M, Assmann G, Breithardt G, Funke H (1999) The LQT syndromes—current status of molecular mechanisms. Z Kardiol 88:245–254

Schwab SG, Eckstein GN, Hallmayer J, Lerer B, Albus M, Borrmann M, Lichtermann D, Ertl MA, Maier W, Wildenauer DB (1997) Evidence suggestive of a locus on chromosome 5q31 contributing to susceptibility for schizophrenia in German and Israeli families by multipoint affected sib-pair linkage analysis. Mol Psychiatry 2:156–160

Wissenbach U, Niemeyer BA, Fixemer T, Schneidewind A, Trost C, Cavalie A, Reus K, Meese E, Bonkhoff H, Flockerzi V (2001) Expression of CaT-like, a novel calcium-selective channel, correlates with the malignancy of prostate cancer. J Biol Chem 276:19461–19468

Wong KF, Siu LL, So CC (2000) Deletion of Xq23 is a recurrent karyotypic abnormality in acute myeloid leukemia. Cancer Genet Cytogenet 122:33–36

Wyszynski DF, Albacha-Hejazi H, Aldirani M, Hammod M, Shkair H, Karam A, Alashkar J, Holmes TN, Pugh EW, Doheny KF, McIntosh I, Beaty TH, Bailey-Wilson JE (2003) A genome-wide scan for loci predisposing to non-syndromic cleft lip with or without cleft palate in two large Syrian families. Am J Med Genet A 123:140–147

Xu C, Dai Y, Lorentzen JC, Dahlman I, Olsson T, Hillert J (2001a) Linkage analysis in multiple sclerosis of chromosomal regions syntenic to experimental autoimmune disease loci. Eur J Hum Genet 9:458–463

Xu J, Meyers DA, Ober C, Blumenthal MN, Mellen B, Barnes KC, King RA, Lester LA, Howard TD, Solway J, Langefeld CD, Beaty TH, Rich SS, Bleecker ER, Cox NJ (2001b) Genomewide screen and identification of gene-gene interactions for asthma-susceptibility loci in three U.S. populations: collaborative study on the genetics of asthma. Am J Hum Genet 68:1437–1446

Xu J, Dimitrov L, Chang BL, et al (2005) A combined genomewide linkage scan of 1,233 families for prostate cancer-susceptibility genes conducted by the international consortium for prostate cancer genetics. Am J Hum Genet 77:219–229

Yu GY, Howell MJ, Roller MJ, Xie TD, Gomez CM (2005) Spinocerebellar ataxia type 26 maps to chromosome 19p13.3 adjacent to SCA6. Ann Neurol 57:349–354

Zandi PP, Willour VL, Huo Y, et al (2003) Genome scan of a second wave of NIMH genetics initiative bipolar pedigrees: chromosomes 2, 11, 13, 14, and X. Am J Med Genet B Neuropsychiatr Genet 119:69–76

Zhang XJ, Chen JJ, Liu JB (2005) The genetic concept of vitiligo. J Dermatol Sci 39:137–146

Zhou X, Tan FK, Wang N, Xiong M, Maghidman S, Reveille JD, Milewicz DM, Chakraborty R, Arnett FC (2003) Genome-wide association study for regions of systemic sclerosis susceptibility in a Choctaw Indian population with high disease prevalence. Arthritis Rheum 48:2585–2592

TRP Channels of the Pancreatic Beta Cell

D. A. Jacobson · L. H. Philipson (✉)

Department of Medicine—Endocrinology Section, University of Chicago,
5841 S Maryland Ave, Chicago IL, , USA
l-philipson@uchicago.edu

1	Introduction	410
2	TRP Channels Found Within the Pancreatic Beta Cell	410
2.1	TRPC1	410
2.2	TRPC4 and TRPC6	410
2.3	TRPM2	412
2.4	TRPM5	412
2.5	TRPV1	412
3	Properties of Neuroendocrine TRP Channels	413
3.1	Nonselective G Protein-Coupled TRPC1 and TRPC4 Cation Channels	413
3.2	Monovalent-Specific TRPM2 and TRPM5 Cation Channels	414
3.3	Nonselective Vanilloid Sensitive TRPV1 Cation Channels	414
4	Beta Cell TRP-Like Channel Activation	414
4.1	Cellular Oxidation	414
4.2	Calcium	415
4.3	Muscarinic Receptors	417
4.4	Glucagon-Like Peptide-1	417
4.5	Capsaicin	418
5	Pharmacological Block of TRP-Like Currents of the Beta Cell	419
5.1	SKF 96365	419
5.2	Trivalents	419
5.3	Capsazepine	419
6	Conclusions	420
	References	420

Abstract Orchestrated ion fluctuations within pancreatic islets regulate hormone secretion and may be essential to processes such as apoptosis. A diverse set of ion channels allows for islet cells to respond to a variety of signals and dynamically regulate hormone secretion and glucose homeostasis (reviewed by Houamed et al. 2004). This chapter focuses on transient receptor potential (TRP)-related channels found within the beta cells of the islet and reviews their roles in both insulin secretion and apoptosis.

Keywords GLP-1 · Islet · Insulin · Nonselective cation channel · SKF 96365 · Pancreatic beta cell

1
Introduction

Insulin secretion occurs from beta cells of pancreatic islets of Langerhans. Exocytosis of insulin-containing granules from beta cells occurs through a complex pathway beginning with glucose uptake from glucose transporters on the beta cell surface. Glucose metabolism leads to changes in the ATP to ADP ratio within the cell. The altered ratio of ATP/ADP causes cellular depolarization through rectification of ATP-sensitive potassium channels (KATP). Depolarization then activates voltage-dependent calcium channels (VDCC), causing oscillatory calcium influx and release of calcium from intracellular stores resulting in insulin release from the cell (reviewed by Houamed et al. 2004). Some of these channels responsible for the ion fluxes in insulin secretion are likely to be TRP channels, perhaps participating in the depolarizing sodium current that leads to activation of voltage-dependent channels, or may also contribute to calcium entry and intracellular regulation of calcium flux.

2
TRP Channels Found Within the Pancreatic Beta Cell

2.1
TRPC1

TRPC1 was the first mammalian TRP channel cloned (Wes et al. 1995) and also the first molecularly identified TRP channel in beta cells (Sakura and Ashcroft 1997; Fig. 1a). Four transcripts for TRPC1, representing different splice variants, are found in RNA from mouse islets and the beta cell line Min6. The beta splice variant is efficiently translated into protein in vitro and shows the highest levels of transcript abundance in beta cells. Although the TRPC1 transcript is expressed in islets, its translation into functional protein has yet to be confirmed. TRPC1 transcripts are also found in brain, heart, kidney, endothelium, ovaries, testis, and smooth muscle cells (Freichel et al. 2005; Montell 2005). TRPC1 functions as a nonselective cation channel in the brain, where it forms heteromeric channels with TRPC3, TRPC4, or TRPC5 that can be activated by $G_{q/11}$ family G proteins; its function in beta cells, however, remains undefined.

2.2
TRPC4 and TRPC6

Together with TRPC1, TRPC4 and TRPC6 transcripts have been identified within the beta cell; however, TRPC6 shows very low expression within neuroendocrine cells (Sakura and Ashcroft 1997; Roe et al. 1998; Fig. 1b). The

TRP Channels of the Pancreatic Beta Cell

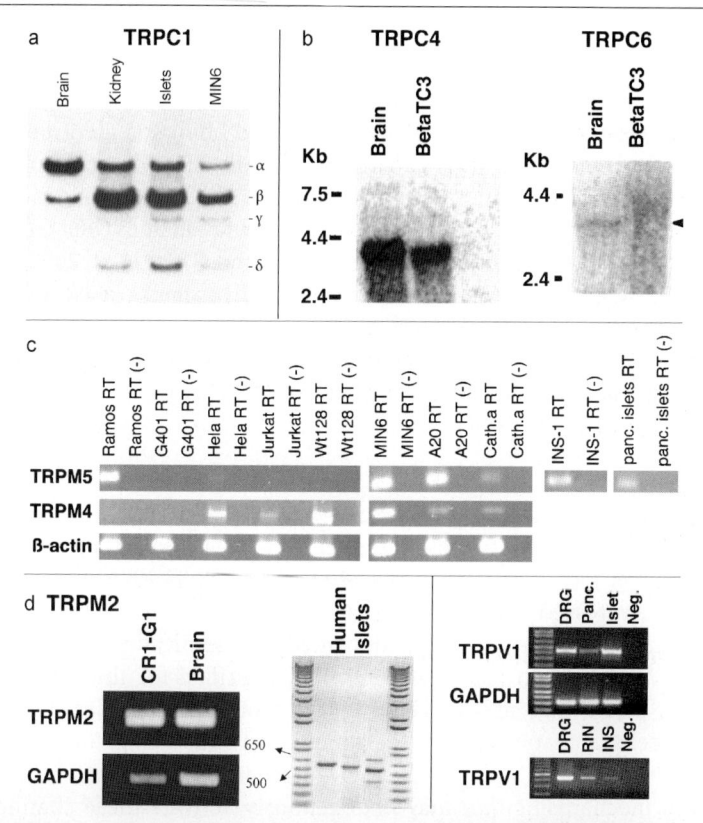

Fig. 1 a–e Expression of TRP channels RNA in pancreatic beta cells. **a** Four variants of Mtrp1were detected by PCR from brain, kidney, islets, and Min6 RNA (Sakura and Ashcroft 1997). **b** Northern blot analysis of RNA from brain and βTC-3 cells with TRPC4 and TRPC6 probes respectively (Roe et al. 1998). **c** RT-PCR performed with species-specific primers for the TRPM4/Trpm4 and the TRPM5/Trpm5 genes (Prawitt et al. 2003). **d** RT-PCR performed on CRG-1 rat insulinoma cells and human islets with TRPM2 specific primers, the human islet RT-PCR was performed with three sets of TRPM2 primers and thus the three differently sized bands (Qian et al. 2002; Inamura et al. 2003). **e** RT-PCR performed with primers specific to TRPV1 on RNA from rat dorsal root ganglion (DRG), whole pancreas (Panc.), rat islets (islet), RINm5F rat insulinoma cells (RIN), and INS1 rat insulinoma cells (INS) (Akiba et al. 2003)

TRPC4 transcript is expressed in brain, lung, testis, and pancreatic islets; however, confirmation of its translation in beta cells has not been confirmed. The endocrine isoforms of TRPC4, cloned by Qian et al. from the mouse beta cell line BetaTC3, contain open reading frames of 974 and 890 amino acids, respectively. TRPC4 is closely related to TRPC1 and can also be activated by $G_{q/11}$ family G proteins and receptor tyrosine kinases. The functions of TRPC4 channels and their ability to form heteromeric complexes with TRPC1 are currently being defined in the islet.

2.3
TRPM2

Along with cloning the TRPC4 channel, Qian and Philipson identified another TRP-like channel within human islets known as LTRPC2 and now termed TRPM2 (Qian et al. 2002; Inamura et al. 2003; Fig. 1d). TRPM2 RNA is expressed in human and mouse beta cells as well as the rat beta cell lines CR1-G1 and RIN-M5, where protein translation has been confirmed (Hara et al. 2002; Inamura et al. 2003). TRPM2 channels are activated by hydrogen peroxide and ADP-ribose, causing nonselective cation entry into the cell. A similar conductance has been identified in rat beta cells and identified as TRPM2 in rat insulinoma cells (Herson et al. 1997, 1999; Hara et al. 2002).

2.4
TRPM5

Another TRPM (melastatin) family member that has recently been identified in beta cells is TRPM5 (Prawitt et al. 2003; Fig. 1c). TRPM5 messenger RNA (mRNA) is expressed in human islets and the rat beta cell line INS-1; however, protein expression has not been confirmed biochemically. The monovalent-specific TRPM5 cation channels were first identified for their roles in taste sensation, where they are activated by calcium and cause depolarization of taste bud cells in response to bitter umami and sweet taste. Similar calcium-activated TRPM5-like currents have been recorded from beta cells, but the G protein mechanisms that may endogenously activate these channels have yet to be defined. TRPM5 channels should be active during calcium-induced insulin secretion and thus may play an important role during insulin secretion.

2.5
TRPV1

Along with the canonical (TRPC) and melastatin (TRPM) channels, a member of the vanilloid family of TRP channels, TRPV1, is also expressed in beta cells. RNA and protein for the TRPV1 channel has been identified in rat islets and the rat beta cell lines RINm5 and INS-1 (Akiba et al. 2003; Fig. 1e). TRPV1 is a nonselective cation channel first cloned for its ability to cause calcium influx in response to the pepper alkaloid capsaicin and it is thus highly expressed in sensory neurons and has also been identified in tissues such as skeletal muscle. Pancreatic activation of TRPV1 has been associated with beta cell insulin secretion (Karlsson et al. 1994; Akiba et al. 2004). However, as TRPV1 is also expressed in pancreatic neurons innervating islets (Karlsson et al. 1994), the exact roles for its endogenous activity during beta cell insulin secretion remain complex.

3
Properties of Neuroendocrine TRP Channels

3.1
Nonselective G Protein-Coupled TRPC1 and TRPC4 Cation Channels

TRPC1 was cloned from human brain in 1995 by sequence similarities to the original *Drosophila* TRP channel (Wes et al. 1995; Zhu et al. 1995). The 793-amino-acid human TRPC1 channel contains similar structural motifs to the 809-amino-acid rodent beta cell TRPC1, including six transmembrane domains, three N-terminal ankyrin repeat domains, and a TRP box in the C-terminus, as well as consensus phosphorylation and glycosylation sequences. TRPC1 has been reported to have multiple interacting domains including inositol triphosphate (IP$_3$) binding and binding to calmodulin, ryanodine receptors, beta-tubulin, the dynamin-like protein MxA, and others, which contribute to its cellular activities (Tang et al. 2001; Sampieri et al. 2005; Bollimuntha et al. 2005; Lussier et al. 2005). Heteromeric complexes between TRPC1, TRPC4, and TRPC5 have been reported, with tetrameric stoichiometry (Hofmann et al. 2002). TRPC1-containing channels are permeable to sodium, calcium, and barium, but not strontium or *N*-methyl-glucamine (Hofmann et al. 2002).

TRPC4 was cloned in 1996 from the bovine adrenal gland and retina by sequence similarity to *Drosophila* TRP. With over 11 identified splice variants, TRPC4 has a great deal of tissue and species-specific variability. The 974-amino-acid alpha splice variant of TRPC4 is expressed in mouse beta cells. Within the canonical TRP channel family, TRPC4 alpha is most closely related to TRPC5 and also shows some similarities to TRPC1. Accordingly, TRPC4 alpha contains similar structural motifs as TRPC1, including six transmembrane domains, a C-terminal TRP box, and an N-terminal ankyrin repeat domain, as well as consensus phosphorylation and glycosylation sequences (Tang et al. 2000). TRPC4 and TRPC5 channels also contain a unique C-terminal PDZ-binding motif allowing interactions with PDZ-containing proteins such as the Na$^+$/H$^+$ exchanger regulatory factor [NHERF (1 and 2)/EBP-50]. Like TRPC1, TRPC4 reportedly interacts with IP$_3$, calmodulin, and other factors to help regulate cellular function (Tang et al. 2001). TRPC4-associating proteins (TAP1, GenBank AF130458; Tap-1A, GenBank AF130459; Tap-1B, GenBank AF130460) have also identified by Qian and Philipson using the beta cell-specific complementary DNA (cDNA) of TRPC4 in a two-hybrid screen; however, the function of these proteins remains unknown. TRPC4 primarily allows calcium flux across the membrane, a response that has been linked to the filling state of calcium stores and G protein-coupled stimulation, but the details remain unclear (Freichel et al. 2005; Montell 2005).

3.2
Monovalent-Specific TRPM2 and TRPM5 Cation Channels

TRPM2 and TRPM5 were cloned in 2001 using human brain expressed sequence tags and in 2000 using brain cDNA libraries, respectively (Nadler et. al. 2001; Prawitt et al. 2000). The TRPM members also have six transmembrane domains, a TRP box in their C-terminal domain, and consensus phosphorylation and glycosylation sequences. TRPM members have a considerably extended N-terminal domain that lacks ankyrin repeats and contains four regions of homology with undefined structural significance (Fleig and Penner 2004). The C-terminal regions of TRPMs are variable and contain important channel modulation domains (Fleig and Penner 2004). The single channel current–voltage relationship is voltage-dependent and linear for TRPM2 as well as TRPM5 and shows outward rectification.

3.3
Nonselective Vanilloid Sensitive TRPV1 Cation Channels

TRPV1 was cloned in 1997 from a dorsal root ganglion-derived cDNA library by David Julius's group (Caterina et al. 1997). Both rat nociceptors and beta cells express an 838-amino-acid TRPV1 ligand-gated calcium-permeant channel. TRPV1 contains six transmembrane domains, three N-terminal ankyrin repeats, and a C-terminal TRP box, as well as multiple phosphorylation and glycosylation sites (Mohapatra and Nau 2003; Bhave et al. 2003). Interactions with many binding partners affect TRPV1 activity including PIP_2 which serves as regulator of the sensitivity of TRPV1 to heat. TRPV1 is an outwardly rectifying primarily calcium permeable channel.

4
Beta Cell TRP-Like Channel Activation

4.1
Cellular Oxidation

Reactive oxygen species (ROS) have been implicated in beta cell destruction in diabetes (Bindokas et al. 2003; Fridlyand and Philipson 2004) and linked to TRPM2 activation (Herson et al. 1997, 1999; Hara et al. 2002; Inamura et al. 2003). TRPM2 in rodent and human beta cells is activated by micromolar concentrations of the ROS hydrogen peroxide (H_2O_2), resulting in calcium entry and cellular depolarization (Herson and Ashford 1997; Herson et al. 1997, 1999; Hara et al. 2002; Inamura et al. 2003). The depolarization of beta cells by H_2O_2 eventually causes a collapse of the membrane potential leading to apoptosis (Herson and Ashford 1997; Herson et al. 1999). Agents that cause

H_2O_2 production in the beta cell such as alloxan have also been linked to TRPM2 activation and diabetes (Herson and Ashford 1997).

H_2O_2-stimulated beta cell TRPM2 currents are composed of three phases consisting of an initial lag phase, dependent on the concentration of H_2O_2, a slowly developing second phase, and a rapid third phase. The change in activation of TRPM2 over time may be due to direct activation of TRPM2 by NAD^+ and a gradual change in the $NAD(P)^+/NAD(P)H$ ratio by H_2O_2 (Hara et al. 2002). The exact mechanism of NAD^+ activation of TRPM2, however, is still uncertain. H_2O_2 can also release adenosine diphosphoribose (ADPR) from mitochondria that activates channel activity through ADPR interactions with the TRPM2 Nudix domain (Perraud et al. 2005). As NAD^+ can be converted into ADPR by CD38 that is expressed in beta cells, it may activate channels through its conversion into ADPR (Furuya et al. 1995). Beta cell TRPM2 has a conductance of around 70 pS and shows a linear current–voltage relationship in response to H_2O_2, $β-NAD^+$, and ADPR (Herson and Ashford 1997; Inamura et al. 2003).

4.2
Calcium

Calcium and ATP play a fundamental role in regulating the secretion of insulin from the beta cell; these signals have also been shown to regulate a TRP-like channel. Calcium activates a nonselective cation current in the insulin-secreting cell line CRI-G1, human insulin producing tumor cells, and rat beta cells (Sturgess et al. 1987a, b; Roe et al. 1998; Fridlyand et al. 2003; Prawitt et al. 2003; Fig. 2a). The resulting calcium-activated TRP-like current is sensitive to ATP and inhibited with K_i of approximately 8 µM (Sturgess et al. 1987a). The current also shows voltage dependence with more activity at positive voltages (Sturgess et al. 1987a; Prawitt et al. 2003). This channel has similar biophysical properties to the recently cloned TRPM5 channel.

TRPM5 is activated by fast changes in calcium causing monovalent cation flux into the cell. The exact concentration of calcium required to activate TRPM5 has varied from the high nanomolar range to greater than 100 µM (Prawitt et al. 2003; Sturgess et al. 1987a). Calcium concentrations in a physiological range, from 300 to 800 nM, are indeed capable of activating a TRPM5-like current in the rat beta cell line INS-1 (Prawitt et al. 2003). The mechanism leading to calcium-induced activation of TRPM5 remains undefined but presumably involves auxiliary subunit interaction modifications or direct effects of calcium on the TRPM5 channel. These observations indicate that calcium-activated TRPM5 activity in the pancreatic beta cell may provide an important depolarizing influence during glucose induced insulin secretion.

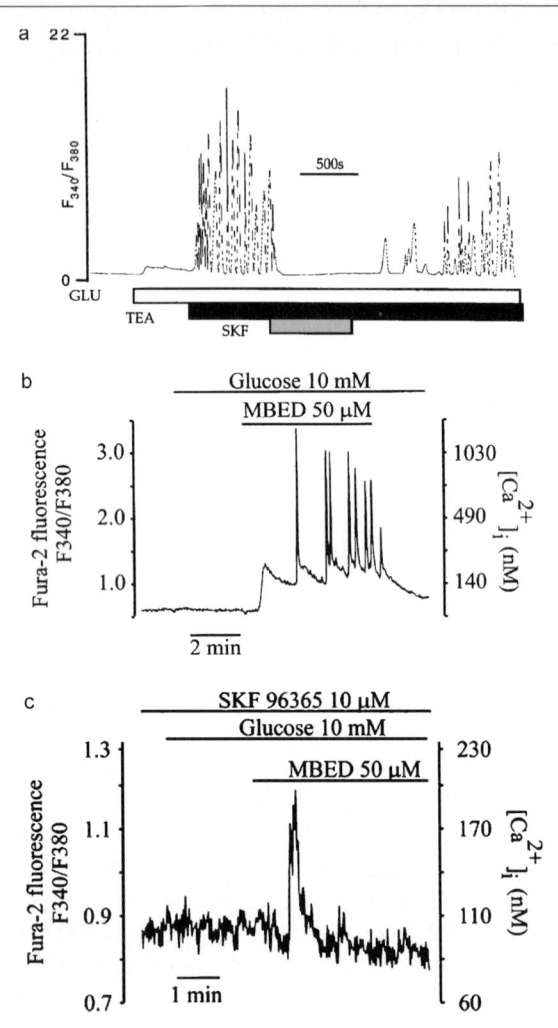

Fig. 2 a–c SKF 96395 inhibits store-depleted calcium entry presumably by inhibiting calcium release-activated TRP-like channels. Calcium was measured using Fura2-loaded cells by determining the change in emission from excitation at 340 nm (F340) to that at 380 nm (F380) (F340/F380). **a** Exposure to the nonspecific TRP channel antagonist SKF 96365 (60 mM; SKF, *gray bar*) ablates calcium oscillations triggered by glucose (1 mM; *open bar*) and TEA (20 mM; *black bar*) (Roe et al. 1998). **b** Activation of ryanodine receptor by 9-methyl 5, 7-dibromoeudistomin D (MBED) resulted in a characteristic pattern of changes in $[Ca^{2+}]_i$. After addition of MBED (50 μM) in the presence 10 mM glucose, there was an initial rapid rise of $[Ca^{2+}]_i$, which was then followed by a plateau. Superimposed on the $[Ca^{2+}]_i$ plateau were a series of large $[Ca^{2+}]_i$ spikes (Gustafsson et al. 2005). **c** MBED (50 μM) was applied in the presence of SKF 96365 (10 μM). This resulted in a transient increase in $[Ca^{2+}]_i$ (Gustafsson et al. 2005)

4.3
Muscarinic Receptors

Acetylcholine (ACh) is released from the intrapancreatic parasympathetic nerve endings during food intake. ACh potentiates insulin secretion through stimulation of muscarinic G protein-coupled receptors (M3) on the pancreatic beta cell. M3 stimulation by acetylcholine can activate TRP-like channels in neurons and beta cells (Miura et al. 1996; Ashcroft and Rorsman 1989; Michel et al. 2005). Activation of TRP-like channels by muscarinic stimulation may contribute to beta cell depolarization and play a role in the insulin-potentiating effects of acetylcholine during feeding.

Two independent TRP-like currents are activated by acetylcholine in the pancreatic beta cell (Mears and Zimliki 2004). These currents can be distinguishable based upon their activation in a calcium store-dependent or -independent manner, respectively (Mears and Zimliki 2004). The calcium store-independent TRP-like current is activated by acetylcholine in a G protein-independent manner causing sodium flux into the cell. The inward sodium-dependent current is small, around 0.77 pA/pF, and inwardly rectifies (Rolland et al. 2002). Interestingly, stimulation of beta cells with similarly sized currents caused beta cell excitability (Rolland et al. 2002).

The calcium store-dependent TRP-like current activated by acetylcholine is also very small, approximately 0.7 pA/pF at −70 mV. Consistent with muscarinic stimulation, emptying endoplasmic reticulum (ER) calcium stores with thapsigargin activates a similarly small inward current (Worley et al. 1994; Bertram et al. 1995; Mears et al. 1997). This has also been demonstrated through release of calcium stores with ryanodine receptor activation, which activates TRP-like calcium entry (Gustafsson et al. 2005; Fig. 2b and c). Accordingly thapsigargin and calcium store depletion cause a depolarizing effect on islets and beta cells that can augment glucose-induced beta cell depolarization and insulin release.

4.4
Glucagon-Like Peptide-1

Glucagon-like peptide (GLP-1) is released from the L cells of the intestine following food intake and causes enhance beta cell insulin secretion, this helps maintain glucose homeostasis during food-induced glucose loads. GLP-1 binds to G protein-coupled receptors on the beta cell causing cyclic AMP (cAMP) production, kinase activity, and ER calcium release. Stimulation with GLP-1 affects beta cell ion flux causing increasing L-type calcium channel activity, reducing ATP-sensitive potassium channel activity, and activating TRP-like nonselective cation channels (Leech and Habener 1997). These changes potentiate insulin release and could play a critical role in glucose homeostasis following food intake, but their exact role remains uncertain.

One of the results of GLP-1 stimulation is increased beta cell permeability to sodium (Miura et al. 2003). Although the molecular identity of the channel responsible for increased sodium flux in response to GLP-1 is undefined, its characteristics resemble a TRP-like channel. The sodium flux is not inhibited by tetrodotoxin, which inhibits voltage-dependent sodium channels, or ouabain, which inhibits the Na^+/K^+ ATPase. However, inhibiting adenylate cyclase inhibits both the sodium influx and the insulinotropic response of GLP-1. Adenylate cyclase activity produces cAMP that can activate protein kinase A (PKA)-dependent and PKA-independent sodium entry in beta cells. These results indicate that sodium entry in response to GLP-1 may result from activation of a TRP-like current, possibly through one already identified molecularly in the beta cell.

4.5
Capsaicin

Capsaicin, the alkaloid of chili peppers that causes the sensation of heat, has been shown to cause beta cell insulin secretion. Dogs treated with a glucose load respond with significantly lower blood glucose values compared with controls when fed concentrated capsaicin solution (Tolan et al. 2001). Similarly, neonatal mice treated with capsaicin to destroy sensory nerves show greater glucose-induced insulin secretion compared to control animals (Karlsson et al. 1994). Rats also respond to systemic capsaicin with increased insulin release (Akiba et al. 2004). Accordingly, the capsaicin-sensitive TRPV1 channels have been identified in both the sensory afferent nerves of the pancreas and in rat beta cells of the islet (Akiba et al. 2004). Although identified in rat beta cells, the effects of capsaicin on the beta cell remain controversial (Carlsson et al. 1996; Akiba et al. 2004). The mechanism for capsaicin-induced increases of insulin release has been shown for the most part to be islet-independent for mice as well as rats through activation of capsaicin-sensitive channels located in pancreatic sensory afferent nerves, which express TRPV1 (Karlsson et al. 1994; Carlsson et al. 1996; Akiba et al. 2004).

Capsaicin interacts with cellular TRPV1 channels leading to nonselective cation flux and depolarization. Mutagenesis studies on the cloned TRPV1 channel have identified multiple distinct intracellular regions that can interact with exogenous capsaicin (Gavva et al. 2004; Jordt and Julius 2002; Jung et al. 2002), and extra cellular activation by capsaicin has also been demonstrated (Vyklicky et al. 2003). Endogenous molecules that affect TRPV1 activation have also been identified, including protons, ATP, bradykinin, leukotriene, *N*-arachidonoyl-dopamine, 12- and 15-(*S*)-hydroperoxyeicosatetraenoic acid, *N*-oleoyldopamine, lipoxygenase products, 5- and 15-(*S*)-hydroxyeicosatetraenoic acids, and *N*-arachidonoyl-ethanolamine (Van Der Stelt et al. 2004). Many of these molecules affect insulin secretion from rodent and human islets; however, their link to TRPV1 is undefined.

5
Pharmacological Block of TRP-Like Currents of the Beta Cell

Although the TRP-like channels of the beta cell are activated by a variety of stimuli, their sensitivity to specific pharmacological block is limited. A limited pharmacology of TRP-like families is emerging and will be an important tool to understanding the role of individual TRP channels in the pancreatic islet.

5.1
SKF 96365

Many beta cell studies of TRP-like channels have used the nonspecific TRP-inhibiting imidazole SKF 96365. SKF 96365 inhibits the store-operated TRP-like calcium influx (Roe et al. 1998; Mears and Zimliki 2004; Gustafsson et al. 2005; Fig. 2a, c) as well as the GLP-1-activated TRP-like conductance (Leech and Habener 1997). Accordingly, glucose-induced calcium oscillations are eliminated with beta cell treatment with SKF 96365, which helps demonstrate the importance of TRP-like channels in glucose-induced insulin secretion. However, the relatively high concentrations of SKF 96365 employed in these studies and its poor specificity makes the interpretation of these studies difficult.

5.2
Trivalents

TRP-like currents, including some within the beta cell, are inhibited by the trivalent cations lanthanum (La^{3+}) and gadolinium (Gd^{3+}). Both the store-operated and the GLP-1-activated TRP-like channels in the beta cell can be blocked with lanthanum and gadolinium (Dyachok and Gylfe 2001; Gustafsson et al. 2005). However, as trivalent cations block many channels, including L-type calcium channels, their use to elucidate the role of TRP-like channels during beta cell insulin secretion is very limited.

5.3
Capsazepine

Capsaicin, the ligand of TRPV1, has been modified with a saturated ring system to form capsazepine, a potent selective inhibitor of the TRPV1 channel, which affects both pancreatic sensory nerves and beta cells. Capsazepine inhibits capsaicin-stimulated insulin secretion from cultured rat beta cells (Akiba et al. 2004) and has also been shown to reduce pain during pancreatitis (Hutter et al. 2005; Wick et al. 2006). Although capsazepine inhibits TRPV1-augmented insulin release in rat beta cells, its role during physiological insulin release may primarily involve sensory nerve channels.

6
Conclusions

Several TRP-like currents, both monovalent cation and nonselective cation, have been biophysically isolated in the pancreatic beta cell. Molecular identification of the TRP-like currents in beta cells is being facilitated through small interfering RNA knockdown and global mouse knockouts. Future studies using animals with beta cell-targeted and developmentally regulated TRP gene disruption in mice will enable more detailed studies of the roles of these genes in glucose metabolism. The development of TRP-specific agonists will also provide further detail of the molecular and biophysical identity of the beta cell TRP channels.

References

Akiba Y, Kato S, Katsube K, Nakamura M, Takeuchi K, Ishii H, Hibi T (2004) Transient receptor potential vanilloid subfamily 1 expressed in pancreatic islet beta cells modulates insulin secretion in rats. Biochem Biophys Res Commun 321:219–225

Amann R, Lembeck F (1986) Capsaicin sensitive afferent neurons from peripheral glucose receptors mediate the insulin-induced increase in adrenal secretion. Naunyn Schmiedebergs Arch Pharmacol 334:71–76

Ashcroft FM, Rorsman P (1989) Electrophysiology of the pancreatic beta-cell. Prog Biophys Mol Biol 54:87–143

Bertram R, Smolen P, Sherman A, Mears D, Atwater I, Martin F, Soria B (1995) A role for calcium release-activated current (CRAC) in cholinergic modulation of electrical activity in pancreatic beta-cells. Biophys J 68:2323–2332

Bhave G, Hu HJ, Glauner KS, Zhu W, Wang H, Brasier DJ, Oxford GS, Gereau RW 4th (2003) Protein kinase C phosphorylation sensitizes but does not activate the capsaicin receptor transient receptor potential vanilloid 1 (TRPV1). Proc Natl Acad Sci U S A 100:12480–12485

Bindokas VP, Kuznetsov A, Sreenan S, Polonsky KS, Roe MW, Philipson LH (2003) Visualizing superoxide production in normal and diabetic rat islets of Langerhans. J Biol Chem 278:9796–9801

Bollimuntha S, Cornatzer E, Singh BB (2005) Plasma membrane localization and function of TRPC1 is dependent on its interaction with beta-tubulin in retinal epithelium cells. Vis Neurosci 22:163–170

Carlsson PO, Sandler S, Jansson L (1996) Influence of the neurotoxin capsaicin on rat pancreatic islets in culture, and on the pancreatic islet blood flow of rats. Eur J Pharmacol 312:75–81

Caterina MJ, Schumacher MA, Tominaga M, Rosen TA, Levine JD, Julius D (1997) The capsaicin receptor: a heat-activated ion channel in the pain pathway. Nature 389:816–824

Chuang HH, Prescott ED, Kong H, Shields S, Jordt SE, Basbaum AI, Chao MV, Julius D (2001) Bradykinin and nerve growth factor release the capsaicin receptor from PtdIns(4,5)P2-mediated inhibition. Nature 411:957–962

Day AL, Maa J, Zerega EC, Richmond AC, Jordan TH, Grady EF, Mulvihill SJ, Bunnett NW, Kirkwood KS (2005) Neutral endopeptidase determines the severity of pancreatitis-associated lung injury. J Surg Res 128:21–27

Dyachok O, Gylfe E (2001) Store-operated influx of Ca(2+) in pancreatic beta-cells exhibits graded dependence on the filling of the endoplasmic reticulum. J Cell Sci 114:2179–2186

Fleig A, Penner R (2004) The TRPM ion channel subfamily: molecular, biophysical and functional features. Trends Pharmacol Sci 25:633–639

Freichel M, Vennekens R, Olausson J, Stolz S, Philipp SE, Weissgerber P, Flockerzi V (2005) Functional role of TRPC proteins in native systems: implications from knockout and knock-down studies. J Physiol 567:59–66

Fridlyand LE, Philipson LH (2004) Does the glucose-dependent insulin secretion mechanism itself cause oxidative stress in pancreatic beta-cells? Diabetes 53:1942–1948

Fridlyand LE, Tamarina N, Philipson LH (2003) Modeling of Ca^{2+} flux in pancreatic beta-cells: role of the plasma membrane and intracellular stores. Am J Physiol Endocrinol Metab 285:E138–154

Fridolf T, Ahrén B (1991) GLP-1(7–36) amide-stimulated insulin secretion in rat islets is sodium-dependent. Biochem Biophys Res Commun 179:701–706

Furuya Y, Takasawa S, Yonekura H, Tanaka T, Takahara J, Okamoto H (1995) Cloning of a cDNA encoding rat bone marrow stromal cell antigen 1 (BST-1) from the islets of Langerhans. Gene 165:329–330

Gavva NR, Klionsky L, Qu Y, Shi L, Tamir R, Edenson S, Zhang TJ, Viswanadhan VN, Toth A, Pearce LV, Vanderah TW, Porreca F, Blumberg PM, Lile J, Sun Y, Wild K, Louis JC, Treanor JJ (2004) Molecular determinants of vanilloid sensitivity in TRPV1. J Biol Chem 279:20283–20295

Gustafsson AJ, Ingelman-Sundberg H, Dzabic M, Awasum J, Nguyen KH, Ostenson CG, Pierro C, Tedeschi P, Woolcott O, Chiounan S, Lund PE, Larsson O, Islam MS (2005) Ryanodine receptor-operated activation of TRP-like channels can trigger critical Ca^{2+} signaling events in pancreatic beta-cells. FASEB J 19:301–323

Hara Y, Wakamori M, Ishii M, Maeno E, Nishida M, Yoshida T, Yamada H, Shimizu S, Mori E, Kudoh J, Shimizu N, Kurose H, Okada Y, Imoto K, Mori Y (2002) LTRPC2 Ca^{2+}-permeable channel activated by changes in redox status confers susceptibility to cell death. Mol Cell 1:163–173

Herson PS, Ashford ML (1997) Activation of a novel non-selective cation channel by alloxan and H2O2 in the rat insulin-secreting cell line CRI-G1. J Physiol 501:59–66

Herson PS, Ashford ML (1999) Reduced glutathione inhibits beta-NAD+-activated non-selective cation currents in the CRI-G1 rat insulin-secreting cell line. J Physiol 514:47–57

Herson PS, Dulock KA, Ashford ML (1997) Characterization of a nicotinamide-adenine dinucleotide-dependent cation channel in the CRI-G1 rat insulinoma cell line. J Physiol 505:65–76

Herson PS, Lee K, Pinnock RD, Hughes J, Ashford ML (1999) Hydrogen peroxide induces intracellular calcium overload by activation of a non-selective cation channel in an insulin-secreting cell line. J Biol Chem 274:833–841

Hofmann T, Schaefer M, Schultz G, Gudermann T (2002) Subunit composition of mammalian transient receptor potential channels in living cells. Proc Natl Acad Sci USA 99:7461–7466

Houamed K, Fu J, Roe MW, Philipson LH (2004) Electrophysiology of the pancreatic beta cell. In: LeRoith DS, Taylor I, Olefsky JM (eds) Diabetes mellitus, 3rd edn. Lippincott Williams and Wilkins, Philadelphia, pp 51–68

Hutter MM, Wick EC, Day AL, Maa J, Zerega EC, Richmond AC, Jordan TH, Grady EF, Mulvihill SJ, Bunnett NW, Kirkwood KS (2005) Transient receptor potential vanilloid (TRPV-1) promotes neurogenic inflammation in the pancreas via activation of the neurokinin-1 receptor (NK-1R). Pancreas 30:260–265

Inamura K, Sano Y, Mochizuki S, Yokoi H, Miyake A, Nozawa K, Kitada C, Matsushime H, Furuichi K (2003) Response to ADP-ribose by activation of TRPM2 in the CRI–G1 insulinoma cell line. J Membr Biol 191:201–207

Jordt SE, Julius D (2002) Molecular basis for species-specific sensitivity to "hot" chili peppers. Cell 108:421–430

Jung J, Lee SY, Hwang SW, Cho H, Shin J, Kang YS, Kim S, Oh U (2002) Agonist recognition sites in the cytosolic tails of vanilloid receptor 1. J Biol Chem 277:44448–44454

Karlsson S, Scheurink AJ, Steffens AB, Ahren B (1994) Involvement of capsaicin-sensitive nerves in regulation of insulin secretion and glucose tolerance in conscious mice. Am J Physiol 267:R1071–R1077

Leech CA, Habener JF (1997) Insulinotropic glucagon-like peptide-1-mediated activation of non-selective cation currents in insulinoma cells is mimicked by maitotoxin. J Biol Chem 272:17987–17993

Lussier MP, Cayouette S, Lepage PK, Bernier CL, Francoeur N, St-Hilaire M, Pinard M, Boulay G (2005) MxA, a member of the dynamin superfamily, interacts with the ankyrin-like repeat domain of TRPC. J Biol Chem 280:19393–19400

Mears D, Zimliki CL (2004) Muscarinic agonists activate Ca^{2+} store-operated and -independent ionic currents in insulin-secreting HIT-T15 cells and mouse pancreatic beta-cells. J Membr Biol 197:59–70

Mears D, Sheppard NF Jr, Atwater I, Rojas E, Bertram R, Sherman A (1997) Evidence that calcium release-activated current mediates the biphasic electrical activity of mouse pancreatic beta-cells. J Membr Biol 155:47–59

Michel FJ, Fortin GD, Martel P, Yeomans J, Trudeau LE (2005) M3-like muscarinic receptors mediate Ca^{2+} influx in rat mesencephalic GABAergic neurones through a protein kinase C-dependent mechanism. Neuropharmacology 48:796–809

Miura Y, Matsui H (2003) Glucagon-like peptide-1 induces a cAMP-dependent increase of [Na+]i associated with insulin secretion in pancreatic beta-cells. Am J Physiol Endocrinol Metab 285:E1001–1009

Miura Y, Gilon P, Henquin JC (1996) Muscarinic stimulation increases Na^+ entry in pancreatic B-cells by a mechanism other than the emptying of intracellular Ca^{2+} pools. Biochem Biophys Res Commun 224:67–73

Mohapatra DP, Nau C (2003) Desensitization of capsaicin-activated currents in the vanilloid receptor TRPV1 is decreased by the cyclic AMP-dependent protein kinase pathway. J Biol Chem 278:50080–50090

Montell C (2005) The TRP superfamily of cation channels. Sci STKE 272:re3

Nadler MJ, Hermosura MC, Inabe K, Perraud AL, Zhu Q, Stokes AJ, Kurosaki T, Kinet JP, Penner R, Scharenberg AM, Fleig A (2001) LTRPC7 is a Mg.ATP-regulated divalent cation channel required for cell viability. Nature 411:590–595

Perraud AL, Takanishi CL, Shen B, Kang S, Smith MK, Schmitz C, Knowles HM, Ferraris D, Li W, Zhang J, Stoddard BL, Scharenberg AM (2005) Accumulation of free ADP-ribose from mitochondria mediates oxidative stress-induced gating of TRPM2 cation channels. J Biol Chem 280:6138–6148

Philipp S, Cavalié A, Freichel M, Wissenbach U, Zimmer S, Trost C, Marquart A, Murakami M, Flockerzi V (1996) A mammalian capacitative calcium entry channel homologous to Drosophila TRP and TRPL. EMBO J 15:6166–6171

Prawitt D, Enklaar T, Klemm G, Gartner B, Spangenberg C, Winterpacht A, Higgins M, Pelletier J, Zabel B (2000) Identification and characterization of MTR1, a novel gene with homology to melastatin (MLSN1) and the trp gene family located in the BWS-WT2 critical region on chromosome 11p15.5 and showing allele-specific expression. Hum Mol Genet 9:203–216

Prawitt D, Monteilh-Zoller MK, Brixel L, Spangenberg C, Zabel B, Fleig A, Penner R (2003) TRPM5 is a transient Ca^{2+}-activated cation channel responding to rapid changes in $[Ca^{2+}]_i$. Proc Natl Acad Sci USA 100:15166–15171

Qian F, Huang P, Ma L, Kuznetsov A, Tamarina N, Philipson LH (2002) TRP genes: candidates for nonselective cation channels and store-operated channels in insulin-secreting cells. Diabetes 51 Suppl 1:S183–189

Roe MW, Worley JF 3rd, Qian F, Tamarina N, Mittal AA, Dralyuk F, Blair NT, Mertz RJ, Philipson LH, Dukes ID (1998) Characterization of a Ca^{2+} release-activated nonselective cation current regulating membrane potential and $[Ca^{2+}]_i$ oscillations in transgenically derived beta-cells. J Biol Chem 273:10402–10410

Rolland JF, Henquin JC, Gilon P (2002) G protein-independent activation of an inward Na(+) current by muscarinic receptors in mouse pancreatic beta-cells. J Biol Chem 277:38373–38380

Sakura H, Ashcroft FM (1997) Identification of four trp1 gene variants murine pancreatic beta-cells. Diabetologia 40:528–532

Sampieri A, Diaz-Munoz M, Antaramian A, Vaca L (2005) The foot structure from the type 1 ryanodine receptor is required for functional coupling to store-operated channels. J Biol Chem 280:24804–24815

Sturgess NC, Carrington CA, Hales CN, Ashford ML (1987a) Calcium and ATP regulate the activity of a non-selective cation channel in a rat insulinoma cell line. Pflugers Arch 409:607–615

Sturgess NC, Carrington CA, Hales CN, Ashford ML (1987b) Nucleotide-sensitive ion channels in human insulin producing tumour cells. Pflugers Arch 410:169–172

Tang J, Lin Y, Zhang Z, Tikunova S, Birnbaumer L, Zhu MX (2001) Identification of common binding sites for calmodulin and inositol 1,4,5-trisphosphate receptors on the carboxyl termini of trp channels. J Biol Chem 276:21303–21310

Tang Y, Tang J, Chen Z, Trost C, Flockerzi V, Li M, Ramesh V, Zhu MX (2000) Association of mammalian trp4 and phospholipase C isozymes with a PDZ domain-containing protein, NHERF. J Biol Chem 275:37559–37564

Tolan I, Ragoobirsingh D, Morrison EY (2001) The effect of capsaicin on blood glucose, plasma insulin levels and insulin binding in dog models. Phytother Res 15:391–394

Van Der Stelt M, Di Marzo V (2004) Endovanilloids. Putative endogenous ligands of transient receptor potential vanilloid 1 channels. Eur J Biochem 271:1827–1834

Vyklicky L, Lyfenko A, Kuffler DP, Vlachova V (2003) Vanilloid receptor TRPV1 is not activated by vanilloids applied intracellularly. Neuroreport 14:1061–1065

Wes PD, Chevesich J, Jeromin A, Rosenberg C, Stetten G, Montell C (1995) TRPC1, a human homolog of a Drosophila store-operated channel. Proc Natl Acad Sci USA 92:9652–9656

Wick EC, Hoge SG, Grahn SW, Kim E, Divino LA, Grady EF, Bunnett NW, Kirkwood KS (2006) Transient receptor potential vanilloid 1, calcitonin gene–related peptide, and substance P mediate nociception in acute pancreatitis. Am J Physiol Gastrointest Liver Physiol 2006 290:G959–G969

Worley JF 3rd, McIntyre MS, Spencer B, Dukes ID (1994) Depletion of intracellular Ca^{2+} stores activates a maitotoxin-sensitive nonselective cationic current in beta-cells. J Biol Chem 269:32055–32058

Yamaguchi H, Matsushita M, Nairn AC, Kuriyan J (2001) Crystal structure of the atypical protein kinase domain of a TRP channel with phosphotransferase activity. Mol Cell 7:1047–1057

Zhu X, Chu PB, Peyton M, Birnbaumer L (1995) Molecular cloning of a widely expressed human homologue for the Drosophila trp gene. FEBS Lett 373:193–198

Zygmunt PM, Petersson J, Andersson DA, Chuang H, Sorgard M, Di Marzo V, Julius D, Hogestatt ED (1999) Vanilloid receptors on sensory nerves mediate the vasodilator action of anandamide. Nature 400:452–457

TRP Channels in Platelet Function

K. S. Authi

Cardiovascular Division, King's College London, New Hunts House, Guy's Campus, London SE1 1UL, UK
kalwant.authi@kcl.ac.uk

1	Introduction	426
2	Ca^{2+} Signalling in Platelets and Megakaryocytes	426
3	Ca^{2+} Channels and TRP Proteins	427
4	Expression of TRPC Genes in Megakaryocytic Cells	428
5	Expression of TRPC Proteins in Platelets and Ca^{2+} Signalling	429
6	Proposed Roles for TRPC1	432
7	Expression and Role for TRPC6	435
8	Expression of Other TRPs and Ca^{2+} Signalling Proteins	438
References		439

Abstract Ca^{2+} entry forms an essential component of platelet activation; however, the mechanisms associated with this process are not understood. Ca^{2+} entry upon receptor activation occurs as a consequence of intracellular store depletion (referred to as store-operated Ca^{2+} entry or SOCE), a direct action of second messengers on cation entry channels or the direct occupancy of a ligand-gated $P2_{X1}$ receptor. The molecular identity of the SOCE channel has yet to be established. Transient receptor potential (TRP) proteins are candidate cation entry channels and are classified into a number of closely related subfamilies including TRPC (canonical), TRPV (vanilloid), TRPM (melastatin), TRPP (polycystin) and TRPML (mucolipins). From the TRPC family, platelets have been shown to express TRPC6 and TRPC1, and are likely to express other TRPC and other TRP members. TRPC6 is suggested to be involved with receptor-activated, diacyl-glycerol-mediated cation entry. TRPC1 has been suggested to be involved with SOCE, though many of the suggested mechanisms remain controversial. As no single TRP channel has the properties described for SOCE in platelets, it is likely that it is composed of a heteromeric association of TRP and related subunits, some of which may be present in intracellular compartments in the resting cell.

Keywords TRPC proteins · Platelets · TRPC1 · TRPC6 · Store-operated Ca^{2+} entry

1
Introduction

Platelets are anucleate cells that play an essential role in haemostasis and thrombosis. At sites of blood vessel injury, platelets adhere to the sub-endothelium and activate, releasing a number of factors that recruit and activate more platelets in addition to activating other vascular cells. These factors include ADP, serotonin and thromboxane A_2 and result in platelet–platelet aggregation and secretion of the contents of their granules, providing a surface for the coagulation reactions [via phosphatidylserine (PS) exposure] that result in the generation of thrombin, and leading to the formation of a haemostatic plug. However, inappropriate activation of platelets in diseased blood vessels, such as those occurring in unstable angina after rupture of an unstable plaque, can lead to myocardial infarction or stroke. Thus, platelets represent important targets for drug intervention. Platelets are formed from megakaryocytes by a complex process that involves the fragmentation of the mature megakaryocyte cytoplasm (Brown et al. 1997). Mature megakaryocytes form less than 0.1% of the cell population of the bone marrow, but the mature cell can be clearly distinguished by its gigantic size compared with other cells in the bone marrow. Megakaryocytes are formed from stem cells during a differentiation process that takes several weeks with one mature megakaryocyte able to produce several thousand platelets. As platelets are anucleate and have little protein synthesising activity, the expression of proteins, be it surface receptors, ion channels or intracellular signalling enzymes, is essentially dictated by their expression in mature megakaryocytes. While the ease with which megakaryocytes can be recognised allows their study at the single cell level, a study of their biochemistry is very difficult because of low numbers. This has led to the use of cell lines isolated from leukaemic patients that represent a megakaryoblastoid phenotype of which the DAMI and human erythroleukemia (HEL) cell lines are a good example. The CHRF-288 cell line is also widely used and represents a more mature megakaryocyte phenotype, as it expresses a higher content of platelet enriched proteins [such as GPIIbIIIa (CD61/CD42b) and GPIb] than the DAMI cell.

2
Ca^{2+} Signalling in Platelets and Megakaryocytes

Recent studies have confirmed an essential role for Ca^{2+} signalling during aggregate formation and thrombus growth in vitro (Nesbitt et al. 2003). Platelet activation (by collagen, thrombin, thromboxane A_2 or ADP) involves the phospholipase C (PLC)-mediated hydrolysis of phosphatidylinositol 4,5-bisphosphate (PIP_2) resulting in the formation of inositol 1,4,5-trisphosphate (IP_3) and 1,2-diacylglycerol (DAG). IP_3 causes the release of Ca^{2+} from intracellular stores via the IP$_3$ receptor (IP$_3$R). Ca^{2+} entry mechanisms from the outside medium are poorly understood but are critical for full activation of the platelet (Nesbitt et al. 2003). A major Ca^{2+} entry pathway results as a consequence of store depletion and is referred to as store-operated Ca^{2+} entry (SOCE) (Putney et al. 2001). A Ca^{2+} release-activated Ca^{2+} current (I_{CRAC}) that represents a specific SOCE current has been demonstrated in the megakaryocytic cell line HEL (Somasundaram et al. 1997) and rat megakaryocytes (Somasundaram and Mahaut-Smith 1994). However, the link between the stores and the plasma membrane (PM) Ca^{2+} entry channel is still poorly defined. Currently favoured mechanisms include the release of a soluble factor (Randriamampita and Tsien 1993) that may involve a Ca^{2+}-independent PLA$_2$ (iPLA$_2$) and calmodulin (Smani et al. 2004), a coupling mechanism involving the IP$_3$R with the PM channel (Irvine 1990) and a model proposing incorporation of a channel or messenger molecule into the plasma membrane (Yao et al. 1999). Rearrangements of the actin cytoskeleton that may affect protein coupling have been shown to block SOCE (Patterson et al. 1999), though the mechanism remains controversial (Lockwich et al. 2001). In platelets, ATP can directly induce Ca^{2+} entry via occupancy of the ligand-gated P2$_{X1}$ receptor (MacKenzie et al. 1996). Other non-SOCE mechanisms include the action of second messengers such as IP$_3$, which has been reported to induce activation of Na$^+$ entry in rat megakaryocytes (Somasundaram and Mahaut-Smith 1995), and a DAG analogue 1-oleoyl-2-acetyl-sn-glycerol (OAG) that is able to induce Ca^{2+} or Ba^{2+} entry in platelets via a possible action on a member of the TRPC channels present in platelets (Hassock et al. 2002). Additionally, phosphatidylinositol 3,4,5-trisphosphate (PIP$_3$), which is the product of phosphatidylinositol 3-kinase (PI-3-K), has been reported to induce Ca^{2+} entry in rabbit platelets and T cells (Lu et al. 1998; Hsu et al. 2000). There are therefore multiple mechanisms of Ca^{2+} (and Na$^+$) entry and the identities of the channels involved and the molecular mechanisms gating them are not established.

3
Ca^{2+} Channels and TRP Proteins

The transient receptor potential (TRP) proteins have been proposed as candidates for SOCE and non-SOCE channels. They are thought to assemble as tetramers to form a channel. There are at least 28 TRP genes in mammalian systems (Clapham 2003), and they have been divided into closely related groups based on sequence identities—the TRPC (canonical), TRPM (melastatin), TRPV (vanilloid), TRPP (polycystin) and TRPML (mucolipin) groups. There is considerable information on the TRPC group, with over-expression studies suggesting most to be receptor-activated (involving PLC), non-selective cation channels. TRPC3, -6 and -7 are the most closely related members, having 75% sequence identity. Heterotetrameric association is known to occur within the TRPC3, -6 and -7 subgroup and TRPC1 is able to complex with -4 and -5 (Hofmann et al. 2002) and may associate with TRPC3 and -7 also (Zagranichnaya et al. 2005). TRPC3, -6 and -7 are known to be activated by DAG independent of protein kinase C (PKC) and independent from store regulation (Hofmann et al. 1999). However, some studies using a variety of tissues have suggested TRPC1, -3, -4 and -5 may be gated by store depletion, though this is controversial.

4
Expression of TRPC Genes in Megakaryocytic Cells

Concerning megakaryocytes and megakaryocytic cell lines, published studies from ourselves (Berg et al. 1997) and others (den Dekker et al. 2001) have shown the expression of messenger RNA (mRNA) for TRPC1, -1A (shorter spliced form), -2, -3, -4 and -6, and for platelets, expression of TRPC1, -1A, -6 and very faintly -4 (TRPC7 was not examined). We have now carried out a further analysis of mRNA expression of TRPC channels in DAMI and CHRF-288 cells prior to studies of knock-down using RNA interference (RNAi) reagents. Excluding TRPC2, which is a pseudogene in humans, in DAMI cells we find good detection of TRPC1 and -3 mRNA using 28 cycles of polymerase chain reaction (PCR), with weak detection of TRPC4 and -6 after 40 cycles of PCR, suggesting predominant expression of TRPC1 and -3 in DAMI cells. In CHRF-288 cells, message for TRPC1 and -3 is again readily detectable at 28 cycles, with that for TRPC6 and -7 detectable at 32 cycles. For TRPC4, 40 cycles are required. In both CHRF-288 and DAMI cells we detected no TRPC5 when tested up to 40 cycles of PCR (K. S. Authi, Y. Shaifta and J.P.T. Ward, unpublished observations). Taken together these observations suggest differentiation-related changes of TRPC expression in megakaryocytes, an increased expression of TRPC6 and -7 with maturation, and potentially an increased role for the latter TRPC channels in the Ca^{2+} signalling characteristics of platelets. This suggestion was mirrored in the earlier study of den Dekker et al. (2001) who reported increased expression of TRPC4 and -6 upon differentiation of stem cells to megakaryocytes using thrombopoietin as the growth factor, with the initial stem cell having detectable levels of only TRPC1A mRNA. It is interesting that TRPC3 mRNA was not detected in the megakaryocytes but, as in our studies, was detected in the cell lines. This may reflect a property of the immortalised cell lines and remains to be established. Further differentiation of stem cells into monocytes did not lead to over-expression of TRPC3, -4 or -6, indicating a lineage-related change. Den Dekker suggested the increased expression of these TRPC channels may relate to the increased SOCE activity of mature megakaryocytes and platelets. In both our work and that of den Dekker, no detection of TRPC5 mRNA was seen in either the megakaryocytes or platelets (or in the cell lines).

5
Expression of TRPC Proteins in Platelets and Ca^{2+} Signalling

Many of the studies published to date regarding the proteins have been from our group and Sage and colleagues, and interestingly have been subject to considerable debate and controversy [see letters to *Blood* (Sage et al. 2002; Authi et al. 2002)]. A great deal of this controversy has arisen from the use of different reagents to identify the TRPC channels present. But before this, it is worth defining some of the characteristics of Ca^{2+} entry in platelets. Figure 1 illustrates important parameters of Ca^{2+} entry after platelets, labelled with Fura2, are stimulated with either a potent platelet agonist—such as thrombin or the sarco-endoplasmic reticulum Ca^{2+} ATPase inhibitor thapsigargin—or a cell-permeable analogue of diacylglycerol, namely OAG, using the Ca^{2+} add back protocol (Hassock et al. 2002). Essentially, store depletion by thapsigargin induces a slow release of the Ca^{2+} from stores, and when Ca^{2+} is added to the medium a rapid and strong entry ensues (Panel b). In our hands, activation of SOCE by thapsigargin is potent at inducing entry of Ca^{2+} but poor at inducing entry of Ba^{2+} when this is substituted for Ca^{2+} in the extracellular medium. This suggests that in human platelets the SOCE channel is selective for Ca^{2+} over Ba^{2+}. If the platelet agonist thrombin is used, this is potent at releasing Ca^{2+} from intracellular stores (by the formation of IP_3) but also effective at inducing the entry of Ca^{2+} and Ba^{2+} (panel c). This suggests that thrombin induces the opening of at least two plasma membrane cation channels. One of these will be the SOCE channel that is selective for Ca^{2+} and another will allow the entry of Ca^{2+} or Ba^{2+}. Figure 1 also shows that OAG induces a slow entry of Ca^{2+} as effectively as Ba^{2+} and thus activates a non-selective cation entry channel (panel a). OAG-induced cation entry occurs without an effect on Ca^{2+} release from intracellular stores (not shown) and is independent of PKC as no effect is observed by inclusion of the PKC inhibitor bisindolylmaleimide I (Bis I). As thrombin leads to the activation of PLC and the formation of DAG, it is likely that endogenous DAG is responsible for the entry of Ba^{2+} by thrombin.

The molecular identification of the channels present is not easy, and many questions regarding the role of TRP channels remain unanswered. In our studies we have used antibodies raised in our laboratory and those of collaborators to identify expression in platelets. In each case the antibody preparations were shown to recognise the corresponding over-expressed TRPC protein and were then tested with platelet membranes. Our studies have shown that there is a low level of TRPC1 and a higher level of TRPC6. TRPC6 was easily detectable using both an in-house-generated antibody and one obtained from Alomone Labs (both to the same epitope sequence); when determined using highly purified fractions of plasma and intracellular membranes prepared by free flow electrophoresis, TRPC6 was found exclusively in the plasma membrane fraction consistent with a role involving cation entry. TRPC1, however, was found predominantly in the intracellular membrane from resting platelets, us-

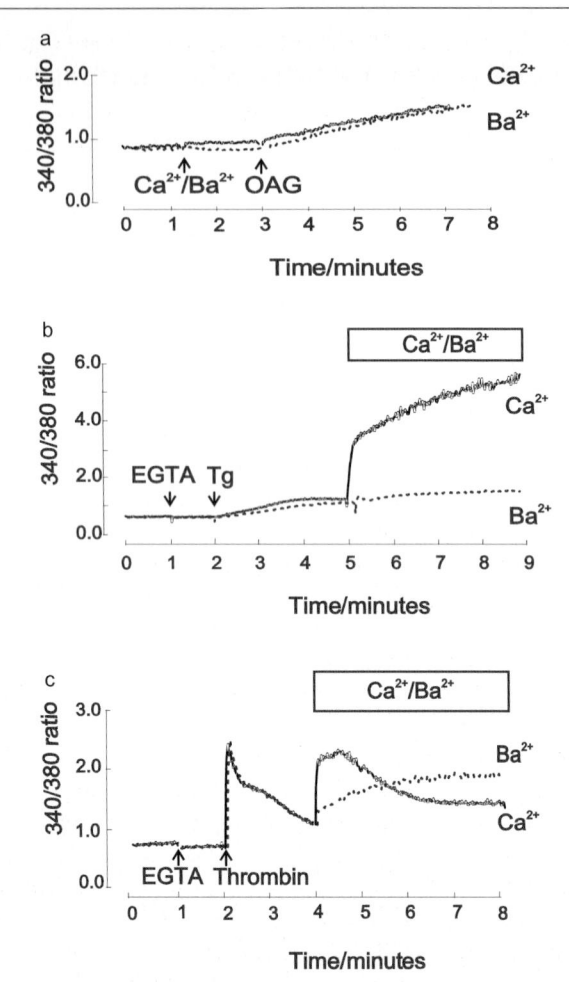

Fig. 1 a–c Characteristics of cation influx in human platelets. Fura2-loaded platelets were incubated at 37°C (additions of the agonists or analogue are marked with *arrows*); the ratio (340/380 nm) of fluorescence increase was measured using extracellular 1 mM Ca^{2+} (*continuous lines*) or Ba^{2+} (*dashed lines*). **a** Responses to 60 µM *OAG*; **b** responses to 3 µM thapsigargin (*Tg*), EGTA was added at 100 µM; **c** responses to 1 U/ml *thrombin*. Other details see text. Taken from Hassock et al. (2002)

ing an antibody raised to the first ankyrin domain of TRPC1 (Ank) and also an anti-*Xenopus* TRP1 antibody that recognises hTRPC1 (Hassock et al. 2002). Using antibodies that were shown to be specific to TRPC3 and one shown to be specific for TRPC4 and -5, we were unable to confirm expression of TRPC3, and TRPC4 or -5 in platelet membranes. This suggests that expression of these isoforms was below the level of detection for these antibodies or that they were not significantly expressed. The absence of TRPC3 detection agrees with the absence of detection of TRPC3 mRNA from platelets and from cultured megakaryocytes (den Dekker et al. 2001), though TRPC3 mRNA is detected in the cell lines DAMI and CHRF-288. Additionally, the absence of TRPC5 also confirms the lack of detection of this isoform at the mRNA level in platelets, megakaryocytes and cell lines. With anti-TRPC4 antibodies we have observed staining of a 250-kDa protein, but the identity of this band is unknown (S.R. Hassock and K.S. Authi, unpublished observations). Studies to determine expression of TRPC7 are under way, and our preliminary studies suggest using an antibody raised to a C-terminal epitope of TRPC7 that recognises mTRPC7 overexpressed in QBI-293 cells that there is expression in platelet membranes (S.R. Hassock and K.S. Authi, unpublished observations).

A number of studies from Sage and colleagues have also reported the presence of TRPC1 in platelets using an antibody obtained from Alomone Labs (Rosado and Sage 2000a, 2001; Rosado et al. 2002), which they suggest is involved with SOCE and couples with the IP$_3$R type II upon store depletion. Very recently they have reported the expression of TRPC3, -4 and -5 to almost similar extents as to TRPC6 and -1, again using antibodies obtained from Alomone Labs (Brownlow and Sage 2005). This reported presence of TRPC3, -4 and -5 may reflect levels that may be undetectable by antibodies used in our laboratory, in which case their roles require elucidation. However, a number of the commercially available antibodies have been subject to considerable criticism; thus, results obtained with these require caution and also reconfirmation with better-characterised antibodies. This is particularly striking with the TRPC1 and -4 polyclonal antibodies from Alomone Labs, and many groups have used these extensively. The manufacturer's Web site (up to December 2005) indicates that the antibody to TRPC1 is raised to a cytoplasmic epitope (residues 557–571) and Western blotting with rat brain membranes reveals protein staining at greater than 250 kDa and approximately 120 kDa. Unfortunately, neither weight is the equivalent of TRPC1. Human TRPC1 migrates at approximately 80 kDa, and so does the endogenous protein from rat brain (Goel et al. 2002). Strikingly, Ong et al. (2002) have shown that the Alomone TRPC1 antibody preparation does not recognise over-expressed hTRPC1 but did recognise proteins of 120 kDa (as stated by the manufacturer) in fractions from mouse liver and brain that remain unidentified. Therefore it remains possible that the 90- to 100-kDa protein recognised in endogenous tissues with this anti-TRPC1 antibody may not be TRPC1, and its identity needs to be confirmed by protein sequencing. Similar criticism has also surfaced regarding the anti-TRPC4 antibody used in the Brownlow and Sage study to show expression of TRPC4 in platelet membranes. Flockerzi et al. (2005) have recently shown that this anti-TRPC4 antibody recognises the appropriate size band in brain microsomes from wild-type mice expressing TRPC4, but it also recognises the same band in brain microsomes prepared from TRPC4$^{-/-}$ mice, where the protein does not exist. It may be that batch difference may explain the discrepancies between laboratories, but use of these reagents alone will continue to generate confusing messages that do not help our understanding of the proteins that regulate Ca^{2+} signalling. Using these antibodies, Brownlow and Sage (2005) show from immunoprecipitation experiments that "TRPC1" associated with "TRPC4" and TRPC5, and that TRPC3 associated with TRPC6 in human platelets. The study lacks any demonstration that the antibody preparations recognise the appropriate positive controls. They reported near equal expression of all these isoforms, whereas mRNA analysis suggests very unequal expression, and in our studies TRPC5, -4 and -3 proteins were not detected with the antibodies used. Verification of this data is therefore needed using better-characterised alternative antibodies.

6
Proposed Roles for TRPC1

Our finding of TRPC1 in intracellular membranes initially surprised us, but there are currently numerous examples of TRPC channels found in membrane systems distributed in the cytoplasm. When over-expressed alone, Strubing et al. (2001) found that TRPC1 did not carry a significant current and Hofmann et al. (2002) showed that it was widely distributed in the cytoplasm. Where expression is successful its cation permeability is similar for Ca^{2+}, Na^+, Cs^+ or Ba^{2+} (Sinkins et al. 1998), and when co-expressed with TRPC5 again it has non-selective cation permeability (Strubing et al. 2001). Thus, by itself or in combination with TRPC5, TRPC1 does not explain the Ca^{2+} selectivity seen in platelet SOCE. However, there is evidence that in a number of cells, including salivary glands and endothelial cells, TRPC1 may be involved with store-mediated Ca^{2+} entry. Its presence in intracellular stores does not rule out its involvement in SOCE, particularly if a population of heteromeric channel complexes exists that is ready to be inserted into the plasma membrane upon signal generation. In our ongoing studies, over-expression of hTRPC1-GFP (green fluorescent protein tagged) into DAMI or CHRF-288 cells reveals expression throughout the cytoplasm and not just localised at the plasma membrane. In avian B cells, Mori et al. (2002) reported that knock-out of TRP1 (avian TRPC1) not only inhibited SOCE but also reduced agonist-induced Ca^{2+} release from stores and IP_3-mediated Ca^{2+} release from the endoplasmic reticulum. Such effects would not be seen if all of the TRP1 was present in the plasma membrane and imply, in line with our finding, that at least a part of the TRPC1 is present in intracellular stores and can interact with the IP_3R. Thus, activation of platelets by surface receptor agonists or store depletion may lead to an insertion of the channel into the plasma membrane.

In other cell systems, TRPC1 has been shown to be capable of binding to a large number of proteins; these include the IP_3R, homer, calmodulin, caveolin, FKBP25, Gq11, PLC, PKC and RhoA, plus others. We have shown that platelets express all three types of IP_3R, with the expression of the type I and type II much greater than the type III receptor (El Daher et al. 2000). The Sage group have reported using human platelets and the Alomone anti-TRPC1 antibody that "TRPC1" couples to the type II IP$_3$R when stores are depleted with either thapsigargin or thrombin treatment; the TRPC1 antibody when applied extracellularly inhibited Ca^{2+} entry or Mn^{2+} entry by approximately 70% (Rosado and Sage 2000a, 2001; Rosado et al. 2002). This suggests that "TRPC1" could account for the majority of Ca^{2+} entry. They suggest a model where the type II IP$_3$R at the intracellular stores specifically couples with "TRPC1" at the plasma membrane when stores are depleted and cite this as evidence for a secretion-like coupling mechanism previously described by Patterson et al. (1999). "TRPC1" is proposed to be only in the plasma membrane, as platelet activation by thapsigargin did not lead to any further increase of antibody binding to intact platelets (Rosado et al. 2002) or an increase of surface biotin labelling of this protein with either thapsigargin or thrombin treatment (Brownlow and Sage 2005). Coupling and SOCE was decreased if platelets were depleted of cholesterol with methyl-β-cyclodextrin (Brownlow et al. 2004), suggesting that "TRPC1" may exist in lipid rafts, as shown in other cells, and was not affected by PKC inhibition (Brownlow and Sage 2003). Further they reported that jasplakinolide, which is known to induce cortical actin assembly and thereby erect a physical barrier, inhibited Ca^{2+} entry and disrupted the coupling of "TRPC1" with the IP$_3$R (Rosado et al. 2002). Surprisingly, xestospongin, an agent that inhibits IP$_3$R function, also inhibited the coupling of the IP$_3$R to "TRPC1", even if store depletion was carried out by thapsigargin (Rosado and Sage 2001). However, and in contrast, coupling induced by thapsigargin was not affected in the presence of 2-aminoethoxydiphenyl borate, which has also been shown to inhibit IP$_3$R and cation entry channels, suggesting that not all inhibitors of IP$_3$R affected this interaction (Diver et al. 2001).

Studies carried out with TRPC1 on platelets in our laboratory differ significantly from those of the Sage group. Incubation of Fura2-loaded human platelets with the Alomone TRPC1 antibody at 10 μg/ml for 10 min did not have any significant effect on Ca^{2+} entry induced by either thapsigargin (200 nM) or 0.25 U/ml thrombin (S.R. Hassock and K.S. Authi, unpublished observations). Again we cannot rule out differences in batches that may lead to loss of antibody activity, but an alternative explanation is that TRPC1 may be present predominantly inside the cell. Thus, the protein band seen in the Sage group studies needs further characterisation. In studies thus far, with our antibodies we have not observed a store depletion-dependent coupling of TRPC1 with the IP$_3$R type II but have seen a constitutive coupling of the two proteins (S.R. Hassock and K.S. Authi, unpublished observations). Studies by Kunzelmann-Marche et al. on HEL cells (which have a similar phenotype to DAMI cells) with the anti-TRPC1 antibody suggest that the antibody had a small significant inhibition of thapsigargin-induced Ca^{2+} entry but also A23187-mediated Ca^{2+} entry, where in the latter case these effects were linked to a similar small inhibition of PS exposure that occurs when pro-coagulant activity is expressed. But surprisingly the antibody also significantly inhibited thapsigargin-mediated Ca^{2+} release from stores, with the authors having to use low concentrations of antibody to "avoid artefactual membrane perturbation" (Kunzelmann-Marche et al. 2002).

Studies from other systems provide an insight into the microenvironment of TRPC1. Ambudkar's group have shown that TRPC1 (using their own in-house-generated antibody) exists in the cholesterol-rich lipid raft domains in the membrane of human salivary gland cells in association with a number of other signalling proteins such as the IP_3R, caveolin 1, plasma membrane Ca^{2+} ATPase and $G\alpha q/11$ (Lockwich et al. 2000). Such an arrangement was also seen with TRPC3 in a stably transfected cell line and where rearrangement of the cytoskeleton with jasplakinolide led to an internalisation of the multimolecular complex containing the IP_3R and TRPC protein rather than a disruption of the association (Lockwich et al. 2001). Studies from Worley's group have reported that the adaptor protein homer plays an important role in the interaction of TRPC1 with IP_3R in human embryonic kidney (HEK)-293 cells and show that store depletion was associated with a disassembly of the complex in contrast to the association reported by the Sage group (Yuan et al. 2003). Homer appears to be important in the association of the IP_3R with TRPC1, and the group reported that a mutated form of homer that doesn't bind TRPC1 leads to a constitutive activation of TRPC1, suggesting that homer may hold TRPC1 in a closed state. Currently it is not known if homer is expressed in platelets. Another cytosolic protein reported to interact with and inhibit TRPC1 is I-mfa (inhibitor of myogenic family a) whose knock-down leads to an enhancement and its over-expression leads to an inhibition of TRPC1-implicated store-mediated currents in CHO-K1 cells (Ma et al. 2003). In microvessel endothelial cells a significant proportion of TRPC1 has been reported to exist in the cytosolic compartment, and its recruitment to the plasma membrane requires the activation of Rho, which increases an interaction of TRPC1 with an IP_3R and its insertion into the plasma membrane (Mehta et al. 2003). The authors reported that inhibition of Rho activation by transfected C3 transferase reduced the association and translocation of TRPC1 to the plasma membrane even if stores were depleted. In a subsequent study they reported that PKC-dependent phosphorylation of TRPC1 was also required to elicit Ca^{2+} entry via TRPC1 and this was also associated with an increase of endothelial cell permeability (Ahmmed et al. 2004). In platelets there is a suggestion that Rho and other small G proteins may play a role in SOCE, and incubation of C3 exoenzyme with platelets for 2 h is reported to cause inhibition of thapsigargin-induced Ca^{2+} entry in addition to use of pharmacological agents that inhibited methylation of these proteins (Rosado and Sage 2000b). Whether these manipulations can alter TRPC1-IP_3R coupling or translocation as is suggested in endothelial cells is not known. But PKC function is thought not to be required for SOCE in platelets, and its inhibition has been reported not to affect the "TRPC1"-IP_3R coupling (Brownlow and Sage 2003).

Phosphorylation of proteins is an effective way of regulating the activity of many proteins. TRPC1 contains consensus sequences for action by various protein kinases such as PKC, PKA (cAMP-dependent protein kinase), PKG (cGMP-), casein kinase II and tyrosine kinases. Platelet stimulatory and inhibitory agents all cause an increase in phosphorylation of a number of protein substrates. Using Western blotting techniques and [^{32}P]-labelled platelets we were unable to show an increase of phosphorylation of TRPC1 by tyrosine kinases, PKC, PKA or PKG under conditions where the activity of the kinases was demonstrated on known substrates after stimulation by appropriate agents (Hassock et al. 2002). However, TRPC1 extraction from human platelets with the Ank antibody did co-extract a number of proteins of molecular weight 250, 120 and 85 kDa that were phosphorylated by either PKA or PKG, and these phosphorylations decreased upon activation of platelets by thrombin. This suggests that the microenvironment around TRPC1 contains proteins that are substrates for PKA and PKG. This is important as, in the circulation, platelets are exposed to either prostacyclin or nitric oxide that will activate PKA and PKG, respectively. These kinases play important regulatory roles in maintaining low cytosolic Ca^{2+} levels. Thus far, however, there are no studies linking TRPC1 activity to platelet function such as shape change, aggregation, secretion or pro-coagulant expression. If TRPC1 is integrally involved with SOCE it is likely that it forms a component rather than the channel itself. Its co-association with TRPC5 or -4 opens the heteromeric channel to augmentation by 100 µM La^{3+} (Strubing et al. 2001; Schaefer et al. 2000). This is in contrast to the reported effects of inhibition of SOCE by La^{3+} and it complicates the proposal that TRPC5 and -4 are associated with TRPC1 in human platelets. The possibility that TRPC1 may exist as a complex with subunits other than TRPC members is worthy of examination. TRPC1 has been shown to bind to polycystin 2 (TRPP2), which itself has channel properties, and the combination may constitute a channel with distinct properties (Tsiokas et al. 1999). TRPP2 is predominantly associated with the endoplasmic reticulum (Koulen et al. 2002) though some is also reported in the plasma membrane (Hanaoka et al. 2000). TRPP2 has also been reported to form a distinct channel when complexed with TRPP1 (polycystin 1) (Hanaoka et al. 2000). Thus, a heteromeric complex comprising TRPC1 with three distinct subunits, some of which may be initially intracellular and translocate to the plasma membrane upon signal generation, may confer the specific properties of SOCE as seen in platelets. Though this remains to be established, it is an interesting hypothesis.

7
Expression and Role for TRPC6

TRPC6 is a non-selective cation channel but there is a fivefold preference for divalent over monovalent cations (Hofmann et al. 1999). In line with the expression of TRPC6 in the platelet plasma membrane, we have shown OAG to cause the entry of Ca^{2+} or Ba^{2+}, independent from store release or PKC activation (Hassock et al. 2002). We have also shown that thrombin stimulated entry of Ba^{2+} is inhibited by the inclusion of the PLC inhibitor U73122 and by a cell-permeable analogue of cAMP that suppresses thrombin-induced PLC activation (S. R. Hassock and K.S. Authi, unpublished observations). Thus, a link between PLC activity, DAG and increase of Ca^{2+} entry is proposed. Further inclusion of the PKC inhibitor Bis I prior to thrombin addition leads to an enhancement of Ba^{2+} entry. The slower kinetics of cation entry by OAG on platelets compared with SOCE-mediated entry suggests a role for TRPC6 during the middle and later stages of platelet activation. Thrombin (but not collagen) stimulation of platelets also results in an increase of TRPC6 translocation to the cytoskeletal fraction. This may be important, as thrombin activation leads to the focussing of a large number of signalling proteins (particularly those that are tyrosine phosphorylated) with cytoskeletal elements that allow enzymes and substrates to be in close association for rapid signalling (Fox 1996). Examples of these include actin-binding protein, α-actinin, actin, myosin, cortactin, talin, spectrin, vinculin, GPIb, GPIIbIIIa, $pp60^{c-src}$, PI-3-K, rap1b and others. TRPC6 has been reported to bind to α-actinin (Goel et al. 2005).

We have shown in platelets that agents that stimulate PKA or PKG lead to TRPC6 phosphorylation and, like TRPC1, TRPC6 also appears to be associated with proteins that are substrates of PKA (Hassock et al. 2002). This localises the channel in an environment that is regulated by cAMP concentrations. Membranes from TRPC6 over-expressing QBI-293 cells incubated with the catalytic subunit of PKA show increased incorporation of phosphate into the protein. Currently the direct consequence of this phosphorylation is not known, as in platelets OAG-stimulated Ca^{2+} or Ba^{2+} entry is unaffected. Two potential sites for PKA-mediated phosphorylation exist at $RRQT^{70}$ and $KKLS^{332}$. Both are in the N-terminal part of the protein which contains a number of ankyrin domains; whether this may alter interaction with cytoskeletal proteins remains to be determined. There are no specific inhibitors of TRPC6 and no specific function-blocking antibody available for testing yet. Analysis of the platelet activities from $TRPC6^{-/-}$ mice may yield valuable information.

There have been recent studies that implicate PIP_3 (the product of PI-3-K) as an activator of TRPC6. PI-3-kinase is heavily expressed in a number of haematopoietic cells, has been described to be important in many functions and has a major role in the activation of PLCγ during platelet stimulation by collagen. TRPC6 contains a YXXM motif that acts as a recognition site for the p85 subunit of PI-3-K. Initially Chen's laboratory showed that cell-permeable analogues of PIP_3 [such as dipalmitoyl-sn-glycerol (DiC16)-PIP_3 or dioctanoyl-sn-glycerol (DiC8)-PIP_3] activated rabbit platelets and caused entry of Ca^{2+} without release from stores (Lu et al. 1998). These studies were extended to PIP_3-activating Ca^{2+} entry in T cells and in rat basophilic leukaemia cells but not in B cells (Hsu et al. 2000; Ching et al. 2001). This suggests that PIP_3 may represent a novel mechanism for store-independent Ca^{2+} entry in some haematopoietic cells. In a recent study using TRPC subunits stably expressed in HEK-293 cells, Chen's group have shown that PIP_3 activates a number of TRPC channels but especially TRPC6 (Tseng et al. 2004).

Because TRPC6 is expressed in platelets we have also examined this activity. As our previous studies suggested that we could use entry of Ba^{2+} to monitor SOCE-independent TRPC6 activity, we examined both activation and entry of cations in Ca^{2+} or Ba^{2+} medium. We have found that exogenously added DiC8- or DiC16-PIP_3 caused aggregation of washed human platelets in a dose-related manner with the DiC8-PI(4,5)P_2 or DiC8-PI(3,4)P_2 either poorly or not effective (S.R. Hassock and K.S. Authi, manuscript in preparation). However, DiC8-PIP_3 was not able to cause the entry of Ca^{2+}, and in only a few preparations (1 in 10) did we observe a small entry of Ba^{2+}. Thus, aggregation of platelets and induction of cation entry could not be correlated. When examined against an effector of PI-3-K, exogenous DiC8-PIP_3 was able to cause activation of Akt (or PKB) in human platelets, suggesting that it did gain entry into the cell and that PKB activation could be related to the aggregation of the cells. In agreement with this effect, an inhibitor of PKB was effective in blocking the aggregation of platelets seen with DiC8-PIP_3. We examined the ability of DiC8-PIP_3 to directly activate transiently expressed TRPC channels in QBI-293 cells. Expression of TRPC proteins was monitored by Western blotting using TRPC-specific antibodies, and the functional activity of hTRPC6, hTRPC3 and mTRPC7 was gauged by monitoring the entry of Ba^{2+} by OAG. Under these conditions we were unable to activate hTRPC6, hTRPC3 or mTRPC7 by DiC8-PIP_3 (S.R. Hassock and K.S. Authi, manuscript in preparation).

The reason for the contrast in data between these findings and those of the Chen group who used stably transfected mTRPC6 is not known, but one possibility may be that stable expression leads to alteration of the expression of PI-3-K effectors or that in our transient expression system an effector protein for PIP_3 may be reduced or missing. We suggest from our data that PIP_3 does not directly activate TRPC6 or its related channels, but we cannot rule out an action of PIP_3 on TRPC channels via another PI-3-K effector. Possible effectors such as PKC isoforms (or others) need to be tested, as these are known to affect TRPC channel activity and a number of the PKC isoforms have been reported to be activated by PIP_3 (Toker and Cantley 1997).

A recent study of TRPC6 gene knock-out in mice resulted in a two- to threefold up-regulation of TRPC3, suggesting a close association of these two members (Dietrich et al. 2005). The mice showed an elevated blood pressure and enhanced agonist-induced contractility in isolated aortic rings that is probably due to the enhanced expression of TRPC3. A number of recent studies have reported that some TRPC channels can associate with other ion-transporting proteins—such as the Na^+/K^+ATPase with TRPC6 and -5 in rat brain (Goel et al. 2005) and the Na^+/Ca^{2+} exchanger with TRPC3 in HEK-293

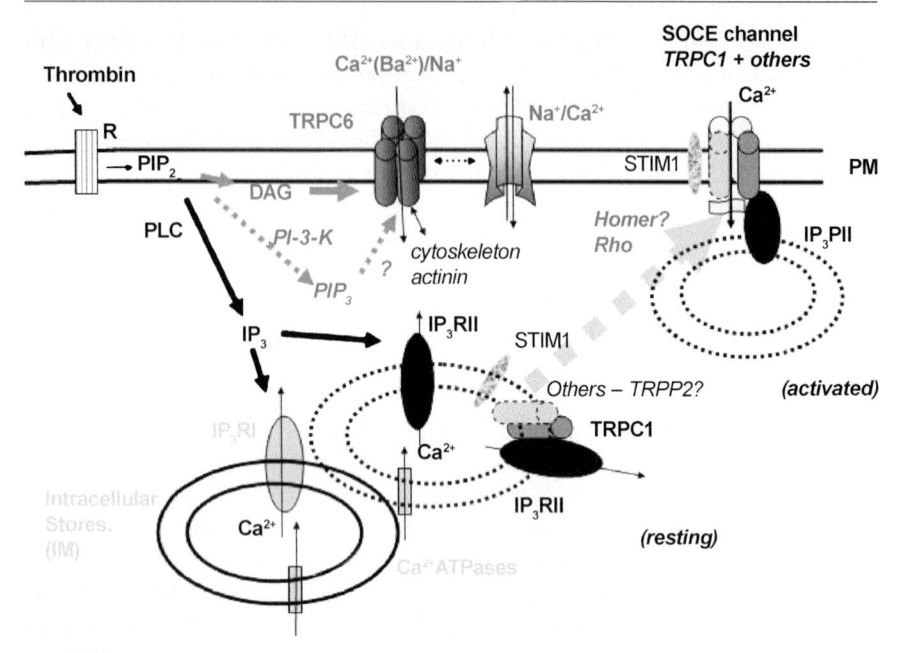

Fig. 2 Model of organisation of TRP proteins in human platelets. For full details see text. Localisation of TRPC7 and other TRPs has yet to be determined

cells (Rosker et al. 2004)—implying a role in Na^+ or K^+ movements. Platelet activation is associated with Na^+ entry (e.g. Roberts et al. 2004), and there is a well-expressed Na^+/Ca^{2+} exchanger in the plasma membrane (Rengasamy et al. 1987; El Daher et al. 2000). Thus, it is possible that TRPC6 may allow the entry of Na^+ which then activates a Na^+/Ca^{2+} exchanger (in the reverse mode) to allow Ca^{2+} entry. A simplified model of the possible organisation of TRP proteins in platelets is depicted in Fig. 2.

8
Expression of Other TRPs and Ca²⁺ Signalling Proteins

Thus far there are no reports of the expression of TRPV or TRPM members in platelets, though it is likely that a number of these subunits are expressed. In an analysis of mRNA detection for TRPM2 and TRPM7 in megakaryocytic cell lines DAMI and CHRF-288, we have observed high expression of TRPM7 in both cell lines and a restricted expression of TRPM2 to the more mature CHRF-288 cell with no detection in the DAMI cells (K.S. Authi and Y. Shaifta, unpublished observations). This is in line with the described ubiquitous expression of TRPM7, which is know to be permeable to Ca^{2+} and Mg^{2+} ions but also to heavy metal divalent cations (Clapham 2003). The channel's suggested ability to sense Mg^{2+}. ATP levels may play a role in metabolic sensing. TRPM2 expression in CHRF-288 cells also suggests its expression in platelets. TRPM2 is a Ca^{2+}-permeable channel that is activated by ADP ribose and is responsive to hydrogen peroxide (H_2O_2). Platelets express CD38 which can generate ADP ribose (Torti et al. 1999), and H_2O_2 is generated as a consequence of platelet activation and has been associated with many oxidative stress-related cellular effects but also Ca^{2+} elevation, inhibition of sequestration, inhibition of extrusion and activation of pp60^{src} (Rosado et al. 2004; Redondo et al. 2004). In a proteomics-based approach using three different strategies to isolate platelet membrane proteins, we have analysed the protein composition of platelet membranes using liquid chromatography linked with mass spectrometry. We have detected expression of TRPM2 and also detected stromal interaction molecule 1 (STIM1), PKD1-like 1 and TRPV3 in addition to other Ca^{2+} signalling proteins and channels (Y.A. Senis, M.G. Tomlinson, Á. García, S. Dumon, V.L. Heath, J. Herbert, S.P. Cobbold, J.C. Spalton, S. Ayman, N. Zitzman, R. Bicknell, J. Frampton, K.S. Authi, A. Martin, M.J. Wakelam, and S.P. Watson, manuscript submitted). Recently STIM1 has been shown by two groups to be the Ca^{2+} sensor, as its knock-down with siRNA inhibited Ca^{2+} entry (Roos et al. 2005; Zhang et al. 2005; Liou et al. 2005). STIM1 can be present in intracellular and plasma membranes and contains an EF hand domain near the N terminus that protrudes either into the lumen of the store or projects outwards to the extracellular medium and binds Ca^{2+}. Store depletion is thought to either induce translocation of STIM1 to the plasma membrane or its re-organisation into punctate regions around the plasma membrane that may allow the regulation of the entry channel. Again, trafficking of proteins from the intracellular compartments to the plasma membrane is stipulated.

In conclusion, platelets have been shown to express TRPC6 as a DAG-activated cation entry channel and a number of other TRP isoforms. The organisation of these subunits into the major Ca^{2+} entry channel and its possible interaction with STIM1 is yet to be established, but may bring about a better understanding of how cation entry occurs into platelets and open up new avenues for therapeutic targeting.

Acknowledgements This work is supported by grants from the British Heart Foundation.

References

Ahmmed GU, Mehta D, Vogel S, Holinstat M, Paria BC, Tiruppathi C, Malik AB (2004) Protein kinase Calpha phosphorylates the TRPC1 channel and regulates store-operated Ca^{2+} entry in endothelial cells. J Biol Chem 279:20941–20949

Authi KS, Hassock S, Zhu MX, Flockerzi V, Trost C (2002) TRPC channels in platelets and Ca^{2+} entry in human platelets. Blood 100:4246–4247

Berg LP, Shamsher MK, El Daher SS, Kakkar VV, Authi KS (1997) Expression of human TRPC genes in the megakaryocytic cell lines MEG01, DAMI and HEL. FEBS Lett 403:83–86

Brown AS, Erusalimsky JD, Martin JF (1997) Megakaryocytopoiesis: the megakaryocyte/ platelet haemostatic axis. In: Bruchlausen FV, Walter U (eds) Antithrombotics. (Handbook of experimental pharmacology, vol. 126) Springer, Berlin Heidelberg New York, pp 3–26

Brownlow SL, Sage SO (2003) Rapid agonist-evoked coupling of type II Ins(1,4,5)P3 receptor with human transient receptor potential (hTRPC1) channels in human platelets. Biochem J 375:697–704

Brownlow SL, Sage SO (2005) Transient receptor potential protein subunit assembly and membrane distribution in human platelets. Thromb Haemost 94:839–845

Brownlow SL, Harper AG, Harper MT, Sage SO (2004) A role for hTRPC1 and lipid raft domains in store-mediated calcium entry in human platelets. Cell Calcium 35:107–113

Ching TT, Hsu AL, Johnson AJ, Chen CS (2001) Phosphoinositide 3-kinase facilitates antigen-stimulated $Ca^{(2+)}$ influx in RBL-2H3 mast cells via a phosphatidylinositol 3,4,5-trisphosphate-sensitive $Ca^{(2+)}$ entry mechanism. J Biol Chem 276:14814–14820

Clapham DE (2003) TRP channels as cellular sensors. Nature 426:517–524

den Dekker E, Molin DG, Breikers G, van Oerle R, Akkerman JW, van Eys GJ, Heemskerk JW (2001) Expression of transient receptor potential mRNA isoforms and $Ca^{(2+)}$ influx in differentiating human stem cells and platelets. Biochim Biophys Acta 1539:243–255

Dietrich A, Mederos YS, Gollasch M, Gross V, Storch U, Dubrovska G, Obst M, Yildirim E, Salanova B, Kalwa H, Essin K, Pinkenburg O, Luft FC, Gudermann T, Birnbaumer L (2005) Increased vascular smooth muscle contractility in $TRPC6^{-/-}$ mice. Mol Cell Biol 25:6980–6989

Diver JM, Sage SO, Rosado JA (2001) The inositol trisphosphate receptor antagonist 2-aminoethoxydiphenylborate (2-APB) blocks Ca^{2+} entry channels in human platelets: cautions for its use in studying Ca^{2+} influx. Cell Calcium 30:323–329

El Daher SS, Patel Y, Siddiqua A, Hassock S, Edmunds S, Maddison B, Patel G, Goulding D, Lupu F, Wojcikiewicz RJ, Authi KS (2000) Distinct localization and function of (1,4,5)IP(3) receptor subtypes and the (1,3,4,5)IP(4) receptor GAP1(IP4BP) in highly purified human platelet membranes. Blood 95:3412–3422

Flockerzi V, Jung C, Aberle T, Meissner M, Freichel M, Philipp SE, Nastainczyk W, Maurer P, Zimmermann R (2005) Specific detection and semi-quantitative analysis of TRPC4 protein expression by antibodies. Pflugers Arch 451:81–86

Fox JE (1996) Studies of proteins associated with the platelet cytoskeleton. In: Watson SP, Authi KS (eds) Platelets, a practical approach. IRL Press, Oxford, pp 217–233

Goel M, Sinkins WG, Schilling WP (2002) Selective association of TRPC channel subunits in rat brain synaptosomes. J Biol Chem 277:48303–48310

Goel M, Sinkins W, Keightley A, Kinter M, Schilling WP (2005) Proteomic analysis of TRPC5- and TRPC6-binding partners reveals interaction with the plasmalemmal $Na^{(+)}/K^{(+)}$-ATPase. Pflugers Arch 451:87–98

Hanaoka K, Qian F, Boletta A, Bhunia AK, Piontek K, Tsiokas L, Sukhatme VP, Guggino WB, Germino GG (2000) Co-assembly of polycystin-1 and -2 produces unique cation-permeable currents. Nature 408:990–994

Hassock SR, Zhu MX, Trost C, Flockerzi V, Authi KS (2002) Expression and role of TRPC proteins in human platelets: evidence that TRPC6 forms the store-independent calcium entry channel. Blood 100:2801–2811

Hofmann T, Obukhov AG, Schaefer M, Harteneck C, Gudermann T, Schultz G (1999) Direct activation of human TRPC6 and TRPC3 channels by diacylglycerol. Nature 397:259–263

Hofmann T, Schaefer M, Schultz G, Gudermann T (2002) Subunit composition of mammalian transient receptor potential channels in living cells. Proc Natl Acad Sci U S A 99:7461–7466

Hsu AL, Ching TT, Sen G, Wang DS, Bondada S, Authi KS, Chen CS (2000) Novel function of phosphoinositide 3-kinase in T cell Ca^{2+} signaling. A phosphatidylinositol 3,4,5-trisphosphate-mediated Ca^{2+} entry mechanism. J Biol Chem 275:16242–16250

Irvine RF (1990) 'Quantal' Ca^{2+} release and the control of Ca^{2+} entry by inositol phosphates—a possible mechanism. FEBS Lett 263:5–9

Koulen P, Cai Y, Geng L, Maeda Y, Nishimura S, Witzgall R, Ehrlich BE, Somlo S (2002) Polycystin-2 is an intracellular calcium release channel. Nat Cell Biol 4:191–197

Kunzelmann-Marche C, Freyssinet JM, Martinez MC (2002) Loss of plasma membrane phospholipid asymmetry requires raft integrity. Role of transient receptor potential channels and ERK pathway. J Biol Chem 277:19876–19881

Liou J, Kim ML, Heo WD, Jones JT, Myers JW, Ferrell JE Jr, Meyer T (2005) STIM is a Ca^{2+} sensor essential for Ca^{2+}-store-depletion-triggered Ca^{2+} influx. Curr Biol 15:1235–1241

Lockwich T, Singh BB, Liu X, Ambudkar IS (2001) Stabilization of cortical actin induces internalization of transient receptor potential 3 (Trp3)-associated caveolar Ca^{2+} signaling complex and loss of Ca^{2+} influx without disruption of Trp3-inositol trisphosphate receptor association. J Biol Chem 276:42401–42408

Lockwich TP, Liu X, Singh BB, Jadlowiec J, Weiland S, Ambudkar IS (2000) Assembly of Trp1 in a signaling complex associated with caveolin-scaffolding lipid raft domains. J Biol Chem 275:11934–11942

Lu PJ, Hsu AL, Wang DS, Chen CS (1998) Phosphatidylinositol 3,4,5-trisphosphate triggers platelet aggregation by activating Ca^{2+} influx. Biochemistry 37:9776–9783

Ma R, Rundle D, Jacks J, Koch M, Downs T, Tsiokas L (2003) Inhibitor of myogenic family, a novel suppressor of store-operated currents through an interaction with TRPC1. J Biol Chem 278:52763–52772

MacKenzie AB, Mahaut-Smith MP, Sage SO (1996) Activation of receptor-operated cation channels via P2X1 not P2T purinoceptors in human platelets. J Biol Chem 271:2879–2881

Mehta D, Ahmmed GU, Paria BC, Holinstat M, Voyno-Yasenetskaya T, Tiruppathi C, Minshall RD, Malik AB (2003) RhoA interaction with inositol 1,4,5-trisphosphate receptor and transient receptor potential channel-1 regulates Ca^{2+} entry. Role in signaling increased endothelial permeability. J Biol Chem 278:33492–33500

Mori Y, Wakamori M, Miyakawa T, Hermosura M, Hara Y, Nishida M, Hirose K, Mizushima A, Kurosaki M, Mori E, Gotoh K, Okada T, Fleig A, Penner R, Iino M, Kurosaki T (2002) Transient receptor potential 1 regulates capacitative $Ca^{(2+)}$ entry and $Ca^{(2+)}$ release from endoplasmic reticulum in B lymphocytes. J Exp Med 195:673–681

Nesbitt WS, Giuliano S, Kulkarni S, Dopheide SM, Harper IS, Jackson SP (2003) Intercellular calcium communication regulates platelet aggregation and thrombus growth. J Cell Biol 160:1151–1161

Ong HL, Chen J, Chataway T, Brereton H, Zhang L, Downs T, Tsiokas L, Barritt G (2002) Specific detection of the endogenous transient receptor potential (TRP)-1 protein in liver and airway smooth muscle cells using immunoprecipitation and Western-blot analysis. Biochem J 364:641–648

Patterson RL, van Rossum DB, Gill DL (1999) Store-operated Ca^{2+} entry: evidence for a secretion-like coupling model. Cell 98:487–499

Putney JW Jr, Broad LM, Braun FJ, Lievremont JP, Bird GS (2001) Mechanisms of capacitative calcium entry. J Cell Sci 114:2223–2229

Randriamampita C, Tsien RY (1993) Emptying of intracellular Ca^{2+} stores releases a novel small messenger that stimulates Ca^{2+} influx. Nature 364:809–814

Redondo PC, Salido GM, Pariente JA, Rosado JA (2004) Dual effect of hydrogen peroxide on store-mediated calcium entry in human platelets. Biochem Pharmacol 67:1065–1076

Rengasamy A, Soura S, Feinberg H (1987) Platelet Ca^{2+} homeostasis: Na^+–Ca^{2+} exchange in plasma membrane vesicles. Thromb Haemost 57:337–340

Roberts DE, McNicol A, Bose R (2004) Mechanism of collagen activation in human platelets. J Biol Chem 279:19421–19430

Roos J, DiGregorio PJ, Yeromin AV, Ohlsen K, Lioudyno M, Zhang S, Safrina O, Kozak JA, Wagner SL, Cahalan MD, Velicelebi G, Stauderman KA (2005) STIM1, an essential and conserved component of store-operated Ca^{2+} channel function. J Cell Biol 169:435–445

Rosado JA, Sage SO (2000a) Coupling between inositol 1,4,5-trisphosphate receptors and human transient receptor potential channel 1 when intracellular Ca^{2+} stores are depleted. Biochem J 350:631–635

Rosado JA, Sage SO (2000b) Farnesylcysteine analogues inhibit store-regulated Ca^{2+} entry in human platelets: evidence for involvement of small GTP-binding proteins and actin cytoskeleton. Biochem J 347:183–192

Rosado JA, Sage SO (2001) Activation of store-mediated calcium entry by secretion-like coupling between the inositol 1,4,5-trisphosphate receptor type II and human transient receptor potential (hTrp1) channels in human platelets. Biochem J 356:191–198

Rosado JA, Brownlow SL, Sage SO (2002) Endogenously expressed Trp1 is involved in store-mediated Ca^{2+} entry by conformational coupling in human platelets. J Biol Chem 277:42157–42163

Rosado JA, Redondo PC, Salido GM, Gomez-Arteta E, Sage SO, Pariente JA (2004) Hydrogen peroxide generation induces pp60src activation in human platelets: evidence for the involvement of this pathway in store-mediated calcium entry. J Biol Chem 279:1665–1675

Rosker C, Graziani A, Lukas M, Eder P, Zhu MX, Romanin C, Groschner K (2004) $Ca^{(2+)}$ signaling by TRPC3 involves $Na^{(+)}$ entry and local coupling to the $Na^{(+)}/Ca^{(2+)}$ exchanger. J Biol Chem 279:13696–13704

Sage SO, Brownlow SL, Rosado JA (2002) TRP channels and calcium entry in human platelets. Blood 100:4245–4246

Schaefer M, Plant TD, Obukhov AG, Hofmann T, Gudermann T, Schultz G (2000) Receptor-mediated regulation of the nonselective cation channels TRPC4 and TRPC5. J Biol Chem 275:17517–17526

Sinkins WG, Estacion M, Schilling WP (1998) Functional expression of TrpC1: a human homologue of the Drosophila Trp channel. Biochem J 331:331–339

Smani T, Zakharov SI, Csutora P, Leno E, Trepakova ES, Bolotina VM (2004) A novel mechanism for the store-operated calcium influx pathway. Nat Cell Biol 6:113–120

Somasundaram B, Mahaut-Smith MP (1994) Three cation influx currents activated by purinergic receptor stimulation in rat megakaryocytes. J Physiol 480:225–231

Somasundaram B, Mahaut-Smith MP (1995) A novel monovalent cation channel activated by inositol trisphosphate in the plasma membrane of rat megakaryocytes. J Biol Chem 270:16638–16644

Somasundaram B, Mason MJ, Mahaut-Smith MP (1997) Thrombin-dependent calcium signalling in single human erythroleukaemia cells. J Physiol 501:485–495

Strubing C, Krapivinsky G, Krapivinsky L, Clapham DE (2001) TRPC1 and TRPC5 form a novel cation channel in mammalian brain. Neuron 29:645–655

Toker A, Cantley LC (1997) Signalling through the lipid products of phosphoinositide-3-OH kinase. Nature 387:673–676

Torti M, Bertoni A, Canobbio I, Sinigaglia F, Balduini C (1999) Hydrolysis of NADP+ by platelet CD38 in the absence of synthesis and degradation of cyclic ADP-ribose 2′-phosphate. FEBS Lett 455:359–363

Tseng PH, Lin HP, Hu H, Wang C, Zhu MX, Chen CS (2004) The canonical transient receptor potential 6 channel as a putative phosphatidylinositol 3,4,5-trisphosphate-sensitive calcium entry system. Biochemistry 43:11701–11708

Tsiokas L, Arnould T, Zhu CW, Kim E, Walz G, Sukhatme VP (1999) Specific association of the gene product of PKD2 with the TRPC1 channel. Proc Natl Acad Sci USA 96:3934–3939

Yao Y, Ferrer-Montiel AV, Montal M, Tsien RY (1999) Activation of store-operated Ca^{2+} current in Xenopus oocytes requires SNAP-25 but not a diffusible messenger. Cell 98:475–485

Yuan JP, Kiselyov K, Shin DM, Chen J, Shcheynikov N, Kang SH, Dehoff MH, Schwarz MK, Seeburg PH, Muallem S, Worley PF (2003) Homer binds TRPC family channels and is required for gating of TRPC1 by IP3 receptors. Cell 114:777–789

Zagranichnaya TK, Wu X, Villereal ML (2005) Endogenous TRPC1, TRPC3, and TRPC7 proteins combine to form native store-operated channels in HEK-293 cells. J Biol Chem 280:29559–29569

Zhang SL, Yu Y, Roos J, Kozak JA, Deerinck TJ, Ellisman MH, Stauderman KA, Cahalan MD (2005) STIM1 is a Ca^{2+} sensor that activates CRAC channels and migrates from the Ca^{2+} store to the plasma membrane. Nature 437:902–905

TRP Channels in Lymphocytes

E. C. Schwarz · M.-J. Wolfs · S. Tonner · A. S. Wenning · A. Quintana · D. Griesemer · M. Hoth (✉)

Institut für Physiologie, Universität des Saarlandes, Gebäude 58, 66421 Homburg/Saar, Germany
markus.hoth@uniklinikum-saarland.de

1	Ca^{2+}/Cation Channels in Lymphocytes	446
2	Expression Pattern of TRP Channels in Lymphocytes	446
3	Ion Channel Properties and Modes of Activation of TRP Channels in Lymphocytes	448
3.1	TRPC1	448
3.2	TRPC3	448
3.3	TRPC7	449
3.4	TRPM2	449
3.5	TRPM4	449
3.6	TRPM7	450
3.7	TRPV6	450
4	Pharmacology	451
5	Biological Relevance and Emerging/Established Biological Roles for TRP Channels in Lymphocytes	452
	References	452

Abstract TRP proteins form ion channels that are activated following receptor stimulation. Several members of the TRP family are likely to be expressed in lymphocytes. However, in many studies, messenger RNA (mRNA) but not protein expression was analyzed and cell lines but not primary human or murine lymphocytes were used. Among the expressed TRP mRNAs are TRPC1, TRPC3, TRPM2, TRPM4, TRPM7, TRPV1, and TRPV2. Regulation of Ca^{2+} entry is a key process for lymphocyte activation, and TRP channels may both increase Ca^{2+} influx (such as TRPC3) or decrease Ca^{2+} influx through membrane depolarization (such as TRPM4). In the future, linking endogenous Ca^{2+}/cation channels in lymphocytes with TRP proteins should lead to a better molecular understanding of lymphocyte activation.

Keywords Lymphocyte · TRP channel · CRAC channel · Cation channel

1
Ca^{2+}/Cation Channels in Lymphocytes

K$^+$ and Ca^{2+}/cation channels are important for lymphocyte activation (Lewis and Cahalan 1995). Ca^{2+}/cation channels provide the Ca^{2+} influx necessary for many signaling processes including gene expression in both B and T cells, whereas K$^+$ channels inhibit membrane depolarization that would otherwise diminish Ca^{2+} entry. Several K$^+$ channels have been cloned in primary human T cells and in the human T cell line Jurkat, and much research has been devoted to their pharmacology because they are promising targets for immunomodulation (Chandy et al. 2004). The recently identified Orai1 protein has been shown to function as a Ca^{2+} channel in lymphocytes and is very likely part of the endogenous Ca^{2+} release-activated Ca^{2+} (CRAC) channel complex (Feske et al. 2006; Prakriya et al. 2006; Yeromin et al. 2006).

Following engagement of the T or B cell receptor and formation of the immunological synapse, several signaling cascades are initiated including the activation of phospholipase C-γ (PLC-γ), which cleaves phosphatidylinositol 4,5 bisphosphate generating the second messengers inositol 1,4,5, trisphosphate (IP$_3$) and diacylglycerol (DAG). IP$_3$ depletes intracellular Ca^{2+} stores, which leads to the opening of store-operated Ca^{2+} (SOC) channels, called Ca^{2+} release-activated Ca^{2+} (CRAC) channels in lymphocytes (Hoth and Penner 1992; Zweifach and Lewis 1993). CRAC channels are the major source for Ca^{2+} influx and thus necessary to activate a variety of transcription factors including nuclear factor-κB (NF-κB), nuclear factor of activated T cells (NFAT), and activator protein-1 (AP-1), which together direct cellular responses such as cytokine secretion, cell proliferation, and cell differentiation (Gallo et al. 2006; Hogan et al. 2003; Lewis 2001; Quintana et al. 2005).

The role of other Ca^{2+}/cation channels in lymphocytes is less clear. It has been proposed that voltage-operated Ca^{2+} channels contribute to Ca^{2+} influx in T cells; additionally, several nonselective cation channels have been described in T and B cells (Gallo et al. 2006; Grafton and Thwaite 2001). Whereas there is no doubt about the existence of messenger RNA (mRNA) from voltage-gated Ca^{2+} channels in lymphocytes, the only paper reporting any electrophysiological data has just been retracted (Flavell et al. 2005).

2
Expression Pattern of TRP Channels in Lymphocytes

Knowledge about the expression pattern of TRP proteins in lymphocytes is a prerequisite to elucidate their role for Ca^{2+} signaling. Most of the studies have been carried out in cell lines used as model for either T cells (human Jurkat T cells) or B cells (chicken DT40 B cells) and only a few data are available from primary lymphocytes. Investigations of TRP protein expression are difficult because of its relatively low abundance and the limited availability of specific and high-affinity antibodies.

Within the TRPC family, Philipp et al. (2003) could detect TRPC3 but not TRPC4 in Jurkat T cells only by immunoprecipitation using self-made anti-TRPC3- or anti-TRPC4-specific antibodies. Andreopoulos et al. (2004) identified TRPC1 (with Santa Cruz antibodies) and TRPC3 (with Alomone antibodies) in Epstein-Barr virus (EBV)-transformed B lymphoblast cell lines from patients with bipolar-I disorder (BD-I, a manic-depressive illness) but not TRPC4 (with Alomone antibodies) or TRPC6 (with Santa Cruz antibodies). In this report, there was, however, an inconsistency between the apparent molecular weight of TRPC3 (123 kDa) and the calculated molecular weight (97.5 kDa). Gamberucci et al. (2002) found TRPC6 in purified plasma membranes of Jurkat T cells and peripheral blood lymphocytes (PBLs).

In all other studies in lymphocytes, only expression of TRP mRNA was analyzed either by Northern blot or reverse transcriptase PCR (RT-PCR) techniques. The expression of TRPC1 and TRPC3 was found to be consistent on protein and mRNA level in T cells (Gamberucci et al. 2002; Philipp et al. 2003) and was also reported in B cells (Vazquez et al. 2003). TRPC4, TRPC5, and TRPC7 are most likely not expressed in T cells (Gamberucci et al. 2002; Philipp et al. 2003; Rao and Kaminski 2006), TRPC2 is a pseudogene in humans (Wissenbach et al. 1998), and expression data of TRPC6, at least in Jurkat T cells, are contradictory [compare Gamberucci et al. (2002) and Rao and Kaminski (2006) to Philipp et al. (2003)]. Expression of TRPC7 mRNA was found in EBV-transformed B lymphoblast cell lines and was reduced in the cell lines from patients with BD-I and BD-II compared to healthy individuals (Yoon et al. 2001).

Expression of TRPV1 was demonstrated in murine dendritic cells by Western blot analysis and RT-PCR (Basu and Srivastava 2005) and TRPV2 was found in murine mast cells (Stokes et al. 2004), but expression in lymphocytes has not been investigated. To our knowledge, no information is available on TRPV3 and TRPV4 in lymphocytes. TRPV5 expression was reported in Jurkat T cells using RT-PCR, and expression of TRPV6 was found in Jurkat T cells using Northern blot and RT-PCR technology (Cui et al. 2002) but not in human leukocytes using Northern blot technology (Wissenbach et al. 2001), which could be a problem of low expression in leukocytes.

Within the TRPM family, TRPM3, TRPM5, and TRPM6 expression have not been reported. Complementary DNA (cDNA) fragments of TRPM1 transcripts were identified in B and T cell lines using a bioinformatics approach (Perraud et al. 2004). TRPM2 mRNA expression was found in Jurkat T cells as well as in human lymphocytes by RT-PCR (Sano et al. 2001), and TRPM4 expression was shown by Western blot in Jurkat T cells as well as in Molt-4 T lymphoblasts (Launay et al. 2004). However, in primary human leukocytes, TRPM4 could not be detected by Northern blot technology (Nilius et al. 2003; Xu et al. 2001). TRPM7 is widely expressed and the knockout of TRPM7 was lethal for DT40 B cells, probably due to its role in Mg^{2+} homeostasis (Nadler et al. 2001).

The expression pattern of the other TRP families (TRPA, TRPP, TRPML, TRPN) has to our knowledge not been investigated in lymphocytes.

3
Ion Channel Properties and Modes of Activation of TRP Channels in Lymphocytes

3.1
TRPC1

At present, TRPC1 may be the strongest candidate of all TRPs to form store-operated nonselective cation channels. From the very beginning of the discovery of human TRPs, TRPC1 has been implicated to form SOC channels (Ambudkar 2005; Liu et al. 2003; Mori et al. 2002; Zhu et al. 1996). However, other modes of activation for TRPC1 have been described [for a review see Ramsey et al. (2005)], which is one example for the lack of consensus regarding the modes of activation of any of the TRPC channels. Targeted disruption of TRPC1 in DT40 chicken B cells reduced SOC activity dramatically, indicating that TRPC1 is an important component of SOC in B cells (Mori et al. 2002). One result of the study by Mori et al. (2002) was, however, unexpected: The authors found no SOC current in 80% of the TRPC1 deficient cells, but normal SOC currents in the other 20%. Another surprising defect of the TRPC1-deficient B cells was the reduced B cell receptor (BCR)-dependent Ca^{2+} mobilization.

3.2
TRPC3

About 10 years ago, Jurkat T cells with reduced CRAC channel activity were generated (Fanger et al. 1995), and several years later damage of the *TRPC3* gene was found in these mutant T cells (Philipp et al. 2003). Overexpressing TRPC3 in the mutants rescued T cell receptor-dependent Ca^{2+} signaling and increased the level of CRAC current amplitudes in patch-clamp recordings (Philipp et al. 2003). In their study, Philipp et al. (2003) found TRPC3 to form highly Ca^{2+}-selective ion channels with biophysical properties similar to endogenous CRAC channels. The high Ca^{2+} selectivity is in clear contrast to all other TRPC3 studies, in which the channels were described as nonselective cation channels. Vazquez et al. (2001) reported, for instance, that TRPC3 expression in DT40 B cells was correlated with an increase in nonselective cation channel activity. The major mode of TRPC3 activation appears to be related to the formation of DAG (Hofmann et al. 1999); however, other stimuli like Ca^{2+} (Zitt et al. 1997) or store-operation (Vazquez et al. 2001) have also been reported to activate TRPC3. Vazquez et al. (2004) recently found Src kinase-mediated tyrosine phosphorylation to be critical for TRPC3 activation. Since Src kinase is one of the most important kinases activated by the T cell receptor, these results support the potential role of TRPC3 for T cell receptor-dependent Ca^{2+} entry. Furthermore, the CRAC channel blocker 3,5-bistrifluoromethyl pyrazole (BTP)2 (Zitt et al. 2004) was recently shown to inhibit specifically TRPC3 and TRPC5 with potencies similar to CRAC channels (He et al. 2005). TRPC3 is thus a candidate to be part of the CRAC channel protein complex.

3.3
TRPC7

A knockout of TRPC7 in DT40 chicken B cells revealed that TRPC7 does not contribute to store-operated Ca^{2+}/cation entry in these cells; however, receptor-activated but store-independent cation entry was basically absent (Lievremont et al. 2005). Similar to its closest relatives, TRPC3 and TRPC6, TRPC7 is probably activated in a DAG-dependent way (Hofmann et al. 1999; Lievremont et al. 2005).

3.4
TRPM2

Adenosine 5′-diphosphoribose (ADPR) activates Ca^{2+}/cation influx through TRPM2 channels in Jurkat T cells (Gasser et al. 2005; Sano et al. 2001). A cytosolic Nudix box sequence motif located within the C-terminus of TRPM2 has been shown to bind ADPR, thereby activating TRPM2 currents (Kuhn and Luckhoff 2004). TRPM2 forms a nonselective cation channel that is believed to be associated with cell death. It is regulated by the intracellular Ca^{2+} concentration with low-level activation at around 100 nM and maximal activation around 600 nM (IC_{50}=340 nM) (McHugh et al. 2003). Recently, Gasser et al. (2005) demonstrated that high concentrations of concanavalin A can elevate endogenous ADPR levels in Jurkat T cells, which in turn activates TRPM2 and subsequent cell death. In this regard it is noteworthy that H_2O_2 can also activate TRPM2, potentially linking reactive oxygen species production with cell death (Wehage et al. 2002).

3.5
TRPM4

TRPM4 is a Ca^{2+}-activated nonselective (CAN) ion channel which has been shown to depolarize the membrane potential following its activation through a rise in $[Ca^{2+}]_i$ (Launay et al. 2002; Nilius et al. 2003). Launay et al. (2004) detected TRPM4-mediated currents in Jurkat T cells by either perfusing free Ca^{2+} concentrations of up to 800 nM into the cells or by stimulation with phytohemagglutinin (PHA). Using an siRNA approach, they showed that downregulation of TRPM4 increased PHA-activated $[Ca^{2+}]_i$ and interleukin (IL)-2 production in Jurkat T cells. It was concluded that TRPM4 activation, through the depolarization of the membrane potential, decreased Ca^{2+} influx and the subsequent $[Ca^{2+}]_i$ rise and IL-2 production. Launay et al. (2004) therefore suggested that TRPM4 may play an important role as negative feedback mechanisms during Ca^{2+} oscillations which are considered to be very important for differential gene expression in T cells (Dolmetsch et al. 1998; Tomida et al. 2003; Utzny et al. 2005).

3.6
TRPM7

TRPM7 is inhibited by intracellular divalent cations (in particular Mg^{2+}) and functions as a divalent-permeable cation channel, which is very likely required for Mg^{2+} homeostasis in many cell types including lymphocytes (Kerschbaum et al. 2003; Monteilh-Zoller et al. 2003; Nadler et al. 2001; Runnels et al. 2001). It is almost certainly responsible for the endogenous Mg^{2+}-inhibited current (MIC), which has been analyzed in great detail in T cells (Hermosura et al. 2002; Kerschbaum et al. 2003; Kozak and Cahalan 2003; Prakriya and Lewis 2003). TRPM7 consists of an ion channel domain and an α-kinase domain, whose role for channel modulation or signaling is controversial (Kozak and Cahalan 2003; Nadler et al. 2001; Runnels et al. 2001). The cellular knockout of TRPM7 in DT40 chicken lymphocytes was found to be lethal, probably because the cellular Mg^{2+} homeostasis was grossly disturbed (Nadler et al. 2001).

3.7
TRPV6

Considering only their pore properties, TRPV5 and TRPV6 are the only channels of the TRP family that are likely candidates to form part of the endogenous CRAC channels in T lymphocytes. Like CRAC channels (Hoth 1995), TRPV5 and TRPV6 have been shown to be highly selective for Ca^{2+} over other cations (Niemeyer et al. 2001; Owsianik et al. 2005; Vennekens et al. 2000). Other biophysical properties, however, were found to be similar but not identical to CRAC channels (Voets et al. 2001). On the basis of its biophysical properties and its proposed activation mode by store depletion, it was postulated that TRPV6 constituted the pore forming unit if not the whole CRAC channel complex (Yue et al. 2001). In support of this, Cui et al. (2002) showed with over expression and antisense strategies that TRPV6 can work as a CRAC-like channel in Jurkat T cells. Subsequent studies with overexpressed TRPV6, however, did not confirm that TRPV6 was related to endogenous CRAC channels. In many of those studies, properties of TRPV6 and CRAC channels were found to be different (Bodding et al. 2002; Voets et al. 2001). In addition, the selective CRAC channel inhibitor BTP2 (Ishikawa et al. 2003; Zitt et al. 2004) had no effect on TRPV6 activity (He et al. 2005; Schwarz et al. 2006). At present, it is unclear if TRPV6 forms functional ion channels in T cells at all.

4
Pharmacology

Most substances interfering with Ca^{2+}/cation channels in lymphocytes or TRP channels present in lymphocytes appear to be rather unspecific (Li et al. 2002). Therefore, we will only discuss the two most promising substances: 2-aminoethoxydiphenyl borate (2-APB), which is one of the best-studied modulators of TRP channel function and, although it is unspecific, may be helpful to define TRP channel function in lymphocytes; the pyrazole derivative BTP2, which is a rather new and promising tool to study the function of endogenous TRP channels in lymphocytes.

2-APB, among other effects, inhibited TRPC5 (Xu et al. 2005), TRPC3, and TRPC6 (Hu et al. 2004; Ma et al. 2000; Thebault et al. 2005; van Rossum et al. 2000) at concentrations much higher than 10 µM. On the other hand, 2-APB worked as an agonist for TRPV1–3 (at concentration as low as 1 µM) but not for TRPV4–6 (Hu et al. 2004; Ramsey et al. 2005). In lymphocytes, 2-APB enhanced CRAC currents up to fivefold at relatively low concentrations (1–5 µM) whereas, after a transient increase, it reduced CRAC currents at concentrations higher than 10 µM (Prakriya and Lewis 2001). Considering only the pharmacological data with 2-APB, one could conclude that CRAC channels could be a multimer consisting of TRPC and TRPV channels.

For TRPC3, this conclusion is supported by recent findings with BTP2. BTP2 has been shown to inhibit CRAC channels and Ca^{2+}-dependent signal transduction in T cells at very low (<100 nM) concentrations (Ishikawa et al. 2003; Zitt et al. 2004). BTP2 was found to be very specific because it did not interfere with any other ion channels tested (Ishikawa et al. 2003; Zitt et al. 2004). The kinetics of inhibition were found to be rather slow (minutes to hours), a finding which cannot easily be reconciled with a classical channel pore blocker. On the other hand, BTP2 did not exhibit its inhibitory effect from the cytoplasmic side of the CRAC channels, as indicated by whole-cell perfusion with the substance (Zitt et al. 2004). Overexpressed TRPC3 and TRPC5 channels were also inhibited by BTP2, with potencies similar to CRAC channels (He et al. 2005), whereas TRPV6 was not susceptible to high BTP2 concentrations (He et al. 2005; Schwarz et al. 2006). Again, from the pharmacological point of view, one could conclude from the correlations between endogenous CRAC and expressed TRP channels that TRPC3 is a good candidate to form part of the CRAC channel. TRPC5 need not be considered because so far there is no evidence that it is expressed in lymphocytes.

5
Biological Relevance and Emerging/Established Biological Roles for TRP Channels in Lymphocytes

As for many cell types so far, the functional role of TRP proteins in lymphocytes is still unclear. Combining all data published in Jurkat T cells, we conclude that TRPM2 is activated by ADPR and related to cell death and that TRPM4 is a Ca^{2+}-activated Na^+-permeable cation channel that may serve to depolarize the membrane potential, thereby reducing Ca^{2+} signals and T cell activation. TRPM7 is almost certainly a housekeeping divalent-permeable cation channel likely to be responsible for Mg^{2+} homeostasis in many tissues including lymphocytes. In DT40 chicken B cells, TRPC7 is clearly working as a receptor–but not store-operated cation channel; its function, however, is still undefined.

The question of which proteins are part of the CRAC channel complex remains a mystery. Genetic and pharmacological evidence point toward a role of TRPC3. This view is, however, not supported by the vast majority of studies in which overexpressed TRPC3 is characterized as a DAG-activated nonselective cation channel, such properties not reconcilable with endogenous CRAC channels in lymphocytes.

We believe that one problem with much of the current TRP channel research is its dependence on overexpression systems. Since it is relatively easy to obtain primary lymphocytes from humans, one should analyze endogenous TRP channels in these cells to finally understand the physiological and pathophysiological roles of TRP channels in mankind.

Acknowledgements Research in our laboratory is funded by the Deutsche Forschungsgemeinschaft (SFB 530, project A3, DFG grant HO 2190/1-1, Graduate College "Cell differentiation and cell proliferation" and Graduate College "Molecular, physiological and pharmacological analysis of cellular membrane transport" to M.H.), the competitive intrauniversity funding HOMFOR (to E.C.S.), and the Central Research Committee Saarbrücken (to D.G.).

References

Ambudkar IS (2006) Ca^{2+} signaling microdomains: platforms for the assembly and regulation of TRPC channels. Trends Pharmacol Sci 27:25–32

Andreopoulos S, Wasserman M, Woo K, Li PP, Warsh JJ (2004) Chronic lithium treatment of B lymphoblasts from bipolar disorder patients reduces transient receptor potential channel 3 levels. Pharmacogenomics J 4:365–373

Basu S, Srivastava P (2005) Immunological role of neuronal receptor vanilloid receptor 1 expressed on dendritic cells. Proc Natl Acad Sci U S A 102:5120–5125

Bodding M, Wissenbach U, Flockerzi V (2002) The recombinant human TRPV6 channel functions as Ca^{2+} sensor in human embryonic kidney and rat basophilic leukemia cells. J Biol Chem 277:36656–36664

Chandy KG, Wulff H, Beeton C, Pennington M, Gutman GA, Cahalan MD (2004) K$^+$ channels as targets for specific immunomodulation. Trends Pharmacol Sci 25:280–289

Cui J, Bian JS, Kagan A, McDonald TV (2002) CaT1 contributes to the stores-operated calcium current in Jurkat T-lymphocytes. J Biol Chem 277:47175–47183

Dolmetsch RE, Xu K, Lewis RS (1998) Calcium oscillations increase the efficiency and specificity of gene expression. Nature 392:933–936

Fanger CM, Hoth M, Crabtree GR, Lewis RS (1995) Characterization of T cell mutants with defects in capacitative calcium entry: genetic evidence for the physiological roles of CRAC channels. J Cell Biol 131:655–667

Feske S, Gwack Y, Prakriya M, Srikanth S, Puppel SH, Tanasa B, Hogan PG, Lewis RS, Daly M, Rao A (2006) A mutation in Orai1 causes immune deficiency by abrogating CRAC channel function. Nature 441:179–185

Flavell RA, Kaczmarek LK, Badou A, Boulpaep EL, Desai R, Basavappa S, Matza D, Peng YQ, Mehal WZ (2005) Retraction. Science 310:1903

Gallo EM, Cante-Barrett K, Crabtree GR (2006) Lymphocyte calcium signaling from membrane to nucleus. Nat Immunol 7:25–32

Gamberucci A, Giurisato E, Pizzo P, Tassi M, Giunti R, McIntosh DP, Benedetti A (2002) Diacylglycerol activates the influx of extracellular cations in T-lymphocytes independently of intracellular calcium-store depletion and possibly involving endogenous TRP6 gene products. Biochem J 364:245–254

Gasser A, Glassmeier G, Fliegert R, Langhorst MF, Meinke S, Hein D, Krueger S, Weber K, Heiner I, Oppenheimer N, Schwarz JR, Guse AH (2006) Activation of T cell calcium influx by the second messenger ADP-ribose. J Biol Chem 281:2489–2496

Grafton G, Thwaite L (2001) Calcium channels in lymphocytes. Immunology 104:119–126

He LP, Hewavitharana T, Soboloff J, Spassova MA, Gill DL (2005) A functional link between store-operated and TRPC channels revealed by the 3,5-bis(trifluoromethyl)pyrazole derivative, BTP2. J Biol Chem 280:10997–11006

Hermosura MC, Monteilh-Zoller MK, Scharenberg AM, Penner R, Fleig A (2002) Dissociation of the store-operated calcium current I(CRAC) and the Mg-nucleotide-regulated metal ion current MagNuM. J Physiol 539:445–458

Hofmann T, Obukhov AG, Schaefer M, Harteneck C, Gudermann T, Schultz G (1999) Direct activation of human TRPC6 and TRPC3 channels by diacylglycerol. Nature 397:259–263

Hogan PG, Chen L, Nardone J, Rao A (2003) Transcriptional regulation by calcium, calcineurin, and NFAT. Genes Dev 17:2205–2232

Hoth M (1995) Calcium and barium permeation through calcium release-activated calcium (CRAC) channels. Pflugers Arch 430:315–322

Hoth M, Penner R (1992) Depletion of intracellular calcium stores activates a calcium current in mast cells. Nature 355:353–356

Hu HZ, Gu Q, Wang C, Colton CK, Tang J, Kinoshita-Kawada M, Lee LY, Wood JD, Zhu MX (2004) 2-Aminoethoxydiphenyl borate is a common activator of TRPV1, TRPV2, and TRPV3. J Biol Chem 279:35741–35748

Ishikawa J, Ohga K, Yoshino T, Takezawa R, Ichikawa A, Kubota H, Yamada T (2003) A pyrazole derivative, YM-58483, potently inhibits store-operated sustained Ca^{2+} influx and IL-2 production in T lymphocytes. J Immunol 170:4441–4449

Kerschbaum HH, Kozak JA, Cahalan MD (2003) Polyvalent cations as permeant probes of MIC and TRPM7 pores. Biophys J 84:2293–2305

Kozak JA, Cahalan MD (2003) MIC channels are inhibited by internal divalent cations but not ATP. Biophys J 84:922–927

Kuhn FJ, Luckhoff A (2004) Sites of the NUDT9-H domain critical for ADP-ribose activation of the cation channel TRPM2. J Biol Chem 279:46431–46437

Launay P, Fleig A, Perraud AL, Scharenberg AM, Penner R, Kinet JP (2002) TRPM4 is a Ca^{2+}-activated nonselective cation channel mediating cell membrane depolarization. Cell 109:397–407

Launay P, Cheng H, Srivatsan S, Penner R, Fleig A, Kinet JP (2004) TRPM4 regulates calcium oscillations after T cell activation. Science 306:1374–1377

Lewis RS (2001) Calcium signaling mechanisms in T lymphocytes. Annu Rev Immunol 19:497–521

Lewis RS, Cahalan MD (1995) Potassium and calcium channels in lymphocytes. Annu Rev Immunol 13:623–653

Li SW, Westwick J, Poll CT (2002) Receptor-operated Ca^{2+} influx channels in leukocytes: a therapeutic target? Trends Pharmacol Sci 23:63–70

Lievremont JP, Numaga T, Vazquez G, Lemonnier L, Hara Y, Mori E, Trebak M, Moss SE, Bird GS, Mori Y, Putney JW Jr (2005) The role of canonical transient receptor potential 7 in B-cell receptor-activated channels. J Biol Chem 280:35346–35351

Liu X, Singh BB, Ambudkar IS (2003) TRPC1 is required for functional store-operated Ca^{2+} channels. Role of acidic amino acid residues in the S5-S6 region. J Biol Chem 278:11337–11343

Ma HT, Patterson RL, van Rossum DB, Birnbaumer L, Mikoshiba K, Gill DL (2000) Requirement of the inositol trisphosphate receptor for activation of store-operated Ca^{2+} channels. Science 287:1647–1651

McHugh D, Flemming R, Xu SZ, Perraud AL, Beech DJ (2003) Critical intracellular Ca^{2+} dependence of transient receptor potential melastatin 2 (TRPM2) cation channel activation. J Biol Chem 278:11002–11006

Monteilh-Zoller MK, Hermosura MC, Nadler MJ, Scharenberg AM, Penner R, Fleig A (2003) TRPM7 provides an ion channel mechanism for cellular entry of trace metal ions. J Gen Physiol 121:49–60

Mori Y, Wakamori M, Miyakawa T, Hermosura M, Hara Y, Nishida M, Hirose K, Mizushima A, Kurosaki M, Mori E, Gotoh K, Okada T, Fleig A, Penner R, Iino M, Kurosaki T (2002) Transient receptor potential 1 regulates capacitative Ca^{2+} entry and Ca^{2+} release from endoplasmic reticulum in B lymphocytes. J Exp Med 195:673–681

Nadler MJ, Hermosura MC, Inabe K, Perraud AL, Zhu Q, Stokes AJ, Kurosaki T, Kinet JP, Penner R, Scharenberg AM, Fleig A (2001) LTRPC7 is a Mg.ATP-regulated divalent cation channel required for cell viability. Nature 411:590–595

Niemeyer BA, Bergs C, Wissenbach U, Flockerzi V, Trost C (2001) Competitive regulation of CaT-like-mediated Ca^{2+} entry by protein kinase C and calmodulin. Proc Natl Acad Sci U S A 98:3600–3605

Nilius B, Prenen J, Droogmans G, Voets T, Vennekens R, Freichel M, Wissenbach U, Flockerzi V (2003) Voltage dependence of the Ca^{2+}-activated cation channel TRPM4. J Biol Chem 278:30813–30820

Owsianik G, Talavera K, Voets T, Nilius B (2005) Permeation and selectivity of TRP channels. Annu Rev Physiol 68:685–717

Perraud AL, Knowles HM, Schmitz C (2004) Novel aspects of signaling and ion-homeostasis regulation in immunocytes. The TRPM ion channels and their potential role in modulating the immune response. Mol Immunol 41:657–673

Philipp S, Strauss B, Hirnet D, Wissenbach U, Mery L, Flockerzi V, Hoth M (2003) TRPC3 mediates T-cell receptor-dependent calcium entry in human T-lymphocytes. J Biol Chem 278:26629–26638

Prakriya M, Feske S, Gwack Y, Srikanth S, Rao A, Hogan PG (2006) Orai1 is an essential pore subunit of the CRAC channel. Nature 443:230–233

Prakriya M, Lewis RS (2001) Potentiation and inhibition of Ca^{2+} release-activated Ca^{2+} channels by 2-aminoethyldiphenyl borate (2-APB) occurs independently of IP(3) receptors. J Physiol 536:3–19

Prakriya M, Lewis RS (2003) CRAC channels: activation, permeation, and the search for a molecular identity. Cell Calcium 33:311–321

Quintana A, Griesemer D, Schwarz EC, Hoth M (2005) Calcium-dependent activation of T-lymphocytes. Pflugers Arch 450:1–12

Ramsey IS, Delling M, Clapham DE (2005) An introduction to TRP channels. Annu Rev Physiol 68:619–647

Rao GK, Kaminski NE (2006) Induction of intracellular calcium elevation by Δ^9-tetrahydrocannabinol in T cells involves TRPC1 channels. J Leukoc Biol 79:202–213

Runnels LW, Yue L, Clapham DE (2001) TRP-PLIK, a bifunctional protein with kinase and ion channel activities. Science 291:1043–1047

Sano Y, Inamura K, Miyake A, Mochizuki S, Yokoi H, Matsushime H, Furuichi K (2001) Immunocyte Ca^{2+} influx system mediated by LTRPC2. Science 293:1327–1330

Schwarz EC, Wissenbach U, Niemeyer BA, Strauss B, Philipp SE, Flockerzi V, Hoth M (2006) TRPV6 potentiates calcium-dependent cell proliferation. Cell Calcium 39:163–173

Stokes AJ, Shimoda LM, Koblan-Huberson M, Adra CN, Turner H (2004) A TRPV2-PKA signaling module for transduction of physical stimuli in mast cells. J Exp Med 200:137–147

Thebault S, Zholos A, Enfissi A, Slomianny C, Dewailly E, Roudbaraki M, Parys J, Prevarskaya N (2005) Receptor-operated Ca^{2+} entry mediated by TRPC3/TRPC6 proteins in rat prostate smooth muscle (PS1) cell line. J Cell Physiol 204:320–328

Tomida T, Hirose K, Takizawa A, Shibasaki F, Iino M (2003) NFAT functions as a working memory of Ca^{2+} signals in decoding Ca^{2+} oscillation. EMBO J 22:3825–3832

Utzny C, Faroudi M, Valitutti S (2005) Frequency encoding of T-cell receptor engagement dynamics in calcium time series. Biophys J 88:1–14

van Rossum DB, Patterson RL, Ma HT, Gill DL (2000) Ca^{2+} entry mediated by store depletion, S-nitrosylation, and TRP3 channels. Comparison of coupling and function. J Biol Chem 275:28562–28568

Vazquez G, Lievremont JP, St JBG, Putney JW Jr (2001) Human Trp3 forms both inositol trisphosphate receptor-dependent and receptor-independent store-operated cation channels in DT40 avian B lymphocytes. Proc Natl Acad Sci U S A 98:11777–11782

Vazquez G, Wedel BJ, Trebak M, St John Bird G, Putney JW Jr (2003) Expression level of the canonical transient receptor potential 3 (TRPC3) channel determines its mechanism of activation. J Biol Chem 278:21649–21654

Vazquez G, Wedel BJ, Kawasaki BT, Bird GS, Putney JW Jr (2004) Obligatory role of Src kinase in the signaling mechanism for TRPC3 cation channels. J Biol Chem 279:40521–40528

Vennekens R, Hoenderop JG, Prenen J, Stuiver M, Willems PH, Droogmans G, Nilius B, Bindels RJ (2000) Permeation and gating properties of the novel epithelial Ca^{2+} channel. J Biol Chem 275:3963–3969

Voets T, Prenen J, Fleig A, Vennekens R, Watanabe H, Hoenderop JG, Bindels RJ, Droogmans G, Penner R, Nilius B (2001) CaT1 and the calcium release-activated calcium channel manifest distinct pore properties. J Biol Chem 276:47767–47770

Wehage E, Eisfeld J, Heiner I, Jungling E, Zitt C, Luckhoff A (2002) Activation of the cation channel long transient receptor potential channel 2 (LTRPC2) by hydrogen peroxide. A splice variant reveals a mode of activation independent of ADP-ribose. J Biol Chem 277:23150–23156

Wissenbach U, Schroth G, Philipp S, Flockerzi V (1998) Structure and mRNA expression of a bovine trp homologue related to mammalian trp2 transcripts. FEBS Lett 429:61–66

Wissenbach U, Niemeyer BA, Fixemer T, Schneidewind A, Trost C, Cavalie A, Reus K, Meese E, Bonkhoff H, Flockerzi V (2001) Expression of CaT-like, a novel calcium-selective channel, correlates with the malignancy of prostate cancer. J Biol Chem 276:19461–19468

Xu SZ, Zeng F, Boulay G, Grimm C, Harteneck C, Beech DJ (2005) Block of TRPC5 channels by 2-aminoethoxydiphenyl borate: a differential, extracellular and voltage-dependent effect. Br J Pharmacol 145:405–414

Xu XZ, Moebius F, Gill DL, Montell C (2001) Regulation of melastatin, a TRP-related protein, through interaction with a cytoplasmic isoform. Proc Natl Acad Sci U S A 98:10692–10697

Yeromin AV, Zhang SL, Jiang W, Yu Y, Safrina O, Cahalan MD (2006) Molecular identification of the CRAC channel by altered ion selectivity in a mutant of Orai. Nature 443:226–229

Yoon IS, Li PP, Siu KP, Kennedy JL, Macciardi F, Cooke RG, Parikh SV, Warsh JJ (2001) Altered TRPC7 gene expression in bipolar-I disorder. Biol Psychiatry 50:620–626

Yue L, Peng JB, Hediger MA, Clapham DE (2001) CaT1 manifests the pore properties of the calcium-release-activated calcium channel. Nature 410:705–709

Zhu X, Jiang M, Peyton M, Boulay G, Hurst R, Stefani E, Birnbaumer L (1996) trp, a novel mammalian gene family essential for agonist-activated capacitative Ca^{2+} entry. Cell 85:661–671

Zitt C, Obukhov AG, Strubing C, Zobel A, Kalkbrenner F, Luckhoff A, Schultz G (1997) Expression of TRPC3 in Chinese hamster ovary cells results in calcium-activated cation currents not related to store depletion. J Cell Biol 138:1333–1341

Zitt C, Strauss B, Schwarz EC, Spaeth N, Rast G, Hatzelmann A, Hoth M (2004) Potent inhibition of Ca^{2+} release-activated Ca^{2+} channels and T-lymphocyte activation by the pyrazole derivative BTP2. J Biol Chem 279:12427–12437

Zweifach A, Lewis RS (1993) Mitogen-regulated Ca^{2+} current of T lymphocytes is activated by depletion of intracellular Ca^{2+} stores. Proc Natl Acad Sci U S A 90:6295–6299

Link Between TRPV Channels and Mast Cell Function

H. Turner (✉) · K. A. del Carmen · A. Stokes

Center for Biomedical Research at Queen's Medical Center, University Tower 811,
1356 Lusitana Street, Honolulu HI, 96813, USA
hturner@queens.org

1	Physiology and Pathophysiology of Mast Cells	458
1.1	Mast Cells Are Mediators of Inflammation	458
1.2	Multiple Classes of Stimuli Activate Mast Cells	458
2	Cation Conductances Observed in Mast Cells	459
2.1	I_{CRAC} Currents Develop After Immunological Stimulation of Mast Cells	460
2.2	Secretagogues Induce Nonselective Cation Conductances in Mast Cells	460
2.3	Physical Stimuli Activate Mast Cells	460
3	TRPV Channels in Mast Cells	461
3.1	Molecular Evidence for Expression of TRPV Cation Channels in Mast Cells	461
3.2	Activation and Roles for TRPV Channels in Mast Cells	461
3.2.1	TRPV1	461
3.2.2	TRPV2	463
3.2.3	TRPV6	464
3.2.4	TRPA1	464
4	Pharmacology of TRPVs in Mast Cells	465
4.1	Pharmacological Activators of Mast Cell TRPVs	465
4.2	Blockers/Inhibitors/Antagonists of Mast Cell TRPVs	465
5	Perspectives	466
	References	466

Abstract Mast cells are tissue-resident immune effector cells. They respond to diverse stimuli by releasing potent biological mediators into the surrounding tissue, and initiating inflammatory responses that promote wound healing and infection clearance. In addition to stimulation via immunological routes, mast cells also respond to polybasic secretagogues and physical stimuli. Each mechanism for mast cell activation relies on the influx of calcium through specific ion channels in the plasma membrane. Recent reports suggest that several calcium-permeant cation channels of the TRPV family are expressed in mast cells. TRPV channels are a family of sensors that receive and react to chemical messengers and physical environmental cues, including thermal, osmotic, and mechanical stimuli. The central premise of this review is that TRPVs transduce physiological and pathophysiological cues that are functionally coupled to calcium signaling and mediator release in mast cells. Inappropriate mast cell activation is at the core of numerous inflammatory pathologies, rendering the mast cell TRPV channels potentially important therapeutic targets.

Keywords Mast cells · Inflammation · Urticaria · Calcium · Ion channels

1
Physiology and Pathophysiology of Mast Cells

1.1
Mast Cells Are Mediators of Inflammation

The term "mastzellen" (lit. well-fed cells) was coined by Paul Ehrlich to describe a densely granular leukocyte type that infiltrates connective tissues and mucosal membranes of the GI, reproductive, and respiratory tracts (Benyon 1989; Galli et al. 2005; Maurer and Metz 2005; Vliagoftis and Befus 2005; Krishnaswamy et al. 2006).

Mast cells respond to a variety of inputs by releasing biologically active mediators (Galli et al. 2005; Yamasaki and Saito 2005; Krishnaswamy et al. 2006). Initially, preformed mediators are released from the mast cell via exocytosis of cytoplasmic granules. Subsequently, de novo synthesis of further mediators occurs by induction of cytoplasmic pathways for synthesis of lipid messengers, and cytokine/chemokine gene transcription. Mast cell activation results in the establishment of an inflammatory site, classically characterized by *rubor* (reddening due to vasodilation), *tumor* (swelling due to edema), *calor* (heat due to intense metabolic activity of infiltrating leukocytes and increased blood flow), and *dolor* (pain caused by the effects of inflammatory mediators on local sensory nerve endings). Tissue remodeling is initiated by the matrix-active proteases secreted from mast cell granules (Noli and Miolo 2001). Overall, mast cell activation promotes clearance of infection and tissue repair.

1.2
Multiple Classes of Stimuli Activate Mast Cells

Mast cells respond to both immunological and nonimmunological stimuli (Church et al. 1991b; Galli et al. 2005; Yamasaki and Saito 2005; Krishnaswamy et al. 2006). Antigenic stimulation of mast cells is best understood via the high-affinity receptor for immunoglobulin (Ig)E (FcεRI). This receptor binds the Fc portion of antigen-specific IgE. FcεRI aggregation initiates numerous signaling pathways, which control the release of proinflammatory mediators and induce transcription of genes encoding cytokines, chemokines, and growth factors. Mast cells also respond to innate challenges, via the Toll-like receptors (Mekori and Metcalfe 2000; Galli et al. 2005; Vliagoftis and Befus 2005).

Three classes of stimuli lead to mast cell activation in the absence of immunological challenge. First, mast cell secretagogues (Church et al. 1991b; Noli and Miolo 2001) are small polybasic molecules that act upon $G\alpha_i$ and are necessary and sufficient to initiate inflammation (Klinker et al. 1995; Chahdi et al. 1998; Lorenz et al. 1998). Secretagogues include the wasp venom constituent mastoparan and the synthetic polyamine 48/80. Second, neurogenic stimuli, specifically nerve growth factor (NGF) and neuropeptides such as substance P and calcitonin gene-related peptide (CGRP), can activate mast cells (Church et al. 1991a; Lewin et al. 1994; Reynier-Rebuffel et al. 1994; Moalem and Tracey 2006). Third, physical stimuli such as mechanical perturbation, temperature elevation, and alterations in osmotic status and pH induce mast cell activation (Black 1985; Greaves 1991; Grabbe 2001). These physical cues reflect the environmental conditions within an inflammatory site, and the ability of mast cells to respond to them may serve as a positive feedback mechanism to amplify proinflammatory responses.

Inappropriate activation of mast cells inevitably results in pathology, which is reviewed extensively elsewhere (Krishnaswamy et al. 2001; Galli and Nakae 2003; Vliagoftis and Befus 2005). Current therapeutic management of mast cell-derived pathologies is mainly symptomatic, using agents that block interactions between inflammatory mediators and their targets, or the IgE-FcεRI interaction. Ion channels in mast cells are attractive targets for therapeutic development, partly because these proteins are highly "druggable" (Hopkins and Groom 2002), and because emerging evidence for coupling of distinct cation channels to distinct upstream stimuli suggests that highly specific therapeutics may be developed. Support for the targeting of ion channels to control mast cell activation can also be derived from the fact that the mast cell stabilizing drug sodium cromoglycate [Nasalcrom (Pfizer, Morris Plains); Opticrom (Aventis, Bridgewater], understood as an effective antiinflammatory since the 1960s, has recently been shown to target mast cell chloride efflux channels (Reinsprecht et al. 1992; Heinke et al. 1995; Alton et al. 1996). Thus, mast cell ion channels are viable targets for antiinflammatories, leaving us with the not inconsiderable tasks of identifying, characterizing, and therapeutically manipulating the channel proteins themselves.

2
Cation Conductances Observed in Mast Cells

The diverse mechanisms for mast cell activation have certain commonalities. Calcium influx from the extracellular milieu is a necessary component of all known activation pathways. Cation channels that specifically mediate calcium entry into mast cells are thus important therapeutic targets. The cation conductances of mast cells have been best characterized in rodent model systems, comprising murine primary bone marrow-derived mast cells (BMMC) or peritoneal mast cells. The need to confirm these data in human mast cells is being addressed (Bradding et al. 2003; Bradding 2005). This review is specifically concerned with the activation-induced conductances for which TRPV channels may be considered as molecular candidates.

2.1
I_{CRAC} Currents Develop After Immunological Stimulation of Mast Cells

FcεRI stimulation of mast cells results activation of a highly calcium-selective cation conductance (Hoth and Penner 1993; Parekh and Penner 1997). Depletion of calcium stores is necessary and sufficient to evoke this calcium release-activated current (I_{CRAC}). TRPC and TRPM channels have been proposed to comprise store-operated conductances (reviewed in Nilius 2003; Penner and Fleig 2004), but no TRP channel exhibits the distinctive features of I_{CRAC}. Studies showing that TRPV6 is a component of I_{CRAC} have now been superceded by the view that TRPV6 and the CRAC channel are distinct at the molecular level (Voets et al. 2001; Yue et al. 2001; Kahr et al. 2004). One intriguing caveat to this distinction arises from recent work that documents dominant interference of CRAC channels by TRPV6 fragments (Cui et al. 2002; Kahr et al. 2004). However, while I_{CRAC} remains of paramount importance for our understanding of antigen responses, its molecular identity remains elusive. The CRAC channel has now been cloned (Yeromin et al. Nature 443:226–229, 2006; Prakriya et al. Nature 443:230–233, 2006; Vig et al. Curr. Biol. Sep 2006).

2.2
Secretagogues Induce Nonselective Cation Conductances in Mast Cells

Secretagogues, such as the polybasic compound 48/80, cause explosive degranulation responses in mast cells. Several studies in rat peritoneal mast cells identified a discrete c48/80-induced cation current (Kuno et al. 1989, 1990; Kuno and Kimura 1992; Fasolato et al. 1993). This $I_{48/80}$ is a nonselective cation conductance (NSCC) carrying Na$^+$ and Ca^{2+} ($P_{Ca}/P_{Na} \sim 1$), with a 50 pS unitary conductance. One study identified a second, large-amplitude, cationic influx in response to secretagogue treatment (Fasolato et al. 1993). Thus, c48/80 induces one or more TRP-like conductances in mast cells. If these conductances are also activated by physiologically occurring secretagogues, then the channel proteins may be important targets for the control of anaphylaxis.

2.3
Physical Stimuli Activate Mast Cells

Direct responses to physical stimuli have been documented in mast cells both in vitro and in vivo (Fredholm and Hagermark 1970; Eggleston et al. 1987; Silber et al. 1988; Church et al. 1991b; Noli and Miolo 2001). Physiologically, this responsiveness may be important in wound healing. Elevated temperature, hypotonicity due to plasma extravasation, and mechanical compression caused by local edema are all features of inflammation. Conversely, mast cell responses to these physical stimuli cause pathologies that derive from inappropriate activation of cutaneous mast cells. Physical urticarias (PU) involve cutaneous wheal-flare reactions that follow dermal exposure to cold, heat, pressure, or UVA light (Daman et al. 1978; Soter and Wasserman 1980; Black 1985; Claudy 2001; Kronauer et al. 2003). Interestingly, two mast cell NSCCs evoked by physical stimuli have been identified.

First, a light-induced NSCC is induced by short exposures of primary mast cells to near-visible UVA light (Mendez and Penner 1998). This cation current is calcium-permeable and present in excised membrane patches. Both electrophysiologically and mechanistically, I_{LiNC} resembles the physically gated TRPV sensors. Second, a temperature-evoked conductance was identified in BMMC (Kuno et al. 1997). Heating from 24°C to 36°C reversibly activated a conductance that was proton- and cation-mediated, and potentiated by intracellular acidification. The Q10 of this conductance was between 6.1 and 9.9, which is similar to that demonstrated for TRPVs, and acid potentiation is a clear property of the heat-evoked TRPV1 current. The recent demonstration of proton permeation by TRPV1 (Hellwig et al. 2004; Vulcu et al. 2004) leads us to the possibility that the current measured by Kuno et al. in 1997 could be mediated by TRPV1.

3
TRPV Channels in Mast Cells

3.1
Molecular Evidence for Expression of TRPV Cation Channels in Mast Cells

The distribution of mRNA for TRPV channels has been addressed by Northern blots, complementary DNA (cDNA) microarrays and reverse transcriptase (RT)-PCR. Our group has been able to RT-PCR transcripts for TRPV1, TRPV2, TRPV4, and TRPV6 from RBL2H3. TRPV3 and TRPV5 mRNA were not detectable. Northern analysis of RBL2H3 and BMMC has confirmed the presence of transcripts and splice variants for TRPV1, TRPV2, and TRPA1 in murine BMMC (H. Turner and A. Stokes, unpublished). Microarray analysis detected significant levels of transcripts for TRPV2 in primary human mast cells (Bradding et al. 2003; Nakajima et al. 2004). Using Western blot analysis of BMMC, RBL2H3, and the human LAD2 (Laboratory of Allergic Diseases cell line 2) mast cells (Kirshenbaum et al. 2003), we have detected protein corresponding to TRPV1, TRPV2, TRPV6, and TRPA1.

3.2
Activation and Roles for TRPV Channels in Mast Cells

3.2.1
TRPV1

Several early studies support the proposition that TRPV1 is expressed in mast cells. Topical capsaicin, substance P, and CGRP, cause histamine and serotonin release from human skin mast cells in vivo (Reynier-Rebuffel et al. 1994;

Forman et al. 2004). However, these agents are also active at sensory neurons, and these studies do not differentiate a direct action of these agents upon dermal mast cells from an indirect, paracrine, process. The latter mechanism (Lewin et al. 1994; Woolf et al. 1996) relies upon the release of NGF from neurons, with NGF then acting as a secretagogue for nearby mast cells. This model allows for sensory neuron activation to cause mast cell activation, which is physiologically important in scenarios such as wound healing. This model removes any logical necessity for direct physical sensors to be present in mast cells. However, emerging molecular evidence for the presence of sensors such as TRPVs in mast cells suggests that parallel stimulation of nerve termini and mast cells may occur.

One of the earliest indications that isolated mast cells responded directly to TRPV activators showed high-affinity binding sites for capsaicin in mast cells (Biro et al. 1998). Capsaicin treatment of some mast cell lines induced degranulation and robust ^{45}Ca uptake and induction of calcium-dependent cytokine gene transcription. Several physiological TRPV1 stimuli are relevant to mast cell activation (Benham et al. 2003; Clapham et al. 2003). Noxious temperatures are encountered during thermal wounding, and there is the potential for the acidification of inflamed tissue to lower the temperature threshold of TRPV1 to physiological levels. TRPV1 is also a direct target for the endocannabinoid anandamide (Smart et al. 2000; Ross 2003). Mast cells produce anandamide, which also ligates the cannabinoid (CB)1 G protein-couples receptor (GPCR) (Facci et al. 1995; Bisogno et al. 1997). Although cannabinoids are widely proposed as antiinflammatories, our recent data suggest that this property is primarily associated with ligands for the CB2 GPCR. CB1-selective ligands, such as anandamide, exhibit a markedly proinflammatory effect on mast cells (Small-Howard et al. 2005). Anandamide-mediated potentiation of mast cell activation occurs over a chronic time course (1–16 h), and involves calcium influx (inhibitable by the TRP blocker ruthenium red) and induction of cytokine transcription (Small-Howard et al. 2005). Release of CB1 ligands, either from sensory neurons or in an autocrine fashion from mast cells, would therefore probably amplify inflammatory responses via TRPV1.

The responsiveness of TRPV1 to GPCR ligands (substance P, CGRP, bradykinin) is highly relevant to inflammation (Suh and Oh 2005). Recent data on a link between the proteinase activated receptors (PAR) and TRPV1 activation is also of interest in the mast cell context (Amadesi et al. 2004; Dai et al. 2004). Mast cells actively secrete PAR ligand proteases (tryptase, thrombin). PAR activation by proteolytic ligands causes the induction of a partial spectrum of inflammatory responses in mast cells. The data from Amadesi et al. now suggest a further downstream effect of protease production by mast cells, PAR-mediated sensitization, and possible activation of TRPV1 (Amadesi et al. 2004; Dai et al. 2004). Since TRPV1 may be located on sensory neurons or on the mast cells themselves, this mechanism provides for autocrine or paracrine intensification of inflammation and associated pain.

3.2.2
TRPV2

TRPV2 is expressed in C-type nociceptors and is activated by temperatures in excess of 49°C (Caterina et al. 1999). Further studies of TRPV2, including those from our laboratory, indicate that this channel is expressed across a range of tissues, including skeletal and cardiac muscle, kidney and liver, and components of the immune system including mast cells, neutrophils, and T cells (Bradding et al. 2003; Clapham et al. 2003; Heiner et al. 2003; Barnhill et al. 2004; Stokes et al. 2004, 2005a).

The expression of TRPV2 in isolated hematopoietic cells suggests that sensing of noxious temperatures is unlikely to be the sole activation mechanism for TRPV2. While heat-mediated calcium influx and secretory responses can be documented in mast cells, it is doubtful that other cell types in which TRPV2 is found would need an intrinsic mechanism to respond to burning temperatures. The resolution to this paradox may be suggested by the observation that TRPV2 acts as an osmosensor in aortic myocytes (Muraki et al. 2003). Plasma extravasation renders an inflammatory site hypotonic, potentially activating or sensitizing TRPV2 on both mast cells and local nerve endings. TRPV2-mediated influx in response to hypoosmolarity may help sustain mast cell activation during an ongoing inflammation.

TRPV2 is not constitutively active at the cell surface in mast cells or in myocytes (Kanzaki et al. 1999; Boels et al. 2001; Stokes et al. 2005a). TRPV2 can be recruited to the cell membrane by either cell-cycle synchronization or treatment with cyclic AMP (cAMP)-mobilizing compounds. Growth factors such as epidermal growth factor (EGF) and NGF have a weak effect on plasma membrane trafficking of TRPV2 in mast cells, although in myocytes the former promotes surface localization of TRPV2. There is an intriguing kinetic difference between the growth factor-induced surface relocalization of TRPV1 and TRPV2. NGF induces rapid membrane insertion of TRPV1 (analogous to the RIViT observed for TRPC5) (Bezzerides et al. 2004; Zhang et al. 2005). In contrast, growth factor or cAMP treatment of TRPV2-expressing cells causes accumulation of V2 in the plasma membrane over a time course of 10 min–1 h (Kanzaki et al. 1999; Boels et al. 2001; Stokes et al. 2004). In myocytes, this relocalization is associated with the appearance of a constitutively active NSCC, indicating that no further signals are required to activate TRPV2 (Kanzaki et al. 1999; Boels et al. 2001). This constitutive NSCC is not observed under similar conditions in the mast cell (H. Turner and A. Stokes, unpublished data). The permissive event for the final step in TRPV2 trafficking appears to be an Asn-linked glycosylation, for which association with the recombinase gene activator (RGA) protein is required (Barnhill et al. 2004; Stokes et al. 2005a).

3.2.3
TRPV6

Several groups have measured CaT1/TRPV6 currents in RBL2H3 (Yue et al. 2001; Kahr et al. 2004). Our recent data confirm that a highly posttranslationally modified TRPV6 protein is present in primary murine and human mast cells. Six distinct mobility forms of TRPV6 migrate around the monomer weight of the channel protein on sodium dodecyl sulfate polyacrylamide gel electrophoresis (SDS-PAGE). Inhibitor analysis suggests that several of these forms are N-linked glycoproteins. Stimulation of mast cells with store-depleting agents or agents that arrest cell cycle lead to striking alterations in the biochemical characteristics of TRPV6, notably altered glycosylation and accumulation of ubiquitinated channel proteins. Future experiments will link these biochemical modifications to alterations in channel functionality.

The role of TRPV6 in mast cells is unknown. It can be activated by depletion of calcium stores, but is not coupled to the intense intracellular calcium release that follows FcεRI stimulation. Recent data suggest that overexpressed TRPV6 contributes calcium signals that can drive cell-cycle progression and proliferation (Schwarz et al. 2006). If this function is conserved for endogenous TRPV6, then we can postulate that its role in mast cells may be to contribute intermittent calcium fluxes that respond to InsP3 transients at certain stages in the cell cycle.

3.2.4
TRPA1

TRPA1 was initially described in human lung fibroblasts (Jaquemar et al. 1999) and later as a cold sensor in nociceptors (Story et al. 2003; Chuang et al. 2004; Montell 2005). Subsequent studies indicate diverse activation mechanisms for TRPA1, including direct and GPCR-mediated activation by pungent compounds, and mechanosensitivity conferred by the ankyrin motifs (Corey et al. 2004; Bautista et al. 2005; Nagata et al. 2005; Xu et al. 2005). We have detected TRPA1 transcripts and protein in human and rodent mast cells (Stokes 2005). Specific activation signals are required to place TRPA1 at the cell surface. Moreover, TRPA1 has a complex pattern of posttranslational modifications (phosphorylations, glycosylations, ligand-induced ubiquitination). Preliminary functional analysis of mast cell TRPA1 indicates that it contributes to a range of GPCR-coupled responses in mast cells, including a key role in the ability of certain cannabinoids to induce proinflammatory calcium signals (Bandell et al. 2004; Jordt et al. 2004; Krause et al. 2005). There is also the potential for TRPA1 to contribute to cold-induced activation of mast cells that manifests in the pathology of cold urticaria (Claudy 2001).

4
Pharmacology of TRPVs in Mast Cells

The study of TRPV channel function in mast cells will be greatly facilitated by specific pharmacological agents. There is a considerable literature on the activation and antagonism of TRPV channels by various small molecules (Clapham et al. 2003; Ross 2003; Bandell et al. 2004; Jordt et al. 2004; Bautista et al. 2005; Calixto et al. 2005; Montell 2005; Suh and Oh 2005).

4.1
Pharmacological Activators of Mast Cell TRPVs

Endogenous ligands for TRPV1 and TRPA1 have been described. The endogenous cannabinoid anandamide is a full agonist of TRPV1 in the mast cell system. Anandamide apparently binds directly to TRPV1, and may share a binding site with capsaicin, which is structurally related to anandamide (Ross 2003; Gavva et al. 2004; Suh and Oh 2005). Given that mast cells constitutively produce anandamide, it will be important to elucidate which of the known TRPV1 stimuli are necessary and sufficient to activate TRPV1, and which are merely sensitizing the channel to the presence of endogenous anandamide.

TRPA1 also responds to cannabinoids, including anandamide and 2-arachidonoylglycerol, and the Δ^9-tetrahydrocannabinol derivative CP-55940 (Jordt et al. 2004; H. Turner, A. Stokes, unpublished data). The structural diversity of the cannabinoids that activate TRPA1 suggests involvement of the GPCR cannabinoid receptors CB1 and CB2 rather than a direct interaction of the cannabinoid with the TRPA1. It seems likely that the pungent compounds that activate TRPA1 (Bandell et al. 2004; Bautista et al. 2005; Calixto et al. 2005; Macpherson et al. 2005; Namer et al. 2005) act through GPCR. The cold mimetic icilin is also a useful tool for TRPA1, since mast cells do not appear to coexpress TRPM8 (Peier et al. 2002; Chuang et al. 2004; Namer et al. 2005).

No specific chemical ligand for TRPV2 has been identified, although 2-aminoethoxydiphenyl borate (2-APB) activates TRPV1, TRPV2, and TRPV3 (Hu et al. 2004; Zimmermann et al. 2005). In mast cells, this profile of activity may allow us to isolate TRPV2-mediated responses using antagonists of TRPV1.

4.2
Blockers/Inhibitors/Antagonists of Mast Cell TRPVs

The TRPV NSCC that are expressed in mast cells are all sensitive to inhibition by ruthenium red and lanthanides, and thus these agents cannot distinguish the function of individual channel species. TRPV1 has a detailed pharmacology of direct inhibitors, many of which are potent, specific, and are based upon the pharmacophores of naturally occurring ligands (Pall and Anderson 2004; Calixto et al. 2005; Suh and Oh 2005). Pharmacological dissection of TRPA1 is likely to be facilitated by the range of compounds that are known to modulate GPCR signaling, especially specific antagonists and inverse agonists of CB1 and CB2 (Rinaldi-Carmona et al. 1994; Rinaldi-Carmona et al. 1998; Howlett et al. 2002). The pharmacological study of TRPV2 and TRPV6 is severely limited, since no specific antagonists have been described.

Therapeutic strategies will absolutely rely upon the eventual identification of specific small-molecule modulators of these channels. However, currently the most important challenge is to dissect the individual functional relevance of each of these channels. Attractive approaches to this problem include (1) isolating BMMC or peritoneal mast cells from mice that are genetically deficient in each channel, and (2) suppression of protein production using small interfering RNA (siRNA) in mast cell lines. Of course, the former is reliant on an available genetic model, but has the key advantage that mast cells can be isolated from multiple tissue contexts.

5
Perspectives

Calcium entry is a key component of mast cell activation. The cation channels that permit calcium entry into mast cells are attractive targets for the therapeutic manipulation of inflammatory responses. Various TRPV cation channels are expressed in mast cells, including TRPV1, TRPV2, TRPV6, and TRPA1. These channels potentially mediate calcium entry in response to diverse stimuli that include changes in the physical environment, immunomodulatory signals, and neurally derived mediators. These initial findings suggest intriguing directions for future studies. Key challenges will include establishing the pattern of TRPV expression in mast cells that reside in distinct tissue contexts. Future work in vivo will elucidate (1) the functional contribution of each TRPV to the initiation of inflammatory responses and the (2) responsiveness of mast cell TRPVs to the environmental conditions that prevail during an ongoing inflammation.

References

Alton EW, Kingsleigh-Smith DJ, Munkonge FM, Smith SN, Lindsay AR, Gruenert DC, Jeffery PK, Norris A, Geddes DM, Williams AJ (1996) Asthma prophylaxis agents alter the function of an airway epithelial chloride channel. Am J Respir Cell Mol Biol 14:380–387

Amadesi S, Nie J, Vergnolle N, Cottrell GS, Grady EF, Trevisani M, Manni C, Geppetti P, McRoberts JA, Ennes H, Davis JB, Mayer EA, Bunnett NW (2004) Protease-activated receptor 2 sensitizes the capsaicin receptor transient receptor potential vanilloid receptor 1 to induce hyperalgesia. J Neurosci 24:4300–4312

Bandell M, Story GM, Hwang SW, Viswanath V, Eid SR, Petrus MJ, Earley TJ, Patapoutian A (2004) Noxious cold ion channel TRPA1 is activated by pungent compounds and bradykinin. Neuron 41:849–857

Barnhill JC, Stokes AJ, Koblan-Huberson M, Shimoda LM, Muraguchi A, Adra CN, Turner H (2004) RGA protein associates with a TRPV ion channel during biosynthesis and trafficking. J Cell Biochem 91:808–820

Bautista DM, Movahed P, Hinman A, Axelsson HE, Sterner O, Hogestatt ED, Julius D, Jordt SE, Zygmunt PM (2005) Pungent products from garlic activate the sensory ion channel TRPA1. Proc Natl Acad Sci U S A 102:12248–12252

Benham CD, Gunthorpe MJ, Davis JB (2003) TRPV channels as temperature sensors. Cell Calcium 33:479–487

Benyon RC (1989) The human skin mast cell. Clin Exp Allergy 19:375–387

Bezzerides VJ, Ramsey IS, Kotecha S, Greka A, Clapham DE (2004) Rapid vesicular translocation and insertion of TRP channels. Nat Cell Biol 6:709–720

Biro T, Maurer M, Modarres S, Lewin NE, Brodie C, Acs G, Acs P, Paus R, Blumberg PM (1998) Characterization of functional vanilloid receptors expressed by mast cells. Blood 91:1332–1340

Bisogno T, Maurelli S, Melck D, De Petrocellis L, Di Marzo V (1997) Biosynthesis, uptake, and degradation of anandamide and palmitoylethanolamide in leukocytes. J Biol Chem 272:3315–3323

Black AK (1985) Mechanical trauma and urticaria. Am J Ind Med 8:297–303

Boels K, Glassmeier G, Herrmann D, Riedel IB, Hampe W, Kojima I, Schwarz JR, Schaller HC (2001) The neuropeptide head activator induces activation and translocation of the growth-factor-regulated $Ca^{(2+)}$-permeable channel GRC. J Cell Sci 114:3599–3606

Bradding P (2005) Mast cell ion channels. Chem Immunol Allergy 87:163–178

Bradding P, Okayama Y, Kambe N, Saito H (2003) Ion channel gene expression in human lung, skin, and cord blood-derived mast cells. J Leukoc Biol 73:614–620

Calixto JB, Kassuya CA, Andre E, Ferreira J (2005) Contribution of natural products to the discovery of the transient receptor potential (TRP) channels family and their functions. Pharmacol Ther 106:179–208

Caterina MJ, Rosen TA, Tominaga M, Brake AJ, Julius D (1999) A capsaicin-receptor homologue with a high threshold for noxious heat. Nature 398:436–441

Chahdi A, Daeffler L, Gies JP, Landry Y (1998) Drugs interacting with G protein alpha subunits: selectivity and perspectives. Fundam Clin Pharmacol 12:121–132

Chuang HH, Neuhausser WM, Julius D (2004) The super-cooling agent icilin reveals a mechanism of coincidence detection by a temperature-sensitive TRP channel. Neuron 43:859–869

Church MK, el-Lati S, Caulfield JP (1991a) Neuropeptide-induced secretion from human skin mast cells. Int Arch Allergy Appl Immunol 94:310–318

Church MK, Okayama Y, el-Lati S (1991b) Mediator secretion from human skin mast cells provoked by immunological and non-immunological stimulation. Skin Pharmacol 4:15–24

Clapham DE, Montell C, Schultz G, Julius D (2003) International Union of Pharmacology. XLIII. Compendium of voltage-gated ion channels: transient receptor potential channels. Pharmacol Rev 55:591–596

Claudy A (2001) Cold urticaria. J Investig Dermatol Symp Proc 6:141–142

Corey DP, Garcia-Anoveros J, Holt JR, Kwan KY, Lin SY, Vollrath MA, Amalfitano A, Cheung EL, Derfler BH, Duggan A, Geleoc GS, Gray PA, Hoffman MP, Rehm HL, Tamasauskas D, Zhang DS (2004) TRPA1 is a candidate for the mechanosensitive transduction channel of vertebrate hair cells. Nature 432:723–730

Cui J, Bian JS, Kagan A, McDonald TV (2002) CaT1 contributes to the stores-operated calcium current in Jurkat T-lymphocytes. J Biol Chem 277:47175–47183

Dai Y, Moriyama T, Higashi T, Togashi K, Kobayashi K, Yamanaka H, Tominaga M, Noguchi K (2004) Proteinase-activated receptor 2-mediated potentiation of transient receptor potential vanilloid subfamily 1 activity reveals a mechanism for proteinase-induced inflammatory pain. J Neurosci 24:4293–4299

Daman L, Lieberman P, Ganier M, Hashimoto K (1978) Localized heat urticaria. J Allergy Clin Immunol 61:273–278

Eggleston PA, Kagey-Sobotka A, Lichtenstein LM (1987) A comparison of the osmotic activation of basophils and human lung mast cells. Am Rev Respir Dis 135:1043–1048

Facci L, Dal Toso R, Romanello S, Buriani A, Skaper SD, Leon A (1995) Mast cells express a peripheral cannabinoid receptor with differential sensitivity to anandamide and palmitoylethanolamide. Proc Natl Acad Sci U S A 92:3376–3380

Fasolato C, Hoth M, Matthews G, Penner R (1993) Ca^{2+} and Mn^{2+} influx through receptor-mediated activation of nonspecific cation channels in mast cells. Proc Natl Acad Sci U S A 90:3068–3072

Forman MF, Brower GL, Janicki JS (2004) Spontaneous histamine secretion during isolation of rat cardiac mast cells. Inflamm Res 53:453–457

Fredholm B, Hagermark O (1970) Studies on histamine release from skin and from peritoneal mast cells of the rat induced by heat. Acta Derm Venereol 50:273–277

Galli SJ, Nakae S (2003) Mast cells to the defense. Nat Immunol 4:1160–1162

Galli SJ, Kalesnikoff J, Grimbaldeston MA, Piliponsky AM, Williams CM, Tsai M (2005) Mast cells as "tunable" effector and immunoregulatory cells: recent advances. Annu Rev Immunol 23:749–786

Gavva NR, Klionsky L, Qu Y, Shi L, Tamir R, Edenson S, Zhang TJ, Viswanadhan VN, Toth A, Pearce LV, Vanderah TW, Porreca F, Blumberg PM, Lile J, Sun Y, Wild K, Louis JC, Treanor JJ (2004) Molecular determinants of vanilloid sensitivity in TRPV1. J Biol Chem 279:20283–20295

Grabbe J (2001) Pathomechanisms in physical urticaria. J Investig Dermatol Symp Proc 6:135–136

Greaves MW (1991) The physical urticarias. Clin Exp Allergy 21:284–289

Heiner I, Eisfeld J, Luckhoff A (2003) Role and regulation of TRP channels in neutrophil granulocytes. Cell Calcium 33:533–540

Heinke S, Szucs G, Norris A, Droogmans G, Nilius B (1995) Inhibition of volume-activated chloride currents in endothelial cells by chromones. Br J Pharmacol 115:1393–1398

Hellwig N, Plant TD, Janson W, Schafer M, Schultz G, Schaefer M (2004) TRPV1 acts as proton channel to induce acidification in nociceptive neurons. J Biol Chem 279:34553–34561

Hopkins AL, Groom CR (2002) The druggable genome. Nat Rev Drug Discov 1:727–730

Hoth M, Penner R (1993) Calcium release-activated calcium current in rat mast cells. J Physiol 465:359–386

Howlett AC, Barth F, Bonner TI, Cabral G, Casellas P, Devane WA, Felder CC, Herkenham M, Mackie K, Martin BR, Mechoulam R, Pertwee RG (2002) International Union of Pharmacology. XXVII. Classification of cannabinoid receptors. Pharmacol Rev 54:161–202

Hu HZ, Gu Q, Wang C, Colton CK, Tang J, Kinoshita-Kawada M, Lee LY, Wood JD, Zhu MX (2004) 2-Aminoethoxydiphenyl borate is a common activator of TRPV1, TRPV2, and TRPV3. J Biol Chem 279:35741–35748

Jaquemar D, Schenker T, Trueb B (1999) An ankyrin-like protein with transmembrane domains is specifically lost after oncogenic transformation of human fibroblasts. J Biol Chem 274:7325–7333

Jordt SE, Bautista DM, Chuang HH, McKemy DD, Zygmunt PM, Hogestatt ED, Meng ID, Julius D (2004) Mustard oils and cannabinoids excite sensory nerve fibres through the TRP channel ANKTM1. Nature 427:260–265

Kahr H, Schindl R, Fritsch R, Heinze B, Hofbauer M, Hack ME, Mortelmaier MA, Groschner K, Peng JB, Takanaga H, Hediger MA, Romanin C (2004) CaT1 knock-down strategies fail to affect CRAC channels in mucosal-type mast cells. J Physiol 557:121–132

Kanzaki M, Zhang YQ, Mashima H, Li L, Shibata H, Kojima I (1999) Translocation of a calcium-permeable cation channel induced by insulin-like growth factor-I. Nat Cell Biol 1:165–170

Kirshenbaum AS, Akin C, Wu Y, Rottem M, Goff JP, Beaven MA, Rao VK, Metcalfe DD (2003) Characterization of novel stem cell factor responsive human mast cell lines LAD 1 and 2 established from a patient with mast cell sarcoma/leukemia; activation following aggregation of FcepsilonRI or FcgammaRI. Leuk Res 27:677–682

Klinker JF, Hageluken A, Grunbaum L, Seifert R (1995) Direct and indirect receptor-independent G-protein activation by cationic-amphiphilic substances. Studies with mast cells, HL-60 human leukemic cells and purified G-proteins. Exp Dermatol 4:231–239

Krause JE, Chenard BL, Cortright DN (2005) Transient receptor potential ion channels as targets for the discovery of pain therapeutics. Curr Opin Investig Drugs 6:48–57

Krishnaswamy G, Kelley J, Johnson D, Youngberg G, Stone W, Huang SK, Bieber J, Chi DS (2001) The human mast cell: functions in physiology and disease. Front Biosci 6:D1109–1127

Krishnaswamy G, Ajitawi O, Chi DS (2006) The human mast cell: an overview. Methods Mol Biol 315:13–34

Kronauer C, Eberlein-Konig B, Ring J, Behrendt H (2003) Influence of UVB, UVA and UVA1 irradiation on histamine release from human basophils and mast cells in vitro in the presence and absence of antioxidants. Photochem Photobiol 77:531–534

Kuno M, Kimura M (1992) Noise of secretagogue-induced inward currents dependent on extracellular calcium in rat mast cells. J Membr Biol 128:53–61

Kuno M, Okada T, Shibata T (1989) A patch-clamp study: secretagogue-induced currents in rat peritoneal mast cells. Am J Physiol 256:C560–568

Kuno M, Kawaguchi J, Mukai M, Nakamura F (1990) PT pretreatment inhibits 48/80-induced activation of $Ca^{(+)}$-permeable channels in rat peritoneal mast cells. Am J Physiol 259:C715–722

Kuno M, Kawawaki J, Nakamura F (1997) A highly temperature-sensitive proton current in mouse bone marrow-derived mast cells. J Gen Physiol 109:731–740

Lewin GR, Rueff A, Mendell LM (1994) Peripheral and central mechanisms of NGF-induced hyperalgesia. Eur J Neurosci 6:1903–1912

Lorenz D, Wiesner B, Zipper J, Winkler A, Krause E, Beyermann M, Lindau M, Bienert M (1998) Mechanism of peptide-induced mast cell degranulation. Translocation and patch-clamp studies. J Gen Physiol 112:577–591

Macpherson LJ, Geierstanger BH, Viswanath V, Bandell M, Eid SR, Hwang S, Patapoutian A (2005) The pungency of garlic: activation of TRPA1 and TRPV1 in response to allicin. Curr Biol 15:929–934

Maurer M, Metz M (2005) The status quo and quo vadis of mast cells. Exp Dermatol 14:923–929

Mekori YA, Metcalfe DD (2000) Mast cells in innate immunity. Immunol Rev 173:131–140

Mendez F, Penner R (1998) Near-visible ultraviolet light induces a novel ubiquitous calcium-permeable cation current in mammalian cell lines. J Physiol 507:365–377

Moalem G, Tracey DJ (2006) Immune and inflammatory mechanisms in neuropathic pain. Brain Res Brain Res Rev 51:240–264

Montell C (2005) The TRP superfamily of cation channels. Sci STKE 2005:re3

Muraki K, Iwata Y, Katanosaka Y, Ito T, Ohya S, Shigekawa M, Imaizumi Y (2003) TRPV2 is a component of osmotically sensitive cation channels in murine aortic myocytes. Circ Res 93:829–838

Nagata K, Duggan A, Kumar G, Garcia-Anoveros J (2005) Nociceptor and hair cell transducer properties of TRPA1, a channel for pain and hearing. J Neurosci 25:4052–4061

Nakajima T, Iikura M, Okayama Y, Matsumoto K, Uchiyama C, Shirakawa T, Yang X, Adra CN, Hirai K, Saito H (2004) Identification of granulocyte subtype-selective receptors and ion channels by using a high-density oligonucleotide probe array. J Allergy Clin Immunol 113:528–535

Namer B, Seifert F, Handwerker HO, Maihofner C (2005) TRPA1 and TRPM8 activation in humans: effects of cinnamaldehyde and menthol. Neuroreport 16:955–959

Nilius B (2003) From TRPs to SOCs, CCEs and CRACs: consensus and controversies. Cell Calcium 33:293–298

Noli C, Miolo A (2001) The mast cell in wound healing. Vet Dermatol 12:303–313

Pall ML, Anderson JH (2004) The vanilloid receptor as a putative target of diverse chemicals in multiple chemical sensitivity. Arch Environ Health 59:363–375

Parekh AB, Penner R (1997) Store depletion and calcium influx. Physiol Rev 77:901–930

Peier AM, Moqrich A, Hergarden AC, Reeve AJ, Andersson DA, Story GM, Earley TJ, Dragoni I, McIntyre P, Bevan S, Patapoutian A (2002) A TRP channel that senses cold stimuli and menthol. Cell 108:705–715

Penner R, Fleig A (2004) Store-operated calcium entry: a tough nut to CRAC. Sci STKE 2004:pe38

Reinsprecht M, Pecht I, Schindler H, Romanin C (1992) Potent block of Cl-channels by antiallergic drugs. Biochem Biophys Res Commun 188:957–963

Reynier-Rebuffel AM, Mathiau P, Callebert J, Dimitriadou V, Farjaudon N, Kacem K, Launay JM, Seylaz J, Abineau P (1994) Substance P, calcitonin gene-related peptide, and capsaicin release serotonin from cerebrovascular mast cells. Am J Physiol 267:R1421–1429

Rinaldi-Carmona M, Barth F, Heaulme M, Shire D, Calandra B, Congy C, Martinez S, Maruani J, Neliat G, Caput D, et al (1994) SR141716A, a potent and selective antagonist of the brain cannabinoid receptor. FEBS Lett 350:240–244

Rinaldi-Carmona M, Barth F, Millan J, Derocq JM, Casellas P, Congy C, Oustric D, Sarran M, Bouaboula M, Calandra B, Portier M, Shire D, Breliere JC, Le Fur GL (1998) SR 144528, the first potent and selective antagonist of the CB2 cannabinoid receptor. J Pharmacol Exp Ther 284:644–650

Ross RA (2003) Anandamide and vanilloid TRPV1 receptors. Br J Pharmacol 140:790–801

Schwarz EC, Wissenbach U, Niemeyer BA, Strauss B, Philipp SE, Flockerzi V, Hoth M (2006) TRPV6 potentiates calcium-dependent cell proliferation. Cell Calcium 39:163–173

Silber G, Proud D, Warner J, Naclerio R, Kagey-Sobotka A, Lichtenstein L, Eggleston P (1988) In vivo release of inflammatory mediators by hyperosmolar solutions. Am Rev Respir Dis 137:606–612

Small-Howard AL, Shimoda LM, Adra CN, Turner H (2005) Anti-inflammatory potential of CB1-mediated cAMP elevation in mast cells. Biochem J 25:25

Smart D, Gunthorpe MJ, Jerman JC, Nasir S, Gray J, Muir AI, Chambers JK, Randall AD, Davis JB (2000) The endogenous lipid anandamide is a full agonist at the human vanilloid receptor (hVR1). Br J Pharmacol 129:227–230

Soter NA, Wasserman SI (1980) Physical urticaria/angioedema: an experimental model of mast cell activation in humans. J Allergy Clin Immunol 66:358–365

Stokes AJ, Shimoda LM, Koblan-Huberson M, Adra CN, Turner H (2004) A TRPV2-PKA signaling module for transduction of physical stimuli in mast cells. J Exp Med 200:137–147

Stokes AJ, Wakano C, Del Carmen KA, Koblan-Huberson M, Turner H (2005a) Formation of a physiological complex between TRPV2 and RGA protein promotes cell surface expression of TRPV2. J Cell Biochem 94:669–683

Stokes AJ, Wakano C, Koblan-Huberson M, Adra CN, Fleig A, Turner H (2005b) TRPA1 is a substrate for de-ubiquitination by the tumor suppressor CYLD. Cell Signal [Epub ahead of print]

Story GM, Peier AM, Reeve AJ, Eid SR, Mosbacher J, Hricik TR, Earley TJ, Hergarden AC, Andersson DA, Hwang SW, McIntyre P, Jegla T, Bevan S, Patapoutian A (2003) ANKTM1, a TRP-like channel expressed in nociceptive neurons, is activated by cold temperatures. Cell 112:819–829

Suh YG, Oh U (2005) Activation and activators of TRPV1 and their pharmaceutical implication. Curr Pharm Des 11:2687–2698

Vliagoftis H, Befus AD (2005) Mast cells at mucosal frontiers. Curr Mol Med 5:573–589

Voets T, Prenen J, Fleig A, Vennekens R, Watanabe H, Hoenderop JG, Bindels RJ, Droogmans G, Penner R, Nilius B (2001) CaT1 and the calcium release-activated calcium channel manifest distinct pore properties. J Biol Chem 276:47767–47770

Vulcu SD, Liewald JF, Gillen C, Rupp J, Nawrath H (2004) Proton conductance of human transient receptor potential-vanilloid type-1 expressed in oocytes of Xenopus laevis and in Chinese hamster ovary cells. Neuroscience 125:861–866

Woolf CJ, Ma QP, Allchorne A, Poole S (1996) Peripheral cell types contributing to the hyperalgesic action of nerve growth factor in inflammation. J Neurosci 16:2716–2723

Xu H, Blair NT, Clapham DE (2005) Camphor activates and strongly desensitizes the transient receptor potential vanilloid subtype 1 channel in a vanilloid-independent mechanism. J Neurosci 25:8924–8937

Yamasaki S, Saito T (2005) Regulation of mast cell activation through FcepsilonRI. Chem Immunol Allergy 87:22–31

Yue L, Peng JB, Hediger MA, Clapham DE (2001) CaT1 manifests the pore properties of the calcium-release-activated calcium channel. Nature 410:705–709

Zhang X, Huang J, McNaughton PA (2005) NGF rapidly increases membrane expression of TRPV1 heat-gated ion channels. EMBO J 24:4211–4223

Zimmermann K, Leffler A, Fischer MM, Messlinger K, Nau C, Reeh PW (2005) The TRPV1/2/3 activator 2-aminoethoxydiphenyl borate sensitizes native nociceptive neurons to heat in wildtype but not TRPV1 deficient mice. Neuroscience 135:1277–1284

TRPV Channels' Role in Osmotransduction and Mechanotransduction

W. Liedtke

Center for Translational Neuroscience, Duke University, Durham NC, 27710, USA
wolfgang@neuro.duke.edu

1	Introduction: The TRPV Subfamily	474
2	Osmo-mechano-TRP TRPV1: Animal Findings	474
3	Osmo-mechano-TRP TRPV2: Tissue Culture Data	476
4	Osmo-mechano-TRP TRPV4: Tissue Culture and Animal Data	476
5	*C. elegans* TRPV Channels and Mechanosensation, Osmosensation	479
5.1	Cloning of the *osm-9* Gene, a Founding Member of the *trpv* Gene Family	479
6	TRPV4 Expression in ASH Rescues *osm-9* Mechanical and Osmotic Deficits	479
7	Outlook for Future Research on TRP Channels	482
References		483

Abstract In signal transduction of metazoan cells, transient receptor potential (TRP) ion channels have been identified that respond to diverse external and internal stimuli, among them osmotic and mechanical stimuli. This chapter will summarize findings on the TRPV subfamily, both its vertebrate and invertebrate members. Of the six mammalian TRPV channels, TRPV1, -V2, and -V4 were demonstrated to function in transduction of osmotic and/or mechanical stimuli. TRPV channels have been found to function in cellular as well as systemic osmotic homeostasis in vertebrates. Invertebrate TRPV channels, five in *Caenorhabditis elegans* and two in *Drosophila*, have been shown to play a role in mechanosensation, such as hearing and proprioception in *Drosophila* and nose touch in *C. elegans*, and in the response to osmotic stimuli in *C. elegans*. In a striking example of evolutionary conservation of function, mammalian TRPV4 has been found to rescue mechanosensory and osmosensory deficits of the TRPV mutant line *osm-9* in *C. elegans*, despite no more than 26% orthology of the respective amino acid sequences.

Keywords TRP · TRPV · Osmotic stimuli · Mechanical stimuli · Osmotransduction · Mechanotransduction

1
Introduction: The TRPV Subfamily

Within the transient receptor potential (TRP) superfamily of ion channels (Cosens and Manning 1969; Montell and Rubin 1989; Wong et al. 1989; Hardie and Minke 1992; Zhu et al. 1995), the TRPV subfamily stepped into the limelight in 1997 (Caterina et al. 1997; Colbert et al. 1997). TRPV2, -V3, and -V4 were each identified by a candidate gene approach (Caterina et al. 1999; Kanzaki et al. 1999; Liedtke et al. 2000; Strotmann et al. 2000; Wissenbach et al. 2000; Peier et al. 2002; Smith et al. 2002; Xu et al. 2002). The latter strategy also led to the identification of four additional *Caenorhabditis elegans ocr* genes (Tobin et al. 2002) and two *Drosophila trpv* genes, Nanchung (NAN) and Inactive (IAV) (Kim et al. 2003; Gong et al. 2004). The TRPV channels can be subgrouped into four branches by sequence comparison:

- TRPV1, -V2, -V3, and -V4, named "thermo-TRPs." Illuminating review articles on thermo-TRPs are available for in-depth reading (Caterina and Julius 1999; Clapham 2003; Tominaga and Caterina 2004; Caterina and Montell 2005; Patapoutian 2005).

- TRPV5 and TRPV6, possibly subserving Ca^{2+} uptake in the kidney and intestine (Hoenderop et al. 1999; Peng et al. 1999; den Dekker et al. 2003; Hoenderop et al. 2003; Peng et al. 2003).

- An invertebrate branch that includes *C. elegans* OSM-9 and *Drosophila* IAV.

- OCR-1 to -4 in *C. elegans* and *Drosophila* NAN.

This chapter will focus on the role of mammalian and invertebrate TRPV channels (with the latter focus on *C. elegans*) in signal transduction in response to osmotic and mechanical stimuli. These "osmo- and mechano-TRPs" (Liedtke and Kim 2005) are TRPV1, -2, -4, OSM-9, OCR-2, NAN, and IAV. Other TRPV channels might join this functional group within the TRP superfamily, which certainly also comprises non-TRPV channels, e.g., TRPA1 (Corey et al. 2004; Nagata et al. 2005) or NompC (Walker et al. 2000).

In this chapter, we will confront the following topics: Do TRPV ion channels function in sensing and transduction of osmotic and mechanical stimuli? Which molecular mechanisms are at play?

2
Osmo-mechano-TRP TRPV1: Animal Findings

In heterologous cellular expression systems, there have not been reports on mechanotransduction by TRPV1. Genetically engineered $trpv1^{-/-}$ mice, which have previously been shown to be devoid of thermal hyperalgesia following inflammation (Caterina et al. 2000; Davis et al. 2000), also displayed an altered response of their bladder to stretch (Birder et al. 2002). TRPV1 could be localized to sensory and autonomic ganglia neurons, and also to urothelial cells. When bladder and urothelial epithelial cells were maintained in culture, their response to mechanical stretch was significantly different from wildtype (wt). Specifically, the TRPV1$^+$ bladders secreted ATP upon stretch, which in turn is known to activate nerve fibers in the bladder. This response to mechanical stimulation was greatly diminished in bladders excised from $trpv1^{-/-}$ mice. It appears likely that this mechanism, operative in mice, also plays a role in human bladder epithelium. Intravesical instillation of TRPV1-activators is used to treat hyperactive bladder syndromes in spinal cord disease (Dinis et al. 2004; Lazzeri et al. 2004; Stein et al. 2004; Apostolidis et al. 2005).

Another instance of an altered response to mechanical stimuli in $trpv1^{-/-}$ mice relates to the response of the jejunum to mechanical stretch (Rong et al. 2004). Afferent jejunal nerve fibers were found to respond with decreased frequency of discharge in $trpv1^{-/-}$ mice than in wt mice. In humans, TRPV1-positive nerve fibers in the rectum were found significantly increased in patients suffering from fecal urgency, a pathologic rectal hypersensitivity in response to mechanical distension (Chan et al. 2003). Expression of TRPV1$^+$ nerve fibers in rectal biopsy samples from these patients was positively correlated with a decreased threshold to mechanical stretch; in addition, the occurrence of TRPV1$^+$ fibers was also correlated with a burning dysesthesia.

Another recent study focused on possible mechanisms of signal transduction in response to mechanical stimuli in blood vessels (Scotland et al. 2004). Elevation of intraluminal mechanical pressure in mesenterial arteries was reported to be associated with generation of 20-hydroxyeicosatetraenoic acid, which in turn activated TRPV1 on C-fibers leading to nerve depolarization and vasoactive neuropeptide release. With respect to nociception, using $trpv1^{-/-}$ mice, $trpv1$ was shown to be involved in inflammatory thermal hyperalgesia, but not inflammatory mechanical hyperalgesia (Caterina and Julius 1999; Gunthorpe et al. 2002). However, a specific and potent blocker of TRPV1 was found to reduce mechanical hyperalgesia in rats (Pomonis et al. 2003). This latter result appears contradictory to the obvious lack of difference between $trpv1^{-/-}$ and wt mice. Either this discrepancy may be due to a species difference between mouse and rat pertaining to signal-transduction by TRPV1 in inflammation-induced mechanical hyperalgesia, or it may be due to the different mechanisms that affect signaling in a $trpv1$ general knockout (with likely compensatory gene regulation) vs a specific temporal pharmacological blocking of the TRPV1 ion channel protein, that most likely participates in a signaling multiplex.

Very recently, in a spectacular finding, Naeini et al. reported that $trpv1^{-/-}$ mice failed to express an N-terminal variant of the $trpv1$ gene in magnocellular neurons of the supraoptic and paraventricular nucleus of the hypothalamus (Naeini et al. 2005). As these neurons are known to secrete vasopressin, the *trpv1*$^{-/-}$ mice were found to have a profound impairment of antidiuretic hormone (ADH) secretion in response to systemic hypertonic stimuli, and their magnocellular neurons did not show an appropriate electrical response to hypertonicity. These findings encouraged Bourque and colleagues to reason that this *trpv1* N-terminal variant, which could not be identified at the molecular level, is very likely involved as (part of) a tonicity sensor of intrinsically osmosensitive magnocellular neurons.

3
Osmo-mechano-TRP TRPV2: Tissue Culture Data

In heterologous cellular systems, TRPV2 was initially described as a temperature-gated ionotropic receptor for stimuli exceeding 52°C (Caterina et al. 1999). Recently, TRPV2 was also demonstrated to respond to hypotonicity and mechanical stimulation (Muraki et al. 2003). Arterial smooth muscle cells from various arteries expressed TRPV2. These myocytes responded to hypotonic stimulation with calcium influx. This activation could be reduced by specific downregulation of TRPV2 protein by an antisense strategy. Heterologously expressed TRPV2 in CHO cells displayed a similar response to hypotonicity. These cells were also subjected to stretch by applying negative pressure to the patch-pipette and by stretching the cell membrane on a mechanical stimulator. Both maneuvers led to Ca^{2+} influx that was dependent on heterologous TRPV2 expression.

4
Osmo-mechano-TRP TRPV4: Tissue Culture and Animal Data

CHO cells responded to hypotonic solution when they were (stably) transfected with TRPV4 (Liedtke et al. 2000). Human embryonic kidney (HEK)-293T cells, when maintained by the same authors, were found to harbor *trpv4* complementary DNA (cDNA), which was cloned from these cells. However, *trpv4* cDNA was not found in other batches of HEK-293T cells, so that this cell line was used as a heterologous expression vehicle by other groups (Strotmann et al. 2000; Wissenbach et al. 2000). Notably, when comparing the two settings it was obvious that the single-channel conductance was different (Liedtke et al. 2000; Strotmann et al. 2000). This underscores the relevance of gene expression in heterologous cellular systems for the functioning of TRPV4 in response to a basic biophysical stimulus. Also, it was found that the sensitivity of TRPV4 could be tuned by warming of the media. Similar results were found in another investigation with mammalian TRPV4 in HEK-293T cells (Gao et al. 2003). In addition, in this investigation the cells were mechanically stretched, without a change of tonicity. At room temperature there was no response upon mechanical stretch; however, at 37°C the isotonic response to stretch resulted in the maximum calcium influx of all conditions tested. In two other investigations, TRPV4 was found to be responsive to changes in temperature (Guler et al. 2002; Watanabe et al. 2002). Temperature change was accomplished by heating the streaming bath solution. Gating was amplified when hypotonic solution was used as streaming bath. In one investigation, temperature stimulation could not activate the TRPV4 channel in cell-detached inside-out patches (Watanabe et al. 2002).

In a recent paper, the ciliary beat frequency of ciliated cells was found to be influenced by gating of TRPV4 (Andrade et al. 2005). In explanted ciliated cells, and also in heterologously transfected HeLa cells, exposing the cells to hyperviscous, isotonic media could activate TRPV4.

Another recent focus in the field of TRP ion channels is intracellular trafficking, posttranslational modification, and subsequent functional modulation. For TRPV4, it was found in heterologous cells (HEK-293T) that N-glycosylation between transmembrane domain 5 and the pore-loop (position 651) decreases osmotic activation via decreased plasma membrane insertion (Xu et al. 2005). Interestingly, N-glycosylation between transmembrane domain 1 and 2 appears to have the same effect on TRPV5, and the antiaging hormone klotho functions as beta-glucuronidase and subsequently activates TRPV5 (Chang et al. 2005). TRPV4 also has been found to play a role in maintenance of cellular osmotic homeostasis. One particular cellular defense mechanism of cellular osmotic homeostasis is regulatory volume change, namely regulatory volume decrease (RVD) in response to hypotonicity and regulatory volume increase (RVI) in response to hypertonicity. In a recent paper, Bereiter-Hahn's group reported that CHO tissue culture cells have a poor RVD which, after transfection with TRPV4, improves significantly (Becker et al. 2005). In another study, Valverde's group published that TRPV4 mediates the cell-swelling induced Ca^{2+} influx into bronchial epithelial cells that triggers RVD via Ca^{2+}-dependent potassium channels (Arniges et al. 2004). This cell swelling response did not work in cystic fibrosis (CFTR) bronchial epithelia, where, on the other hand, TRPV4 could be activated by 4-alpha-PDD, leading to Ca^{2+} influx. In yet another investigation, Ambudkar and colleagues found the concerted interaction of aquaporin 5 (AQP-5) with TRPV4 in hypotonic swelling-induced RVD of salivary gland epithelia (Liu et al. 2006). These findings shed light on mechanisms operative in secretory epithelia that underlie watery secretion based on a concerted interaction of TRPV4 and AQP-5.

In $trpv4^{-/-}$ mice, the response to mechanical stimulation is diminished (Liedtke and Friedman 2003; Suzuki et al. 2003). In mice, the absence of TRPV4, which in wt animals could be shown to be expressed in sensory ganglia (Liedtke and Friedman 2003) and in subcutaneous nerve fibers and keratinocytes of skin (Delany et al. 2001; Guler et al. 2002), the threshold to noxious mechanical stimulation was significantly elevated. This result was obtained using two standard tests, the Randall-Selitto test, which applies mechanical pressure by squeezing the paw, and an automated von-Frey test, which applies mechanical stimulation from underneath the hind-paw, leading to withdrawal (Liedtke and Friedman 2003). When rats were sensitized with taxol, their sensitivity to noxious mechanical stimuli was strikingly lowered as a result of the taxol-induced neuropathy (Dina et al. 2001; Dina et al. 2004). When these rats were treated intrathecally with the TRPV4-specific antisense oligonucleotide, the threshold for mechanical painful stimuli went up (Alessandri-Haber et al. 2004). This clearly suggests a role for TRPV4 in mediating hyperalgesia in response to mechanical stimuli in an animal model for neuropathic pain. Last, but not least, with respect to mechanotransduction in $trpv4^{-/-}$ mice, they do not show any sign of inner ear dysfunction including deafness (Liedtke and Friedman 2003), which has to be viewed against the expression pattern of *trpv4* in the inner ear (Liedtke et al. 2000; Mutai and Heller 2003). Messenger RNA (mRNA) of *trpv4* could be demonstrated in the secretory epithelia of the stria vascularis and in neurosensory inner ear hair-cells of both rodents and birds.

When stressed with systemic hypertonicity, $trpv4^{-/-}$ mice did not counter-regulate their systemic tonicity as efficiently as wt littermates (Liedtke and Friedman 2003). Their drinking was diminished, and systemic tonicity was significantly higher. Continuous infusion of the ADH analog dDAVP led to systemic hypotonicity, whereas renal water readsorption capacity was not changed in both genotypes. Antidiuretic hormone synthesis in response to osmotic stimulation was reduced in $trpv4^{-/-}$ mice. Hypertonic stress led to reduced expression of c-FOS$^+$ cells in the sensory circumventricular organ, OVLT (organum vasculosum laminae terminalis), indicative of impaired osmotic activation. These findings in $trpv4^{-/-}$ mice point toward a deficit in central osmotic sensing. Thus, TRPV4 is necessary for the maintenance of osmotic equilibrium in mammals. It is conceivable that TRPV4 acts as an osmotic sensor in the CNS. The impaired osmotic regulation in $trpv4^{-/-}$ mice reported in the author's paper differs from that published in another report. While our experiments showed that $trpv4^{-/-}$ mice secrete lower amounts of ADH in response to hypertonic stimuli, the results from Mizuno et al. (2003) suggest that there is an increased ADH response to water deprivation and subsequent systemic administration of propylene glycol. The reasons for this discrepancy are not obvious. In our study, a blunted ADH response and diminished cFOS response in the OVLT in $trpv4^{-/-}$ mice upon systemic hypertonicity suggests, as one possibility, an activation of TRPV4$^+$ sensory cells in the OVLT by hypertonicity.

In a recent publication, Allessandri-Haber et al. demonstrate that hypertonic subcutaneous solution leads to pain-related behavior in wt mice, which is not present in $trpv4^{-/-}$ mice (Alessandri-Haber et al. 2005). When sensitizing nociceptors with prostaglandin E2, the pain-related responses to hypertonic stimulation became more frequent, and were greatly diminished in $trpv4^{-/-}$ mice. These in vivo data could not be recapitulated in acutely dissociated dorsal root ganglia (DRG) neurons upon stimulation with hypertonicity and subsequent calcium imaging. As a whole, this study indicates differences in the response of mice to noxious tonicity stimuli depending on the presence or absence of TRPV4. Yet at the level of a critical transducer cell, namely the DRG sensory neuron, only hypotonicity led to a rise of intracellular calcium that was dependent on the presence of TRPV4.

5
C. elegans TRPV Channels and Mechanosensation, Osmosensation

5.1
Cloning of the *osm-9* Gene, a Founding Member of the *trpv* Gene Family

As mentioned in the introduction, the *osm-9* mutant was first reported in 1997 (Colbert et al. 1997). The forward genetics screen in *C. elegans* applied a confinement assay with a high-molar osmotically active substance. The *osm-9* mutants did not respect this barrier, and the mutated gene was found to be a TRP channel. On closer analysis, *osm-9* mutants did not respond to aversive tonicity stimuli, they did not respond to mechanical stimuli to their "nose," and they did not respond to (aversive) odorants. The OSM-9 channel protein was expressed in amphid sensory neurons, constituting the worm's cellular substrate of exteroceptive sensing of chemical, osmotic, and mechanical stimuli. The OSM-9 channel protein was expressed in the sensory cilia of the AWC and ASH amphid sensory neurons. Bilateral laser ablation of the ASH neuron has been shown to lead to a deficit in osmotic, nose touch and olfactory avoidance (Kaplan and Horvitz 1993). ASH has therefore been termed the "nociceptive" neuron (Bargmann and Kaplan 1998). Next, four additional TRPV channels from *C. elegans* were cloned, named OCR-1 to -4 (Tobin et al. 2002). Of these four channels, only OCR-2 was expressed in ASH. The *ocr-2* mutant phenotype was virtually identical to the *osm-9* phenotype with respect to worm "nociception," and there was genetic evidence that the two channels were necessary for the proper intracellular trafficking of each other in sensory neurons, indicating an OSM-9/OCR-2 interaction. When expressing the mammalian capsaicin receptor TRPV1 in the ASH sensory neurons, neither *osm-9* nor *ocr-2* mutants could be rescued for any of their deficits, but *osm-9 ash:trpv1* transgenic worms displayed a strong avoidance to capsaicin, which normal worms do not respond to.

6
TRPV4 Expression in ASH Rescues *osm-9* Mechanical and Osmotic Deficits

In order to understand better sensory transduction in *C. elegans* after the identification of the OSM-9 mutant, TRPV4 was transgenically targeted to

ASH amphid neurons of *osm-9* mutants. Surprisingly, TRPV4 expression in *C. elegans* ASH rescued *osm-9* mutants' defects in avoidance of hyperosmotic stimuli and nose touch (Liedtke et al. 2003). However, mammalian TRPV4 did not rescue the odorant avoidance defect of *osm-9*, suggesting that this specific function of TRPV channels differs between vertebrate and invertebrate. This basic finding of the rescue experiments in *osm-9 ash:trpv4* worms has important implication for mechanisms of signal transduction (see Fig. 1).

TRPV4 appeared to be integrated into the normal ASH sensory neuron signaling apparatus, since the transgene failed to rescue these deficits in other *C. elegans* mutants defective in osmosensation and mechanosensation (including OCR-2, bespeaking of the specificity of the observed response). A point mutation in the pore-loop of TRPV4, M680K, markedly reduced rescue, indicating that TRPV4 very likely functions as a transducing ion channel. In an attempt to recapitulate the properties of the mammalian channel in the nociceptive behavior of the worm, it was found that the sensitivity for osmotic stimuli and

Fig. 1 a–c Signal transduction in sensory (nerve) cells in response to odorant (**a**), osmotic (**b**), and mechanical (**c**) stimuli. **a** The odorant activates the TRPV ion channel via a G-protein coupled receptor mechanism. Such a mechanism is operative in the ASH sensory neuron of *C. elegans* in response to, e.g., 8-octanone, a repulsive odorant. Intracellular signaling cascades downstream of the G-protein coupled receptor activate the TRPV channel—OSM-9 or OCR-2. Calcium influx through the TRPV channel serves as an amplification mechanism, which is required for this signaling pathway to lead to the stereotypical withdrawal response. **b** This drawing represents two possibilities of how tonicity signaling could work. In one alternative scenario, depicted on the *right* side, the TRPV channel functions downstream of a yet unknown osmotic stimulus transduction mechanism, which is directly activated by a change in tonicity. This is conceptually related to what is depicted in (**a**). Intracellular signaling via phosphorylation (dephosphorylation)-dependent pathways activates the TRPV channel. For heterologous cellular expression systems, two groups have obtained data, contradictory in their details, that suggest phosphorylation of TRPV4 to be of relevance (Vriens et al. 2003; Xu et al. 2003). On the *left* side of the representation, note another scenario where the TRPV channel is at the top of the signaling cascade, i.e., it is directly activated by a change in tonicity, which in turn leads to an altered mechanical tension of the cytoplasmic membrane via volume change. Note that the two alternatives need not be mutually exclusive. Apart from phosphorylation of the TRPV channel, which could possibly be of relevance in vivo, a direct physical linkage of the TRPV channel to the cytoskeleton, the extracellular matrix, and the lipids of the plasma membrane adjacent to the channel has to be entertained. **c** This drawing represents two possibilities of how mechanotransduction could work. Here, depicted on the *right* side, an unknown mechanotransduction channel responds directly to the mechanical stimulus with calcium influx. This activity and the subsequent signal transduction are modulated more indirectly by the TRPV channel, which acts on the unknown transduction channel, onto the biophysical properties of the membrane, and via other, yet unknown intracellular signaling mechanisms. The *left* side depicts another alternative. Here, the TRPV channel is the mechanotransducer itself, i.e., it is activated directly via mechanical stimulation. (From Liedtke and Kim 2005, with permission)

the effect of temperature on the avoidance responses of *osm-9 ash:trpv4* worms more closely resembled the known properties of mammalian TRPV4 than that of normal worms. TRPV4 does not rescue the odorant avoidance deficit of *osm-9* mutant worms, where G-protein coupled receptors function as odorant sensors, and TRPV4 did not function downstream of other known mutations that affect nose touch and osmotic avoidance in *C. elegans*.

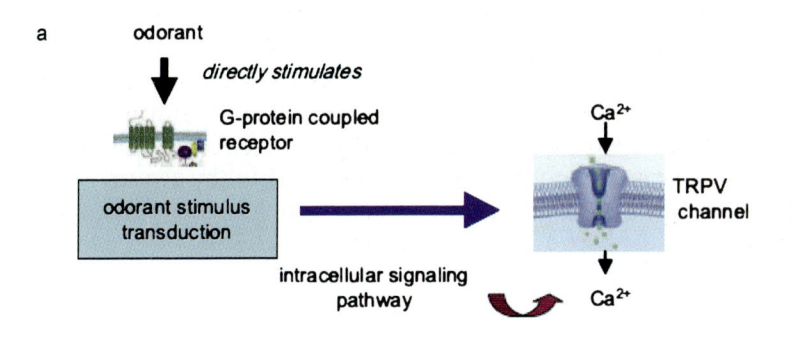

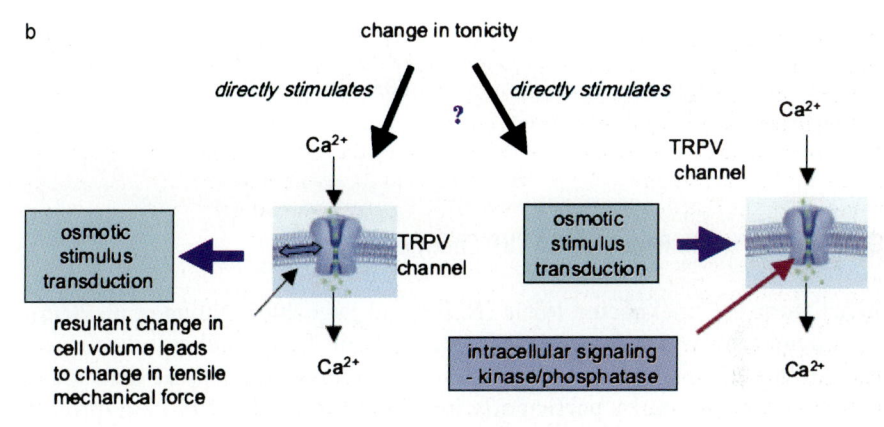

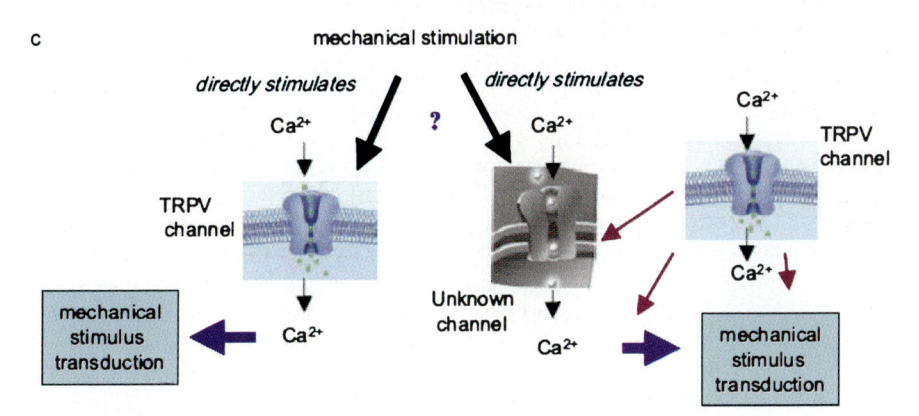

In aggregate, these data suggest that mammalian TRPV4 was functioning as the osmotic and mechanical sensor or at least a component of it. It should be realized that TRPV4 was expressed only in ASH, a single sensory neuron, where the mammalian protein, with a similarity to OSM-9 of approximately 25%, was %routed %correctly to the ASH sensory cilia, a distance of about 100 µm. The rescue was specific (not for OCR-2, not by mammalian TRPV1), and it respected genetically defined pathways.

This study prompts stimulating questions. Whereas TRPV4 restores responsiveness to hyper-osmotic stimuli in *C. elegans osm-9* mutants, it is only gated by hypo-osmotic stimuli in transfected mammalian cells. The reasons for this discrepancy are not understood. Related to that study, it was recently reported that TRPV2 could rescue one particular deficit of the *ocr-2* mutant, namely the dramatic downregulation of serotonin biosynthesis in the sensory ADF neuron, but mammalian TRPV2, unlike TRPV4 directing behavior in *osm-9*, did not complement the lack of osmotic avoidance reaction of *ocr-2* (Zhang et al. 2004; Sokolchik et al. 2005). Common to these two investigations is the conservation of TRPV signaling across phyla that were separated by several hundred million years of molecular evolution, in view of the low sequence homology.

In reference to the *Drosophila* TRPV channels, NAN and IAV, the reader is directed to original papers (Kim et al. 2003; Gong et al. 2004) and reviews (Vriens et al. 2004; Liedtke and Kim 2005).

7
Outlook for Future Research on TRP Channels

Apart from the unexpected turns (Nilius and Sage 2005; Nilius et al. 2005), the obvious topic for future investigation is the functional significance of protein–protein interactions of TRP(V) ion channels with yet-to-be-discovered interaction partners. [A particularly interesting example of protein–protein interactions of TRPV4 splice-variants from airway epithelia was reported by Arniges et al. (2005)]. In addition, there is the obvious potential of using TRP channels as therapeutical targets (Gudermann and Flockerzi 2005; Wissenbach et al. 2004).

Acknowledgements The author was supported by a K08 career development award of the National Institutes of Mental Health, by funding from the Whitehall Foundation (Palm Springs, FL), American Federation for Aging Research (New York, NY), the Klingenstein Fund (New York, NY), and by Duke University (Durham, NC).

References

Alessandri-Haber N, Dina OA, Yeh JJ, Parada CA, Reichling DB, Levine JD (2004) Transient receptor potential vanilloid 4 is essential in chemotherapy-induced neuropathic pain in the rat. J Neurosci 24:4444–4452

Alessandri-Haber N, Joseph E, Dina OA, Liedtke W, Levine JD (2005) TRPV4 mediates pain-related behavior induced by mild hypertonic stimuli in the presence of inflammatory mediator. Pain 118:70–79

Allessandri-Haber N, Yeh J, Boyd AE, Parada CA, Chen X, Reichling DB, Levine JD (2003) Hypotonicity induces TRPV4-mediated nociception in rat. Neuron 39:497–511

Andrade YN, Fernandes J, Vazquez E, Fernandez-Fernandez JM, Arniges M, Sanchez TM, Villalon M, Valverde MA (2005) TRPV4 channel is involved in the coupling of fluid viscosity changes to epithelial ciliary activity. J Cell Biol 168:869–874

Apostolidis A, Brady CM, Yiangou Y, Davis J, Fowler CJ, Anand P (2005) Capsaicin receptor TRPV1 in urothelium of neurogenic human bladders and effect of intravesical resiniferatoxin. Urology 65:400–405

Arniges M, Vazquez E, Fernandez-Fernandez JM, Valverde MA (2004) Swelling-activated Ca^{2+} entry via TRPV4 channel is defective in cystic fibrosis airway epithelia. J Biol Chem 279:54062–54068

Arniges M, Fernandez-Fernandez JM, Albrecht N, Schaefer M, Valverde MA (2005) Human TRPV4 channel splice variants revealed a key role of ankyrin domains in multimerization and trafficking. J Biol Chem 281:8990

Bargmann CI, Kaplan JM (1998) Signal transduction in the Caenorhabditis elegans nervous system. Annu Rev Neurosci 21:279–308

Becker D, Blase C, Bereiter-Hahn J, Jendrach M (2005) TRPV4 exhibits a functional role in cell-volume regulation. J Cell Sci 118:2435–2440

Birder LA, Nakamura Y, Kiss S, Nealen ML, Barrick S, Kanai AJ, Wang E, Ruiz G, De Groat WC, Apodaca G, Watkins S, Caterina MJ (2002) Altered urinary bladder function in mice lacking the vanilloid receptor TRPV1. Nat Neurosci 5:856–860

Caterina MJ, Julius D (1999) Sense and specificity: a molecular identity for nociceptors. Curr Opin Neurobiol 9:525–530

Caterina MJ, Montell C (2005) Take a TRP to beat the heat. Genes Dev 19:415–418

Caterina MJ, Schumacher MA, Tominaga M, Rosen TA, Levine JD, Julius D (1997) The capsaicin receptor: a heat-activated ion channel in the pain pathway. Nature 389:816–824

Caterina MJ, Rosen TA, Tominaga M, Brake AJ, Julius D (1999) A capsaicin-receptor homologue with a high threshold for noxious heat. Nature 398:436–441

Caterina MJ, Leffler A, Malmberg AB, Martin WJ, Trafton J, Petersen-Zeitz KR, Koltzenburg M, Basbaum AI, Julius D (2000) Impaired nociception and pain sensation in mice lacking the capsaicin receptor. Science 288:306–313

Chan CL, Facer P, Davis JB, Smith GD, Egerton J, Bountra C, Williams NS, Anand P (2003) Sensory fibres expressing capsaicin receptor TRPV1 in patients with rectal hypersensitivity and faecal urgency. Lancet 361:385–391

Chang Q, Hoefs S, van der Kemp AW, Topala CN, Bindels RJ, Hoenderop JG (2005) The beta-glucuronidase klotho hydrolyzes and activates the TRPV5 channel. Science 310:490–493

Clapham DE (2003) TRP channels as cellular sensors. Nature 426:517–524

Colbert HA, Smith TL, Bargmann CI (1997) OSM-9, a novel protein with structural similarity to channels, is required for olfaction, mechanosensation, and olfactory adaptation in Caenorhabditis elegans. J Neurosci 17:8259–8269

Corey DP, Garcia-Anoveros J, Holt JR, Kwan KY, Lin SY, Vollrath MA, Amalfitano A, Cheung EL, Derfler BH, Duggan A, Geleoc GS, Gray PA, Hoffman MP, Rehm HL, Tamasauskas D, Zhang DS (2004) TRPA1 is a candidate for the mechanosensitive transduction channel of vertebrate hair cells. Nature 432:723–730

Cosens DJ, Manning A (1969) Abnormal electroretinogram from a Drosophila mutant. Nature 224:285–287

Davis JB, Gray J, Gunthorpe MJ, Hatcher JP, Davey PT, Overend P, Harries MH, Latcham J, Clapham C, Atkinson K, Hughes SA, Rance K, Grau E, Harper AJ, Pugh PL, Rogers DC, Bingham S, Randall A, Sheardown SA (2000) Vanilloid receptor-1 is essential for inflammatory thermal hyperalgesia. Nature 405:183–187

Delany NS, Hurle M, Facer P, Alnadaf T, Plumpton C, Kinghorn I, See CG, Costigan M, Anand P, Woolf CJ, Crowther D, Sanseau P, Tate SN (2001) Identification and characterization of a novel human vanilloid receptor-like protein, VRL-2. Physiol Genomics 4:165–174

den Dekker E, Hoenderop JG, Nilius B, Bindels RJ (2003) The epithelial calcium channels, TRPV5 and TRPV6: from identification towards regulation. Cell Calcium 33:497–507

Dina OA, Chen X, Reichling D, Levine JD (2001) Role of protein kinase Cepsilon and protein kinase A in a model of paclitaxel-induced painful peripheral neuropathy in the rat. Neuroscience 108:507–515

Dinis P, Charrua A, Avelino A, Yaqoob M, Bevan S, Nagy I, Cruz F (2004) Anandamide-evoked activation of vanilloid receptor 1 contributes to the development of bladder hyperreflexia and nociceptive transmission to spinal dorsal horn neurons in cystitis. J Neurosci 24:11253–11263

Gao X, Wu L, O'Neil RG (2003) Temperature-modulated diversity of TRPV4 channel gating: activation by physical stresses and phorbol ester derivatives through protein kinase C-dependent and -independent pathways. J Biol Chem 278:27129–27137

Gong Z, Son W, Chung YD, Kim J, Shin DW, McClung CA, Lee Y, Lee HW, Chang DJ, Kaang BK, Cho H, Oh U, Hirsh J, Kernan MJ, Kim C (2004) Two interdependent TRPV channel subunits, inactive and Nanchung, mediate hearing in Drosophila. J Neurosci 24:9059–9066

Gudermann T, Flockerzi V (2005) TRP channels as new pharmacological targets. Naunyn Schmiedebergs Arch Pharmacol 371:241–244

Guler AD, Lee H, Iida T, Shimizu I, Tominaga M, Caterina M (2002) Heat-evoked activation of the ion channel, TRPV4. J Neurosci 22:6408–6414

Gunthorpe MJ, Benham CD, Randall A, Davis JB (2002) The diversity in the vanilloid (TRPV) receptor family of ion channels. Trends Pharmacol Sci 23:183–191

Hardie RC, Minke B (1992) The trp gene is essential for a light-activated Ca^{2+} channel in Drosophila photoreceptors. Neuron 8:643–651

Hoenderop JG, van der Kemp AW, Hartog A, van de Graaf SF, van Os CH, Willems PH, Bindels RJ (1999) Molecular identification of the apical Ca^{2+} channel in 1, 25-dihydroxyvitamin D3-responsive epithelia. J Biol Chem 274:8375–8378

Hoenderop JG, Nilius B, Bindels RJ (2003) Epithelial calcium channels: from identification to function and regulation. Pflugers Arch 446:304–308

Kanzaki M, Zhang YQ, Mashima H, Li L, Shibata H, Kojima I (1999) Translocation of a calcium-permeable cation channel induced by insulin-like growth factor-I. Nat Cell Biol 1:165–170

Kaplan JM, Horvitz HR (1993) A dual mechanosensory and chemosensory neuron in Caenorhabditis elegans. Proc Natl Acad Sci U S A 90:2227–2231

Kim J, Chung YD, Park DY, Choi S, Shin DW, Soh H, Lee HW, Son W, Yim J, Park CS, Kernan MJ, Kim C (2003) A TRPV family ion channel required for hearing in Drosophila. Nature 424:81–84

Lazzeri M, Vannucchi MG, Zardo C, Spinelli M, Beneforti P, Turini D, Faussone-Pellegrini MS (2004) Immunohistochemical evidence of vanilloid receptor 1 in normal human urinary bladder. Eur Urol 46:792–798

Lee H, Iida T, Mizuno A, Suzuki M, Caterina MJ (2005) Altered thermal selection behavior in mice lacking transient receptor potential vanilloid 4. J Neurosci 25:1304–1310

Liedtke W, Friedman JM (2003) Abnormal osmotic regulation in trpv4$^{-/-}$ mice. Proc Natl Acad Sci U S A 100:13698–13703

Liedtke W, Kim C (2005) Functionality of the TRPV subfamily of TRP ion channels: add mechano-TRP and osmo-TRP to the lexicon! Cell Mol Life Sci 62:2985–3001

Liedtke W, Choe Y, Marti-Renom MA, Bell AM, Denis CS, Sali A, Hudspeth AJ, Friedman JM, Heller S (2000) Vanilloid receptor-related osmotically activated channel (VR-OAC), a candidate vertebrate osmoreceptor. Cell 103:525–535

Liedtke W, Tobin DM, Bargmann CI, Friedman JM (2003) Mammalian TRPV4 (VR-OAC) directs behavioral responses to osmotic and mechanical stimuli in C. elegans. Proc Natl Acad Sci U S A 100:14531–14536

Liu X, Bandyopadhyay B, Nakamoto T, Singh B, Liedtke W, Melvin JE, Ambudkar I (2006) A role for AQP5 in activation of TRPV4 by hypotonicity: concerted involvement of AQP5 and TRPV4 in regulation of cell volume recovery. J Biol Chem 281:15485–15495

Martinac B, Hamill OP (2002) Gramicidin A channels switch between stretch activation and stretch inactivation depending on bilayer thickness. Proc Natl Acad Sci U S A 99:4308–4312

Mizuno A, Matsumoto N, Imai M, Suzuki M (2003) Impaired osmotic sensation in mice lacking TRPV4. Am J Physiol Cell Physiol 285:C96–101

Montell C, Rubin GM (1989) Molecular characterization of the Drosophila trp locus: a putative integral membrane protein required for phototransduction. Neuron 2:1313–1323

Muraki K, Iwata Y, Katanosaka Y, Ito T, Ohya S, Shigekawa M, Imaizumi Y (2003) TRPV2 is a component of osmotically sensitive cation channels in murine aortic myocytes. Circ Res 93:829–838

Mutai H, Heller S (2003) Vertebrate and invertebrate TRPV-like mechanoreceptors. Cell Calcium 33:471–478

Naeini RS, Witty MF, Seguela P, Bourque CW (2005) An N-terminal variant of Trpv1 channel is required for osmosensory transduction. Nat Neurosci 9:93–98

Nagata K, Duggan A, Kumar G, Garcia-Anoveros J (2005) Nociceptor and hair cell transducer properties of TRPA1, a channel for pain and hearing. J Neurosci 25:4052–4061

Nilius B, Sage SO (2005) TRP channels: novel gating properties and physiological functions. J Physiol 567:33–34

Nilius B, Voets T, Peters J (2005) TRP channels in disease. Sci STKE 2005:re8

Patapoutian A (2005) TRP channels and thermosensation. chem senses 30 Suppl 1:i193–i194

Peier AM, Reeve AJ, Andersson DA, Moqrich A, Earley TJ, Hergarden AC, Story GM, Colley S, Hogenesch JB, McIntyre P, Bevan S, Patapoutian A (2002) A heat-sensitive TRP channel expressed in keratinocytes. Science 296:2046–2049

Peng JB, Chen XZ, Berger UV, Vassilev PM, Tsukaguchi H, Brown EM, Hediger MA (1999) Molecular cloning and characterization of a channel-like transporter mediating intestinal calcium absorption. J Biol Chem 274:22739–22746

Peng JB, Brown EM, Hediger MA (2003) Epithelial Ca^{2+} entry channels: transcellular Ca^{2+} transport and beyond. J Physiol 551:729–740

Pomonis JD, Harrison JE, Mark L, Bristol DR, Valenzano KJ, Walker K (2003) N-(4-Tertiarybutylphenyl)-4-(3-cholorphyridin-2-yl)tetrahydropyrazine-1(2H)-carbox-amide (BCTC), a novel, orally effective vanilloid receptor 1 antagonist with analgesic properties: II. in vivo characterization in rat models of inflammatory and neuropathic pain. J Pharmacol Exp Ther 306:387–393

Rong W, Hillsley K, Davis JB, Hicks G, Winchester WJ, Grundy D (2004) Jejunal afferent nerve sensitivity in wild-type and TRPV1 knockout mice. J Physiol 560:867–881

Scotland RS, Chauhan S, Davis C, De Felipe C, Hunt S, Kabir J, Kotsonis P, Oh U, Ahluwalia A (2004) Vanilloid receptor TRPV1, sensory C-fibers, and vascular autoregulation: a novel mechanism involved in myogenic constriction. Circ Res 95:1027–1034

Smith GD, Gunthorpe MJ, Kelsell RE, Hayes PD, Reilly P, Facer P, Wright JE, Jerman JC, Walhin JP, Ooi L, Egerton J, Charles KJ, Smart D, Randall AD, Anand P, Davis JB (2002) TRPV3 is a temperature-sensitive vanilloid receptor-like protein. Nature 418:186–190

Sokolchik I, Tanabe T, Baldi PF, Sze JY (2005) Polymodal sensory function of the Caenorhabditis elegans OCR-2 channel arises from distinct intrinsic determinants within the protein and is selectively conserved in mammalian TRPV proteins. J Neurosci 25:1015–1023

Stein RJ, Santos S, Nagatomi J, Hayashi Y, Minnery BS, Xavier M, Patel AS, Nelson JB, Futrell WJ, Yoshimura N, Chancellor MB, De Miguel F (2004) Cool (TRPM8) and hot (TRPV1) receptors in the bladder and male genital tract. J Urol 172:1175–1178

Strotmann R, Harteneck C, Nunnenmacher K, Schultz G, Plant TD (2000) OTRPC4, a nonselective cation channel that confers sensitivity to extracellular osmolarity. Nat Cell Biol 2:695–702

Suzuki M, Mizuno A, Kodaira K, Imai M (2003) Impaired pressure sensation in mice lacking TRPV4. J Biol Chem 278:22664–22668

Tian W, Salanova M, Xu H, Lindsley JN, Oyama TT, Anderson S, Bachmann S, Cohen DM (2004) Renal expression of osmotically responsive cation channel TRPV4 is restricted to water-impermeant nephron segments. Am J Physiol Renal Physiol 287:F17–24

Tobin D, Madsen DM, Kahn-Kirby A, Peckol E, Moulder G, Barstead R, Maricq AV, Bargmann CI (2002) Combinatorial expression of TRPV channel proteins defines their sensory functions and subcellular localization in C. elegans neurons. Neuron 35:307–318

Todaka H, Taniguchi J, Satoh J, Mizuno A, Suzuki M (2004) Warm temperature-sensitive transient receptor potential vanilloid 4 (TRPV4) plays an essential role in thermal hyperalgesia. J Biol Chem 279:35133–35138

Tominaga M, Caterina MJ (2004) Thermosensation and pain. J Neurobiol 61:3–12

Vriens J, Watanabe H, Janssens A, Droogmans G, Voets T, Nilius B (2003) Cell swelling, heat, and chemical agonists use distinct pathways for the activation of the cation channel TRPV4. Proc Natl Acad Sci U S A 101:396–401

Vriens J, Owsianik G, Voets T, Droogmans G, Nilius B (2004) Invertebrate TRP proteins as functional models for mammalian channels. Pflugers Arch 449:213–226

Walker RG, Willingham AT, Zuker CS (2000) A Drosophila mechanosensory transduction channel. Science 287:2229–2234

Watanabe H, Vriens J, Suh SH, Benham CD, Droogmans G, Nilius B (2002) Heat-evoked activation of TRPV4 channels in a HEK293 cell expression system and in native mouse aorta endothelial cells. J Biol Chem 277:47044–47051

Wissenbach U, Bodding M, Freichel M, Flockerzi V (2000) Trp12, a novel Trp related protein from kidney. FEBS Lett 485:127–134

Wissenbach U, Niemeyer BA, Flockerzi V (2004) TRP channels as potential drug targets. Biol Cell 96:47–54

Wong F, Schaefer EL, Roop BC, LaMendola JN, Johnson-Seaton D, Shao D (1989) Proper function of the Drosophila trp gene product during pupal development is important for normal visual transduction in the adult. Neuron 3:81–94

Xu H, Ramsey IS, Kotecha SA, Moran MM, Chong JA, Lawson D, Ge P, Lilly J, Silos-Santiago I, Xie Y, DiStefano PS, Curtis R, Clapham DE (2002) TRPV3 is a calcium-permeable temperature-sensitive cation channel. Nature 418:181–186

Xu H, Zhao H, Tian W, Yoshida K, Roullet JB, Cohen DM (2003) Regulation of a transient receptor potential (TRP) channel by tyrosine phosphorylation. SRC family kinase-dependent tyrosine phosphorylation of TRPV4 on TYR-253 mediates its response to hypotonic stress. J Biol Chem 278:11520–11527

Xu H, Fu Y, Tian W, Cohen DM (2005) Glycosylation of the osmoresponsive transient receptor potential channel TRPV4 on Asn-651 influences membrane trafficking. Am J Physiol Renal Physiol 290:F1103–F1109

Xu XZ, Li HS, Guggino WB, Montell C (1997) Coassembly of TRP and TRPL produces a distinct store-operated conductance. Cell 89:1155–1164

Xu XZ, Chien F, Butler A, Salkoff L, Montell C (2000) TRPgamma, a Drosophila TRP-related subunit, forms a regulated cation channel with TRPL. Neuron 26:647–657

Zhang S, Sokolchik I, Blanco G, Sze JY (2004) Caenorhabditis elegans TRPV ion channel regulates 5HT biosynthesis in chemosensory neurons. Development 131:1629–1638

Zhu X, Chu PB, Peyton M, Birnbaumer L (1995) Molecular cloning of a widely expressed human homologue for the Drosophila trp gene. FEBS Lett 373:193–198

Nociception and TRP Channels

M. Tominaga

Section of Cell Signaling, Okazaki Institute for Integrative Bioscience, National Institutes of Natural Sciences, 444-8787 Okazaki, Japan
tominaga@nips.ac.jp

1	Capsaicin Receptor TRPV1	490
1.1	TRPV1 and Inflammatory Pain	493
1.2	TRPV1 and Visceral Pain	494
1.3	TRPV1 as a Target of Analgesics	495
2	TRPV2	495
3	TRPV3	496
4	TRPV4	497
5	TRPM8	497
6	TRPA1	499
References		500

Abstract Pain is initiated when noxious stimuli excite the peripheral terminals of specialized primary afferent neurons called nociceptors. Many molecules are involved in conversion of the noxious stimuli to the electrical signals in the nociceptor endings. Among them, TRP channels play important roles in detecting noxious stimuli.

Keywords Noci-TRPs · Capsaicin · Analgesic · Thermo-TRPs · Heat perception · Cold perception

Pain is initiated when noxious thermal, mechanical, or chemical stimuli excite the peripheral terminals of specialized primary afferent neurons called nociceptors, most of which are small-diameter, unmyelinated C fibers (Wood and Perl 1999; Woolf and Salter 2000; Scholz and Woolf 2002). These fibers transmit this information to the central nervous systems, leading to the sensation of pain. Many kinds of ionotropic and metabotropic receptors are known to be involved in the process (Caterina and Julius 1999; McCleskey and Gold 1999; Julius and Basbaum 2001). The most important function of peripheral nociceptors is action potential generation, which is then transmitted as nociceptive information. Therefore, most nociceptors are cation channels, including TRP channels, P2X receptors, and ASIC channels whose opening causes cation influx leading to depolarization necessary for activation of voltage-gated

Table 1 Comparison of the six TRP channels thought to be involved in nociception: noci-TPRs

	Activators	Blockers
TRPV1	Voltage, vanilloids, heat ($\geq 43°C$), proton, anandamide, oleoyletahnolamide, NADA, OLDA, camphor, allicin	Capsazepine, ruthenium red, iodo-resiniferatoxin, BCTC, PIP$_2$
TRPV2	Heat ($\geq 52°C$), 2-APB, mechanical stimulus	Ruthenium red, La^{3+}
TRPV3	Heat ($\geq 32°C-39°C$), camphor, carvacrol, eugenol, thymol, 2-APB	Ruthenium red
TRPV4	Heat ($\geq 27°C-35°C$), 4a-PDD, EETs, hypotonic stimulus, mechanical stimulus	Ruthenium red, Gd^{3+}, La^{3+}
TRPM8	Cold ($<25°C-28°C$), menthol, icilin, PIP$_2$	BCTC
TRPA1	Cold ($<17°C$), allyl isothiocyanate, cinnamaldehyde, carvacrol, allicin, Δ^9-tetrahydrocannabinol, mechanical stimulus	Ruthenium red, camphor, amiloride, Gd^{3+}, La^{3+}

NADA, *N*-arachidonoyl dopamine; OLDA, *N*-oleoyl dopamine; APB, 2-aminoethoxy-diphenyl borate; BCTC, *N*-(4-tertiarybutylphenyl)-4-(3-cholorphyridin-2-yl)tetra-hydropyrazine-1(2*H*)-carbox-amide; 4αPDD, 4-phorbol 12,13 didecanoate; EETs, epoxyeicosatrienoic acids

Na$^+$ channels (Fig. 1). Several kinds of TRP channels are reportedly expressed exclusively in subsets of sensory neurons, suggesting their involvement in detecting external stimuli (Table 1).

1
Capsaicin Receptor TRPV1

Capsaicin is known to depolarize nociceptors and increase their cytosolic free-Ca^{2+} concentration (Szallasi and Blumberg 1999). This property allowed Julius and colleagues to isolate the gene encoding the capsaicin receptor by using a Ca^{2+} imaging-based expression strategy in 1997 (Caterina et al. 1997). The cloned receptor was designated a vanilloid receptor subtype-1 (VR1) because a vanilloid moiety constitutes an essential chemical component of both capsaicin and its ultra potent analog, resiniferatoxin (Caterina and Julius 2001). The capsaicin receptor VR1 is now called TRPV1 as the first member of the TRPV subfamily of ion channels. A functional analysis using a patch-clamp technique revealed that heterologously expressed TRPV1 exhibits membrane current activation upon capsaicin application whose properties were almost identical to those observed in native dorsal root ganglion (DRG) neurons,

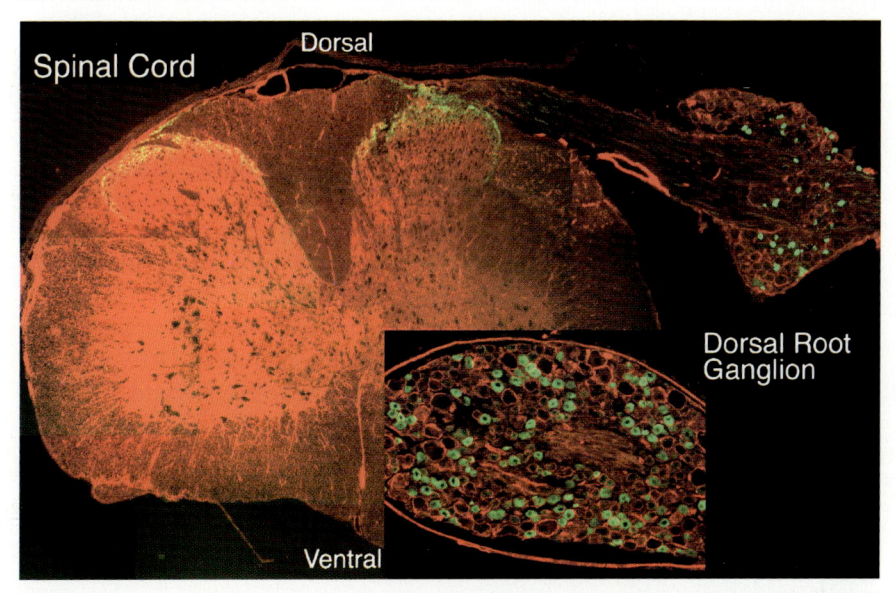

Fig. 1 Expression of TRPV1 proteins in rat lumbar spinal cord (indicated in *green*) detected by anti-TRPV1 antibody. TRPV1 expression in dorsal root ganglion (DRG) neurons, superficial layers of dorsal horn, and nerve fibers is observed. TRPV1 is expressed in predominantly in small-diameter neurons, probably cell bodies of unmyelinated C fibers

suggesting that TRPV1 protein consists of a homomeric tetramer, although Fas-associated factor 1 was reported to bind to TRPV1 as a regulatory factor (Kim et al. 2006). Upon application of capsaicin to excised membrane patches, clear single channel openings were observed, indicating that no cytosolic second messengers were necessary for TRPV1 activation by capsaicin. Furthermore, it became clear that capsaicin-activated currents showed nonselective cation permeability with an outwardly rectifying current–voltage (I–V) relationship and that Ca^{2+} was found to be about ten times more permeable than Na^+ which is effective for release of substance P (SP) or calcitonin gene-related peptide (CGRP), a phenomenon called neurogenic inflammation.

It is known that heat stimulus also induces pain sensation through the activation of polymodal nociceptors. TRPV1 was found to be activated by heat with a threshold of about 43°C, a temperature causing pain responses in vivo, suggesting that TRPV1 is involved in the detection of painful heat. Tissue acidification induced in inflammation and ischemia is also thought to exacerbate or cause pain. Extracellular acidification (pH ∼6.4) potentiated both capsaicin- and heat-evoked TRPV1 currents by increasing agonist potency without altering efficacy (Tominaga et al. 1998). When the proton concentration was increased to a level of pH 6.0, TRPV1 could be activated by the proton concentration itself at room temperature. The concentration range needed for activating TRPV1 is well attainable in local acidosis associated with tissue injury. Both heat- and proton-evoked TRPV1 currents were observed in excised membrane patches, indicating that the activation process does not involve cytosolic second messengers. Thus, TRPV1 can be activated by different stimuli causing pain in vivo. While capsaicin is an exogenous ligand for TRPV1, it remains possible that there exist endogenous, pain-producing, chemical regulators for this channel, other than protons. Several candidates for such endogenous ligands have been identified (Table 1), including anandamide, lipoxygenase products, oleoylethanolamide, N-arachidonoyl dopamine (NADA) and N-oleoyl dopamine (OLDA) (Zygmunt et al. 1999; Hwang et al. 2000; Huang et al. 2002; Ahern 2003; Chu et al. 2003). A number of these compounds are released during tissue injury and therefore may have a role in TRPV1 activation during inflammatory pain and hyperalgesia (De Petrocellis and Di Marzo 2005). Anandamide acts both at cannabinoid CB1 receptors and TRPV1, and has been shown to act as an intracellular messenger that would provide an indirect means for metabotropic receptor amplification of TRPV1-mediated Ca^{2+} influx (van der Stelt et al. 2005). Further, camphor isolated from the wood of the camphor laurel tree has been recently found to activate TRPV1 (Xu et al. 2005).

Expression of TRPV1 was observed in the terminals of primary afferent neurons projecting to the superficial layers of the spinal cord dorsal horn and the trigeminal nucleus caudalis in addition to primary afferent neurons' cell bodies, although peripheral nerve endings are the main parts in which TRPV1 detects noxious stimuli (Fig. 1; Tominaga et al. 1998). TRPV1 immunoreactivity was also observed in the nucleus of the solitary tract and area postrema, which receive vagal projections from visceral organs through the nodose ganglion. These observations, plus the fact that TRPV1 protein was detected in nerve terminals, indicates that TRPV1 is expressed in both central and peripheral termini of small-diameter sensory neurons. In the central nervous system, TRPV1 expression was observed in the brain (Mezey et al. 2000), although TRPV1 function in the brain is not clear. However, the fact that NADA, which is present predominantly in dopaminergic neurons in the brain, can activate TRPV1 and that capsaicin or anandamide enhances neuronal excitability or influences synaptic transmission might suggest that TRPV1 is associated with synaptic plasticity in the central nervous system (Marinelli et al. 2003). TRPV1 in the spinal cord (in the presynaptic terminals) might also be involved in synaptic transmission. It has been reported that inflammation or tissue injury induced increase in the number of unmyelinated C fibers expressing TRPV1 (Carlton and Coggeshall 2001). Furthermore, increase of TRPV1 expression induced by inflammation was predominantly observed in myelinated Aδ fibers compared to C fibers (Amaya et al. 2003). Nerve growth factor (NGF) induced increase of TRPV1 messenger RNA (mRNA) and release of CGRP with capsaicin treatment in primary cultured DRG neurons (Winston et al. 2001). Activation of p38 mitogen-activated protein kinase (MAPK) by NGF was found to enhance the translocation of TRPV1 proteins from cell bodies in DRG to sensory nerve endings (Ji et al. 2002). Furthermore, p38 MAPK inhibitor reduced inflammatory hyperalgesia. Thus, in addition to the phosphorylation of TRPV1 described in the following section, increase of TRPV1 expression in the sensory nerve endings seems to be involved in the development of hyperalgesia as well.

Electrophysiological analyses in a heterologous expression system revealed the importance of TRPV1 in detecting noxious thermal and chemical stimuli. To determine whether TRPV1 really contributes to the detection of these noxious stimuli in vivo, mice lacking this protein were generated and analyzed for nociceptive function (Caterina et al. 2000; Davis et al. 2000). Sensory neurons from mice lacking TRPV1 were deficient in their responses to each of the reported noxious stimuli: capsaicin, protons, and heat. Consistent with this observation, behavioral responses to capsaicin were absent and responses to acute thermal stimuli were diminished in these mice. In contrast, TRPV1 knockout mice showed normal physiological and behavioral responses to noxious mechanical stimuli, implying the existence of other mechanisms for the detection of such stimuli. The most prominent feature of the knockout mouse thermosensory phenotype was a virtual absence of thermal hypersensitivity in the setting of inflammation.

These findings indicate that TRPV1 is essential for selective modalities of pain sensation and for tissue injury-induced thermal hyperalgesia. Recent studies have provided further support for TRPV1 involvement in inflammatory pain and have demonstrated that the participation of TRPV1 in pain sensation may also extend to neuropathic pain, mechanical allodynia, and mechanical hyperalgesia. These conclusions are based on enhanced expression of TRPV1 in sensory neurons in the context of these conditions, as well as behavioral effects of TRPV1 antagonism with capsazepine, a competitive antagonist of TRPV1.

1.1
TRPV1 and Inflammatory Pain

One important aspect of TRPV1 regulation that has received considerable attention concerns the mechanisms by which the inflammatory mediators in damaged tissues sensitize TRPV1 to its chemical and physical stimuli. The fact that one of the most impressive phenotypes observed in TRPV1-deficient mice was attenuation of thermal hyperalgesia may be related to this aspect. Whereas protons act directly on TRPV1, others influence TRPV1 indirectly such as through receptors with intrinsic Tyr kinase activity, G protein-coupled receptors, or receptors coupled to the JAK/STAT (Janus kinase/signal transducers and activators of transcription) signaling pathway. Like other ion channels, TRPV1 can be phosphorylated by several kinases including protein kinase (PK)A (Bhave et al. 2002; Rathee et al. 2002), PKC (Premkumar and Ahern 2000; Tominaga et al. 2001), Ca^{2+}/calmodulin (CaM)-dependent kinase II

(CaM kinase II) (Jung et al. 2004), and Src kinase (Jin et al. 2004; Zhang et al. 2005). There has been extensive work demonstrating that activation of a PKA-dependent pathway by inflammatory mediators such as prostaglandins influences capsaicin- or heat-mediated actions in sensory neurons, probably by acting on TRPV1 (Lopshire and Nicol 1998; Distler et al. 2003). These results suggest that PKA plays a pivotal role in the development of hyperalgesia and inflammation by inflammatory mediators. PKC-dependent phosphorylation of TRPV1 occurs downstream of activation of Gq-coupled receptors by several inflammatory mediators including ATP, bradykinin, prostaglandins, and trypsin or tryptase (Sugiura et al. 2002; Moriyama et al. 2003; Dai et al. 2004; Moriyama et al. 2005). PKC-dependent phosphorylation of TRPV1 caused not only potentiation of capsaicin- or proton-evoked responses but it also reduced the temperature threshold for TRPV1 activation so that normally nonpainful temperatures in the range of normal body temperature were capable of activating TRPV1. In this way, activation of TRPV1 at normal body temperature could lead to the sensation of pain. These phenomena were also confirmed in native sensory neurons. Direct phosphorylation of TRPV1 by PKC was proved biochemically, and two target Ser residues were identified (Numazaki et al. 2002).

A reduction of direct PIP_2 (phosphatidylinositol-4,5-bisphosphate)-mediated TRPV1 inhibition and generation of lipoxygenase derived products through phospholipase A $(PLA)_2$ activation are also reported to be involved in the potentiation of TRPV1 activity following Gq-coupled receptor activation (Chuang et al. 2001). Both PKC-dependent and independent potentiation mechanisms might occur at downstream of Gq-coupled receptor activation. Inhibition of calcineurin inhibits desensitization of TRPV1 (Docherty et al. 1996), indicating that a phosphorylation/dephosphorylation process is important for TRPV1 activity. Indeed, CaM kinase II was reported to control TRPV1 activity upon phosphorylation of TRPV1 by regulating capsaicin binding (Jung et al. 2004). Thus, phosphorylation of TRPV1 by three different kinases seems to control TRPV1 activity through the dynamic balance between the phosphorylation and dephosphorylation.

1.2
TRPV1 and Visceral Pain

TRPV1 is also expressed in visceral afferent neurons in the gastrointestinal tract (Ward et al. 2003). TRPV1-expressing fibers appear to be predominantly of spinal origin although a few vagal TRPV1-positive fibers exist in the stomach, suggesting the involvement of TRPV1 in visceral nociception in normal and diseased conditions. In fact, TRPV1 expression has been reported to be increased in inflammatory diseases of the gastrointestinal tract and in patients with rectal hypersensitivity (Yiangou et al. 2001; Chan et al. 2003). A main stimulus causing visceral pain is a mechanical one that is not an effective stimulus for TRPV1 in vitro. What activates TRPV1 in the sensory neurons in the gastrointestinal tract? Protons or candidate lipid ligands described above can activate TRPV1. Alternatively, body temperature can activate TRPV1 through PKC-mediated phosphorylation in the inflammation. Indeed, serotonin, existing extensively in the gastrointestinal tract, was found to facilitate TRPV1 activity in DRG neurons innervating the colon. Analyses using TRPV1-deficient mice will provide more information about TRPV1 function in the gastrointestinal tract.

1.3
TRPV1 as a Target of Analgesics

Capsaicin not only causes pain, but also seems to exhibit analgesic properties, particularly when used to treat pain associated with diabetic neuropathies or rheumatoid arthritis (Szallasi and Blumberg 1999). This paradoxical effect may relate to the ability of capsaicin to desensitize nociceptive terminals to capsaicin, as well as to other noxious stimuli following prolonged exposure. At the molecular level, an extracellular Ca^{2+}-dependent reduction of TRPV1 responsiveness upon continuous vanilloid exposure (electrophysiological desensitization) may partially underlie this phenomenon, although physical damage to the nerve terminal or depletion of CGRP or SP probably contributes to this effect as well. Ca^{2+}- and voltage-dependent desensitization of capsaicin-activated currents has also been observed in rat DRG neurons. This inactivation of nociceptive neurons by capsaicin has generated extensive research on the possible therapeutic effectiveness of capsaicin as a clinical analgesic tool. Desensitization to capsaicin is a complex process with varying kinetic components: a fast component that appears to depend on Ca^{2+} influx through TRPV1 and a slow component that does not. Calcineurin inhibitors reduce TRPV1 desensitization (the slow component), indicating the involvement of a Ca^{2+}-dependent phosphorylation/dephosphorylation process (Docherty et al. 1996). In addition, PKA-dependent phosphorylation of TRPV1 has been reported to mediate the slow component of TRPV1 desensitization (Bhave et al. 2002). CaM has also been reported to be involved in Ca^{2+}-dependent desensitization of TRPV1 through its binding to the cytosolic domains of TRPV1 (Numazaki et al. 2003; Rosenbaum et al. 2004).

2
TRPV2

TRPV2 might be a potential candidate for the receptor detecting high heat stimulus responsible for the residual high temperature-evoked nociceptive responses observed in TRPV1-deficient mice (Table 1). TRPV2, with about 50% identity to TRPV1, was found to be activated by high temperatures with a threshold of approximately 52°C (Caterina et al. 1999). TRPV2 currents showed similar properties to those of TRPV1, such as an outwardly rectifying I–V relationship and relatively high Ca^{2+} permeability. Intense TRPV2 immunoreactivity was observed in medium- to large-diameter cells in rat DRG neurons, and very few of the TRPV2-positive cells stained with the isolectin IB4 or with SP antibody (Lewinter et al. 2004). However, about one-third of the TRPV2-positive cells co-stained with antibody against CGRP. Many of the TRPV2-positive neurons co-stained with a marker for myelinated neurons. Dense TRPV2 immunoreactivity in the spinal cord was found in lamina I, inner lamina II, and laminae III/IV. This is consistent with the expression of TRPV2 in myelinated nociceptors that target laminae I and inner lamina II and in non-nociceptive Aβ fibers that target laminae III/IV.

Aδ mechano- and heat-sensitive (AMH) neurons in monkey are medium- to large-diameter, lightly myelinated neurons that fall into two groups: type I has a heat threshold of approximately 53°C, and type II is activated at 43°C (Treede et al. 1995). The TRPV2 localization data and the residual high temperature-evoked responses observed in the TRPV1-deficient mice suggest that TRPV2 expression might account for the high thermal threshold ascribed to type I AMH nociceptors. Temperatures activating TRPV2 are more harmful to our body than those activating TRPV1. Therefore, expression of TRPV2 in the myelinated sensory fibers seems reasonable because Aδ fibers can transmit the nociceptive information much faster than C fibers, which exclusively express TRPV1. It has been recently reported that expression of TRPV2 is increased after inflammation, suggesting the involvement of TRPV2 in the development of inflammation-induced hyperalgesia to noxious high temperature stimuli (Shimosato et al. 2005). The function of TRPV2 in Aβ fibers is not known. Studies using mice lacking TRPV2 will tell us how important TRPV2 is for detecting noxious thermal stimulus.

TRPV2 transcript and protein were found not only in sensory neurons but also in motoneurons and in many nonneuronal tissues that are unlikely to be exposed to temperature above 50°C. These results indicate that TRPV2 undoubtedly contributes to numerous functions in addition to nociceptive processing.

3
TRPV3

TRPV3 has been found to be activated by warm temperatures (∼34°C–38°C), and to be expressed in several tissues such as keratinocytes, sensory neurons, and brain (Peier et al. 2002b; Smith et al. 2002; Xu et al. 2002). Involvement of TRPV3 in nociception was suggested from the several observations that it is transcribed from a gene adjacent to *TRPV1* in the chromosome 17p13, is co-expressed with TRPV1 in DRG neurons, is able to associate with TRPV1 when heterologously expressed, and TRPV3 currents are sensitized upon repeated heat stimuli. Mice lacking TRPV3 showed strong deficits in response to innocuous and noxious heat, suggesting the involvement of TRPV3 in acute thermal nociception, although thermal hyperalgesia in the context of inflammation seems not to involve TRPV3 (Moqrich et al. 2005). TRPV3 was also shown to be chemesthetic receptor in oral and nasal epithelium and to be a molecular target for plant-derived skin sensitizer (Table 1), because TRPV3 was activated by carvacrol, eugenol, and thymol, major ingredient of oregano, savory, and thyme, respectively (Xu et al. 2006).

4
TRPV4

TRPV4 has been found to be activated by warm temperatures (~27°C–35°C) in addition to hypotonic stimulus in heterologous expression systems (Guler et al. 2002; Watanabe et al. 2002), and to be expressed in multiple tissues, including, among others, sensory and hypothalamic neurons and keratinocytes. TRPV4 is also known to be activated by epoxyeicosatrienoic acids (EETs), the cytochrome P450 epoxygenase products (Table 1), suggesting the TRPV4 involvement in controlling vasoconstriction since EETs cause vasodilation and account for endothelium-derived hyperpolarizing factor activity in some vascular beds. Several approaches, including the knockdown of TRPV4 with gene-disruption or antisense oligonucleotides, have led to reports that this protein is involved in mechanical stimulus- and hypotonicity-induced nociception in rodents at baseline or following hypersensitivity induced by prostaglandin injection or taxol neurotoxicity (Alessandri-Haber et al. 2003; Suzuki et al. 2003; Alessandri-Haber et al. 2004). Both TRPV3 and TRPV4 have been reported to be involved in thermosensation by keratinocytes, based on studies of wildtype and TRPV3- or TRPV4-deficient keratinocytes (Chung et al. 2004). However, there is no physiological evidence yet that TRPV4 is involved in thermal nociception.

5
TRPM8

A distinct class of cold-sensitive fibers has been described as polymodal nociceptors, responding to noxious cold, heat, and pinch (Campero et al. 1996) although an early study suggested a distinction between cold sensation and painful sensation (Klement and Arndt 1992). The cooling sensation of menthol, a chemical agent found in mint, is well established, and both cooling and menthol was suggested to be transduced through a nonselective cation channel in DRG neurons (Reid and Flonta 2001). Two groups independently cloned and characterized a cold receptor, TRPM8, which can also be activated by menthol (McKemy et al. 2002; Peier et al. 2002a). In heterologous expression systems, TRPM8 could be activated by menthol or by cooling, with an activation temperature of approximately 28°C, and the cold-activated currents were increased in magnitude down to 8°C, thus spanning both innocuous cool and noxious cold temperatures. The TRPM8-mediated cold- and menthol-induced currents were reminiscent of those observed in native sensory neurons regarding electrophysiological properties, suggesting that TRPM8 protein consists of a homomeric tetramer such as TRPV1. There also appears to be interaction between effective stimuli for TRPM8, in that subthreshold concentrations of menthol increased the temperature threshold for TRPM8 activation from 25°C to 30°C. This is reminiscent of TRPV1, whose activation temperature is reduced under mildly acidic conditions that do not open TRPV1 alone. Patch-clamp analysis revealed that TRPM8 is a nonselective cation channel with relatively high Ca^{2+} permeability and that TRPM8 shows an outwardly rectifying I–V relationship similar to TRPV1.

TRPM8 could alternatively be activated by other cooling compounds, such as menthone, eucalyptol and icilin (Table 1). However, icilin can activate TRPM8 only in the presence of extracellular Ca^{2+} (McKemy et al. 2002). Further studies clarified that TRPM8 acted as a coincidence detector of icilin and Ca^{2+}, and mapped a critical amino acid residue required for icilin activity to a single glutamine in the third transmembrane domain, although the glutamine residue does not seem to be involved in menthol and cold sensitivity (Chuang et al. 2004), suggesting the distinct mechanisms for TRPM8 activation depend on ligands. Interestingly, the corresponding region was mapped as capsaicin-binding site in TRPV1 (Jordt and Julius 2002), suggesting a conserved mechanism for ligand activation of the two thermosensitive channels. In terms of menthol sensitivity, two regions were identified (Bandell et al. 2006) from a high-throughput random mutagenesis screening: the second transmembrane domain and C-terminal TRP domain of TRPM8.

TRPM8 activity was reported to be regulated by several factors. Lowering of the intracellular pH to below 7.0 was able to completely block TRPM8 currents elicited by either cold or icilin, but not menthol (Andersson et al. 2004). However, it is not clear whether the pH dependence of TRPM8 can be related to acidification observed in inflammation or tissue injury. TRPM8 is also regulated by PIP_2. Interestingly, however, PIP_2 was found to activate TRPM8, unlike the action on TRPV1 (Liu and Qin 2005; Rohacs et al. 2005). Furthermore, PIP_2-interacting sites were mapped to conserved positive residues in the highly conserved proximal C-terminal TRP domain, suggesting that regulation by PIP_2 is a common feature of members of the TRP channel family. PKC-mediated phosphorylation was reported to downregulate TRPM8 (Premkumar et al. 2005). In the context of inflammation, it is expected that PIP_2 hydrolysis is accelerated and PKC is activated downstream of Gq-coupled receptor activation, both of which cause inactivation of TRPM8. Therefore, these regulation mechanisms suggest that inflammatory thermal hyperalgesia mediated by TRPV1 could be further aggravated by downregulation of TRPM8, since activation of TRPM8 provided the much-needed cool sensation. If this is the case, TRPM8 and TRPV1 located at the peripheral terminals (and perhaps at the central terminals in the spinal cord as well) could play a significant role in controlling the gain of painful input.

TRPM8 is expressed in a subset of DRG and TG neurons that can be classified as small-diameter C fiber sensory neurons (less than 15% of small-diameter neurons), consistent with the population of excitable cells shown to be cold- and menthol-sensitive (McKemy et al. 2002; Peier et al. 2002a). Interestingly, however, TRPM8 is not coexpressed with nociceptor markers such as TRPV1, neurofilament, and CGRP, indicating that TRPM8 is not expressed in a class of afferents historically considered to be nociceptors. Whether TRPM8 is involved in cold nociception remains to be clarified.

6
TRPA1

TRPA1 was reported as a distantly related TRP channel that is activated by cold, with a lower activation threshold as compared to TRPM8 (Story et al. 2003). In heterologous expression systems, TRPA1 was activated by cold stimuli with an activation temperature of about 17°C, which is close to the reported noxious cold threshold. This finding led to the suggestion that TRPA1 is involved in cold nociception. Indeed, TRPA1 has been found to be involved in cold hyperalgesia caused by inflammation or nerve injury via activation of MAPK (Obata et al. 2005). Patch-clamp analysis of TRPA1 revealed cationic permeability and an outwardly rectifying I–V relationship, although cold-evoked current activation has not been clearly shown, and one group reported no TRPA1-mediated cold-evoked responses in heterologous expression system (Jordt et al. 2004). Whether TRPA1 is activated directly by cold remains to be elucidated.

TRPA1 was also found to be activated by pungent isothiocyanate compounds such as those found in wasabi, horseradish, and mustard oil, cinnamonaldehyde, a main pungent ingredient of cinnamon, or allicin, a pungent ingredient of garlic (Bandell et al. 2004; Jordt et al. 2004; Macpherson et al. 2005; Table 1). TRPA1 was reported to be a target of bradykinin signaling; an endogenous pro-algesic agent is released during inflammation and acts via the Gq-coupled bradykinin receptor B2R in vitro (Bandell et al. 2004), emphasizing the importance of TRPA1 in inflammation. TRPA1 activity was found to be inhibited by camphor, an agonist for TRPV1 and TRPV3. It might be one of the reasons why camphor is commonly applied to the skin for analgesic and counterirritant properties (Burkhart and Burkhart 2003).

Unlike TRPM8, TRPA1 is specifically expressed in a subset of sensory neurons that express the nociceptive markers CGRP and SP. Furthermore, TRPA1 is frequently coexpressed with TRPV1, raising the possibility that TRPA1 and TRPV1 mediate the function of a class of polymodal nociceptors. Such a coexpression might also explain the paradoxical hot sensation experienced when one is exposed to a very cold stimulus. Interestingly, strong TRPA1 expression was also found in hair cell in the inner ear (Corey et al. 2004). Furthermore, TRPA1-mediated current properties were similar to those reported for mechanotransduction channels in hair cells (Nagata et al. 2005), suggesting the possibility that TRPA1 is a promising candidate for the mechanosensitive transduction channel of hair cells in the inner ear.

Thus, TRPA1 was assumed to have abilities to detect a wide variety of chemical and physical stimuli, and studies using TRPA1-deficient mice have been awaited. Two independent analyses have shown that TRPA1 is required in mice for sensing endogenous pain-causing substances, pungent natural compounds, mechanical stimuli, and environmental irritants (acrolein), although the cold-sensing phenotype is different between the two studies (Bautista et al. 2006; Kwan et al. 2006). Interestingly, both studies showed no significant role of TRPA1 in hearing. One group observed marked deficits in sensation of acute noxious cold and mechanical sensation in mice lacking TRPA1, whereas these deficits were not detected by another group (Bautista et al. 2006; Kwan et al. 2006). The threshold of mechanically induced pain after bradykinin-mediated inflammation was higher in TRPA1-deficient mice. Although direct mechanical activation of TRPA1 has not yet been demonstrated, knockout mice data strongly suggest that TRPA1 mediates responses to high-threshold mechanical stimuli in peripheral nociceptors. Further analyses of the mice are clearly necessary to clarify the contribution of TRPA1 to the detection of nociceptive stimuli, especially of cold. Regardless, the knockout mice studies have shown that TRPA1 is required for sensing various noxious stimuli and make TRPA1 a potential target for the development of new analgesics.

References

Ahern GP (2003) Activation of TRPV1 by the satiety factor oleoylethanolamide. J Biol Chem 278:30429–30434

Alessandri-Haber N, Yeh JJ, Boyd AE, Parada CA, Chen X, Reichling DB, Levine JD (2003) Hypotonicity induces TRPV4-mediated nociception in rat. Neuron 39:497–511

Alessandri-Haber N, Dina OA, Yeh JJ, Parada CA, Reichling DB, Levine JD (2004) Transient receptor potential vanilloid 4 is essential in chemotherapy-induced neuropathic pain in the rat. J Neurosci 24:4444–4452

Amaya F, Oh-hashi K, Naruse Y, Iijima N, Ueda M, Shimosato G, Tominaga M, Tanaka Y, Tanaka M (2003) Local inflammation increases vanilloid receptor 1 expression within distinct subgroups of DRG neurons. Brain Res 963:190–196

Andersson DA, Chase HW, Bevan S (2004) TRPM8 activation by menthol, icilin, and cold is differentially modulated by intracellular pH. J Neurosci 24:5364–5369

Bandell M, Story GM, Hwang SW, Viswanath V, Eid SR, Petrus MJ, Earley TJ, Patapoutian A (2004) Noxious cold ion channel TRPA1 is activated by pungent compounds and bradykinin. Neuron 41:849–857

Bandell M, Dubin AE, Petrus MJ, Orth A, Mathur J, Hwang SW, Patapoutian A (2006) High-throughput random mutagenesis screen reveals TRPM8 residues specifically required for activation by menthol. Nat Neurosci 9:493–500

Bautista DM, Jordt SE, Nikai T, Tsuruda PR, Read AJ, Poblete J, Yamoah EN, Basbaum AI, Julius D (2006) TRPA1 mediates the inflammatory actions of environmental irritants and proalgesic agents. Cell 124:1269–1282

Bhave G, Zhu W, Wang H, Brasier DJ, Oxford GS, Gereau RWt (2002) cAMP-dependent protein kinase regulates desensitization of the capsaicin receptor (VR1) by direct phosphorylation. Neuron 35:721–731

Burkhart CG, Burkhart HR (2003) Contact irritant dermatitis and anti-pruritic agents: the need to address the itch. J Drugs Dermatol 2:143–146

Campero M, Serra J, Ochoa JL (1996) C-polymodal nociceptors activated by noxious low temperature in human skin. J Physiol 497:565–572

Carlton SM, Coggeshall RE (2001) Peripheral capsaicin receptors increase in the inflamed rat hindpaw: a possible mechanism for peripheral sensitization. Neurosci Lett 310:53–56

Caterina MJ, Julius D (1999) Sense and specificity: a molecular identity for nociceptors. Curr Opin Neurobiol 9:525–530

Caterina MJ, Julius D (2001) The vanilloid receptor: a molecular gateway to the pain pathway. Annu Rev Neurosci 24:487–517

Caterina MJ, Schumacher MA, Tominaga M, Rosen TA, Levine JD, Julius D (1997) The capsaicin receptor: a heat-activated ion channel in the pain pathway. Nature 389:816–824

Caterina MJ, Rosen TA, Tominaga M, Brake AJ, Julius D (1999) A capsaicin-receptor homologue with a high threshold for noxious heat. Nature 398:436–441

Caterina MJ, Leffler A, Malmberg AB, Martin WJ, Trafton J, Petersen-Zeitz KR, Koltzenburg M, Basbaum AI, Julius D (2000) Impaired nociception and pain sensation in mice lacking the capsaicin receptor. Science 288:306–313

Chan CL, Facer P, Davis JB, Smith GD, Egerton J, Bountra C, Williams NS, Anand P (2003) Sensory fibres expressing capsaicin receptor TRPV1 in patients with rectal hypersensitivity and faecal urgency. Lancet 361:385–391

Chu CJ, Huang SM, De Petrocellis L, Bisogno T, Ewing SA, Miller JD, Zipkin RE, Daddario N, Appendino G, Di Marzo V, Walker JM (2003) N-Oleoyldopamine, a novel endogenous capsaicin-like lipid that produces hyperalgesia. J Biol Chem 278:13633–13639

Chuang HH, Prescott ED, Kong H, Shields S, Jordt SE, Basbaum AI, Chao MV, Julius D (2001) Bradykinin and nerve growth factor release the capsaicin receptor from PtdIns(4,5)P2-mediated inhibition. Nature 411:957–962

Chuang HH, Neuhausser WM, Julius D (2004) The super-cooling agent icilin reveals a mechanism of coincidence detection by a temperature-sensitive TRP channel. Neuron 43:859–869

Chung MK, Lee H, Mizuno A, Suzuki M, Caterina MJ (2004) TRPV3 and TRPV4 mediate warmth-evoked currents in primary mouse keratinocytes. J Biol Chem 279:21569–21575

Corey DP, Garcia-Anoveros J, Holt JR, Kwan KY, Lin SY, Vollrath MA, Amalfitano A, Cheung EL, Derfler BH, Duggan A, Geleoc GS, Gray PA, Hoffman MP, Rehm HL, Tamasauskas D, Zhang DS (2004) TRPA1 is a candidate for the mechanosensitive transduction channel of vertebrate hair cells. Nature 432:723–730

Dai Y, Moriyama T, Higashi T, Togashi K, Kobayashi K, Yamanaka H, Tominaga M, Noguchi K (2004) Proteinase-activated receptor 2-mediated potentiation of transient receptor potential vanilloid subfamily 1 activity reveals a mechanism for proteinase-induced inflammatory pain. J Neurosci 24:4293–4299

Davis JB, Gray J, Gunthorpe MJ, Hatcher JP, Davey PT, Overend P, Harries MH, Latcham J, Clapham C, Atkinson K, Hughes SA, Rance K, Grau E, Harper AJ, Pugh PL, Rogers DC, Bingham S, Randall A, Sheardown SA (2000) Vanilloid receptor-1 is essential for inflammatory thermal hyperalgesia. Nature 405:183–187

De Petrocellis L, Di Marzo V (2005) Lipids as regulators of the activity of transient receptor potential type V1 (TRPV1) channels. Life Sci 77:1651–1666

Distler C, Rathee PK, Lips KS, Obreja O, Neuhuber W, Kress M (2003) Fast Ca^{2+}-induced potentiation of heat-activated ionic currents requires cAMP/PKA signaling and functional AKAP anchoring. J Neurophysiol 89:2499–2505

Docherty RJ, Yeats JC, Bevan S, Boddeke HW (1996) Inhibition of calcineurin inhibits the desensitization of capsaicin-evoked currents in cultured dorsal root ganglion neurones from adult rats. Pflugers Arch 431:828–837

Guler AD, Lee H, Iida T, Shimizu I, Tominaga M, Caterina M (2002) Heat-evoked activation of the ion channel, TRPV4. J Neurosci 22:6408–6414

Huang SM, Bisogno T, Trevisani M, Al-Hayani A, De Petrocellis L, Fezza F, Tognetto M, Petros TJ, Krey JF, Chu CJ, Miller JD, Davies SN, Geppetti P, Walker JM, Di Marzo V (2002) An endogenous capsaicin-like substance with high potency at recombinant and native vanilloid VR1 receptors. Proc Natl Acad Sci U S A 99:8400–8405

Hwang SW, Cho H, Kwak J, Lee SY, Kang CJ, Jung J, Cho S, Min KH, Suh YG, Kim D, Oh U (2000) Direct activation of capsaicin receptors by products of lipoxygenases: endogenous capsaicin-like substances. Proc Natl Acad Sci U S A 97:6155–6160

Ji RR, Samad TA, Jin SX, Schmoll R, Woolf CJ (2002) p38 MAPK activation by NGF in primary sensory neurons after inflammation increases TRPV1 levels and maintains heat hyperalgesia. Neuron 36:57–68

Jin X, Morsy N, Winston J, Pasricha PJ, Garrett K, Akbarali HI (2004) Modulation of TRPV1 by nonreceptor tyrosine kinase, c-Src kinase. Am J Physiol Cell Physiol 287:C558–563

Jordt SE, Julius D (2002) Molecular basis for species-specific sensitivity to "hot" chili peppers. Cell 108:421–430

Jordt SE, Bautista DM, Chuang HH, McKemy DD, Zygmunt PM, Hogestatt ED, Meng ID, Julius D (2004) Mustard oils and cannabinoids excite sensory nerve fibres through the TRP channel ANKTM1. Nature 427:260–265

Julius D, Basbaum AI (2001) Molecular mechanisms of nociception. Nature 413:203–210

Jung J, Shin JS, Lee SY, Hwang SW, Koo J, Cho H, Oh U (2004) Phosphorylation of vanilloid receptor 1 by Ca^{2+}/calmodulin-dependent kinase II regulates its vanilloid binding. J Biol Chem 279:7048–7054

Kim S, Kang C, Shin CY, Hwang SW, Yang YD, Shim WS, Park MY, Kim E, Kim M, Kim BM, Cho H, Shin Y, Oh U (2006) TRPV1 recapitulates native capsaicin receptor in sensory neurons in association with Fas-associated factor 1. J Neurosci 26:2403–2412

Klement W, Arndt JO (1992) The role of nociceptors of cutaneous veins in the mediation of cold pain in man. J Physiol 449:73–83

Kwan KY, Allchorne AJ, Vollrath MA, Christensen AP, Zhang DS, Woolf CJ, Corey DP (2006) TRPA1 contributes to cold, mechanical, and chemical nociception but is not essential for hair-cell transduction. Neuron 50:277–289

Lewinter RD, Skinner K, Julius D, Basbaum AI (2004) Immunoreactive TRPV-2 (VRL-1), a capsaicin receptor homolog, in the spinal cord of the rat. J Comp Neurol 470:400–408

Liu B, Qin F (2005) Functional control of cold- and menthol-sensitive TRPM8 ion channels by phosphatidylinositol 4,5-bisphosphate. J Neurosci 25:1674–1681

Lopshire JC, Nicol GD (1998) The cAMP transduction cascade mediates the prostaglandin E2 enhancement of the capsaicin-elicited current in rat sensory neurons: whole-cell and single-channel studies. J Neurosci 18:6081–6092

Macpherson LJ, Geierstanger BH, Viswanath V, Bandell M, Eid SR, Hwang S, Patapoutian A (2005) The pungency of garlic: activation of TRPA1 and TRPV1 in response to allicin. Curr Biol 15:929–934

Marinelli S, Di Marzo V, Berretta N, Matias I, Maccarrone M, Bernardi G, Mercuri NB (2003) Presynaptic facilitation of glutamatergic synapses to dopaminergic neurons of the rat substantia nigra by endogenous stimulation of vanilloid receptors. J Neurosci 23:3136–3144

McCleskey EW, Gold MS (1999) Ion channels of nociception. Annu Rev Physiol 61:835–856

McKemy DD, Neuhausser WM, Julius D (2002) Identification of a cold receptor reveals a general role for TRP channels in thermosensation. Nature 416:52–58

Mezey E, Toth ZE, Cortright DN, Arzubi MK, Krause JE, Elde R, Guo A, Blumberg PM, Szallasi A (2000) Distribution of mRNA for vanilloid receptor subtype 1 (VR1), and VR1-like immunoreactivity, in the central nervous system of the rat and human. Proc Natl Acad Sci U S A 97:3655–3660

Moqrich A, Hwang SW, Earley TJ, Petrus MJ, Murray AN, Spencer KS, Andahazy M, Story GM, Patapoutian A (2005) Impaired thermosensation in mice lacking TRPV3, a heat and camphor sensor in the skin. Science 307:1468–1472

Moriyama T, Iida T, Kobayashi K, Higashi T, Fukuoka T, Tsumura H, Leon C, Suzuki N, Inoue K, Gachet C, Noguchi K, Tominaga M (2003) Possible involvement of P2Y2 metabotropic receptors in ATP-induced transient receptor potential vanilloid receptor 1-mediated thermal hypersensitivity. J Neurosci 23:6058–6062

Moriyama T, Higashi T, Togashi K, Iida T, Segi E, Sugimoto Y, Tominaga T, Narumiya S, Tominaga M (2005) Sensitization of TRPV1 by EP1 and IP reveals peripheral nociceptive mechanism of prostaglandins. Mol Pain 1:3–12

Nagata K, Duggan A, Kumar G, Garcia-Anoveros J (2005) Nociceptor and hair cell transducer properties of TRPA1, a channel for pain and hearing. J Neurosci 25:4052–4061

Numazaki M, Tominaga T, Toyooka H, Tominaga M (2002) Direct phosphorylation of capsaicin receptor VR1 by protein kinase Cepsilon and identification of two target serine residues. J Biol Chem 277:13375–13378

Numazaki M, Tominaga T, Takeuchi K, Murayama N, Toyooka H, Tominaga M (2003) Structural determinant of TRPV1 desensitization interacts with calmodulin. Proc Natl Acad Sci U S A 100:8002–8006

Obata K, Katsura H, Mizushima T, Yamanaka H, Kobayashi K, Dai Y, Fukuoka T, Tokunaga A, Tominaga M, Noguchi K (2005) TRPA1 induced in sensory neurons contributes to cold hyperalgesia after inflammation and nerve injury. J Clin Invest 115:2393–2401

Peier AM, Moqrich A, Hergarden AC, Reeve AJ, Andersson DA, Story GM, Earley TJ, Dragoni I, McIntyre P, Bevan S, Patapoutian A (2002a) A TRP channel that senses cold stimuli and menthol. Cell 108:705–715

Peier AM, Reeve AJ, Andersson DA, Moqrich A, Earley TJ, Hergarden AC, Story GM, Colley S, Hogenesch JB, McIntyre P, Bevan S, Patapoutian A (2002b) A heat-sensitive TRP channel expressed in keratinocytes. Science 296:2046–2049

Premkumar LS, Ahern GP (2000) Induction of vanilloid receptor channel activity by protein kinase C. Nature 408:985–990

Premkumar LS, Raisinghani M, Pingle SC, Long C, Pimentel F (2005) Downregulation of transient receptor potential melastatin 8 by protein kinase C-mediated dephosphorylation. J Neurosci 25:11322–11329

Rathee PK, Distler C, Obreja O, Neuhuber W, Wang GK, Wang SY, Nau C, Kress M (2002) PKA/AKAP/VR-1 module: a common link of gs-mediated signaling to thermal hyperalgesia. J Neurosci 22:4740–4745

Reid G, Flonta ML (2001) Physiology. Cold current in thermoreceptive neurons. Nature 413:480

Rohacs T, Lopes CM, Michailidis I, Logothetis DE (2005) PI(4,5)P2 regulates the activation and desensitization of TRPM8 channels through the TRP domain. Nat Neurosci 8:626–634

Rosenbaum T, Gordon-Shaag A, Munari M, Gordon SE (2004) Ca^{2+}/calmodulin modulates TRPV1 activation by capsaicin. J Gen Physiol 123:53–62

Scholz J, Woolf CJ (2002) Can we conquer pain? Nat Neurosci 5 Suppl:1062–1067

Shimosato G, Amaya F, Ueda M, Tanaka Y, Decosterd I, Tanaka M (2005) Peripheral inflammation induces up-regulation of TRPV2 expression in rat DRG. Pain 119:225–232

Smith GD, Gunthorpe MJ, Kelsell RE, Hayes PD, Reilly P, Facer P, Wright JE, Jerman JC, Walhin JP, Ooi L, Egerton J, Charles KJ, Smart D, Randall AD, Anand P, Davis JB (2002) TRPV3 is a temperature-sensitive vanilloid receptor-like protein. Nature 418:186–190

Story GM, Peier AM, Reeve AJ, Eid SR, Mosbacher J, Hricik TR, Earley TJ, Hergarden AC, Andersson DA, Hwang SW, McIntyre P, Jegla T, Bevan S, Patapoutian A (2003) ANKTM1, a TRP-like channel expressed in nociceptive neurons, is activated by cold temperatures. Cell 112:819–829

Sugiura T, Tominaga M, Katsuya H, Mizumura K (2002) Bradykinin lowers the threshold temperature for heat activation of vanilloid receptor 1. J Neurophysiol 88:544–548

Suzuki M, Mizuno A, Kodaira K, Imai M (2003) Impaired pressure sensation in mice lacking TRPV4. J Biol Chem 278:22664–22668

Szallasi A, Blumberg PM (1999) Vanilloid (capsaicin) receptors and mechanisms. Pharmacol Rev 51:159–212

Tominaga M, Caterina MJ, Malmberg AB, Rosen TA, Gilbert H, Skinner K, Raumann BE, Basbaum AI, Julius D (1998) The cloned capsaicin receptor integrates multiple pain-producing stimuli. Neuron 21:531–543

Tominaga M, Wada M, Masu M (2001) Potentiation of capsaicin receptor activity by metabotropic ATP receptors as a possible mechanism for ATP-evoked pain and hyperalgesia. Proc Natl Acad Sci U S A 98:6951–6956

Treede RD, Meyer RA, Raja SN, Campbell JN (1995) Evidence for two different heat transduction mechanisms in nociceptive primary afferents innervating monkey skin. J Physiol 483:747–758

van der Stelt M, Trevisani M, Vellani V, De Petrocellis L, Schiano Moriello A, Campi B, McNaughton P, Geppetti P, Di Marzo V (2005) Anandamide acts as an intracellular messenger amplifying Ca^{2+} influx via TRPV1 channels. EMBO J 24:3026–3037

Ward SM, Bayguinov J, Won KJ, Grundy D, Berthoud HR (2003) Distribution of the vanilloid receptor (VR1) in the gastrointestinal tract. J Comp Neurol 465:121–135

Watanabe H, Vriens J, Suh SH, Benham CD, Droogmans G, Nilius B (2002) Heat-evoked activation of TRPV4 channels in a HEK293 cell expression system and in native mouse aorta endothelial cells. J Biol Chem 277:47044–47051

Winston J, Toma H, Shenoy M, Pasricha PJ (2001) Nerve growth factor regulates VR-1 mRNA levels in cultures of adult dorsal root ganglion neurons. Pain 89:181–186

Wood JN, Perl ER (1999) Pain. Curr Opin Genet Dev 9:328–332

Woolf CJ, Salter MW (2000) Neuronal plasticity: increasing the gain in pain. Science 288:1765–1769
Xu H, Ramsey IS, Kotecha SA, Moran MM, Chong JA, Lawson D, Ge P, Lilly J, Silos-Santiago I, Xie Y, DiStefano PS, Curtis R, Clapham DE (2002) TRPV3 is a calcium-permeable temperature-sensitive cation channel. Nature 418:181–186
Xu H, Blair NT, Clapham DE (2005) Camphor activates and strongly desensitizes the transient receptor potential vanilloid subtype 1 channel in a vanilloid-independent mechanism. J Neurosci 25:8924–8937
Xu H, Delling M, Jun JC, Clapham DE (2006) Oregano, thyme and clove-derived flavors and skin sensitizers activate specific TRP channels. Nat Neurosci 9:628–635
Yiangou Y, Facer P, Dyer NH, Chan CL, Knowles C, Williams NS, Anand P (2001) Vanilloid receptor 1 immunoreactivity in inflamed human bowel. Lancet 357:1338–1339
Zhang X, Huang J, McNaughton PA (2005) NGF rapidly increases membrane expression of TRPV1 heat-gated ion channels. EMBO J 24:4211–4223
Zygmunt PM, Petersson J, Andersson DA, Chuang H, Sorgard M, Di Marzo V, Julius D, Hogestatt ED (1999) Vanilloid receptors on sensory nerves mediate the vasodilator action of anandamide. Nature 400:452–457

Regulation of TRP Ion Channels by Phosphatidylinositol-4,5-Bisphosphate

F. Qin

Department of Physiology and Biophysics, State University of New York at Buffalo, Buffalo NY, , USA
qin@buffalo.edu

1	Introduction	510
2	PIP$_2$ and Its Metabolism	512
3	Interactions of PIP$_2$ with Proteins	513
4	TRPC Channels	514
5	TRPV Channels	515
6	TRPM Channels	518
7	TRPA1	520
8	TRP Channels as PLC-Activating Receptors	521
9	Perspectives	521
	References	522

Abstract Phosphatidylinositol-4,5-bisphosphate (PIP$_2$) has emerged as a versatile regulator of TRP ion channels. In many cases, the regulation involves interactions of channel proteins with the lipid itself independent of its hydrolysis products. The functions of the regulation mediated by such interactions are diverse. Some TRP channels absolutely require PIP$_2$ for functioning, while others are inhibited. A change of gating is common to all, endowing the lipid a role for modulation of the sensitivity of the channels to their physiological stimuli. The activation of TRP channels may also influence cellular PIP$_2$ levels via the influx of Ca^{2+} through these channels. Depletion of PIP$_2$ in the plasma membrane occurs upon activation of TRPV1, TRPM8, and possibly TRPM4/5 in heterologous expression systems, whereas resynthesis of PIP$_2$ requires Ca^{2+} entry through the TRP/TRPL channels in *Drosophila* photoreceptors. These developments concerning PIP$_2$ regulation of TRP channels reinforce the significance of the PLC signaling cascade in TRP channel function, and provide further perspectives for understanding the physiological roles of these ubiquitous and often enigmatic channels.

Keywords TRP channels · Regulation · PIP$_2$ · Lipid · Gating · Nociceptor · Thermoreceptor

1
Introduction

Lipids in plasma membranes can have functions far beyond forming permeability barriers. Among them are phosphatidylinositides, a minority group constituting less than 5% of the total membrane lipids (Fig. 1). The versatile biological roles of phosphoinositides were first appreciated when the inositol 1,4,5-triphosphate (IP$_3$) signaling system was discovered (Berridge and Irvine 1984). Stimulation of the phospholipase C (PLC)-activating receptors causes hydrolysis of phosphatidylinositol 4,5-bisphosphate (PIP$_2$), a most abundant variant of inositide lipids in the inner leaflet of membranes. Two products are generated: DAG and IP$_3$, both of which are essential signaling messengers. IP$_3$ diffuses into cytosols to activate Ca^{2+} release from stores, whereas DAG remains on membranes to recruit and activate cytoplasmic proteins such as protein kinase (PKC). Numerous surface receptors exploit the inositide signaling pathways to convey their messages into cells.

Aside from production of second messengers, PIP$_2$ itself may interact with many cellular components. With one end hydrophobic and the other charged, the lipid serves as an ideal adapter to anchor cytoplasmic proteins to plasma membranes. Actin-binding proteins are among the first shown to be associated with PIP$_2$ (Lassing and Lindberg 1985). The finding of the pleckstrin homology

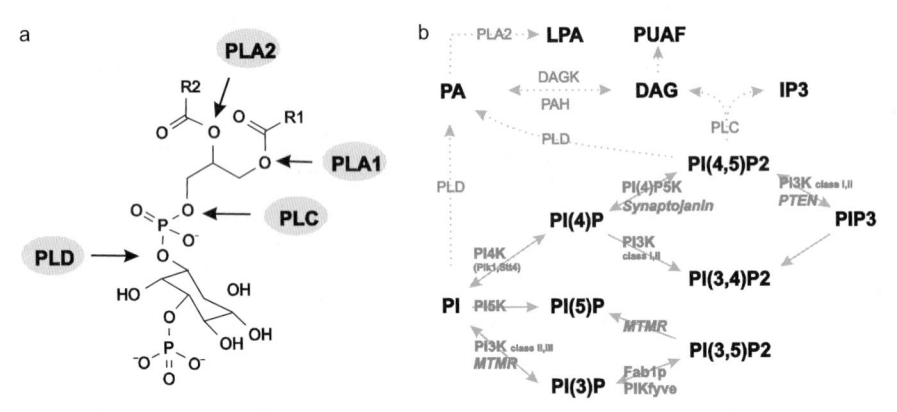

Fig. 1 a,b Phosphoinositides. **a** The structure of a phosphorylated derivative of the phospholipid phosphatidylinositol (phosphatidylinositol 3-monophosphate). The R groups are long-chain fatty acids and for natural phosphoinositides, R1 is typically stearoyl, and R2 is arachidonoyl diacylglycerol. The *arrows* indicate the positions for hydrolysis by phospholipases A1, A2, C, and D. **b** Phosphoinositide metabolism. The major phosphorylation and dephosphorylation pathways are shown. Also shown are lipases and associated reactions. *DAG*, diacylglycerols; *DAGK*, diacylglycerol kinase; *MTMR*, myotubularin-related proteins; *PI3K*, phosphoinositide 3-kinase; *PI4K*, phosphoinositide 4-kinase; *PI5K*, phosphoinositide 5-kinase; *PIKfyve*, PI(3)P 5-kinase; *PI(4)P5P*, PI(4)P 5-kinase; *PLA2*, cytosolic phospholipase A2; *PLD*, phospholipase D; *PUFA*, polyunsaturated fatty acid

domain and other related protein modules now brings more than 100 proteins potentially interacting with PIP$_2$ (Lemmon and Ferguson 2000) and has led to the development of powerful probes for detection and visualization of the subcellular localization and translocation of the lipid within cells (Varnai and Balla 1998; Hirose et al. 1999). To date, PIP$_2$ has been shown to regulate such diverse processes as cell signaling, vesicle trafficking, cytoskeleton remodeling, and so on (Fig. 2).

PIP$_2$ as a direct signaling molecule is not limited to interacting only with intracellular proteins for membrane targeting. Proteins that are located in the membrane are also subject to its attack. Pioneering work in this regard has been done on the Na$^+$/Ca^{2+} exchangers and the ATP-sensitive K$_{ATP}$ channels (Hilgemann and Ball 1996; Huang et al. 1998). Transmembrane proteins such as ion channels generally contain large intracellular regions, which may inevitably expose polar patches on the surface. Thus, it makes sense intuitively that interactions may take place between PIP$_2$ and these proteins, especially in view of their close proximity. However, it comes as a surprise that such interactions are not merely regulatory but often obligatory for channel functions. Recent studies have revealed an increasing number of ion channels interacting with

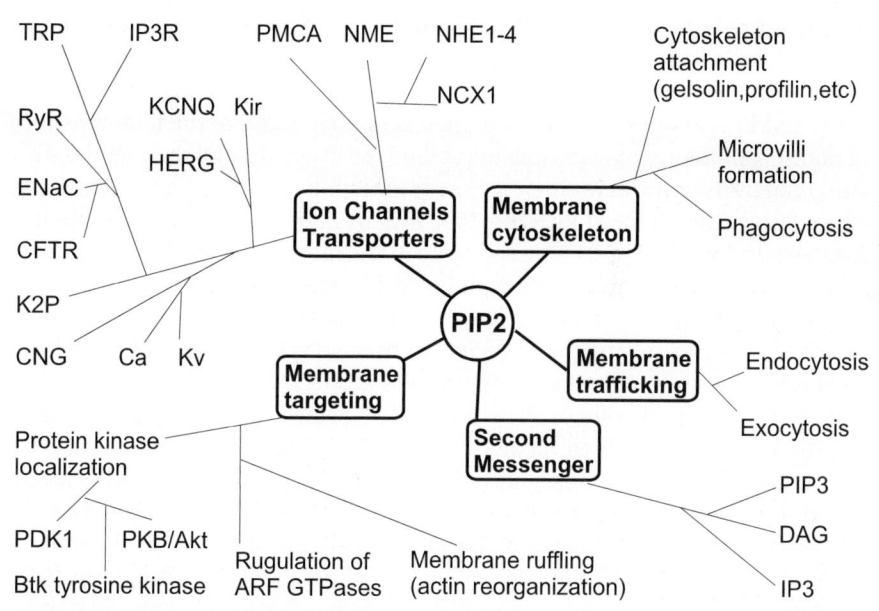

Fig. 2 Some functions of PIP$_2$ in the plasma membrane. PIP$_2$ is involved in modulation of a host of transmembrane proteins including transporters and ion channels, production of second messengers in response to receptor stimulation, clathrin coating and uncoating of vesicles in endocytosis and exocytosis, anchoring of actin binding proteins for membrane attachment to cytoskeleton, localization of protein kinases, regulation of membrane ruffling, and so on

PIP$_2$, and the members of the TRP family are among the latest. Several excellent reviews are available with detailed coverage of PIP$_2$-regulated ion channels (Hilgemann et al. 2001; Suh and Hille 2005) and the biophysical principles of the interactions between PIP$_2$ and proteins (McLaughlin et al. 2002). Here we will focus on the recent advances concerning TRP channels.

2
PIP$_2$ and Its Metabolism

Contrary to their rich functions, phosphoinositides amount to only a small fraction of total membrane lipids. Phosphatidylinositol (PI) is the precursor of all phosphoinositides. The lipid can be reversibly phosphorylated at one or a combination of the hydroxyl groups at positions 3, 4, or 5 of the inositol ring, resulting in eight forms of phosphoinositides, comprising the nonphosphorylated, three singly phosphorylated, three doubly phosphorylated, and one triply phosphorylated, all of which exist in the biological membranes (Hinchliffe and Irvine 1997).

The different forms of phosphoinositides are maintained at dramatically different levels in mammalian cells under basal conditions (Cullen et al. 2001). PIP$_2$ comprises 1%–5% of the phospholipids in the plasma membrane. Phosphatidylinositol and phosphatidylinositol 4-monophosphate [PI(4)P] are two other major components, where PI(4)P exists in roughly equal amounts as PIP$_2$, and PI is about 10–20 times higher. Less than 0.25% of the total inositol-containing lipids are phosphorylated at the 3-position, though their levels (e.g., PIP$_3$) can transiently increase enormously in response to receptor stimulations. The relative abundance of PIP$_2$ together with its high number of charges in the head group ($\sim$3 at pH 7.0) renders it one of the most active lipids in cells.

Enzymes responsible for phosphoinositide metabolism include lipid kinases, phosphomonoesterases, and phosphodiesterases (Fig. 1). Different forms of phosphoinositides can be interconverted via lipid kinases and phosphatases by adding or removing phosphates at specific positions. Two kinases involved in the synthesis of PIP$_2$ are PI(4)K, which phosphorylates PI into PI(4)P, and PIP(5)K, which further converts PI(4)P into PIP$_2$ (Tolias and Carpenter 2000). Pharmacological inhibition of PI(4)K downregulates PIP$_2$ levels in many cells. For a long time this was considered the only route of PIP$_2$ synthesis. But recently an alternative pathway involving PIP(4)K has been reported (Rameh et al. 1997). Another lipid kinase, PI(3)K, converts PIP$_2$ into PIP$_3$ (Cantley 2002). The degradation of PIP$_2$ can also occur via phosphodiesterase cleavage by PLCs or dephosphorylation by phosphomonoesterases. The latter likely involves those of type II of the inositol 5-phosphatase family, which removes 5-phosphate from the inositol ring (Woscholski and Parker 1997; Majerus et al. 1999). Among the many kinases and phosphatases involved in PIP$_2$ metabolism, PLC and PI(3)K appear to be most susceptible to regulation

(Rhee 2001; Cantley 2002). The production of PIP_2 is generally slow, in the order of minutes, and requires a high concentration of ATP in the millimolar range as opposed to the submicromolar range for protein kinases (Hilgemann 1997; Varnai and Balla 1998). The degradation of PIP_2 is relatively fast; PLC-catalyzed hydrolysis of PIP_2 proceeds on a time course of seconds (Varnai and Balla 1998).

3
Interactions of PIP_2 with Proteins

There are two types of interactions between PIP_2 and proteins: stereospecific (structured) and electrostatic (unstructured). The structured interactions involve specialized protein domains such as PH, PX, FYVE, and ENTH (Lemmon and Ferguson 2000). The three-dimensional structures of a number of PH domains have been determined (Rebecchi and Scarlata 1998). Despite a low level of homology in sequence, the core structure of these domains is highly conserved, which consists of two sets of β-sheets stacked orthogonally with one comprising three and the other four strands. One side of the structure has a positively charged surface, forming a cleft at the positions of the variable loops. The region has been suggested as a binding site for negatively charged membrane surfaces. In the structure of $PLC\delta_1$ PH domain in complex with IP_3, the inositide ring binds to the cleft and makes extensive hydrogen bonds with the positively charged residues around the loops. It is noted that the PH domain from $PLC\delta_1$ is among the few exceptions in the family of PH domains for its stereospecific and high-affinity binding of PIP_2. The majority of PH domains bind phosphoinositides only weakly. As an extreme, the Bkt-PH domain binds specifically to PIP_3 instead of PIP_2.

Without stereospecific binding, electrostatic interactions alone can also produce high-affinity binding between PIP_2 and proteins. One well-characterized example is the effector domain of the myristoylated alanine-rich C kinase substrate (MARCKS) (McLaughlin et al. 2002). Consisting of a cluster of basic residues interspaced by aromatic ones, the peptide can be absorbed to a $PC/PD/PIP_2$ membrane and laterally sequesters PIP_2, even when the phosphatidylserine (PS) is present in excess of PIP_2. Consistent with electrostatic interactions, the peptide binds $PI(3,4)P_2$ and $PI(4,5)P_2$ with the same affinity, and also binds with high affinity to membranes containing physiological concentrations of monovalent acidic lipids. An important feature for the unstructured interaction is that the stoichiometry of the binding is not limited to 1:1. Theoretical calculations support that a single MARCK, despite charge screening by counterions, can sequester three PIP_2 molecules (Wang et al. 2004). A high concentration of such peptides can therefore potentially sequester PIP_2 into discrete pools with high local concentrations in the plasma membrane. Also characteristic of the unstructured interactions is the seemingly required aromatic residues embedded between basic residues. Hydrophilic peptides lacking aromatic residues generally fail to produce significant sequestration of PIP$_2$ in the model membranes. It is thought that the hydrophobic penetration of the aromatics into the acyl chain region of the membrane is necessary to help pull the adjacent basic residues close to or within the polar head group region (Gambhir et al. 2004).

4
TRPC Channels

The regulation of TRPC channels by PIP$_2$, either by the lipid itself or through its metabolic products, manifests in the involvement of PLC for activation of these channels (Harteneck et al. 2000; Vazquez et al. 2004). As early as when the founding member of the TRP family, the *Drosophila* TRP channel, was cloned (Montell and Rubin 1989), PLCβ was found essential for light-induced currents in fly photoreceptors (Bloomquist et al. 1988). The signaling pathways that link PLC activation and TRP/TRPL opening have been enigmatic. In view of the significance of inositide signaling in Ca^{2+} release, earlier studies have focused on store depletion as an activation mechanism (Scott and Zuker 1998; Montell and Rubin 1989). The model has since been challenged on multiple grounds, and the more recent studies of the mammalian homologs have further fueled the debate (Vazquez et al. 2004). Regardless of the controversy on the role of store depletion, the available evidence is generally consistent in that all TRPC channels are activated following receptor-mediated stimulation of PLCs. The PIP$_2$-derived lipid product, DAG, or its polyunsaturated fatty acid (PUFA) metabolites, has been shown to activate several TRPC channels, including dTRP, dTRPL, TRPC1, and the TRPC3, TRPC6, TRPC7 group (Hardie 2003). In most cases, the activation occurs in a membrane-delimited, PKC-independent manner at physiologically relevant concentrations.

That neither store depletion nor lipid products of PIP$_2$ are capable of fully duplicating the trp phenotype in *Drosophila* phototransduction has recently led to investigations of possible involvement of PIP$_2$ itself. TRPL channels expressed in heterologous systems are inhibited by application of exogenous PIP$_2$ to excised membrane patches (Estacion et al. 2001). The inhibition is reversible, is specific to PIP$_2$, and was not observed with PI, phosphatidylcholine (PC), PS, or phosphatidylethanolamine (PE). PIP$_2$ is also reported to suppress light-induced responses of another invertebrate (molluscan) photoreceptor (del Pilar Gomez and Nasi 2005). Additionally, TRP and TRPL channels in the native *Drosophila* photoreceptors are activated by metabolic stress or reduction of cellular ATP (Agam et al. 2000), which could cause depletion of PIP$_2$.

More direct insights into the role of PIP$_2$ in the *Drosophila* trp phenotype have been gained recently by tracking dynamic changes of PIP$_2$ levels in vivo (Hardie et al. 2001, 2004). A functional assay was cleverly developed using transgenic flies in which a human inward rectifying K$^+$ channel (Kir2.1) was genetically targeted to photoreceptors. All inward rectifiers have their functions dependent on phosphoinositides, but Kir 2.1 shows a particularly high specificity and affinity toward PIP$_2$ binding. The currents generated by the expressed Kir2.1 channels thus provide a readout of the PIP$_2$ levels in the plasma membrane. The results indicate that the depletion of PIP$_2$ or impaired resynthesis is correlated with the trp phenotype. Under normal conditions, PIP$_2$ levels were maintained even during saturating illumination, but when TRP channels were blocked, microvillar PIP$_2$ was rapidly depleted by even moderate intensities. As a model, it was suggested that the Ca^{2+} influx from TRP channels is required to maintain cellular PIP$_2$ levels by inhibition of background PLC activity. Further quantitative measurements corroborate a high level of in vivo PLC activity, equivalent to approximately 50% depletion of total PIP$_2$ min^{-1} basal and in excess of 100% s^{-1} after activation by light, and a relatively slow resynthesis rate with a half-time of 1 min (Hardie et al. 2004). According to the model, the TRP/TRPL channels are not only responsible for transduction of lights into electrical signals, but also provide a feedback mechanism to switch off PLCs immediately after the transduction.

5
TRPV Channels

TRPV1 is among the first TRP channels shown to be regulated by PIP$_2$ directly. As a pain receptor, the channel responds to a wide range of noxious stimuli, including proinflammatory substances, which are released upon tissue injury and inflammation. While some of the agents (e.g., protons) act on the channel directly, others work through intracellular signaling cascades. Examples of the latter are bradykinin and NGF, two principal components of the so-called "inflammatory soup." TRPV1-deficient mice do not develop thermal hypersensitivity in response to NGF or bradykinin treatment (Chuang et al. 2001), suggesting that TRPV1 is an essential downstream target for modulation by these proalgesic mediators.

A signaling pathway downstream of both NGF and bradykinin receptor activation is the PLC system. Bradykinin activates the BK2 receptors on nociceptive neurons, which are coupled to PLC via G proteins. Earlier studies in sensory neurons show that treatment with bradykinin increases both the amplitude and sensitivity of heat-activated currents in these cells (Cesare and McNaughton 1996). The effect has been ascribed to receptor phosphorylation by the ε-isoform of PKC. Consistent with this proposition, direct phosphorylation of TRPV1 is observed in vitro, and PKC-ε-knockout mice show attenuated thermal and mechanical hyperalgesia (Khasar et al. 1999).

The PKC-dependent mechanism, however, does not account for all aspects of PLC-mediated actions on TRPV1 (Chuang et al. 2001). Application of a recombinant PI-specific PLC (PI-PLC), for example, potentiates TRPV1 responses in the absence of ATP. Some of the common PKC inhibitors are ineffective in blocking the sensitization of the channel induced by inflammatory agents. Anti-PIP$_2$ monoclonal antibody, on the other hand, stimulates basal activity and enhances capsaicin responses when applied to the cytoplasmic side in excised inside-out membrane patches containing TRPV1. These data are consistent with a direct role for PIP$_2$ and suggest that the bradykinin-induced potentiation may partly arise from the breakdown of PIP$_2$ itself. The PLC pathway is also found essential to NGF-mediated potentiation of TRPV1. Only the mutation that disrupts the coupling between TrkA and PLC is found effective in abrogating NGF effects, whereas the mutation that uncouples TrkA and mitogen-activeated protein kinase (MAPK) is ineffectual. As an alternative to the phosphorylation model, the release of the channel from tonic inhibition of PIP$_2$ binding has been suggested as a common mechanism underlying the hypersensitivity of nociceptors in response to NGF or bradykinin treatment.

Mutagenesis experiments have mapped the site of PIP$_2$ binding to a region in the distal C-terminal of TRPV1 (Prescott and Julius 2003). The region is rich in positively charged residues. Mutations that are expected to weaken PIP$_2$–TRPV1 interactions reduce the thresholds for chemical or thermal stimuli. Truncation of the region renders the channel unresponsive to PLC-coupled receptor activation. Replacement of the region with the PIP$_2$ binding site of an inward-rectifying K$^+$ channel preserves PIP$_2$ sensitivity. In a recent intriguing study based on sequence analyses, the PIP$_2$ binding region of TRPV1 has been suggested as a partial PH domain (van Rossum et al. 2005). As shown with TRPC3, such a PH domain fragment may interact with a cognate fragment in another protein (e.g., PLCγ) to form a functional intermolecular PH domain. Whether TRPV1 binds PIP$_2$ independently or through this novel mode of protein–protein interaction remains to be elucidated.

It is of note that NGF can activate a plethora of intracellular signaling pathways. Other models for potentiation of TRPV1 by NGF have also been proposed. Previous findings in cultured sensory neurons have implicated PKA (Shu and Mendell 2001). A more recent study suggests PI-3 kinase as a predominant player, which activates Src kinase leading to tyrosine phosphorylation of TRPV1 (Zhang et al. 2005a). Tyrosine phosphorylation can serve as a trafficking signal for membrane proteins. Indeed, tyrosine phosphorylation of TRPV1 at an N-terminal residue enhanced its expression on surface membranes.

While the study of the sensitization of TRPV1 leads to the finding of PIP$_2$ inhibition of the channel, the study of the desensitization process reveals another prominent role of the lipid in the recovery of the channel from desensitization (Liu et al. 2005). Capsaicin responses of TRPV1 show strong desensitization when extracellular Ca^{2+} is present. The desensitization is peculiarly long-lasting and often irreversible under typical patch-clamp experimental conditions. The fact that topical capsaicin and other naturally occurring pungent compounds can be used as an analgesic for temporary pain relief has been thought to depend on the presence of such long refractory states of the channel (Szallasi and Blumberg 1999). The first hint for the involvement of PIP_2 in the recovery of TRPV1 comes from the finding that a high concentration of ATP can confer a complete recovery of a fully desensitized capsaicin response over a time course of several minutes. This profile of ATP concentration dependence and kinetics is reminiscent of that of PIP_2 resynthesis after depletion. Indeed, the manipulation of protein kinases, either stimulation or inhibition, produces no effect, while the inhibition of PIP_2 synthesis or the stimulation of its hydrolysis impedes recovery of channel activity. When the Kir 2.1 channel was coexpressed to monitor membrane PIP_2 levels, the activation of TRPV1 by capsaicin causes substantial suppression of Kir2.1 currents, and this suppressed activity recovers concurrently with the desensitized capsaicin response of TRPV1. These data suggest that the activation of TRPV1 results in potent depletion of PIP_2, presumably due to the Ca^{2+} influx, and that PIP_2 is required to recover the surface activity of the channel after desensitization. The slow replenishment of PIP_2 thus provides an explanation for the long-lasting nature of the desensitization of capsaicin responses of TRPV1. How PIP_2 mediates the recovery process remains unknown, but its binding at the distal C-terminal of the channel is not required because mutant TRPV1 channels lacking the region show no defect in the recovery process.

TRPV5 is the other member that has been shown to be regulated by PIP_2 in the TRPV subfamily (Lee et al. 2005). Within the phylogenetic tree of TRP channels, TRPV5 is most homologous to TRPV6, and the two channels are distinct for their high selectivity to Ca^{2+}. Also a prominent feature of these channels is their complex dependences on intracellular Mg^{2+} ions (Nilius et al. 2001; Hoenderop et al. 2003; Voets et al. 2003; Yeh et al. 2003; Lee et al. 2005). Three effects of Mg^{2+} can be distinguished: a fast voltage-dependent pore block, a slow reversible inhibition on a time course of tens of seconds, and an even slower irreversible inhibition with a time course of minutes. This last slow, irreversible inhibition has been attributed to reduction of PIP_2 levels in the membranes. Mg^{2+} has been implicated in activation of lipid phosphatases for depletion of PIP_2 in excised membrane patches. Many channels that are sensitive to PIP_2 also show a strong dependence on a physiological level of intracellular Mg^{2+}. On excised membrane patches containing TRPV5, application of Mg^{2+} solutions causes progressive rundown of currents, whereas application of exogenous PIP_2 recovers the rundown activity. The binding of PIP_2 to the channel also alters the slow reversible inhibition by Mg^{2+}, raising the possibility that PIP_2 may regulate the channel in vivo by modulating its sensitivity to intracellular Mg^{2+}. Whole-cell experiments confirm many of the observations in excised patches, with one notable exception: the Mg^{2+}-induced rundown becomes negligible in whole-cell conditions despite prolonged, direct perfusion of intracellular milieu. This observation helps explain one puzzle: some channels run down rapidly in patches

a fractional activity in excised membrane patches vs whole cells (Reid and Flonta 2002), raising concerns on whether the apparent thermal sensitivity is intrinsic to the channel itself or other regulatory sources. In an effort to understand the mechanism of the rundown of channel activity in excised patches, it was found that a cocktail solution containing fluoride, vanadate, and pyrophosphate could stabilize both menthol- and cold-evoked currents, whereas Mg^{2+}-containing solutions dramatically increased the rundown rate (Liu and Qin 2005). These agents are known to have effects on lipid phosphatases (Huang et al. 1998), thus implicating a role for PIP_2 in the rundown process. Indeed, manipulations of the endogenous PIP_2 levels produced predictable effects on channel activity. PI(4)P was also able to partially restore the rundown activity, albeit less effectively. The successful prevention of the rundown of the channel in excised patches then argues that the cold sensitivity of TRPM8 is a property of the channel itself. Interestingly, the cold and menthol responses of the channel were found to have significantly different rundown rates, suggesting that the rundown process is highly temperature-dependent (Liu and Qin 2005). Another study by Rohacs et al. (2005) has further identified the TRP box as a site for PIP_2 binding in TRPM8 and the depletion of PIP_2 as a mechanism for Ca^{2+}-dependent desensitization of the channel (see above). The TRP box, rich in positively charged residues, lies proximal to the C-terminal, a region that has been seen in other channels for PIP_2 binding. Mutations of several residues in the region produced profound effects on PIP_2 sensitivity of TRPM8. The TRP box is one of the few structural features conserved throughout the TRP family. Given PIP_2 sensitivity as a widespread property of these channels, it is tempting to speculate whether PIP_2 binding represents the function of this well-conserved domain. Preliminary study of TRPV5 is indeed consistent with the hypothesis (Rohacs et al. 2005; Lee et al. 2005).

7
TRPA1

TRPA1 (ANKTM1) is coexpressed with TRPV1 in nociceptive neurons. Bradykinin, an inflammatory peptide known to stimulate PLC activity in nociceptors, activates TRPA1 in cultured dorsal root ganglion (DRG) neurons (Bandell et al. 2004). Stimulation of coexpressed M1 muscarinic acetylcholine receptors also elicits a current in heterologous expression systems (Jordt et al. 2004). Further evidence implicating a role for PIP_2 in regulation of TRPA1 comes from the recent findings that the channel is a likely candidate for mechanotransduction in hair cells (Corey et al. 2004; Nagata et al. 2005), a function that has been shown to require PIP_2 (Hirono et al. 2004). Whether PIP_2 modulates TRPA1 via direct interactions or through its hydrolysis products has not been determined.

8
TRP Channels as PLC-Activating Receptors

In the cascade of cell signaling, ion channels are often considered as a passive end target. In the case of the regulation of TRP channels by PIP_2, for example, the signal flows from hormone receptors to PLCs, which then convey the message to channels. The communication is unidirectional. Recent developments in the PIP_2 regulation of TRP channels and others have revealed that the channel may play a more active role. The activity of *Drosophila* TRP/TRPL channels is essential to maintaining PIP_2 levels for phototransduction (Hardie et al. 2001). The activation of TRPV1 causes depletion of PIP_2 in the plasma membrane (Liu et al. 2005). Hydrolysis of PIP_2 is found concomitant to activation of TRPM8 (Rohacs et al. 2005). In these cases, the channel functions both as an electric transducer and a PLC-activating (inhibiting) receptor.

The signal that links the channel activation and PIP_2 metabolism is likely the Ca^{2+} influx through these channels. The enzymes that deplete PIP_2 are still enigmatic. The translocation of the PLCδ-PH domain resulting from TRPM8 activation suggests the involvement of PI-specific PLCs. Ca^{2+} activates all isoforms of PLC in vitro (Rhee 2001). But the activation is far less certain in living cells except for PLCδ (Horowitz et al. 2005), which is more sensitive to Ca^{2+} and can be stimulated by an increase of $[Ca^{2+}]$ within the physiological range. Colocalization of the channel and the enzyme will be essential for an optimal local $[Ca^{2+}]$. Although PLCs are leading candidates, other enzymes may be involved too, especially those whose products can synergize or activate PLCs. Examples include phospholipase D (PLD) and phospholipase A_2 (PLA_2). Ca^{2+} activation has been reported for some isoforms of both enzymes (Exton 1997; Dennis 1994). Thus it is possible that the influx of Ca^{2+} leads to the generation of multiple cofactors/activators that converge on PLC for its maximal stimulation.

9
Perspectives

Recent development has defined PIP_2 as an inescapable regulator for TRP channels. Present work has been mostly based on heterologous expression systems. Precise functions of the regulation await further corroboration in more relevant physiological contexts. The widespread roles of PIP_2 in the functions of TRP channels reinforce the significance of the PLC/TRP signaling cascade. The pathway is likely one of the earliest mechanism evolved for sensory transduction, as exemplified by *Drosophila* phototransduction and mammalian taste sensation. Even for channels like TRPV1, which has evolved with other direct activators, the pathway is still retained for fine-tuning of channel responses. Generality of the pathway as a physiological mechanism for TRP functions merits attention in future investigations. The finding that TRP channel activity may cause PIP$_2$ depletion is intriguing and endows these channels with possible cellular functions beyond membrane depolarization. Much remains to be learned about the effectiveness of the depletion of PIP$_2$ and its physiological consequences.

The stage is also set for rigorous investigations of mechanisms underlying the interaction between PIP$_2$ and TRP channels. At issue is the nature and structural basis of the interaction, the effects on channel gating, and the interplay with other stimuli. A recurring question that has been raised by many researchers is: How can a single lipid play so many roles? The recent findings on TRP channels seem to only further aggravate the puzzle. Channels that are regulated by PIP$_2$ in the family are highly diverse in their structure and function. It will be interesting to see whether a common mechanism can emerge out of this diversity.

References

Agam K, von Campenhausen M, Levy S, Ben-Ami HC, Cook B, Kirschfeld K, Minke B (2000) Metabolic stress reversibly activates the Drosophila light-sensitive channels TRP and TRPL in vivo. J Neurosci 20:5748–5755

Bandell M, Story GM, Hwang SW, Viswanath V, Eid SR, Petrus MJ, Earley TJ, Patapoutian A (2004) Noxious cold ion channel TRPA1 is activated by pungent compounds and bradykinin. Neuron 41:849–857

Berridge MJ, Irvine RF (1984) Inositol trisphosphate, a novel second messenger in cellular signal transduction. Nature 312:315–321

Bloomquist BT, Shortridge RD, Schneuwly S, Perdew M, Montell C, Steller H, Rubin G, Pak WL (1988) Isolation of a putative phospholipase C gene of Drosophila, norpA, and its role in phototransduction. Cell 54:723–733

Cantley LC (2002) The phosphoinositide 3-kinase pathway. Science 296:1655–1657

Cesare P, McNaughton P (1996) A novel heat-activated current in nociceptive neurons and its sensitization by bradykinin. Proc Natl Acad Sci U S A 93:15435–15439

Chuang HH, Prescott ED, Kong H, Shields S, Jordt SE, Basbaum AI, Chao MV, Julius D (2001) Bradykinin and nerve growth factor release the capsaicin receptor from PtdIns(4,5)P2-mediated inhibition. Nature 411:957–962

Corey DP, Garcia-Anoveros J, Holt JR, Kwan KY, Lin SY, Vollrath MA, Amalfitano A, Cheung EL, Derfler BH, Duggan A, Geleoc GS, Gray PA, Hoffman MP, Rehm HL, Tamasauskas D, Zhang DS (2004) TRPA1 is a candidate for the mechanosensitive transduction channel of vertebrate hair cells. Nature 432:723–730

Cullen PJ, Cozier GE, Banting G, Mellor H (2001) Modular phosphoinositide-binding domains—their role in signalling and membrane trafficking. Curr Biol 11:R882–R893

del Pilar Gomez M, Nasi E (2005) A direct signaling role for phosphatidylinositol 4,5-bisphosphate (PIP2) in the visual excitation process of microvillar receptors. J Biol Chem 280:16784–16789

Dennis EA (1994) Diversity of group types, regulation, and function of phospholipase A2. J Biol Chem 269:13057–13060

Estacion M, Sinkins WG, Schilling WP (2001) Regulation of Drosophila transient receptor potential-like (TrpL) channels by phospholipase C-dependent mechanisms. J Physiol 530:1–19

Exton JH (1997) Phospholipase D: enzymology, mechanisms of regulation, and function. Physiol Rev 77:303–320

Gambhir A, Hangyas-Mihalyne G, Zaitseva I, Cafiso DS, Wang J, Murray D, Pentyala SN, Smith SO, McLaughlin S (2004) Electrostatic sequestration of PIP2 on phospholipid membranes by basic/aromatic regions of proteins. Biophys J 86:2188–2207

Hardie RC (2003) Regulation of TRP channels via lipid second messengers. Annu Rev Physiol 65:735–759

Hardie RC, Raghu P, Moore S, Juusola M, Baines RA, Sweeney ST (2001) Calcium influx via TRP channels is required to maintain PIP2 levels in Drosophila photoreceptors. Neuron 30:149–159

Hardie RC, Gu Y, Martin F, Sweeney ST, Raghu P (2004) In vivo light-induced and basal phospholipase C activity in Drosophila photoreceptors measured with genetically targeted phosphatidylinositol 4,5-bisphosphate-sensitive ion channels (Kir2.1). J Biol Chem 279:47773–47782

Harteneck C, Plant TD, Schultz G (2000) From worm to man: three subfamilies of TRP channels. Trends Neurosci 23:159–166

Hilgemann DW (1997) Cytoplasmic ATP-dependent regulation of ion transporters and channels: mechanisms and messengers. Annu Rev Physiol 59:193–220

Hilgemann DW, Ball R (1996) Regulation of cardiac Na$^+$,Ca^{2+} exchange and KATP potassium channels by PIP2. Science 273 5277:956–959

Hilgemann DW, Feng S, Nasuhoglu C (2001) The complex and intriguing lives of PIP2 with ion channels and transporters. Sci STKE 2001:RE19

Hinchliffe K, Irvine R (1997) Inositol lipid pathways turn turtle. Nature 390:123–124

Hirono M, Denis CS, Richardson GP, Gillespie PG (2004) Hair cells require phosphatidylinositol 4,5-bisphosphate for mechanical transduction and adaptation. Neuron 44:309–320

Hirose K, Kadowaki S, Tanabe M, Takeshima H, Iino M (1999) Spatiotemporal dynamics of inositol 1,4,5-trisphosphate that underlies complex Ca^{2+} mobilization patterns. Science 284:1527–1530

Hoenderop JG, Voets T, Hoefs S, Weidema F, Prenen J, Nilius B, Bindels RJ (2003) Homo- and heterotetrameric architecture of the epithelial Ca^{2+} channels TRPV5 and TRPV6. EMBO J 22:776–785

Horowitz LF, Hirdes W, Suh BC, Hilgemann DW, Mackie K, Hille B (2005) Phospholipase C in living cells: activation, inhibition, Ca^{2+} requirement, and regulation of M current. J Gen Physiol 126:243–262

Huang CL, Feng S, Hilgemann DW (1998) Direct activation of inward rectifier potassium channels by PIP2 and its stabilization by Gbetagamma. Nature 3919:803–806

Jordt SE, Bautista DM, Chuang HH, McKemy DD, Zygmunt PM, Hogestatt ED, Meng ID, Julius D (2004) Mustard oils and cannabinoids excite sensory nerve fibres through the TRP channel ANKTM1. Nature 427:260–265

Khasar SG, McCarter G, Levine JD (1999) Epinephrine produces a beta-adrenergic receptor-mediated mechanical hyperalgesia and in vitro sensitization of rat nociceptors. J Neurophysiol 81:1104–1112

Lassing I, Lindberg U (1985) Specific interaction between phosphatidylinositol 4,5-bisphosphate and profilactin. Nature 314:472–474

Lee J, Cha SK, Sun TJ, Huang CL (2005) PIP2 activates TRPV5 and releases its inhibition by intracellular Mg^{2+}. J Gen Physiol 126:439–451

Lemmon MA, Ferguson KM (2000) Signal-dependent membrane targeting by pleckstrin homology (PH) domains. Biochem J 350:1–18

Liu B, Qin F (2005) Functional control of cold- and menthol-sensitive TRPM8 ion channels by phosphatidylinositol 4,5-bisphosphate. J Neurosci 25:1674–1681

Liu B, Zhang C, Qin F (2005) Functional recovery from desensitization of vanilloid receptor TRPV1 requires resynthesis of phosphatidylinositol 4,5-bisphosphate. J Neurosci 25:4835–4843

Liu D, Liman ER (2003) Intracellular Ca^{2+} and the phospholipid PIP2 regulate the taste transduction ion channel TRPM5. Proc Natl Acad Sci U S A 100:15160–15165

Majerus PW, Kisseleva MV, Norris FA (1999) The role of phosphatases in inositol signaling reactions. J Biol Chem 274:10669–10672

McLaughlin S, Wang J, Gambhir A, Murray D (2002) PIP(2) and proteins: interactions, organization, and information flow. Annu Rev Biophys Biomol Struct 31:151–175

Montell C, Rubin GM (1989) Molecular characterization of the Drosophila trp locus: a putative integral membrane protein required for phototransduction. Neuron 2:1313–1323

Nagata K, Duggan A, Kumar G, Garcia-Anoveros J (2005) Nociceptor and hair cell transducer properties of TRPA1, a channel for pain and hearing. J Neurosci 25:4052–4061

Nilius B, Vennekens R, Prenen J, Hoenderop JG, Droogmans G, Bindels RJ (2001) The single pore residue Asp542 determines Ca^{2+} permeation and Mg^{2+} block of the epithelial Ca^{2+} channel. J Biol Chem 276:1020–1025

Prescott ED, Julius D (2003) A modular PIP2 binding site as a determinant of capsaicin receptor sensitivity. Science 300:1284–1288

Rameh LE, Tolias KF, Duckworth BC, Cantley LC (1997) A new pathway for synthesis of phosphatidylinositol-4,5-bisphosphate. Nature 390:192–196

Rebecchi MJ, Scarlata S (1998) Pleckstrin homology domains: a common fold with diverse functions. Annu Rev Biophys Biomol Struct 27:503–528

Reid G, Flonta ML (2002) Ion channels activated by cold and menthol in cultured rat dorsal root ganglion neurones. Neurosci Lett 324:164–168

Rhee SG (2001) Regulation of phosphoinositide-specific phospholipase C. Annu Rev Biochem 70:281–312

Rohacs T, Lopes CM, Michailidis I, Logothetis DE (2005) PI(4,5)P2 regulates the activation and desensitization of TRPM8 channels through the TRP domain. Nat Neurosci 8:626–634

Runnels LW, Yue L, Clapham DE (2002) The TRPM7 channel is inactivated by PIP(2) hydrolysis. Nat Cell Biol 4:329–336

Scott K, Zuker C (1998) TRP, TRPL and trouble in photoreceptor cells. Curr Opin Neurobiol 8:383–388

Shu X, Mendell LM (2001) Acute sensitization by NGF of the response of small-diameter sensory neurons to capsaicin. J Neurophysiol 86:2931–2938

Suh BC, Hille B (2005) Regulation of ion channels by phosphatidylinositol 4,5-bisphosphate. Curr Opin Neurobiol 15:370–378

Szallasi A, Blumberg PM (1999) Vanilloid (capsaicin) receptors and mechanisms. Pharmacol Rev 51:159–212

Tolias KF, Carpenter CL (2000) Enzymes involved in the synthesis of PI(4,5)P2 and their regulation: PI kinases and PIP kinases. In: Cockcroft S (ed) Biology of phosphoinositides. Oxford University Press, Oxford

van Rossum DB, Patterson RL, Sharma S, Barrow RK, Kornberg M, Gill DL, Snyder SH (2005) Phospholipase Cgamma1 controls surface expression of TRPC3 through an intermolecular PH domain. Nature 434:99–104

Varnai P, Balla T (1998) Visualization of phosphoinositides that bind pleckstrin homology domains: calcium- and agonist-induced dynamic changes and relationship to myo-[3H]inositol-labeled phosphoinositide pools. J Cell Biol 143:501–510

Vazquez G, Wedel BJ, Aziz O, Trebak M, Putney JW Jr (2004) The mammalian TRPC cation channels. Biochim Biophys Acta 1742:21–36

Voets T, Janssens A, Prenen J, Droogmans G, Nilius B (2003) Mg^{2+}-dependent gating and strong inward rectification of the cation channel TRPV6. J Gen Physiol 121:245–260

Wang J, Gambhir A, McLaughlin S, Murray D (2004) A computational model for the electrostatic sequestration of PI(4,5)P2 by membrane-adsorbed basic peptides. Biophys J 86:1969–1986

Woscholski R, Parker PJ (1997) Inositol lipid 5-phosphatases—traffic signals and signal traffic. Trends Biochem Sci 22:427–431

Yeh BI, Sun TJ, Lee JZ, Chen HH, Huang CL (2003) Mechanism and molecular determinant for regulation of rabbit transient receptor potential type 5 (TRPV5) channel by extracellular pH. J Biol Chem 278:51044–51052

Zhang X, Huang J, McNaughton PA (2005a) NGF rapidly increases membrane expression of TRPV1 heat-gated ion channels. EMBO J 24:4211–4223

Zhang Z, Okawa H, Wang Y, Liman ER (2005b) Phosphatidylinositol 4,5-bisphosphate rescues TRPM4 channels from desensitization. J Biol Chem 280:39185–39192

TRPC, cGMP-Dependent Protein Kinases and Cytosolic Ca^{2+}

X. Yao

Department of Physiology, Faculty of Medicine, The Chinese University of Hong Kong, Hong Kong China
yao2068@cuhk.edu.hk

1	Introduction	528
2	PKG: Structure and Activation Mechanism	528
3	The Regulation of TRPC by PKG	529
3.1	PKG Phosphorylation Sites in TRPC Proteins	529
3.2	Inhibition of TRPC3 Activity by PKG	531
3.3	Involvement of PKA?	532
3.4	PKC Inhibition of TRPC Is Partly Mediated by PKG	533
4	PKG-Mediated Negative Feedback Control of [Ca^{2+}]$_i$	534
5	Functional Role of PKG-Mediated Regulation of TRPC	536
6	Concluding Remarks	537
References		537

Abstract Ca^{2+}, nitric oxide (NO), and protein kinase G (PKG) are important signaling molecules that play pivotal roles in many physiological processes such as vascular tone control, platelet activation, and synaptic plasticity. TRPC channels allow Ca^{2+} influx, thus contributing to the production of NO, which subsequently stimulates PKG. It has been demonstrated that PKG can phosphorylate human TRPC3 at Thr-11 and Ser-263 and that this phosphorylation inactivates TRPC3. These two PKG phosphorylation sites, Thr-11 and Ser-263 in human TRPC3, are conserved in other members of the TRPC3/6/7 subfamily, suggesting that PKG may also phosphorylate TRPC6 and TRPC7. In addition, protein kinase C (PKC) also inactivates TRPC3, partly through activating PKG. The PKG-mediated inhibition of TRPC channels may provide a feedback control for the fine tuning of [Ca^{2+}]$_i$ levels and protect the cells from the detrimental effects of excessive [Ca^{2+}]$_i$ and/or NO.

Keywords Protein kinase G · TRPC channel · Cytosolic Ca^{2+} · Protein kinase C · Nitric oxide

1
Introduction

There are seven members of the TRPC channel family, which can be divided into four subfamilies: TRPC1, TRPC2, TRPC4/5, and TRPC3/6/7 (Montell 2005; Minke and Cook 2002). Functional channels are believed to be composed of four subunits. Each subunit contains the predicted topology of six transmembrane segments with three or four ankyrin repeats at the NH$_2$ terminal domain and a proline-rich sequence in the COOH-terminal domain (Minke and Cook 2002). The charged residues in the putative fourth transmembrane segment helix, which usually underlie voltage gating in voltage-gated channels, are not present in TRPC. The physiological roles of TRP channels are diverse. TRPC channels mediate Ca^{2+} influx through capacitative and/or noncapacitative Ca^{2+} entry mechanisms (Montell 2005; Minke and Cook 2002; Parekh and Putney 2005). This TRPC-mediated Ca^{2+} influx has important roles in various functions such as regulation of vascular tone and vascular permeability (Freichel et al. 2001; Tiruppathi et al. 2002), guidance of neuronal growth cones in the developing nervous system (Li et al. 2005; Greka et al. 2003), and sperm fertilization (Jungnickel et al. 2001).

A high degree of diversity exists in the modes of activation and regulation of TRP channels. Physical stimuli (such as mechanical forces) and chemical stimuli (such as hormones, neurotransmitters, and growth factors) activate different TRP channels by various mechanisms including alteration of cellular diacylglycerol levels, changes in cytosolic Ca^{2+} ([Ca^{2+}]$_i$) levels, depletion of Ca^{2+} stores, and various protein–protein interactions that occur either between different TRP isoforms or when TRP channel proteins interact with other specific proteins (Montell 2005). The activity of TRPC is also regulated by protein phosphorylation. For example, calmodulin (CaM)-kinase II activates TRPC6 (Shi et al. 2004), Src family nonreceptor tyrosine kinases activate both TRPC3 (Vazquez et al. 2004) and TRPC6 (Hisatsune et al. 2004), and protein kinase (PK)C inhibits TRPC3, -4, -5, -6, and -7 (Trebak et al. 2005; Venkatachalam et al. 2003; Zhang et al. 2001; Kwan et al. 2006). Recent evidence indicates that the activity of several TRPC isoforms can be inhibited by cyclic guanosine monophosphate (cGMP)-dependent protein kinase 1α (PKG-1α). PKG is a key component in the nitric oxide (NO)-cGMP-PKG signaling pathway. The regulation of TRPC channels by PKG confers the channels with additional properties, which enable TRPC to participate in NO-cGMP-PKG-mediated functional changes.

2
PKG: Structure and Activation Mechanism

PKG belongs to the family of serine/threonine kinases that are activated by cGMP. Two PKG genes, coding for PKG type I (PKG-I) and type II (PKG-II), have been identified in mammals. The PKG-I gene encodes two alternative spliced isoforms, PKG-1α and PKG-1β, which differ in their amino termini by approximately 100 amino acids. The PKG-1α isoform is approximately tenfold more sensitive to cGMP activation than PKG-1β. Both PKG-I and PKG-II are homodimers of subunits with a molecular weight of approximately 75 kDa and approximately 85 kDa, respectively. Each subunit comprises three major functional domains (Fig. 1), which include a regulatory domain that binds to cGMP, a catalytic domain catalyzing phosphorylation reactions, and an amino terminal domain mediating dimerization as well as being responsible for the suppression of kinase activity in the absence of cGMP (Feil et al. 2005). The regulatory domain contains two tandem cGMP-binding sites that interact allosterically and bind cGMP with high and low affinity. Occupation of both binding sites induces a conformational change that releases inhibition of the catalytic core by the amino terminal domain, subsequently allowing the phosphorylation of serine/threonine residues in the target proteins (Fig. 1; Feil et al. 2003). Autophosphorylation initiated by the binding of low cGMP concentration to high-affinity sites may precede activation of heterophosphorylation. The binding increases the spontaneous activity of PKG-I and PKG-II. Autophosphorylation of mammalian PKGs occurs at multiple sites, with at least six such sites in PKG-1α, in the N-terminal (~100) amino acids of the regulatory domain (Hofmann 2005). When cGMP dissociates from the active form of the enzymes and is hydrolyzed by cGMP phosphodiesterases, the inactive conformation of the enzymes is reestablished. The PKGs differ in their subcellular localizations: PKG-I is principally found in the cytoplasm, while PKG-II is anchored to the plasma membrane by its myristoylated amino terminus (Feil et al. 2003).

3
The Regulation of TRPC by PKG

3.1
PKG Phosphorylation Sites in TRPC Proteins

PKGs recognize the phosphorylation site motif of $(R/K)_{2-3}$-X-S/T, where X is any amino acid. All seven isoforms of TRPC contain this PKG phosphorylation site motif (Pearson and Kemp 1991). Based upon results from in vitro phosphorylation experiments with human TRPC3 proteins that were purified either from human embryonic kidney (HEK) cells that overexpress TRPC3 or a rabbit reticulocyte in vitro translation system that expresses TRPC3, TRPC3 can be phosphorylated by bovine recombinant PKG-1α (Kwan et al. 2004). In the presence of KT5823, a potent and highly selective PKG-1α inhibitor, the recombinant PKG-1α was unable to phosphorylate TRPC3 (Kwan et al. 2004), substantiating the argument that TRPC3 is a substrate for PKG-1α. Point mutations were constructed on each of the three potential phosphorylation sites on

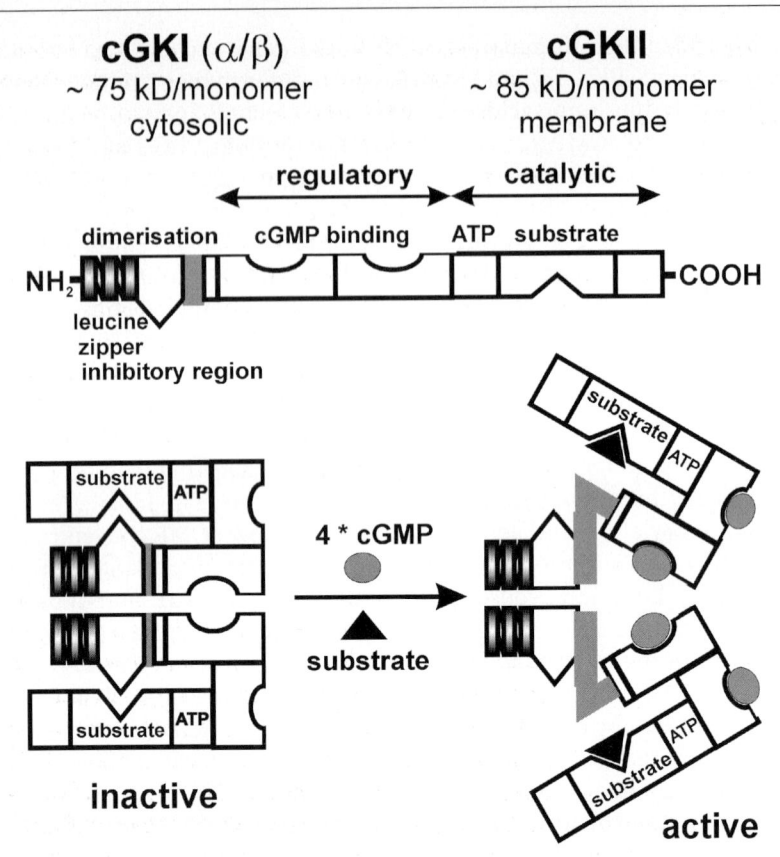

Fig. 1 Structural and biochemical characteristics of homodimeric PKG type I (PKG-I) and type II (PKG-II). The functional domains within each PKG-I subunit are illustrated. Binding of four cGMP molecules is required for a conformational change from the inactive to the active state. (Courtesy of Dr. Thomas Kleppisch; Feil et al. 2005)

human TRPC3 in an attempt to identify the PKG phosphorylation sites (Kwan et al. 2004). Mutations at Thr-11 and Ser-263, two of three potential PKG phosphorylation sites, significantly reduced the PKG1α-mediated phosphorylation on human TRPC3 proteins, demonstrating that Thr-11 and Ser-263 are the real PKG phosphorylation sites (Kwan et al. 2004). These two PKG phosphorylation sites in human TRPC3, (-RRxT^{11}- and -YRKLS^{263}MQC-), are both conserved in human TRPC6 and -7 (Table 1), which are the two TRPC members most closely related to TRPC3. Therefore, it is likely that human TRPC6 and -7 are also the substrates for PKGs. Cross-species sequence alignment reveals that one of the PKG phosphorylation sites in human TRPC3, -YRKLS^{263}MQC-, is conserved in rat and mouse TRPC3 (Table 2). However, it is not known if PKGs can phosphorylate other human TRPC proteins, such as TRPC1, -4, and -5. The two PKG phosphorylation sites of human TRPC3, Thr-11 and Ser-263, are not

Table 1 Conservation of PKG phosphorylation sites in the TRPC3/6/7 subfamily. The putative phosphorylated residues and their positions in corresponding TRPC proteins are labeled in bold italics

	Isoforms	Amino acid sequence alignment
Position 1	Human TRPC3	RRM*T^{11}*VMREKGRR
	Human TRPC6	RRQ*T^{70}*VLREKGRR
	Human TRPC7	RRH*T^{15}*TLREKGRR
Position 2	Human TRPC3	YRKL*S^{263}*MQCKD
	Human TRPC6	YKKL*S^{322}*MQCKD
	Human TRPC7	YRKL*S^{267}*MQCKD

Table 2 Conservation of a PKG phosphorylation site on TRPC3 across different species. The putative phosphorylated residues and their positions in corresponding TRPC proteins are labeled in bold italics

Isoforms	Species	Amino acid sequence alignment
TRPC3	Human	YRKL*S^{263}*MQCKD
	Rat	YRKL*S^{251}*MQCKD
	Mouse	YRKL*S^{263}*MQCKD

conserved in human TRPC1, -4, or -5. On the other hand, human TRPC1, -4, and -5 contain other potential PKG phosphorylation sites, one for TRPC4 and three for TRPC1 and -5. Future experiments are needed to learn if these TRPC isoforms are the substrates for PKGs.

3.2
Inhibition of TRPC3 Activity by PKG

TRPC is activated either by agonist binding to plasma membrane G protein-coupled receptors, leading to the activation of phospholipase C-β, or by binding of growth factors to tyrosine kinase-linked receptors, resulting in the activation of phospholipase C-γ (Minke and Cook 2002; Parekh and Putney 2005). Both phospholipase C-β and phospholipase C-γ hydrolyze the lipid precursor phosphatidylinositol-4,5-bisphosphate to yield inositol 1,4,5-trisphosphate (InsP$_3$) and diacylglycerol (Minke and Cook 2002; Parekh and Putney 2005). Diacylglycerol can activate TRPC3. The activation of phospholipase C also results in the InsP$_3$-mediated Ca^{2+} release from intracellular Ca^{2+} stores. The resultant reduction of [Ca^{2+}]$_i$ level in the stores may stimulate TRPC3 via the capacitative mechanism at least under certain conditions (Kiselyov et al. 1998; Vazquez et al. 2003). On the other hand, the activity of TRPC3 can be inhibited by 8-bromocGMP (8-BrcGMP), a membrane-permeant analog of cGMP. In the presence of KT5823, 8-BrcGMP has no inhibitory effect on TRPC3. Furthermore, disruptions of two PKG phosphorylation sites on TRPC3 by point mutations T11A and S263Q drastically reduce the inhibitory action of 8-BrcGMP on TRPC3 (Kwan et al. 2004). Taken together, it is clear that PKG directly phosphorylates human TRPC3 at Thr-11 and Ser-263, and as a consequence reduces the activity of the channels.

3.3
Involvement of PKA?

Cyclic AMP (cAMP)-dependent PKAs exists in two major forms, PKA-I and PKA-II, constituting a ubiquitous enzyme family present in all eukaryotic cells. PKA is structurally similar to PKG. The PKA also contains two regulatory domains capable of binding to cAMP, and two catalytic domains responsible for the phosphorylation reactions. However, in PKA the regulatory domain and catalytic domain are located in separate subunits, but in PKG these two domains are fused. PKA and PKG recognize a similar substrate sequence (Schwede et al. 2000). The PKG phosphorylation motif $(R/K)_{2-3}$-X-S/T is also a recognition site for PKA (Wang and Robinson 1997; Pearson and Kemp 1991). However, there are some subtle differences in the substrate specificity between PKG and PKA. Some PKG substrates, such as the vasodilator-stimulated phosphoprotein and phosphodiesterase 5, are relatively poor substrates for PKA (Wang and Robinson 1997; Corbin et al. 2000). A key relevant question here is whether the two PKG phosphorylation sites in human TRPC3, Thr-11 and Ser-263, are also the phosphorylation targets of PKA. Functional studies have demonstrated that 8-BrcAMP, a membrane-permeant activator of PKA, failed to inhibit TRPC3 in TRPC3-overexpressing HEK cells, suggesting that Thr-11 and Ser-263 may not be good substrates for PKA (Kwan et al. 2004).

PKG is predominantly activated by cGMP, but PKA is predominantly activated by cAMP. PKA contains a key alanine in its cAMP-binding sites; in contrast, PKG contains a threonine in this position. This amino acid variation accounts for the cyclic nucleotide specificity and underlies evolutionary divergence of cGMP and cAMP effects in animals. The K_a value, which is the concentration of cyclic nucleotides required for half-maximal stimulation of the kinase, for PKG-1α and cGMP is approximately 110 nM, and the corresponding K_a value for PKA and cAMP is approximately 90 nM (Corbin et al. 1986; Ogreid et al. 1985). However, due to the structural similarities of PKA and PKG, PKA may be cross-activated by cGMP and likewise PKG may be cross-activated by cAMP, if the concentration of cyclic nucleotides reaches the 10- to 100-μM range (Corbin et al. 1986; Ogreid et al. 1985). This could cause experimental problems. For example, membrane-permeant analogs of cyclic nucleotides, such as 8-BrcGMP and Bt$_2$-cGMP, are frequently used to stimulate PKG in order to study the involvement of PKG in cellular functions. However, due to their poor membrane permeability and low specificity, these analogs are frequently used at 0.1-10-mM concentrations in experiments with intact cells (Schwede et al. 2000). Under these conditions, cross-activation of PKA by a cGMP-analog should be a concern because the actual intracellular concentration of these cGMP-analogs is unknown and difficult to estimate. Additional evidence is therefore needed to verify the involvement of PKG in these situations. In such experiments, PKG-specific inhibitors could provide useful tools. There are several classes of highly selective PKG inhibitors available, which include: Rp-stereoisomers of cGMP-phosphorothioates that prevent activation of PKGs at the cGMP-binding site of the regulatory domain (Schwede et al. 2000); KT5823 that interferes the ATP binding site of the catalytic domain (Gadbois et al. 1992); and peptide inhibitor DT-2/DT-3 that interferes with the interaction of the catalytic domain and its substrates (Dostmann et al. 2000).

3.4
PKC Inhibition of TRPC Is Partly Mediated by PKG

PKC is another kinase that can regulate TRPC channels. PKC inhibits the activity of multiple TRPC channels, including members of the TRP3/6/7 and TRPC4/5 subfamilies. It was postulated that this inhibition may represent a negative feedback mechanism for the fine control of $[Ca^{2+}]_i$. In this negative feedback pathway, the activation of TRPC results in Ca^{2+} entry. A rise in $[Ca^{2+}]_i$, together with an elevated diacylglycerol level, stimulates PKC activity, which feeds back to inactivate multiple TRPC isoforms, thus limiting a further rise in $[Ca^{2+}]_i$ (Venkatachalam et al. 2003). Extensive studies have been conducted to examine the detailed mechanisms of PKC inhibition of TRPC3. Current evidence suggests that PKC inhibits TRPC3 channels by two independent pathways. (1) PKC directly phosphorylates Ser-712 in human TRPC3 proteins, resulting in channel inactivation (Trebak et al. 2005); (2) PKC phosphorylates and activates PKG-1α. Activated PKG-1α then phosphorylates Thr-11 and Ser-263 in human TRPC3 proteins, causing channel inactivation (Kwan et al. 2006). The relative importance of these two pathways may depend on the cellular level of PKG-1α. In a cell type that has no or low levels of PKG-1α activity, PKC would inhibit TRPC3 predominantly by the direct PKC phosphorylation of Ser-712. In cells with moderate or high PKG-1α activity, PKC could inhibit TRPC channels by two mechanisms, directly by PKC phosphorylation of Ser-712 and indirectly through PKG-1α. In latter case, the PKG-mediated indirect pathway may have an equivalent or a slightly greater contribution toward the overall PKC inhibition than the direct PKC effect on TRPC3. For example, it was demonstrated that in wildtype HEK-293 cells, PKC inhibited TRPC3 principally by the direct PKC phosphorylation of Ser-712, but in PKG-1α-overexpressing HEK cells, the PKG-mediated indirect pathway made a large contribution toward the overall PKC inhibition of TRPC3 (Kwan et al. 2006).

The detailed mechanisms of PKC and PKG inhibition have been elucidated only with TRPC3. However, the amino acid sequence alignment of different TRPC isoforms suggests that PKC and PKG may inhibit other TRPC isoforms through similar mechanisms. As mentioned, Thr-11 and Ser-263 of human TRPC3 are conserved in two other members of the TRPC3/6/7 subfamily. Furthermore, Ser-712 in human TRPC3, which is located within a consensus PKC phosphorylation sequence of -PS^{712}PKS-, is conserved in two TRPC subfamilies, TRPC3/6/7 and TRPC4/5, suggesting that the same site could be responsible for PKC-induced inhibition of TRPC4–7. However, the response of TRPC1 to PKC is very different from that of other TRPC isoforms. In contrast to other TRPC channels, TRPC1 is activated by direct PKC phosphorylation (Ahmmed et al. 2004).

4
PKG-Mediated Negative Feedback Control of [Ca^{2+}]$_i$

Signaling cascades initiated by NO play important roles in diverse physiological and pathological processes. NO is produced by various nitric oxide synthases (NOS), including the endothelial type, eNOS, neuronal type, nNOS, and inducible type, iNOS. nNOS and eNOS are dependent upon increased [Ca^{2+}]$_i$ to activate calmodulin, which binds to NOS and activates the enzyme. An elevated NO level activates guanylyl cyclases, which are present either in a soluble or a particulate form. Guanylyl cyclases catalyze the production of cGMP, which activates multiple pathways by binding to a number of receptor proteins, among them the PKGs. PKG can inhibit TRPC channels. Thus a loop of negative feedback regulation can be postulated. In this negative feedback loop, a rise in [Ca^{2+}]$_i$ provides a signal to reduce Ca^{2+} influx through a signal transduction pathway involving NOS-NO-cGMP-PKG and TRPC (Fig. 2). The inhibition of TRPC by PKGs is an important step in this negative feedback mechanism, which may serve to protect the cells from the detrimental effects of excessive [Ca^{2+}]$_i$ and/or NO.

The above-mentioned negative feedback mechanism involving Ca^{2+}-NOS-NO-cGMP-PKG-TRPC is only one among other feedback pathways that also serve to limit [Ca^{2+}]$_i$ and NO levels. Several other established feedback pathways include:

- An increase in [Ca^{2+}]$_i$ may rapidly inactivate some Ca^{2+} influx channels (Zweifach and Lewis 1995).

- An increase in [Ca^{2+}]$_i$ may inhibit inwardly rectifying K$^+$ channels (K$_{ir}$), reducing the driving force for Ca^{2+} influx into nonexcitable cells, such as

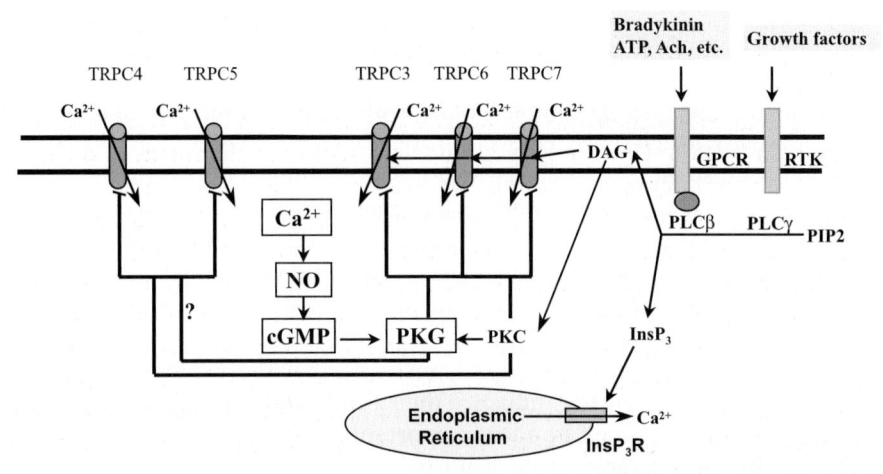

Fig. 2 PKC- and PKG-mediated regulation of TRPC channels. Agonists bind to either G protein-coupled receptors (*GPCR*) or tyrosine kinase-linked receptors (*RTK*) and activate phospholipase C (*PLC*)-β or PLC-γ, producing diacylglycerol (*DAG*) and InsP$_3$. InsP$_3$ binds to its receptor (*InsP$_3$R*), resulting in Ca^{2+} release from intracellular Ca^{2+} stores and activating TRP via capacitative Ca^{2+} entry. DAG activates some TRP channels such as TRPC3, -6, and -7 by non-capacitative mechanism. DAG also activates PKC, which inhibits TRPC3-7. Increased [Ca^{2+}]$_i$ also activates PKG by the NO/cGMP pathway. PKG then inactivates TRPC3, -6, and -7 by a negative feedback mechanism. PKC can inhibit TRPC by direct phosphorylation of TRPC and by directly activating PKG, which then inhibits TRPC3, -6, and -7. (↑), Stimulatory processes; (⊤), inhibitory processes

vascular endothelial cells, through a pathway that involves Ca^{2+}-NOS-NO-cGMP-K$_{ir}$ (Shimoda et al. 2002).

– An increase in [Ca^{2+}]$_i$ may activate additional Ca^{2+} removal mechanisms by multiple PKG-mediated pathways. For example:
a. PKG may enhance Ca^{2+} sequestration into intracellular Ca^{2+} stores by stimulating the sarcoplasmic/endoplasmic reticulum Ca^{2+}-ATPase (Yao and Huang 2003).
b. PKG may inhibit voltage-gated Ca^{2+} channels in smooth muscle cells (Lincoln et al. 2001).
c. PKG may inhibit mechanosensitive Ca^{2+} influx channels in vascular endothelial cells (Yao et al. 2000).
d. PKG may activate Ca^{2+}-sensitive K$^+$ channels, causing membrane hyperpolarization, which reduces Ca^{2+} influx through voltage-gated Ca^{2+} channels in vascular smooth muscle cells (Lincoln et al. 2001).
e. PKG may decrease InsP$_3$-induced Ca^{2+} release from intracellular Ca^{2+} stores (Lincoln et al. 2001).

5
Functional Role of PKG-Mediated Regulation of TRPC

pkg is not a housekeeping gene and its level of expression varies greatly in different tissues and cell types. PKG-I is present in high concentration (>0.1 µM) in smooth muscles, platelets, cerebellum, hippocampus, dorsal root ganglia, neuromuscular endplates, and kidney (Hofmann 2005). Its activity can also be found in vascular endothelium and cardiomyocytes (Feil et al. 2003). PKG-II is expressed in several brain nuclei, intestinal mucosa, kidney, adrenal cortex, chondrocytes, and lung (Hofmann 2005). TRPC channels also display differential expression patterns in different tissues and cell types (Montell 2005). Therefore, it is reasonable to postulate that the expression level of PKG and TRPC may greatly influence their functional importance. The PKG inhibition of TRPC3 may be more important in the cell types where both PKG and PKG-inhibitable TRPC isoform(s) are highly expressed. Indeed, it has been demonstrated that NO, cGMP, and PKG can inhibit Ca^{2+} influx in various cell types that express both PKG and TRPC channels. These include vascular endothelial cells (Kwan et al. 2000; Yao et al. 2000), neurons (Viso-Leon et al. 2004), vascular smooth muscle cells (Lincoln et al. 2001; Cohen et al. 1999), platelets (Trepakova et al. 1999), and glial cells (Yoshioka et al. 2000).

Ca^{2+}, NO, and PKG are signaling molecules that have pivotal roles in many physiological processes. TRPC channels allow Ca^{2+} influx, and the activity of some TRPC channels is negatively regulated by PKG. These unique properties of TRPC channels may enable TRPC to play a key role in the following functional changes:

- In vascular tone control, Ca^{2+} entry into endothelial cells activates eNOS, resulting in NO production. NO then diffuses to the underlying smooth muscle cells, which activates PKG and its downstream targets, resulting a reduction of $[Ca^{2+}]_i$ in smooth muscle cells and subsequent vascular dilation. NO may also activate PKG of endothelial cells. The activated PKG reduces the $[Ca^{2+}]_i$ and NO levels in endothelial cells by negative feedback mechanisms (Yao and Huang 2003). Multiple isoforms of TRPC are expressed in vascular endothelial cells and smooth muscle cells. It is likely that these TRPC channels serve as the downstream targets for PKG. Indeed, it has been shown that PKG may inhibit TRPC channels to reduce the $[Ca^{2+}]_i$ level in vascular cells (Kwan et al. 2006).

- There is evidence that Ca^{2+} influx through a number of TRPC isoforms, including TRPC1, -4, and -6, contributes to the increased vascular permeability in response to the inflammatory mediators thrombin and histamine. In this process, Ca^{2+} influx through TRPC channels activates multiple downstream pathways, among them the eNOS-cGMP-PKG pathway, resulting in an increased vascular permeability (Yuan 2003). It is possible that PKG may inhibit TRPC in a negative feedback manner to limit Ca^{2+} entry into endothelial cells, thus reducing further vascular leakage.

- Ca^{2+}, NO, and PKG are important signaling molecules involved in hippocampal long-term potentiation and long-term depression, which underlie learning and memory. Recent evidence indicates that TRPC channels are also involved in synaptic plasticity. In this process, stimulation of group I metabotropic glutamate receptors activates several TRPC isoforms. The resultant changes in synaptic currents may modulate synaptic plasticity (Strubing et al. 2001; Kim et al. 2003).

- In platelets, stimulatory agents increase the $[Ca^{2+}]_i$ level. Conversely, NO, cGMP, PKG, and PKC all reduce the $[Ca^{2+}]_i$ level, contributing to the inhibition of platelet activation (Haslam et al. 1999; Tertyshnikova et al. 1998; Zavoico et al. 1985). TRPC6 appears to be a crucial Ca^{2+} influx channel involved in platelet activation (Hassock et al. 2002). It is possible that PKG and PKC may inhibit TRPC6 in platelets, leading to the inhibition of the platelet activation.

6
Concluding Remarks

Signaling cascades involving Ca^{2+}-NOS-NO-cGMP-PKG have important roles in diverse physiological and pathological processes. TRPC channels allow Ca^{2+} influx, and the activity of several TRPC channels is inhibited by PKG. PKC can modulate TRPC channels and part of this PKC action is mediated via PKGs (Kwan et al. 2006). Taken together, PKG-mediated inhibition of TRPC channels may serve to finely regulate $[Ca^{2+}]_i$, NO, and PKC levels, and to protect cells from the detrimental effects (oxidative stress, apoptotic cell death, and uncontrolled cell proliferation) caused by excessive NO, $[Ca^{2+}]_i$, and PKC activity (Yao and Huang 2003).

References

Ahmmed GU, Mehta D, Vogel S, Holinstat M, Paria BC, Tiruppathi C, Malik AB (2004) Protein kinase Cα phosphorylates the TRPC1 channel and regulates store-operated Ca^{2+} entry in endothelial cells. J Biol Chem 279:20941–20949

Cohen RA, Weisbrod RM, Gericke M, Yaghoubi M, Bierl C, Bolotina VM (1999) Mechanism of nitric oxide-induced vasodilation: refilling of intracellular stores by sarcoplasmic reticulum Ca^{2+} ATPase and inhibition of store-operated Ca^{2+} influx. Circ Res 84:210–219

Corbin JD, Ogreid D, Miller JP, Suva RH, Jastorff B, Doskeland SO (1986) Studies of cGMP analog specificity and function of the two intrasubunit binding sites of cGMP-dependent protein kinase. J Biol Chem 261:1208–1214

Corbin JD, Turko IV, Beasley A, Francis SH (2000) Phosphorylation of phosphodiesterase-5 by cyclic nucleotide-dependent protein kinase alters its catalytic and allosteric cGMP-binding activities. Eur J Biochem 267:2760–2767

Dostmann WR, Taylor MS, Nickl CK, Brayden JE, Frank R, Tegge WT (2000) Highly specific, membrane-permeant peptide blockers of cGMP-dependent protein kinase Iα inhibit NO-induced cerebral dilation. Proc Natl Acad Sci USA 97:14772–14777

Feil R, Lohmann SM, de Jonge H, Walter U, Hofmann F (2003) Cyclic GMP-dependent protein kinases and the cardiovascular system. Insight from genetically modified mice. Circ Res 93:907–916

Feil R, Hofmann F, Kleppisch T (2005) Function of cGMP-dependent protein kinases in the nervous system. Rev Neurosci 16:23–42

Freichel M, Suh SH, Pfeifer A, Schweig U, Trost C, Weissgerber P, Biel M, Philipp S, Freise D, Droogmans G, Hofmann F, Flockerzi V, Nilius B (2001) Lack of an endothelial store-operated Ca^{2+} current impairs agonist-dependent vasorelaxation in $TRP4^{-/-}$ mice. Nat Cell Biol 3:121–127

Gadbois DM, Crissman HA, Tobey RA, Bradbury EM (1992) Multiple kinase arrest points in the G1 phase of nontransformed mammalian cells are absent in transformed cells. Proc Natl Acad Sci USA 89:8626–8630

Greka A, Navarro B, Oancea E, Duggan A, Clapham DE (2003) TRPC5 is a regulator of hippocampal neurite length and growth cone morphology. Nat Neurosci 6:837–845

Haslam RJ, Dickinson NT, Jang EK (1999) Cyclic nucleotides and phosphodiesterases in platelets. Thromb Haemost 82:412–423

Hassock SR, Zhu MX, Trost C, Flockerzi V, Authi KS (2002) Expression and role of TRPC proteins in human platelets: evidence that TRPC6 forms the store-independent calcium entry channel. Blood 100:2801–2811

Hisatsune C, Kuroda Y, Nakamura K, Inoue T, Nakamura T, Michikawa T, Mizutani A, Mikoshiba K (2004) Regulation of TRPC6 channel activity by tyrosine phosphorylation. J Biol Chem 279:18887–18894

Hofmann F (2005) The biology of cyclic cGMP-dependent protein kinases. J Biol Chem 280:1–4

Jungnickel MK, Marrero H, Birnbaumer L, Lemos JR, Florman HM (2001) Trp2 regulates entry of Ca into mouse sperm triggered by egg ZP3. Nat Cell Biol 3:499–502

Kim SJ, Kim YS, Yuan JP, Petralia RS, Worley PF, Linden DJ (2003) Activation of the TRPC1 cation channel by metabotropic glutamate receptor mGluR1. Nature 426:285–291

Kiselyov K, Xu X, Mozhayeva G, Kuo T, Pessah I, Mignery G, Zhu X, Birnbaumer L, Muallem S (1998) Functional interaction between InsP3 receptors and store-operated Htrp3 channels. Nature 396:478–482

Kwan HY, Huang Y, Yao X (2000) Store-operated calcium entry in vascular endothelial cells is inhibited by cGMP via a protein kinase G-dependent mechanism. J Biol Chem 275:6758–6763

Kwan HY, Huang Y, Yao X (2004) Regulation of canonical transient receptor potential isoform 3 (TRPC3) channel by protein kinase G. Proc Natl Acad Sci USA 101:2625–2630

Kwan HY, Huang Y, Yao X (2006) Protein kinase C can inhibit TRPC3 channels indirectly via stimulating protein kinase G. J Cell Physiol 207:315–321

Li Y, Jia YC, Cui K, Li N, Zheng ZY, Wang YZ, Yuan XB (2005) Essential role of TRPC channels in the guidance of nerve growth cones by brain-derived neurotrophic factor. Nature 434:894–898

Lincoln TM, Dey N, Sellak H (2001) Invited review: cGMP-dependent protein kinase signaling mechanisms in smooth muscle: from the regulation of tone to gene expression. J Appl Physiol 91:1421–1430

Minke B, Cook B (2002) TRP channel proteins and signal transduction. Physiol Rev 82:429–472

Montell C (2005) The TRP superfamily of cation channels. Sci STKE 272:re3

Ogreid D, Ekanger R, Suva RH, Miller JP, Sturm P, Corbin JD, Doskeland SO (1985) Activation of protein kinase isozymes by cyclic nucleotide analogs used singly or in combination. Principles for optimizing the isozyme specificity of analog combinations. Eur J Biochem 150:219–227

Parekh AB, Putney Jr JW (2005) Store depletion and calcium influx. Physiol Rev 85:757–810

Pearson RB, Kemp BE (1991) Protein kinase phosphorylation site sequences and consensus specificity motif: tabulations. Methods Enzymol 200:62–81

Schwede F, Maronde E, Genieser H, Jastorff B (2000) Cyclic nucleotide analogs as biochemical tools and prospective drugs. Pharmacol Ther 87:199–226

Shi J, Mori E, Mori Y, Mori M, Li J, Ito Y, Inoue R (2004) Multiple regulation by calcium of murine homologues of transient receptor potential proteins TRPC6 and TRPC7 expressed in HEK293 cells. J Physiol 561:415–432

Shimoda LA, Welsh LE, Pearse DB (2002) Inhibition of inwardly-rectifying K channels by cGMP in pulmonary vascular endothelial cells. Am J Physiol Lung Cell Mol Physiol 283:L297–L304

Strubing C, Krapivinsky G, Krapivinsky L, Clapham DE (2001) TRPC1 and TRPC5 form a novel cation channel in mammalian brain. Neuron 29:645–655

Tertyshnikova S, Yan XW, Fein A (1998) cGMP inhibits IP3-induced Ca^{2+} release in intact rat megakaryocytes via cGMP- and cAMP-dependent protein kinase. J Physiol 512:89–96

Tiruppathi C, Freichel M, Vogel SM, Paria BC, Mehta D, Flockerzi V, Malik AB (2002) Impairment of store-operated Ca^{2+} entry in TRPC4(−/−) mice interferes with increase in lung microvascular permeability. Circ Res 91:70–76

Trebak M, Hempel N, Wedel BJ, Smyth JT, Bird GSJ, Putney Jr JW (2005) Negative regulation of TRPC3 channels by protein kinase C-mediated phosphorylation of serine 712. Mol Pharmacol 67:558–563

Trepakova ES, Cohen RA, Bolotina VM (1999) Nitric oxide inhibits capacitative cation influx in human platelets by promoting sarcoplasmic/endoplasmic reticulum Ca^{2+}-ATPase-dependent refilling of Ca^{2+} stores. Circ Res 84:201–209

Vazquez G, Wedel BJ, Trebak M, Bird GS, Putney JW Jr (2003) Expression level of the canonical transient receptor potential 3 (TRPC3) channel determines its mechanism of activation. J Biol Chem 278:21649–21654

Vazquez G, Wedel BJ, Kawasaki BT, Bird GS, Putney JW Jr (2004) Obligatory role of Src kinase in the signaling mechanism for TRPC3 cation channels. J Biol Chem 279:40521–40528

Venkatachalam K, Zheng F, Gill DL (2003) Regulation of canonical transient receptor potential (TRPC) channel function by diacylglycerol and protein kinase C. J Biol Chem 278:29031–29040

Viso-Leon MC, Ripoll C, Nadal A (2004) Oestradiol rapidly inhibits Ca^{2+} signals in ciliary neurons through classical oestrogen receptors in cytoplasm. Pflugers Arch 449:33–41

Wang X, Robinson PJ (1997) Cyclic GMP-dependent protein kinase and cellular signaling in the nervous system. J Neurochem 68:443–456

Yao X, Huang Y (2003) From nitric oxide to endothelial cytosolic Ca^{2+}: a negative feedback control. Trends Pharmacol Sci 24:263–266

Yao X, Kwan HY, Chan FL, Chan NW, Huang Y (2000) A protein kinase G-sensitive channel mediates flow-induced Ca^{2+} entry into vascular endothelial cells. FASEB J 14:932–938

Yoshioka A, Yamaya Y, Saiki S, Kanemoto M, Hirose G, Pleasure D (2000) Cyclic GMP/cyclic GMP-dependent protein kinase system prevents excitotoxicity in an immortalized oligodendroglial cell line. J Neurochem 74:633–640

Yuan SY (2003) Protein kinase signaling in the modulation for microvascular permeability. Vascul Pharmacol 39:213–223

Zavoico GB, Halenda SP, Sha'afi RI, Feinstein MB (1985) Phorbol myristate acetate inhibits thrombin-stimulated Ca^{2+} mobilization and phosphatidylinositol 4,5-bisphosphate hydrolysis in human platelets. Proc Natl Acad Sci USA 82:3859–3862

Zhang L, Saffen D (2001) Muscarinic acetylcholine receptor regulation of TRP6 Ca^{2+} channel isoforms. Molecular structures and functional characterization. J Biol Chem 276:13331–13339

Zweifach A, Lewis R (1995) Rapid inactivation of depletion-activated calcium current (ICRAC) due to local calcium feedback. J Gen Physiol 105:209–226

Trafficking of TRP Channels: Determinants of Channel Function

I. S. Ambudkar

Secretory Physiology Section, NIH, Building 10, Room 1N-113, Bethesda MD, 20892, USA
indu.ambudkar@nih.gov

1	Introduction	542
2	Overview of Trafficking Mechanisms That Determine Surface Expression of Proteins	542
3	Mechanisms Regulating TRP Channel Trafficking	544
3.1	TRPC Subfamily	544
3.1.1	TRPC1	545
3.1.2	TRPC3	546
3.1.3	TRPC4 and TRPC5	547
3.1.4	TRPC6	548
3.1.5	Invertebrate TRPCs	549
3.2	TRPV Subfamily	549
3.3	TRPM Subfamily	550
3.4	TRPP and TRPML Subfamilies	550
4	Emerging Commonalities Underlying TRP Channel Trafficking	551
4.1	Assembly of TRP Channel Complexes	551
4.2	Role of Cytoskeletal Proteins in TRP Trafficking	552
4.3	Vesicular Trafficking Mechanisms	553
5	Conclusion	553
	References	554

Abstract Transient receptor potential (TRP) channels are members of a relatively newly described family of cation channels that display a wide range of properties and mechanisms of activation. The exact physiological function and regulation of most of these channels have not yet been conclusively determined. Studies over the past decade have revealed important features of the channels that contribute to their function. These include homomeric interactions between TRP monomers, selective heteromeric interactions within members of the same subfamily, interactions of TRPs with accessory proteins and assembly into macromolecular signaling complexes, and regulation within functionally distinct cellular microdomains. Further, distinct constitutive and regulated vesicular trafficking mechanisms have a critical role not only in controlling the surface expression of TRP channels but also their activation in response to stimuli. A number of cellular components such as cytoskeletal and scaffolding proteins also contribute to TRP channel trafficking. Thus, mechanisms involved in the assembly and trafficking of TRP channels control their plasma membrane expression and critically impact their function and regulation.

Keywords TRP channels · Receptors · Sensory · Vesicular trafficking · Microdomains · Signaling · Ca^{2+} · Cations

1
Introduction

The transient receptor potential (TRP) proteins constitute a superfamily of cation channels with remarkably diverse properties, mode of regulation, and physiological functions (extensively reviewed in Minke and Cook 2002; Montell et al. 2002; Montell 2005; Pedersen et al. 2005). Despite the recent intense focus on TRP channels, conclusive data regarding the regulation and exact physiological function of most of these channels are still lacking. TRPs generate channels by homomeric or heteromeric interactions between members of the same subfamily (Schilling and Goel 2004; Schaefer 2005; Montell 2005). Further, TRP proteins interact with accessory proteins to form large multi-protein signaling complexes, the components of which likely regulate their gating as well as their localization and expression within specialized plasma membrane domains (Ambudkar 2004, 2005; Kiselyov et al. 2005; Montell 2005). Thus, unique structural features of different TRPs, their auxiliary proteins, and their trafficking mechanisms are critical determinants of their function and regulation. This chapter will be directed toward the mechanisms that control plasma membrane expression and trafficking of TRP channels and how these processes affect their regulation and function.

2
Overview of Trafficking Mechanisms That Determine Surface Expression of Proteins

Irrespective of the mode of activation of various TRP channels, their function ultimately depends on their insertion into a specific plasma membrane location where they can be regulated by upstream signaling mechanisms and in turn regulate downstream processes that control cellular function. Thus, localization of TRP channels is critical not only for their regulation but also for the downstream cellular function that is controlled via channel activity. This close link between the cellular localization of channels and the function(s) they regulate in different tissues is slowly emerging.

The function of plasma membrane ion channels can be regulated by controlling the level of the protein in the surface membrane (Royle and Murrell-

Lagnado 2003). While this can be achieved on a relatively slow time-scale via synthesis and degradation of the channel components, plasma membrane expression of a protein ultimately depends on constitutive and regulated vesicular trafficking mechanisms that control their transport to, and insertion into, the plasma membrane as well as their internalization (Fig. 1). These trafficking processes can effectively regulate the surface expression of the channels; e.g., a slow insertion and fast retrieval will retain most of the protein in an intracellular compartment and vice versa. Further, sorting of the proteins into recycling vesicles provides a readily available pool from which they can be recruited to the plasma membrane by exocytosis and recovered back into the pool by endocytosis. Such recycling vesicles are often localized near the target membrane to facilitate rapid insertion and retrieval of the protein (Royle and Murrell-Lagnado 2003). Surface expression can also be increased by retention of the protein via interaction with scaffolding proteins either during or after insertion into the plasma membrane (Altschuler et al. 2003). Some common examples of these include interactions with PDZ or WD domain-containing proteins, direct or indirect interaction with cytoskeletal components and microtubules, and as described recently, interaction with membrane lipids such as phosphatidylinositol 4,5-bisphosphate (PIP_2) (Altschuler et al. 2003; Royle and Murrell-Lagnado 2003; Symons and Rusk 2003; Patterson et al. 2005). The constitutive vesicle trafficking pathway can be regulated to either increase or decrease its level in the surface membrane. Alternatively, stimulation can result in activation of a different trafficking mechanism and plasma membrane insertion of a different pool of the channel.

Irrespective of the mode of trafficking, surface expression of the channel requires vesicle fusion, via exocytosis, while endocytotic mechanisms achieve internalization. Soluble N-ethylmaleimide-sensitive factor attachment receptor protein (SNARE), as well as components of specific plasma membrane domains such as clathrin coat proteins, caveolin, and PIP_2, are involved in the internalization processes (Brown and London 1998; Czech 2003; Jahn et al. 2003; Laude and Prior 2004). Thus, regulation of vesicular exocytosis and endocytosis provides efficient control of the number of channels as well as the duration of their expression in the surface membrane. Although there is relatively little information about TRP channel trafficking, it is rapidly emerging as a key mechanism for both constitutive and regulated control of channel function. An important endpoint of trafficking is routing of TRP channels to specific cellular locations and their compartmentalization into segregated signaling complexes (Ambudkar 2005; Kiselyov 2005; Montell 2005). In polarized cells, such as those from endothelium, kidney, gastrointestinal tissues, as well as pancreatic and salivary glands, proteins are routed to specific plasma membrane regions of the cell (Altschuler et al. 2003; Hoenderop and Bindels 2005). For example, TRPC3 has been recently shown to be present in apical membranes of polarized epithelial cells, TRPC6 is found in both apical and basolateral membranes, while TRPC1 is primarily found in the basolateral membrane (Singh et al. 2001;

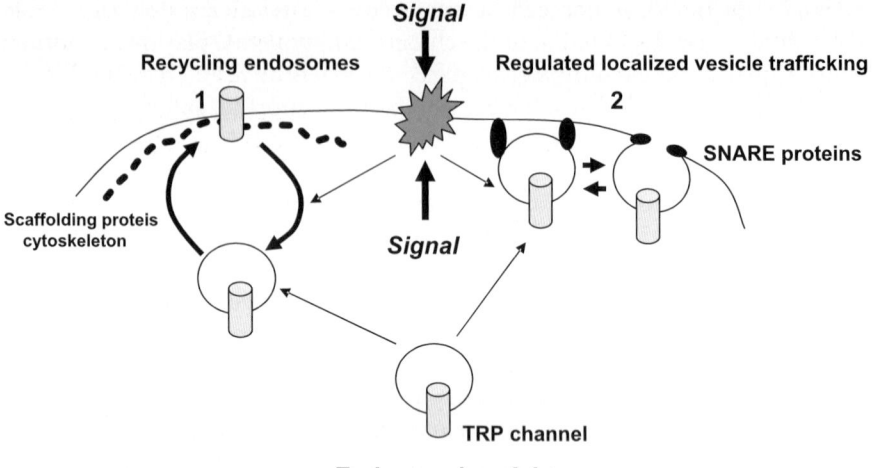

Fig. 1 Vesicle trafficking mechanisms determining plasma membrane expression of TRP channels: The figure illustrates (*1*) recycling vesicle trafficking as well as (*2*) regulated vesicle fusion via a "kiss and run" mechanism as two possible mechanisms for regulating plasma membrane expression of TRP channels. Both intracellular and extracellular signals can modulate these trafficking processes. In addition, vesicle trafficking depends on interactions of channel and vesicle components with scaffolding and cytoskeletal proteins. Vesicle fusion and internalization are mediated by the interaction of SNARE proteins that are localized in the vesicles and in the plasma membrane. It is proposed that such trafficking results in the assembly and regulation of TRP channels in specific cellular microdomains

Liu et al. 2000; Bandyopadhyay et al. 2005; Goel et al. 2005). Similar polarized trafficking governs localization of proteins' neuronal cells (Moran et al. 2004). Routing signals that are inherent within the sequence of the protein most likely determine the specific cellular localization of TRPC channels (Ma and Jan 2002). However, such signals are still largely unknown for most plasma membrane proteins and there is little information for TRP channel routing. Table 1 summarizes the primary stimuli for activation of different TRP subfamilies and highlights documented mechanisms that determine plasma membrane expression of the channels (discussed in more detail below).

3
Mechanisms Regulating TRP Channel Trafficking

3.1
TRPC Subfamily

While the exact role of TRPC channels in agonist-stimulated Ca^{2+} signaling mechanisms is not yet completely understood (Putney et al. 2005; Spassova et al.

Table 1 Function and regulation of TRP channel function. The table lists different TRP subfamily members and highlights current knowledge regarding their mechanism of activation as well as factors involved in regulating their plasma membrane expression, both of which are key determinants of channel activity

TRP subfamily	Channel proteins	Regulation	Plasma membrane expression
TRPC	C1–C7	PIP$_2$ hydrolysis, store-depletion	Multimerization, accessory and scaffolding proteins, constitutive and regulated trafficking
TRPV	V1–V3	Moderate to high heat, ligands, PIP$_2$, pH	Protein kinase, regulated vesicular trafficking
	V4	Osmolarity, phorbol ester	Multimerization
	V5, V6	Spontaneously active, [Ca^{2+}] (?)	Constitutive trafficking, accessory scaffold proteins
TRPM	M1	?	Multimerization/ regulated trafficking (?)
	M2	ADP ribose, NAD, H$_2$O$_2$	Not known
	M4, M5	Ca^{2+}, voltage	Not known
	M6, M7	Mg, Mg-ATP	Multimerization
	M8	Cold temperature, ligands	Not known
TRPA		Cold, ligands, mechanical	Not known
TRPN		Mechanical	Not known
TRP	P1, P2	Ca^{2+}, pH	Multimerization
	P3	Ca^{2+}	Not known

2005), there is increasing evidence that mammalian TRPC proteins interact with each other selectively and with a number of signaling and scaffolding proteins to generate distinct channels (Ambudkar 2005; Montell 2005). Recent data suggest that specific constitutive and regulated trafficking mechanisms contribute toward plasma membrane expression of TRPCs and their function.

3.1.1
TRPC1

TRPC1 is currently the strongest candidate component of store-operated Ca^{2+} entry channels (Beech 2005; Ambudkar et al. 2004). Plasma membrane expression of TRPC1 depends on its interaction with other TRPCs (e.g., TRPC4) in some types of cells (Hofmann et al. 2002). Localization of TRPC1 in the plasma membrane depends on the status of microtubules through an interaction with β-tubulin in retinal cells (Bollimuntha 2005). An interaction between Rho and TRPC1 determines regulation of channel localization by remodeling of the cytoskeleton (Mehta 2002). TRPC1 is assembled in a signaling complex that is localized in distinct plasma membrane lipid domains called lipid raft domains (Brownlow et al. 2004). In some cells, the channel interacts with a protein component of this domain, caveolin-1 (Lockwich 2000; Brazer 2003; Ambudkar 2004, 2005). Importantly, caveolar domains have been demonstrated to be centers for the assembly and regulation of Ca^{2+} signaling mechanisms (Brown and London 1998; Isshiki and Anderson 2003; Brownlow et al. 2004). Phospholipase C (PLC), $G_{\alpha q/11}$, and PIP_2, which are key molecular components in the activation of TRPC channels, are also localized in the same domain. Thus, it has been suggested that regulation of TRPC1 is coordinated by early events in the Ca^{2+} signaling cascade within this domain. Consistent with this, disruption of caveolar lipid raft domains decreases thapsigargin-stimulated Ca^{2+} entry in salivary gland cells (Lockwich et al. 2000; Brazer et al. 2003) and in platelets (Brownlow et al. 2004) and vascular smooth muscle cells (Beech 2005). Interestingly, expression of mutant caveolins lacking the lipid binding domain or TRPC1 lacking the caveolin-binding domain causes mislocalization of TRPC1 (Brazer et al. 2003). Thus, caveolin 1 has a role in plasma membrane expression of TRPC1, although it is not clear whether TRPC1 trafficking to, or its retention in, the plasma membrane depends on the interaction with caveolin.

Regulated TRPC1 trafficking to the membrane occurs in endothelial cells (Mehta et al. 2003) following stimulation with thrombin. Under these conditions, TRPC1 is assembled in a complex with inositol 1,4,5-trisphosphate receptor (IP_3R) and RhoA. Assembly of this complex and translocation of the channel to the plasma membrane is dependent on actin polymerization and mediated via its interaction of TRPC1 with RhoA. It is interesting to note that lipid raft domains are enriched in cytoskeletal and vesicle trafficking proteins as well as ezrin, which anchors the cytoskeleton to the plasma membrane. Additionally, PIP_2, which also scaffolds actin, is found in lipid rafts. Thus, further studies are required to determine the interplay between cytoskeleton and lipid rafts domains in regulating the assembly, localization, and function of TRPC1.

3.1.2
TRPC3

TRPC3 can form both store-independent and store-dependent channels (Putney 2005; Spassova et al. 2005). TRPC3 localization in the plasma membrane depends on the status of the cytoskeleton. Actin polymerization induces internalization of the channel, although actin depolymerization does not affect plasma membrane localization of TRPC3 (Lockwich 2001; Itagaki 2004). The calmodulin (CaM)/IP_3 receptor-binding (CIRB) region of TRPC3 is also involved in targeting the channel to the plasma membrane (Wedel et al. 2003). Whether TRPC3 trafficking or surface expression of the channel depends on its interaction with IP$_3$R and CaM is not yet clear. TRPC3 interaction with PLCγ regulates its trafficking and cell surface expression (van Rossum et al. 2005; Patterson et al. 2005). Interestingly, PLCγ facilitates interaction/anchoring of TRPC3 with the plasma membrane lipid, PIP$_2$. The PLCγ–TRPC3 interaction does not appear to be required for trafficking of TRPC3 to the plasma membrane or its activation, but rather for channel retention in the membrane. A TRPC3 mutant that cannot bind to PIP$_2$, but retains PLCγ binding, does not appear to reach the membrane. Thus, interaction with PIP$_2$ promotes surface expression of TRPC3. These studies demonstrate that both PLCγ and PIP$_2$ are important regulatory components of TRPC3. However, whether the agonist-induced decrease in PIP$_2$ levels directly affects TRPC3 trafficking is not yet known.

TRPC3 is dynamically regulated in response to muscarinic receptor stimulation (Singh et al. 2004). The protein is localized in mobile intracellular vesicles and interacts with key proteins involved in vesicular trafficking. Carbachol stimulation increases plasma membrane expression of TRPC3, this translocation is abolished by hydrolysis of vesicle-associated membrane protein (VAMP)2 by tetanus toxin treatment. Thus, fusion of intracellular vesicles is involved in regulated translocation of TRPC3 to the membrane. Whether the TRPC3–PLCγ–PIP$_2$ interaction is also involved in this translocation has not yet been determined. Regulated recruitment of TRPC3 is independent of changes in intracellular [Ca^{2+}]. Constitutive plasma membrane expression of TRPC3 is also achieved by vesicular trafficking and VAMP2-dependent membrane fusion events. Distinct from the regulated mechanism, it appears to be mediated via a process that is dependent on Ca^{2+}. The molecular components of these different mechanisms have not yet been identified. While TRPC3 does display spontaneous activity (Putney 2005), it is unclear whether active channels are inserted into the plasma membrane or whether the channels are activated following insertion.

3.1.3
TRPC4 and TRPC5

These channels interact with the PDZ-domain proteins Na$^+$/H$^+$ exchanger regulatory factor (NHERF) and ZO1 via their C-terminal PDZ-interacting domains (Tang et al. 2000). The interaction with NHERF mediates association of TRPC4 and TRPC5 with PLBβ and regulates their surface expression (Mery et al. 2002). The exact role of the TRPC4-interaction with PLCβ has not yet been determined. Translocation of TRPC4 to the plasma membrane in response to epidermal growth factor (EGF) stimulation has been recently reported (Odell et al. 2005). This interesting study showed that EGF stimulation of cells leads to activation of protein tyrosine kinases, fyn being the dominant kinase, which leads to phosphorylation of two tyrosine sites on the TRPC4 C-terminus. This increases the interaction of TRPC4 with NHERF as well as its exocytotic insertion into the plasma membrane. Thus, the dynamic interplay between EGF, tyrosine kinases, TRPC4, and NHERF coordinates plasma membrane expression and activation of TRPC4 channels. Interestingly, TRPC4 is also associated with caveolae (Torihashi et al. 2002) where growth factor receptor signaling proteins as well as NHERF binding proteins, such as ezrin, are localized. Involvement of vesicular trafficking mechanisms in TRPC4 activation has not yet been described.

TRPC5 is trafficked to specific sites in hippocampal neurons (Greka et al. 2003). TRPC1+TRPC5 heteromers are localized in the neurites, whereas TRPC5 homomers are found in the growth cones. Trafficking of TRPC5 to the growth cone is mediated via binding to the exocyst component protein stathmin 2 and Ca^{2+} entry via the channel inhibits extension of growth cones. Notably, SNARE proteins are found to be associated with the TRPC5-trafficking complex. Further, TRPC5 channels are localized in vesicles. Rapid trafficking of these vesicles to the plasma membrane is activated in response to stimulation of hippocampal and other neuronal cells with EGF and nerve growth factor (NGF), with relatively weaker response to brain-derived neurotrophic factor (BDNF) and insulin-like growth factor-1 (IGF-1). Incorporation of the channel into the plasma membrane also involves PI3-kinase (PI3K) and the GTPase, Rac1, as well as PI4P-5 kinase (Bezzerides et al. 2004). TRPC5-containing vesicles appear to be retained in a subplasma membrane region from where they are rapidly recruited to the membrane. The internalization mechanism is not yet known. Thus, both TRPC4 and TRPC5 are regulated in response to stimulation of tyrosine kinase receptors by various growth hormones.

3.1.4
TRPC6

Although there are no reported mechanisms for regulation of constitutive plasma membrane expression of TRPC6, the protein is translocated to the plasma membrane upon stimulation with muscarinic agonists by a Ca^{2+}-independent mechanism (Cayouette 2004). The tyrosine kinase, fyn, has been identified as an accessory protein for TRPC6 and suggested to regulate channel activation (Hisatsune et al. 2004). However, it is not clear whether phosphorylation has a role in this translocation. Although the involvement of SNARE proteins in TRPC6 trafficking has not yet been reported, a member of the dynamin family, MxA, interacts with TRPC6 (Lussier et al. 2005). While dynamin can potentially exert an effect on channel trafficking, i.e., internalization, the role of MxA in TRPC6 trafficking has not been directly demonstrated. In addition, recruitment of several proteins—including protein kinase C (PKC) and the muscarinic receptor to TRPC6—following stimulation with carbachol has been reported (Kim and Saffen 2005). It is important to recognize that both TRPC3 and TRPC6 are regulated by PIP_2 hydrolysis and undergo translocation to the plasma membrane in response to agonist stimulation. Furthermore, these TRPCs interact with each other to form heteromeric proteins. However, any direct impact of TRP multimerization on TRPC3/C6 trafficking is not yet known. It is also not clear whether trafficking of homomeric channels differ from that of heteromeric channels. This might be an important consideration for all TRPCs.

3.1.5
Invertebrate TRPCs

Regulated translocation also occurs in the case of invertebrate TRPC proteins. *Drosophila* TRPL shuttles from the microvillar membranes to the cell bodies in response to light, and this movement functions in long-term adaptation (Bahner et al. 2002; Montell 2005). The TRPC protein in worms, TRP-3, is expressed in spermatids in intracellular vesicles; upon sperm activation it is detected in the plasma membrane (Xu and Sternberg 2003). This translocation is correlated with an increase in store-operated calcium entry (SOCE), suggesting that regulated exocytosis is involved in Ca^{2+} influx.

3.2
TRPV Subfamily

TRPV channels are typically sensory in function and are activated by heat, changes in osmolarity, odorants, and mechanical stimuli (detailed descriptions of the function of these channels are provided in other chapters of this book). There is relatively less information regarding the mechanisms involved plasma membrane expression of TRPV channels, although regulated trafficking was first described for TRPV2. TRPV1 and TRPV2 are dynamically translocated to the plasma membrane following cell stimulation (Morenilla-Palao et al. 2004). This trafficking has been suggested to contribute to thermal inflammatory hyperalgesia. TRPV1 interacts with the vesicular SNARE proteins snapin and synaptotagmin IX, which participate in channel exocytosis and colocalize with the proteins in intracellular vesicular compartments in neuronal cells. PKC promotes insertion of the channel into the plasma membrane via a mechanism regulated by SNARE proteins. Metabotropic glutamate receptor stimulation also regulates the same PKC-dependent trafficking mechanism. Recently, it was shown that insulin and IGF-1 also induce translocation of TRPV1 (Van Buren et al. 2005). The stimulation is mediated via activation of PI3K and PKC. Thus, PKC activation appears to be a central mechanism for regulating plasma membrane expression and function of TRPV1. TRPV1 trafficking in response to temperature sensation has not yet been reported.

Mouse TRPV2 is also trafficked from intracellular pools to the plasma membrane in response to stimulation with either IGF-1 or neuropeptide head activator in insulinoma cells and neuroendocrine cells (Boels et al. 2001; Kanzaki et al. 2001). Additionally, regulated trafficking of TRPV2 also occurs in mouse mast cells (Stokes et al. 2004, 2005). It is interesting to note that TRPV2 forms a complex with PKA and a protein-kinase adaptor protein. This interaction could facilitate the trafficking of TRPV2 to the plasma membrane by cyclic AMP (cAMP) mobilizing stimuli. An interaction between TRPV2 and the recombinase gene activator (RGA), a four-transmembrane protein, potentiates TRPV2 surface localization but does not participate in the final exocytotic step (Stokes et al. 2005). Whether all these observations reflect the same basic mechanism for TRPV2 trafficking needs to be established.

As mentioned in Sect. 3, accessory proteins and interactions between TRP monomers can have a critical role in the trafficking and regulation of TRP channels. Recent studies demonstrate that alternatively spliced variants of TRPV4 with deletions in the N-terminal region show defective trafficking and are retained in the endoplasmic reticulum (ER) (Arniges et al. 2005). How differences in protein structure of these TRP variants affect their localization is not known. TRPV5 and TRPV6 interact with the S-100–annexin 2 complex, which acts as a scaffold for the channels in the plasma membrane. Notably, annexin 2 also binds membrane lipids (van de Graf et al. 2003). Whether the S-100-TRPV5/6 interaction is also regulated by any stimuli is not known. Presently, TRPV5 and TRPV6 have been suggested to form spontaneously active channels. Thus, any factors that affect their plasma membrane expression can directly control their function.

3.3
TRPM Subfamily

There are eight mammalian TRPM proteins. TRPM proteins are also encoded in the *Drosophila* and *Caenorhabditis elegans* genomes, and in worms (Montell 2005). Studies with TRPM1 suggest that it is regulated by trafficking. TRPM1-S, an alternatively spliced variant of TRPM1 that lacks the transmembrane and C-terminal region, associates with TRPM1 and suppresses the activity of the channel by inhibiting its translocation to the plasma membrane (Xu et al. 2001). Thus, TRPM1 activation potentially involves regulation of its plasma membrane expression, although such a mechanism has not yet been directly described. TRPM6 is capable of efficiently heteromultimerizing with TRPM7, and this interaction promotes translocation of TRPM6 from an intracellular compartment to the plasma membrane (Chubanov et al. 2004). Mutations in the channel affect assembly of the channels as well as trafficking.

3.4
TRPP and TRPML Subfamilies

TRPP and TRPML exhibit high homology with each other but low similarity to the other TRPs. There is little information about trafficking mechanisms involved in regulation of these TRP proteins (Kottgen and Walz 2005a). Coexpression of TRPP1 along with TRPP2 induces translocation of TRPP2 to the plasma membrane and generation of a Ca^{2+}-permeable nonselective cation conductance (Tsiokas et al. 1997; Neill et al. 2004). TRPPs may be situated on primary cilia, which protrude from the surface of most vertebrate cells, to facilitate the detection of extracellular signaling molecules or to sense changes in fluid flow, osmolarity, or other mechanical stress. How the trafficking of these channels to this specific cellular locale is regulated is not known. The adaptor proteins PACS1 and PACS2 have been reported to control ER-plasma membrane trafficking of TRPP2. This mechanism depends on the phosphorylation of TRPP2 (Kottgen et al. 2005b). The main TRPML subfamily member, TRPML1 (mucolipidin 1; ML1), is responsible for the lysosomal storage disorder mucolipidosis type IV (Sun et al. 2005). The subcellular distribution of TRPML proteins is largely unresolved. TRPML1 has been proposed to be a lysosomal-associated cation channel (Soyombo et al. 2005). However, heterologous expression in *Xenopus* oocytes demonstrated translocation of TRPML1 to the plasma membrane upon treatment with the Ca^{2+} ionophore ionomycin (LaPlante et al. 2004). Whether such translocation also occurs in the case of endogenous TRPML is not known.

4
Emerging Commonalities Underlying TRP Channel Trafficking

Currently available data suggest some marked commonalities in the underlying trafficking mechanisms. Although some of these mechanisms are likely to be more basic mechanisms for plasma membrane proteins in general, they currently form the basis of our understanding of TRP channel trafficking.

4.1
Assembly of TRP Channel Complexes

Homomeric and heteromeric interactions between TRPs can give rise to a large number of channels with diverse properties. The type of channel generated in cells is dictated by the specific physiological function they serve in a particular cell. For example, TRPC5 homomers are found in growth cones, and C1+C5 heteromers are localized in the neurites (Greka et al. 2003), the former affects growth cone extension. While dominant routing signals in the channel proteins can dictate their cellular locale, TRP proteins interact with a number of signaling, scaffolding, and trafficking proteins to form distinct multiprotein complexes. Typically, TRP channels appear to be associated with (1) signaling proteins such as plasma membrane receptors, PLCs, PKCs and other protein kinases, and CaM; (2) ion channels and transporters; (3) scaffolding proteins such as INAD, HOMER, and NHERF; (4) cytoskeletal/microtubule components (NINAC, ezrin, tubulin, Rho-GTPase, S-100, annexin 2, 4.1, and NHERF);

and (5) vesicular trafficking proteins such as plasma membrane and vesicle SNARES, small molecular GTPases, PI3 and PIP$_4$ kinases, dynamin, and caveolin (Montell 2005; Ambudkar 2005; Kiselyov 2005). Nonprotein components such as PIP$_2$, DAG, and plasma membrane lipid raft domains also appear to be important in TRPC channel function and regulation (Patterson et al. 2005; Ambudkar 2004, 2005). In addition to regulating the specificity and rate of TRP channel activation, interactions between components of the TRP-signalplex also determine trafficking of the channel to specific cellular locations and retention of the protein in the plasma membrane, as well as its inactivation. A good example of the critical role of the signalplex in channel function is the *Drosophila* TRP (Montell 2005) where TRP and INAD form the core complex required for assembly and retention of the channel in the rhabdomeres, as well as rapid activation and inactivation. In addition, there is regulated recruitment and trafficking of channel components such as TRPL and TRPγ.

4.2
Role of Cytoskeletal Proteins in TRP Trafficking

Interaction of TRPs with components of the cytoskeleton and microtubules serves a stable anchor for retaining the channels in specific plasma membrane locations (Minke and Cook 2002; Montell 2005). Further, dynamic interactions with cytoskeletal components or microtubules can provide a system for rapid or slow trafficking of channel to the membrane. Regulated changes in the assembly of the cytoskeleton can also contribute to the trafficking of TRP channels. These include localized remodeling, which can either promote or limit access of the vesicles to the plasma membrane. Cytoskeletal interactions can also be involved in bringing required molecular components that are involved in regulating their function or trafficking in close proximity to TRP channels. TRPC4 and TRPC5 interaction with NHERF links the channels to the cytoskeleton via ezrin, which interacts with NHERF as well as the cytoskeleton (Tang et al. 2000; Odell et al. 2005). Myosin light chain kinase regulates TRPC5 function (Kim et al. 2005; Shimizu et al. 2005), although it is not clear whether this involves direct effects on TRPC5 or is mediated via regulation of the cytoskeleton. Other examples of such interactions with the cytoskeleton involve 4.1 protein interaction with TRPC4 (Cioffi et al. 2005) and annexin 2–S100-interaction with TRPV5 and TRPV6 (van de Graf 2004). Dynamic recruitment of proteins that regulate the status of the cytoskeleton has been reported in the case of TRPC1 (Mehta 2003). Actin polymerizing agents cause internalization of TRPC3 (Lockwich et al. 2001). This might be a result of their association with caveolar plasma membrane lipid rafts, which have been shown to be internalized when ezrin is phosphorylated, resulting in polymerization of cortical actin. Importantly, cytoskeletal components also interact with PIP$_2$, and thus changes in PIP$_2$ levels significantly affect the status of the underlying cytoskeleton. Vesicular trafficking, including internalization and exocytosis, is regulated via compartmentalized changes in plasma membrane PIP$_2$-cytoskeletal interactions. Thus, both protein and nonprotein components, such as plasma membrane phosphoinositides, are likely to contribute to the regulation of TRP channels.

4.3
Vesicular Trafficking Mechanisms

It is important to emphasize the emerging role of vesicular trafficking mechanisms in regulating plasma membrane expression of TRP channels. Such mechanisms impact both constitutive and regulated trafficking of TRP channels. Thus, it is significant that a number of proteins involved in vesicular trafficking are auxiliary proteins of TRP channels. These include exocyst proteins, such as stathmin; SNARE proteins such as VAMP2, syntaxin, SNAP, and NSF; and small molecular GTP-binding proteins that are involved in vesicular trafficking such as Rac1 (TRPC5), rab11 (TRPV5 and TRPV6), and Rho-GTPase (TRPC1). Other modulators of vesicular trafficking that have been associated with TRPs include PI3K, caveolin, PLCγ, and likely PIP$_2$ itself (Ambudkar et al. 2004; Bezzerides et al. 2004; Patterson et al. 2005). While generation of PIP$_2$ is connected with protein insertion into the plasma membrane, recruitment of PIP kinases and PIP$_2$-binding proteins initiates membrane retrieval (Czech 2003). Further vesicular trafficking proteins as well as cytoskeletal attachment to the plasma membrane appears to be localized in the same microdomain. Since PIP$_2$ is relatively enriched in lipid raft domains, it is interesting to speculate that these plasma membrane lipid microdomains coordinate trafficking as well as regulation of TRP channels (Ambudkar 2004).

The type of molecular components that have thus far been identified in TRP-signalplexes are quite consistent with the involvement of specialized microdomains, local rearrangement of cytoskeleton or microtubules, and localized trafficking of vesicles. TRPC5 trafficking involves a "kiss-and-run" mechanism that involves rapid insertion and retrieval of vesicles to regulate the level of the channel in the plasma membrane (Bezzerides et al. 2004). The channel basically recycles into a subplasma membrane vesicular pool from which it can be recruited following stimulation. TRPC3 is also trafficked to the plasma membrane region and retained in a subplasma membrane pool (Singh et al. 2004). Thus, intracellular recycling vesicles significantly contribute to the regulation of TRP channel trafficking. The exact route and proteins involved in the delivery of TRP proteins from the ER to the plasma membrane should be investigated.

5
Conclusion

TRP channels represent a relatively large family of distinct ion channels. Members within TRP subfamilies form homomeric or heteromeric channels that differ in their function and regulation. There are sufficient data to demonstrate that TRP channels are localized and regulated within signaling microdomains via both protein and nonprotein components. It is also increasingly evident that constitutive and regulated trafficking processes contribute to the plasma membrane expression and regulation of TRP channels. Homomeric or heteromeric interaction between TRP monomers and their interaction with auxiliary proteins determine the expression of the channels in the surface membrane or their movement to the plasma membrane. Critical mechanisms governing TRP channel trafficking involve cytoskeletal remodeling, vesicular trafficking, and interaction with scaffolding and cytoskeletal proteins. A major question that needs to be addressed is how cells coordinate the precise temporal and spatial constraints involved in the trafficking of distinct TRP channels under resting conditions and following stimulation.

References

Altschuler Y, Hodson C, Milgram SL (2003) The apical compartment: trafficking pathways, regulators, and scaffolding proteins. Curr Opin Cell Biol 15:423–429

Ambudkar IS (2004) Cellular domains that contribute to Ca^{2+} entry events. Sci STKE 2004:pe29

Ambudkar IS (2005) Ca(2+) signaling microdomains: platforms for the assembly and regulation of TRPC channels. Trends Pharmacol Sci 27:25–32

Ambudkar IS, Brazer SC, Liu X, et al (2004) Plasma membrane localization of TRP channels: role of caveolar lipid rafts. Novartis Found Symp 258:63–70

Arniges M, Fernandez-Fernandez JM, Albrecht N, et al (2005) Human TRPV4 channel splice variants revealed a key role of ankyrin domains in multimerization and trafficking. J Biol Chem 281:8990

Bahner M, Frechter S, Da Silva N, Minke B, Paulsen R, Huber A (2002) Light-regulated subcellular translocation of Drosophila TRPL channels induces long-term adaptation and modifies the light-induced current. Neuron 34:83–93

Bandyopadhyay BC, Swaim WD, Liu X, et al (2005) Apical localization of a functional TRPC3/TRPC6-Ca^{2+}-signaling complex in polarized epithelial cells. Role in apical Ca^{2+} influx. J Biol Chem 280:12908–12916

Beech DJ (2005) TRPC1: store-operated channel and more. Pflugers Arch 451:53–60

Bezzerides V, Ramsey IS, Kotecha S, et al (2004) Rapid vesicular translocation and insertion of TRP channels. Nat Cell Biol 6:709–720

Boels K, Glassmeier G, Herrmann D, et al (2001) The neuropeptide head activator induces activation and translocation of the growth-factor-regulated Ca^{2+}-permeable channel GRC. J Cell Sci 114:3599–3606

Bollimuntha S, Cornatzer E, Singh BB (2005) Plasma membrane localization and function of TRPC1 is dependent on its interaction with β tubulin in retinal epithelial cells. Vis Neurosci 22:163–170

Brazer SC, Singh BB, Liu X, et al (2003) Caveolin-1 contributes to assembly of store-operated Ca^{2+} influx channels by regulating plasma membrane localization of TRPC1. J Biol Chem 278:27208–27215

Brown DA, London E (1998) Functions of lipid rafts in biological membranes. Annu Rev Cell Dev Biol 14:111–136

Brownlow SL, Harper AG, Harper MT, et al (2004) A role for hTRPC1 and lipid rafts domains in store-operated calcium entry in human platelets. Cell Calcium 35:107–113

Cayouette S, Lussier MP, Mathieu L, et al (2004) Exocytotic insertion of TRPC6 channel into the plasma membrane upon Gq protein-coupled receptor activation. J Biol Chem 279:7241–7246

Chubanov V, Waldegger S, Mederosy Schnitzeler M, et al (2004) Disruption of TRPM6/ TRPM7 complex formation by a mutation in the TRPM6 gene causes hypomagnesemia with secondary hypocalcemia. Proc Natl Acad Sci U S A 101:2894–2899

Cioffi DL, Wu S, Alexeyev M, et al (2005) Activation of the endothelial store-operated ISOC Ca^{2+} channel requires interaction of protein 4.1 with TRPC4. Circ Res 97:1164–1172

Czech MP (2003) Dynamics of phosphoinositides in membrane retrieval and insertion. Annu Rev Physiol 65:791–815

Goel M, Sinkins WG, Zuo CD, et al (2005) Identification and localization of TRPC channels in rat kidney. Am J Physiol Renal Physiol 290:F1241–F1252

Greka A, Navarro B, Oancea E, et al (2003) TRPC5 is a regulator of hippocampal neurite length and growth cone morphology. Nat Neurosci 6:837–845

Hisatsune C, Kuroda Y, Nakamura K, et al (2004) Regulation of TRPC6 activity by tyrosine phosphorylation. J Biol Chem 279:18887–18894

Hoenderop JG, Bindels RJ (2005) Epithelial Ca^{2+} and Mg^{2+} channels in health and disease. J Am Soc Nephrol 16:15–26

Hofmann T, Schaefer M, Schultz G, et al (2002) Subunit composition of mammalian transient receptor potential channels in living cells. Proc Natl Acad Sci U S A 99:7461–7466

Isshiki M, Anderson RGW (2003) Function of caveolae in Ca^{2+} entry and Ca^{2+}-dependent signal transduction. Traffic 4:717–723

Itagaki K, Kannan KB, Singh BB, et al (2004) Cytoskeletal reorganization internalizes multiple transient receptor potential channels and blocks calcium entry into human neutrophils. J Immunol 172:601–607

Jahn R, Lang T, Sudhof TC (2003) Membrane fusion. Cell 112:519–533

Kim JY, Saffen D (2005) Activation of M1 muscarinic receptors stimulates the formation of a multiprotein complex centered on TRPC6 channels. J Biol Chem 280:32035–32047

Kim MT, Kim BJ, Lee JH, et al (2005) Involvement of calmodulin and myosin light chain kinase in the activation of mTRPC5 expressed in HEK cells. Am J Physiol Cell Physiol 290:C1031–C1040

Kiselyov K, Kim JY, Zeng W, et al (2005) Protein-protein interaction and functionTRPC channels. Pflugers Arch 451:116–124

Kottgen M, Walz G (2005) Subcellular localization and trafficking of polycystins. Pflugers Arch 451:286–293

Kottgen M, Benzing T, Simmen T, et al (2005) Trafficking of TRPP by PACS protein represents a novel mechanisms of mechanisms of ion channel regulation. EMBO J 24:705–716

LaPlante JM, Ye CP, Quinn SJ, et al (2004) Functional links between mucolipin and Ca^{2+}-dependent membrane trafficking in mucolipidosis IV. Biochem Biophys Res Commun 322:1384–1391

Laud AJ, Prior IA (2004) Plasma membrane microdomains: organization, function, and trafficking. Mol Membr Biol 21:193–205

Liu X, Wang W, Singh BB, et al (2000) Trp1, a candidate protein for the store-operated Ca(2+) influx mechanism in salivary gland cells. J Biol Chem 275:3403–3411

Lockwich TP, Liu X, Singh BB, et al (2000) Assembly of Trp1 in a signaling complex associated with caveolin-scaffolding lipid raft domains. J Biol Chem 275:11934–11942

Lockwich TP, Singh BB, Liu X, et al (2001) Stabilization of cortical actin induces internalization of transient receptor potential 3 (Trp3)-associated caveolar Ca^{2+} signaling complex and loss of Ca^{2+} influx without disruption of Trp3-inositol trisphosphate receptor association. J Biol Chem 276:42401–42408

Lussier MP, Cayoutte S, Lepage PK, et al (2005) MxA, a member of the dynamic superfamily interacts with the anykyrin-like repeat domain of TRPC. J Biol Chem 280:19393–193400

M Kanzaki, Zhang YO, Mashima H, et al (1999) Translocation of a calcium-permeable cation channel induced by insulin-like growth factor-I. Nat Cell Biol 1:165–170

Ma D, Jan L (2002) ER transport signals and trafficking of potassium channels and receptors. Curr Opin Neurobiol 12:287–292

Mehta D, Ahmed GU, Paria BC, et al (2003) RhoA interaction with inositol 1,4,5-trisphosphate receptor and transient receptor potential channel-1 regulates Ca^{2+} entry. Role in signaling increased endothelial permeability. J Biol Chem 278:33492–33500

Mery L, Strauss B, Dufour JF, et al (2002) The PDZ-interacting domain of TRPC4 controls its localization and surface expression in HEK293 cells. J Cell Sci 115:3497–3508

Minke B, Cook B (2002) TRP channel proteins and signal transduction. Physiol Rev 82:429–472

Montell C (2005) The TRP superfamily of cation channels. Sci STKE 2005:re0

Montell C, Birnbaumer L, Flockerzi V, et al (2002) A unified nomenclature for the superfamily of TRP cation channels. Mol Cell 9:229–231

Moran MM, Xu H, Clapham DE (2004) TRP ion channels in the nervous system. Curr Opin Neurobiol 14:362–369

Morenilla-Palao C, Planells-Cases R, Garcia-Sanz N, et al (2004) Regulated exocytosis contributes to protein kinase C potentiation of vanilloid receptor activity. J Biol Chem 279:25665–25672

Neill T, Moy GW, Vacquier VD, et al (2004) Polycystin-2 associates with the polycystin-1 homolog, suREJ3, and localizes to the acrosomal region of sea urchin spermatozoa. Mol Reprod Dev 67:472–477

Odell AF, Scott JL, Van Helden DI, et al (2005) Epidermal growth factor induces tyrosine phosphorylation membrane insertion and activation of transient receptor potential 4. J Biol Chem 280:37974–37984

Pattersen RL, VanRossum DB, Nikolaidis N, et al (2005) Phospholipase C gamma: diverse roles in receptor-mediated calcium signaling. Trends Biochem Sci 30:688–697

Pedersen SF, Owsianik G, Nilius B (2005) TRP channels: an overview. Cell Calcium 38:233–252

Putney JW (2005) Physiological mechanisms of TRPC activation. Pflugers Arch 451:29–34

Royle SJ, Murrell-Lagnado RD (2003) Constitutive cycling: a general mechanism to regulate cell surface proteins. Bioessays 25:39–46

Schaefer M (2005) Homo- and heteromeric assembly of TRP channel subunits. Pflugers Arch 451:35–42

Schilling WP, Goel M (2004) Mammalian TRPC channel subunit assembly. Novartis Found Symp 258:18–30

Shimizu S, Yoshida T, Wakamori M, et al (2005) Calmodulin dependent myosin light chain kinase is essential for activation of TRPC5 channels expressed in HEK293 cells. J Physiol 570:219–235

Singh BB, Zheg C, Liu X, et al (2001) Trp1-dependent enhancement of salivary gland fluid secretion: role of store-operated calcium entry. FASEB J 15:1652–1654

Singh BB, Lockwich TP, Liu X, et al (2004) VAMP-2-dependent exocytosis regulates plasma membrane insertion of TRPC3 channels and contributes to agonist-stimulated Ca^{2+} influx. Mol Cell 15:635–646

Soyombo AA, Tjon-Kon-Sang S, Rbaibi Y, et al (2005) TRP-ML1 regulates lysosomal pH and lysosomal lipid hydrolytic activity. J Biol Chem 281:7294–7301

Spassova MA, Soboloff J, He LP, et al (2004) Calcium entry mediated by SOCs and TRP channels: variations and enigma. Biochim Biophys Acta 1742:9–20

Stokes AJ, Shimoda LM, Koblan-Huberson M, et al (2004) A TRPV2-PKA signaling module for transduction of physical stimuli in mast cells. J Exp Med 200:137–147

Stokes AJ, Wakano C, Del Carmen KA, et al (2005) Formation of a physiological complex between TRPV2 and RGA protein promotes cell surface expression of TRPV2. J Cell Biochem 94:669–683

Sun M, Goldin E, Stahl S, et al (2000) Mucolipidosis type IV is caused by mutations in a gene encoding a novel transient receptor potential channel. Hum Mol Genet 9:2471–2478

Symons M, Rusk N (2003) Control of vesicular trafficking by Rho GTPases. Curr Biol 13:409–418

Tang Y, Tang J, Chen Z, et al (2000) Association of mammalian Trp4 and phospholipase C isozymes with a PDZ domain-containing protein, NHERF. J Biol Chem 275:37559–37564

Torihashi S, Fujimoto T, Trost C, Nakayama S (2002) Calcium oscillation linked to pacemaking of interstitial cells of Cajal: requirement of calcium influx and localization of TRP4 in caveolae. J Biol Chem 277:19191–19197

Tsiokas L, Kim E, Arnold T, et al (1997) Homo- and heterodimeric interactions between the gene products of PKD1 and PKD2. Proc Natl Acad Sci USA 94:6965–6970

Van Buren JJ, Bhat S, Rotello R, et al (2005) Sensitization and translocation of TRPV1 by insulin and IGF-1. Mol Pain 1:17–28

van de Graaf SF, Hoenderop J, Gkika D, et al (2003) Functional expression of the epithelial Ca^{2+} channels (TRPV5 and TRPV6) requires association of the S100A10-annexin 2 complex. EMBO J 22:1478–1487

van Rossum DB, patterson RL, Sharma S, et al (2005) Phospholipase Cgamma1 controls surface expression of TRPC3 through an intermolecular PH domain. Nature 434:99–104

Wedel BJ, Vazquez G, McKay RR, et al (2003) A calmodulin/IP3 receptor binding region targets TRPC3 to the plasma membrane in a calmodulin/IP3 receptor-independent process. J Biol Chem 278:25758–25765

Xu XZ, Sternberg PW (2003) A C. elegans sperm TRP protein required for sperm-egg interactions during fertilization. Cell 114:285–297

Xu XZ, Moebius F, Gill DL, et al (2001) Regulation of melastatin, a TRP-related protein, through interaction with a cytoplasmic isoform. Proc Natl Acad Sci USA 98:10692–110697

TRPC Channels: Interacting Proteins

K. Kiselyov[1] · D. M. Shin[2] · J.-Y. Kim[3] · J. P. Yuan[3] · S. Muallem[3] (✉)

[1] Department of Biological Sciences University of Pittsburgh, 15260, Pittsburgh PA, , USA

[2] Department of Oral Biology, Oral Science Research Center, Brain Korea 21 Project for Medical Science, Yonsei University College of Dentistry, 120-752 Seoul, Korea

[3] Department of Physiology, University of Texas Southwestern Medical Center at Dallas, Dallas TX, 75390, USA
Shmuel.Muallem@UTSouthwestern.edu

1	Activation Mechanisms and Permeation Properties of TRPC Channels in Recombinant and Native Systems	560
1.1	TRPC Channels in Receptor-Activated Ca^{2+} Influx	560
1.2	Multimerization of TRPC Isoforms	560
1.3	Interaction of TRPC Channels with ER Resident Proteins	561
2	Interaction of TRPC Channels with Auxiliary Proteins	562
2.1	Regulatory Proteins	563
2.1.1	Protein Kinases as Regulators of TRPC Channels Activity	563
2.1.2	Calmodulin	564
2.2	Proteins Involved in the PM Targeting and Insertion of TRPC Channels	564
3	TRPC Channels and Scaffolding Proteins	565
3.1	Near-Membrane Scaffolds	565
3.2	Scaffolds that Mediate PM/ER Communication	566
3.2.1	Junctate	566
3.2.2	Homer	567
References		568

Abstract TRP channels, in particular the TRPC and TRPV subfamilies, have emerged as important constituents of the receptor-activated Ca^{2+} influx mechanism triggered by hormones, growth factors, and neurotransmitters through activation of phospholipase C (PLC). Several TRPC channels are also activated by passive depletion of endoplasmic reticulum (ER) Ca^{2+}. Although in several studies the native TRP channels faithfully reproduce the respective recombinant channels, more often the properties of Ca^{2+} entry and/or the store-operated current are strikingly different from that of the TRP channels expressed in the same cells. The present review aims to discuss this disparity in the context of interaction of TRPC channels with auxiliary proteins that may alter the permeation and regulation of TRPC channels.

Keywords TRP channels · Receptor-induced Ca^{2+} influx · Scaffold proteins · Homer · Junctate

Abbreviations

ER	Endoplasmic reticulum
IP$_3$	Inositol (1,4,5)-trisphosphate
IP$_3$R	IP$_3$ receptor
PLC	Phospholipase C
PM	Plasma membrane
RyR	Ryanodine receptor
TRP	Transient receptor potential (channels)

1
Activation Mechanisms and Permeation Properties of TRPC Channels in Recombinant and Native Systems

1.1
TRPC Channels in Receptor-Activated Ca^{2+} Influx

Ca^{2+} signaling mediates cell functions that occur on time scales of microseconds to several days (Berridge 2001; Dolmetsch et al. 1998; Orrenius et al. 2003). The Ca^{2+} signal evoked by plasma membrane (PM)-localized receptors involves Ca^{2+} release from internal stores, mostly the endoplasmic reticulum (ER), which is followed by Ca^{2+} influx across the PM (Parekh and Putney 2005; Putney et al. 2001). The molecular identity of the receptor-activated Ca^{2+} influx channels is not known with certainty. A large body of evidence implicates a member of the transient receptor potential (TRP) channels in receptor-activated Ca^{2+} influx. The evidence that supports a role of TRPC and TRPV channels in receptor-operated Ca^{2+} influx comes from three sources: (1) selective knockdown using small interfering RNA (siRNA) or antisense DNA probes (Dietrich et al. 2005; Rao et al. 2005; Soboloff et al. 2005; Wu et al. 2000, 2002, 2004; Zagranichnaya et al. 2005), (2) pharmacological assays (Boulay et al. 1997; Fasolato and Nilius 1998; Liu et al. 2005a; Thebault et al. 2005), and (3) direct demonstrations of TRPC/V-like activity in several cell types (Dietrich et al. 2005; Freichel et al. 2001; Greka et al. 2003; Kim et al. 2003; Kriz 2005; Liu et al. 2000, 2003; Paria et al. 2004; Reiser et al. 2005; Sweeney et al. 2002; Thebault et al. 2005; Wang and Poo 2005; Winn et al. 2005). These results suggest a role of TRPC channels in the receptor-operated Ca^{2+} influx and indicate that several of these channels may be activated by store depletion.

1.2
Multimerization of TRPC Isoforms

Many PM- and ER-resident channels, including Ca^{2+} channels, retain their selectivity and regulatory mechanism when expressed as recombinant proteins. This is not the case for several TRPC channels whose permeation and activation mechanisms are often very different depending on the expression systems and even on the level of the expression. Thus, a recombinant TRPC3 behaves as a large conductance, nonselective cation channel, which can be activated by the diacylglycerol (DAG) analog 1-oleoyl-2-acetyl-*sn*-glycerol (OAG) (Hofmann et al. 1999). By contrast, TRPC3 knockdown in human embryonic kidney (HEK)-293 cells suppresses receptor- and store depletion-induced Ca^{2+} influx, the latter of which displays high Ca^{2+} selectivity and is not activated by OAG (Wu et al. 2000). Furthermore, TRPC3 behaves as an OAG-activated, store-independent channel when expressed at high levels (Trebak et al. 2002; Vazquez et al. 2001), but as partially (Trebak et al. 2002; Vazquez et al. 2001) or largely (Kiselyov et al. 1998) store-activated when expressed at low levels.

A significant contributing factor for the discrepancies between permeability and activation mechanism of native and expressed TRPC channels is likely the multimerization of several TRPC subunits expressed in the same cells. Interaction between the native TRPC isoforms is evident, as downregulation of TRPC isoforms reduces Ca^{2+} influx to the same extent as knockdown of a combination or all isoforms expressed in a given cell (Wu et al. 2000; Zagranichnaya et al. 2005). TRPC1, -4, and -5 can interact with each other as well as TRPC3, -6, and -7 in native cells (Goel et al. 2002; Liu et al. 2005b; Schilling and Goel 2004) and expression systems (Strubing et al. 2001). Biochemical and functional interactions between TRPC1 and TRPC5 were shown in hippocampal neurons (Goel et al. 2002; Strubing et al. 2001), whereas similar interactions between TRPC1 and TRPC3 were reported in salivary gland cell lines (Liu et al. 2005b). Coexpression of TRPC1 and TRPC5 (Strubing et al. 2001) and TRPC1 and TRPC4 in HEK-293 cells (Gudermann et al. 2004) resulted in a novel channel conductance. The differential activation of TRPC6 by agonist stimulation and OAG in A75r5 cells was attributed to activation of TRPC6 alone by OAG and activation of multiple isoforms (TRPC1, -4, and -6) by agonist stimulation (Soboloff et al. 2005).

1.3
Interaction of TRPC Channels with ER Resident Proteins

Another mechanism that may account for the apparent dependence of the activation pattern of recombinant TRPCs on the expression condition is interaction of TRPC channels with ER proteins. Without exception, all TRPC channels interact with inositol 1,4,5-trisphosphate receptors (IP$_3$Rs), as was demonstrated by coimmunoprecipitation (Kiselyov et al. 1999; Yuan et al. 2003) and pull-down (Tang et al. 2001; Zhang et al. 2001) assays. Interaction of TRPC1 and TRPC6 with IP$_3$Rs was used as a readout of the composition of the bradykinin B2 and muscarinic M1 signaling complexes and to probe proximity of IP$_3$Rs to the G protein-coupled receptors (GPCRs) (Delmas et al. 2002). The interaction between TRPCs and IP$_3$Rs is mediated by the N-terminal domain of the IP$_3$R and the C-termini and possibly the N-termini of the TRPCs (Tang et al. 2001; Zhang et al. 2001). The IP$_3$R and TRPC binding sites were identified and shown to participate in TRPC channels gating (Boulay et al. 1999; Zhang et al. 2001). TRPC3 and TRPC1 were also shown to interact with RyRs and the site of interaction was localized at the foot structure of the RyRs (Kiselyov et al. 2000; Sampieri et al. 2005).

A new regulator of Ca^{2+} influx channels discovered recently is stromal interaction molecule (STIM)1 (Liou et al. 2005; Roos et al. 2005; Zhang et al. 2005). STIM1 is an ER resident protein that functions as a sensor of ER Ca^{2+} content and translocates toward the PM in response to store depletion (Spassova et al. 2006; Zhang et al. 2005). At the PM region it assembles into punctae and stimulates store-operated Ca^{2+} channels (SOCs) and calcium release-activated current (I_{CRAC}) (Liou et al. 2005; Roos et al. 2005; Spassova et al. 2006; Zhang et al. 2005). If STIM1 interacts with TRPC channels, it may participate in the communication between the ER and PM through its interaction and activation of the TRPC channels.

It should be noted that there is still a significant disagreement concerning the physiological implication of the interaction between IP$_3$Rs/RyRs and TRPC channels. Thus, while several studies (Boulay et al. 1999; Delmas et al. 2002), including ours (Kiselyov et al. 1998; Yuan et al. 2003), demonstrate that IP$_3$Rs are required or at least are involved in activation of TRPC3, others concluded that DAG can specifically activate TRPC3 (but not TRPC1) in the absence of IP$_3$Rs (Ma et al. 2001; Vazquez et al. 2001; Venkatachalam et al. 2001). Of course these findings are not mutually exclusive. It is possible that regulation of TRPC channels by IP$_3$Rs is conditional and requires auxiliary or scaffolding proteins that enhance and modulate the interaction between the proteins to allow the faithful regulation of TRPC channels by the IP$_3$Rs/RyRs.

2
Interaction of TRPC Channels with Auxiliary Proteins

TRPC channels interact directly with several proteins that regulate their activity and with several scaffolding proteins that assemble them within Ca^{2+}-signaling complexes. As illustrated in Fig. 1, interactions of TRPC channels with different proteins can be divided into three categories: (1) proteins that directly interact with TRPC channels at near-PM domain; (2) proteins that mediate insertion and retrieval of TRPC channels from the PM; and (3) scaffolding proteins that mediate TRPC channel–ER coupling. The chapter by G.Y Rychkov and G.J. Barritt (in this volume) is focused on proteins that appear to interact only with TRPC channels and modulate their activity.

TRPC Channels: Interacting Proteins 563

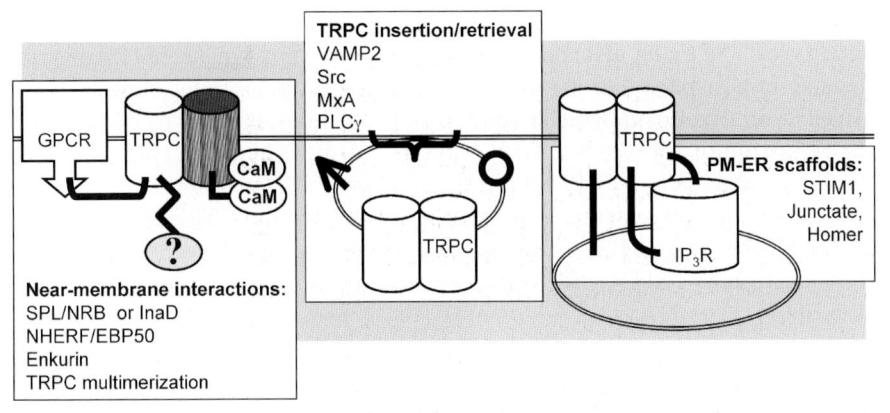

Fig. 1 A schematic of the key scaffolding and auxiliary proteins that interact with TRPC channels in the near-PM domain (*left*), mediate insertion and retrieval of TRPC channels from the PM (*middle*), and mediate the coupling between TRPC channels and the ER coupling (*right*). The scaffolding and auxiliary proteins are identified by *thick lines*. Each panel lists the proteins mediating the specific interactions. The binding partner(s) of NHERF/EBP50 and the other scaffolds are unknown and are marked by a *question mark*. The role of SPL/NRB and MxA is only partially established. Although VAMP2 and PLCγ were shown to interact with TRPC3, for simplicity the direct interaction is not shown in the model. The *left panel* shows (*left to right*) the interaction of GPCR with TRPC channels mediated by InaD (Chevesich et al. 1997; Tsunoda et al. 1997) or SPL/NRB (Wang 2005), and that interaction is aided by TRPC-NHERF complexes (Obukhov and Nowycky 2004; Tang et al. 2000), thus facilitating TRPC multimerization (Schilling and Goel 2004; Strubing et al. 2001), which is regulated by direct (Zhu 2005) or Enkurin-mediated TRPC channel interaction with calmodulin (*CaM*) (Sutton et al. 2004). The *middle panel* shows (*left to right*) the role of VAMP2 (*arrow*) in PM insertion of TRPC channels (Singh et al. 2004), which may be related to TRPC tyrosine phosphorylation that promotes insertion of TRPC channels into the PM (Bezzerides et al. 2004; Hisatsune et al. 2004). The role of MxA in membrane fission and TRPC channels withdrawal (*circle*) is inferred from its homology with dynamins; PLCγ regulates the PIP_2-dependent (*brackets*) steady-state level of TRPC channels' expression at the PM (van Rossum et al. 2005). The *right panel* shows (*left to right*) PM-ER coupling by STIM1 (Feske et al. 2005; Liou et al. 2005; Roos et al. 2005; Zhang et al. 2005) and TRPC-IP_3R complexes formed and stabilized by junctate and Homer (Hong et al. 2001; Treves et al. 2004; Yuan et al. 2003)

2.1
Regulatory Proteins

2.1.1
Protein Kinases as Regulators of TRPC Channels Activity

Phosphorylation and dephosphorylation are well-known regulatory mechanisms of ion channel activity, and similar regulation was reported for several TRPC channels. Protein kinase C (PKC) was shown to inhibit the activity of TRPC3 (Venkatachalam et al. 2003), TRPC4, (Venkatachalam et al. 2004),

TRPC5 (Zhu et al. 2005), and TRPC6 (Venkatachalam et al. 2003, 2004). The activity of TRPC6 is also regulated by a calmodulin kinase II (CaMK II)-dependent phosphorylation (Shi et al. 2004), although it is not yet known if the kinase directly phosphorylates the channel. Src kinases seem to be required for activation of TRPC3 and TRPC6 by the M3 muscarinic receptor and epidermal growth factor (EGF) receptor, respectively (Hisatsune et al. 2004). The myosin light chain kinase is involved in the activation of TRPC5 (Kim et al. 2005; Shimizu et al. 2005). Tyrosine phosphorylation regulates EGF-mediated translocation of TRPC4 and TRPC5 to the PM (Bezzerides et al. 2004; Odell et al. 2005).

2.1.2
Calmodulin

Since TRPC channels are at the "epicenter" of Ca^{2+} signaling, it is not surprising that, like all other Ca^{2+} transporters, they are subject to regulatory feedback loops mediated by Ca^{2+} itself. The Ca^{2+}-binding protein CaM is the prime conveyor of the cellular Ca^{2+} signal and has been shown to regulate several Ca^{2+}-signaling proteins, including Ca^{2+} pumps and channels (Bezprozvanny 2005; Bosanac et al. 2005; Catterall 2000; Guerini et al. 2005). Regulation by CaM was reported for the *Drosophila* TRP and TRPL, which are negatively regulated by CaM (Scott et al. 1997). All the mammalian TRPC channels contain at least one CaM binding site at their C-terminus (Tang et al. 2001; Zhu 2005). Interestingly, the CaM binding site at the C-terminus of TRPC channels overlaps with the IP_3R binding site, and IP_3R and CaM appear to compete for binding to this site: binding of IP_3R activates, whereas binding of CaM inhibits, the TRPC channels (Zhang et al. 2001). CaM also interacts with TRPC1 and mediates the Ca^{2+}-dependent inactivation of the channel (Singh et al. 2002).

Another form of regulation of TRPC channels by CaM is by indirect interaction mediated by the protein enkurin (Sutton et al. 2004). Enkurin binds CaM, phosphatidylinositol-3-kinase (PI3K) and TRPC1, TRPC2, and TRPC5 (Sutton et al. 2004) and may thus function as an adaptor for interaction of other CaM-regulated proteins with TRPC channels.

2.2
Proteins Involved in the PM Targeting and Insertion of TRPC Channels

A novel form of regulation of TRPC channels that has been recently elucidated is the translocation and insertion of TRPC into the PM. This process is regulated by tyrosine phosphorylation in the case of TRPC4 and TRPC5 and by a yet unknown mechanism in the case of TRPC3 and TRPC6. TRPC3 was shown to interact with vesicle-associated membrane protein 2 (VAMP2) and αsoluble *N*-ethylmaleimide-sensitive factor-associated protein (αSNAP), which mediate membrane fusion (Jahn et al. 2003), and interaction with VAMP2 is required for fusion of TRPC3-containing vesicles with the PM (Singh et al. 2004). The role of αSNAP is not known. It may dissociate the soluble N-ethylmaleimide-sensitive factor attachment protein receptor (SNARE) complexes after vesicle fusion or prevent fusion of the TRPC3-containing vesicles with membranes other than the PM. It will be interesting to determine which step in the assembly and translocation or fusion of the TRPC3-containing vesicle is inhibited by disrupting the interaction of TRPC3 with the SNARE proteins.

An unexpected candidate for lipid-mediated fusion of TRPC3 with the PM is phospholipase C (PLC)γ (Patterson et al. 2002; van Rossum et al. 2005), which contributes half of its compound PH domain to form a lipid-binding sequence in TRPC3. This mechanism seems to regulate the long-term steady-state level of TRPC3 at the PM (van Rossum et al. 2005). Withdrawal from, rather than insertion of, the channels into the PM may mediate long-term modulation of TRPC channel expression at the PM. In this respect, it is interesting that TRPC1–6 interact with MxA, a member of the dynamin superfamily (Lussier et al. 2005a). Dynamins are involved in endocytosis and membrane fission (Praefcke and McMahon 2004). Coexpression of TRPC6 with MxA increases activation of TRPC6 by stimulation of GPCRs and by OAG (Lussier et al. 2005b). Is it possible that MxA interferes with endocytosis of TRPC6 to increase its steady-state level at the PM.

3
TRPC Channels and Scaffolding Proteins

3.1
Near-Membrane Scaffolds

InaD is the prototypical near-membrane scaffold that assembles TRP channels and other Ca^{2+}-signaling proteins into signaling complexes to facilitate communication between them and thus signal transduction (Chevesich et al. 1997; Tsunoda et al. 1997). The role of the mammalian InaD homolog is not known. Further, the regulation of TRPC channels may not be as straightforward as in the *Drosophila* photoreceptor since no TRPC channel has been shown to function in a pathway that is dedicated to one task. It is likely that several scaffolds interact with specific TRPC channels depending on the cellular context and environment.

TRPC4 and TRPC5 have PDZ-binding ligands in their extreme C-termini and interact with the bivalent scaffolding protein NHERF/EBP50 (Na^+/H^+ exchanger regulatory factor–ezrin/moesin/radixin-binding phosphoprotein 50), which also contains an ezrin binding domain (Mery et al. 2002; Obukhov and Nowycky 2004; Tang et al. 2000). Disruption of the interaction with NHERF/EBP50 inhibits TRPC4 and delays activation of TRPC5 (Obukhov and Nowycky 2004). Since TRPC4 and TRPC5 are also regulated by PM insertion, it would be interesting to determine whether and how interaction with PDZ-containing scaffolds affects PM insertion and retention of these channels.

Spinophilin (SPL) and neurabin (NRB) are homologous scaffolds that have actin and protein phosphatase 1 binding domains, a PDZ domain, and three (SPL) or four (NRB) coiled-coil domains (Allen et al. 1997; McAvoy et al. 1999; Nakanishi et al. 1997; Satoh et al. 1998). It was recently reported that SPL and NRB bind TRPC5 and TRPC6 (Goel et al. 2005). SPL also binds the third intracellular loop of GPCRs and the regulators of G proteins signaling (RGS) proteins (Wang 2005). Therefore SPL and NRB may bind multiple Ca^{2+}-signaling proteins to coordinate the activity of the biochemical component of the Ca^{2+}-signaling complex that generates IP_3 production and consequently the TRPC channel activity.

Another potential TRPC scaffold is caveolin. Spatial segregation of TRPC channels, such as TRPC1, into caveolae was suggested to be important for their regulation. TRPC1 binds to caveolin (Brazer et al. 2003), and truncation of the ankyrin repeats at the N-terminus of TRPC1 and dominant-negative caveolin inhibit targeting of TRPC1 to the PM (Brazer et al. 2003). How this interaction results in targeting and retention of TRPC1 in PM caveolae is not known.

3.2
Scaffolds that Mediate PM/ER Communication

As mentioned above, all TRPC channels interact with IP_3Rs, and sequences that mediate direct interaction between the proteins have been identified (Boulay et al. 1999; Tang et al. 2001). IP_3R was suggested to play a role in activation and/or gating of TRPC channels (Boulay et al. 1999; Kiselyov et al. 1998; Tang et al. 2001; Zhang et al. 2001). The question that arises is whether the strength of the interaction between TRPC channels and IP_3Rs is regulated. The evidence for such a regulation by the scaffolds junctate and Homer, as well as the potential role of STIM1 in PM/ER communication, is discussed below.

3.2.1
Junctate

Junctate binds IP_3Rs and TRPC3 to form the tertiary complex IP_3R–junctate–TRPC3 (Stamboulian et al. 2005; Treves et al. 2000, 2004). Junctate resides in the ER, where it binds to IP_3R, while its N-terminal domain binds TRPC3. Overexpression of junctate increases agonist-induced and store depletion-induced Ca^{2+} influx, and its downregulation dissociates the IP_3R–junctate–TRPC3 complex and decreases activation of TRPC3 (Treves et al. 2004). A recent report showed that junctate also binds TRPC5 and TRPC2 but not the ubiquitously expressed TRPC1 present in the same cells (Stamboulian et al. 2005). This highlights the complex interaction of TRPC channels with scaffolding proteins. It is possible that in highly specialized cells, such as sperm in which the TRPC5/junctate interaction takes place, TRPC5 function is highly specialized. Similar exclusive interaction with a subset of scaffolding proteins may serve to confer specific function to other TRPC channels.

3.2.2
Homer

Homer proteins are coded by three genes, Homer1, -2, and -3, and bind many Ca^{2+} signaling proteins to regulate Ca^{2+} signaling (Brakeman et al. 1997; Fagni et al. 2002; Olson et al. 2005; Stiber et al. 2005; Tu et al. 1998; Xiao et al. 1998; Yamamoto et al. 2005). The family of Homer proteins was identified with the finding of Homer1a as an immediate early gene product that is upregulated in response to neuronal stress like seizure (Brakeman et al. 1997). Full-length Homer proteins contain an Ena/VASP homology (EVH) domain that binds proline-rich sequences flanked by phenylalanine (PPXXF and PPXF), and coiled-coil and leucine zipper domains, which multimerize Homer proteins to generate a lattice that is used to assemble and regulate Ca^{2+}-signaling complexes (Fagni et al. 2002; Xiao et al. 2000). The complex may include metabotropic glutamate receptor, IP_3R, plasma membrane Ca^{2+} ATPase (PMCA) pump, sarcoplasmic/endoplasmic Ca^{2+} ATPase (SERCA) pump, and ER luminal Ca^{2+} binding proteins (Sala et al. 2005). Homer1a lacks the coiled-coil domain and cannot multimerize, and thus Homer1a dissociates Homer-assembled complexes and inhibits Ca^{2+} signaling.

All TRPC and several other TRP channels contain a highly conserved proline-rich sequence PPPF at their C-terminus. TRPC1 also contains the Homer binding proline-rich motif PXXP in its N-terminus (Yuan et al. 2003). Homer mediates the interaction and gating of TRPC channels by IP_3Rs. Disabling the TRPC/Homer binding by mutations of the Homer binding sites in TRPC1 or by infusing the cells with Homer1a dissociates the TRPC1/Homer/IP_3R complex to make TRPC1 constitutively active (Yuan et al. 2003). Moreover, deletion of Homer1 in mice, but not of Homer2 and Homer3, increases the spontaneous Ca^{2+} influx in native cells (Yuan et al. 2003).

The Homer proteins appear to actively regulate the strength and timing of the cellular Ca^{2+}-signaling memory. Expression of the uncoupling Homer1a is a dynamic process. Changes in Homer1 expression have been reported in response to neuronal activity (Kane et al. 2005), methamphetamine stimulation (Fujiyama et al. 2003), pain (Miletic et al. 2005), ER dysfunction (Paschen and Mengesdorf 2003), and animal behavior (Ehrengruber et al. 2004), and expression of Homer1 displays circadian changes (Nelson et al. 2004). It is thus conceivable that the multimerizing Homer proteins enhance the interaction between the Ca^{2+}-signaling proteins to strengthen the Ca^{2+} signal, while the uncoupling homer dissociates the signaling complexes to reduce signaling strength. Long-term changes in signaling activity may be achieved by changes in the expression ratio of multimerizing/uncoupling Homer1. Although challenging, it should be informative to establish the operation of such a mechanism.

References

Allen PB, Ouimet CC, Greengard P (1997) Spinophilin, a novel protein phosphatase 1 binding protein localized to dendritic spines. Proc Natl Acad Sci U S A 94:9956

Berridge MJ (2001) The versatility and complexity of calcium signalling. Novartis Found Symp 239:52

Bezprozvanny I (2005) The inositol 1,4,5-trisphosphate receptors. Cell Calcium 38:261

Bezzerides VJ, Ramsey IS, Kotecha S, Greka A, Clapham DE (2004) Rapid vesicular translocation and insertion of TRP channels. Nat Cell Biol 6:709

Bosanac I, Yamazaki H, Matsu-Ura T, Michikawa T, Ikura K, Mikoshiba M (2005) Crystal structure of the ligand binding suppressor domain of type 1 inositol 1,4,5-trisphosphate receptor. Mol Cell 17:193

Boulay G, Zhu X, Peyton M, Jiang M, Hurst R, Stefani E, Birnbaumer L (1997) Cloning and expression of a novel mammalian homolog of Drosophila transient receptor potential (Trp) involved in calcium entry secondary to activation of receptors coupled by the Gq class of G protein. J Biol Chem 272:29672

Boulay G, Brown DM, Qin N, Jiang M, Dietrich A, Zhu MX, Chen Z, Birnbaumer M, Mikoshiba K, Birnbaumer L (1999) Modulation of $Ca^{(2+)}$ entry by polypeptides of the inositol 1,4,5-trisphosphate receptor (IP3R) that bind transient receptor potential (TRP): evidence for roles of TRP and IP3R in store depletion-activated $Ca^{(2+)}$ entry. Proc Natl Acad Sci U S A 96:14955

Brakeman PR, Lanahan AA, O'Brien R, Roche K, Barnes CA, Huganir RL, Worley PF (1997) Homer: a protein that selectively binds metabotropic glutamate receptors. Nature 386:284

Brazer SC, Singh BB, Liu X, Swaim W, Ambudkar IS (2003) Caveolin-1 contributes to assembly of store-operated Ca^{2+} influx channels by regulating plasma membrane localization of TRPC1. J Biol Chem 278:27208

Catterall WA (2000) Structure and regulation of voltage-gated Ca^{2+} channels. Annu Rev Cell Dev Biol 16:521

Chevesich J, Kreuz AJ, Montell C (1997) Requirement for the PDZ domain protein, INAD, for localization of the TRP store-operated channel to a signaling complex. Neuron 18:95

Delmas P, Wanaverbecq N, Abogadie FC, Mistry M, Brown DA (2002) Signaling microdomains define the specificity of receptor-mediated InsP(3) pathways in neurons. Neuron 34:209

Dietrich A, Mederos YSM, Gollasch M, Gross V, Storch U, Dubrovska G, Obst M, Yildirim E, Salanova B, Kalwa H, Essin K, Pinkenburg O, Luft FC, Gudermann T, Birnbaumer L (2005) Increased vascular smooth muscle contractility in $TRPC6^{-/-}$ mice. Mol Cell Biol 25:6980

Dolmetsch RE, Xu K, Lewis RS (1998) Calcium oscillations increase the efficiency and specificity of gene expression. Nature 392:933

Ehrengruber MU, Kato A, Inokuchi K, Hennou S (2004) Homer/Vesl proteins and their roles in CNS neurons. Mol Neurobiol 29:213

Fagni L, Worley PF, Ango F (2002) Homer as both a scaffold and transduction molecule. Sci STKE 2002:RE8

Fasolato C, Nilius B (1998) Store depletion triggers the calcium release-activated calcium current (ICRAC) in macrovascular endothelial cells: a comparison with. Jurkat and embryonic kidney cell lines. Pflugers Arch 436:69

Feske S, Prakriya M, Rao A, Lewis RS (2005) A severe defect in CRAC Ca^{2+} channel activation and altered K^+ channel gating in T cells from immunodeficient patients. J Exp Med 202:651

Freichel M, Suh SH, Pfeifer A, Schweig U, Trost C, Weissgerber P, Biel M, Philipp S, Freise D, Droogmans G, Hofmann F, Flockerzi V, Nilius B (2001) Lack of an endothelial store-operated Ca^{2+} current impairs agonist-dependent vasorelaxation in $TRP4^{-/-}$ mice. Nat Cell Biol 3:121

Fujiyama K, Kajii Y, Hiraoka S, Nishikawa T (2003) Differential regulation by stimulants of neocortical expression of mrt1, arc, and homer1a mRNA in the rats treated with repeated methamphetamine. Synapse 49:143

Goel M, Sinkins WG, Schilling WP (2002) Selective association of TRPC channel subunits in rat brain synaptosomes. J Biol Chem 277:48303

Goel M, Sinkins W, Keightley A, Kinter M, Schilling WP (2005) Proteomic analysis of TRPC5- and TRPC6-binding partners reveals interaction with the plasmalemmal $Na^{(+)}/K^{(+)}$-ATPase. Pflugers Arch 451:87

Greka A, Navarro B, Oancea E, Duggan A, Clapham DE (2003) TRPC5 is a regulator of hippocampal neurite length and growth cone morphology. Nat Neurosci 6:837

Gudermann T, Hofmann T, Mederos y Schnitzler M, Dietrich A (2004) Activation, subunit composition and physiological relevance of DAG-sensitive TRPC proteins. Novartis Found Symp 258:103

Guerini D, Coletto L, Carafoli E (2005) Exporting calcium from cells. Cell Calcium 38:281

Hisatsune C, Kuroda Y, Nakamura K, Inoue T, Nakamura T, Michikawa T, Mizutani A, Mikoshiba K (2004) Regulation of TRPC6 channel activity by tyrosine phosphorylation. J Biol Chem 279:18887

Hofmann T, Obukhov AG, Schaefer M, Harteneck C, Gudermann T, Schultz G (1999) Direct activation of human TRPC6 and TRPC3 channels by diacylglycerol. Nature 397:259

Hong CS, Kwak YG, Ji JH, Chae SW, Kim do H (2001) Molecular cloning and characterization of mouse cardiac junctate isoforms. Biochem Biophys Res Commun 289:882

Jahn R, Lang T, Sudhof TC (2003) Membrane fusion. Cell 112:519

Kane JK, Hwang Y, Konu O, Loughlin SE, Leslie FM, Li MD (2005) Regulation of Homer and group I metabotropic glutamate receptors by nicotine. Eur J Neurosci 21:1145

Kim MT, Kim BJ, Lee JH, Kwon SC, Yeon DS, Yang DK, So I, Kim KW (2005) Involvement of calmodulin and myosin light chain kinase in the activation of mTRPC5 expressed in HEK cells. Am J Physiol Cell Physiol 290:C1031–C1040

Kim SJ, Kim YS, Yuan JP, Petralia RS, Worley PF, Linden DJ (2003) Activation of the TRPC1 cation channel by metabotropic glutamate receptor mGluR1. Nature 426:285

Kiselyov K, Xu X, Mozhayeva G, Kuo T, Pessah I, Mignery G, Zhu X, Birnbaumer L, Muallem S (1998) Functional interaction between InsP3 receptors and store-operated Htrp3 channels. Nature 396:478

Kiselyov K, Mignery GA, Zhu MX, Muallem S (1999) The N-terminal domain of the IP3 receptor gates store-operated hTrp3 channels. Mol Cell 4:423

Kiselyov KI, Shin DM, Wang Y, Pessah IN, Allen PD, Muallem S (2000) Gating of store-operated channels by conformational coupling to ryanodine receptors. Mol Cell 6:421

Kriz W (2005) TRPC6—a new podocyte gene involved in focal segmental glomerulosclerosis. Trends Mol Med 11:527

Liou J, Kim ML, Do Heo W, Jones JT, Myers JW, Ferrell JE Jr, Meyer T (2005) STIM is a $Ca^{(2+)}$ sensor essential for $Ca^{(2+)}$-store-depletion-triggered $Ca^{(2+)}$ influx. Curr Biol 15:1235

Liu D, Scholze A, Zhu Z, Kreutz R, Wehland-von-Trebra M, Zidek W, Tepel M (2005a) Increased transient receptor potential channel TRPC3 expression in spontaneously hypertensive rats. Am J Hypertens 18:1503

Liu X, Wang W, Singh BB, Lockwich T, Jadlowiec J, O'Connell B, Wellner R, Zhu MX, Ambudkar IS (2000) Trp1, a candidate protein for the store-operated Ca$^{(2+)}$ influx mechanism in salivary gland cells. J Biol Chem 275:3403

Liu X, Singh BB, Ambudkar IS (2003) TRPC1 is required for functional store-operated Ca^{2+} channels. Role of acidic amino acid residues in the S5-S6 region. J Biol Chem 278:11337

Liu X, Bandyopadhyay BC, Singh BB, Groschner K, Ambudkar IS (2005b) Molecular analysis of a store-operated and 2-acetyl-sn-glycerol-sensitive non-selective cation channel. Heteromeric assembly of TRPC1–TRPC3. J Biol Chem 280:21600

Lussier MP, Cayouette S, Lepage PK, Bernier CL, Francoeur N, St-Hilaire M, Pinard M, Boulay G (2005a) MxA, a member of the dynamin superfamily, interacts with the ankyrin-like repeat domain of TRPC. J Biol Chem 280:19393–19400

Lussier MP, Cayouette S, Lepage PK, Bernier CL, Francoeur N, St-Hilaire M, Pinard M, Boulay G (2005b) MxA, a member of the dynamin superfamily, interacts with the ankyrin-like repeat domain of TRPC. J Biol Chem 280:19393

Ma HT, Venkatachalam K, Li HS, Montell C, Kurosaki T, Patterson RL, Gill DL (2001) Assessment of the role of the inositol 1,4,5-trisphosphate receptor in the activation of transient receptor potential channels and store-operated Ca^{2+} entry channels. J Biol Chem 276:18888

McAvoy T, Allen PB, Obaishi H, Nakanishi H, Takai Y, Greengard P, Nairn AC, Hemmings HC Jr (1999) Regulation of neurabin I interaction with protein phosphatase 1 by phosphorylation. Biochemistry 38:12943

Mery L, Strauss B, Dufour JF, Krause KH, Hoth M (2002) The PDZ-interacting domain of TRPC4 controls its localization and surface expression in HEK293 cells. J Cell Sci 115:3497

Miletic G, Miyabe T, Gebhardt KJ, Miletic V (2005) Increased levels of Homer1b/c and Shank1a in the post-synaptic density of spinal dorsal horn neurons are associated with neuropathic pain in rats. Neurosci Lett 386:189

Nakanishi H, Obaishi H, Satoh A, Wada M, Mandai K, Satoh K, Nishioka H, Matsuura Y, Mizoguchi A, Takai Y (1997) Neurabin: a novel neural tissue-specific actin filament-binding protein involved in neurite formation. J Cell Biol 139:951

Nelson SE, Duricka DL, Campbell K, Churchill L, Krueger JM (2004) Homer1a and 1bc levels in the rat somatosensory cortex vary with the time of day and sleep loss. Neurosci Lett 367:105

Obukhov AG, Nowycky MC (2004) TRPC5 activation kinetics are modulated by the scaffolding protein ezrin/radixin/moesin-binding phosphoprotein-50 (EBP50). J Cell Physiol 201:227

Odell AF, Scott JL, Van Helden DF (2005) Epidermal growth factor induces tyrosine phosphorylation, membrane insertion, and activation of transient receptor potential channel 4. J Biol Chem 280:37974

Olson PA, Tkatch T, Hernandez-Lopez S, Ulrich S, Ilijic E, Mugnaini E, Zhang H, Bezprozvanny I, Surmeier DJ (2005) G-protein-coupled receptor modulation of striatal CaV1.3 L-type Ca^{2+} channels is dependent on a Shank-binding domain. J Neurosci 25:1050

Orrenius S, Zhivotovsky B, Nicotera P (2003) Regulation of cell death: the calcium-apoptosis link. Nat Rev Mol Cell Biol 4:552

Parekh AB, Putney JW Jr (2005) Store-operated calcium channels. Physiol Rev 85:757

Paria BC, Vogel SM, Ahmmed GU, Alamgir S, Shroff J, Malik AB, Tiruppathi C (2004) Tumor necrosis factor-alpha-induced TRPC1 expression amplifies store-operated Ca^{2+} influx and endothelial permeability. Am J Physiol Lung Cell Mol Physiol 287:L1303

Paschen W, Mengesdorf T (2003) Conditions associated with ER dysfunction activate homer 1a expression. J Neurochem 86:1108

Patterson RL, van Rossum DB, Ford DL, Hurt KJ, Bae SS, Suh PG, Kurosaki T, Snyder SH, Gill DL (2002) Phospholipase C-gamma is required for agonist-induced Ca^{2+} entry. Cell 111:529

Praefcke GJ, McMahon HT (2004) The dynamin superfamily: universal membrane tubulation and fission molecules? Nat Rev Mol Cell Biol 5:133

Putney JW Jr, Broad LM, Braun FJ, Lievremont JP, Bird GS (2001) Mechanisms of capacitative calcium entry. J Cell Sci 114:2223

Rao JN, Platoshyn O, Golovina VA, Liu L, Zou T, Marasa BS, Turner DJ, J XJY, Wang JY (2005) TRPC1 functions as a store-operated Ca^{2+} channel in intestinal epithelial cells and regulates early mucosal restitution after wounding. Am J Physiol Gastrointest Liver Physiol 290:G782–G792

Reiser J, Polu KR, Moller CC, Kenlan P, Altintas MM, Wei C, Faul C, Herbert S, Villegas I, Avila-Casado C, McGee M, Sugimoto H, Brown D, Kalluri R, Mundel P, Smith PL, Clapham DE, Pollak MR (2005) TRPC6 is a glomerular slit diaphragm-associated channel required for normal renal function. Nat Genet 37:739

Roos J, DiGregorio PJ, Yeromin AV, Ohlsen K, Lioudyno M, Zhang S, Safrina O, Kozak JA, Wagner SL, Cahalan MD, Velicelebi G, Stauderman KA (2005) STIM1, an essential and conserved component of store-operated Ca^{2+} channel function. J Cell Biol 169:435

Sala C, Roussignol G, Meldolesi J, Fagni L (2005) Key role of the postsynaptic density scaffold proteins Shank and Homer in the functional architecture of Ca^{2+} homeostasis at dendritic spines in hippocampal neurons. J Neurosci 25:4587

Sampieri A, Diaz-Munoz M, Antaramian A, Vaca L (2005) The foot structure from the type 1 ryanodine receptor is required for functional coupling to store-operated channels. J Biol Chem 280:24804

Satoh A, Nakanishi H, Obaishi H, Wada M, Takahashi K, Satoh K, Hirao K, Nishioka H, Hata Y, Mizoguchi A, Takai Y (1998) Neurabin-II/spinophilin. An actin filament-binding protein with one pdz domain localized at cadherin-based cell-cell adhesion sites. J Biol Chem 273:3470

Schilling WP, Goel M (2004) Mammalian TRPC channel subunit assembly. Novartis Found Symp 258:18

Scott K, Sun Y, Beckingham K, Zuker CS (1997) Calmodulin regulation of Drosophila light-activated channels and receptor function mediates termination of the light response in vivo. Cell 91:375

Shi J, Mori E, Mori Y, Mori M, Li J, Ito Y, Inoue R (2004) Multiple regulation by calcium of murine homologues of transient receptor potential proteins TRPC6 and TRPC7 expressed in HEK293 cells. J Physiol 561:415

Shimizu S, Yoshida T, Wakamori M, Ishii M, Okada T, Takahashi M, Seto M, Sakurada K, Kiuchi Y, Mori Y (2005) Ca^{2+}/calmodulin dependent myosin light chain kinase is essential for activation of TRPC5 channels expressed in HEK293 cells. J Physiol 570:219–235

Singh BB, Liu X, Tang J, Zhu MX, Ambudkar IS (2002) Calmodulin regulates $Ca^{(2+)}$-dependent feedback inhibition of store-operated $Ca^{(2+)}$ influx by interaction with a site in the C terminus of TrpC1. Mol Cell 9:739

Singh BB, Lockwich TP, Bandyopadhyay BC, Liu X, Bollimuntha S, Brazer SC, Combs C, Das S, Leenders AG, Sheng ZH, Knepper MA, Ambudkar SV, Ambudkar IS (2004) VAMP2-dependent exocytosis regulates plasma membrane insertion of TRPC3 channels and contributes to agonist-stimulated Ca^{2+} influx. Mol Cell 15:635

Soboloff J, Spassova M, Xu W, He LP, Cuesta N, Gill DL (2005) Role of endogenous TRPC6 channels in Ca^{2+} signal generation in A7r5 smooth muscle cells. J Biol Chem 280:39786

Spassova MA, Soboloff J, He LP, Xu W, Dziadek MA, Gill DL (2006) STIM1 has a plasma membrane role in the activation of store-operated Ca^{2+} channels. Proc Natl Acad Sci U S A 103:4040

Stamboulian S, Moutin MJ, Treves S, Pochon N, Grunwald D, Zorzato F, De Waard M, Ronjat M, Arnoult C (2005) Junctate, an inositol 1,4,5-triphosphate receptor associated protein, is present in rodent sperm and binds TRPC2 and TRPC5 but not TRPC1 channels. Dev Biol 286:326

Stiber JA, Tabatabaei N, Hawkins AF, Hawke T, Worley PF, Williams RS, Rosenberg P (2005) Homer modulates NFAT-dependent signaling during muscle differentiation. Dev Biol 287:213

Strubing C, Krapivinsky G, Krapivinsky L, Clapham DE (2001) TRPC1 and TRPC5 form a novel cation channel in mammalian brain. Neuron 29:645

Sutton KA, Jungnickel MK, Wang Y, Cullen K, Lambert S, Florman HM (2004) Enkurin is a novel calmodulin and TRPC channel binding protein in sperm. Dev Biol 274:426

Sweeney M, Yu Y, Platoshyn O, Zhang S, McDaniel SS, Yuan JX (2002) Inhibition of endogenous TRP1 decreases capacitative Ca^{2+} entry and attenuates pulmonary artery smooth muscle cell proliferation. Am J Physiol Lung Cell Mol Physiol 283:L144

Tang J, Lin Y, Zhang Z, Tikunova S, Birnbaumer L, Zhu MX (2001) Identification of common binding sites for calmodulin and inositol 1,4,5-trisphosphate receptors on the carboxyl termini of trp channels. J Biol Chem 276:21303

Tang Y, Tang J, Chen Z, Trost C, Flockerzi V, Li M, Ramesh V, Zhu MX (2000) Association of mammalian trp4 and phospholipase C isozymes with a PDZ domain-containing protein NHERF. J Biol Chem 275:37559

Thebault S, Zholos A, Enfissi A, Slomianny C, Dewailly E, Roudbaraki M, Parys J, Prevarskaya N (2005) Receptor–operated $Ca^{(2+)}$ entry mediated by TRPC3/TRPC6 proteins in rat prostate smooth muscle (PS1) cell line. J Cell Physiol 290:C1060–C1066

Trebak M, Bird GS, McKay RR, Putney JW Jr (2002) Comparison of human TRPC3 channels in receptor-activated and store-operated modes. Differential sensitivity to channel blockers suggests fundamental differences in channel composition. J Biol Chem 277:21617

Treves S, Feriotto G, Moccagatta L, Gambari R, Zorzato F (2000) Molecular cloning, expression, functional characterization, chromosomal localization, and gene structure of junctate, a novel integral calcium binding protein of sarco(endo)plasmic reticulum membrane. J Biol Chem 275:39555

Treves S, Franzini-Armstrong C, Moccagatta L, Arnoult C, Grasso C, Schrum A, Ducreux S, Zhu MX, Mikoshiba K, Girard T, Smida-Rezgui S, Ronjat M, Zorzato F (2004) Junctate is a key element in calcium entry induced by activation of InsP3 receptors and/or calcium store depletion. J Cell Biol 166:537

Tsunoda S, Sierralta J, Sun Y, Bodner R, Suzuki E, Becker A, Socolich M, Zuker CS (1997) A multivalent PDZ-domain protein assembles signalling complexes in a G-protein-coupled cascade. Nature 388:243

Tu JC, Xiao B, Yuan JP, Lanahan AA, Leoffert K, Li M, Linden DJ, Worley PF (1998) Homer binds a novel proline-rich motif and links group 1 metabotropic glutamate receptors with IP3 receptors. Neuron 21:717

van Rossum DB, Patterson RL, Sharma S, Barrow RK, Kornberg M, Gill DL, Snyder SH (2005) Phospholipase Cgamma1 controls surface expression of TRPC3 through an intermolecular PH domain. Nature 434:99

Vazquez G, Lievremont JP, St JBG, Putney JW Jr (2001) Human Trp3 forms both inositol trisphosphate receptor-dependent and receptor-independent store-operated cation channels in DT40 avian B lymphocytes. Proc Natl Acad Sci U S A 98:11777

Venkatachalam K, Ma HT, Ford DL, Gill DL (2001) Expression of functional receptor-coupled TRPC3 channels in DT40 triple receptor InsP3 knockout cells. J Biol Chem 276:33980

Venkatachalam K, Zheng F, Gill DL (2003) Regulation of canonical transient receptor potential (TRPC) channel function by diacylglycerol and protein kinase C. J Biol Chem 278:29031

Venkatachalam K, Zheng F, Gill DL (2004) Control of TRPC and store-operated channels by protein kinase C. Novartis Found Symp 258:172

Wang GX, Poo MM (2005) Requirement of TRPC channels in netrin-1-induced chemotropic turning of nerve growth cones. Nature 434:898

Wang X, Zeng W, Soyombo AA, Tang W, Ross EM, Barnes AP, Milgram SL, Penninger JM, Allen PB, Greengard P, Muallem S (2005) Spinophilin regulates $Ca^{(2+)}$ signalling by binding the N-terminal domain of RGS2 and the third intracellular loop of G-protein-coupled receptors. Nat Cell Biol 7:405–411

Winn MP, Conlon PJ, Lynn KL, Farrington MK, Creazzo T, Hawkins AF, Daskalakis N, Kwan SY, Ebersviller S, Burchette JL, Pericak-Vance MA, Howell DN, Vance JM, Rosenberg PB (2005) A mutation in the TRPC6 cation channel causes familial focal segmental glomerulosclerosis. Science 308:1801

Wu X, Babnigg G, Villereal ML (2000) Functional significance of human trp1 and trp3 in store-operated $Ca^{(2+)}$ entry in HEK-293 cells. Am J Physiol Cell Physiol 278:C526

Wu X, Babnigg G, Zagranichnaya T, Villereal ML (2002) The role of endogenous human Trp4 in regulating carbachol-induced calcium oscillations in HEK-293 cells. J Biol Chem 277:13597

Wu X, Zagranichnaya TK, Gurda GT, Eves EM, Villereal ML (2004) A TRPC1/TRPC3-mediated increase in store-operated calcium entry is required for differentiation of H19-7 hippocampal neuronal cells. J Biol Chem 279:43392

Xiao B, Tu JC, Petralia RS, Yuan JP, Doan A, Breder CD, Ruggiero A, Lanahan AA, Wenthold RJ, Worley PF (1998) Homer regulates the association of group 1 metabotropic glutamate receptors with multivalent complexes of homer-related synaptic proteins. Neuron 21:707

Xiao B, Tu JC, Worley PF (2000) Homer: a link between neural activity and glutamate receptor function. Curr Opin Neurobiol 10:370

Yamamoto K, Sakagami Y, Sugiura S, Inokuchi K, Shimohama S, Kato N (2005) Homer 1a enhances spike-induced calcium influx via L-type calcium channels in neocortex pyramidal cells. Eur J Neurosci 22:1338

Yuan JP, Kiselyov K, Shin DM, Chen J, Shcheynikov N, Kang SH, Dehoff MH, Schwarz MK, Seeburg PH, Muallem S, Worley PF (2003) Homer binds TRPC family channels and is required for gating of TRPC1 by IP3 receptors. Cell 114:777

Zagranichnaya TK, Wu X, Villereal ML (2005) Endogenous TRPC1, TRPC3, and TRPC7 proteins combine to form native store-operated channels in HEK-293 cells. J Biol Chem 280:29559

Zhang SL, Yu Y, Roos J, Kozak JA, Deerinck TJ, Ellisman MH, Stauderman KA, Cahalan MD (2005) STIM1 is a Ca^{2+} sensor that activates CRAC channels and migrates from the Ca^{2+} store to the plasma membrane. Nature 437:902

Zhang Z, Tang J, Tikunova S, Johnson JD, Chen Z, Qin N, Dietrich A, Stefani E, Birnbaumer L, Zhu MX (2001) Activation of Trp3 by inositol 1,4,5-trisphosphate receptors through displacement of inhibitory calmodulin from a common binding domain. Proc Natl Acad Sci U S A 98:3168

Zhu MH, Chae M, Kim HJ, Lee YM, Kim MJ, Jin NG, Yang DK, So I, Kim KW (2005) Desensitization of canonical transient receptor potential channel 5 by protein kinase C. Am J Physiol Cell Physiol 289:C591

Zhu MX (2005) Multiple roles of calmodulin and other $Ca^{(2+)}$-binding proteins in the functional regulation of TRP channels. Pflugers Arch 451:105

/n# TRPC Channels: Integrators of Multiple Cellular Signals

J. Soboloff[1] · M. Spassova[1] · T. Hewavitharana[1] · L.-P. He[1] · P. Luncsford[1] · W. Xu[1] · K. Venkatachalam[2] · D. van Rossum[3] · R. L. Patterson[4] · D. L. Gill[1] (✉)

[1]Department of Biochemistry and Molecular Biology,
University of Maryland School of Medicine, 108 North Greene Street,
Baltimore MD, 21201, USA
dgill@umaryland.edu

[2]Department of Biological Chemistry, Johns Hopkins School of Medicine, ,
Baltimore MD, 21205, USA

[3]Department of Neuroscience, Johns Hopkins School of Medicine, Baltimore MD, 21205, USA

[4]Department of Biology, Pennsylvania State University, University Park, PA, 16801, USA

1	Introduction	576
2	TRPC Ion Channel Properties and Cellular Role	577
3	Modes of Activation: Are TRPC Channels SOCs?	577
4	Role of Phospholipase C in TRPC Function	579
5	TRPC Channel Modification by Protein Kinase C	580
6	Role of the InsP3 Receptor in TRPC Activation	582
7	Pharmacology of TRPC Channels	583
8	Biological Relevance and Emerging Roles for TRPC Channels	585
References		587

Abstract TRPC channels are ubiquitously expressed among cell types and mediate signals in response to phospholipase C (PLC)-coupled receptors. TRPC channels function as integrators of multiple signals resulting from receptor-induced PLC activation, which catalyzes the breakdown of phosphatidylinositol 4,5-bisphosphate (PIP_2) to produce inositol 1,4,5-trisphosphate ($InsP_3$) and diacylglycerol (DAG). $InsP_3$ depletes Ca^{2+} stores and TRPC3 channels can be activated by store-depletion. $InsP_3$ also activates the $InsP_3$ receptor, which may undergo direct interactions with the TRPC3 channel, perhaps mediating store-dependence. The other PLC product, DAG, has a direct non-PKC-dependent activating role on TRPC3 channels likely by direct binding. DAG also has profound effects on the TRPC3 channel through PKC. Thus PKC is a powerful inhibitor of most TRPC channels and DAG is a dual regulator of the TRPC3 channel. PLC-mediated DAG results in rapid channel opening followed later by a slower DAG-induced PKC-mediated deactivation of the channel. The decreased level of PIP$_2$ from PLC activation also has an important modifying action on TRPC3 channels. Thus, the TRPC3 channel and PLCγ form an intermolecular PH domain that has high specificity for binding PIP$_2$. This interaction allows the channel to be retained within the plasma membrane, a further operational control factor for TRPC3. As nonselective cation channels, TRPC channel opening results in the entry of both Na$^+$ and Ca^{2+} ions. Thus, while they may mediate Ca^{2+} entry signals, TRPC channels are also powerful modifiers of membrane potential.

Keywords Calcium signaling · TRPC channels · Store-operated channels

1
Introduction

Ca^{2+} signals in response to receptors mediate and control countless cellular functions ranging from short-term responses such as secretion and contraction to longer term control of growth, cell division, and apoptosis. TRPC channels are ubiquitously expressed among cell types and mediate important signals in response to phospholipase C (PLC)-coupled receptors. TRPC channels function as integrators of multiple signals resulting from receptor-induced PLC activation, which catalyzes the breakdown of phosphatidylinositol 4,5-bisphosphate (PIP$_2$) to produce inositol 1,4,5-trisphosphate (InsP$_3$) and diacylglycerol (DAG). InsP$_3$ depletes Ca^{2+} stores and evidence indicates TRPC3 channels are store-operated. InsP$_3$ also activates the InsP$_3$ receptor, which is able to undergo direct interactions with the TRPC3 channel, perhaps mediating store-dependence. The other PLC product, DAG, has a direct non-protein kinase C (non-PKC)-dependent activating role on TRPC3 channels likely as a direct result of binding to the TRPC3 molecule itself. DAG also has profound effects on the TRPC3 channel through PKC. Thus PKC is a powerful inhibitor of most TRPC channels and DAG is a dual regulator of the TRPC3 channel. PLC-mediated DAG results in rapid channel opening followed later by a slower DAG-induced PKC-mediated deactivation of the channel. The decreased level of PIP$_2$ as a result of PLC activation also has an important modifying action on TRPC3 channels. Thus, the TRPC3 channel and PLCγ form an intermolecular pleckstrin homology (PH) domain that has high specificity for binding PIP$_2$. This interaction with the TRPC3 channel allows the channel to be retained within the plasma membrane, a further operational control factor determining channel function in the plasma membrane. As nonspecific cation channels, TRPC channel opening results in the entry of both Na$^+$ and Ca^{2+} ions. Thus, while they may mediate Ca^{2+} entry signals, TRPC channels are also powerful modifiers of membrane potential. Here we discuss recent information on the control and significance of the operation of TRPC channels' coupling properties, TRPC pharmacology, and new insights regarding the physiological roles of these cation channels.

2
TRPC Ion Channel Properties and Cellular Role

Although often studied in the context of Ca^{2+} entry, all of the TRPC channels are nonselective cation channels, with no selective preference for Ca^{2+} over other divalent or monovalent cations (Clapham 2003; Spassova et al. 2004). This is in distinct contrast to known store-operated Ca^{2+} channels (SOCs), which show remarkable selectivity for Ca^{2+} (Parekh and Penner 1997; Parekh and Putney 2005). Thus, Sr^{2+}, which has similar properties to the Ca^{2+} ion including the ability to alter the fluorescence properties of indicator dyes such as fura-2, enters cells poorly in response to store-depletion. However, Sr^{2+} readily enter cells transfected with TRPC channels in response to either receptor activation or increases in DAG concentrations (Ma et al. 2000; Venkatachalam et al. 2003; He et al. 2005). While this has provided a useful tool to distinguish between TRPC channels and SOCs, it does not address the physiological role of TRPC channels. Considering the relatively small magnitude of the Ca^{2+} ion difference across the plasma membrane compared with that for Na^+, it is likely that the opening of TRPC channels has a greater effect upon membrane potential than on Ca^{2+} concentration. Recent studies provide strong support for this concept (Welsh et al. 2002; Beech et al. 2004; Beech 2005; Reading et al. 2005; Soboloff et al. 2005). After careful analysis of the expression patterns of TRPC channels in A7r5 vascular smooth muscle cells using real-time PCR and Western blot, we established that TRPC6 was the only member of the DAG-responsive TRPC3/6/7 subfamily of channels expressed in this system. Hence, responses to the soluble DAG analog 1-oleoyl-2-acetyl-*sn*-glycerol (OAG) appear primarily to be mediated by TRPC6. However, TRPC6 short interfering RNA (siRNA), which decreased TRPC6 levels without significant compensation by other channels, decreased the OAG-induced current, but not OAG-induced Ca^{2+} entry (Soboloff et al. 2005). This was apparently because the Ca^{2+} entry was primarily mediated by voltage-operated Ca^{2+} channels rather than TRPC6 channels. Moreover, we determined that, even after TRPC6 knockdown, the response of residual TRPC6 channels was more than sufficient to depolarize the cells. Hence, the primary role of TRPC6, at least in this system, was depolarization rather than Ca^{2+} entry.

3
Modes of Activation: Are TRPC Channels SOCs?

There is controversy concerning the role of TRP channels in mediating store-operated Ca^{2+} entry. TRP channels appear ubiquitously among cell types, and multiple TRP channel subtypes are expressed in most cells (Venkatachalam et al. 2002; Clapham 2003). One problem with assessing the physiological role of TRP channels is that functional analysis of overexpressed channels is usually against a background of endogenous SOCs and TRP channels. Similarly, reduction of TRP channel expression by antisense, siRNA, or knockout approaches can be confused by the functional overlap and redundancy with multiple endogenous TRP channel subtypes. One unifying characteristic among the *Drosophila* TRP and most members of the mammalian TRPC family, is their response to receptor-activated PLC (Montell 2001). Thus, these channels can be classified as "receptor-operated" channels (ROCs). However, PLC activation and concomitant InsP$_3$ production ensures at least some Ca^{2+} store release in most cell types. Therefore, the question of whether the TRPC channels function as SOCs as opposed to ROCs is difficult to answer, and there is obvious overlap in these modes of entry. The "hallmark" that has been used to define store-operated Ca^{2+} entry is a response to store-depletion alone using Ca^{2+} pump blockers such as thapsigargin, or ionophores such as ionomycin (Parekh and Penner 1997). However, store-emptying with these agents is not a physiological process per se. Thus, physiologically, stores are released following PLC-coupled receptor activation, and the consequences of PLC activation are not limited to store-release alone.

There seems little disagreement that members of the quite diverse TRP family of channels do not yet fulfill the criteria of being authentic Ca^{2+} release-activated Ca^{2+} (CRAC) channels. The Ca^{2+}-selective TRPV5 and TRPV6 channels may represent the closest match to characteristics of CRAC channels (Yue et al. 2001), but it is clear that there are a number of channel distinctions from CRAC channels (Voets et al. 2001; Kahr et al. 2004). Whether or not exogenously expressed TRP channels operate in a "store-dependent" mode is controversial, with studies claiming a number of TRPC, TRPM, and TRPV channels are activated when stores are emptied, and as many studies suggesting they are not (extensively reviewed in Venkatachalam et al. 2002; Spassova et al. 2004).

This abundance of apparently conflicting information on the coupling of TRPC channels to store-emptying is overwhelming at first sight. However, in heterologous expression systems in which the TRP channels are expressed alongside endogenous SOCs, the determination of whether TRP channels are directly activated by store-depletion as opposed to modifying or being modified by existing SOCs is difficult. Thus, it appears very hard to truly assess the TRPC channel phenotype in overexpression systems. In certain cell types, the high expression levels of specific TRPC channel subtypes have allowed more definitive analysis of endogenously expressed TRPC channels. However, in such cases the channels are generally not store-operated. For example, TRPC3 channels are highly expressed in pontine neurons and allow passage of Na$^+$ and Ca^{2+} ions in response to PLC-coupled receptors, but are not activated by store-emptying (Li et al. 1999). We have made similar observations in vascular smooth muscle cells regarding the mechanism of activation of TRPC6 (Soboloff et al. 2005). However, the variation in the observed activation of TRPC channels may result from their assembly into multimeric structures (Montell 2001). Thus, it was originally observed that coexpression of the *Drosophila*

TRP and TRPL channels in mammalian cells resulted in a store-operated phenotype, whereas when they were singly expressed the channels were store-independent (Xu et al. 1997). A heteromer comprising TRPC1, TRPC3, and TRPC7 was recently reported to mediate SOC activity in human embryonic kidney (HEK)-293 cells (Zagranichnaya et al. 2005); however, the entry of Ba^{2+} mediated by this combination of TRPC channels suggests a divergence from endogenous SOCs. The overall conclusion from many studies is that it is possible that TRPC channels may be activated in response to store depletion, but it is unproved whether they actually contribute to endogenously operating SOCs or CRAC channels.

The questionable physiological relevance of artificially emptying stores (Gill and Patterson 2004) and the uncertainty of overexpressing TRP channels makes it difficult to draw conclusions on their store-dependence. Moreover, since many of the TRP channels are widely expressed, it is difficult to assess whether using knockout or knockdown approaches has led to a definitive identification of the function of natively expressed TRP channels (Gill and Patterson 2004). In some cases, Ca^{2+} signaling has been affected by such deletion approaches, but curiously, the "lost" currents do not correspond to those attributed to the overexpression of the corresponding channel. The properties of channels may be influenced not only by heteromeric makeup, but also by any one of the many proteins suggested to be in tight proximity in the molecular domains or "signalplexes" (Montell et al. 2002a). When overexpressed, particularly by transient overexpression, which yields far higher channel levels than endogenous expression, it is very likely that channels are not assembled and organized as in native cells. We should also consider the importance of other classes of "receptor-activated" or "store-dependent" channels that, unlike CRAC, are not selective for Ca^{2+} ions. This is an increasingly voiced perspective (Gill and Patterson 2004), and it is likely that nonselective cation channels play crucial roles in mediating depolarization responses and hence contributing to the modification of a number of other voltage-dependent channels or electrogenic transporters (for example, the K^+ channels, Cl^- channels, or the Na^+-Ca^{2+} exchanger).

4
Role of Phospholipase C in TRPC Function

In the fly retina, TRP channels appear to be organized by the PDZ-containing scaffold protein INAD, and they are likely in intimate contact with the rhodopsin-triggered phototransduction machinery (Montell 1999). Less is known about the organization of TRPC channels in mammalian cells, but there are indications that the PDZ-containing Na^+/H^+ exchanger regulatory factor (NHERF) protein interacts with TRPC channels (Tang et al. 2000) and may organize a spatial relationship with G protein-coupled receptors (GPCRs), PLC, and the cytoskeleton in an analogous fashion (Suh et al. 2001).

It is likely that many different structural and adaptor proteins contribute to such a coupling complex. In recent studies, we determined that PLC-γ plays an adaptor-role independent of its enzymic activity in the coupling between receptors and the activation of Ca^{2+} entry (Patterson et al. 2002). Thus, experiments revealed that knocking down of PLC-γ1 or PLC-γ2 isoforms prevented GPCRs (which activate PLC-β) from activating Ca^{2+} entry, even though there was still $InsP_3$-mediated Ca^{2+} release. In contrast, thapsigargin (TG)-induced Ca^{2+} entry was not affected. Using DT40 B cells in which PLC-γ2 (the only isoform in B cells) was knocked out, GPCR-induced Ca^{2+} entry was prevented without an effect on purely store-operated Ca^{2+} entry induced by TG. The B cell receptor (BCR) in these cells activates the enzymic activity of PLC-γ2, and as expected, the $InsP_3$-mediated Ca^{2+} release is restored in the PLC-γ2-knockout cells by expression of wildtype PLC-γ2 but not by the lipase-inactive mutant. Whereas the PLC-γ2 knockout cells were devoid of Ca^{2+} entry in response to the muscarinic receptor (a GPCR), both wildtype and LIM PLC-γ2 restored Ca^{2+} entry in response to activation of the muscarinic receptor. The results indicate that PLC-γ2 plays a permissive role that is independent of its enzymic activity in mediating receptor- but not TG-induced Ca^{2+} release. We have since determined that this occurs as a result of intermolecular PH domains that form when PLC-γ and TRPC channels interact (van Rossum et al. 2005).

Hence, whereas both TRPC channels and PLC-γ lack complete PH domains along with the corresponding capacity to bind to lipids (in particular, PIP_2), they can interact with each other, causing the formation of a functional PH domain containing portions of each molecule. The resulting increase in affinity for lipids in the plasma membrane causes the TRPC channels to be retained at the plasma membrane, rather than recycled into the cell. Therefore, PLC-γ plays a dual role (1) mediating production of DAG and $InsP_3$, and (2) providing a permissive adaptor that is required for GPCRs to activate Ca^{2+} entry channels. The implication from this is that PLC-γ allows coupling to channels without necessarily requiring release of Ca^{2+} from stores.

5
TRPC Channel Modification by Protein Kinase C

For the TRPC family of proteins, it seems that their primary mechanism of activation is through PLC activation, making them functionally analogous to their fly homologs. While $InsP_3$ clearly causes Ca^{2+} release, the other product of PLC, DAG, has a direct role in TRPC activation. The closely related TRPC3, TRPC6, and TRPC7 subgroup of channels, therefore, is directly activated by DAG and its cell-permeant analogs (Hofmann et al. 1999; Okada et al. 1999; Ma et al. 2000; Venkatachalam et al. 2001; Zitt et al. 2002). This action of DAG is independent of PKC (Hofmann et al. 1999; Okada et al. 1999) and may be mediated by an N-terminal domain in the TRPC6 channel (Zhang and Saffen 2001). However, the response to DAG is subgroup-specific. Thus, overexpressed TRPC1, TRPC4, and TRPC5 channels do not respond to DAG (Hofmann et al. 1999; Schaefer et al. 2000; Venkatachalam et al. 2003). DAG also has an important dual role in controlling TRPC channels (Venkatachalam et al. 2003; Venkatachalam et al. 2004). Hence, while not activating TRPC4 or TRPC5 channels, DAG in fact has a profound inhibitory effect on both channels.

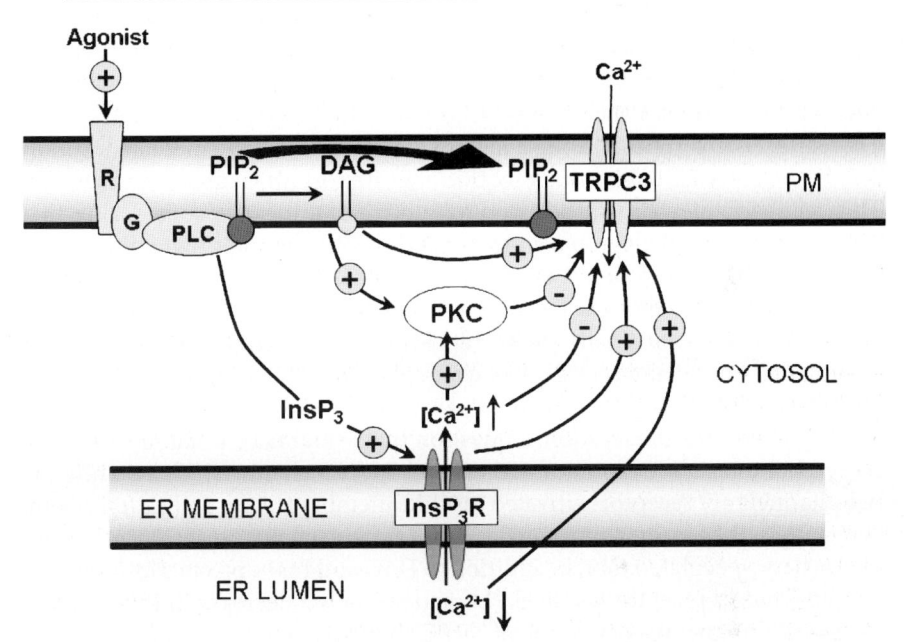

Fig. 1 Model depicting the different signals that activate or modify the TRPC3 channel (from Soboloff et al. 2006). PLC has multiple effects on the channel. Thus, PLC enzymatically breaks down PIP_2 into DAG and $InsP_3$. $InsP_3$ depletes stores of Ca^{2+} and much information indicates TRPC3 channels are store-operated. $InsP_3$ also activates the $InsP_3$ receptor, which is able to undergo direct interactions with the TRPC3 channel, perhaps mediating store-dependence. The other PLC product, DAG, has a direct non-PKC-dependent activating role on TRPC3 channels likely as a direct result of binding to the TRPC3 molecule itself. DAG also has profound effects of the TRPC3 channel through PKC. PKC is therefore a powerful inhibitor of most TRPC channels and DAG is a dual regulator of the TRPC3 channel. PLC-mediated DAG results in rapid channel opening followed later by a slower PKC-mediated deactivation of the channel. The decreased level of PIP_2 level as a result of PLC activation also has an important modifying action on TRPC3 channels. Hence, the TRPC3 channel and PLCγ form an intermolecular PH domain that has high specificity for binding PIP_2. This interaction with the TRPC3 channel allows the channel to be retained within the plasma membrane, a further operational control factor determining channel function in the plasma membrane. Details of these mechanisms are given in the text

Moreover, this inhibitory effect is clearly mediated through PKC (in contrast to the stimulatory action on TRPC3 channels). The effect was seen not only with exogenously added DAG. Thus, we could modify endogenously produced DAG by manipulating both DAG kinase and DAG lipase the two major metabolizing enzymes controlling turnover of DAG (Venkatachalam et al. 2003).

We were also able to show that this same PKC-mediated inhibition functions on TRPC3 channels that are activated by DAG. Treatment with the PKC activator, PMA, blocks the stimulation by subsequent addition of DAG. The action of PMA, however, is completely prevented by GF 109203X, a reliable aminoalkyl bisindolylmaleimide PKC blocker. In this case, the action of DAG is not only rapid, but longer lasting, indicating that endogenously activated PKC likely results in the fast turnoff of TRPC3 channels. With TRPC3 channels, therefore, the single molecule has two opposing actions on the same channel. DAG clearly activates the channel through a PKC-independent mechanism. Thereafter, there appears to be a slower DAG-induced PKC-mediated inhibition of the channel, suggesting that this is an important feedback mechanism. This feedback mechanism takes on yet added significance considering that the Ca^{2+} entering the cytoplasm through the TRP channels will enhance the activation of PKC and hence the turning off of the channel. Interestingly, we were able to show that Ca^{2+} entry induced by receptors in nontransfected HEK-293 cells was partially (50%) blocked by PKC activation (Fig. 1), suggesting that endogenous TRPC-like channel activity may account for some of the entry.

Despite all this information, a question that remains is: What mediates the receptor-induced activation of TRPC4 and TRPC5 channels? Thus, while the two channels are receptor-activated just like members of the TRPC3/6/7 group of channels, they are not activated by DAG (Venkatachalam et al. 2002). We and others have speculated that, in addition to DAG and InsP$_3$ production, another consequence of receptor-induced PLC activation is a decrease in PIP$_2$. Hence, the possibility that the activation of TRPC channels may also reflect decreased PIP$_2$ levels is an area of investigation worthy of considerable attention.

6
Role of the InsP$_3$ Receptor in TRPC Activation

There have been a number of studies indicating that TRPC3 channels can specifically interact with InsP$_3$ receptors. InsP$_3$R constructs containing the N-terminal InsP$_3$ binding domain were sufficient to couple to TRPC3 channels and render them activatable by InsP$_3$ (Kiselyov et al. 1999). It was revealed that a relatively short segment within the C-terminus of TRPC3 was required for interaction with the InsP$_3$R, and two specific InsP$_3$R sequences close to the InsP$_3$ binding domain interacted with TRPC3 (Boulay et al. 1999). Studies have shown that InsP$_3$ receptors interact with an inhibitory calmodulin-binding site on TRPC3, displacing calmodulin and thereby activating the channel (Tang et al. 2001; Zhang et al. 2001).

While the evidence for TRPC–InsP$_3$R interactions is quite compelling, the question of the role of InsP$_3$Rs in physiological store-operated Ca^{2+} entry is less clear. The actions of the two InsP$_3$R antagonists, 2-aminoethoxydiphenyl borate (2-APB) and xestospongin C, to block both endogenous store-operated Ca^{2+} entry and receptor-induced TRPC3 activation was considered evidence for a mediating role of InsP$_3$Rs in the activation of both channels (Ma et al. 2000). More compelling, however, were studies revealing that in the DT40 B lymphocyte triple InsP$_3$R knockout variant line, store-operated Ca^{2+} entry was entirely identical to that observed in wildtype cells (Sugawara et al. 1997; Broad et al. 2001; Ma et al. 2001, 2002; Venkatachalam et al. 2001). Exhaustive searching for the presence of InsP$_3$Rs in these knockout cells by examination of transcripts, full-length proteins or fragments thereof, InsP$_3$-binding activity, or physiological InsP$_3$R-mediated Ca^{2+} release in intact or permeabilized cells confirmed in all cases the absence of all InsP$_3$Rs in these cells (Sugawara et al. 1997; Broad et al. 2001; Ma et al. 2001; Venkatachalam et al. 2001). Moreover, the blocking action of 2-APB on capacitative Ca^{2+} entry (CCE) was unaltered by InsP$_3$R elimination (Broad et al. 2001; Ma et al. 2001; Ma et al. 2002). These and further studies indicated that its action was either upon the channel (Bakowski et al. 2001; Broad et al. 2001; Prakriya and Lewis 2001) or, more interestingly, upon the coupling machinery for SOCs (Ma et al. 2002; Schindl et al. 2002). However, TRPC3 channels expressed rapidly and with high efficiency in DT40 cells were shown to be activated by PLC-coupled receptors independently of InsP$_3$R expression and independently of store-emptying (Venkatachalam et al. 2001). On the other hand, it was observed that TRPC3 channels expressed more slowly and at lower efficiency in the same cells were partially-dependent on InsP$_3$R expression and wholly-dependent on store-emptying (Vazquez et al. 2001).

While apparently enigmatic, the coupling phenotype could rest on expression conditions (Venkatachalam et al. 2002; Vazquez et al. 2003; Putney 2004). Cells may have a limited quantity of the machinery required for store-coupling (be it junctions, InsP$_3$Rs, or linking proteins); thereby the level of expression of channels determines whether store-coupling occurs (Venkatachalam et al. 2002; Putney 2004).

7
Pharmacology of TRPC Channels

Recent studies have revealed the 3,5-bistrifluoromethyl pyrazole derivative BTP2 inhibits SOC/CRAC channels with approximately 100-fold greater potency than econazole or SKF (Ishikawa et al. 2003; Zitt et al. 2004; He et al. 2005). This has since been followed with the discovery that BTP2 also inhibits

TRPC channels with a similar potency (He et al. 2005). However, BTP2's actions are not nonspecific. It does not affect the properties of K^+ channels or voltage-gated Ca^{2+} channels, nor does it alter the Ca^{2+} permeability of mitochondria or ER (Ishikawa et al. 2003; He et al. 2005). Moreover, TRPV6 channels, with some properties similar to SOCs, are unaffected by BTP2 (He et al. 2005). Although the mechanism of this inhibition is not clear, we recently established that it decreases the open probability of the TRPC3 channels, without affecting amplitude, using single channel measurements of TRPC3 (He et al. 2005). Hence, rather than inhibiting conductance, BTP2 may target the process of opening or closing (or both the opening and closing) of the channel.

The observation that BTP2 appears to selectively inhibit SOC and TRPC channels indicates that TRPC channels share some common properties with SOC channels. This action is not limited to the TRPC3/6/7 channel family, since BTP2 also blocked TRPC5 with similar sensitivity as TRPC3 (He et al. 2005). While BTP2 does not appear to discriminate between TRPC channels, recent studies have revealed some specific pharmacological modifiers of TRPC channels. For example, flufenamate, a somewhat nonspecific cation channel blocker, reversibly enhances TRPC6 channel activity (Inoue et al. 2001), while inhibiting all other members of the TRPC family. Although this is the only published example of an agent that is specific for a single TRPC channel type, other agents are somewhat effective in differentiating between different TRPC subfamilies. Thus, Hisatsune et al. (2004) revealed a requirement for TRPC6 phosphorylation by Fyn, a member of the Src family of protein tyrosine kinases (PTKs). Specifically, Fyn enhanced DAG- and epidermal growth factor-induced TRPC6 activation by increasing Ca^{2+} conductance (Hisatsune et al. 2004). In a separate study, Vazquez et al. (2004) showed that the Src kinase phosphorylates TRPC3, and that the Src kinase inhibitor 4-amino-5-(4-chlorophenyl)-7-(t-butyl)pyrazolo[3,4-d]pyrimidine (PP2) inhibits TRPC3 channel activation. Importantly, they also found that TRPC5 was not sensitive to PP2. Hence, it appears that this Src kinase sensitivity is unique to the TRPC3/6/7 subfamily of channels.

The lanthanides La^{3+} and Gd^{3+} also have differential effectiveness across TRPC channel function. Whereas SOC-mediated Ca^{2+} entry is sensitive to concentrations as lower than 1 µM, TRPC3/6/7 channels are only modified at 10- to 50-fold higher levels (Ma et al. 2000; Putney 2001). In contrast, TRPC4 and -5 are activated by lanthanides (Jung et al. 2003). Single channel measurements revealed that this activation resulted from a substantial increase in the open probability of TRPC5 channels due to three negatively charged residues close to the extracellular mouth of the channel pore (Jung et al. 2003). These residues are absent in other channel members, providing a clear explanation for why this response is unique to the TRPC4/5 subfamily (Jung et al. 2003). Based upon these studies, while TRPC channels have some similarities in their activation and channel properties, they appear to have some unique pharmacological properties.

8
Biological Relevance and Emerging Roles for TRPC Channels

Although the TRPC channels are similar proteins, both in their modes of activation (PLC-mediated) and channel properties (slow nonselective cation channels), differences in the expression patterns of the different proteins along with the phenotype of a number of different knockout models have revealed distinct biological roles for the different channels. Perhaps the clearest example of this is TRPC2. A pseudogene in humans and higher primates (Liman and Innan 2003), this protein is primarily expressed in the vomeronasal organ, where it mediates the pheromone response. Elimination of this gene leads to a clear change in behavioral responses of mice, indicating that TRPC2 is required for this biological function (Leypold et al. 2002; Stowers et al. 2002). In contrast, the physiological impact of elimination of the TRPC6 gene, which has a well-established role in vascular smooth muscle contraction (Inoue et al. 2001; Jung et al. 2002; Welsh et al. 2002; Soboloff et al. 2005), has been far less straightforward, due primarily to compensation by expression of TRPC3, a closely related channel (Dietrich et al. 2005). However, since, unlike TRPC6, TRPC3 exhibits significant constitutive activity, there was a significant increase in blood pressure and agonist-induced contractility. Hence, despite the considerable homology that exists between different TRPC channels, TRPC channels have distinct and likely irreplaceable biological functions.

Whereas knockout of TRPC6 in the whole animal was complicated by the upregulation of related proteins, TRPC6 RNA interference (RNAi) was achieved in cultured cells without significant compensation by other channels. Thus, in a recent study Soboloff et al. (2005), targeted RNAi was combined with a rigorous assessment of both message and protein, to provide new information on the presence and function of *endogenously* expressed TRPC6 channels in A7r5 smooth muscle cells. TRPC6 knockdown experiments revealed that an OAG-activated nonselective cation current with a current–voltage relationship close to that of known TRPC6 channels was substantially reduced (Soboloff et al. 2005). This reduction in current mirrored the reduction in TRPC6 protein. However, the corresponding TRPC6-mediated OAG-dependent entry of Ca^{2+} was not significantly altered by TRPC6 knockdown. Yet the OAG-induced Ca^{2+} entry was almost completely inhibited by L-type Ca^{2+} channel blockers, indicating Ca^{2+} was entering through L-type voltage-dependent Ca^{2+} channels. However, pharmacological characterization of this current revealed TRPC6-like characteristics. Thus, it was sensitive to inhibitors of Src kinase and it was strongly inhibited by activation of PKC known to inhibit TRPC channels (Venkatachalam et al. 2002, 2003). The explanation for these results is that the TRPC6 channel, as a nonselective cation channel, is predominantly mediating the entry of Na^+ as opposed to Ca^{2+} ions, resulting in depolarization and the opening of L-type channels. Hence the TRPC6 channel is a mediator between

PLC-generated DAG and the activation of Ca^{2+} entry through L-type channels. A scheme depicting this signaling process is shown in Fig. 2.

Calculations reveal that even 90% reduction of TRPC6 channels would still allow depolarization sufficient to activate L-type channels. Thus, under conditions of RNAi resulting in approximately 90% reduction of TRPC6 protein and current carried by TRPC6 channel, there was still substantial depolarization-mediated activation of Ca^{2+} entry through L-type channels (Soboloff et al. 2005). The function of TRPC channels mediating depolarization and activation of L-type channels has also been indicated in other studies. Thus, in cerebral arteries, TRPC6 antisense treatment reduced pressure-induced depolarization and arterial constriction, suggesting that TRPC6 channels are activated as a result of pressure and may play an important role in the control of myogenic tone (Welsh et al. 2002). Recently, the TRPC3 channel, which is also expressed in cerebral arteries, was shown to mediate purinergic receptor-induced depolarization and contraction (Reading et al. 2005). Thus, members of the TRPC3/6/7 subfamily of nonselective cation channels may play an important role in the control of smooth muscle cell membrane potential to effect control over voltage-operated Ca^{2+} entry and muscle contraction.

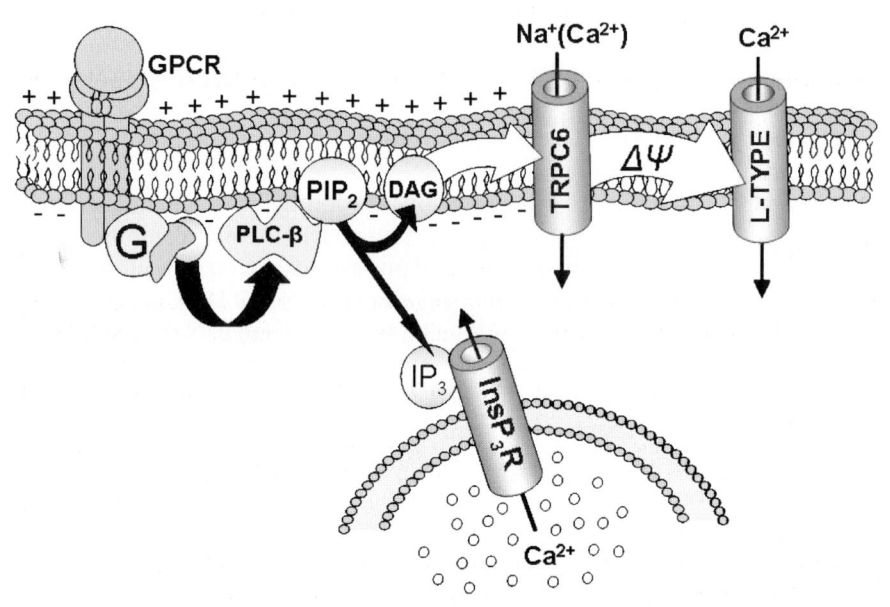

Fig. 2 Model to depict the role of TRPC6 channels in the coupling between receptor-induced PLC stimulation and L-type Ca^{2+} channel activation (from Soboloff et al. 2006). The G protein-coupled receptor (*GPCR*) activates PLC-β via G protein (*G*) and results in the formation of DAG and InsP$_3$. While InsP$_3$ induces Ca^{2+} release from stores, DAG activates the nonselective cation channel TRPC6. The predominant entry of Na$^+$ ions (in addition to Ca^{2+} ions) results in depolarization of the membrane and the activation of voltage-sensitive L-type Ca^{2+} channels

Overall, rather than a group of putative SOCs, TRPC channels should be viewed as a group of receptor-operated nonselective cation channels, with distinct cell type-specific roles. Hence, while the low basal activity of TRPC6 is a required characteristic in vascular smooth muscle, the leakiness of TRPC3, dangerous in the circulatory system, likely serves an important functional role in pontine neurons, where it is highly expressed. In addition, it has become increasingly clear that the nonselective cation properties of TRPC channels are important in their biological function. As regulators of membrane potential, their activation has an impact on many different cellular functions including Ca^{2+} entry. Recent information reveals the TRPC1 channel as a required component of the mechanosensitive cation channel in *Xenopus* oocytes (Maroto et al. 2005), indicating it opens in response to shape or pressure changes. Earlier, the TRPA1 channel was suggested to be the mechanosensitive channel that transduces cation flux in cochlear hair cells (Corey et al. 2004). Based on the profound actions of DAG and PIP_2 on TRPC channels, we are considering a model in which TRPC channels are primarily "lipid sensors." We consider that their sensing of membrane stretch may be another manifestation of this lipid-sensing role. Indeed, it is possible that on a broader scale, many other TRP channel subtypes may be sensing mechanical and thermal changes by detecting changes in the local protein-lipid environment.

Acknowledgements This work was supported by NIH grants HL55426 and AI058173, and the Interdisciplinary Training Program in Muscle Biology, University of Maryland School of Medicine.

References

Bakowski D, Glitsch MD, Parekh AB (2001) An examination of the secretion-like coupling model for the activation of the Ca^{2+} release-activated Ca^{2+} current ICRAC in RBL-1 cells. J Physiol 532:55–71

Beech DJ (2005) Emerging functions of 10 types of TRP cationic channel in vascular smooth muscle. Clin Exp Pharmacol Physiol 32:597–603

Beech DJ, Muraki K, Flemming R (2004) Non-selective cationic channels of smooth muscle and the mammalian homologues of Drosophila TRP. J Physiol 559:685–706

Berridge MJ (2001) The versatility and complexity of calcium signalling. Novartis Found Symp 239:52–159

Berridge MJ (2002) The endoplasmic reticulum: a multifunctional signaling organelle. Cell Calcium 32:235–249

Berridge MJ, Bootman MD, Roderick HL (2003) Calcium signalling: dynamics, homeostasis and remodelling. Nat Rev Mol Cell Biol 4:517–529

Boulay G, Brown DM, Qin N, Jiang M, Dietrich A, Zhu MX, Chen Z, Birnbaumer M, Mikoshiba K, Birnbaumer L (1999) Modulation of Ca^{2+} entry by polypeptides of the inositol 1,4,5-trisphosphate receptor (IP3R) that bind transient receptor potential (TRP): evidence for roles of TRP and IP3R in store depletion-activated Ca^{2+} entry. Proc Natl Acad Sci U S A 96:14955–14960

Broad LM, Braun FJ, Lievremont JP, Bird GS, Kurosaki T, Putney JW Jr (2001) Role of the phospholipase C-inositol 1,4,5-trisphosphate pathway in calcium release-activated calcium current and capacitative calcium entry. J Biol Chem 276:15945–15952

Clapham DE (2003) TRP channels as cellular sensors. Nature 426:517–524

Corey DP, Garcia-Anoveros J, Holt JR, Kwan KY, Lin SY, Vollrath MA, Amalfitano A, Cheung EL, Derfler BH, Duggan A, Geleoc GS, Gray PA, Hoffman MP, Rehm HL, Tamasauskas D, Zhang DS (2004) TRPA1 is a candidate for the mechanosensitive transduction channel of vertebrate hair cells. Nature 432:723–730

Dietrich A, Mederos YSM, Gollasch M, Gross V, Storch U, Dubrovska G, Obst M, Yildirim E, Salanova B, Kalwa H, Essin K, Pinkenburg O, Luft FC, Gudermann T, Birnbaumer L (2005) Increased vascular smooth muscle contractility in TRPC6$^{-/-}$ mice. Mol Cell Biol 25:6980–6989

Gill DL, Patterson RL (2004) Toward a consensus on the operation of receptor-induced calcium entry signals. Sci STKE 2004:pe39

He LP, Hewavitharana T, Soboloff J, Spassova MA, Gill DL (2005) A functional link between store-operated and TRPC channels revealed by the 3,5-bis(trifluoromethyl)pyrazole derivative, BTP2. J Biol Chem 280:10997–11006

Hisatsune C, Kuroda Y, Nakamura K, Inoue T, Nakamura T, Michikawa T, Mizutani A, Mikoshiba K (2004) Regulation of TRPC6 channel activity by tyrosine phosphorylation. J Biol Chem 279:18887–18894

Hofmann T, Obukhov AG, Schaefer M, Harteneck C, Gudermann T, Schultz G (1999) Direct activation of human TRPC6 and TRPC3 channels by diacylglycerol. Nature 397:259–263

Inoue R, Okada T, Onoue H, Hara Y, Shimizu S, Naitoh S, Ito Y, Mori Y (2001) The transient receptor potential protein homologue TRP6 is the essential component of vascular alpha(1)-adrenoceptor-activated Ca^{2+}-permeable cation channel. Circ Res 88:325–332

Ishikawa J, Ohga K, Yoshino T, Takezawa R, Ichikawa A, Kubota H, Yamada T (2003) A pyrazole derivative, YM-58483, potently inhibits store-operated sustained Ca^{2+} influx and IL-2 production in T lymphocytes. J Immunol 170:4441–4449

Jung S, Strotmann R, Schultz G, Plant TD (2002) TRPC6 is a candidate channel involved in receptor-stimulated cation currents in A7r5 smooth muscle cells. Am J Physiol Cell Physiol 282:C347–359

Jung S, Muhle A, Schaefer M, Strotmann R, Schultz G, Plant TD (2003) Lanthanides potentiate TRPC5 currents by an action at extracellular sites close to the pore mouth. J Biol Chem 278:3562–3571

Kahr H, Schindl R, Fritsch R, Heinze B, Hofbauer M, Hack ME, Mortelmaier MA, Groschner K, Peng JB, Takanaga H, Hediger MA, Romanin C (2004) CaT1 knock-down strategies fail to affect CRAC channels in mucosal-type mast cells. J Physiol 557:121–132

Kiselyov K, Mignery GA, Zhu MX, Muallem S (1999) The N-terminal domain of the IP3 receptor gates store-operated hTrp3 channels. Mol Cell 4:423–429

Leypold BG, Yu CR, Leinders-Zufall T, Kim MM, Zufall F, Axel R (2002) Altered sexual and social behaviors in trp2 mutant mice. Proc Natl Acad Sci U S A 99:6376–6381

Li HS, Xu XZ, Montell C (1999) Activation of a TRPC3-dependent cation current through the neurotrophin BDNF. Neuron 24:261–273

Liman ER, Innan H (2003) Relaxed selective pressure on an essential component of pheromone transduction in primate evolution. Proc Natl Acad Sci U S A 100:3328–3332

Ma HT, Patterson RL, van Rossum DB, Birnbaumer L, Mikoshiba K, Gill DL (2000) Requirement of the inositol trisphosphate receptor for activation of store-operated Ca^{2+} channels. Science 287:1647–1651

Ma HT, Venkatachalam K, Li HS, Montell C, Kurosaki T, Patterson RL, Gill DL (2001) Assessment of the role of the inositol 1,4,5-trisphosphate receptor in the activation of transient receptor potential channels and store-operated Ca^{2+} entry channels. J Biol Chem 276:18888–18896

Ma HT, Venkatachalam K, Parys JB, Gill DL (2002) Modification of store-operated channel coupling and inositol trisphosphate receptor function by 2-aminoethoxydiphenyl borate in DT40 lymphocytes. J Biol Chem 277:6915–6922

Maroto R, Raso A, Wood TG, Kurosky A, Martinac B, Hamill OP (2005) TRPC1 forms the stretch-activated cation channel in vertebrate cells. Nat Cell Biol 7:179–185

Montell C (1999) Visual transduction in Drosophila. Annu Rev Cell Dev Biol 15:231–268

Montell C (2001) Physiology, phylogeny, and functions of the TRP superfamily of cation channels. Sci STKE 2001:RE1

Montell C, Birnbaumer L, Flockerzi V (2002a) The TRP channels, a remarkably functional family. Cell 108:595–598

Montell C, Birnbaumer L, Flockerzi V, Bindels RJ, Bruford EA, Caterina MJ, Clapham DE, Harteneck C, Heller S, Julius D, Kojima I, Mori Y, Penner R, Prawitt D, Scharenberg AM, Schultz G, Shimizu N, Zhu MX (2002b) A unified nomenclature for the superfamily of TRP cation channels. Mol Cell 9:229–231

Okada T, Inoue R, Yamazaki K, Maeda A, Kurosaki T, Yamakuni T, Tanaka I, Shimizu S, Ikenaka K, Imoto K, Mori Y (1999) Molecular and functional characterization of a novel mouse transient receptor potential protein homologue TRP7. Ca^{2+}-permeable cation channel that is constitutively activated and enhanced by stimulation of G protein-coupled receptor. J Biol Chem 274:27359–27370

Parekh AB, Penner R (1997) Store depletion and calcium influx. Physiol Rev 77:901–930

Parekh AB, Putney JW Jr (2005) Store-operated calcium channels. Physiol Rev 85:757–810

Patterson RL, van Rossum DB, Ford DL, Hurt KJ, Bae SS, Suh PG, Kurosaki T, Snyder SH, Gill DL (2002) Phospholipase C-gamma is required for agonist-induced Ca^{2+} entry. Cell 111:529–541

Prakriya M, Lewis RS (2001) Potentiation and inhibition of Ca^{2+} release-activated Ca^{2+} channels by 2-aminoethyldiphenyl borate (2-APB) occurs independently of IP3 receptors. J Physiol 536:3–19

Putney JW Jr (2001) Pharmacology of capacitative calcium entry. Mol Interv 1:84–94

Putney JW Jr (2004) The enigmatic TRPCs: multifunctional cation channels. Trends Cell Biol 14:282–286

Putney JW Jr, Broad LM, Braun FJ, Lievremont JP, Bird GS (2001) Mechanisms of capacitative calcium entry. J Cell Sci 114:2223–2229

Reading SA, Earley S, Waldron BJ, Welsh DG, Brayden JE (2005) TRPC3 mediates pyrimidine receptor-induced depolarization of cerebral arteries. Am J Physiol Heart Circ Physiol 288:H2055–2061

Schaefer ML, Wong ST, Wozniak DF, Muglia LM, Liauw JA, Zhuo M, Nardi A, Hartman RE, Vogt SK, Luedke CE, Storm DR, Muglia LJ (2000) Altered stress-induced anxiety in adenylyl cyclase type VIII-deficient mice. J Neurosci 20:4809–4820

Schindl R, Kahr H, Graz I, Groschner K, Romanin C (2002) Store depletion-activated CaT1 currents in rat basophilic leukemia mast cells are inhibited by 2-aminoethoxydiphenyl borate. Evidence for a regulatory component that controls activation of both CaT1 and CRAC (Ca^{2+} release-activated Ca^{2+} channel) channels. J Biol Chem 277:26950–26958

Soboloff J, Spassova M, Xu W, He LP, Cuesta N, Gill DL (2005) Role of endogenous TRPC6 channels in Ca^{2+} signal generation in A7r5 smooth muscle cells. J Biol Chem 280:39786–39794

Spassova MA, Soboloff J, He LP, Hewavitharana T, Xu W, Venkatachalam K, van Rossum DB, Patterson RL, Gill DL (2004) Calcium entry mediated by SOCs and TRP channels: variations and enigma. Biochim Biophys Acta 1742:9–20

Stowers L, Holy TE, Meister M, Dulac C, Koentges G (2002) Loss of sex discrimination and male-male aggression in mice deficient for TRP2. Science 295:1493–1500

Sugawara H, Kurosaki M, Takata M, Kurosaki T (1997) Genetic evidence for involvement of type 1, type 2 and type 3 inositol 1,4,5-trisphosphate receptors in signal transduction through the B-cell antigen receptor. EMBO J 16:3078–3088

Suh PG, Hwang JI, Ryu SH, Donowitz M, Kim JH (2001) The roles of PDZ-containing proteins in PLC-beta-mediated signaling. Biochem Biophys Res Commun 288:1–7

Tang J, Lin Y, Zhang Z, Tikunova S, Birnbaumer L, Zhu MX (2001) Identification of common binding sites for calmodulin and inositol 1,4,5-trisphosphate receptors on the carboxyl termini of trp channels. J Biol Chem 276:21303–21310

Tang Y, Tang J, Chen Z, Trost C, Flockerzi V, Li M, Ramesh V, Zhu MX (2000) Association of mammalian trp4 and phospholipase C isozymes with a PDZ domain-containing protein, NHERF. J Biol Chem 275:37559–37564

van Rossum DB, Patterson RL, Sharma S, Barrow RK, Kornberg M, Gill DL, Snyder SH (2005) Phospholipase Cgamma1 controls surface expression of TRPC3 through an intermolecular PH domain. Nature 434:99–104

Vazquez G, Lievremont JP, St JBG, Putney JW Jr (2001) Human Trp3 forms both inositol trisphosphate receptor-dependent and receptor-independent store-operated cation channels in DT40 avian B lymphocytes. Proc Natl Acad Sci U S A 98:11777–11782

Vazquez G, Wedel BJ, Trebak M, St John Bird G, Putney JW Jr (2003) Expression level of the canonical transient receptor potential 3 (TRPC3) channel determines its mechanism of activation. J Biol Chem 278:21649–21654

Vazquez G, Wedel BJ, Kawasaki BT, Bird GS, Putney JW Jr (2004) Obligatory role of Src kinase in the signaling mechanism for TRPC3 cation channels. J Biol Chem 279:40521–40528

Venkatachalam K, Ma HT, Ford DL, Gill DL (2001) Expression of functional receptor-coupled TRPC3 channels in DT40 triple receptor InsP3 knockout cells. J Biol Chem 276:33980–33985

Venkatachalam K, van Rossum DB, Patterson RL, Ma HT, Gill DL (2002) The cellular and molecular basis of store-operated calcium entry. Nat Cell Biol 4:E263–272

Venkatachalam K, Zheng F, Gill DL (2003) Regulation of canonical transient receptor potential (TRPC) channel function by diacylglycerol and protein kinase C. J Biol Chem 278:29031–29040

Venkatachalam K, Zheng F, Gill DL (2004) Control of TRPC and store-operated channels by protein kinase C. Novartis Found Symp 258:172–185

Voets T, Prenen J, Fleig A, Vennekens R, Watanabe H, Hoenderop JG, Bindels RJ, Droogmans G, Penner R, Nilius B (2001) CaT1 and the calcium release-activated calcium channel manifest distinct pore properties. J Biol Chem 276:47767–47770

Welsh DG, Morielli AD, Nelson MT, Brayden JE (2002) Transient receptor potential channels regulate myogenic tone of resistance arteries. Circ Res 90:248–250

Xu XZ, Li HS, Guggino WB, Montell C (1997) Coassembly of TRP and TRPL produces a distinct store-operated conductance. Cell 89:1155–1164

Yue L, Peng JB, Hediger MA, Clapham DE (2001) CaT1 manifests the pore properties of the calcium-release-activated calcium channel. Nature 410:705–709

Zagranichnaya TK, Wu X, Villereal ML (2005) Endogenous TRPC1, TRPC3, and TRPC7 proteins combine to form native store-operated channels in HEK-293 cells. J Biol Chem 280:29559–29569

Zhang L, Saffen D (2001) Muscarinic acetylcholine receptor regulation of TRP6 Ca^{2+} channel isoforms. Molecular structures and functional characterization. J Biol Chem 276:13331–13339

Zhang Z, Tang J, Tikunova S, Johnson JD, Chen Z, Qin N, Dietrich A, Stefani E, Birnbaumer L, Zhu MX (2001) Activation of Trp3 by inositol 1,4,5-trisphosphate receptors through displacement of inhibitory calmodulin from a common binding domain. Proc Natl Acad Sci U S A 98:3168–3173

Zitt C, Halaszovich CR, Luckhoff A (2002) The TRP family of cation channels: probing and advancing the concepts on receptor-activated calcium entry. Prog Neurobiol 66:243–264

Zitt C, Strauss B, Schwarz EC, Spaeth N, Rast G, Hatzelmann A, Hoth M (2004) Potent inhibition of Ca^{2+} release-activated Ca^{2+} channels and T-lymphocyte activation by the pyrazole derivative BTP2. J Biol Chem 279:12427–12437

Phospholipase C-Coupled Receptors and Activation of TRPC Channels

M. Trebak · L. Lemonnier · J. T. Smyth · G. Vazquez · J. W. Putney Jr. (✉)

Laboratory of Signal Transduction, Department of Health and Human Services, National Institute of Environmental Health Sciences—NIH, Research Triangle Park, PO Box 12233, NC, 27709, USA
putney@niehs.nih.gov

1	Calcium Entry Pathways	594
2	Capacitative Calcium Entry	594
2.1	Non-Capacitative Ca^{2+} Entry	594
3	Role of TRPC Channels in CCE and NCCE	595
4	Modes of Activation of TRPC Channels	599
4.1	Store Depletion	599
4.2	Non-Store-Operated Mechanisms	601
4.2.1	TRPC1	601
4.2.2	TRPC2	602
4.2.3	TRPC3/6/7	602
4.2.4	TRPC4/5	603
4.3	Role of Channel Trafficking to the Plasma Membrane	604
5	Biological Relevance of TRPC Channels	605
	References	607

Abstract The canonical transient receptor potential (TRPC) cation channels are mammalian homologs of the photoreceptor channel TRP in *Drosophila melanogaster*. All seven TRPCs (TRPC1 through TRPC7) can be activated through Gq/11 receptors or receptor tyrosine kinase (RTK) by mechanisms downstream of phospholipase C. The last decade saw a rapidly growing interest in understanding the role of TRPC channels in calcium entry pathways as well as in understanding the signal(s) responsible for TRPC activation. TRPC channels have been proposed to be activated by a variety of signals including store depletion, membrane lipids, and vesicular insertion into the plasma membrane. Here we discuss recent developments in the mode of activation as well as the pharmacological and electrophysiological properties of this important and ubiquitous family of cation channels.

Keywords Ion channels · TRPC channels · Nonselective cation channel · Store-operated channel · Phospholipase C

1
Calcium Entry Pathways

Elevation of intracellular free Ca^{2+} upon receptor stimulation is essential for cell survival and function. By means of Ca^{2+}-permeable channels, cells carefully use Ca^{2+} to transduce signals in response to a wide variety of stimuli such as hormones, growth factors, and antigens (Berridge et al. 2000; Clapham 1995). The dysfunction of this Ca^{2+} homeostasis can result in diseases such as immune deficiency, hypertension, neurodegenerative diseases, and cancer.

2
Capacitative Calcium Entry

Gq/11 or receptor tyrosine kinase (RTK) receptor stimulation activates a phospholipid-specific phospholipase C (PLC). PLC action on phosphatidylinositol 4,5-bisphosphate (PIP_2) produces two second messengers, inositol 1,4,5-trisphosphate (IP_3) and diacylglycerol (DAG). IP_3 binds to its receptor (IP_3R) on the membrane of the endoplasmic reticulum (ER), inducing release of ER Ca^{2+} stores (Berridge 1993). The fall of the Ca^{2+} concentration within the lumen of the ER by the action of IP_3 on the IP_3R in some manner activates a concomitant entry of Ca^{2+} across the plasma membrane. This ubiquitous Ca^{2+} entry pathway is known as capacitative Ca^{2+} entry (CCE) and the channels mediating this entry are called store-operated channels (SOC) (Barritt 1999; Berridge 1995; Parekh and Putney 2005; Putney 1986, 1997). SOC channels can be activated by any procedure that empties ER stores. For instance, exposure of cells to sarcoplasmic/ER Ca^{2+}-ATPase (SERCA) pump inhibitors such as thapsigargin, which prevent the pump from refilling the stores, activates SOC channels. One of the characteristics of SOC channels is Ca^{2+}-dependent inactivation, whereby a rise of intracellular Ca^{2+}, generally due to Ca^{2+} entering through the channels, exerts a negative feedback on the channels (Parekh and Putney 2005; Zweifach and Lewis 1995a; Zweifach and Lewis 1995b). The molecular identity of SOC channels and the signaling mechanism(s) linking internal Ca^{2+} store depletion to this Ca^{2+} entry remain elusive and have been the subject of intensive investigations in recent years.

2.1
Non-Capacitative Ca^{2+} Entry

A large number of pathways are now known to contribute to the generation of Ca^{2+} signals in cells following receptor activation. The interplay between these different pathways generates a diverse array of Ca^{2+} signals, which are needed to ensure specificity of the information transmitted to the nucleus.

Although CCE is clearly a major means of Ca^{2+} entry into nonexcitable cells, there are other, no less important, modes of regulated Ca^{2+} entry across the plasma membrane. All of these routes of Ca^{2+} entry into cells that do not depend on the state of filling of internal Ca^{2+} stores are often referred to as non-capacitative Ca^{2+} entry pathways (NCCE).

It is now clearly established that both limbs of the PLC pathway (IP_3 and DAG) are involved in Ca^{2+} signaling at the plasma membrane. In addition to its known role as a PKC activator, DAG acts, in some instances, as a Ca^{2+} channel activator (Chakrabarti and Kumar 2000; Gamberucci et al. 2002; Hofmann et al. 1999; Lievremont et al. 2005b; Trebak et al. 2003a), and also serves as a source of arachidonic acid (AA), another important player in Ca^{2+} entry regulation (Broad et al. 1999; Shuttleworth 1999; Shuttleworth et al. 2004). IP_3 was also shown to act directly on channels at the plasma membrane (presumably IP_3R channels) to induce Ca^{2+} entry in olfactory neurons (Fadool and Ache 1992; Okada et al. 1994) and B cells (Vazquez et al. 2002). Phosphoinositide 3-kinase (PI3K) activation results in the production of phosphatidylinositol 3,4,5-trisphosphate (PIP_3), a second messenger that was recently implicated in stimulating a Ca^{2+} entry pathway in lymphocytes (Hsu et al. 2000). The second messengers, cyclic ADP-ribose (cADPr) and cyclic GMP (cGMP) were implicated in another Ca^{2+} entry pathway (Guse et al. 1999; Podesta et al. 2000; Sadighi Akha et al. 1996). The rise of intracellular Ca^{2+} can also activate Ca^{2+}-sensitive entry channels at the plasma membrane (Barritt 1999). Although the exact signaling mechanisms controlling NCCE channels as well as the identities of these channels are not yet defined, there is considerable evidence that those intracellular messengers produced downstream of receptor stimulation (i.e., DAG, PIP_3, IP_3, AA, cADPr, cGMP) can play important roles in channel regulation.

3
Role of TRPC Channels in CCE and NCCE

Given the importance of PLC-generated calcium signals in nonexcitable cells, considerable effort has been devoted to identifying the molecular basis of CCE and NCCE channels and their exact mechanisms of activation. Members of the transient receptor potential (TRP) superfamily of channels are considered potential candidates for PLC-dependent calcium entry in many cells types (Benham et al. 2002; Birnbaumer et al. 1996; Clapham et al. 2002; Harteneck et al. 2000; Minke and Cook 2002; Montell et al. 2002; Vazquez et al. 2004; Zitt et al. 2002).

TRP proteins constitute a large superfamily of ion channels that are expressed in many different tissues and cell types (Birnbaumer et al. 1996; Minke and Cook 2002; Montell et al. 2002). The TRP superfamily can be divided into three major families: (1) six channels homologous to the vanilloid receptor, called TRPV, that are activated by a variety of signals, including vanilloid compounds such as capsaicin, noxious signals, hypotonic cell swelling, and heat (Gunthorpe et al. 2002; Hoenderop et al. 2003; Nilius et al. 2004). Also included in the TRPV family are two channels (TRPV5 and -6) that are highly Ca^{2+} selective and are involved in epithelial Ca^{2+} transport (Hoenderop et al. 2003); (2) eight melastatin-related TRP channels (TRPM), with diverse functional properties including Ca^{2+} and Mg^{2+} influx, modulating membrane potential, and sensing cold in sensory neurons (Fleig and Penner 2004; Perraud et al. 2004); and (3) the canonical TRP (TRPC) channels, so named because they are structurally closest to the founding family member, *Drosophila* TRP (Montell et al. 2002; Vazquez et al. 2004). The seven TRPCs share in common the property of activation through PLC-coupled receptors and are therefore proposed components of native receptor-regulated channels in different cell types (Vazquez et al. 2004; Zitt et al. 2002). The TRPC family can be divided into four subfamilies: TRPC1, TRPC2, TRPC3/6/7, and TRPC4/5 (or, TRPC1 can be included with TRPC4 and -5). TRPC2 is a pseudogene in humans but encodes a functional channel in most other mammals. TRPC3, -6, and -7 form a subfamily sharing 70%–80% amino acid identity (Trebak et al. 2003b). TRPC4 and -5 are also closely related and form another subfamily within the TRPC family.

All members of the TRPC family are proposed to share a common transmembrane topology, in which six predicted transmembrane domains (TM1–TM6) separate the cytoplasmic N- and C-termini (Vannier et al. 1998; see also Dohke et al. 2004). This model predicts a putative pore region between TM5 and TM6. A few studies have examined the effects of point mutations within the pore region and extracellular loops that line the pore (Hofmann et al. 2002; Jung et al. 2003; Liu et al. 2003; Strubing et al. 2003); however, a systematic analysis of the predicted pore region of TRPC family members has not yet been performed. Other salient structural features of TRPCs include: on the N-terminus, three to four ankyrin repeats, a predicted coiled-coil region, and a putative caveolin binding region; on the C-terminus, a conserved 6-amino-acid sequence, the TRP signature motif (EWKFAR), a highly conserved proline-rich motif, the CIRB (*c*almodulin/IP$_3$ *r*eceptor *b*inding) region and a predicted coiled-coil region. TRPC4 and -5 have an extended C-terminus that contains a PDZ-binding motif. The reader is referred to a recent review for a thorough discussion of the structure-function relationships for TRPCs (Vazquez et al. 2004). Table 1 provides a summary of the pharmacological and electrophysiological properties of TRPC channels, based on measurements with ectopically expressed channels. The different modes of activation proposed for individual TRPC channels are also depicted in Table 1 and are discussed in detail below.

Table 1 Properties of TRPC channels

	TRPC1	TRPC2	TRPC3	TRPC4	TRPC5	TRPC6	TRPC7
Expression pattern in human	Ubiquitous (1)	Pseudogene in human (2); highly expressed in vomeronasal organ, sperm and brain (mice) (3, 4)	Primarily brain (5)	Brain, endothelial cells, adrenal gland, prostate, uterus, heart (6)	Brain (7)	Brain, lung (8), smooth muscle (mice) (9, 10); kidney (11)	Kidney, pituitary, brain (12)
Rectification	Slightly inwardly rectifying (13, 14)	Linear (15)	Outwardly rectifying (16)	Inwardly rectifying (6); doubly rectifying (17)	Inwardly rectifying (18, 19); doubly rectifying (17, 20)	Doubly rectifying (21, 22)	Doubly rectifying (23)
Selectivity (pCa/pNa)	Nonselective (13, 14)	2.7 (15)	1.6 (24)	7.7 (6); 1.05 (17)	14.3 (18); 1.79 (17)	5 (21)	1.9–5.9 (23)
Single channel conductance	16 pS (13)	42 pS (15)	17 pS (25); 23 pS (24); 60–66 pS (25, 26, 27)	41 pS (17); 30 pS (28)	47.6 pS (29); 63 pS (17); 38 pS (30)	35 pS (21); 46.6 pS (22); 34.4 pS (34)	49.3 pS (34)

Table 1 continued

	TRPC1	TRPC2	TRPC3	TRPC4	TRPC5	TRPC6	TRPC7
Activation mechanisms	Store depletion (5, 13); OAG, PLC (14); stretch (35)	Store depletion (36); DAG (15)	Store depletion (5, 25, 38, 42); OAG (21, 16)	GTPγS (6, 17); store depletion (6); PLC (17); La^{3+} (17)	IP$_3$, store depletion (7); activation of IP$_3$ receptor (31); GTPγS (17); La^{3+} (17, 30); Gd^{3+} (22); PLC (32); extracellular Ca^{2+} (17, 33)	OAG (21, 22); 20-HETE (38); PIP$_3$ (39)	OAG (23, 40); store depletion (40)
Pharmacological inhibitors	Lanthanides (13, 41); extracellular Ca^{2+} (14)	2-APB (15)	Lanthanides (24, 27); SKF 96265 (25); 2-APB (43); flufenamic acid (9); phorbol esters (16)	10 μM Gd^{3+} (44), but see (17) in which 100 μM activated both TRPC4 and TRPC5	SK&F96365 (18); OAG/DAG (32); 2-APB (45)	Lanthanides (8, 22); SK&F96365 (46); 2-APB (45, 47); phorbol esters (34, 48), but see (49)	SK&F96365 (23); 2-APB (47); flufenamic acid (9); phorbol esters (23, 34)

20-HETE, 20-hydroxyeicosatetraenoic acid; see the text for other abbreviation definitions 1, Zhu et al. 1995; 2, Wes et al. 1995; 3, Liman et al. 1999; 4, Jungnickel et al. 2001; 5, Zhu et al. 1996; 6, Philipp et al. 1996; 7, Philipp et al. 1998; 8, Boulay et al. 1997; 9, Inoue et al. 2001; 10, Dietrich et al. 2005; 11, Reiser et al. 2005; 12, Winn et al. 2005; 13, Riccio et al. 2002a; 14, Zitt et al. 1996; 14, Lintschinger et al. 2000; 15, Lucas et al. 2003; 16, Trebak et al. 2003a; 17, Schaefer et al. 2000; 18, Okada et al. 1998; 19, Philipp et al. 1998; 20, Strübing et al. 2001; 21, Hofmann et al. 1999; 22, Jung et al. 2003; 23, Okada et al. 1999; 24, Kamouchi et al. 1999; 25, Kiselyov et al. 1998; 26, Zitt et al. 1997; 27, Hurst et al. 1998; 28, Schaefer et al. 2002; 29, Yamada et al. 2000; 30, Strübing et al. 2001; 31, Kanki et al. 2001; 32, Venkatachalam et al. 2003; 33, Zeng et al. 2004; 34, Shi et al. 2004; 35, Maroto et al. 2005; 36, Vannier et al. 1999; 37, Vazquez et al. 2001; 38, Basora et al. 2003; 39, Tseng et al. 2004; 40, Lievremont et al. 2004; 41, Sinkins et al. 1998; 42, Vazquez et al. 2001; 43, Ma et al. 2000; 44, McKay et al. 2000; 45, Xu et al. 2005; 46, Estacion et al. 2004; 47, Lievremont et al. 2005a; 48, Kim and Saffen 2005; 49, Estacion et al. 2004

4
Modes of Activation of TRPC Channels

4.1
Store Depletion

The heterogeneity in both cation selectivity and pharmacological profile of SOCs in different cell types (see Parekh and Putney 2005, and references therein) suggests that these channels may be formed from more than a single gene product. The recognition of the product of the *Drosophila trp* gene as a light-sensitive Ca^{2+} permeable channel responsible for Ca^{2+}-entry upon light-dependent activation of PLC (Montell 1999) originally suggested mammalian TRP proteins as molecular candidates for SOCs (Hardie and Minke 1993). As a result, several studies attempting to elucidate the molecular nature of SOCs initially focused on mammalian TRP channels, particularly those from the TRPC family. Most members of the TRPC family, either endogenous or ectopically expressed in cell lines, have been experimentally implicated in store-operated cation entry in one or more study.

A considerable amount of experimental evidence supports the notion that TRPC1 or TRPC4 (or both) may form or function as components of SOC channels. Overexpression of TRPC1 in many different cell lines results in channels with different degrees of Ca^{2+} selectivity (Liu et al. 2000; Zitt et al. 1996) that can be activated by either agonist stimulation or pharmacological depletion of stores with the SERCA pump inhibitor thapsigargin. TRPC1 antisense constructs have in general proved to be efficient in reducing SOC entry in a variety of cell types (Brough et al. 2001; Liu et al. 2000; Sweeney et al. 2002; Tomita et al. 1998; Wu et al. 2000), and genetic disruption of the TRPC1 gene reduced the highly Ca^{2+} selective, electrophysiological archetypical store-operated current, I_{CRAC} (calcium release-activated current), in the avian DT40 B lymphocyte cell line (Mori et al. 2002). In vascular smooth muscle cells and human platelets, extracellular application of anti-TRPC1 antibodies directed against extracellular loops of TRPC1 significantly reduced SOC entry (Rosado et al. 2002; Xu and Beech 2001). Similarly, ectopic expression of TRPC4 in human embryonic kidney (HEK)-293 and CHO cells increased SOC entry induced by either PLC-dependent release of Ca^{2+} from ER or pharmacological depletion of stores with thapsigargin (Philipp et al. 1996; Warnat et al. 1999). In addition, I_{crac} currents are enhanced by overexpression of TRPC4 (Warnat et al. 1999).

As for TRPC1, the use of antisense oligonucleotides against TRPC4 markedly reduced I_{crac}-like currents in an adrenal cortical cell line and store-operated calcium entry in mouse mesangial cells (Philipp et al. 2000; Wang et al. 2004). Additionally, SOC entry is significantly impaired in vascular endothelium from TRPC4 knockout (TRPC4$^{-/-}$) mice (Freichel et al. 2001; Tiruppathi et al. 2002). It should be noted, however, that the role of TRPC4 in SOC entry is not as well defined as for TRPC1. In contrast with the above-mentioned findings, ectopic expression of mouse TRPC4 in diverse cell lines was found to form PLC-regulated cation channels functioning independently of store depletion (Obukhov and Nowycky 2002; Schaefer et al. 2000). Also, in HEK-293 cells, overexpression of human TRPC4 failed to produce agonist- or store depletion-regulated channels, resulting in constitutively active cation channels (McKay et al. 2000). In addition, in another study, stable expression of a TRPC4 antisense construct did not alter store-operated entry (Wu et al. 2002).

Following the initial cloning almost simultaneously by two different groups, heterologous expression of mouse TRPC5 in cell lines yielded conflicting results as to its regulation by store depletion (Okada et al. 1998; Philipp et al. 1998), and in most instances channel activation seems to be dependent upon receptor-stimulation of PLC, in a manner not related to store-depletion (Kanki et al. 2001; Schaefer et al. 2000). However, recent data from Beech and colleagues showed that at the single-cell level TRPC5 could be activated by a multiplicity of signals, including store depletion (Zeng et al. 2004).

As mentioned above, TRPC2 is a pseudogene in humans, but seems to be functional in mouse, rat, and other mammalian species (Liman and Innan 2003; Vazquez et al. 2004). Two splice variants of mouse TRPC2 (TRPC2a and TRPC2b) enhanced SOC in COS cells (Vannier et al. 1999), whereas two additional splice variants (TRPC2α and TRPC2β) cloned by a different laboratory (Hofmann et al. 2000) did not show any measurable function upon heterologous expression. In sperm and CHO cells, evidence suggests that TRPC2 might be part of endogenous SOCs (Gailly and Colson-Van Schoor 2001; Jungnickel et al. 2001).

After TRPC1, TRPC3, a member of the TRPC3/6/7 subfamily, is probably the next most extensively investigated TRPC family member. Yet its role in CCE remains controversial. A recurrent cause of misinterpretation of the ability of TRPC3 to function in a store-operated mode is the fact that ectopic expression of TRPC3 in cell lines results in a channel with significant constitutive activity (McKay et al. 2000; Preuß et al. 1997; Zhu et al. 1996, 1998; discussed in Trebak et al. 2003b). Indeed, most available evidence supports the notion that TRPC3 is a receptor-activated channel not activated by store depletion (Hofmann et al. 1999; Ma et al. 2002; McKay et al. 2000; Trebak et al. 2003a; Zitt et al. 1997). Recently, however, a study from Vazquez et al. showed that ectopic expression of TRPC3 in the avian B lymphocyte DT40 cell line results in a channel with the ability to function in either receptor- or store-operated modes, as determined by the channel expression level (Vazquez et al. 2001, 2003; discussed in Putney 2004). Store-operation was observed only at very low channel expression levels, whereas conditions that resulted in higher levels of TRPC3 protein expression gave rise to the receptor-regulated mode of TRPC3 activation (Vazquez et al. 2003; Venkatachalam et al. 2001). In addition, TRPC3 in its SOC mode showed pharmacological properties reminiscent of those exhibited by native SOCs, i.e., inhibition by low concentrations of Lanthanides in the micromolar range (Trebak et al. 2002). Interestingly, a more recent study showed that TRPC7, another member from the TRPC3/6/7 family, might also be able to function in either store-operated or receptor-activated mode depending upon expression conditions (Lievremont et al. 2004). This may provide a plausible explanation for previous conflicting results regarding regulation of this channel (Okada et al. 1999; Riccio et al. 2002a). In this case, the channels could be activated by store depletion only when stably expressed. Thus, the potential impact of channel expression conditions on channel regulation should be considered in studies aimed at evaluating the role of TRPCs in CCE, or in any functional mode for that matter. Although the impact of channel expression in regulation of TRPC6 has not been addressed, the majority of studies have concluded that ectopic expression of both mouse and human TRPC6 in cell lines results in a nonselective cation channel whose activation is linked to PLC-derived products but independent of store depletion (Boulay et al. 1997; Hofmann et al. 1999; Inoue et al. 2001; reviewed in Trebak et al. 2003b).

4.2
Non-Store-Operated Mechanisms

While there is considerable evidence for TRPC channels functioning as SOCs, in no case has an expressed TRPC served to recapitulate the electrophysiological behavior of the archetypical I_{crac}. Thus, it is thought that if these channels do function physiologically as SOCs, they more likely function as subunits of SOCs that show less Ca^{2+} selective behavior (Parekh and Putney 2005; Putney 2004). [However, as mentioned above, one study demonstrated that knockout of TRPC1 reduced the activity of a Ca^{2+}-selective I_{crac}-like current in DT40 B-cells (Mori et al. 2002).] On the other hand, there is an increasing list of naturally occurring currents which are apparently not store-operated and do appear to have TRPC-like electrophysiological behavior. This section will review the current understanding and ideas on the regulation of TRPC channels in their non-store-operated modes.

4.2.1
TRPC1

As discussed in the preceding section, a considerable number of publications have implicated TRPC1 in the function of store-operated channels, based largely on knockdown strategies. Actually, many laboratories have found it difficult to obtain functional channels when TRPC1 is expressed alone, finding that it did not traffic to the plasma membrane properly unless accompanied in a heterotetramer with TRPC4 or -5 (Hofmann et al. 2002). Thus, the mechanism of activation of TRPC1 in its non-store-operated mode may be assumed to be similar to that which is involved in the activation of TRPC4 and -5, a problem without a definitive answer as yet, as discussed below. However, more than one report has indicated that TRPC1 can also associate with TRPC3 (Lintschinger et al. 2000; Liu et al. 2005; Strubing et al. 2003), and in such a conformation the channels can apparently be activated by diacylglycerols (Lintschinger et al. 2000; Liu et al. 2005) (as for TRPC3/6/7, as discussed below).

4.2.2
TRPC2

Very little data are available on TRPC2, and its mechanism of activation is largely unknown. The suggestion that it may in some instances function as a SOC was discussed above. One report (Lucas et al. 2003) identified a native Ca^{2+}-permeable channel in mouse vomeronasal neuron dendrites that is activated by DAG, independently of Ca^{2+} and PKC. Ablation of the TRPC2 gene in mice caused a severe deficit in the DAG-activated channel. These authors concluded that TRPC2 encodes a principal subunit of this DAG-activated channel and that the primary electrical response to pheromones depends on DAG but not IP_3, Ca^{2+} stores, or arachidonic acid. Unfortunately, to date, the effect of DAG on ectopically expressed TRPC2 channels has not been explored.

4.2.3
TRPC3/6/7

When members of the TRPC3, -6, and -7 subfamily are ectopically expressed in HEK-293 or CHO cell lines they have been shown to be activated by membrane permeant DAG analogs such as 1-oleoyl-2-acetyl-*sn*-glycerol (OAG), but not by monoacylglycerols (reviewed by Trebak et al. 2003b). These channels can also be activated by procedures that increase intracellular DAG such as DAG lipase and DAG kinase inhibitors. Therefore, DAG activation provides a plausible mechanism of activation of these channels via PLC-coupled receptors.

The activation of TRPC3-mediated cation entry by DAG has been confirmed by several groups using different expression systems (Halaszovich et al. 2000; Hofmann et al. 1999; Ma et al. 2000; McKay et al. 2000; Trebak et al. 2003a; Vazquez et al. 2001; Venkatachalam et al. 2001). The effects of DAG on members of the TRPC3/6/7 subfamily are not mediated by PKC. Indeed, phorbol esters inhibit OAG-mediated activation of TRPC3/6/7 subfamily members, suggesting a negative regulation of these channels by PKC (Okada et al. 1999; Trebak et al. 2003a, 2005; Zhang and Saffen 2001). Trebak et al. (2005) showed that PKC exerts a negative regulation on TRPC3 channels via phosphorylation on serine 712. Subsequently, Kim and Saffen (2005) reported that the equivalent residue in TRPC6, serine 714, is involved in PKC-mediated phosphorylation of TRPC6 channels (however, see Estacion et al. 2004). There are also reports of negative regulation of TRPC3 by cyclic GMP-dependent kinase, both directly and by an action downstream of PKC (Kwan et al. 2004, 2005).

Ma et al. (2000) showed that TRPC3 expressed in HEK-293 cells could be activated after exogenous application of OAG or a DAG lipase inhibitor

(RHC80267) to the cells. These authors also reported that OAG could activate TRPC3 in the presence of 2-aminoethoxydiphenyl borate (2-APB), a presumed IP$_3$R inhibitor, whereas ATP failed to activate TRPC3 in the presence of 2-APB. The study concluded that DAG activates TRPC3 by a mechanism distinct from activation by IP$_3$R. However, Trebak et al. (2003a) reported that IP$_3$ does not activate TRPC3; rather the mechanism of activation by PLC -linked receptors seemed to result from DAG. Most recently, Lievremont et al. (2005a) showed that 2-APB is incapable of significant inhibition of IP$_3$ receptors when applied to intact cells and that it partially inhibits divalent cation entry in HEK-293 cells expressing TRPC3, TRPC6, or TRPC7 whether these channels were activated by a muscarinic agonist or by OAG. Therefore, the effect of 2-APB on TRPC3 channels is more likely due to a direct action on the channels themselves, unrelated to IP$_3$R, as was demonstrated for native SOCs (Braun et al. 2001).

When ectopically expressed in cells, TRPC6 and TRPC7 function as receptor-regulated cation channels (Hofmann et al. 1999; Okada et al. 1999). As is the case for all members of the TRPC family, PLC activity is required for TRPC6 function since the PLC inhibitor, U73122, blocked agonist activation of TRPC6 (Hofmann et al. 1999). The same authors showed that no G proteins, thapsigargin, ionomycin, PIP$_2$, or IP$_3$ activated TRPC6 and that DAG analogs such as OAG were efficient at activating TRPC6, providing a possible activation mechanism of TRPC6 via receptor stimulation. Okada et al. showed that TRPC7 can be activated by OAG and the DAG lipase inhibitor RHC80267 (Okada et al. 1999). One study (Tseng et al. 2004) reported that PIP$_3$, a lipid product of PI3K, induced Ca^{2+} entry via a non-capacitative pathway in many cell types and that TRPC6 is a component of this calcium entry pathway.

4.2.4
TRPC4/5

While it is clear that in their NCCE mode TRPC4 and -5 are activated downstream of PLC, the precise signal for activation of these channels is not known. When TRPC4 and -5 proteins were exogenously expressed in HEK-293 cells, they were shown to be insensitive to DAG analogs and to IP$_3$ (Schaefer et al. 2000). These authors reported that receptor-mediated activation of TRPC4 and -5 currents was unaffected by infusion of heparin through the patch pipette, arguing against the involvement of IP$_3$ receptors in TRPC4/5 activation. In another study (Venkatachalam et al. 2003), receptor activation of TRPC5 could occur in DT-40 B cells lacking all three IP$_3$ receptor subtypes. Strübing et al. (2001) showed that homomeric TRPC4 and -5, as well as heteromers formed between TRPC1 and -5, were not activated by infusion of IP$_3$ into cells through the patch pipette, but responded to subsequent muscarinic receptor activation. One possibility that remains to be tested is that PLC activates TRPC4/5 through PIP$_2$ hydrolysis by relieving an inhibitory effect on the channels by PIP$_2$. Such a mechanism has been described for one member of the TRP superfamily,

TRPV1, where PIP$_2$ appears to act as an inhibitor of the channels (Chuang et al. 2001).

4.3
Role of Channel Trafficking to the Plasma Membrane

A seemingly obvious, yet critical aspect of ion channel activation is the requirement that the channels must be properly inserted within the plasma membrane to function. But beyond this basic need for channel trafficking to the plasma membrane is the idea that *regulated* transport of TRPC channels to the plasma membrane is a requisite step in the activation of channels via G protein-coupled receptors. In this section, we will summarize results from the current literature on TRPC channel transport as it relates to their activation by PLC-coupled receptors.

An interaction with members of the soluble N-ethylmaleimide-sensitive factor attachment protein receptor (SNARE) complex, which is involved in vesicle attachment and exocytosis, has been demonstrated for TRPC3 (Singh et al. 2004) and TRPC5 (Greka et al. 2003). In the case of TRPC3, it was shown that inhibition of vesicular fusion with either tetanus toxin or brefeldin A significantly reduced the levels of surface-expressed TRPC3 and, as expected, these drugs also significantly inhibited TRPC3 channel activity (Singh et al. 2004). Thus, an interaction of TRPC3 with the SNARE complex appears to be important for maintenance of sufficient levels of channel proteins in the plasma membrane. These authors further showed, by surface biotinylation labeling, that in HEK-293 cells overexpressing TRPC3, treatment with carbachol significantly increased the amount of surface-expressed TRPC3 protein, and this regulated trafficking was likewise sensitive to inhibition by tetanus toxin. Importantly, this agonist-induced insertion of TRPC3 into the plasma membrane was not a consequence of agonist-induced Ca^{2+} release, since it also occurred in the presence of BAPTA. Increases in surface expression of TRPC1 (Mehta et al. 2003) and TRPC6 (Cayouette et al. 2004) in response to G protein-coupled receptor stimulation have also been reported. In addition, the work with TRPC6 showed that surface expression of the channel could be increased by treatment of cells with thapsigargin or BAPTA, leading the authors to suggest a role for TRPC6 functioning as a store-operated channel (Cayouette et al. 2004). Alternatively, this result may raise questions at to whether trafficking is specifically induced by agonist stimulation or is a result of pharmacological interference with cellular Ca^{2+} homeostasis.

In none of the studies that demonstrated agonist-induced insertion of TRPC channels was it clearly demonstrated that membrane insertion constitutes the major mechanism of channel activation; in fact, a recent study demonstrated that increased surface expression of TRPC5 induced by bradykinin is likely not a direct effector of TRPC5 channel activity induced by the agonist (Ordaz et al. 2005). Thus, it remains possible that regulated transport of TRPC channels to the plasma membrane may serve as a secondary adaptive mechanism, whereby channel density in the membrane is increased with the result being a greater availability of channels for activation, as opposed to a primary mechanism of channel activation. Adaptation by means of regulation of TRP channel surface expression has been demonstrated for *Drosophila* TRPL, where it was shown that dark adaptation of the fly visual response involves an increase in surface expression of TRPL (Bahner et al. 2002), as well as for the *Caenorhabditis elegans* TRP3, which translocates to the plasma membrane of sperm during sperm activation in preparation for fertilization (Xu and Sternberg 2003). More recently it was shown using a combination of total internal reflection fluorescence (TIRF) microscopy and surface biotinylation labeling that TRPC5 overexpressed in HEK-293 cells traffics to the plasma membrane in response to treatment with epidermal growth factor (EGF) (Bezzerides et al. 2004). An increased La^{3+}-induced whole-cell current accompanied this EGF effect, although EGF treatment alone did not induce a current. The authors further showed that surface expression of native TRPC5 in primary hippocampal neurons was also increased by growth factors, and that this increase may be associated with an inhibitory effect on neurite length. Thus, this study provides further evidence for a physiological paradigm whereby regulated increases in surface expression of TRPC channels may prime or enhance channel activation events. Furthermore, growth factors may be important mediators of TRPC surface expression in a variety of systems, since it was also shown that TRPC4 traffics to the plasma membrane in response to EGF treatment (Odell et al. 2005).

5
Biological Relevance of TRPC Channels

In nonexcitable cells, Ca^{2+} entry modulates, either directly or indirectly, an extraordinary variety of cellular functions (Berridge et al. 2000; Putney 1997). Within that context, CCE and/or NCCE Ca^{2+} entry in response to activation of plasma membrane PLC-coupled receptors plays a critical role in both short- and long-term regulation of Ca^{2+}-dependent processes and requires tight regulation of the channels involved. Although the molecular identity of SOCs and non-SOCs is not yet clearly defined, and despite some conflicting reports in the literature, members of the TRPC family of cation channels continue to emerge as good candidates for both types of Ca^{2+} entry pathways. In fact, there is experimental evidence for involvement of each member of the TRPC family in both CCE and NCCE entry in different cell types, whether endogenously or ectopically expressed. An important property shared by all TRPC members is that, regardless of the mode of operation, they are activated by PLC-derived signals, an early and common event in signaling through either G protein-coupled- or RTK receptors in diverse physiological situations. One of the more significant examples available of the physiological relevance of TRPC channels is the potential involvement of TRPC1, -4, and -6 in regulation of vascular function. TRPC6 seems to be a significant component of endogenous nonselective cation channels in smooth muscle cells (Inoue et al. 2001) where it may play an important role in regulation of vascular tone. Interestingly, in studies with the vascular smooth muscle cell line A7r5, Soboloff et al. (2005) demonstrated that TRPC6 was an important mediator of endogenous, agonist-generated Ca^{2+} signals and that the majority of Ca^{2+} entry likely resulted from activation of L-type Ca^{2+} channels, secondary to TRPC6-induced membrane depolarization. Paradoxically, in TRPC6 knockout mice, vascular reactivity was increased (Dietrich et al. 2005). This was attributed to compensatory upregulation of the more permeable channel, TRPC3, underscoring the difficulty in interpreting the often complex phenotypes obtained with knockout of a single gene. TRPC6 is also abundant in the lung where it may take part in modulation of mucus secretion (Nilius et al. 2005). Finally, two recent studies have demonstrated that mutations in TRPC6 appear to underlie certain progressive and often fatal kidney diseases (Reiser et al. 2005; Winn et al. 2005).

Functional studies using mouse knockout models have recently provided a potential connection between TRPC1- and TRPC4-mediated SOC entry and regulation of endothelial permeability. Increased expression of TRPC1 in mouse lung endothelial cells promotes an augmentation of both thrombin-induced Ca^{2+} entry and SOC currents (Paria et al. 2004). Aortic endothelial cells derived from TRPC4 knockout (TRPC4$^{-/-}$) mice lack SOC currents induced by either receptor activation or pharmacological depletion of Ca^{2+} stores, and this strongly correlates with impaired agonist-dependent vasorelaxation (Freichel et al. 2001). Also, in lung endothelial cells derived from these TRPC4$^{-/-}$ mice, thrombin receptor-dependent activation of both CCE and lung microvascular permeability is significantly reduced compared to wildtype cells (Tiruppathi et al. 2002).

As discussed in Sect. 4.1, recent evidence suggests that changes in the channel expression level within the same cell type may switch the coupling mechanism underlying channel activation. This introduces a novel variable into the equation governing channel modulation and function: a channel protein functioning in two distinct ways depending on its expression level. It remains to be explored how these changes are regulated, what factors are implicated, and what impact channel expression might have on either physiological or pathological states. In a different context, understanding the impact of TRPC channel expression conditions and environment might help to clarify existing discrepancies in the literature that hamper the definition of the mode of regulation for some of these channels.

Acknowledgements Drs. Elaine Gay and Paige Adams read the manuscript and provided helpful comments.

References

Bahner M, Frechter S, Da SN, Minke B, Paulsen R, Huber A (2002) Light-regulated subcellular translocation of Drosophila TRPL channels induces long-term adaptation and modifies the light-induced current. Neuron 34:83–93

Barritt GJ (1999) Receptor-activated Ca^{2+} inflow in animal cells: a variety of pathways tailored to meet different intracellular Ca^{2+} signalling requirements. Biochem J 337:153–169

Basora N, Boulay G, Bilodeau L, Rousseau E, Payet MD (2003) 20-Hydroxyeicosatetraenoic acid (20-HETE) activates mouse TRPC6 channels expressed in HEK293 cells. J Biol Chem 278:31709–31716

Benham CD, Davis JB, Randall AD (2002) Vanilloid and TRP channels: a family of lipid-gated cation channels. Neuropharmacology 42:873–888

Berridge MJ (1993) Inositol trisphosphate and calcium signalling. Nature 361:315–325

Berridge MJ (1995) Capacitative calcium entry. Biochem J 312:1–11

Berridge MJ, Lipp P, Bootman MD (2000) The versatility and universality of calcium signalling. Nat Rev Mol Cell Biol 1:11–21

Bezzerides VJ, Ramsey IS, Kotecha S, Greka A, Clapham DE (2004) Rapid vesicular translocation and insertion of TRP channels. Nat Cell Biol 6:709–720

Birnbaumer L, Zhu X, Jiang M, Boulay G, Peyton M, Vannier B, Brown D, Platano D, Sadeghi H, Stefani E, Birnbaumer M (1996) On the molecular basis and regulation of cellular capacitative calcium entry: roles for Trp proteins. Proc Natl Acad Sci USA 93:15195–15202

Boulay G, Zhu X, Peyton M, Jiang M, Hurst R, Stefani E, Birnbaumer L (1997) Cloning and expression of a novel mammalian homolog of Drosophila transient receptor potential (Trp) involved in calcium entry secondary to activation of receptors coupled by the Gq class of G protein. J Biol Chem 272:29672–29680

Braun FJ, Broad LM, Armstrong DL, Putney JW Jr (2001) Stable activation of single CRAC-channels in divalent cation-free solutions. J Biol Chem 276:1063–1070

Broad LM, Cannon TR, Taylor CW (1999) A non-capacitative pathway activated by arachidonic acid is the major Ca^{2+} entry mechanism in rat A7r5 smooth muscle cells stimulated with low concentrations of vasopressin. J Physiol (Lond) 517:121–134

Brough GH, Wu S, Cioffi D, Moore TM, Li M, Dean N, Stevens T (2001) Contribution of endogenously expressed Trp1 to a Ca^{2+}-selective, store-operated Ca^{2+} entry pathway. FASEB J 15:1727–1738

Cayouette S, Lussier MP, Mathieu EL, Bousquet SM, Boulay G (2004) Exocytotic insertion of TRPC6 channel into the plasma membrane upon Gq protein-coupled receptor activation. J Biol Chem 279:7241–7246

Chakrabarti R, Kumar S (2000) Diacylglycerol mediates the T-cell receptor-driven Ca^{2+} influx in T cells by a novel mechanism independent of protein kinase C activation. J Cell Biochem 78:222–230

Chuang HH, Prescott ED, Kong H, Shields S, Jordt SE, Basbaum AI, Chao MV, Julius D (2001) Bradykinin and nerve growth factor release the capsaicin receptor from PtdIns(4,5)P2-mediated inhibition. Nature 411:957–962

Clapham DE (1995) Calcium signaling. Cell 80:259–268

Clapham DE, Runnels LW, Strübing C (2002) The TRP ion channel family. Nat Rev Neurosci 2:387–396

Dietrich A, Mederos YS, Gollasch M, Gross V, Storch U, Dubrovska G, Obst M, Yildirim E, Salanova B, Kalwa H, Essin K, Pinkenburg O, Luft FC, Gudermann T, Birnbaumer L (2005) Increased vascular smooth muscle contractility in TRPC6$^{-/-}$ mice. Mol Cell Biol 25:6980–6989

Dohke Y, Oh YS, Ambudkar IS, Turner RJ (2004) Biogenesis and topology of the transient receptor potential Ca^{2+} channel TRPC1. J Biol Chem 279:12242–12248

Estacion M, Li S, Sinkins WG, Gosling M, Bahra P, Poll C, Westwick J, Schilling WP (2004) Activation of human TRPC6 channels by receptor stimulation. J Biol Chem 279:22047–22056

Fadool DA, Ache BW (1992) Plasma membrane inositol 1,4,5-trisphosphate-activated channels mediate signal transduction in lobster olfactory receptor neurons. Neuron 9:907–918

Fleig A, Penner R (2004) The TRPM ion channel subfamily: molecular, biophysical and functional features. Trends Pharmacol Sci 25:633–639

Freichel M, Suh SH, Pfeifer A, Schweig U, Trost C, Weißgerber P, Biel M, Philipp S, Freise D, Droogmans G, Hofmann F, Flockerzi V, Nilius B (2001) Lack of an endothelial store-operated Ca^{2+} current impairs agonist-dependent vasorelaxation in TRP4$^{-/-}$ mice. Nat Cell Biol 3:121–127

Gailly P, Colson-Van Schoor M (2001) Involvement of trp-2 protein in store-operated influx of calcium in fibroblasts. Cell Calcium 30:157–165

Gamberucci A, Giurisato E, Pizzo P, Tassi M, Giunti R, McIntosh DP, Benedetti A (2002) Diacylglycerol activates the influx of extracellular cations in T-lymphocytes independently of intracellular calcium-store depletion and possibly involving endogenous TRP6 gene products. Biochem J 364:245–254

Greka A, Navarro B, Oancea E, Duggan A, Clapham DE (2003) TRPC5 is a regulator of hippocampal neurite length and growth cone morphology. Nat Neurosci 6:837–845

Gunthorpe MJ, Benham CD, Randall A, Davis JB (2002) The diversity in the vanilloid (TRPV) receptor family of ion channels. Trends Pharmacol Sci 23:183–191

Guse AH, da Silva CP, Berg I, Skapenko AL, Weber K, Heyer P, Hohenegger M, Ashamu GA, Schulze-Koops H, Potter BVL, Mayr GW (1999) Regulation of calcium signalling in T lymphocytes by the second messenger cyclic ADP-ribose. Nature 398:70–73

Halaszovich CR, Zitt C, Jüngling E, Lückhoff A (2000) Inhibition of TRP3 by lanthanides. Block from the cytosolic side of the plasma membrane. J Biol Chem 275:37423–37428

Hardie RC, Minke B (1993) Novel Ca^{2+} channels underlying transduction in Drosophila photoreceptors: implications for phosphoinositide-mediated Ca^{2+} mobilization. Trends Neurosci 16:371–376

Harteneck C, Plant TD, Schultz G (2000) From worm to man: three subfamilies of TRP channels. Trends Neurosci 23:159–166

Hoenderop JG, Nilius B, Bindels RJ (2003) Epithelial calcium channels: from identification to function and regulation. Pflugers Arch 446:304–308

Hofmann T, Obukhov AG, Schaefer M, Harteneck C, Gudermann T, Schultz G (1999) Direct activation of human TRPC6 and TRPC3 channels by diacylglycerol. Nature 397:259–262

Hofmann T, Schaefer M, Schultz G, Gudermann T (2000) Cloning, expression and subcellular localization of two splice variants of mouse transient receptor potential channel 2. Biochem J 351:115–122

Hofmann T, Schaefer M, Schultz G, Gudermann T (2002) Subunit composition of mammalian transient receptor potential channels in living cells. Proc Natl Acad Sci USA 99:7461–7466

Hsu AL, Ching TT, Sem G, Wang DS, Bondada S, Authi KS, Chen CS (2000) Novel function of phosphoinositide 3-kinase in T cell signaling. A phosphatidylinositol 3,4,5-trisphosphate-mediated Ca^{2+} entry mechanism. J Biol Chem 275:16242–16250

Hurst RS, Zhu X, Boulay G, Birnbaumer L, Stefani E (1998) Ionic currents underlying HTRP3 mediated agonist-dependent Ca^{2+} influx in stably transfected HEK293 cells. FEBS Lett 422:333–338

Inoue R, Okada T, Onoue H, Hara Y, Shimizu S, Naitoh S, Ito Y, Mori Y (2001) The transient receptor potential protein homologue TRP6 is the essential component of vascular α1-adrenoceptor-activated Ca^{2+}-permeable cation channel. Circ Res 88:325–332

Jung S, Mühle A, Schaefer M, Strotmann R, Schultz G, Plant TD (2003) Lanthanides potentiate TRPC5 currents by an action at extracellular sites close to the pore mouth. J Biol Chem 278:3562–3571

Jungnickel MK, Marreo H, Birnbaumer L, Lémos JR, Florman HM (2001) Trp2 regulates entry of Ca^{2+} into mouse sperm triggered by egg ZP3. Nat Cell Biol 3:499–502

Kamouchi M, Philipp S, Flockerzi V, Wissenbach U, Mamin A, Raeymaekers L, Eggermont J, Droogmans G, Nilius B (1999) Properties of heterologously expressed hTRP3 channels in bovine pulmonary artery endothelial cells. J Physiol (Lond) 518:345–358

Kanki H, Kinoshita M, Akaike A, Satoh M, Mori Y, Kaneko S (2001) Activation of inositol 1,4,5-trisphosphate receptor is essential for the opening of mouse TRP5 channel. Mol Pharmacol 60:989–998

Kim JY, Saffen D (2005) Activation of M1 muscarinic acetylcholine receptors stimulates the formation of a multiprotein complex centered on TRPC6 channels. J Biol Chem 280:32035–32047

Kiselyov K, Xu X, Mozhayeva G, Kuo T, Pessah I, Mignery G, Zhu X, Birnbaumer L, Muallem S (1998) Functional interaction between InsP3 receptors and store-operated Htrp3 channels. Nature 396:478–482

Kwan HY, Huang Y, Yao X (2004) Regulation of canonical transient receptor potential isoform 3 (TRPC3) channel by protein kinase G. Proc Natl Acad Sci U S A 101:2625–2630

Kwan HY, Huang Y, Yao X (2005) Protein kinase C can inhibit TRPC3 channels indirectly via stimulating protein kinase G. J Cell Physiol 207:315–321

Lievremont JP, Bird GS, Putney JW Jr (2004) Canonical transient receptor potential TRPC7 can function as both a receptor- and store-operated channel in HEK-293 cells. Am J Physiol Cell Physiol 287:C1709–C1716

Lievremont JP, Bird GS, Putney JW Jr (2005a) Mechanism of inhibition of TRPC cation channels by 2-aminoethoxydiphenylborane. Mol Pharmacol 68:758–762

Lievremont JP, Numaga T, Vazquez G, Lemonnier L, Hara Y, Mori E, Trebak M, Moss SE, Bird GS, Mori Y, Putney JW Jr (2005b) The role of canonical transient receptor potential 7 in B-cell receptor-activated channels. J Biol Chem 280:35346–35351

Liman ER, Innan H (2003) Relaxed selective pressure on an essential component of pheromone transduction in primate evolution. Proc Natl Acad Sci U S A 100:3328–3332

Liman ER, Corey DP, Dulac C (1999) TRP2: a candidate transduction channel for mammalian pheromone sensory signaling. Proc Natl Acad Sci USA 96:5791–5796

Lintschinger B, Balzer-Geldsetzer M, Baskaran T, Graier WF, Romanin C, Zhu MX, Groschner K (2000) Coassembly of Trp1 and Trp3 proteins generates diacylglycerol- and Ca^{2+}-sensitive cation channels. J Biol Chem 275:27799–27805

Liu X, Wang W, Singh BB, Lockwich T, Jadlowiec J, O'Connell B, Wellner R, Zhu MX, Ambudkar IS (2000) Trp1, a candidate protein for the store-operated Ca^{2+} influx mechanism in salivary gland cells. J Biol Chem 275:3403–3411

Liu X, Singh BB, Ambudkar IS (2003) TRPC1 is required for functional store-operated Ca^{2+} channels. Role of acidic amino acid residues in the S5-S6 region. J Biol Chem 278:11337–11343

Liu X, Bandyopadhyay BC, Singh BB, Groschner K, Ambudkar IS (2005) Molecular analysis of a store-operated and 2-Acetyl-sn-glycerol-sensitive non-selective cation channel: heteromeric assembly of TRPC1-TRPC3. J Biol Chem 280:21600–21606

Lucas P, Ukhanov K, Leinders-Zufall T, Zufall F (2003) A diacylglycerol-gated cation channel in vomeronasal neuron dendrites is impaired in TRPC2 mutant mice: mechanism of pheromone transduction. Neuron 40:551–561

Ma HT, Patterson RL, van Rossum DB, Birnbaumer L, Mikoshiba K, Gill DL (2000) Requirement of the inositol trisphosphate receptor for activation of store-operated Ca^{2+} channels. Science 287:1647–1651

Ma HT, Venkatachalam K, Parys JB, Gill DL (2002) Modification of store-operated channel coupling and inositol trisphosphate receptor function by 2-aminoethoxydiphenyl borate in DT40 lymphocytes. J Biol Chem 277:6915–6922

Maroto R, Raso A, Wood TG, Kurosky A, Martinac B, Hamill OP (2005) TRPC1 forms the stretch-activated cation channel in vertebrate cells. Nat Cell Biol 7:179–185

McKay RR, Szymeczek-Seay CL, Lièvremont JP, Bird GS, Zitt C, Jüngling E, Lückhoff A, Putney JW Jr (2000) Cloning and expression of the human transient receptor potential 4 (TRP4) gene: localization and functional expression of human TRP4 and TRP3. Biochem J 351:735–746

Mehta D, Ahmmed GU, Paria B, Holinstat M, Voyno-Yasenetskaya T, Tiruppathi C, Minshall RD, Malik AB (2003) RhoA interaction with inositol 1,4,5-triphosphate receptor and transient receptor potential channel-1 regulates Ca^{2+} entry. Role in signaling increased endothelial permeability. J Biol Chem 278:33492–33500

Minke B, Cook B (2002) TRP channel proteins and signal transduction. Physiol Rev 82:429–472

Montell C (1999) Visual transduction in Drosophila. Annu Rev Cell Dev Biol 15:231–268

Montell C, Birnbaumer L, Flockerzi V (2002) The TRP channels, a remarkably functional family. Cell 108:595–598

Mori Y, Wakamori M, Miyakawa T, Hermosura M, Hara Y, Nishida M, Hirose K, Mizushima A, Kurosaki M, Mori E, Gotoh K, Okada T, Fleig A, Penner R, Iino M, Kurosaki T (2002) Transient receptor potential 1 regulates capacitative Ca^{2+} entry and Ca^{2+} release from endoplasmic reticulum in B lymphocytes. J Exp Med 195:673–681

Nilius B, Vriens J, Prenen J, Droogmans G, Voets T (2004) TRPV4 calcium entry channel: a paradigm for gating diversity. Am J Physiol Cell Physiol 286:C195–C205

Nilius B, Voets T, Peters J (2005) TRP channels in disease. Sci STKE 2005:re8

Obukhov AG, Nowycky MC (2002) TRPC4 can be activated by G-protein-coupled receptors and provides sufficient Ca^{2+} to trigger exocytosis in neuroendocrine cells. J Biol Chem 277:16172–16178

Odell AF, Scott JL, Van Helden DF (2005) Epidermal growth factor induces tyrosine phosphorylation, membrane insertion, and activation of transient receptor potential channel 4. J Biol Chem 280:37974–37987

Okada T, Shimizu S, Wakamori M, Maeda A, Kurosaki T, Takada N, Imoto K, Mori Y (1998) Molecular cloning and functional characterization of a novel receptor-activated TRP Ca^{2+} channel from mouse brain. J Biol Chem 273:10279–10287

Okada T, Inoue R, Yamazaki K, Maeda A, Kurosaki T, Yamakuni T, Tanaka I, Shimizu S, Ikenaka K, Imoto K, Mori Y (1999) Molecular and functional characterization of a novel mouse transient receptor potential protein homologue TRP7. Ca^{2+}-permeable cation channel that is constitutively activated and enhanced by stimulation of G protein-coupled receptor. J Biol Chem 274:27359–27370

Okada Y, Teeter JH, Restrepo D (1994) Inositol 1,4,5-trisphosphate-gated conductance in isolated rat olfactory neurons. J Neurophysiol 71:595–602

Ordaz B, Tang J, Xiao R, Salgado A, Sampieri A, Zhu MX, Vaca L (2005) Calmodulin and calcium interplay in the modulation of TRPC5 channel activity: identification of a novel C-terminal domain for calcium/calmodulin-mediated facilitation. J Biol Chem 280:30788–30796

Parekh AB, Putney JW Jr (2005) Store-operated calcium channels. Physiol Rev 85:757–810

Paria BC, Vogel SM, Ahmmed GU, Alamgir S, Shroff J, Malik AB, Tiruppathi C (2004) Tumor necrosis factor-alpha-induced TRPC1 expression amplifies store-operated Ca^{2+} influx and endothelial permeability. Am J Physiol Lung Cell Mol Physiol 287:L1303–L1313

Perraud AL, Knowles HM, Schmitz C (2004) Novel aspects of signaling and ion-homeostasis regulation in immunocytes. The TRPM ion channels and their potential role in modulating the immune response. Mol Immunol 41:657–673

Philipp S, Cavalié A, Freichel M, Wissenbach U, Zimmer S, Trost C, Marguart A, Murakami M, Flockerzi V (1996) A mammalian capacitative calcium entry channel homologous to Drosophila TRP and TRPL. EMBO J 15:6166–6171

Philipp S, Hambrecht J, Braslavski L, Schroth G, Freichel M, Murakami M, Cavalié A, Flockerzi V (1998) A novel capacitative calcium entry channel expressed in excitable cells. EMBO J 17:4274–4282

Philipp S, Trost C, Warnat J, Rautmann J, Himmerkus N, Schroth G, Kretz O, Nastainczyk W, Cavalié A, Hoth M, Flockerzi V (2000) Trp4 (CCE1) protein is part of native calcium release-activated Ca^{2+}-like channels in adrenal cells. J Biol Chem 275:23965–23972

Podesta M, Zocchi E, Pitto A, Usai C, Franco L, Bruzzone S, Guida L, Bacigalupo A, Scadden DT, Walseth TF, De Flora A, Daga A (2000) Extracellular cyclic ADP-ribose increases intracellular free calcium concentration and stimulates proliferation of human hemopoietic progenitors. FASEB J 14:680–690

Preuß KD, Nöller JK, Krause E, Göbel A, Schulz I (1997) Expression and characterization of a trpl homolog from rat. Biochem Biophys Res Commun 240:167–172

Putney JW Jr (1986) A model for receptor-regulated calcium entry. Cell Calcium 7:1–12

Putney JW Jr (1997) Capacitative calcium entry. Landes Biomedical Publishing, Austin

Putney JW Jr (2004) The enigmatic TRPCs: multifunctional cation channels. Trends Cell Biol 14:282–286

Reiser J, Polu KR, Moller CC, Kenlan P, Altintas MM, Wei C, Faul C, Herbert S, Villegas I, vila-Casado C, McGee M, Sugimoto H, Brown D, Kalluri R, Mundel P, Smith PL, Clapham DE, Pollak MR (2005) TRPC6 is a glomerular slit diaphragm-associated channel required for normal renal function. Nat Genet 37:739–744

Riccio A, Mattei C, Kelsell RE, Medhurst AD, Calver AR, Randall AD, Davis JB, Benham CD, Pangalos MN (2002a) Cloning and functional expression of human short TRP7, a candidate protein for store-operated Ca^{2+} influx. J Biol Chem 277:12302–12309

Riccio A, Medhurst AD, Mattei C, Kelsell RE, Calver AR, Randall AD, Benham CD, Pangalos MN (2002b) mRNA distribution analysis of human TRPC family in CNS and peripheral tissues. Mol Brain Res 109:95–104

Rosado JA, Brownlow SL, Sage SO (2002) Endogenously expressed Trp1 is involved in store-mediated Ca^{2+} entry by conformational coupling in human platelets. J Biol Chem 277:42157–42163

Sadighi Akha AA, Willmott NJ, Brickley K, Dolphin AC, Galione A, Hunt SV (1996) Anti-Ig-induced c alcium influx in rat B lymphocytes mediated by cGMP through a dihydropyridine-sensitive channel. J Biol Chem 271:7297–7300

Schaefer M, Plant TD, Obukhov AG, Hofmann T, Gudermann T, Schultz G (2000) Receptor-mediated regulation of the nonselective cation channels TRPC4 and TRPC5. J Biol Chem 275:17517–17526

Schaefer M, Plant TD, Stresow N, Albrecht N, Schultz G (2002) Functional differences between TRPC4 splice variants. J Biol Chem 277:3752–3759

Shi J, Mori E, Mori Y, Mori M, Li J, Ito Y, Inoue R (2004) Multiple regulation by calcium of murine homologues of transient receptor potential proteins TRPC6 and TRPC7 expressed in HEK293 cells. J Physiol 561:415–432

Shuttleworth TJ (1999) What drives calcium entry during $[Ca^{2+}]_i$ oscillations? Challenging the capacitative model. Cell Calcium 25:237–246

Shuttleworth TJ, Thompson JL, Mignen O (2004) ARC channels: a novel pathway for receptor-activated calcium entry. J Appl Physiol 19:355–361

Singh BB, Lockwich TP, Bandyopadhyay BC, Liu X, Bollimuntha S, Brazer SC, Combs C, Das S, Leenders AG, Sheng ZH, Knepper MA, Ambudkar SV, Ambudkar IS (2004) VAMP2-dependent exocytosis regulates plasma membrane insertion of TRPC3 channels and contributes to agonist-stimulated Ca(2+) influx. Mol Cell 15:635–646

Sinkins WG, Estacion M, Schilling WP (1998) Functional expression of TrpC1: a human homologue of the Drosophila Trp channel. Biochem J 331:331–339

Soboloff J, Spassova M, Xu W, He LP, Cuesta N, Gill DL (2005) Role of endogenous TRPC6 channels in Ca^{2+} signal generation in A7r5 smooth muscle cells. J Biol Chem 280:39786–39794

Strubing C, Krapivinsky G, Krapivinsky L, Clapham DE (2003) Formation of novel TRPC channels by complex subunit interactions in embryonic brain. J Biol Chem 278:39014–39019

Strübing C, Krapivinsky G, Krapivinsky L, Clapham DE (2001) TRPC1 and TRPC5 form a novel cation channel in mammalian brain. Neuron 29:645–655

Sweeney M, Yu Y, Platoshyn O, Zhang S, McDaniel SS, Yuan JXJ (2002) Inhibition of endogenous TRP1 decreases capacitative Ca^{2+} entry and attenuates pulmonary artery smooth muscle cell proliferation. Am J Physiol 283:L144–L155

Tiruppathi C, Freichel M, Vogel SM, Paria BC, Mehta D, Flockerzi V, Malik AB (2002) Impairment of store-operated Ca^{2+} entry in TRPC4(−/−) mice interferes with increase in lung microvascular permeability. Circ Res 91:70–76

Tomita Y, Kaneko S, Funayama M, Kondo H, Satoh M, Akaike A (1998) Intracellular Ca^{2+} store-operated influx of Ca^{2+} through TRP-R, a rat homolog of TRP, expressed in Xenopus oocytes. Neurosci Lett 248:195–198

Trebak M, Bird GS, McKay RR, Putney JW Jr (2002) Comparison of human TRPC3 channels in receptor-activated and store-operated modes. Differential sensitivity to channel blockers suggests fundamental differences in channel composition. J Biol Chem 277:21617–21623

Trebak M, Bird GS, McKay RR, Birnbaumer L, Putney JW Jr (2003a) Signaling mechanism for receptor-activated TRPC3 channels. J Biol Chem 278:16244–16252

Trebak M, Vazquez G, Bird GS, Putney JW Jr (2003b) The TRPC3/6/7 subfamily of cation channels. Cell Calcium 33:451–461

Trebak M, Hempel N, Wedel BJ, Smyth JT, Bird GS, Putney JW Jr (2005) Negative regulation of TRPC3 channels by protein kinase C-mediated phosphorylation of serine 712. Mol Pharmacol 67:558–563

Tseng PH, Lin HP, Hu H, Wang C, Zhu MX, Chen CS (2004) The canonical transient receptor potential 6 channel as a putative phosphatidylinositol 3,4,5-trisphosphate-sensitive calcium entry system. Biochemistry 43:11701–11708

Vannier B, Zhu X, Brown D, Birnbaumer L (1998) The membrane topology of human transient receptor potential 3 as inferred from glycosylation-scanning mutagenesis and epitope immunocytochemistry. J Biol Chem 273:8675–8679

Vannier B, Peyton M, Boulay G, Brown D, Qin N, Jiang M, Zhu X, Birnbaumer L (1999) Mouse trp2, the homologue of the human trpc2 pseudogene, encodes mTrp2, a store depletion-activated capacitative Ca^{2+} channel. Proc Natl Acad Sci U S A 96:2060–2064

Vazquez G, Lièvremont JP, Bird GS, Putney JW Jr (2001) Human Trp3 forms both inositol trisphosphate receptor-dependent and receptor-independent store-operated cation channels in DT40 avian B-lymphocytes. Proc Natl Acad Sci USA 98:11777–11782

Vazquez G, Wedel BJ, Bird GS, Joseph SK, Putney JW Jr (2002) An inositol 1,4,5-trisphosphate receptor-dependent cation entry pathway in DT40 B lymphocytes. EMBO J 21:4531–4538

Vazquez G, Wedel BJ, Trebak M, Bird GS, Putney JW Jr (2003) Expression level of TRPC3 channel determines its mechanism of activation. J Biol Chem 278:21649–21654

Vazquez G, Wedel BJ, Aziz O, Trebak M, Putney JW Jr (2004) The mammalian TRPC cation channels. Biochim Biophys Acta 1742:21–36

Venkatachalam K, Ma HT, Ford DL, Gill DL (2001) Expression of functional receptor-coupled TRPC3 channels in DT40 triple receptor InsP3 knockout cells. J Biol Chem 276:33980–33985

Venkatachalam K, Zheng F, Gill DL (2003) Regulation of canonical transient receptor potential (TRPC) channel function by diacylglycerol and protein kinase C. J Biol Chem 278:29031–29040

Wang J, Shimoda LA, Sylvester JT (2004) Capacitative calcium entry and TRPC channel proteins are expressed in rat distal pulmonary arterial smooth muscle. Am J Physiol Lung Cell Mol Physiol 286:L848–L858

Warnat J, Philipp S, Zimmer S, Flockerzi V, Cavalié A (1999) Phenotype of a recombinant store-operated channel: highly selective permeation of Ca^{2+}. J Physiol (Lond) 518:631–638

Wes PD, Chevesich J, Jeromin A, Rosenberg C, Stetten G, Montell C (1995) TRPC1, a human homolog of a Drosophila store-operated channel. Proc Natl Acad Sci USA 92:9652–9656

Winn MP, Conlon PJ, Lynn KL, Farrington MK, Creazzo T, Hawkins AF, Daskalakis N, Kwan SY, Ebersviller S, Burchette JL, Pericak-Vance MA, Howell DN, Vance JM, Rosenberg PB (2005) A mutation in the TRPC6 cation channel causes familial focal segmental glomerulosclerosis. Science 308:1801–1804

Wu X, Babnigg G, Villereal ML (2000) Functional significance of human trp1 and trp3 in store-operated Ca^{2+} entry in HEK-293 cells. Am J Physiol 278:C526–C536

Wu X, Babnigg G, Zagranichnaya T, Villereal ML (2002) The role of endogenous human Trp4 in regulating carbachol-induced calcium oscillations in HEK-293 cells. J Biol Chem 277:13597–13608

Xu SZ, Beech DJ (2001) TrpC1 is a membrane-spanning subunit of store-operated Ca(2+) channels in native vascular smooth muscle cells. Circ Res 88:84–87

Xu SZ, Zeng F, Boulay G, Grimm C, Harteneck C, Beech DJ (2005) Block of TRPC5 channels by 2-aminoethoxydiphenyl borate: a differential, extracellular and voltage-dependent effect. Br J Pharmacol 145:405–414

Xu XZ, Sternberg PW (2003) A C. elegans sperm TRP protein required for sperm-egg interactions during fertilization. Cell 114:285–297

Yamada H, Wakamori M, Hara Y, Takahashi Y, Konishi K, Imoto K, Mori Y (2000) Spontaneous single-channel activity of neuronal TRP5 channel recombinantly expressed in HEK293 cells. Neurosci Lett 285:111–114

Zeng F, Xu SZ, Jackson PK, McHugh D, Kumar B, Fountain SJ, Beech DJ (2004) Human TRPC5 channel activated by a multiplicity of signals in a single cell. J Physiol 559:739–750

Zhang L, Saffen D (2001) Muscarinic acetylcholine receptor regulation of TRP6 Ca^{2+} channel isoforms. J Biol Chem 276:13331–13339

Zhu X, Chu PB, Peyton M, Birnbaumer L (1995) Molecular cloning of a widely expressed human homologue for the Drosophila trp gene. FEBS Lett 373:193–198

Zhu X, Jiang M, Peyton M, Boulay G, Hurst R, Stefani E, Birnbaumer L (1996) trp, a novel mammalian gene family essential for agonist-activated capacitative Ca^{2+} entry. Cell 85:661–671

Zhu X, Jiang M, Birnbaumer L (1998) Receptor-activated Ca^{2+} influx via human Trp3 stably expressed in human embryonic kidney (HEK)293 cells. Evidence for a non-capacitative calcium entry. J Biol Chem 273:133–142

Zitt C, Zobel A, Obukhov AG, Harteneck C, Kalkbrenner F, Lückhoff A, Schultz G (1996) Cloning and functional expression of a human Ca^{2+}-permeable cation channel activated by calcium store depletion. Neuron 16:1189–1196

Zitt C, Obukhov AG, Strübing C, Zobel A, Kalkbrenner F, Lückhoff A, Schultz G (1997) Expression of TRPC3 in Chinese hamster ovary cells results in calcium-activated cation currents not related to store depletion. J Cell Biol 138:1333–1341

Zitt C, Halaszovich CR, Lückhoff A (2002) The TRP family of cation channels: probing and advancing the concepts on receptor-activated calcium entry. Prog Neurobiol 66:243–264

Zweifach A, Lewis RS (1995a) Rapid inactivation of depletion-activated calcium current (ICRAC) due to local calcium feedback. J Gen Physiol 105:209–226

Zweifach A, Lewis RS (1995b) Slow calcium-dependent inactivation of depletion-activated calcium current. J Biol Chem 270:14445–14451

Subject Index

A-fiber 349
A-G-35 352
acetazolamide 306
acetylcholine 417
acidic nociception 12
acrolein 350, 353
ADP-glucose 241
agonist 355
AITC 355
allicin 8, 350, 352, 353
alloxan 245, 248
allyl isothiocyanate 9, 350
alternative splicing 56, 254
amiloride 148, 354
AMP 241, 243
anandamide 8, 161
anesthetics 183
ankyrin 348, 354
ankyrin repeat motifs 23
ankyrin repeats 4
annexin I 304
antagonist 354
2-APB (2-aminoethoxydiphenylborate) 8, 24, 44, 87, 148, 305, 451
2-APB analog 178, 179, 182
– dimethyl 2APB 179
– diphenhydramine 179, 181
– DPBA 178–180, 182
– DPTHF 178–182, 184
apoptosis 248
arachidonic acid 36, 243
2 arachidonoyl-glycerol (2-AG) 8
arachidonylethanolamide (AEA) 8
ATP 241, 369
auditory 350, 354
auditory function 13

Basic fibroblast growth factor receptor-1 30

behaviour 1
– gender-specific 1
biological functions 4
blood pressure 11
bradykinin (BK) 165, 350, 353
brain 78, 88
BTP2 451

C fiber 349, 491, 499
(Ca^{2+}+Mg^{2+})ATP-ase Ca^{2+} pump (SERCA) 30, 35–37, 40, 43, 44
Ca^{2+} 2, 53
– entry 6
– gradients 77
– influx 2
– influx factor, CIF 38
– permeation 82
– re-absorption 12
– release 6
Ca^{2+} release-activated Ca^{2+} channel 9, 446
Ca^{2+} signalling 77, 80, 82
Ca^{2+}-activated K^+ channel 305, 353
cADPR 241, 247
Caenorhabditis elegans 2, 348
calcitonin gene-related peptide (CGRP) 166
calmodulin (CaM) 30, 83, 84, 147, 159, 198, 257
– TRPV4 198
calmodulin and IP_3 receptor binding site (CIRB) 83, 84, 87, 93, 96
CaMKII 158
camphor 8
capacitative Ca^{2+} entry 6
capsaicin (see also vanilloids) 8, 157, 158, 160–164, 166, 167, 175–177, 183, 490
capsazepine 8, 9, 164, 175, 177, 178, 183

cardiomyopathy 12
carvacrol 8
catalase 245
cation permeation 81
- regulation 162
- selectivity 4
caveolae 84, 86
caveolin 84, 86
caveolin-1 30, 84
CD38 247
cell death 242, 305
cell-cycle control 6
central nervous system (CNS) 144
cerebrospinal fluid 266
cGMP 84, 86
CGRP 166, 349
channel 1
- activated 356
- activation 355
- amplitude 355
- block 354, 356
- blocking 355
- conductance 355, 356
- gating 355
- inactivation 355, 356
- permeation 356
- pore 4
- potentiation 355, 356
channel blocker 148
- 2-Aminoethoxydiphenyl borate (2-APB) 8, 24, 44, 87, 148, 305, 451
- Amiloride 148, 354
- Gd^{3+} 8, 82, 87, 102, 103, 148, 354
- La^{3+} 8, 87, 99, 102–104, 148
chanzyme 240
chemosensation 370
chemotaxis 248
chemotherapeutic regime 350
cholesterol 86
choroid plexus 258
chromosomal location 11, 365
cibacron blue 247
cinnamaldehyde 9, 350
Ciona intestinalis 2
CIRB 83, 84, 87, 93, 96
CMR1 330
concavalin A 243
constitutive activity 146
corpora quadrigemina 349

CRAC (Ca^{2+} release-activated Ca^{2+}) channels 9, 37, 45, 446
CRACM 9, 10
crystal structures 2
current–voltage relationship 147
cyclic nucleotide-gated channel 37
cyclophosphamide 350
cystic kidney disease 363, 364
cytoskeleton 366, 369

DAG 10, 64, 68, 70, 82, 87, 146
DAG-induced current 66
Danio rerio 2
DBHQ (2,5-di(*tert*-butyl)-1,4-hydroquinone) 35
D-erythro-sphingosine 263
dentate gyrus 258
development 146
dexamethasone 307
diabetes mellitus 242
diacylglycerol (DAG) 10, 64, 68, 70, 82, 87, 146
differentiation 146
dimethylthiourea 245
diseases 248
- Alzheimer 248
- bipolar disorder 249
divalent cations 162
dominant-negative mutant 80, 81
dorsal root ganglia (DRG) 156, 164, 176–178, 349, 490, 499
Drosophila melanogaster 2
DT40 147

EF hand 10
EGF 369
elongation factor-2 kinase 300
endocannabinoid lipids 8
endocytosis 83
endoplasmic reticulum (ER) 43, 84
- Ca^{2+} homeostasis 149
endothelial permeability 1, 11
5',6'epoxyeicosatrienoic acid (5',6'-EET) 8
eucalyptol 9, 330
eugenol 8
evolution 348
exocytosis 83

fatty acid 183

FKBP12   84
flufenamic acid   244
fluorescence resonance energy transfer (FRET)   80, 302
fMLP   247
focal segmental glomerular sclerosis   15
fruit fly   1
Fugu rubripes   2

G protein-coupled receptors   53, 63
$G_{i2}\alpha$ 63
$G_o\alpha$ 63
– β2 microglobulin   63
– accessory olfactory bulb   63
– main olfactory bulb   62
– major histocompatibility complex (MHC1b)   63
– pheromone   62
GABA-release   11
gadolinium (Gd^{3+})   8, 82, 87, 102, 103, 148, 354
ganglia   349, 350, 354
– dorsal root ganglia   156, 164, 176–178, 349, 490, 499
– jugular   177
– nodose ganglia   156, 177, 349
– trigeminal ganglia   156, 349, 353
garlic (allicin)   9
gastrointestinal muscles   105
gender specific sexual behavior   68
gender-specific social behaviour   11
genetic arrangement   144
gentamicin   354
geraniol   9
glucagon-like peptide   417
glutamate metabotropic receptor 1α   30, 105
glutathione   242
glycosylation   348
– N-linked   348
GPCR   10
Gq/11 α   30

hair cells   13, 350, 354, 355
hearing loss   12
heart   78, 79, 88
heat   159, 160, 162, 164, 165, 176, 491, 495
HEK293 cell   78, 80, 84, 86, 87, 146

20-HETE   166
5-(S)-HETE   8
Homer   30, 41
12-(S)-HPETE   8
12-HPETE   161
15-(S)-HPETE   8
human disease   6
hydrogen peroxide   414
hydroxycitronellal   9
hyperalgesia   353, 493, 499
hypercalciuria   12, 307
hypocalcaemia   300
hypomagnesaemia   300
hypomagnesaemia with secondary hypocalcaemia   300
hypomagnesia   15
– TRPM6   15
hypotonic   263

icilin   9, 330, 332, 336, 352, 353
identity   144
ifosfamide   350
IgE-mediated Ca^{2+} entry   12
IGF-I   8
inflammation   156, 162, 165–168, 350, 353
inflammatory pain   13
Inhibitor of myogenic family (I-mfa)   30
inner ear   349, 350
inositol 1,4,5-trisphosphate (IP_3)   10, 54, 63, 83, 87, 147
– receptor (IP_3R)   30, 36, 41, 83, 84, 87
insulin   308
ionic selectivity   261
irritants   350
isothiocyanates   9, 350, 353

jugular ganglia   177
junctate   566

K^+ channels   2
– voltage-gated   2
keratinocyte   12, 178
kidney   258
kinocilia   350
Krd   14

La^{3+} 8, 87, 99, 102–104, 148
latrunculin   40

leukotriene B4   8
linalool   9
lipid   86, 87
lipid kinase   512
– PI(3)K   512
– PI(4)K   512
– PIP(5)K   512
lipid mediator   82
lung   78
lung vascular endothelial cells (LEC)   103, 104
lymphocytes   445
– B cells   447
– DT40 B cells   446
– Jurkat T cells   446
– T cells   447

male–male aggression   11
mannitol   245
mast cells   12
mechanical   350
– activation   354
mechanosensation   5, 364, 369
mechanosensitive cation channel   42
mechanosensory   349
mechanotransducing   354
mechanotransduction   354, 355
melastatin   238
membrane-spanning   79
menthol   9, 238, 330, 332, 354, 497
mesencephalon   349
metabolic acidosis   306
metabolic alkalosis   306
metabotropic glutamate receptor   30, 105
Mg^{2+} 99, 102
MicroRNA   256
mitogen-activated protein kinase (MAPK)   492
mouse aortic endothelial cell (MAEC)   101–104
mucolipidosis   15
– TRPML1   15
mucolipins   3
– TRPMLs   3
mustard oil   352
mutation   11
– spontaneous   11
– type   11

MxA (member of dynamin superfamily) β-Tubulin   30

N-arachidonoyl dopamine (NADA)   161
$(Na^+ + K^+)$ ATP-ase   45
NAADP   247
NCX1   88
nerve growth factor (NGF)   163
nerves   350
– sciatic   350
NGF   165, 167
nociceptive   352
nociceptor   349, 353
nodose ganglia   156, 177, 349
non-excitable cells   5, 77
Nudix box   238, 240, 241, 243

odorant receptors   64
– TRPC2$^{-/-}$ mice   66
oleoylethanolamide (OEA)   161, 168
olfaction   5
olfactory nucleotide (cAMP)-gated channel   63
olfactory transduction   63
olvanil   8
oocyte   302
Orai   9, 10, 446
orthologue   145
osmosensation   5
outward rectification   99, 100, 102
oxidative stress   246

p-loop   3, 348
pain   156, 162, 165–167, 350, 353, 489
– inflammatory pain   493
– visceral pain   494
pain hypersensitivity   13
pain perception   1, 5
PARP   242, 248
PDZ domains   97
2-pentenal   353
permeation   148, 260
pheromone   68
– perception   11
4α-PDD (4α-Phorbol 12,13-didecanoate)   8
phosphatidylinositol 4,5-bisphosphate (PIP_2)   4, 9, 10, 158, 162, 304, 337, 494, 498, 512, 513

– interaction   513
phosphoinositide   2, 512
– PI(4)P   512
– PIP$_2$ 512
phospholipase A2   36
phospholipase C (PLC)   1, 10, 30, 68, 80, 82–84, 86, 87, 146, 162, 353, 510
– γ1   84
– β   30
phosphorylation   158, 162, 163
– c-Src kinase   163
– Ca^{2+}/calmodulin-dependent kinase II (CaMKII)   163
– PKA   159, 163
– PKC   161, 163
– PKG   84, 86
– Src kinase   163
phototransduction   1, 2
piperine   8
PKC   30, 84, 86, 494, 498
PKD1   2
– polycystin-1, PC1   2
PKD2   2, 15
– TRPP2, PC2   2, 15
PKD2L1   2
– TRPP3, polycystin L, PCL   2
PKD2L2   2
– TRPP5   2
Plasma membrane (Ca^{2+}+Mg^{2+}) ATP-ase (PMCA)   30
PLC   1, 10, 30, 68, 80, 82–84, 86, 87, 146, 162, 353, 510
polyamines   159
– putrescine   159
– spermidine   159
– spermine   159, 162
polycystic kidney   14
polycystic kidney disease   15, 370
– PC2   2, 15
– PKD2   2, 15
polycystin-1L1   2
polycystin-1L2   2
polycystin-2 (PC2, PKD2)
– cation channel   366–369
– conductivity   367, 368
– expression pattern   365
– interacting   370
– interacting proteins   366, 369
– mutant proteins   367–369
– permeability ratios   367, 368
– promoter   364
polycystin-REJ   2
polycystins   3
– TRPPs   2, 3
pore
– ion-conducting   4, 259, 261, 354
pore helix   98, 99
pore loop   3
pressure sensitivity   12
primary afferent neurons   165
primary afferent nociceptors   13
primary cilium   364, 366, 369, 370
primitive node   370
proalgesic   350, 353
prognostic marker   6
proliferation   305
prostate   329, 339
prostate cancer   6, 15
– TRPM8   6, 15
– TRPV6   6, 15
protein domain   5, 513
– ankyrin repeat   4, 5, 23, 348, 354
– coiled-coil   4, 5
– ENTH   513
– FYVE   513
– kinase   5
– PH   513
– PLCδ1 PH   513
– PX   513
– TRP   4, 5
protein kinase A (PKA)   158, 159, 163
protein kinase C (PKC)   146, 158, 161, 163
protein kinase G (PKG)   84, 86
proton   157–159, 162–164, 167, 176
pseudogene   2, 348
pungent   350, 353
pyrophosphatase   238, 241

receptor-operated channel   149
redox signals   77
relative permeability   101
resiniferatoxin   8
retinal pigment epithelium   258
RhoA   30, 40
ribose-5-phosphate   241
ruthenium red (RR)   8, 9, 159, 164, 175–178, 354
ryanodine receptors   35

seizure 308
SERCA (Sarcoplasmic-endoplasmic ($Ca^{2+}+Mg^{2+}$) ATP-ase) 30, 35–37, 40, 43, 44
serine/threonine protein kinase 301
sex discrimination 11
single channel 240
– conductance 240
– inside-out 241
– outside-out 358
situs inversus 370
SK&F-96365 8, 24, 44, 87, 265
SOC 6, 24, 35, 36, 38, 43, 45, 46, 146
spermine 8
splice variants 53, 94, 144, 239
– TRPC4α 94, 96, 97
– TRPC4α 98
– TRPC4β 94, 96–98
STIM, stromal interaction molecule 9, 10
store-operated Ca^{2+} channels (SOCs) 24
store-operated Ca^{2+} current 304
store-operated calcium entry (SOCE) 174, 180, 181
store-operated channel (SOC) 6, 24, 35, 36, 38, 43, 45, 46, 146
streptozotocin 308
stretch 43
stretch activation pathway 38
substance P 166, 349
substantia nigra 178
supporting cells 350
supraoptic nucleus 157

taste 5, 13
temperature 160, 161
tetany 308
Δ^9-Tetrahydrocannabinol 9
tetrameric channel 80
– stoichiometry 80
– subunit composition 80
thapsigargin 35, 146
thermal hyperalgesia 11, 12
thermosensation 1, 5, 12
thermoTRP 5, 8, 330, 332
thyme 8
TIRFM 83, 84
TM5–TM6 linker 4
TPRC1

– patch clamp 33
transgenic mouse models 11
– phenotype 11
transient receptor potential, see TRP 1
transmembrane domains 2, 3, 80
tremor 308
trigeminal 353
trigeminal ganglia 156, 349, 353
TrkA 349
TrkB 349
TrkC 349
TRP 1, 15, 53, 54
– disease 15
trp 1
TRP box 4, 94
TRP channel pharmacology 4
TRP channels 2, 3, 381
– TRPA1 347
– TRPC1 23
– TRPC2 53
– TRPC3 77
– TRPC4 93
– TRPC5 109
– TRPC6 125
– TRPC7 143
– TRPM2 237
– TRPM3 253
– TRPM4 269
– TRPM5 287
– TRPM6 300
– TRPM7 313
– TRPM8 329
– TRPP2 363
– TRPV1 156, 490
– TRPV2 173, 495
– TRPV3 173
– TRPV4 189, 497
– TRPV5 207
– TRPV6 221
TRP domain 4
trp locus 1
TRP-p8 330
TRPA1 2, 3, 9, 13, 353, 355
TRPC 9, 44, 54, 81, 82, 174, 183, 566
– interaction with caveolin 566
– interaction with IP_3R 566
– interaction with junctate 566
– interaction with spinophilin and neurabin 566
– TRPC1 174

Subject Index   621

- TRPC3   174, 182
- TRPC5   174, 182, 183
- TRPC6   174, 175
- TRPC7   174
TRPC1   11, 23–26, 29, 31, 33–39, 41–44, 46, 80–82, 174
- 1,4,5-trisphosphate receptor   29
- 2-aminoethoxydiphenyl borate (2APB)   44
- ankyrin-like repeats   26
- antibody   29, 31, 36, 38, 44, 46
- brain-derived neurotrophic factor   46
- calmodulin   24, 26, 29, 43, 44
- caveolin-1   24, 26, 29, 33
- cytoskeleton   38
- endoplasmic reticulum   32
- ER   38
- Gd^{3+} 33, 44
- Glycosylation   31
- Golgi apparatus   32
- Homer   24, 29
- $InsP_3$ receptor   24
- $InsP_3$ receptor (IP_3R)   26
- IP_3R 38, 43
- IP_3R1 39
- IP_3R2 37, 38
- La^{3+} 44, 46
- lipid rafts   33
- myelin-associated glycoprotein   46
- Na^+ 33, 45
- Na^+ 44
- nerve growth cones   45
- netrin 1   46
- patch-clamp recording   42
- phospholipase C   24
- pore   34
- pore-forming domain   26
- protein kinase C   43
- SERCA 2B   39
- SKF96365   46
- small interfering RNAs   36
- SOCs   25, 37
- splice variants   25
TRPC2   11, 26, 29, 53, 56, 58, 60, 62, 64
- acrosome   58
- acrosome reaction   53, 61
- ankyrin   58
- calmodulin   58
- diacylglycerol (DAG)   54
- enkurin   58

- gender specific sexual behavoir   53
- gene   2
- junctate   58
- oocyte   58
- pheromones   54
- phospholipase C   54
- sperm   58
- store operated Ca^{2+} entry   55
- $TRPC2^{-/-}$ mice   62
- V1R   53
- V2R   53
- vomeronasal sensory organ (VNO)   53
TRPC3   11, 26, 29, 34, 566
- basal activity   129
- TRPC3/6 heteromers   126, 128, 134
TRPC3 mutant   80
TRPC4   11, 26, 28, 32, 34, 80, 94
TRPC4 knock-out   6
TRPC5   11, 26, 28, 32, 34, 80, 83, 566
TRPC6   6, 11, 26, 29, 80, 82, 126
- Bayliss effect   133
- blood cells   136
- brain   128, 137
- Ca^{2+} and Ca^{2+}/Calmodulin   131
- DAG activation   130
- glycosylation   126
- heteromultimerization   126
- human   126
- immune cells   136
- kidney   127, 128, 135
- mutants   135
- MxA interaction   127
- olfactory system   136
- pharmacology   132
- phosphorylation   131
- presenilin 2 interaction   137
- prostate   137
- ROC   130
- rods   137
- smooth muscle cells   128, 133
- splice variants   128
- translocation   131
- TRPC3/6 heteromers   126, 128, 134
- $TRPC6^{-/-}$ 134
TRPC6 knock-out   6
TRPC7   11, 26, 29, 80, 82
TRPL   1
TRPM   2, 9
TRPM1   6, 12, 15, 256

TRPM1 to TRPM8   2
TRPM2   12, 174
TRPM3   4, 12, 174, 254
TRPM4   4, 8, 12
TRPM5   4, 8, 13
TRPM6   13
TRPM7   13, 15, 174
TRPM8   4, 9, 13, 175, 352, 353
TRPML1   2, 14
TRPML2   2, 14
TRPML3   2, 14
TRPN1   2, 3, 348
TRPP   2
– TRPP2 (PC2, PKD2)   2, 14, 15, 174
TRPP3 (PKD2L1)   14
TRPP5 (PKD2L2)   14
TRPV   9, 174, 175
TRPV1   4, 8, 11, 353
TRPV2   8, 12
TRPV3   8, 12
TRPV4   8, 12, 189–198, 200–202
– EETs   196
– expression pattern   191
– gene and protein   190
– ion channel properties   192
– ion selectivity   194
– lipid messengers   196
– mechano-/osmosensation   200
– mechanosensitivity   196
– modes of TRPV4 activation   194
– osmosensitivity   195
– pharmacology   199
– phorbol ester derivatives   197
– systemic osmoregulation   200
– temperature sensitivity   197
– thermosensation   201
– vascular regulation   201
TRPV5   4, 12
TRPV6   4, 12
– 7q33–q34   222
– ankyrin repeats   225
– calcium selectivity   228
– econazole   228
– klotho   227
– pancreas   224
– placenta   224
– proliferation   230
– ruthenium red (RR)   228
– S100 annexin A2   227
– small intestine   224
– tumor staging   230

utricle   354

V1R   63, 66
V2R   63, 66
VAMP   84, 86
vanilloids (see also capsaicin)   11, 158, 161–166
– olvanil   161
– resiniferatoxin (RTX)   161
– RTX   164, 165
varitint-waddler   14
vasopressin   369
vasorelaxation   6, 11
vesicle fusion   84
vision   5
vitamin D_3 229
VNO   64, 66
voltage-gated Ca^{2+} channel   36, 87, 88
– CaV1.2   88
voltage-gated K^+ channels   3, 80
voltage-sensitivity   160, 161

white blood cells   243
– granulocytes   241, 243, 247, 248
– T lymphocytes   249

Xenopus laevis   42, 46
xestospongin C   38

yeast   2
yvc1   2

zebrafish   354
zinc-finger domain   301

Printing: Krips bv, Meppel
Binding: Stürtz, Würzburg